Climate Change and Chemicals

About the Authors

Golam Kibria, Ph. D. (Australia)

Dr. Golam Kibria is currently a Senior Environmental Scientist with the Australia's Largest Rural Water Authority (Goulburn – Murray Rural Water Authority, Victoria, Australia) since 2001. His current research activities includes among others assessing the impacts of pollutants on aquatic biodiversity; use of new and cutting edge technology such as passive sampling technology to monitor non-polar, polar micro-pollutants in waterways (Rivers, channels, creeks); assessing microbial risk using traditional, biochemical and molecular (DNA fingerprinting) techniques; assessing risks of pesticides/herbicides/heavy metals to human, livestock, flora and fauna and drinking water; potential for greenhouse gas emissions from reservoirs; ecotoxicological studies with native fish, climate change impacts on water resources and agriculture and Australian bush fire impacts. He is currently conducting a wide range of environmental and human health related research in collaboration with the Australia's most reputable and key scientific organizations such as the RMIT University, University of Melbourne, Commonwealth Scientific and Industrial Research Organization (CSIRO, Australia), Department of Primary Industries (Werribee and Queenscliff Research Centre, Melbourne, Australia), the Centre for Ecotoxicology (Department of Environment & Climate Change, NSW, Australia). Dr Kibria is an Adjunct Professor with the City University of Hong Kong, China and a Senior Associate with the RMIT University, where he coordinates cooperative research and training with the above Universities. He is also associated with supervising both undergraduate and post-graduate students as part of research collaboration and examining Ph.D dissertations. He was awarded an Australian Research Scholarship which enabled him to complete a degree of the Doctorate of Philosophy (Ph. D.) in Environmental Management from the Victoria University of Technology, Australia. He has also completed a Master of Science degree from the U.K through a British Council Award. Dr Kibria received the following international prizes: Runner up prize for scientific review (ICLARM, Philippines) and an Award for paper/poster presentation on Artificial mussel technology (Australasian Society for Ecotoxicology, Australia). Previously, he worked in different countries under different capacities as follows: Visiting Scientist with the University of Melbourne, Lecturer/Researcher with the Spencer Institute/Lincoln Marine Science Centre, South Australia, Specialist with the United Nations Development Program in Zimbabwe and Senior Scientist, DoF, Bangladesh. He already published 30 peer-reviewed papers in national and international journals, 15 peer-reviewed technical reports and 24 popular scientific articles and one book on issues relevant to environment and human health. He is a member of the Australasian Society for Ecotoxicology.

A. K. Yousuf Haroon, Dr. Sci. (Norway)

Dr. A. K. Yousuf Haroon is currently National Adviser, Strengthening Institutional Capacity of DoF Project, DoF-Danida, Bangladesh since 2008. His current research activities and field of interests include fish, sharks-rays & phytoplankton taxonomy; limnology, hydrobiology & wetland ecology; warm water aquaculture, and captive breeding of carps, minnows and catfishes; feeding strategies, niches, selectivity, & daily ration of fishes; fish population dynamics, stock assessment of natural depressions & impact assessment of flood control structures on fisheries and their mitigative measures and integrated rice-fish/shrimp farming and its synergistic. Previously, Dr. Yousuf served the Department of Fisheries (DoF), Bangladesh as an ecologist; Department of Biology, University of Basrah, Iraq as a Teaching Assistant; Master Plan Organization of the National Water Plan Project, Bangladesh Water Development Board as a Fisheries Consultant and the BFRI as Senior, Principal and Chief Scientist. He completed the Doctorate of Science (Dr. Sci.) in Tropical Aquaculture from the world famous school of fisheries, the Institute of Fisheries and Marine Biology, University of Bergen, Norway. He has to his credits more than 46 peer-reviewed research papers published in national and international journals, 30 technical reports, peer-reviewed proceeding's papers, and 11 popular scientific articles relevant to limnology-hydrobiology, phytoplankton taxonomy and various aspects of Bangladesh fisheries. He is an active member of the Network of Tropical Aquaculture and Fisheries Professionals (NTAFP), Penang, Malaysia and life-member of the Asian Fisheries Society, WorldFish Center, Penang, Malaysia. He is also associated with the prestigious Asiatic Society of Bangladesh in the Encyclopedia of Flora and Fauna Project (contributing author to marine bony fishes and elasmobranchs) of Bangladesh.

Dayanthi Nugegoda, Ph. D. (U. K.)

Dr. Dayanthi Nugegoda is currently a Professor of Ecotoxicology with the RMIT University, Melbourne, Australia since 1997. Professor Dayanthi Nugegoda leads the Ecotoxicology and Environmental Biology research team within the School of Applied Sciences at RMIT University. She is a core member of the College of Science Engineering and Technology, Environmental Change research strength group. Her current research includes Ecotoxicology and Environmental Biology with focuses on aquatic ecotoxicology, pesticides, trace metals, endocrine disrupting chemicals (EDCs), cyanobacterial toxins and their effects on fish and invertebrates. Her other research interests include biomonitoring of environmental contamination using fish and invertebrates in water and earthworms in soil; development of toxicity tests using Australian species; use of enzymes and hormones as biomarkers of environmental exposure to xenobiotics; and monitoring the effects of dry land salinity on freshwater ecosystems and bioremediation of metal contaminated and salinised sites and minimising nutrient pollution from aquaculture and evaluating water quality with a focus on reclaimed and recycled water. Professor Nugegoda is deeply committed to international issues being an Alumnus of the University of Colombo, Sri Lanka and the University of London, U.K. and having work experience in Sri Lanka, the UK, Germany, China, Singapore and Vietnam. She received the following awards, scholarships and fellowships : Visiting Fellowship, Flinders Research Centre for Coastal & Catchment Environments, Flinders University; RMIT Faculty of Life Sciences: Research Supervision Award; Visiting Research Fellowships from the GSF National Research Centre for Environment and Health, Munich Germany to collaborate on the research project "Biomarkers of pollution in fish from contaminated sites in the Yangtze River region, China" with the Institute of Ecological Chemistry; Drapers Company/Queen Mary College Award for Postgraduate Study, University of London; Open Commonwealth Scholarship awarded by the Commonwealth Scholarship Commission of the U.K.; Fulbright-Hayes Graduate Student Award for study in the U.S. and Scholarship for the best results in the First Examination in Natural Sciences, University of Colombo, Sri Lanka. She published a total of 68 refereed journal articles, 1 book chapter and 11 refereed publications in Conference proceedings. Professor Nugegoda supervised several Ph. D (11)., M. Sc. (2), Honours (22) theses at, RMIT University, and Victoria University, Australia, in addition to previous research supervision at The National University of Singapore and The University of Colombo, Sri Lanka. She is a Member of the Sri Lanka Association for the Advancement of Science; Member of the Asian Fisheries Society; Member of the Australasian Society for Ecotoxicology; Elected Representative for the State of Victoria; Member of the Scientific Panel of the Victorian Coastal Council.; Member of the Society for Ecotoxicology and Chemistry (SETAC), USA; Member of the Australian Society of Limnology and RMIT representative on the Council of the Queenscliff Marine Consortium.

Mr. Gavin Rose, B. A. (Hons., organic and analytical chemistry; Australia)

Gavin Rose is currently a Senior Research Scientist in Organic Chemistry with the Department of Primary Industries (Werribee Chemistry Centre, Melbourne, Australia) since 2001. Gavin is a member of the prestigious Royal Australian Chemical Institute and a Chartered Chemist. He has extensive experience in providing analytical chemistry services to agricultural industries with extensive experience as a professional residue chemist in several government laboratories encompassing pesticides and antibiotic veterinary drugs. Mr. Gavin has expertises in Quality Assurance/Quality control and has worked closely with Australia's National Association of Testing Authorities (NATA) as a laboratory assessor and signatory. He has been a leader of implementation of ISO 17025 (NATA) accreditation at two government laboratories and had led implementation of OECD Good Laboratory Practice recognition at DPI Werribee laboratory. He has been an invited speaker on residue testing and its quality assurance at international conferences. The group Gavin leads at DPI Werribee is a national leader in application of tandem mass spectrometry techniques to pesticide residue sample analysis for a range of insecticides, fungicides and herbicides. He has co-supervised post-graduate students at RMIT University and Murdoch University investigating organic residue and flavour methods in food and environmental samples. Mr. Gavin also has experience in conducting and management of field studies of pesticide applications. Recently he has collaborated with a range of catchment management, irrigation and land management authorities in application of passive sampling techniques to environmental samples.

Climate Change and Chemicals

Environmental & Biological Aspects

Golam Kibria, *Ph.D (Australia)*
Senior Environmental Scientist
Goulburn Murray Rural Water Authority, Victoria, Australia

A. K. Yousuf Haroon, *Dr. Sci. (Norway)*
National Adviser, Strengthening Institutional Capacity of DoF Project
DoF-Daniada, Dhaka, Bangladesh

Dayanthi Nugegoda, *Ph.D (U.K)*
Professor of Ecotoxicology
RMIT University, Bundoora, Melbourne, Australia

Gavin Rose, *B.A (Hons) (Australia)*
Senior Research Scientist
Future Farming Systems Research Division, DPI, Werribee, Victoria, Australia

New India Publishing Agency
Pitam Pura, New Delhi-110 088

ISBN : 978-9380235-30-1

Citation
Kibria, G., Yousuf Haroon, A. K., Nugegoda, D. and Rose, G. 2010. Climate change and chemicals - environmental and biological aspects. New India Publishing Agency (NIPA), Delhi, India.

For comments about the book (*email address*)
kibria@froggy.com.au, dharoon@ymail.com, dayanthi.nugegoda@rmit.edu.au, gavchrisrose@hotmail.com

First edition : 2010

Cover Layout and Planning : Golam Kibria

Declaration 1

"The author's current official title and current official address included in the book in good faith and intention. However, the views expressed in the book are of the authors only and doesn't reflect the views of the organization/countries etc. They presently belong to or belonged to in the past"

Declaration 2

"All figures/diagrams/tables included in the book were acknowledged in the reference section of each chapter as well as at the bottom of each figure/diagram/table and were used in good faith and intention of promoting science in the world in particular in the third world countries such as Asia-Pacific regions where information are not readily available or lacking on climate change and chemical impacts".

Declaration 3

"This publication may be of assistance to you but the authors or the publisher do not guarantee that the publication is without flaw of any kind or wholly appropriate for your particular purposes and therefore disclaims all liability for any error, loss or other consequences which may arise from relying on any information in this book"

Published by
Sumit Pal Jain *for*
New India Publishing Agency
101, Vikas Surya Plaza, CU Block, L.S.C. Mkt., Pitam Pura, New Delhi- 110 088, (India)
Phone: 011-27341717, Fax: 011-27341616, Mobile : 09717133558
E-mail: newindiapublishingagency@gmail.com, Web: www.bookfactoryindia.com

Composed and Designed by NIPA

Preface

Climate change refers to any change in climate over time, whether due to natural variability or as a result of human activity. The United Nations Framework Convention on Climate Change (UNFCCC) defines 'climate change' as 'Change of climate which is attributed directly or indirectly to human activity that alters the composition of the global atmosphere and which is in addition to natural climate variability observed over comparable time periods'.

Our climate is changing. These changes are due to two main reasons. The first and foremost is due to emissions of green house gases (GHGs) to atmosphere. The principal GHGs are carbon dioxide, methane and nitrous oxide. Human activities such as burning of fossil fuels, and agricultural activities (growing rice, application of nitrogen fertilisers, livestock rearing) have increased the amount of GHGs in the atmosphere in recent times. The enhanced greenhouse effect (an increase in the concentrations of GHGs) has resulted in a change in weather and climate of our planet. Some of these effects are rising of surface and ocean temperatures, rising of sea levels, widespread melting of glaciers and snows, rising of atmospheric CO_2 concentrations and extreme weather events (intense drought, intense precipitation, heat waves). The likely negative impacts of climate change or GHG effects are acidification of oceans, changes in river hydrology, changes in abundance and distribution of biodiversity, coral bleaching, increased incidence of pests and diseases (malaria, dengue), harmful algal blooms and increase in the toxicity of common harmful pollutants and more heat waves, etc. Despite several negative impacts, there are a few beneficial aspects of enhanced GHGs effects, for example, higher atmospheric concentrations of CO_2 (or Carbon dioxide fertilisation) may enhance the productivity of seagrasses, mangroves and crop production including rice, wheat, soybeans (so called C_3 crops). Higher temperatures may accelerate the growth of temperate fish. Part 1 of the book (chapter 1-5) provides an introduction to GHGs and its relationships with climate change and likely impacts of climate change on freshwater resources, agriculture and livestock, fisheries and aquatic ecosystems and human health.

Secondly, human activities resulting from agricultural, industrial and mining activities release chemicals and change our environment through contamination of air, land and water. Pesticides used to control pest plants and animals may enter into the environment via spray drift or drainage-discharge or run-off. Water contaminated with pesticides could be unfit for human consumption, stock drinking, irrigation, food processing and fish farming. Arsenic is released into the environment as natural or industrial processes. Arsenic contamination of groundwater has been reported from 20 countries including Bangladesh, India (West Bengal), where drinking water as well as vegetables and rice have been contaminated. Heavy metals' pollution can harm aquatic organisms through lethal and sub-lethal effects and can reduce or eliminate species from an ecosystem through increased disease or mortality or decreased fecundity. During combustion processes (power generation, bush fire) dioxins are released into the environment. Dioxins are some of the most toxic chemicals so far known to science and have the risk of cancer in adult humans. Sewage effluents can contain endocrine disrupting chemicals (EDCs) such as Estrogens EE2, and E2 which are likely to cause greatest risk to environment and wildlife at a dose as low as 0.1-1.0 ng/l. Furthermore, other EDCs - nonylphenol (NP) has been detected in human blood, breast milk, urine and adipose tissue and in seafood confirming persistent and bio-accumulative nature of NP. Exposures to high levels of other EDCs - phthalates have reported to cause miscarriages and pregnancy complications in women and are the leading cause of reproductive disorders in humans including reduced sperm counts, histological changes in testes, and reduced fertility. Pharmaceuticals (human and veterinary medicines) have been detected worldwide in surface waters including drinking water. The decline in vulture population in Bangladesh, India and Pakistan is reported to be associated with renal failure related to the accumulation of pharmaceutical diclofenac used in livestock industry. Biotoxins are naturally occurring highly toxic compounds produced by freshwater cyanobacteria and marine algae (reported from all over the world), these toxins can kill a wide range of organisms including birds, cattle, chicken, ducks, fish and human. Arsenic, heavy metals, pesticides, dioxins, EDCs, human and veterinary drugs and biotoxins are harmful chemicals, which are persistent, bio-accumulative as well as toxic. Most of these chemicals can be transferred to human via food chains or drinking water and could be of health hazard, if consumed. Part 2 of the book (chapter 6-13) highlights the physico-chemical properties, potential sources, national and international guidelines, world and regional reviews and the impacts of these harmful chemicals on wildlife, environment and human health.

The book '**Climate change and chemicals – Environmental and Biological Aspects**' reviewed and summarised research results and information from both developed and developing countries including Asia-Pacific, Australasia and other parts of the world. It was intended to document complex and technical information in most simplified manner by providing more illustrations in the form of tables and figures. The book would be a useful resource for academic and research institutes, health professionals, environmental scientists, chemists, chemical engineers, water engineers/scientists, fisheries biologists, agriculturists, ecotoxicologists and government planners and regulators. To benefit a wide range of readers the book has been written in simple and plain English, with appropriate acronyms and glossary and the key facts or summary given at the beginning of each chapter.

Comments and criticisms are welcome, we would be grateful to receive any comments from the readers since it would help us to improve the next edition. Last but not the least; our effort would be fruitful if the book helps to improve the human and environmental health for which the book has been written.

Golam Kibria	*A. K. Yousuf Haroon*	*Dayanthi Nugegoda*	*Gavin Rose*
Australia	Bangladesh	Australia	Australia
<kibria@froggy.com.au>	<dharoon@ymail.com>	<dayanthi.nugegoda@rmit.edu.au>	<gavchrisrose@hotmail.com>

Contents

Part-1 : Climate Change Impacts

1 Climate Change and Greenhouse Gases

Key Facts

- The United Nations Framework Convention on Climate Change (UNFCCC) defines 'climate change' as a 'Change of climate which is attributed directly or indirectly to human activity that alters the composition of the global atmosphere and which is in addition to natural climate variability observed over comparable time periods'.
- The main green house gases (GHG) that have caused the global climate change or global warming are carbon dioxide (CO_2), methane (CH_4) and nitrous oxide (NO_2).
- CO_2 is the principal anthropogenic greenhouse gas that affects the Earths radiative balance. It is the reference gas against which other GHGs are measured and therefore has a Global Warming Potential (GWP) of 1. CH_4 and N_2O have a GWP of 21 and 310 respectively.
- CO_2 is responsible for most of the enhanced green house effect of more than 50%, CH_4 about 10 to 15% and NO_2 about 4%.
- The global atmospheric concentrations of carbon-dioxide, methane and nitrous oxide have increased since 1750, CO_2 increased from a pre-industrial value of 280 ppm to 379 ppm in 2005, CH_4 increased from 715 ppb to 1774 ppb and NO_2 increased from 270 ppb to 319 ppb in 2005.
- 21 countries including western developed countries and large developing countries are responsible for 80% of global CO_2 emissions including USA (22.2%), China and Taiwan (18.4%), Russia (5.6%), India (4.9%), Japan (4.6%), Germany (3.1%), Canada (2.3%), UK (2.2%), South Korea (1.7%), Italy (1.7%), Mexico (1.6%), South Africa (1.6%), Iran (1.6%), Indonesia (1.4%), France (1.4%), etc.
- Recent observed global GHG emission impacts include an increase in global air and ocean temperatures; melting of snow and ice, and rising of global average sea levels, etc.
- Projections for global temperature rise are in the range of 1.4-5.8 °C; global average sea level rise is in the range of 0.26-0.59 m (high scenario) and a doubling of global carbon dioxide by 2100.
- Use of renewable energy sources (wind power, solar power, hydropower and biofuels) and improvement in technologies (fuel efficient vehicles, solar design for heating and cooling of buildings) would help to reduce a significant amount of GHG emissions. Replantation of plants or reforestation could further reduce the amount of carbon dioxide from the atmosphere.

Contents

List of Tables

List of Figures

List of Boxes

Acronyms and Glossary

Atmosphere: The gaseous envelope surrounding the earth. The dry atmosphere consists of almost entirely nitrogen (78.1% volume mixing ratio) and oxygen (20.9%), together with trace gases including argon (0.93%), carbon dioxide (0.035%) and ozone. In addition, the atmosphere contains water vapour (highly variable) and clouds and aerosols.

Biofuels: Liquid or gaseous fuels derived from biomass and used to power vehicle engines. They can be manufactured from crops, trees, animal fat or the biodegradable fraction of waste.

Carbon credits: Carbon credits are a key component of national and international CO_2 emissions trading schemes that has been implemented to mitigate global warming. They provide a way to reduce greenhouse effect emissions on an industrial scale by capping total annual emissions and letting the market assign a monetary value to any shortfall through trading. Credits can be exchanged between businesses or bought and sold in international markets at the prevailing market price. Credits can be used to finance carbon reduction schemes between trading partners and around the world.

Carbon-dioxide A naturally occurring gas and also a by-product of burning fossil fuels and biomass, as well as land-use changes and other industrial processes. It is the principal anthropogenic greenhouse gas that affects the Earth's radiative balance. It is the reference gas against which other greenhouse gases are measured and therefore has a Global Warming Potential of 1.

Carbon sequestration: The process of increasing the carbon content of a reservoir/pool other than the atmosphere.

Climate: Average weather or mean and variability of relevant quantities over a period of time ranging from months to thousands or millions of years.

Climate change: Climate change refers to any change in climate over time, whether due to natural variability or as a result of human activity. The United Nations Framework Convention on Climate Change (UNFCCC) defines 'climate change' as 'Change of climate which is attributed directly or indirectly to human activity that alters the composition of the global atmosphere and which is in addition to natural climate variability observed over comparable time periods'.

Climate model: A numerical representation of the climate system based on the physical, chemical and biological properties of its components, their interactions and feedback processes, and accounting for all or some of its known properties.

Climate prediction: A climate prediction or climate forecast is the result of an attempt to produce an estimate of the actual evolution of the climate in the future (seasonal, inter-annual, or long-term time scale).

Climate projection: A projection of the response of the climate system to emission or concentration scenarios of greenhouse gases and aerosols, or radiative forcing scenarios, often based upon simulations by climate model. Climate projections are critically dependent on the emissions/concentration/ radiative forcing scenario used, which are based on highly uncertain assumptions of future socio-economic and technological development.

Climate system: The climate system is defined by the dynamics and interactions of five major components: atmosphere, hydrosphere, cryosphere (snow, ice and permafrost), land surface, and biosphere (land and ocean). Climate system dynamics are driven by both internal and external forcing, such as volcanic eruptions, solar variations, or human-induced modifications to the planetary radiative balance (via anthropogenic emissions of green house gases).

Climate variability: Climate variability refers to variations in the mean state and other statistics (such as standard deviation, occurrence of extremes) of the climate on all temporal and spatial scales beyond that of individual weather events. Variability may be due to natural internal processes within the climate system (internal variability), or to variations in natural or anthropogenic external forcing (external variability).

Emission trading: Emissions trading (or emission trading) is an administrative approach used to control pollution by providing economic incentives for achieving reductions in the emissions of pollutants. It is sometimes called cap and trade.

Fossil fuel: A primary fuel consisting of the fossilised organic material derived from plants and animals which lived in the remote geological past; the main fossil fuels are coal, petroleum and natural gas.

Glacier: A large body of ice moving slowly down a slope or valley or spreading outward on a land surface.

Global warming : potential (GWP) Global warming potential (GWP) is the warming effect of each green house gas on the climate relative to that of carbon dioxide.

Greenhouse gases or : GHGs Those gases in the atmosphere which absorb and emit radiation at specific wavelengths within the spectrum of infrared radiation emitted by the Earth's surface, the atmosphere and clouds. Water vapour (H_2O), carbon dioxide (CO_2), nitrous oxide (N_2O), methane (CH_4) and ozone (O_3) are the primary greenhouse gases in the Earths atmosphere. Moreover, there are a number of entirely human-made gases in the atmosphere, such as the halocarbons and others dealt with under the Montreal and Kyoto Protocols [sulphur hexafluoride (SF_6), hydroflurocarbons (HFCs) and perfluorocarbons (PFCs)].

Greenhouse effect: The process in which the absorption of infrared radiation by the atmosphere warms the Earth. In common parlance, the term 'greenhouse effect' may be used to refer either to the natural greenhouse effect, due to naturally occurring greenhouse gases, or to the enhanced (anthropogenic) greenhouse effect, which results from gases emitted as a result of human activities.

IPCC: Intergovernmental Panel on Climate Change (IPCC). The IPCC is a scientific intergovernmental body set up by the World Meteorological Organization (WMO) and by the United Nations Environment Programme (UNEP).

Photovoltaic (PV) cell: An electric device that converts solar energy directly into electricity without any moving parts (apart from the electrons).

ppb: Parts per billion (1 billion = 1,000 million).

ppm: Parts per million (1 million = 100,000); is the number of greenhouse gas molecules to the total number of molecules of dry air.

ppt: Parts per trillion.

Sequestration: See carbon sequestration.

TWh: Terawatt-hour 3.6×10^{15} J.

MWH: Megawatt.

UNFCC: United Nations Framework Convention on Climate Change.

1.1: Introduction

The atmosphere consists of a number of green house gases (GHGs) that trap heat emitted from the ground and the oceans. These naturally occurring greenhouse gases such as water vapour, carbon-dioxide and methane that keep the surface of our planet warm enough to sustain us. Without heat-trapping greenhouse gases the surface would have an average temperature of -18^{0}C rather than current average of 15^{0}C. Over the 20th century, the earth's surface has increased in temperature by 0.76^{0}C, of which most of warming occurred between 1976 and 2000 (Figure 1.1). Scientists believe it is likely for a further rise of earth's temperatures by another 1.4 to 5.8^{0}C by the end of 21st century. This predicted increase of temperatures is generally known as 'climate change' or 'global warming'. The UNFCC defines 'climate change' as a change of climate which is attributed directly or indirectly to human activity that alters the composition of the global atmosphere and which is in addition to natural climate variability observed over a comparable time periods.

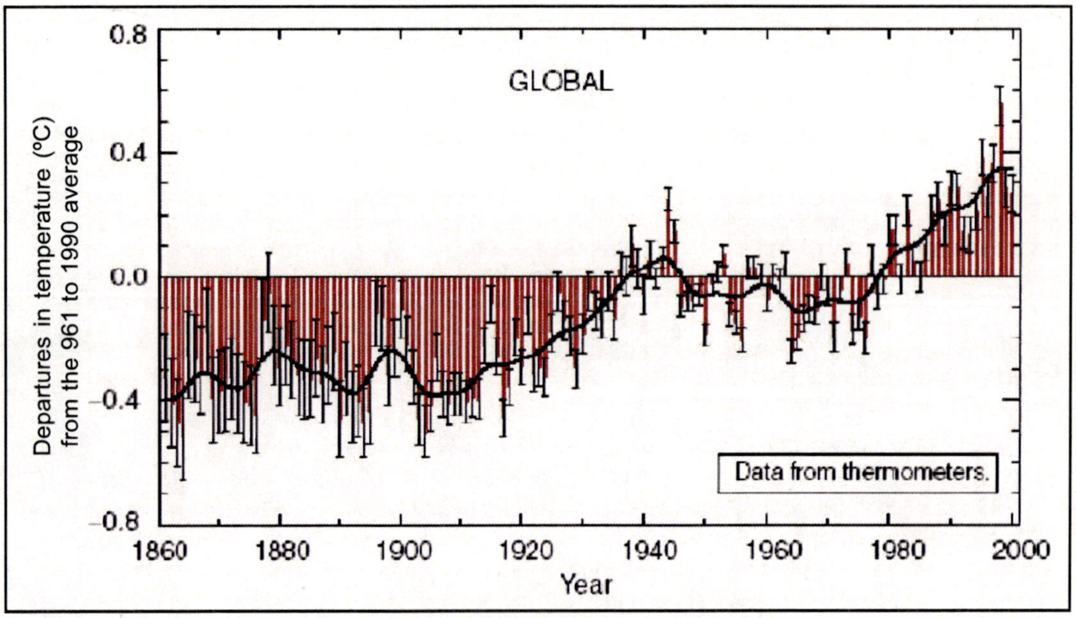

Figure 1.1 : Variations of the Earth's surface temperature for the past 140 years *[IPCC 2001; Note temperature increase from 1980 to 2000]*
http:/www.grida.no/publications/other/ipcc_tar/?src=/climate/ipcc_tar/wg1/figspm-1.htm

1.2: Causes of climate change

Climate change is caused by an increase in the concentration of the GHGs due to human activities. These greenhouse gases are carbon dioxide (CO_2), methane (CH_4), nitrous oxide (N_2O) and hydrofluorocarbon (HFCs). Higher concentrations of GHGs will lead to increased heat trapping, producing enhanced green house effect. The result is likely to cause a changing weather and climate of our planet. The enhanced greenhouse effect is results of the following human activities as listed in Table 1.1.

1.3: Changes in greenhouse gases concentrations

The global atmospheric concentrations of carbon-dioxide, methane and nitrous oxide have increased since 1750. Carbon-dioxide has increased from a pre-industrial value of 280 ppm to 379 ppm in 2005 (due mainly to fossil fuel use and CO_2 is responsible for at least half of the enhanced green house effect), methane has increased from 715 ppb to 1774 ppb (mainly due to agriculture and fossil fuel use) and nitrous oxide has increased from 270 ppb to 319 ppb in 2005 (mainly due to agriculture) (see Figure 1.2 and Table 1.1).

1.4: Global warming potential

Global warming potential (GWP) is the warming effect of each green house gas on the climate relative to that of carbon-dioxide. Carbon-dioxide was chosen as the reference gas by the Intergovernmental Panel on Climate Change (IPCC) (Table 1.1). Each greenhouse gas has different potential to trap heat. The global warming potential for carbon-dioxide is 1 and for methane is 21, it means that methane has a GWP 21 times greater than carbon-dioxide.

1.5: Global emissions

Rich countries and large developing countries account for most of global carbon-dioxide emissions (USA 22.2%; China 18.4%, India 4.9%, Japan 4.6%, European union 14.7% (see Table 1.2) with the United States - the world's top emitter of greenhouse gases - leaving a carbon footprint that is nearly 70 times higher than in Bangladesh. Bangladesh accounts for only 0.1% carbon dioxide emissions. India contributes around 4.9% of the global greenhouse gas emissions and is regarded as one of the world's top polluters like China. But it has resisted international pressure to make

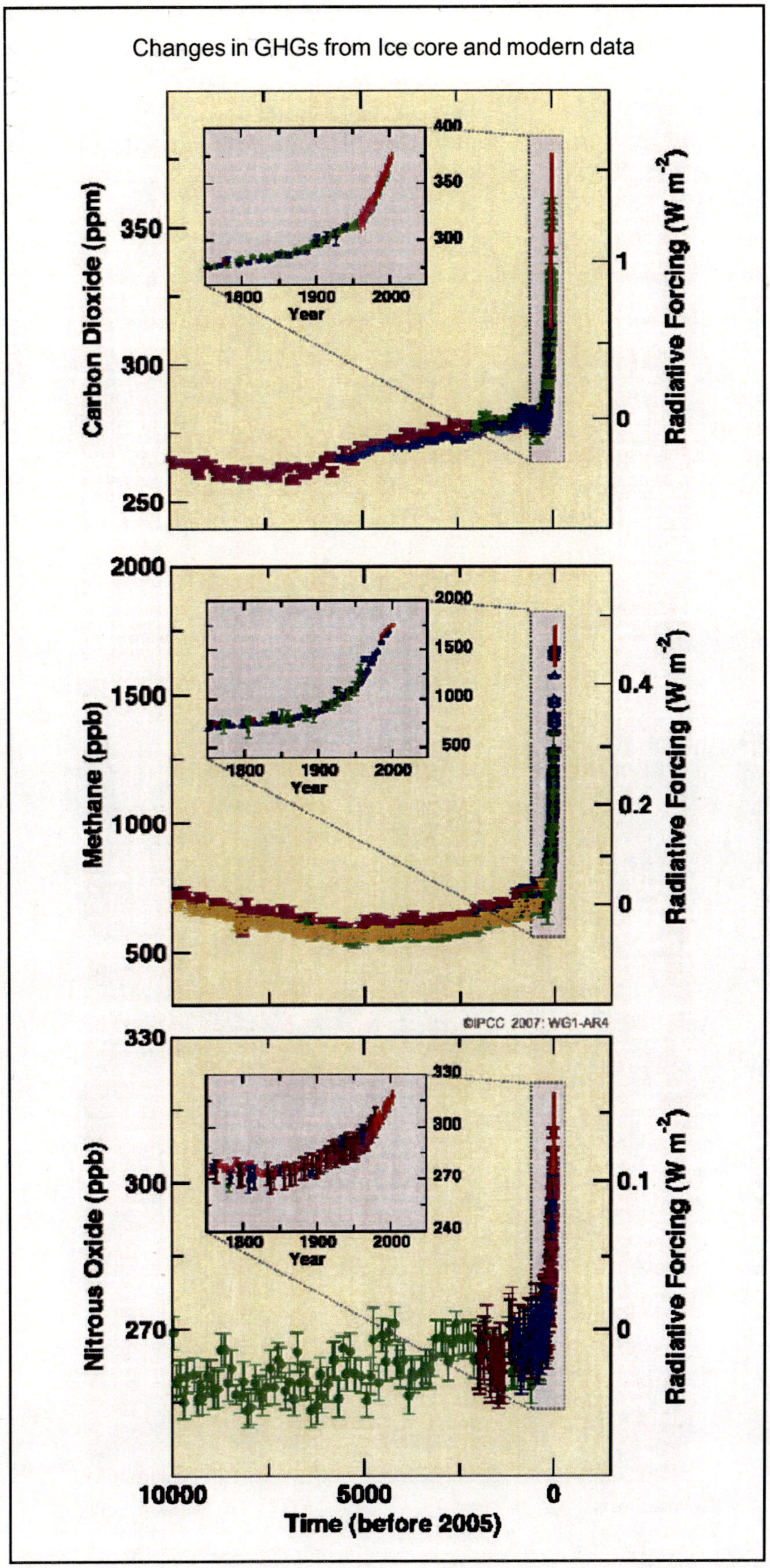

Figure 1.2: *Atmospheric concentrations of carbon dioxide, methane and nitrous oxide over the last 10,000 years (large panels) and since 1759 inset panels) [IPCC, 2007a].*

commitments to limit emissions. India along with China, is not included for targeted emission cuts under the 1997 Kyoto Protocol (see Box 1.1), the only global agreement that sets specific targets for reducing greenhouse gas emissions (AFP, New Delhi 08 Feb. 2008). At the Bali Climate Change Conference in December 2007, Europe proposed that major global emitters, including emerging markets such as China and India would commit to binding targets of reducing greenhouse gas emissions by 15-30% within 2020 and 60-80% by 2050 (Huq 2008). Australia generates only 1.2% of global GHGs emissions; however, Australia is one of the world's biggest polluters on a per capita basis.

Table 1.1: ***Examples of greenhouse gases that are affected by human activities*** *[source IPCC 2007a and IPCC 2007b].*

Enhanced greenhouse gases	Sources	Pre-industrial concentration	Current concentration	Global warming potential (GWP) compared to CO_2
Carbon-dioxide (CO_2) [*Atmospheric Lifetime (5-220 years*]	**a.** burning of fossil fuels (e.g. coal, oil and natural gas) for cars, electricity generation, heating of homes; **b.** clearing of land and burning of vegetation; [*CO_2 is responsible for at least half of the enhanced green house effect*]	280 ppm	379 ppm (2005)	1
Methane (CH_4) [*Atmospheric Lifetime 12 years]*	**a.** irrigated agriculture; ruminant animals (cows, sheep); **b.** rice paddies; **c.** decomposition of animal waste; **d.** burning of wastes, vegetation, wood and forest fire, coal mines, oil and gas fields (natural gas is composed of 95% CH_4), garbage in landfills; **e.** decomposition of biomass in dams and reservoirs (anaerobic conditions) (significant). [*Methane contributes about 10 to 15% of the enhanced green house effect]*	715 ppb	1723 ppb (1990) 1774 ppb (2005)	21
Nitrous oxide (N_2O) [*Atmospheric Lifetime 114 years]*	**a.** motor vehicles (automobile exhaust), **b.** use of nitrogen based fertilizers, **c.** dispose of human and animal waste in sewage treatment plants; [*Nitrous oxide contributes about 4% of the enhanced green house effect]*	270 ppb	319 ppb (2005)	310
Hydrofluorocarbons (HFCs-23) [*Atmospheric Lifetime (260 years)]*	**a.** refrigerants, **b.** plastic industry	0 ppt	20 ppt	140-11,700
Perflurocarbons [*Atmospheric Lifetime (> 50,000 years)]*		40 ppt	80 ppt	6,500-9,500
Sulfur hexafluoride		0	5 ppt	23,900

Table 1.2. ***List of countries against carbon dioxide emissions*** *(data of 2004 emissions). [http://en.wikipedia.org/wiki/List_of_countries_by_carbon_dioxide_emissions]*

Rank	Country	Annual CO_2 emissions (in thousands metric tonnes)	Percent age of total emissions	Rank	Country	Annual CO_2 emissions (in thousands metric tonnes)	Percentage of total emissions
	World	**27,245,758**	**100%**				
1	United States	6,049,435	22.2 %	102	Ghana	7,190	<0.1 %
				103	Latvia	7,098	<0.1 %
				104	Bolivia	6,973	<0.1 %
				105	Cyprus	6,750	<0.1 %
				106	Costa Rica	6,405	<0.1 %
2	China and Taiwan	5,010,170	18.4 %	107	El Salvador	6,167	<0.1 %
				108	Kyrgyzstan	5,727	<0.1 %
				109	Panama	5,661	<0.1 %
-	European Union	4,001,222	14.7 %	110	Uruguay	5,477	<0.1 %
3	Russia	1,524,993	5.6 %	111	Equatorial Guinea	5,426	<0.1 %
4	India	1,342,962	4.9 %	112	Côte d'Ivoire	5,162	<0.1 %
5	Japan	1,257,963	4.6 %	113	Tajikistan	5,004	<0.1 %
6	Germany	860,522	3.1 %	114	Senegal	4,993	<0.1 %
7	Canada	639,403	2.3 %	115	Tanzania	4,352	<0.1 %
8	United Kingdom	587,261	2.2 %	116	Botswana	4,301	<0.1 %
9	South Korea	465,643	1.7 %	117	Paraguay	4,180	<0.1 %
10	Italy	449,948	1.7 %	118	Netherlands Antilles	4,088	<0.1 %
11	Mexico	438,022	1.6 %	119	Nicaragua	4,007	<0.1 %
12	South Africa	437,032	1.6 %	120	Georgia	3,912	<0.1 %
13	Iran	433,571	1.6 %	121	Cameroon	3,839	<0.1 %
14	Indonesia	378,250	1.4 %	122	Albania	3,674	<0.1 %
15	France	373,693	1.4 %	123	Armenia	3,648	<0.1 %
16	Brazil	331,795	1.2 %	124	Republic of the Congo	3,542	<0.1 %
17	Spain	330,497	1.2 %	125	Mauritius	3,197	<0.1 %
18	Ukraine	330,039	1.2 %	126	Nepal	3,043	<0.1 %
19	Australia	326,757	1.2 %	127	Madagascar	2,731	<0.1 %
20	Saudi Arabia	308,393	1.1 %	128	New Caledonia	2,577	<0.1 %
21	Poland	307,238	1.1 %	129	Mauritania	2,555	<0.1 %
22	Thailand	268,082	1.0 %	130	Namibia	2,471	<0.1 %
23	Turkey	226,125	0.8 %	131	Malta	2,453	<0.1 %
24	Kazakhstan	200,278	0.7 %	132	Papua New Guinea	2,449	<0.1 %
25	Algeria	194,001	0.7 %	133	Benin	2,387	<0.1 %
26	Malaysia	177,584	0.7 %	134	Togo	2,310	<0.1 %
27	Venezuela	172,623	0.6 %	135	Zambia	2,288	<0.1 %
28	Egypt	158,237	0.6 %	136	Suriname	2,284	<0.1 %
29	United Arab Emirates	149,188	0.5 %	137	Réunion	2,277	<0.1 %
30	Netherlands	142,061	0.5 %	138	Iceland	2,229	<0.1 %
31	Argentina	141,786	0.5 %	139	Macau	2,207	<0.1 %
32	Uzbekistan	137,907	0.5 %	140	Mozambique	2,167	<0.1 %
33	Pakistan	125,669	0.5 %	141	Aruba	2,156	<0.1 %
34	Czech Republic	116,991	0.4 %	142	Democratic Republic of the Congo	2,104	<0.1 %
35	Nigeria	114,025	0.4 %	143	Bahamas	2,009	<0.1 %
36	Belgium	100,716	0.4 %	144	Uganda	1,826	<0.1 %
37	Kuwait	99,364	0.4 %	145	Haiti	1,756	<0.1 %

Contd...

38	Vietnam	98,663	0.4 %	146	Guadeloupe	1,734	<0.1 %
39	Greece	96,695	0.4 %	147	Guyana	1,445	<0.1 %
40	Romania	90,425	0.3 %	148	Gabon	1,371	<0.1 %
42	Iraq	81,652	0.3 %	149	Guinea	1,338	<0.1 %
43	Philippines	80,512	0.3 %	150	Martinique	1,291	<0.1 %
44	North Korea	79,111	0.3 %	151	Laos	1,280	<0.1 %
45	Israel	71,247	0.3 %	152	Barbados	1,269	<0.1 %
46	Austria	69,846	0.3 %	153	Niger	1,214	<0.1 %
47	Syria	68,420	0.3 %	154	Burkina Faso	1,096	<0.1 %
48	Finland	65,799	0.2 %	155	Fiji	1,071	<0.1 %
49	Belarus	64,890	0.2 %	156	Malawi	1,045	<0.1 %
50	Chile	62,418	0.2 %	157	French Guiana	1,005	<0.1 %
51	Libya	59,914	0.2 %	158	Sierra Leone	994	<0.1 %
52	Portugal	58,906	0.2 %	159	Swaziland	957	<0.1 %
53	Hungary	57,183	0.2 %	160	Belize	792	<0.1 %
54	Colombia	53,634	0.2 %	161	Eritrea	755	<0.1 %
55	Serbia and Montenegro	53,322	0.2 %	162	Maldives	726	<0.1 %
56	Sweden	53,033	0.2 %	163	Afghanistan	693	<0.1 %
57	Denmark	52,956	0.2 %	164	French Polynesia	671	<0.1 %
58	Qatar	52,904	0.2 %	165	Faroe Islands	660	<0.1 %
59	Singapore	52,252	0.2 %	166	Palestinian Authority	649	<0.1 %
60	Norway	43,149	0.2 %	167	Greenland	572	<0.1 %
61	Ireland	42,353	0.2 %	168	Rwanda	572	<0.1 %
62	Turkmenistan	41,726	0.2 %	169	Mali	565	<0.1 %
63	Morocco	41,169	0.2 %	170	Bermuda	550	<0.1 %
64	Switzerland	40,457	0.2 %	171	Seychelles	546	<0.1 %
65	Hong Kong	37,411	0.1 %	172	Cambodia	535	<0.1 %
66	Bangladesh	37,165	0.1 %	173	Liberia	470	<0.1 %
67	Slovakia	36,289	0.1 %	174	Bhutan	414	<0.1 %
68	Trinidad and Tobago	32,557	0.1 %	175	Antigua and Barbuda	414	<0.1 %
69	New Zealand	31,570	0.1 %	176	Gibraltar	374	<0.1 %
70	Peru	31,493	0.1 %	177	Saint Lucia	367	<0.1 %
71	Azerbaijan	31,365	0.1 %	178	Djibouti	367	<0.1 %
72	Oman	30,899	0.1 %	179	Cayman Islands	312	<0.1 %
73	Ecuador	29,268	0.1 %	180	Gambia	286	<0.1 %
74	Cuba	25,818	0.1 %	181	Cape Verde	275	<0.1 %
75	Croatia	23,501	0.1 %	182	Guinea-Bissau	271	<0.1 %
77	Tunisia	22,885	0.1 %	183	Central African Republic	253	<0.1 %
78	Yemen	21,114	0.1 %	184	Western Sahara (SADR)	238	<0.1 %
79	Dominican Republic	19,640	0.1 %	185	Palau	238	<0.1 %
80	Estonia	18,944	0.1 %	186	Burundi	220	<0.1 %
81	Bahrain	16,949	0.1 %	187	Grenada	216	<0.1 %
82	Jordan	16,465	0.1 %	188	Saint Vincent and the Grenadines	198	<0.1 %
83	Lebanon	16,263	0.1 %	189	Timor-Leste	176	<0.1 %
84	Slovenia	16,212	0.1 %	190	Solomon Islands	176	<0.1 %
85	Bosnia and Herzegovina	15,596	0.1 %	191	Samoa	150	<0.1 %

Contd...

86	Lithuania	13,309	0.1 %	192	Nauru	143	<0.1 %
87	Guatemala	12,220	0.1 %	193	Saint Kitts and Nevis	125	<0.1 %
88	Sri Lanka	11,534	<0.1 %	194	Chad	125	<0.1 %
89	Luxembourg	11,277	<0.1 %	195	Tonga	117	<0.1 %
90	Jamaica	10,592	<0.1 %	196	Dominica	106	<0.1 %
91	Kenya	10,588	<0.1 %	197	São Tomé and Príncipe	92	<0.1 %
92	Zimbabwe	10,559	<0.1 %	198	Vanuatu	88	<0.1 %
93	Macedonia	10,420	<0.1 %	199	Comoros	88	<0.1 %
94	Sudan	10,372	<0.1 %	200	British Virgin Islands	84	<0.1 %
95	Myanmar	9,760	<0.1 %	201	Saint Pierre and Miquelon	62	<0.1 %
96	Brunei	8,810	<0.1 %	202	Montserrat	62	<0.1 %
97	Mongolia	8,553	<0.1 %	203	Falkland Islands	44	<0.1 %
98	Ethiopia	7,982	<0.1 %	204	Kiribati	29	<0.1 %
99	Angola	7,897	<0.1 %	205	Cook Islands	29	<0.1 %
100	Moldova	7,685	<0.1 %	206	Saint Helena	11	<0.1 %
101	Honduras	7,615	<0.1 %	207	Niue	4	<0.1 %

Box 1.1 : *What is Kyoto protocol?* *[http://en.wikipedia.org/wiki/Kyoto_Protocol].*

The Kyoto Protocol is a protocol to the United Nations Framework Convention on Climate Change (UNFCCC or FCCC), aimed at combating global warming. The treaty is intended to achieve "stabilization of greenhouse gas concentrations" in the atmosphere at a level that would prevent dangerous anthropogenic interference with the climate system. The Kyoto Protocol establishes legally binding commitments for the reduction of four greenhouse gases (carbon dioxide, methane, nitrous oxide, sulfur hexafluoride), and two groups of gases (hydrofluorocarbons and perfluorocarbons. As of 2009, 187 parties have ratified the protocol (see Figure below) which was initially adopted for use on 11 December 1997 in Kyoto, Japan and entered into force on 16 February 2005. Under Kyoto Protocol, industrialized countries agreed to reduce their collective GHG emissions by 5.2% compared to the year 1990. It will reduce global GHG emissions by requiring developed countries to meet national targets for GHG emissions over the five year period from 2008 to 2012. The Protocol recognises that developed countries have a responsibility to take the lead in international action because they are responsible for most of the world's past emissions. Each developed country's target was negotiated and agreed internationally. Under the Protocol, developed countries are required to take domestic action to reduce GHG emissions.

*Participation in the Kyoto Protocol, as of June 2009, where **dark green** indicates the countries that have signed and ratified the treaty, **grey colour** represenets those countries which have not yet decided and **red colour** represents those countries which have no intention to ratify. The United States (U.S.), although a signatory to the Kyoto Protocol, has neither ratified nor withdrawn from the Protocol [Figure sourced from http://en.wikipedia.org/wiki/Kyoto_Protocol; 20 December 2009].*

Box 1.2 : COP 15 Accord *[http://unfccc.int/resource/docs/2009/cop15/eng/l07.pdf; http://en.cop15.dk/news/view+news?newsid=3006; http://en.wikipedia.org/wiki/United_Nations_Framework_Convention_on_Climate_Change#2010_-_COP_16.2FMOP_6.2C_Mexico].*

The United Nations Climate Change Conference (15th Conference of the Parties or COP 15) to the United Nations Framework Convention on Climate Change) was held in Copenhagen, Denmark, between 7 December and 18 December, where 192 countries attended. The Copenhagen Accord was drafted by the US, China, India, Brazil and South Africa on 18 December 2009 (see Figure below), The ***Copenhagen Accord*** *is the document in which the delegates at the United Nations Climate Change Conference agreed to "take note of". The document recognised that climate change is one of the greatest challenges of the present and that actions should be taken to keep any temperature increases to below 2°C. The document is not legally binding and does not contain any legally binding commitments for reducing CO_2 emissions. The main points of COP 15 are highlighted below :*

(a) to keep global temperature rise below 2 °C;
(b) pledging funds to help poor countries adapt to climate change such as least developed countries, small island developing states and Africa;
(c) to reduce deforestation in return for cash from developed countries.

The COP 15 Accord/agreement was reached after a meeting among President Barack Obama (USA) and Premier Wen Jiabao (China), Prime Minister Manmohan Singh (India), President Luiz Inacio Lula da Silva (Brazil) and President Jacob Zuma (South Africa).
(photo source: http://en.cop15.dk/news/view+news?newsid=3064; Scanpix/New York Times).

1.6: Observations of recent climate changes

According to IPCC (IPCC 2007a and 2007b), the enhanced GHGs have already caused an increase in global surface temperature, a rise of sea level and melting of snow cover (Figure 1.3). The main changes observed during the recent times are summarised below:

- *Increases in global average air/land and ocean temperatures, the eleven of the last twelve years (1995-2006) were among the warmest years; the total temperature increase from 1850-1899 to 2001-2005 is 0.76^{0}C; the average temperature of the global ocean has increased up to depths of 3,000 m (note: oceans absorb more than 80% of the heat added to the climate system).*
- *Global warming cause the oceans to warm and expand, inducing a rise in the sea levels and it was observed that global sea level rose at an average of 1.8 mm per year, and the rate was faster during 1993 to 2003 (see Table 1.3).*
- *Mountain glaciers and snow cover have declined on an average in both the hemispheres; widespread decreases in glaciers and ice caps have caused an additional contribution to sea level rise (see Table 1.3) (also see chapter 2.5.3 impacts of Himalayas' glaciers melt on fresh water resources).*

- *Widespread changes in precipitation amounts, ocean salinity, wind patterns, extreme weather patterns (droughts, heavy precipitation, heat waves and the intensity of tropical cyclones) (see Table 1.4).*

Table 1.3: ***Observed rates of sea level rise and estimated contributions from different sources during 1961-2003*** *[IPCC 2007a].*

Source of sea level rise	Rate of sea level rise (mm/year) during 1961-2003	Rate of sea level rise (mm/year)1993-2003
Thermal expansion	0.42	1.6
Glaciers and ice caps	0.50	0.77
Greenland ice sheet	0.05	0.21
Antarctic ice sheet	0.14	0.21
Sum of total climate contributions to sea level rise (estimated)	1.1	2.8
Observed total sea level rise	1.8	3.1

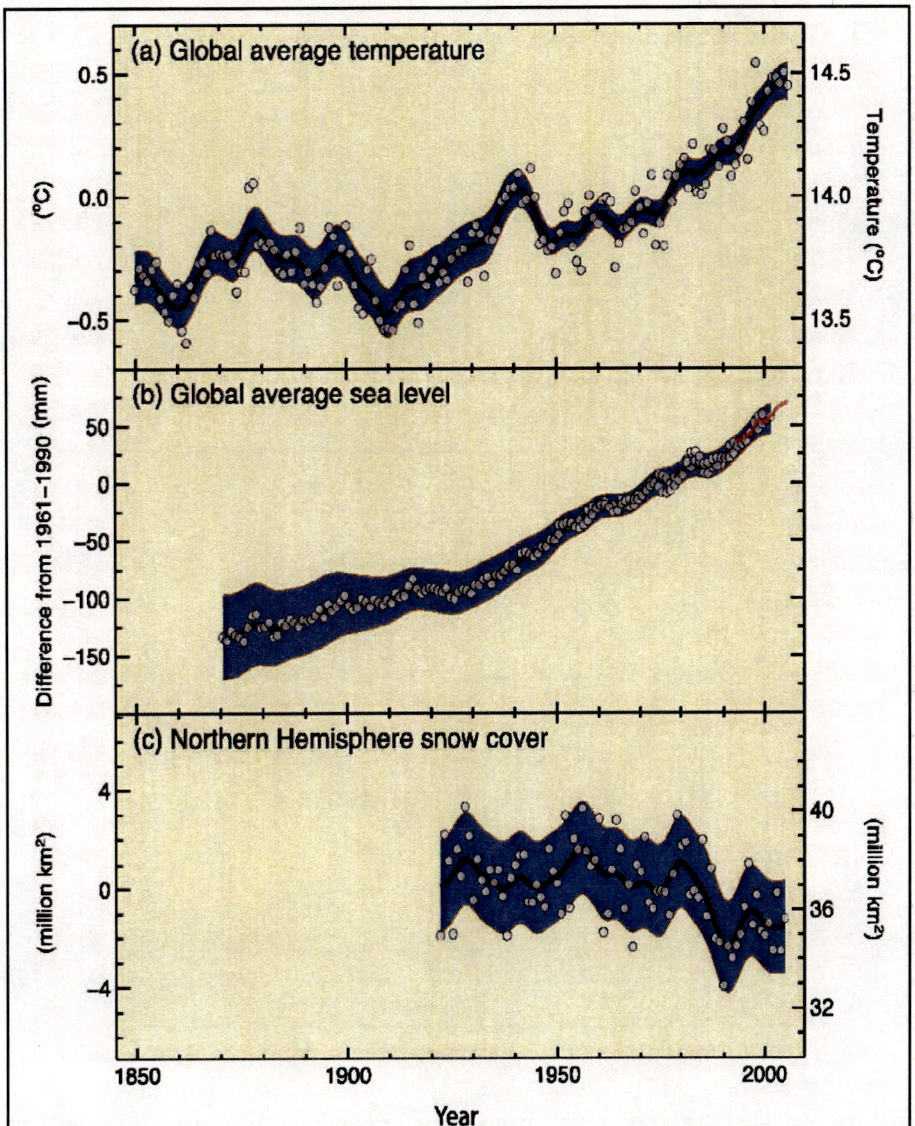

Figure 1.3: ***Observed changes in global surface temperature, sea level and snow cover*** *[IPCC 2007a].*

1.7: Projections for future climate changes

The climate change model projected the following changes:

1.7.1: Projections for temperatures, sea level, precipitation, glaciers and cyclones [based on IPCC 2001; IPCC 2007a, 2007b, 2007c].

- *A global warming of about 0.2°C per decade is predicted, of which, warming is expected to be greatest over land and most at northern high latitudes, and least over the southern oceans and parts of the North Atlantic Ocean.*
- *The global average warming could be in the range from 1.4° C to 5.8° C by 2100 (see Figure 1.4).*
- *The global average sea level rise could be in the range of 0.18-0.38 m (low GHG emissions scenario) and 0.26-0.59 m (high GHG emissions scenario).*
- *A doubling of global atmospheric CO_2 may occur by the end of 2100 (compared to pre-industrial levels of 280 ppm) (Figure 1.4).*

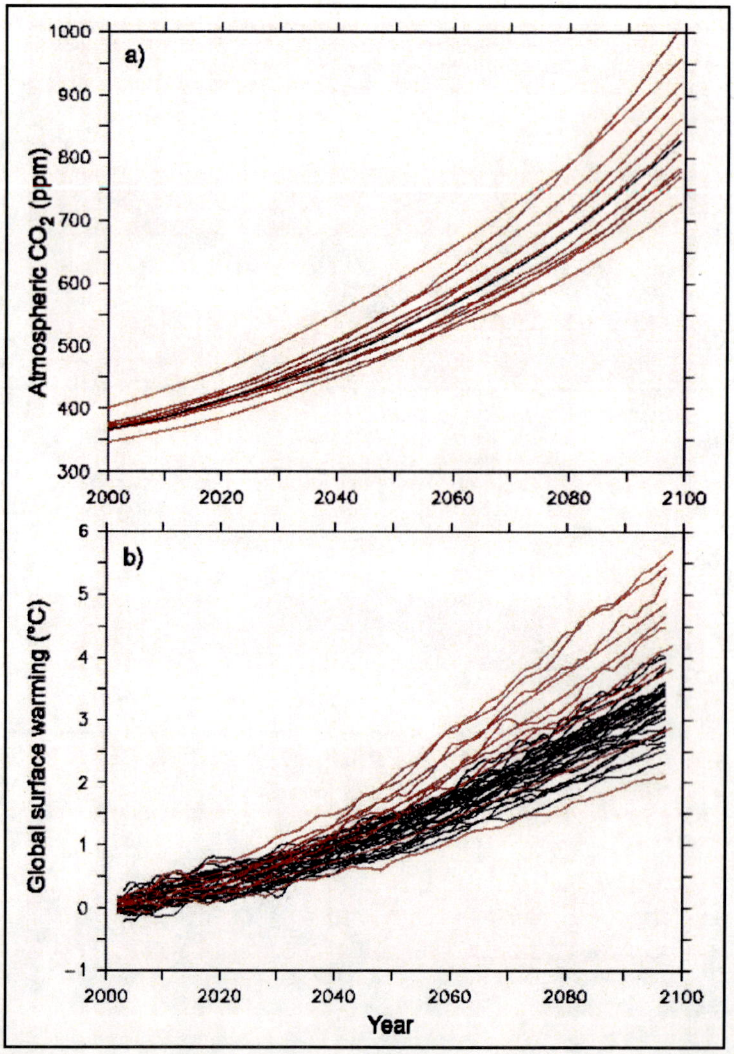

Figure 1.4: ***Climate model predications for global temperature change relative to atmospheric CO_2 concentrations from 2000 to 2100.***

a. atmospheric CO_2 concentration change; b. global average surface temperature change; Red line = simulated models and black line = various IPCC models *[See Meehl et al. 2007, page 790, figure 10.20 for further details. http://ipcc-wgi.vcar.edu/wg1/wg1-report.html].*

- *The snow cover is expected to contract and contraction of the Greenland ice sheet is projected to continue to contribute sea level rise after 2100.*
- *The sea ice is projected to shrink in both the Arctic and Antarctic regions.*
- *Hot extremes, heat waves and heavy precipitation events will continue to become more frequent (see Table 1.4 for future projections).*
- *The amount of precipitation are very likely to increase in high altitudes and decrease in most sub-tropical land regions (Table 1.4).*
- *Future tropical cyclones (typhoons and hurricanes) will become more intense with large peak wind speeds and heavier precipitation.*

Table 1.4: ***Projections of extreme weather and climate events*** *[IPCC 2007b].*

Phenomenon and direction of trend	Likelihood of trend occurred in the 20th century (post 1960)	Likelihood of future trend based on projections for 21st century
Warmer and fewer cold days and nights over most land areas	*Very likely*	*Virtually certain*
Warmer and more frequent hot days and nights over most land areas	*Very likely*	*Virtually certain*
Warm spells/heat waves: frequency increases over most land areas	*Likely*	*Very likely*
Heavy precipitation events, frequency increases over most areas	*Likely*	*Very like*
Areas affected by droughts increases	*Likely in many regions since 1970s*	*Likely*
Intense tropical cyclone activity increases	*Likely in some regions since 1970s*	*Likely*
Increased incidence of extreme high sea level	*Likely*	*Likely*

Virtually certain = > 99% probability of occurrence; Very likely = >90% probability; Likely = >66% probability.

1.7.2: *Other projections*

According to Australian Government (2005), Department of Climate Change [*http://www.climatechange.gov.au/impacts/settlements.html*], the future increases in temperatures may cause an increase in peak demand for electricity in summer. Extreme temperatures are likely to have impacts on the production and transmission of energy. For example, higher temperatures are likely to affect the transmission efficiency of power lines. Moreover, higher water temperatures (combined with reduced water availability) will result in decreased cooling capacity for thermo-electric power generations. Extreme temperature events (heat waves) could cause disruptions to transport - through damage to transport infrastructure (e.g. road pavement and rails). There is likely that maintenance costs of transport infrastructure will increase since materials are needed to be replaced more often. The decrease in the longevity of exterior materials of buildings and infrastructure will lead to increased maintenance and replacement costs.

1.8: Possible measures to reduce greenhouse gas emissions

Industries are the main energy consumer and CO_2 emitter. Improving energy efficiency in industries such as paper, pulp, cement, ammonia, ethylene, iron and steel, other chemical industries (petro-chemical, pharmaceuticals) could bring benefit to global environment and help to reduce energy consumption. Many countries have introduced and accepted emissions trading and emissions reduction targets as a policy measure to decrease GHG emissions by the 2008-2012 time-frames (see Table 1.5). Improving energy efficiency through proper energy efficiency policy could significantly make energy savings and reduce CO_2 emissions, for example, mandatory utilization of energy saving bulbs in many countries is a good example of energy saving. Other measures that can be implemented to reduce GHG emissions are discussed below (section 1.8.1 to 1.8.4)

Table 1.5: ***Emission reduction targets for some selected countries.*** *[http://www.climatechange.gov.au/whitepaperfactsheets/pubs/fact-sheets-all.pdf].*

Country	2020 targets	2020 per capita reduction	2050 targets
Australia	5-15% below 2000 levels (4-14% below 1990 levels)	7-34% below 2000 levels (34-41% below 1990 levels)	60% reduction below 2000 levels (60% below 1990 levels)
European Union	20-30% below 1990 levels	24-34% below 1990 levels	60-80% below 1990 levels
United Kingdom	26-32% below 1990 levels	33-39% below 1990 levels	80% below 1990 levels
USA (proposal, president- elect Obama)	Return to 1990 levels	25% below 1990 levels	80% below 1990 levels

1.8.1: Power generation options

Energy sectors are the main contributors of greenhouse gas emissions. For example, in 2005 Australia's net greenhouse gas emissions were 522.2 million tonnes (Mt) CO_2 - equivalent, of which the energy sectors comprised 74.9% (391.0 Mt CO_2 - equivalent) followed by agriculture (16.8%) (see Table 1.6). The world emissions of GHGs also show that energy sector (electricity, transport) is the highest GHG emitter equivalent to 39% (see Table 1.7).

Table: 1.6. ***Australia's main greenhouse gas emissions by sectors*** *[Australian Government 2005].*
Mt = Million tonnes; CO_2-e = Carbon dioxide-equivalent; NA = not available

Green house gases	CO_2	CH_4	N_2O	CO_2 -e
1. All Energy (combustion + fugitive)				
Stationary	277.2 Mt (73.5%)	0.1 Mt (1.1%)	0.003 Mt (4.0%)	279.4 Mt (53.5%)
Transport	78.3 Mt (20.8%)	0.03 Mt (0.5%)	0.005 (6.2%)	80.4 Mt (15.4%)
Fugitive emissions from fuel	5.6 Mt (1.5%)	1.2 Mt (22.5%)	0.0001 (0.1%)	31.2 Mt (6.0%)
2. Agriculture	NA	3.2 Mt (58.9%)	0.07 Mt (84.2%)	87.9 Mt (16.8%)
3. Industrial processes	23 Mt (6.1%)	0.003 Mt (0.1%)	0.0001 Mt (0.1%)	29.5 Mt (5.6%)
4. Waste	0.03 Mt (0.01%)	0.8 Mt (14.4%)	0.002 Mt (2.3%)	17 Mt (3.3%)
5. Land use, land use change and forestry	- 6.8 Mt (-1.8%)	0.1 Mt (2.5%)	0.002 Mt (3.1%)	-3.2 Mt (-0.6%)
Total net emissions	**377.4 Mt**	**5.4 Mt**	**0.08 Mt**	**522.2 Mt**

Using more environmentally friendly energy source such as renewable energy could reduce significant amount of GHG emissions to atmosphere. These renewable energy sources are wind power, hydro, solar and biofuels which do not produce many GHGs and other pollutants as fossil fuels such coal, oil or diesel does (see Table 1.8).

Table: 1.7: ***Energy consumption, power generation and greenhouse emissions in Australia and in the world*** *[source The Age, 4 July 2008, page 4 insight].*

I: Energy consumption	Sector	Percent		Sector	Percent
Australia	Oil/petroleum	35.2%	World	Oil/petroleum	35.0%
	Black coal	29.0%		Coal	25.3%
	Brown coal	12.4%		Gas	20.7%
	Gas	18.6%		Renewables/waste	10.0%
	Renewables	4.8%		Nuclear	6.3%
				Hydro	2.2%
				other	0.5%
II: Power generation	**Sector**	**Percent**		**Sector**	**Percent**
Australia	Black coal	54.5%	World	Coal	40.3%
	Brown coal	21.1%		Gas	19.7%
	Gas	15.0%		Hydro	16.0%
	Hydro	6.4%		Nuclear	15.2%
	Oil	1.8%		Oil	6.6%
	Other (wind, biomass etc.)	1.2%		Other (wind, biomass etc.	2.2%
III: Emissions	**Sector**	**Percent**		**Sector**	**Percent**
(CO_2-equivalent) Australia	Electricity generation	34.4%	World	Electricity generation	25.9%
	Agriculture	15.6%		Industry	19.4%
	Other stationary energy	15.5%		Land use (forestry clearing)	17.4%
	Transport	13.7%		Agriculture	13.5%
	Land use	6.9%		Transport	13.1%
	Industrial processes	4.9%		Waste	2.8%
	Waste	2.9%		Other	7.9%
	Other	6.1%			

Table 1.8: ***Different electricity options with respect to potential environmental impacts*** *[Greenhouse emissions from main power generation options (kt eq CO_2/TWH) in North America for existing technology; www.hydroquebec.com/environment/climate].*

Electricity generation technology	Greenhouse emissions	Greenhouse emissions of energy options (kt eq CO_2/ TWH)	Air pollutants	Waste	Other comments
Bituminous coal/ conventional coal	High	974	High	Yes	Moderate or high cooling water demand
'Clean coal' with CO_2	Medium		Low	Yes	Increased coal consumption per MWH
Heavy oil	High	778	High	Yes	Moderate cooling water demand
Diesel	High	778		Yes	Moderate cooling water demand
Natural gas	Medium	511	Low	No	Moderate cooling water demand
Biomass plantation/ bio-energy	Low	118	Low	Yes	Fertilizers for energy crops; burning biomass creates pollution
Nuclear	None	15	None	Radioactive	High cooling water demand
Hydro reservoir	Low	15	None	No	Can impair fish migration; flooding; inexpensive
Solar-photovoltaic	None	13	None	No	High energy consumption during manufacture; requires large space
Wind power	None	9	None	No	Bird/bat kills; inexpensive
Hydro run-of-river	None	1	None	No	Can interfere with recreational activity
Geothermal power	Low		None	Yes	Odour
Tidal current power	None		None	No	Other impacts unknown
Wave power	None		None	No	Other impacts unknown

The merits of each kind of renewable energy are summarised below:

- **Wind power** : This involves using the energy of wind to turn blades of windmills to produce electricity; it produces no GHGs during operation. Areas where winds are stronger and more constant, such as offshore and high altitudes are preferred locations for windmill farms.
- **Water or hydro power** : Hydroelectricity generation involves using flowing water to turn turbines to generate electricity. Water is about thousand times denser than air and a slow flowing stream of water, or moderate sea swell can yield considerable amount of energy. Hydroelectricity is an example of water power.
- **Solar energy** : This involves using solar cells (photovoltaic cells or concentrated solar power) to convert the sun rays into electricity. It is regarded as being one of the cleanest forms of energy available. Solar energy can be used to heat buildings through passive solar design of houses (called environmentally friendly house or eco-house), heating food stuffs through solar ovens, heat water or air for domestic hot water and heat and cool air through use of solar chimneys.
- **Biofuels/biomass generation** : It involves generation of electricity from organic sources such as plant material or methane gas from rubbish tips. Agriculturally produced biomass fuels, such as biodiesel, ethanol and bagasse (a by-product of sugar cane cultivation) can be burned in internal combustion engines or boilers. Liquid biofuel or bioalcohol (e.g. ethanol or bio oil such as biodiesel or vegetable oil) can be used in modern diesel vehicles.

1.8.2: Clean coal technology

Coal is the most abundant and widely distributed fossil fuel source in the world. In Australia alone, burning coal produces about 9 billion tonnes of carbon-dioxide each year, of which 70% from the power generation. New 'clean coal' technologies can be utilised which could reduce amount of GHGs released during burning of coals. Clean coal technologies include washing of coals which reduces emissions of ash and sulphur dioxide when coal is burned. Adding of electrostatic precipitators and fabric filters can remove 99% of the fly ash from the flue gases (flue gas desulfurisation reduces the output of sulphur dioxide to the atmosphere by upto 97%).

1.8.3: Storage and sequestration of carbon dioxide

Carbon dioxide can be pumped deep underground and stored in geological formations such as empty gas reservoirs. It may also be possible to collect carbon dioxide from power stations and other major sources and pump it directly into the deep ocean as ocean storage. In this process CO_2 is injected deep into the ocean, where major portion of it dissolves (undergoes rapid biochemical reactions with the water) and small portion is deposited on the ocean floor as 'lake of dense CO_2'. However, such sequestration process may turn ocean waters more acidic (Galib 2008) and thereby may impact on oceanic biota (see chapter 4 on impacts of higher CO_2 on corals and other calcareous organisms). In fact, ocean plays an important role in regulating or absorbing the amount of CO_2 in the atmosphere. For example, more than 80% of anthropogenic CO_2 emissions are absorbed naturally. This is a slow process of exchange between the atmosphere and the ocean's surface water. CO_2 entering the deep waters of the oceans are removed from interaction with the atmosphere for periods of hundreds of years.

1.8.4: Biomass as a sink

Replanting or generation of new forests could reduce the amount of carbon-dioxide since trees are a carbon sink as it absorb carbon-dioxide, incorporating carbon into wood and leaves. For example, one hectare of new forest can absorb about 25 tonnes of carbon-dioxide per year; however, trees and other plants are regarded as a carbon sink only when they are growing. Phytoplankton and aquatic plants in oceans are generally absorbing 20 to 40% of global carbon-dioxide emissions. Whether addition of nutrients in ocean can accelerate the carbon-dioxide absorption rate (through marine biological production) needs further investigations. A list of available technologies to reduce GHG emissions are listed in Table 1.9.

Table 1.9 : *Key mitigation technologies available for reduction of Greenhouse gas emissions from different sectors* [IPCC 2007c].

Sector	Key mitigation technologies and practices currently commercially available
Energy supply	-Improved supply and distribution efficiency -Fuel switching from coal to gas -Nuclear power -Renewable heat and power (hydropower, solar, wind, geothermal and bioenergy) -Combined heat and power -Early applications of carbon capture and storage
Transport	-More fuel efficient vehicles -Hybrid vehicles -Cleaner diesel vehicles -Compressed natural gas (CNG) run vehicles -Biofuels -Shifts from road transport to rail and public transport systems -Non-motorised transport (cycling, walking)
Buildings	-Efficient lighting and day lighting -More efficient electrical appliances, heating and cooling devices -Improved cooked stoves -Improved insulation -Passive and active solar design for heating and cooling -Alternative refrigeration fluids, recovery and recycle of fluorinated gases
Industry	-More efficient end-use electrical equipment -Heat and power recovery -Material cycling and substitution -Control of non-CO_2 gas emissions -A wide array of process-specific technologies
Agriculture	-Improved crop and grazing land management to increase soil carbon storage -Restoration of cultivated peat soils and degraded lands -Improved rice cultivation techniques and livestock and manure management to reduce CH_4 emissions -Improved nitrogen fertilizer application techniques to reduce N_2O emissions -Dedicated energy crops to replace fossil fuel use -Improved energy efficiency
Forestry/forests	-Afforestation -Reforestation -Forest management -Reduced deforestation -Harvested wood product management -Use of forestry products for bioenergy to replace fossil fuel use
Waste management	-Landfill methane recovery -Waste incineration with energy recovery -Composting of organic waste -Controlled wastewater treatment -Recycling and waste minimization

1.9: Conclusion

Climate change is one of the greatest economic, social, and environmental challenges. Climate change is the result of changes in our weather patterns because of an increase in the Earth's average temperature caused by increases in greenhouse gases in the atmosphere. The principal GHG concentrations that have increased over the industrial period are carbon dioxide (CO_2), methane (CH_4), nitrous oxide (N_2O), and chlorofluorocarbons (CFCs). Changes in the climate system that are confidently predicted by IPCC in response to increases in GHGs include increases in mean land and sea temperature, rising sea level and melting of snows and glaciers. The impacts of climate change can be negative or positive or intermediate. Chapters 2 to 5 discusses the environmental and biological impacts of projected climate change on different sectors as listed below:

- ***Chapter 2: Climate change impacts on worlds' freshwater resources including Asia-Pacific and Australasia*** (vulnerability of water resources, projections of future impacts on freshwater resources, impacts of snows and glaciers melting, impacts of sea level rise, climate change impacts on shared water resources, case studies with Asia and Australia and adaptation measures).
- ***Chapter 3: Climate change impacts on worlds' agriculture and livestock including Asia-Pacific and Australasia*** (factors relating to future agricultural production including effects due to CO_2, higher temperatures, sea level rise, pests and diseases and case studies with Asia and Australasia and adaptation measures).
- ***Chapter 4: Climate change impacts on world's fisheries and aquatic ecosystems including Asia Pacific and Australasia*** (effects on freshwater and marine ecosystems including rising temperatures, CO_2 concentrations and sea levels, case studies with Asia and Australia and adaptation measures).
- ***Chapter 5: Climate change impacts on global human health including Asia-Pacific and Australasia*** (recent observations of climate change related health effects, extreme weather events and future projections, vector borne diseases, other climate change related health effects, case studies with Asia and Australia and adaptation measures).

1.10: References

Australian Government 2005. *Australia's National Greenhouse Accounts*. The Australia's Government submission to the UN Framework convention on climate change April 2007. National Inventory Report 2005-volume 1. 192p.

Galib, M. 2008. Can the ocean slow global warming? The Daily Star, Bangladesh. Sub-editorial. 08 February, 2008.

Huq, M. 2008. Japan aspires to become a low carbon society. The Daily Star, Bangladesh. Sub-editorial, 08 February, 2008.

IPCC 2001. Intergovernmental Panel on Climate Change, Working Group II. 2001. *Climate change 2001. Impacts, Adaptation and Vulnerability*. IPCC, Third Assessment report, Cambridge, UK: Cambridge University Press

IPCC 2007a. Summary for the policy makers. In: *Climate change 2007. The physical science basis.* Contributing of working group I to the Fourth Assessment Report of the Intergovernmental Panel on Climate Change. Solomon, S., Qin, D., Manning, M., Chen, Z., Marquis, M., Averyt, K. B., Tignor, M and Miller, H. L. (eds.). Cambridge University Press, Cambridge, United Kingdom and New York, USA. 996 pp.

IPCC 2007b. *Climate change 2007: Impacts, adaptation and vulnerability*. Contribution of working group II to the Fourth Assessment Report of the Intergovernmental Panel on Climate Change. Parry, M. L., Canziani, O. F., Palutikop, J. P., van der Linden, P. J., and Hanson, C. E. (eds.). Cambridge University Press, Cambridge, United Kingdom and New York, USA, 976 pp. (http://www.ipcc.ch)

IPCC 2007c. Summary for policy maker. In: Climate change 2007: Mitigation. Contribution of working group III to the Fourth Assessment Report of the Intergovernmental Panel on Climate Change. Metz, B., Davidson, O. R., Bosch, P. R., Dave R., Meyer L. A (eds.). Page 10, Table SPM 3. Cambridge University Press, Cambridge, United Kingdom and New York, USA.

Meehl, G. A., Stocker, T. F., Collins, W. D., Friedlingstein, P., Gaye, A. T., Gregory, J. M., Kitoh, A., Knutti, R., Murphy, J. M., Noda, A., Raper, S. C. B., Watterson, I. G., Weaver, A. J and Zhao, Z-C. 2007. Global climate change projections. Chapter 10 (747-845). In: *Climate change 2007: The physical science basis. Contribution of working group I to the fourth assessment report of the intergovernmental panel on climate change* [Solomon, S., D. Qin., M. Manning, Z. Chen., M. Marquis, K. B. Averyt., M. Tignor and H. L. Miller. (eds). Cambridge University Press, Cambridge, United Kingdom and New York, USA.

Websites

www.hydroquebec.com/environment/climate
http://www.canelect.ca/en/pdf/handbook.pdf
http://hydroquebec.com/enviornment/climate
http://www.grida.no/climate/vital/20.htm (potential impacts of climate change)
http://www.climatechange.gov.au/whitepaper/factsheets/pubs/fact-sheets-all.pdf

2 Climate Change Impacts on Freshwater Resources

Key facts

- Though water is the most widely occurring substance on earth, freshwater accounts for only 2.5% whereas the rest of water is saltwater. The agriculture sector uses more than 75% of freshwater available.
- The current observed climate change impacts on water resources include increased precipitation at high latitudes and temperate areas, decreasing precipitation at low latitudes; decreasing snow covers, and glaciers in most regions; decreasing flows in some river basins; floods and droughts in some countries and significant decreases in lake and river size in Africa.
- A number of climate models projected that a doubling of atmospheric concentrations of CO_2 would increase global precipitation by at least 5%. The precipitation is expected to increase in some areas, such as, at high latitudes and in the tropics (south east monsoon regions and over the tropical Pacific), whereas the precipitation is likely to decrease in the sub-tropics (e.g. much of North Africa and northern Sahara).
- The future impacts on freshwater resources include an increase in flooded areas in Bangladesh, a decrease in freshwater resources in small islands, a decrease in stream flows in the USA and a decline in hydropower stations in North Europe.
- Where intense rainfall is expected to increase, pollutants such as pesticides, heavy metals, fertilizers and organic matters will be increasingly washed from soils to water bodies causing water quality problems.
- Melting of Himalayan glaciers may cause major changes in freshwater flow regimes and would have a dramatic impact on drinking water supplies, biodiversity, hydropower, industry, agriculture and others. About 2.4 billion people in Asia depend on the summer melt water of the Himalayan glacier, therefore, rivers depending on Himalayas' glaciers such as the Indus, Ganges, Brahmmaputra, Syr Darya, Amu Darya could face water shortages or river run-off.
- The rise of sea level would have influence on freshwater resources, water quality of shallow groundwater and surface water could thus be affected due to salt water intrusion in Bangladesh, Mekong and Nile deltas for example. It is also projected that more than 25% of the Netherlands drinking water facilities (about 200 drinking water facilities) are estimated to be threatened by salt water intrusion from coastal areas.
- Climate change would likely to lead disputes and tensions among the water shared countries.
- In case of Australia, it has been projected that most of southern and eastern Australia (major food growing regions) will have reduced inflows to water storages, reduced stream flows in major catchments, reduced water availability for rain-fed agriculture and reduced recharge of groundwater. The occurrence and severity of drought events are projected to increase in north-eastern and Southern Australia.
- Water savings program, reuse and recycling of water, research to develop more efficient irrigation technology and preventing or minimising water pollution are some of the measures to augment future demand for freshwater resources.

Contents

List of Tables

List of Figures

Acronyms and Glossary

Adaptation: Adjustment in natural or human systems in response to actual or expected climatic stimuli or their effects, which moderates harm or exploits beneficial opportunities.

Afforestation: Direct human induced conversion of land that has not been forested for a period of at least 50 years to forested land through, planting, seeding and /or the human-induced promotion of natural seed sources.

Avalanches: An avalanche is a flow of snow down a mountainside. Avalanches are among the biggest dangers in the mountains for both life and property.

Boreal: Located or growing in Northern or mountainous regions.

Cryosphere: The component of the climate system consisting of all snow, ice and permafrost on and beneath the surface of the earth and ocean.

Evapotranspiration: It is the process in which water is drawn from the soil into plant roots, and then is evaporated or transpired from leaves and other plant surfaces into the air.

Glaciers: A large body of ice; large masses of ice that is formed by the compaction and recrystallization of snow under freezing conditions; glaciers often move down slope or outward in all directions because of the stress of their own weight; they may be stagnant or retreating under warming conditions.

Moraine: An accumulation of earth and stones deposited by a glacier.

Mushing: Mushing is a general term for a sport or transport method powered by dogs, and includes carting, pulka, scootering, sled dog racing, skijoring, freighting, and weight pulling. More specifically, it implies the use of one or more dogs to pull a sled on snow. The term is thought to come from the French word *marche*, or go, run, the command to the team to commence pulling.

Permafrost: Perennially frozen ground that occurs wherever the temperature remains below 0^0C for several years.

Sahel: The Sahel is the boundary zone in Africa between the Sahara to the north and the more fertile region to the south, known as the Sudan. The Sahel runs 3862 km from the Atlantic Ocean in the west to the Red Sea in the east, in a belt that varies from several hundred to a thousand kilometers (620 miles) in width, covering an area of 3,053,200 square kilometers. It is a transitional ecoregion of semi-arid grasslands, savannas, steppes, and thorn shrublands lying between the wooded Sudanian savanna to the south and the Sahara to the north. The countries of the Sahel today include Senegal, Mauritania, Mali, Burkina Faso, Niger, Nigeria, Chad, Sudan, and Eritrea.Over the history of Africa the region has been home to some of the most advanced kingdoms benefiting from trade across the desert. Collectively these states are known as the Sahelian kingdoms.

Sea level rise: An increase in the mean level of the ocean.

Sink: Any process, activity, or mechanism that removes greenhouse gas, an aerosol, or a precursor of a greenhouse gas (Chapter 1).

Snow-pack: A seasonal accumulation of slow-melting snow.

Snowmobile: A snowmobile, or a snowmachine, is a land vehicle for travel on snow that is commonly propelled by a continuous track or tracks at the rear and steered by skis at the front.

Snowshoes: Snowshoes, sometimes colloquially referred to as webs are footwear for walking over snow.

Stream flow: Water flow within a river channel.

Rangeland: Unmanaged grassland, shrub lands, savannas, tundra.

2.1: Introduction

Freshwater can be defined as water that has a dissolved salt content of less than 0.5 parts per thousand or 0.5 ppt (seawater has dissolved salt content of 35 ppt). Though water is the most widely occurring substance on earth, freshwater accounts for only 2.5% whereas, the rest of water is saltwater. Freshwater is the most precious resource on the earth as it is vital for all living organisms and major ecosystems (about 9,000-25,000 plants and animal species depend on freshwater ecosystems).

Freshwater is used for human consumption (drinking), cooking, cleaning of clothes and in home gardens. It is used to grow agriculture food, livestock and fish farming. The other use is in industry (as a means of removing wastes and by-products), in hydroelectric power generations, fire fighting, transportation and healthy functioning of nature. Freshwater is also required for many recreational activities including fishing and swimming. It has also some aesthetic functions. The agriculture sector is by far the biggest user of freshwater in Asia, Africa, Middle East, South America and Oceania (where more than 75% of freshwater is used for irrigation), whereas, in Europe and North America it is mostly used in industry (Table 2.1).

Table 2.1: ***World's freshwater resources and its usage (selected countries)*** *[Source: Original from FAO 2007 - http://www.fao.org/waicent/faoinfo/agrcultut/agl/w/aquastat/water_res/index.htm and cited in Earth Trends data tables: http:///earthtrends.wri.org/searchable__db/; a = dependency ratio is the percentage of total renewable water resources originating outside the country; b = the average daily industrial discharge of organic water pollutants in metric tons between 1997 and 2000].*

Country	Actual renewable water resources			Water withdrawals					Industrial water pollution (mt/day) (b)
						Sectoral Withdrawals (percent), 2000			
	Total (km^3)	Per capita (m^3 per person)	Dependency ratio (a)	Total (million m^3 2000)	Per capita (m^3 per person 2000)	Agriculture	Industry	Domestic	
Asia									
Afghanistan	65	2,608	15	23,261	1,087	98	0	2	-
Bangladesh	1,211	8,089	91	79,394	576	96	1	3	273
Bhutan	95	40,860	0	420	204	95	1	4	-
Cambodia	476	32,876	75	4,091	311	98	1	2	-
China	2,830	2,206	1	630,289	494	68	26	7	7,060
India	1,897	1,754	34	645,837	635	86	5	8	1,605
Indonesia	2,838	12,749	0	82,773	391	91	1	8	662
Japan	430	3,365	0	88,432	696	62	18	20	1,379
Korea, Rep	70	1,454	7	18,590	397	48	16	36	299
Malaysia	580	23,316	0	9,016	392	62	21	17	154
Myanmar/ Burma	1,046	20,870	16	33,224	699	98	1	1	4
Nepal	210	8,171	6	10,177	433	96	1	3	-
Pakistan	223	1,415	76	169,384	1,187	96	2	2	-
Philippines	479	5,884	0	28,520	377	74	9	17	202
Sri Lanka	50	2,602	0	12,604	678	92	5	4	-
Thailand	410	6,459	49	87,065	1,429	95	2	2	-

Contd...

Turkey	229	3,171	1	37,519	550	74	11	15	175
Vietnam	891	10,085	59	71,392	914	68	24	8	-
Europe									
Bulgaria	21	2,721	1	10,498	1,296	19	78	3	106
Czech Rep	13	1,286	0	2,566	250	2	57	41	-
Denmark	6	1,116	0	1,267	238	42	26	32	84
France	204	3,3711	12	39,959	674	10	74	16	280
Germany	154	1,866	31	47,052	572	20	68	12	788
Greece	74	6,764	22	7,759	712	81	3	16	57
Hungary	104	10,579	94	7,641	763	32	59	9	147
Iceland	170	582,192	0	153	543	0	66	34	-
Ireland	52	13,003	6	,129	296	0	77	3	42
Italy	191	3,336	5	44,372	771	45	37	18	495
Norway	382	83,919	0	2,185	489	10	67	23	427
Portugal	69	6,821	45	1,263	1,125	78	12	10	124
Spain	112	2,711	0	35,635	874	68	19	13	357
Sweden	174	19,581	2	2,965	335	9	54	37	101
Switzerland	54	7,468	24	2,571	359	2	74	24	-
United Kingdom	147	2,474	1	9,541	163	3	75	22	583
Middle East and Africa									
Angola	184	13,070	0	343	28	61	16	22	-
Botswana	14	8,022	80	140	81	43	19	38	-
Central African Rep	144	36,912	2	22	6	4	19	77	-
Chad	43	4,857	65	234	30	80	1	19	-
Congo	832	217,915	73	39	11	10	30	59	-
Cote de 'Ivory	81	4,794	5	931	59	65	12	23	12
Egypt	58	794	97	68,653	1,013	78	14	8	202
Ethiopia	110	1,519	0	2,648	40	93	6	1	21
Gambia	164	121,392	0	128	102	40	11	48	
Ghana	53	2,489	43	520	27	48	15	37	-
Guinea	31	20,156	48	110	81	91	1	9	-
Guinea Bissau	31	20,516	48	110	81	91	1	9	-
Iran	138	1,970	7	72,877	1,097	91	2	7	-
Iraq	75	2,917	53	42,702	1,839	92	5	3	-
Israel	2	255	55	2,041	338	63	7	31	-
Jordan	1	157	23	1,016	202	75	4	21	16
Kenya	30	932	33	1,576	52	64	6	30	51
Lebanon	4	1,189	1	1,372	394	67	1	33	15
Lesotho	3	1,678	0	54	30	19	41	40	3
Libya	1	106	0	4,811	919	89	3	8	-
Madagascar	337	18,826	0	14,970	937	96	2	3	-

Contd...

Malawi	17	1,401	7	1,005	88	81	5	15	-
Morocco	29	934	0	12,758	438	90	2	8	88
Mozambique	11	3,826	96	1,689	642	88	3	9	-
Namibia	18	8,921	66	268	142	63	5	33	-
Niger	34	2,710	90	2,187	204	95	1	4	-
Nigeria	286	2,252	23	8,004	70	69	10	21	-
Oman	1	337	0	1,350	518	91	2	7	5
Rwanda	5	613	0	76	10	39	14	48	-
Saudi Arabia	2	96	0	17,320	782	89	1	10	-
Senegal	39	3,811	33	1,591	169	90	4	6	8
Somalia	160	30,960	0	380	86	93	2	5	-
South Africa	50	1,106	10	15,306	348	73	10	17	235
Sudan	65	1,879	77	37,314	1,187	97	1	3	-
Syria	26	1,441	80	19,947	1,205	95	2	3	15
Tanzania	91	2,416	10	1,996	57	93	1	6	34
Yemen	4	198	0	6,631	368	95	1	4	-
Zambia	20	1,547	30	2,612	207	86	5	10	-
Zimbabwe	20	1,547	30	2,612	207	86	5	10	-
North America									
Canada	2,902	91,419	2	45,974	1,494	12	69	20	300
United States	3,069	10,333	8	479,293	1,682	41	46	13	2,433
Central, South America and Caribbean									
Argentina	814	20,914	66	29,072	764	74	9	16	-
Brazil	8,233	45,573	34	59,298	345	62	18	20	-
Chile	922	57,639	4	12,539	824	64	25	11	71
Costa Rica	112	26,447	0	2,677	681	53	17	29	34
Cuba	383,365	0	8,204	732	69	12	19	-	
Ecuador	432	32,747	0	16,980	1,367	82	5	12	83
Mexico	457	4,357	11	78,219	791	77	5	17	291
Panama	148	46,579	0	824	279	28	5	66	12
Uruguay	139	40,419	58	3,146	941	96	1	2	21
Oceania									
Australia	492	24,708	0	23,932	1,250	75	10	15	94
Fiji	29	33,707	0	69	85	78	11	11	6
New Zealand	327	83,760	0	2,111	558	42	9	49	48

World's Freshwater resources is under increasing pressure mainly due to rising of human population, growing concentrations of human habitation in urban areas, demands for more food production. The requirements for protection of freshwater ecosystems are highlighted below:

- Population growth: an increase in world's population means greater demand for water.
- Population concentration: population is becoming increasingly concentrated in large cities where water use is different (generally much more) than in a rural environment.
- Expansion of irrigation: future increase of world's population will lead to more water usage by agriculture.

- Environmental requirements: increasing demands for biodiversity and environment protection will put additional pressure on water resource use.

The further pressure on this finite resource is due to the effects of climate change, which may cause a change in world's rainfall patterns and possible intrusion of saline seawater into inland freshwater. The impacts of climate change will bring varying consequences for the availability of the freshwater around the world (see Table 2.2).

2.2: Vulnerability of water resources

Climate change will increase water system dynamics through temperature changes, changes in precipitation patterns (more extreme precipitation events), evaporation and water quality and water storage in ice packs. Water system dependent economical stakeholders, such as drinking water companies have to cope with consequences of climate change, e.g. floods and water storages in river systems, upcoming of brackish groundwater, saltwater intrusion, increasing peak demands and microbiological activity due to temperature rise (Meuleman *et al.* 2007). The likely impacts of climate change on water resources are given in Table 2.2.

Table 2.2: ***Likely impacts of climate change on water resources*** *[IPCC 2007a].*

Likelihood	Effects on water resources
Virtually certain (warmer and fewer cold days and nights; more frequent hot days)	• effects on water resources relying on snow melt • effects on some water supply
Very likely (Warm spells/heat waves)	• increased water demand • increased water quality problems (e.g. algal blooms)
Very likely (increases of heavy precipitation events)	• adverse effects on surface and ground water quality • contamination of water supply with chemicals and pathogens
Likely (increases of drought events)	• wide spread water stress
Likely (increases of tropical cyclone)	• disruption of public water supply due to power outages
Likely (sea level rise)	• decreased freshwater availability due to saltwater intrusion

2.3: Observed trends on water resources

The current observed climate change related impacts on world's water resources are summarised below (see also Table 2.3 A and 2.3 B):

- increased precipitation at high latitudes and temperate areas and decreasing trends at low latitudes.
- decreasing snow covers, and glaciers in most regions.
- earlier spring flows in North America and Europe.
- decreasing flows in some river basins.
- floods and droughts in some countries (in Bangladesh, for example, three extreme floods have occurred in the last two decades and in recent times, more intense droughts have been reported from Australia, semi-arid and semi-humid regions, western USA and southern Canada).
- significant decreases in lake and river size in Africa and China (declining precipitation has been linked for the shrinking of Lake Chad in Africa and Huanghe river in China) .

Table 2.3: ***Climate related observed trends of various components of the global freshwater system (2.3.A) and examples of current vulnerability of world's freshwater resources (2.3.B)*** *[based on Kundzewicz et al. 2007].*

2.3.A: Climate-related observed trends of worlds freshwater resources	
Components	**Climate-related trends**
Precipitation	- Increasing over land north of 30^{0}N over the period 1901-2005. - Decreasing over land between 10^{0}S and 30^{0}N after 1970. - Increasing intensity of precipitation.
Cryosphere	
Snow cover	- Decreasing in most regions, especially in spring.
Glacier	- Decreasing almost everywhere.
Permafrost	- Thawing between 0.02 m/yr (Alaska) and 0.4 m/yr (Tibetan plateau).
Surface waters	
Stream flow	- Increasing in Eurasia Arctic, significant increases or decreases in some river basins. - Earlier spring peak flows and increased winter base flows in Northern America and Eurasia.
Evapotranspiration	- Increased actual evapotranspiration in some areas.
Lakes	- Warming, significant increases or decreases of some lake levels and reduction in ice cover.
Groundwater	- No evidence for ubiquitous climate-related trend.
Floods and droughts	
Flood	- No evidence of climate - related trend, but flood damages are increasing.
Droughts	- Intensified droughts in some drier regions since 1970s.
Water quality	- No evidence of climate related trend.
Erosion and sediment	- No evidence of climate related trend.
Irrigation water	- No evidence of climate related trend.
2.3.B: Current vulnerability of worlds freshwater resources	
Country	**Examples of current vulnerability**
Asia	- Flood disasters in Bangladesh, where more than 70% of the country inundated in 1998. - Health problems due to arsenic in groundwater in India and Bangladesh (see chapter 7 on arsenic). - Huanghe river in China has temporarily run dry due to precipitation decreases and irrigation use.
Australia	- Damage to aquatic ecosystems due to decreased stream flow and increased salinity in Murray-Darling basin.
Africa	- Area of Lake Chad declining. - Rural water supply affected by extended dry season in Benin.
Europe	- Damage to riparian ecosystems due to flood protection along the Elbe River in Germany.
North America	- Multi-year droughts in USA and Southern Canada.
South America	- Land subsistence and land slides in Mexico city. - Water supply affected by shrinking glaciers in Andes. - Water supply reduced by erosion and sedimentation in reservoirs.

2.4: Projections for future impacts on freshwater resources

[based on Kundzewicz et al. 2007].

A number of climate models projected that a doubling of atmospheric concentrations of CO_2 would increase global precipitation by at least 5%. However, it is also projected that future rising of temperature is likely to increase evaporation as well. The precipitation is expected to increase in some areas, such as, at high latitudes and in the tropics (south east monsoon region and over the tropical Pacific), whereas the precipitation is likely to decrease in the sub-tropics (e.g. much of North Africa and northern Sahara) (see Figure 2.1). Groundwater reserve will steadily shrink in areas receiving reduced precipitation. In areas where evaporation rates are expected to increase it would lead to reduced river and stream run-off. In contrast, areas receiving higher precipitation would likely have higher lake and river levels and may experience floods. The future impacts on freshwater resources are summarised in Figure 2.1 which forecasts an increase in flooded areas in Bangladesh, a decreases in freshwater resources in small islands, a decrease in steam flows in the USA and a decline in hydropower stations in North Europe due to reduced annual river run-off. The other impacts of climate change on water resources are highlighted in the following sections (2.4.1 to 2.4.4):

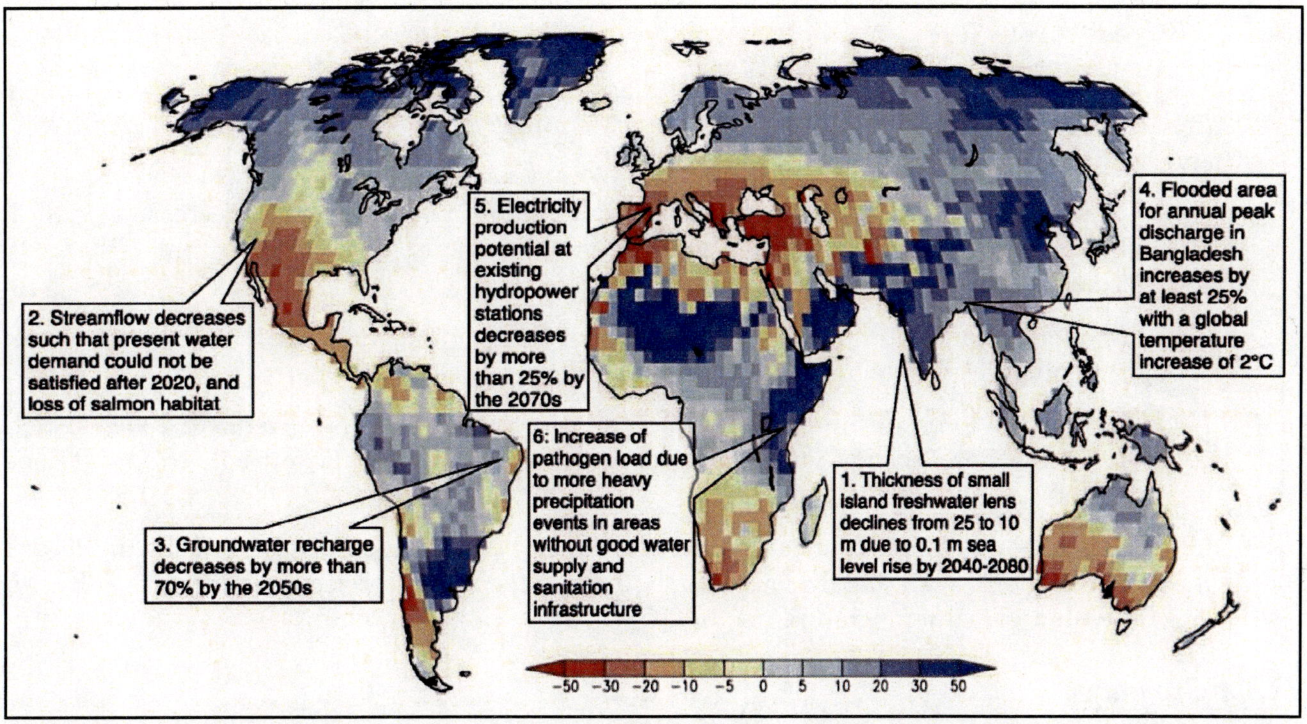

Figure 2.1: ***Map of future climate change impacts on freshwater representing the mean change of annual run-off, in percentage, between present (1981-2000) and 2081-2100) (blue = increased runoff; red = decreased run-off)*** *[see Kundzewicz et al. 2007].*

2.4.1: Surface waters

- Global mean run-off has been simulated to increase by 5-17% due to climate change related increase of precipitation at mid to high latitudes.
- Changes in the seasonality of river flows where much of the precipitation currently falls as snow (European Alps, the Himalayas, Western North America). In many cases, peak flow would occur at least a month earlier and an increase in winter flows and a decrease in summer flows.
- Projected higher temperatures would increase glacial melt in the Hindu Kush-Himalaya (Asia), and South American Andes (more than 1/6th of worlds population relies on snow melt water

from glaciers and seasonal snow packs for their water supply) (see section 2.5 for more effects of melting of ice and glaciers).

- The likelihood of precipitation falling as rain rather than snow would increase in areas with temperatures near to 0°C due to rise of temperature.
- In many rain dominated catchments, the seasonal flows may increase, with higher flows in the peak flow season and either lower flow or extended dry periods during the low flow season.
- Little changes will be in the timing of peak or low flows in rain dominated catchments.
- Changes in lake levels will depend on evaporation loss and precipitation inputs.
- By 2015, the ice cover duration on the rivers in Siberia would be shorter by 15 to 27 days and maximum ice cover would be thinner by 20 to 40%.
- The climate change has the potential to exacerbate considerably the reductions in run-off caused by afforestation in Australia.
- The increased evapotranspiration and sea level rise, would cause saltwater intrusion into aquifer (see section 2.6) and freshwater groundwater affecting groundwater resources.

2.4.2: *Floods and droughts*

- The frequency and risk of floods (river floods, flash floods, urban floods, sewer floods) and droughts (meteorological drought, hydrological drought) is projected to increase due to warmer climate and its increased climate variability.
- An increase of drought in low latitudes and mid latitudes in summer and a decrease in summer precipitation in southern Europe has been projected (an increase of drought from 1-3% for the present day to 30% by the 2090).
- Flood frequencies are expected to rise in northern Europe and north-eastern Europe, whereas drought frequencies will increase in southern and south-eastern Europe.
- Flood is projected to occur more frequently in some areas where it is currently a 100 year flood event (large uncertainty) and in temperate regions the snow melt that contribute to spring floods are projected to decrease.
- Up to 20% of worlds' population is likely to be affected by increased flood hazards by the 2080s in the course of global warming. In Bangladesh, the flooded area is projected to increase at least by 23-29% with a global temperature rise of 2°C.

2.4.3: *Groundwater*

- The demand for groundwater is likely to increase due to global increase of water use and a decline in surface water and reduced summer flows in snow-dominated basins.
- In high latitudes, thawing of permafrost will cause changes in groundwater level and quality.
- Climate change will affect groundwater recharge, in particular, recharge will decrease by more than 70% in north-eastern Brazil, south-west Africa and southern rim of Mediterranean sea, whereas recharge will increase by more than 30% by 2050s in Sahel (Africa), the Near East, northern China, Siberia and western USA.
- In England, in summer, groundwater recharge and stream flows are projected to decrease by as much as 50%, which may lead to water quality problems and groundwater withdrawal restrictions.
- In higher latitudes, the sensitivity of groundwater and run-off to increasing temperature is greater because of increasing biomass and leaf area index (improved growth conditions and increased evapotranspiration).

2.4.4: Water quality

Higher water temperatures and variations in run-off are likely to cause adverse changes in water quality affecting human health; ecosystems and water use (for example, decline in rivers and lake levels will lead to the resuspension of bottom sediments and liberating compounds which could have negative effects on water supplies).

- Projected intense rainfall in some regions will increase the amount of suspended solids and pollutants (nutrients, sediments) in lakes and reservoirs.
- Higher surface water temperature will promote harmful algal blooms such as cyanobacteria. An increase in bacteria and fungi content and density in lakes and reservoirs will lead to increasing water treatment costs (see chapter 13 for the effects of freshwater biotoxins).
- Higher water temperatures will enhance the transfer of volatile and semi-volatile compounds (ammonia, mercury, pesticides and dioxins) from surface water bodies to the atmosphere (see chapter 8, 9 and 10 respectively for the effects of mercury (heavy metals), pesticides and dioxins on human, wildlife and environment). In addition, where intense rainfall is expected to increase, pollutants such as pesticides, heavy metals, fertilizers and organic matters will be increasingly washed from soils to water bodies.
- The shrinking of summer sea ice may allow some of the persistent organic pollutants (POPs), long trapped under sea ice to evaporate into the atmosphere and spread further around the Polar Regions and some chemicals trapped in glaciers or permafrost may get washed out by the process of melting.
- Rivers and lakes are expected to be more acidic due to increasing amount of CO_2.
- Toxicity of contaminants (POPs such as atrazine, endosulfan) may be enhanced with increasing temperatures affecting fish and other wildlife living in the aquatic environment (see chapter 4, section 4.2.1).
- The decreasing stream flows in some regions will increase salinity in estuaries and inland rivers. It is projected that salt concentration in the Murray-Darling Basin (Australia) will increase by 13 - 19% by 2050 and 21-72% by 2100.
- Due to sea level rise, groundwater salinisation will very likely increase.
- Water borne diseases (e.g. *Cryptosporidium* outbreaks) are likely to increase due to extreme rainfall and thereby will increase water treatment costs (see chapter 5 for climate change impacts on human health).
- Areas where amounts of surface and groundwater recharge are projected to decrease, water quality will likely to decrease due to lower dilution.
- Increased rainfall amounts and intensities will lead to greater rates of erosion and sediment transport, an increase in erosion in the order of 20 to 60% is projected in case of Saxony, Germany and by 15 to 100% in areas where precipitation increases.

In summary, the reduction of storage in snow pack and decreasing river run-off would impact both the water availability and quality (see Figure 2.1 and Table 2.3). The impacts of climate change would induce a decrease in river run-off causing an increasing concentrations of pollutants (e.g. pesticides, heavy metals; chapters 8 and 9 discusses the impacts of these two chemicals on environment and human health respectively) thereby would increase the water treatment costs. Decrease in river run-off accompanied by predicted increase in surface temperature would increase the risk of toxin producing algal blooms (see chapter 13).

2.5: Impacts of snows and glaciers melting

[based on Bajracharya et al. 2007; IPCC 2007a and 2007b].

Rising temperatures will cause a decline in snow cover, lake and sea ice, glaciers, and permafrost. It is estimated that 40% of the world's population could be affected by loss of snow and glaciers. Seasonal snow cover is the main source of run-off in the dry season in many mountain regions, on which over a billion people depends for water supplies for various purposes including domestic, agricultural and industrial uses as well as for hydro-electric power generation. Snow is also important for winter sports, tourism, and recreation (skiing, snowmobiling, dog mushing, sledging, snowshoeing, etc. are important for local economies).

2.5.1: *Snow cover*

Satellite monitoring reveals that since 1960s, snow cover in the Northern hemisphere has been decreasing by 1.3% per decade. Analyses of results suggest that a 2°C temperature rise could reduce temperate snow cover by over 20%. Permafrost or frozen ground is important for the stability of buildings and infrastructures. Large quantities of GHGs such as methane could be released if widespread thawing of permafrost takes place. The permafrost in China is expected to decline by 30-50% during this century.

2.5.2: *Sea ice*

Sea ice is important in relation to ocean circulations, for the food chain, wildlife (polar bears), walruses as well as fisheries. Coastal Arctic indigenous people are inextricably linked with sea ice. Sea ice in the North Sea has decreased by 2.5% per decade in March and close to nine percent in September over the past quarter century.

2.5.3: *Himalayan glaciers*

Many glaciers around the world are receding in response to climate change forming new glacier lakes from melting. One of the clear evidence is that the Himalayan glaciers have been melting (see Figure 2.2) at an unprecedented rate in recent decades (retreating at rates ranging from 10-60 m/yr i.e. receding faster than any other part of the world). For example, 30.2 km long Gangotri glacier, Nepal (Figure 2.2) was receding at an average of 7.3 m every year (1842-1935), and 23 m/yr (between 1985 and 2001). Glaciers' environment is especially sensitive to impacts of climate change since temperature changes are pronounced at higher altitudes.

Himalayan glaciers are a freshwater reserve, which provide the headwaters for nine major rivers. Melting of glaciers may cause major changes in freshwater flow regimes and would have a dramatic impact on drinking water supplies, biodiversity, hydropower, industry, agriculture and others. About 2.4 billion people in Asia depend on the summer melt water of the Himalayan glacier (which is the largest concentrations of glaciers outside the polar region). Rivers depending on the Himalaya's glaciers such as the Indus, Ganges, Brahmmaputra, Syr Darya and Amu Darya could face water shortages or river run-off. Irrigation, industry and power generation could be disrupted as a result of water shortage. Other consequences of the Himalayan glacier retreat are the formation of glacier lakes at the terminal moraine. This would cause an increasing potential threat of glacier lake outburst floods that would have devastating consequences on lives and properties of downstream of glacier lakes.

2.6: Impacts of sea level rise

Antarctic and Greenland ice sheets contain about 98-99% of worlds freshwater ice. The meting of ice sheets with those of mountain glaciers, ice caps and thermal expansion of the oceans led to a sea level rise of 18 cm between 1961 and 2003 and 31 cm between 1993 and 2003 (see Chapter 1, Table 1.3). One meter sea level rise (without adaptation measures) would expose some 145 million people to flooding in the small islands, mega deltas of the Ganges-Brahmmaputra, the Mekong and the Nile in Africa (see Figure 2.3). Bangladesh would be most affected due to sea level rise (about 16% land area would be affected for a 1.5 m sea level rise (*see <http://www.grida.no/climate/vital/33.htm>*).

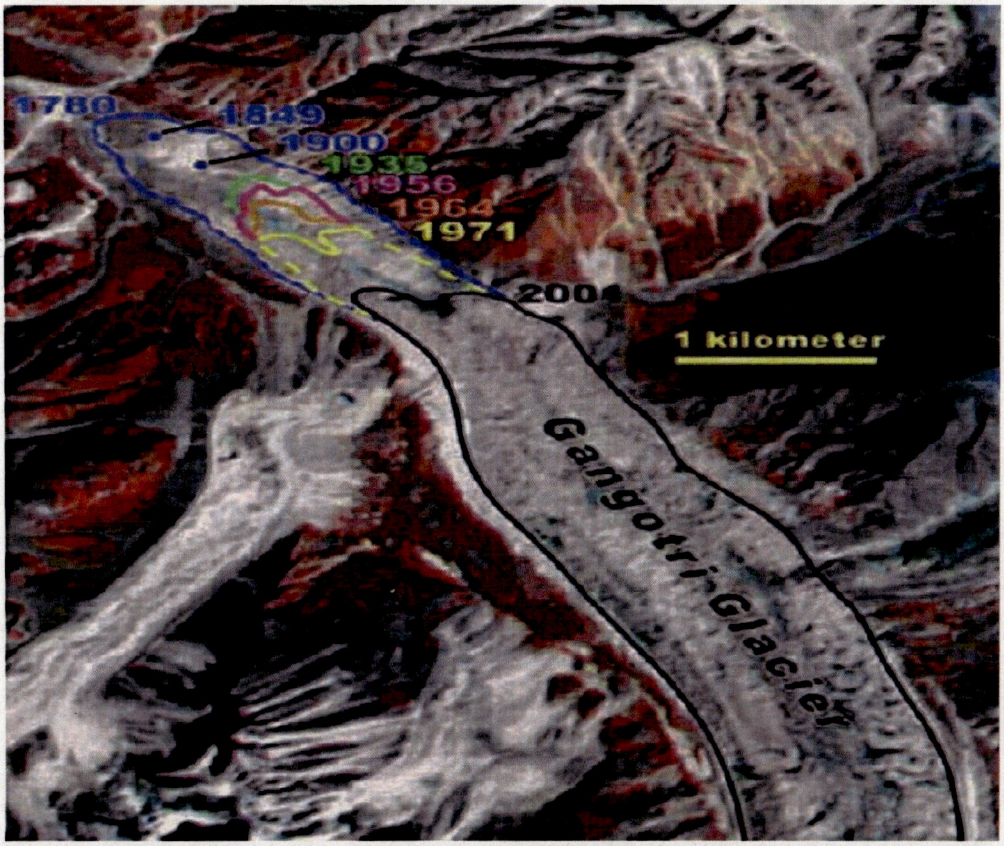

Figure 2.2: ***Retreat of the Gangotri glacier terminus in Nepal during the last 220 years from 1780*** *[Bajracharya et al. 2007; and NASA EROS Data Centre, 2001; http://www.icimod/org/index.php/search/publication/169].*

Figure 2.3: ***Relative vulnerability of coastal deltas in the world and potential population displacement scenario to 2050 due to sea level rise*** *[extreme = >1 million; high =1 million to 50,000; medium = 50,000-5,000; Nicholls, et al. 2007, http://www.ipcc.ch/pdf/assessment-report/ar4-wg2-chapter6.pdf adapted from Ericson et al. 2006].*

The sea level rise would have influence on freshwater resources due to sea water intrusion (salinization) into surface waters, coastal aquifers, coastal river systems, estuaries and inundation of inland coastal zones. Such impacts would be higher for developing countries with a high proportion of coastal low lands (see Figure 2.3). Water quality of shallow groundwater and surface water could thus be affected due to salt water intrusion in the Ganges-Brahmmaputra (Bangladesh), Mekong and Nile deltas for example. It is also projected that more than 25% of the Netherlands drinking water facilities (about 200 drinking water facilities) are estimated to be threatened by salt water intrusion from coastal areas.

2.7: Climate change impacts on shared water resources

According to Gleick (1992), the greatest threat from climate change is the increase in evaporative losses and water demands caused by high average temperatures. The long-term water shortfall due to climate change would likely to lead disputes and tensions among water sharing nations. This will likely worsen disparities in water use for irrigation, development and sanitation within water shared nations.

Many freshwater resources and rivers are shared by two or more nations (e.g. the Indus, Ganges and Brahmmaputra in Southern Asia, and the Nile, Jordan and Euphrates in the Middle East, and the Colorado and Rio Grande in North America). Demands due to climate change and growing populations for agriculture and economic development for water resources is expected to grow in countries sharing water with other nations or where bulk of water originates from outside or where there are water shortages. There are a number of nations who are highly dependent on water resources originating outside of their borders or under control of other nations (Table 2.4). For example, more than 95% of Egypt's water (the run-off from the Nile River) originates outside Egypt. Bangladesh, Iraq and Mexico are some of the down stream nations; these nations would suffer much if climate changes reduce water flows, since the stronger upper (upstream) nations are in a position to dominate the use of remaining water.

Table 2.4: *Countries whose 70% water originates outside* [FAO 2007].

Country	% total renewable water resources originating outside the country
Azerbaijan	73
Bangladesh	91
Botswana	80
Cambodia	75
Congo	73
Egypt	97
Hungary	94
Kuwait	100
Mauritania	96
Moldova rep.	91
Netherlands	88
Niger	90
Pakistan	76
Paraguay	72
Romania	80
Syrian Arab Rep	80
Turkmenistan	97
Uzbekistan	77

2.8: Case studies - Asia-Pacific and Australasia Region

General

Different climate models projected that global average precipitation (rain or snow) and evaporation will increase during the 21st century. The rainfall is expected to be higher over the northern mid- to high latitudes and Antarctica in winter. At lower latitudes, models simulate both regional increases and decreases over land areas. Extreme events such as droughts and floods could have dominating affects of climate change on water resources. Water quality would degrade with rising pollution, temperatures and reduced stream flows. Demand for water is likely to increase due to expected higher evaporative losses in irrigation, higher population and economic growth. In short, climate change will increase uncertainty in water management (see section 2.4).

2.8.1: Asia

[based on Cruz et al. 2007]

The impacts of climate change on water resources in Asia will be positive in some areas and negative in others. It is projected that more than a billion people could be affected by a decline in the availability of freshwater, particularly in large river basins by 2050.

- Climate change will cause melting of glaciers in the Himalayas thereby increasing flooding to downstream countries such as India and Bangladesh.
- The river flows in Asian countries is expected to decrease as glaciers recede.
- The thawing volume and speed of snow cover in spring in North-West China and Mongolia is projected to accelerate and the thawing time could advance, which will increase water sources but may cause some floods in spring.
- By 2050, freshwater availability in Central, South, East and South-East Asia (in particular large river basins) are projected to decrease.
- In some parts of Russia, climate change could significantly alter the variability of river run-off which may cause extremely low run-off events much more frequently in the crop growing regions of the south-west.
- In North China, irrigation from surface and groundwater sources will meet only 70% of the water required for agriculture production due to climate change and increasing demand.
- Changes in run-off of river basins could have a significant effect on the power output of hydropower generating countries such as Tajikistan.
- Sea level rise could cause saltwater intrusions into groundwater and surface water (thereby increase salinity) in India, China and Bangladesh (see also section 2.6). Increasing frequency and intensity of droughts in the catchments areas will led to more serious and frequent saltwater intrusions in the estuary.

The other impacts of climate change on water resources in Asian region is summarised in Table 2.5.

Table 2.5: ***Climate change impacts on water resources in Asia*** *[Cruz et al. 2007].*

Country	Projections
China	- Glacier melt could affect water supplies to a quarter of a billion people.
Himalayan (Hindu-Kush region)	- Glacier melt could affect water supplies to a half a billion people (Himalayas region) affecting agriculture production in downstream countries of South Asia.
India, China and Bangladesh	- Susceptible to increase salinity of groundwater as well as surface waters along the coast.
Iraq	- Euphrates and Tigris Rivers may be affected due to alteration of river flows.
Lebanon	- Annual net usable water resources is projected to decrease by 15% with a rise of 1.2^{0}C and doubling of CO_2. - Rivers flows would increase in winter and decrease in spring.
Mekong delta	- Increasing flooding risks during wet season and increasing water shortage in dry season.
North China	- Irrigation from surface and groundwater sources will meet only 70% of the water requirement.
North-West China, Mongolia	- Thawing process could accelerate which will increase water sources and may cause spring floods but could cause water shortages in winter.
Russia	- Frequent low river run-off in crop growing areas situated in southwest.
Tajikistan	- Changes in run-off of river basins could significantly affect output of hydropower generation.

2.8.2: *Australia and New Zealand*

[based on Hennessy et al. 2007].

In Australia, increases in temperature and reduced rainfall (Table 2.6) has been projected in most of the southern and eastern Australia (major food growing regions) which will cause reduced inflows to water storages, reduced stream flows in major catchments, reduced water available for rain-fed agriculture and reduced recharge of groundwater. Increases in the occurrence and severity of drought events are projected in the north-eastern and Southern Australia. Over the last ten years (1997-2005), the inflows to the Murray River has been almost 50% less (average 5,936 GL) compared to pre-1997 period average inflows of 11,631 GL (see Figure 2.4). Simulated models projected a decrease in winter-spring rainfall (June-November) and increase in evaporation over the southern half of the Australian continent.

Table 2.6: ***Projected changes in average temperature and rainfall in Australia relative to 1990*** *[Hennessy et.al. 2007 cited from Suppiah et al. 2007].*

Temperature change (^{0}C)	2020	2050	2080
0 to 400 km inland of coast	+0.1 to 1.0	+0.3 to 2.7	+0.4 to 5.4
400 to 800 km inland	+0.2 to 1.3	+0.5 to 3.4	+0.8 to 6.7
Central Australia	+0.2 to 1.5	+0.5 to 4.0	+0.8 to 8.0
Rainfall change (%)	**2020**	**2050**	**2080**
Within 400 km of western and southern coasts	-15 to 0	-40 to 0	-80 to 0
Sub-tropics (20-28^{0}S)	-10 to +5	-27 to +13	-54 to +27
Northern NSW, Tasmania and central Northern Territory	-5 to +10	-13 to +27	-27 to +54
Central South Australia, southern NSW and North Latitude 20^{0}S, except central NT	-5 to +5	-13 to +13	-27 to +27
Inland Queensland	-10 to +10	-27 to +27	-54 to +54

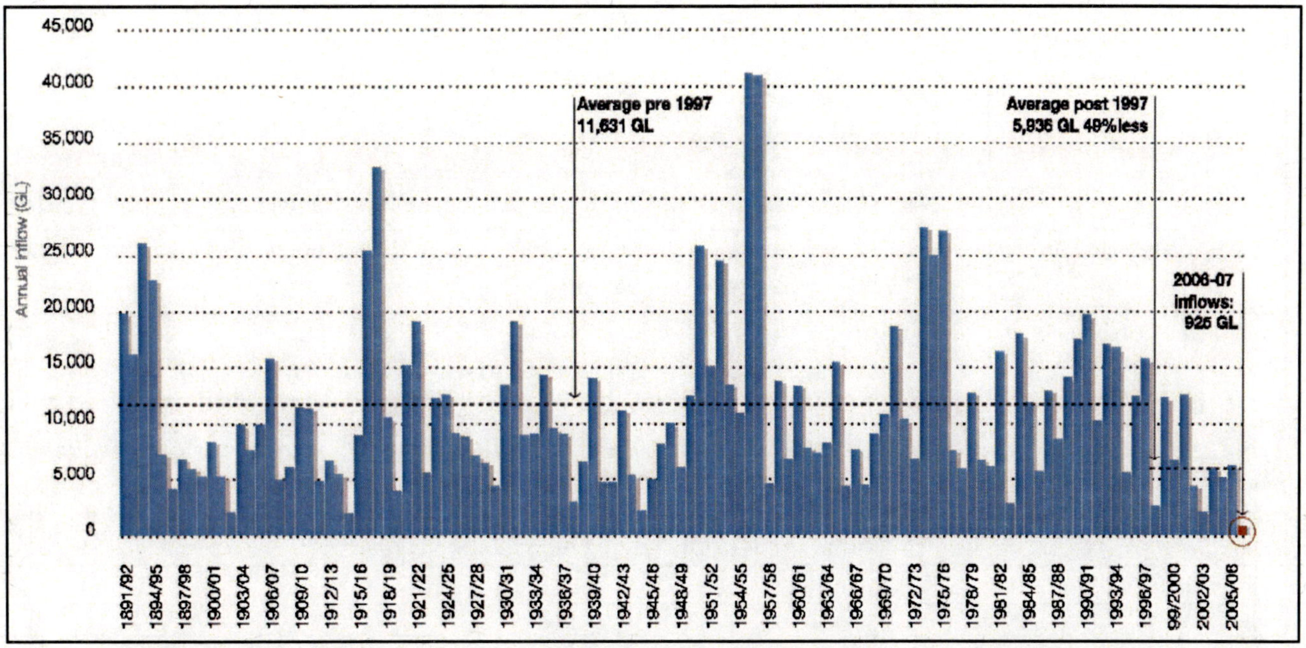

Figure 2.4: ***River Murray, Australia inflows during 1891-2007*** *[State Water Report 2004/05].*

2.8.2.1: Water security in Australia

The Murray Darling Basin is the Australia's largest river basin covering an area of around 1 million square kilometres. The basin accounts for 70% of irrigated crops and pastures (around 40% of Australia's gross value of agricultural production). Based on climate modelled scenario, the annual stream flow in the basin is projected to fall by 10-25% by 2050 and 16-48% by 2100 (Table 2.7). The reduction in water availability would have severe consequences on irrigated agriculture (see Table 2.8), for example dairy production could be affected up to 94%. There is a 50% chance by 2020 of the average salinity of the lower Murray River, in Australia may exceed the 800 EC (electrical conductivity) threshold set for desirable drinking and irrigation water standard. Toxic algal blooms are likely to become frequent and to last longer due to climate change (due to rise of temperature and decrease of river flow). They can pose a threat to human health and can kill fish and livestock (see Chapter 13 for the effects on freshwater biotoxins).

In New Zealand, annual flow from larger rivers with headwaters in Southern Alps has been projected to increase, which is likely to provide more water for hydro-electric generation during winter peak demand period. Industries dependent on irrigation may experience negative effects due to lower water availability in spring and summer. By 2030, water security problems are projected to intensify in southern and eastern Australia and in New Zealand, (in Northland and some eastern regions).

Table 2.7: ***Future projections of Australia and New Zealand's water resources*** *[Hennessy et al. 2007].*

	Impacts
Water security	- Reduction in water supply for irrigation, cities, industry and riverine environment in those areas where stream flow declines.
Stream flow	- 2030: North-east coast - change in annual run-off by – 5 to +15%. South-east - a decline of run-off up to 20%. Tasmania – a decline of run-off up to ±10%. South Australia (Gulf of St. Vincent) - a decline of run-off up to 25%. South West - decline of up to -25% to +10%. Most Eastern Australia- decline of annual run-off by 6-8%.

Contd...

	NSW (Burrendong dam) – inflows change by 10 to - 15%. Victoria – run-off in 29 catchments declines by 0-45%. - 2050: Murray Darling Basin - stream flow may drop by 10-25%. Melbourne – stream flow may decline by 7-35%. - 2070: NSW (Burrendong dam) – inflows may change by + 5 to - 35%. - 2100: Stream flow in the Murray Darling Basin may drop by 16-48%.
Water quality (salinity)	- 2050: Salinity changes in the Murray Darling Basin may be in the range of – 6% to + 19%. - 2100: Salinity changes in the Murray Darling Basin may be in the range of – 16% to + 72%.
Identified hot spots	- Southern and Eastern Australia. - Northland and parts of eastern lowlands of New Zealand.

Table 2.8: ***Change in irrigated land use in the Murray Darling Basin, Australia (from a 20% reduction in water availability)*** *[Australian Commodities, 2008; http://www.abareconomics.cominteractive/08ac_march/htm/a5.htm].*

	Horticulture (%)	Cotton (%)	Rice/grain (%)	Dairy (%)
Queensland	-3.2	-18.7	33.1*	-68.2*
Northern NSW	-1.2	-19.1	-3.9	-3.2
Riverina	-1.2	-16.2*	-18.2	-2.3
Western NSW	-1.7	-21.7	12.9	-31.1*
Eastern Victoria	-3.1		-52.2*	-18.0
Western Victoria	-1.6		-41.7*	-16.8
South Australia	-3.1		-100.0*	-93.7
Total	**- 2.2**	**-19.2**	**- 16.9**	**- 13.8**

* represent less than 3% of total land use for that activity in the basin.

2.8.2.2: Water adaptation strategies in Australia

To tackle water shortages and water crisis resulting from climate change, a wide range of water adaptation strategies has been implemented or proposed in Australia, these include water savings, water recycling, seepage loss prevention, new pipe lines and desalination (see Table 2.9).

Table 2.9: ***Examples of government adaptation strategies to cope with current and future water shortages in Australia*** *[Hennessy et al. 2007; G-MW web site 2008].*

Government	Strategy	Investment
Australia	Drought aid payments to rural communities	2001-2006; US $1.7 billion
Australia	National water initiative supported by Australian water fund	2004-2009; US $ 1.5 billion
Australia	Murray-Darling Basin water agreement	2004-2009; US $ 0.4 billion
Victoria	Melbourne's Eastern treatment plant to supply recycled water	By 2012; US $153 million
Victoria	New pipeline from Bendigo to Ballarat, water recycling, interconnections between dams, reducing channel seepage, conservation measures	By 2015; US $153 million
Victoria	Wimmera Mallee pipeline replacing open irrigation channels	By 2010; US $376 million
Victoria	G-MWs Modernisation project: (Shepparton Modernisation- 52 GL of water savings) and the Central Goulburn 1, 2, 3 and 4 Modernization (15.2 GL water savings) through automated regulator gates, upgrading Dethridge Meter, channel remediation works including plastic and clay lining and pipelining and remodeling (through reducing leaks and seepage losses)	A$ 188.2 million (Shepparton); A$ 41.8 million (Central Goulburn)

Contd...

Victoria	G-MWs Water tight 2020 - to save 400,000 ML of water from within the Goulburn-Murray Water irrigation system by the year 2020 (channel automation, more accurate metering, channel lining, piping and system reconfiguration)	
Victoria	Water desalination project (Wonthaggi, Melbourne)-150 GL by reverse osmosis	AS$ 4.9 billion
Victoria	75 GL water diversion from Goulburn River to Sugarloaf, Melbourne	
NSW	NSW water savings fund supports projects which save or recycle water in Sydney	US$ 98 million for round 3 plus more than US $ 25 million to 68 other projects
Queensland	QLD water plan 2005-2010 to improve water use efficiency and quality, recycling, drought preparedness, new water pricing	US $ 182 million for water infrastructure in south-east QLD, and US$ 302 million for other infrastructure programs
South Australia	Conservation and development of Adelaide's water resources to 2025	
Western Australia	State water strategy (2003), state water plan, WA doubled water supply and Perth desalination plan	US $ 500 million spent plus $ 290 million for desalination plant
Australia	National Water Plan (Federal Government)	US$ 8 billion

2.8.3: Other countries

Africa

Drought would intensify in Africa as a result of climate change causing shortages and rising demand for water. This would result 'water scarcity' in Africa. Shortages of water supplies will likely severely affect the agricultural production.

Antarctica

The frozen continent contains almost 90% of the planet's freshwater. However, both temperatures and glaciers have remained relatively constant for the last 50 years; therefore no prediction has been projected for this continent.

Europe

Researchers believe that Europe's northern regions will have greater precipitation and will be more suitable for agriculture production in the future. Whereas southern Europe will experience hotter summers, less precipitation, more droughts with reduced water availability.

Latin America

The disappearance of glaciers and predicted changes in precipitation patterns could reduce water availability for different purposes.

North America

Climate change (rising temperature) is projected to diminish the mountain snow packs, thereby increase winter floods, reduced summer water flows, thus alter the seasonal availability of water.

Small island states

Rising sea levels could threat freshwater resources in small islands.

2.9: Adaptation measures

There are many adaptation measures that could be applied to minimise the impacts of climate change on water resources. These include modernisation of existing irrigation schemes, recycling of water in water stressed countries (arid and semi-arid) and promotion of better practices for wiser use of water in agriculture including recycling of waste water. Recycled water can be used to create or enhance wetlands and riparian habitats. Reduction of water wastage and leakages should be the top priority in countries or regions where substantial decline in precipitation and rise of temperature has been projected. The Australian Government has taken many adaptation measures to cope with climate change (see Table 2.9), for example, desalinisation schemes are being considered in some Australian cities (Perth, Adelaide, Melbourne, Sydney) and installation of rain water tanks is being encouraged through incentive policies and rebates.

2.10: Conclusion

Climate change increases water system dynamics through temperature changes, changes in precipitation patterns, evaporation and water quality and water storage in ice packs. Water supply authorities have to cope up with the consequences of climate change, for example, droughts, floods, water shortages in river systems, salt water intrusions, increasing concentrations of pollutants, increasing demand for water (high evaporation and drought) and microbiological activities due to temperature.

Apart from the impacts of climate change on water resources, increasing human population and human activities will cause more freshwater demand in the future. By 2025, it is predicted that agriculture sector will require about 1.2 times more water, industry by 1.5 times and domestic consumption by 1.8 times. World's freshwater resource needs protection and conservation for its long-term sustainability. This may be achieved through water saving programs, reuse and recycling of water (where feasible) and research to develop more efficient irrigation technology. The other measures to cope with climate change may include the following:

- Building of new reservoirs/large dams to stock rain and run-off water
- Reuse of waste water after treatment
- Desalination to increase water supply in semi-arid and arid regions
- Efficient use of irrigation water
- Innovative technology to treat micro-pollutants such as endocrine disrupting chemicals (EDCs), and pharmaceuticals (see chapter 11 for EDCs and chapter 12 for pharmaceuticals) from waste water, which cannot be removed by current available treatment technology.

2.11: References

Australian Commodities 2008. Murray Darling Basin-economic implications of water scarcity. March. 15(1): 265-281. *http://www.abareconomics.cominteractive/08ac_march/htm/a5.htm*

Bajracharya, S. R., Mool, P. K. and Shrestha, B. R. 2007. Impacts of climate change on Himalayan glaciers and glacial lakes-Case studies on GLOF and associated hazards in Nepal and Bhutan. International Centre for Integrated Mountain Development (ICIMDO) and United Nations Environment Program. 119 pp. http://books/icimod/org/index.php/search/publication/169

Cruz, R. V., Harasawa, H., Lal, M., Wu, S., Anokhin, Y., Punsalmaa, B., Honda, Y., Jafari, M., Li, C. and Huu Ninh, N. 2007. Asia. *Climate change 2007: Impacts, Adaptation and Vulnerability. Contribution of Working Group II to the Fourth Assessment Report to the Intergovernmental Panel on Climate Change.* Parry, M. L., Canziani O. F., Palutikof J. P., van der Linden P.J. and Hanson C. E. (eds.). Cambridge University Press, Cambridge, U.K. pp. 469-506 (chapter 10). http://www.ipcc.ch/pdf/assessment-report/ar4/wg2/ar4-wg2-chapter6.pdf.

Ericson, J. P., Vorosmarty, C. J., Dingman, S. L., Ward, L.G and Meybeck, M. 2006. Effective sea-level rise and deltas: causes of change and human dimension implications. *Global Planet Change* 50: 63-82.

FAO of the UN 2007. http://earthtrends.wri.org/gsearth.php?kw=freshwater+resources+2005&action+results (world resources institute) original data from FAO, population divn. of the Dept. of Economics and Social Affairs of the UN Secretariat, World Bank.

Gleick, P. H. 1992. Effects of climate change on shared freshwater resources. Chapter 9. In: *Confronting climate change: Risks, implications and responses.* Irving M. Mintzer (eds.). Stockholm Environment Institute, Cambridge University Press.

Hennessy, K., Fitzharris, B., Bates, B. C., Harvey, N., Howden, S. M., Hughes, L., Salinger, J., and Warrick, R. 2007. Australia and New Zealand. *Climate change 2007: Impacts, Adaptation and Vulnerability. Contribution of working group II to the Fourth Assessment Report of the Intergovernmental Panel on Climate Change.* Parry, M. L., Canziani, O. F., Palutikof, J. P., van der Linden, P. J.and Hanson, C. E. (eds.). Cambridge University Press, Cambridge, UK, chapter 11: pp. 507-540.

IPCC 2007a. *Climate change 2007: Impacts, adaptation and vulnerability.* Working group II to the fourth assessment report of the intergovernmental panel on climate change. (Parry, M. L., Canziani, O. F., Palutikop, J. P., van der Linden, P. J., and Hanson, C. E. (eds.) Cambridge University Press, Cambridge, United Kingdom and New York, USA, 976 pp. (http://www.ipcc.ch).

IPCC 2007b. *Climate change 2007: The physical science basis.* Working group II to the fourth assessment report of the intergovernmental panel on climate change. Parry, M. L., Canziani, O. F., Palutikop, J. P., van der Linden, P. J., and Hanson, C. E. (eds.). Cambridge University Press, Cambridge, United Kingdom and New York, USA, 976p. (http://www.ipcc.ch).

Kundzewicz, Z. W., Mata, L. J., Arnell, N. W., Doll, P., Kabat, P., Jimenez, B., Miller, K. A., Oki, T., Sen, Z. and Shiklomanov, I. A. 2007. Chapter-3. Freshwater resources and their management. Climate change 2007: Impacts, adaptation and vulnerability. Contribution of working group II to the Fourth Assessment Report of the Intergovernmental Panel on climate Change. Parry., M. L., Canziani, O. F., Palutikof, J. P., van der Linden, P. J. and Hanson, C. E. (eds.). Cambridge University Press, U.K. pp. 173-210.

Meuleman, A. F. M. and Cirkel, G. and Zwolsman, G. J. J. 2007. When climate change is a fact! Adaptive strategies for drinking water production in a changing natural environment. *Water Science & Technology.* 56(4): 137-144.

Nicholls, R. J., Wong, P. P., Burkett, V. R., Codignotto, J.O., Hay, J. E., McLean, R. F., Ragoonaden, S and Woodroffe, C. D. 2007. Coastal systems and low-lying areas. *Climate change 2007. Adaptation and Vulnerability. Contribution of Working Group II to the Fourth Assessment Report to the intergovernmental panel on Climate Change.* M. L. Parry., Canziani, O. F., Palutikof, J. P., van der Linden, P. J. and Hanson, C. E. (eds.). Cambridge University Press, U.K. pp. 173-210. http://www.ipcc.ch/pdf/assessment-report/ar4/wg2/ar4-wg2-chapter6.pdf.

Sate Water Report. 2004/05: Water supply, drought and climate change. Our Water Our Future. The Next Stage of the Governments Water Plan. Department of Sustainability and Environment (DSE), Victoria, Australia.

Suppiah, R., Hennessy, K. J., Wheton, P .H., McInnes, K. Macadam, I, Bathols, J.and Rickets, J. 2007. Australian climate change projections derived from simulations performed for the IPCC 4th Assessment Report. Aust. Meteorol. Mag. (accepted).

Websites

http://www.fao.org/waicent/faoinnfo/agricul/agl/w/quastat/water_res/index.htm

UNEP (2007). Melting ice-A hot topic? New UNEW reports shows how hot it's getting.

http://www.unep.org/documents.multilingual/default.print.asp?documentID=512&articleID

http://www.cs.ntu.edu/homepge/jmitroy/sid101/uncc/fs104.html (The impact of climate change on water resources, information unit of climate change) accessed on 12 December 2007

http://www.grida.no.climate/vital/33.htm (potential impact of sea-level rise on Bangladesh)

http://www.planetark.com/dailyewsstory.cfm/newsid/40844/story.htm (regional impacts of climate change-UN draft on water resources). Accessed on 3 December 2007)

www.un.org/climatechange (The consequences for the future. Fact sheet-Climate change 24 September 2007) http://www.ipcc.ch/pdf/assessment-report/ar4/wg2/ar4-wg2-chapter3.pdf

http://www.g-mwater.com.au

□□□

3 Climate Change Impacts on Agriculture and Livestock

Key facts

- Agriculture is the major land use across the globe. Agriculture production (plant growth) is highly dependent on climate since crop growth is influenced by solar radiation, temperature and precipitation. Agriculture is also sensitive to climate variability and weather extremes (droughts, floods, severe storms).
- The yields of C_3 crops (rice, wheat, and soybean) is expected to increase (10-20%) with predicted doubling of CO_2 concentrations compared to C_4 crops (maize, sorghum, millet, sugarcane) since C_3 plants respond readily/positively to increased CO_2 levels. Therefore, CO_2 enrichment may benefit temperate and humid tropical agriculture compared to semi-arid tropics.
- The predicted rise of temperature could impact crop and horticulture yield if it exceeds optimal temperature range. For example, low latitudes crops are being grown near the limits of temperature tolerance and global warming may put them to higher stress.
- On a world wide basis, global warming would benefit the mid to high latitude zones for agriculture (for a small change of temperature of 1-3^0 C), whereas low latitudes, semi-arid and tropical areas will have much reduced crop and livestock yields. However, adaptation measures may reduce such impacts.
- The impact of climate change on potential crop production (e.g. based on doubled, 2xCO_2 and general circulation model, GCM) reveals that Africa and Latin America are more likely to have negative consequences. Whereas for Europe, United States, Canada and Asia the results range from severe negative effects to large potential increase in yield.
- In the middle and higher latitudes, global warming would extend the length of the potential growing season and will allow earlier planting of crops in the spring, earlier maturation and harvesting. Crop producing areas may expand pole ward in countries such as Canada and Russia. In tropical areas, mountain environment may be more favourable than the flat lands for major crops.
- The demand for irrigation is expected to rise in a warmer climate, causing increased competition between different water users. Globally, by 2080, the irrigation requirement for crops is estimated to increase to + 5% to 20% (due to net transpiration loss, longer growing seasons and evaporative loss) of which larger impacts will be in developing regions. The intensified evaporation will increase the hazard of salt accumulation in the soil.
- A warmer climate would enable more tropical pests, parasites and pathogen species to spread northwards and southwards. Climate change models projected that bluetongue, which mostly affects sheep, and occasionally goat and deer, would spread from the tropics to mid-latitudes. It would also increase the ability of some pest species to survive under warmer conditions. The possible increases in pest infestations would require greater use of pesticides to control them.
- With higher temperatures and overall decreased water supplies, beef-cattle, and sheep are likely to experience increased incidences of stress related deaths, the other climate change related effects on livestock are reduction in animal weight, reproduction rates and milk yields.
- Adaptive measures that may be taken include development of crop varieties that are adaptive to the greater tolerance to stresses, drought, heat and flood; altering the timing or location of cropping activities, altering fertilizer rates to maintain grain or fruit quality consistent with prevailing climate and altering amounts and timing of irrigation and other water management.

Contents

Acronyms and Glossary

Adaptation: Adjusting practices.

Bluetongue: A viral disease of sheep and cattle that is transmitted by biting midges or flies.

Broadcare crops: The word broadcare is an Australian term used to describe land suitable for farms practicing large-scale crop (agriculture) operations. The key crops in this category are as follows: Oilseeds - Canola, sunflowers; Winter and summer cereals - Wheat, barley, oats, triticale, sorghum, maize, millets; Pulses - Lupins, chickpeas, faba beans.

C_3 plants: Plants that produce three-carbon compounds during photosynthesis, including most trees and agricultural crops. C_3 species respond more to increased CO_2. C_3 plants include cotton, rice, wheat, barley, soybeans, sunflower, potatoes, most leguminous, woody plants, most horticultural crops (including vegetables) and many weeds.

C_4 plants: Plants mainly of tropical origin and widely grown in Africa, which produce a four-carbon compound during photosynthesis, including many grasses, pasture and agriculturally important crops like maize, sugarcane, millet and sorghum, halophytes (salt tolerant plants), many tall tropical grasses, forage and weed species. C_4 plants respond better to higher temperature.

Carbon-dioxide (CO_2): A colourless, odourless gas that occurs naturally and is also emitted by fossil fuel combustion and land clearing.

Carbon-dioxide (CO_2) fertilisation : Increasing plant growth or yield by elevated concentrations of atmospheric carbon-dioxide. A plants photosynthesis mechanism determines its sensitivity to carbon- dioxide fertilisation. Most tress, most agricultural crops such as, wheat and rice and most cold climate plants are more sensitive to changes in atmospheric carbon dioxide than other plants such as tropical grasses and crops including maize and sugar cane.

CGIAR: The Consultative Group on International Agricultural Research (CGIAR), established in 1971, is a strategic partnership of countries, international and regional organizations and private foundations supporting the work of 15 international Centres. In collaboration with national agricultural research systems, civil society and the private sector, the CGIAR fosters sustainable agricultural growth through high-quality science aimed at benefiting the poor through stronger food security, better human nutrition and health, higher incomes and improved management of natural resources.

Climate change: Climate change refers to any change in climate over time, whether due to natural variability or as a result of human activity. The United Nations Framework Convention on Climate Change (UNFCCC) defines 'climate change' as 'Change of climate which is attributed directly or indirectly to human activity that alters the composition of the global atmosphere and which is in addition to natural climate variability observed over comparable time periods'.

Cultivar: A plant that has been developed and maintained by cultivation as a result of agricultural or horticultural practices.

Drought: Meteorological drought (when atmospheric conditions result in low rainfall); hydrological drought (prolonged moisture deficits that affect surface or sub-surface water supply; and agricultural drought (short-term dryness in the surface soil layers (root-zone) at a critical time in the growing season.

El Niño: A warm event which brings drought. In general, El Niño indicates drier than average climate (like the drought)

La Niña: A cool event which brings rainfall. In general, La Niña indicates wetter than average climate (like the flooding)

IRRI: International Rice Research Institute, HQ in Metro Manila 1301, Philippines.

Extreme event: Refers to meteorological conditions that are rare for a particular place and / or item. Such as intense storm or heat wave.

Photosynthesis The process by which plants convert water and carbon dioxide into carbohydrates, using sunlight as the source of energy with the aid of chlorophyll.

Transpiration: The process by which water absorbed by plants (usually through the roots) is evaporated into the atmosphere from the plant surface (principally from the leaves).

3.1: Introduction

Agriculture is the major land use across the globe (see Table 3.1). World agriculture land occupied 5,023 Mha in 2002, of which most area was under pasture (69%) and crop land occupied 28%. In developing countries, agriculture is the largest supporter of livelihoods of rural people. Agriculture production (plant growth) is highly dependent on climate since crop growth is influenced by solar radiation, temperature, and precipitation. Agriculture is also sensitive to climate variability and weather extremes (droughts, floods, severe storms). A significant fraction of the world's food is grown in rain-fed tropics, where climate variability (e.g. El Niño events) could play an important role in determining crop's productivity and food production. It is estimated that >80% of total agriculture land and close to 100% pasture land is rain-fed.

Table 3.1: ***Global use of land for agriculture in 2001-02*** *[Smith et al. 2007].*

	Million hectares
1. World	
Agriculture land	5,023
Arable land	1,405
Permanent crops	130
Permanent pasture	3,488
2. Developed countries	
Agriculture land	1,838
Arable land	613
Permanent crops	24
Permanent pasture	1,202
2. Developing countries	
Agriculture land	3,184
Arable land	792
Permanent crops	106
Permanent pasture	2,286

Climate change is projected to have an effect on local agriculture and the net result could be harmful (e.g. frequent droughts, salinization of agriculture land) or beneficial, (e.g. enhance CO_2 and higher yield, longer growing season, increased precipitation). Though climate change due to GHG emissions could cause several effects on agriculture production but agriculture sector itself is responsible for global greenhouse gas emissions contributing about 25% carbon (due to deforestation), 50% methane (due to rice and livestock production), and 75% nitrous oxide (due to use of nitrogen fertilisers) (Tubiello *et al.* 2007).

3.2: Factors relating to future agricultural production

Several factors are directly related to climate change and agricultural productivity including:

- Rising atmospheric concentrations of CO_2,
- Increase of average temperature,
- Precipitation, water availability, irrigation,
- Change in climate variability and extreme events,
- Rises in sea level,
- Pests and diseases.

3.2.1: *Enhanced CO_2 or CO_2 fertilisation effect on crop yield*

[based on Parry 1990; Rosenzweig and Hillel 1995; Easterling et al. 2007; Tubiello et al. 2007; Hatfield et al. 2008].

Plants grow through the well-known process of photosynthesis, where CO_2 is vital for photosynthesis. There is evidence that increases in CO_2 would increase the rate of plant growth. However, the photosynthetic mechanisms of different crop plants differ in response to increasing CO_2. For example, C_3 plants (mid-latitude crops such as wheat, rice and soybean), respond readily/positively to increase CO_2 levels (see Figure 3.1 for wheat). However, the responses to C_4 plants which follow different pathway are less responsive to enriched CO_2 levels (Figure 3.1 for maize). The low-latitude crops such as maize, corn, sorghum, sugarcane and millet, many pastures and forage grasses are C_4 plants. Therefore,

CO_2 enrichment will benefit temperate and humid tropical agriculture compared to semi-arid tropics. Experimental results showed that yields of C_3 crops may be increased by 10% to 20% with doubling of atmospheric CO_2 (or $2xCO_2$ scenarios) whereas yield in C_4 crops (maize, sugarcane, millet) may increase 0 to 10% (maize and sugarcane) (see Table 3.2).

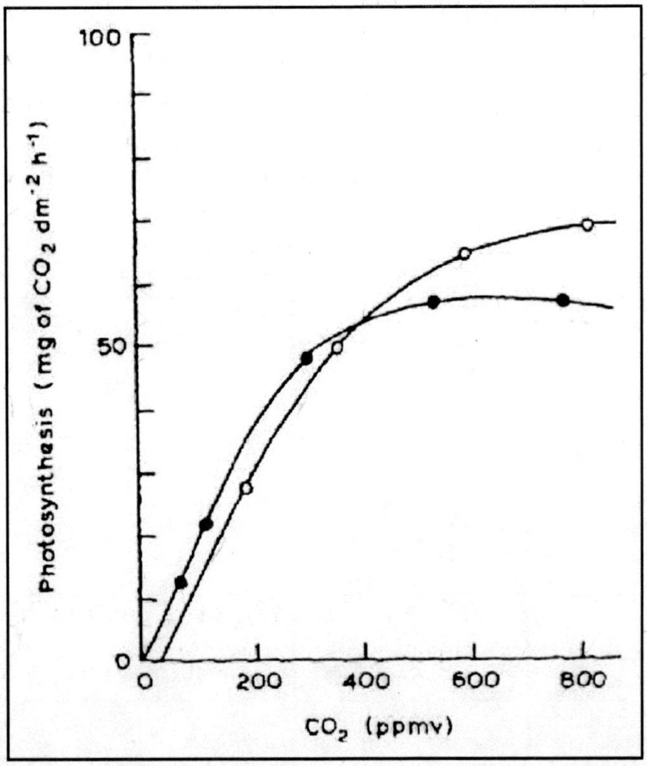

Figure 3.1: ***Typical photosynthetic response of C_3 and C_4 plants (C_3, empty circle and C_4, filled circle) to CO_2.*** *Net photosynthesis of wheat (C_3) is about 70 mg of CO_2 dm^2/h compared with maize (C_4) (about 55 mg of CO_2 dm^2/h) for equivalent light intensity (0.4 cal cm^2/min). Maize (C_4 crop) is saturated at a lower CO_2 concentration (c. 450 ppmv) than wheat (c. 850 ppmv) [Cited in Parry 1990 original from Akita and Moss 1973].*

Table 3.2: ***Effects of doubling of atmospheric CO_2 for various crops and plants (increase of CO_2 concentrations from 300 ppm to 660 ppm or 350 ppm to 700 ppm)*** *[Various sources compiled by Easterling et al. 2007; Hatfield et al. 2008].*

Plants	Leaf photosynthesis	Yield	Comments
C_3 plants	30-50	10-20	Responsive to CO_2
C_4 plants	10-25	0-10	Less responsive to CO_2
Bean (C_3)	50	27	Responsive to CO_2
Cotton (C_3 non-legume)	33	44	Responsive to CO_2
Peanut (C_3 legume)	27	30	Responsive to CO_2
Rice (C_3 grass)	36	30	Responsive to CO_2
Soybean (C_3 legume)	39	34-38	Responsive to CO_2
Wheat (C_3 grass)	35	31	Responsive to CO_2
Corn (C_4)	3	4	Less responsive to CO_2
Maize (C_4 grass)	4.4		Less responsive to CO_2
Sorghum (C_4 grass)	9	8	Less responsive to CO_2
Pasture grasses and legumes	~ +10 % and +20%		
Forest tree (biomass)		0-30%	Higher yield in young trees and no response to mature natural forest

The impact of climate change on potential crop production (e.g. based on doubled, 2xCO_2, and general circulation model, GCM) reveals that Africa and Latin America are more likely to have negative consequences. Whereas for Europe, United States, Canada and Asia the results range from severe negative effects to large potential increase in yield (Table 3.3). The impact on rice yields in Asian countries ranged from negative to positive (-3 to +28 for India; + 2 to +27% for Malaysia; -14% to + 14% for the Philippines). The impacts on wheat response in Australia ranged around - 34% to + 65%. In summary, increasing atmospheric concentrations of CO_2 can act as a fertilizer and enhance the growth of some crops such as wheat, rice and soybeans. Therefore, future CO_2 levels may favour C_3 plants over C_4.

Table 3.3: ***Regional crop yields for 2 x CO_2, General Circulation Model (GCM) equilibrium climates*** *[John Rielly-http://www.fao.org/docrep/W5183E0c.htm].*

Region	Crop	Yield impact (%)	Countries studied/comments
Africa	1. Maize	1. -65 to +6	1. Egypt, Kenya, South Africa, Zimbabwe. With CO_2 effect, range across sites and climate scenarios.
	2. Millet	2. -79 to -63	2. Senegal. Carrying capacity fell to 11-38%.
	3. Biomass	3. decrease	3. South Africa; agro zones shifts.
Asia (South)	1. Rice 2. Maize 3. Wheat	1. -22 to +28 2. -65 to +67 3. -61 to +67	1, 2, 3. Bangladesh, India, Philippines, Thailand, Indonesia, Malaysia, Myanmar. Range over GCM scenarios and sites; with CO_2 effect, some studies consider adaptation.
Asia (China and Taiwan)	1. Rice	1. -78 to + 28	1. Includes rain fed and irrigated rice. Positive effects in NE and NW China, negative in most of the country. Genetic variation provides scope for adaptation
Asia (other)	1. Rice	1. -45 to + 30	1. Japan and South Korea. Range in across GCM scenarios. Generally positive in northern Japan; negative in South.
Australia and New Zealand	1. Pasture 2. Wheat	1. -1 to + 35 2. -41 to + 65	1. Australia, New Zealand. Regional variation 2. Australia and Japan. Wide variation, depending on cultivar.
Europe	1. Maize	1. -30 to increase	1. France, U.K, N. Europe, with adaptation, CO_2 effect. Longer growing season; irrigation efficiency loss; northward shift.
	2. Wheat	2. increase or decrease	2. France, U.K, N. Europe, with adaptation, CO_2 effect. Longer growing season; irrigation efficiency loss; northward shift, greater pest damage.
	3. Vegetables	3. increase	3. Lower risk of crop failure.
Former Soviet Union	1. Wheat grain	1. -19 to 41 -14 to + 13	1. Range is across GCM scenarios, with or without the CO_2 effect.
Latin America	1. Maize	1. -61 to increase	1. Argentina, Brazil, Chile, Mexico. Range is across GCM scenarios, with or without CO_2 effect.
	2. Wheat	2. -50 to -5	2. Argentina, Uruguay, Brazil. Range is across GCM scenarios, with or without CO_2 effect.
	3. Soybean	3. -10 to +40	3. Brazil. Range is across GCM scenarios, with or without CO_2 effect.
North America	1. Maize	1. -55 to +62	1. USA and Canada. Range across GCM scenarios.
	2. Wheat	2. -100 to +234	2. Sites with/without CO_2 effect.
	3. Soybean	3. -96 to +58	3. USA. Less sever or increase in yield when CO_2 effect and adaptation considered.

3.2.2: Higher temperatures

3.2.2.1: Temperature and plant response

[based on Hatfield et al 2008].

Each crop has a specific temperature range at which vegetative and reproductive growth will proceed at the optimal rate (Table 3.4) and exposures to extremely high temperature during these phases can impact growth and yield. Higher temperatures during the reproductive stage of development can affect pollen viability, fertilization and grain or fruit formation. Even a short-term exposure of high temperatures during the pollination stage of initial grain or fruit set will reduce yield potential. Therefore, the temperatures that fall above specific thresholds at critical times during development can have significant impact on crop yields. For most perennial and temperate fruit and nut crops, winter temperatures (winter chilling) play a significant role in productivity for optimum flowering and fruit set. Horticultural crops in general are highly sensitive to minor environmental stresses related to temperatures outside the optimal ranges, which negatively affect visual and flavour quality. Therefore, the predicted rise of temperature could impact crop and horticulture yield (Table 3.4) if it exceeds optimal temperature range. Low latitude crops are being grown near the limits of temperature tolerance and global warming may put them to higher stress. The effect of high temperature could also accelerate in hastened maturation resulting low yield in crops.

Table 3.4: ***Optimum temperatures (ºC) for selected crops (a. failure temperature at which grain yield falls to zero; b. frost sensitivity: + = sensitive to weak frost; - = relatively insensitive; () = uncertain)*** *[Hatfield et al. 2008].*

1. Crop	Optimum temperature range for reproductive (grain) yield	Failure temperature for reproductive yield (a)
Maize	18-22	35
Soybean (*Glycine max*)	22-44	39
Wheat (*Triticum aestivum*)	15	34
Rice (*Oryza sativa*)	23-26	35-36
Sorghum (*Sorghum bicolour*)	25	35
Cotton (*Gossypium hirsutum*)	25-26	35
Peanut (*Arachis hypogea*)	20-26	39
Tomato (*Lycopersicon esculentum*)	22-25	30
2. Horticulture	**Optimum temperature range for reproductive (grain) yield**	**Frost sensitivity (b)**
Water melon	25-27	+
Okra	25-27	+
Melon	25-27	+
Sweet potato	25-27	+
Cucumber	20-25	+
Pepper	20-25	+
Sweet corn	20-25	+
Snap bean	20-25	+
Tomato	20-25	+
Onion	20-25	-
Garlic	20-25	-
Turnip	18-25	-
Pea	18-15	()
Potato	16-25	+
Lettuce	16-25	(+)
Cabbage	16-18(25)	-
Broccoli	16-25 (25)	-
Spinach	16-18(25)	-

3.2.2.2: Temperature and agriculture productivity
[based on Parry 1990; Easterling et al. 2007; IPCC 2007].

In the middle and higher latitudes, global warming would extend the length of the potential growing season. This would allow earlier planting of crops in the spring, earlier maturation and harvesting. For example, in Europe the growing season is estimated to lengthen by about 10 days per degree Celsius. There is a possibility of completing two or more crops during the same season. Crop producing areas may expand pole ward in countries such as Canada and Russia. As result, it is virtually certain that crop yields will increase in colder environments (Table 3.5). In tropical areas, mountain environment may be more favourable than the flat lands for major crops. Tropical crops are likely to move upwards with increasing global temperature.

Table 3.5: ***Summary of possible impacts of climate change on agriculture sector*** *[IPCC, 2007].*

Likelihood	Agriculture, forestry and ecosystems
Virtually certain (warmer and fewer cold days and nights; more frequent hot days)	• increased yields in colder environments; • decreased yields in warmer environments • increased insect outbreaks causing more pesticide use
Very likely (Warm spells/heat waves)	• reduced yields in warmer regions due to heat stress • increase of wild fire danger events
Very likely (increases of heavy precipitation events)	• damage to crops • soil erosion • water logging of soils
Likely (increases of droughts events)	• land degradation • lower yields/crop damage and failure • increased livestock deaths • increased risk of wildfire
Likely (increases of tropical cyclone)	• damage to crops • uprooting of trees • damage to coral reefs
Likely (sea level rise)	• salinisation of irrigation water, estuaries and freshwater systems

Climate modelling studies indicated a small beneficial effects on major cereals yield (rice, wheat) in temperate (mid to high latitude) due to small change of temperature (1-3^0 C), by contrast, in tropical or low latitude regions, model indicated negative yield impacts for the major cereals even with moderate temperature increases (1-2^0 C) (see Figure 3.2). The results in Figure 3.2 demonstrate that there is a potential for global food production increase with increases in local average temperature over a range of 1-3^0 C, but above this range the yield may decrease. However, adaptation measures proposed in section 3.4 may reduce the impacts.

On a world wide basis, global warming would benefit the mid to high latitudes zones for agriculture, whereas low latitudes, semi-arid and tropical areas will have much reduced crop and livestock yields (see Table 3.6, Figure 3.3).

3.2.3: Precipitation, water availability, irrigation
[based Rosenzweig and Hillel 1995; Tubiello et al. 2007].

According to IPCC, the precipitation will increase in high latitudes (chapter 2, Figure 2.1) and decrease in most sub-tropical land regions in the future. Climate change will modify rainfall, evaporation, run-off, and soil moisture storage. Increased evaporation from soil and accelerated transpiration in plants will cause moisture stress. The demand for irrigation is expected to rise in a warmer climate, causing increased competition between different water users. Globally, by 2080, the irrigation requirement for crops is estimated to increase to + 5% to 20% (due to net transpiration loss, longer growing seasons and evaporative loss), of which larger impacts will be in developing regions. The intensified evaporation will increase the hazards of salt accumulation in the soil.

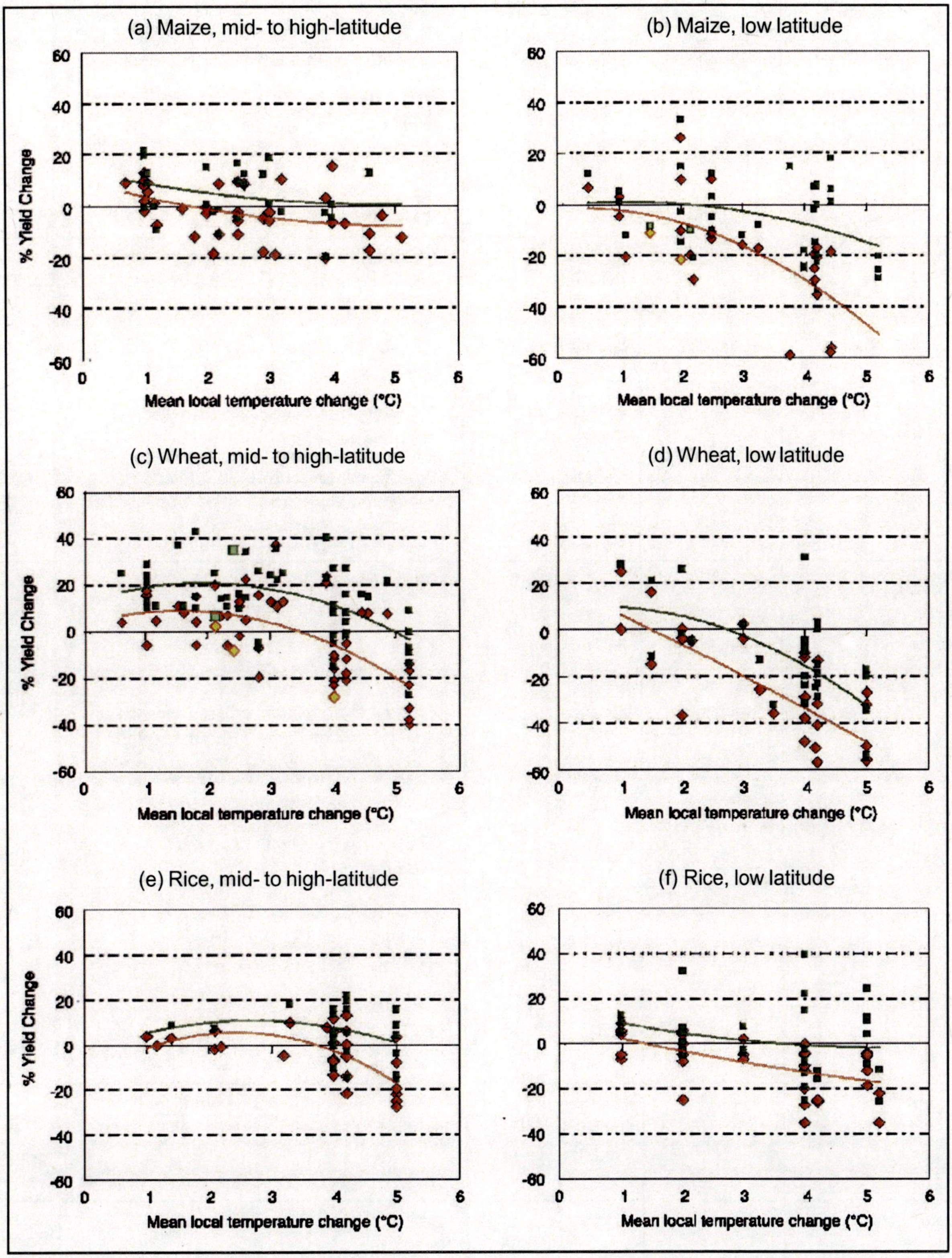

Figure 3.2: ***Sensitivity of cereal yields to climate change for maize, wheat and rice against mean local temperature changes.*** *Red dots = responses without adaptation; dark green dots = responses with adaptation. Results derived from 69 published studies at multiple simulation sites [see Easterling et al. 2007, page 286 for further details. http://www.ipcc.ch/pdf/assessment-report/ar4/wg2/ar4-wg2-chapter5.pdf].*

Table 3.6: ***Future warming increment effects on world's agriculture and livestock (based on climatic regions)*** *[adapted from Easterling et al. 2007) +ve = positive; -ve = negative; uc = unchanged].*

Climatic region	Temperature change	Sub-sector	Finding	Effects
Mid to high-latitudes	+1 to 2^0 C	Food crops	- Cold limitation alleviated for all crops.	+ve
			- Adaptation of maize and wheat increases yields 10-15%; Rice yield-no change; regional variation is high.	
	+2 to 3^0 C	Food crops	- Adaptation increases all crops above base line levels.	+ve
Temperate	+1 to 2^0 C	Pastures and livestock	- Cold limitation alleviated for pastures; seasonal increased frequency of heat stress for livestock.	+ve/-ve
	+2 to 3^0 C	Food crops	- Adaptation increases all crops above baseline yield.	+ve
		Pastures and livestock	- Moderate production loss in swine and confined cattle.	+ve
		Fisheries	- Positive effects on trout in winter, negative in summer.	+ve/-ve
	+3 to 5^0 C	Forestry	- Increase in fire hazard and insect damage.	-ve
Low latitudes	+1 to 2^0 C	Food crops	- Wheat and maize yields reduced below base line levels; rice unchanged.	-ve/ uc
	+2 to 3^0 C	Food crops	- Yields (all crops) above base line with adaptation but drops without adaptation.	+ ve & -ve
	+3 to +5^0 C	Food crops	- Yields (all crops) above base line with adaptation but drops without adaptation.	+ ve/-ve
		Food crops	- Maize and wheat yields reduced below baseline regardless of adaptation.	-ve
			- Adaptation maintains rice yields at baseline levels.	
Semi-arid	+1 to 2^0 C	Pastures and livestock	- No increase in net primary productivity; increased heat stress for livestock.	-ve
	+2 to 3^0 C	Pastures and livestock	- Reduction in animal weight and pasture yield; increased heat stress for livestock.	-ve
	+3 to +5^0 C	Pastures and livestock	- Reduction in animal weight & pasture yield; increased heat stress for livestock.	-ve
Tropical	+3 to +5^0 C	Pastures and livestock and Forestry	- Strong production loss in swine and confined cattle	-ve
			- Massive Amazonian deforestation possible.	-ve
Global	+1 to 2^0 C	Prices	- Agriculture prices could rise -10 to -30%.	+ve
	+2 to 3^0 C	Prices	- Agriculture prices could rise -10 to +20%.	+ve/ -ve
	+3 to +5^0 C	Prices and trade	- Agriculture prices could rise +10 to +40%.	-ve
			- Cereal imports of developing counters to increase by 10-40%.	-ve

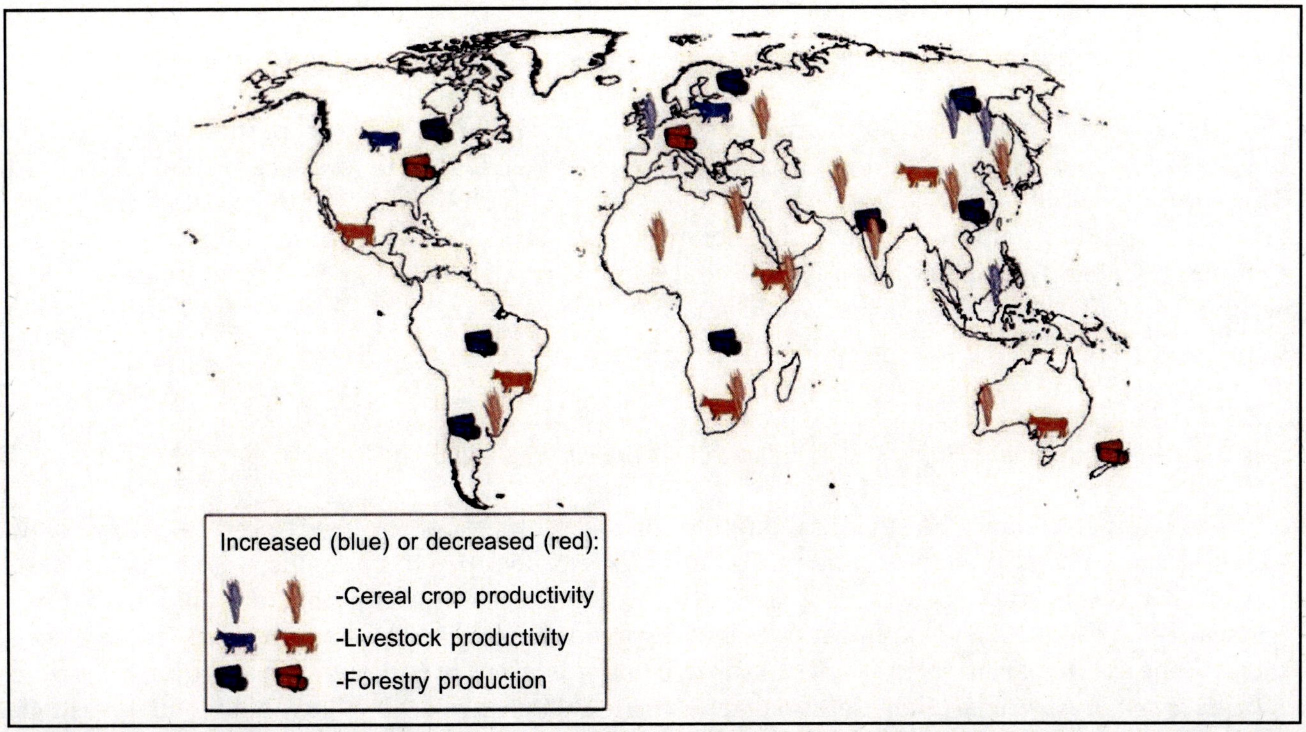

Figure 3.3: ***Major impacts of climate change on crop and livestock yields, and forestry production by 2050 (without adaptation).*** *[Note: Decreasing agricultural productivity in Australia, Africa, Asia and south America; source Easterling et al. 2007 page 30 . http://www.ipcc.ch/pdf/assessment-report/ar4/wg2/ar4-wg2-chapter5.pdf].*

3.2.4: Extreme events and climate variability

[based on Slingo et al. 2005].

Extreme events such as heat spell, heavy storms and drought and heavy precipitation are likely to increase (see Table 3.5) which would disrupt crop production and may cause production losses. For example, frequent droughts would not only reduce water supplies but also increase the demand for water needed for plant transpiration. Frequent droughts would also cause heat stress and water stress during the life cycle of crops affecting the yield. Certain varieties of crops which are grown near their limits of maximum temperature tolerance, such as rice in Southern Asia, heat spells can be particularly detrimental. A greater incidence of summer drought is expected in central and Southern Europe. Short-term natural extremes (storms, floods, El Nino southern oscillation) would also cause important effects on crop, pasture and forest production (Table 3.7).

Table 3.7: ***Examples of cyclones and sea level rise on certain agriculture crops*** *[adapted from Nicholls et al. 2007].*

Climate change	Examples
Extreme climate events (cyclone - floods and destruction)	- Negative effects on crop yields. - ***India:*** Negative effects on coconut production. - ***Australia:*** Negative effects on sugar cane and bananas (e.g. Cyclone Larry, March 2006). - ***Bangladesh:*** Negative effects on rice yield (e.g. Cyclone Sidr, November 2007). - ***Myanmar (Burma):*** Negative effects on rice yield (e.g. Cyclone Nargis, May 2008).
Rising sea level	- Negative impacts on coastal agriculture due to inundation. - ***Mekong delta:*** Significant changes to number of rice crops possible under 20-40 cm of sea level rise. - ***Benin & Côte d' Ivoire*****:** Inundation and soil salinisation could affect palm oil and coconut production. - ***Kenya*****:** Inundation and soil salinisation could affect mangoes, cashew nuts and coconuts.

3.2.5: Sea level rise

[based on Rosenzweig and Hillel 1995; Nicholls et al. 2007].

Global warming will lead to thermal expansion of seawater, along with partial melting of land-based glaciers and sea-ice, resulting in sea level rise from 0.26 to 0.59 m (see section 1.6 in Chapter 1). This would result in inundation of farmland threatening agriculture in low-lying coastal areas. It will also make coastal groundwater saltier. Agriculture farming in low-lying areas of Egypt, Bangladesh, Indonesia, China, the Netherlands, Florida and some island states could be affected (see chapter 2: section 2.4.2). The rising sea-levels would inundate farmland and make coastal groundwater saltier (Table 3.7).

3.2.6: Pests and diseases

[based on Rosenzweig and Hillel 1995; Pittock 2003; Tubiello et al. 2007].

The warmer climates are more favourable for the proliferation of insect pests. Also the longer growing season would enable some insects such as grasshopper to complete a greater number of reproductive cycles during the spring, summer and autumn. A warmer climate would enable more tropical pest, parasites and pathogen species to spread northwards and southwards. It would also increase the ability of some pest species to survive under warmer conditions such as fruit fly, *Bactrocera (Dacus) tryoni* in Australia. Likewise, simulated climate change increased vulnerability of the Australian beef industry to the cattle tick (*Boophilus microplus*). Increased climate extremes may promote plant disease and pest outbreaks. Climate change models projected that bluetongue, which mostly affects sheep, and occasionally goat and deer, would spread from the tropics to mid-latitudes. The possible increases in pest infestations would require greater use of pesticides to control them. Projected increases in CO_2 could accelerate growth of some invasive plants since they may tolerate a wide range of environmental conditions and may be more successful in a warming world because they can migrate and establish in new sites more rapidly than native plants.

3.2.7: Soil fertility and erosion

[based on Parry 1990].

Higher air temperatures are likely to speed up the natural decomposition of organic matters and increase the rates of other soil processes that affect fertility. This will require additional application of fertilizer to counteract the processes. The additional use of chemicals may impact water and air quality.

3.3: Case studies - Asia-Pacific and Australasia Region

3.3.1: Asia

Asia is the most populus continent, where total population in 2002 was 3,902 million, of which 61% people lives in rural areas and 38.5% lives within 100 km of the coast. The production of rice, maize and wheat has declined in recent times in many parts of Asia due to increasing water stress from increasing temperature and reduction in the number of rainy days. It is likely that temperature as well as rainfall will increase in North Asia (Table 3.8). The effects of climate change on Asian agriculture are listed below:

- Most warming is projected to be at high latitudes (North Asia) and Central Asia and West Asia and least rapid in South Asia, and stronger over South Asia and East Asia (see Table 3.8). Recent modelling experiments suggest that the warming would also be significant in the Himalayan highlands including the Tibetan Plateau.
- Studies conducted indicate an increase in annual precipitation in most Asia, of which the relative increase being largest in North Asia, East Asia and the Tibetan plateau (Table 3.8).

Table 3.8: ***Projected changes in surface air temperature and precipitation in different sub-regions of Asia**** *[Cruz et al. 2007].*

Subregions and Temperature change (°C)	2010-2039	2040-2069	2070-2099
North Asia	1.69-2.94	3.13-6.65	4.00-10.45
Central Asia	1.52-1.89	2.58-4.42	3.42-7.50
West Asia	1.06-1.55	2.0-3.7	2.7-6.3
Tibetan Plateau	1.49-2.05	2.74-4.44	3.73-7.62
East Asia	1.24-1.82	2.24-4.18	3.00-6.95
South Asia	0.54-1.17	0.88-3.16	1.56-5.44
South-East Asia	0.72-0.92	1.30-2.32	1.87-2.02
Subregions and Precipitation change (%)	**2010-2039**	**2040-2069**	**2070-2099**
North Asia	+7 to +16	+ 8 to + 35	+ 10 to + 59
Central Asia	-5 to +5	-7 to + 8	-13 to + 10
West Asia	-4 to + 18	-9 to + 27	-25 to + 52
Tibetan Plateau	+ 4 to +14	+5 to +21	+7 to + 31
East Asia	0 to + 6	+2 to + 13	+4 to +21
South Asia	-3 to +7	0 to + 26	-16 to + 31
South-East Asia	-2 to +1	-1 to +3	+1 to +12

* ***North Asia*** = Russia, Mongolia; ***Central Asia*** = North-West China; ***West Asia*** = Middle East; ***East Asia*** = China, Japan, Korea; ***South Asia*** = India, Nepal, Pakistan, Bangladesh, Sri Lanka; and ***South-East Asia*** = Indonesia, Philippines, Thailand, and Vietnam

- Crop yields could increase up to 20% in East and South East Asia and a decrease up to 30% in Central and South Asia (Table 3.9).
- Combined influence of 2 x CO_2 fertilisation, thermal stress and water scarcity may cause a 3.8% decline of rice production in Asia.
- Rice and wheat yields in Bangladesh and wheat and maize yield in India is projected to drop (Table 3.9).

Table 3.9: ***Projections of climate change impacts on Asian agriculture*** *[based on Cruz et al. 2007].*

Sub-regions or country	Yield
General crop yields	increase by 20% in East and South Asia decrease by 30% in Central and South Asia
Cereal production in South Asia	decline by 4-10%
Rice yield in China (rain fed)	drop by 5-12% with a rise of 2°C temperature
Rice yield in Asia	decline by 3.8%
Rice yield-Bangladesh	drop by 8% by 2050
Wheat yield-Bangladesh	drop by 32% by 2050
Wheat and maize yield in India	decrease by 2-5% for a rise of 0.5- 1.5 °C

3.3.2: *Australia and New Zealand*

[based on Hennessy et al. 2007].

In Australia, wheat production in south Western Australia is projected to have significant reduction in yields by 2070 compared to a moderate increase in yields in north-eastern Australia. Climate change is likely to change land use in southern Australia where growing of crop may be non-viable (at dry margins) if rainfall is reduced substantially as projected (see Table 2.6 of Chapter 2 for future rainfall pattern in Australia). The other effects of climate change on Australia and New Zealand's agriculture are given in Table 3.10.

In Australia, temperate fruits and nuts is projected to be negatively affected by warmer conditions (see Table 2.6 in Chapter 2 for projected temperature rise in Australia) because temperate fruits require winter chill or vernalisation. Similarly, in New Zealand, kiwi fruit are likely to increase vegetative growth at the expanse of fruit growth and quality can be affected due to warmer summer temperatures. Also, production of kiwi fruit varieties may be uneconomic in Northland (NZ) by 2050 because of lack of winter chilling. Moreover, some orchards will be affected which are reliant on irrigation. Earlier ripening and reductions in grape quality and value is projected in viticulture in Australia and New Zealand. On the other hand, warming will allow alternative varieties or grapes to be grown. Some tropical pests such as Queensland fruit fly *B. tryoni* could be a threat to southern Australia (see Table 3.10 for other effects of climate change on horticulture and other crops).

Table 3.10: *Likely effect of climate change on Australia and New Zealand's agriculture* [*Pittock 2003; GoV 2007; Hennessy et al. 2007*].

Country	Sectors	Climate change effects
Australia	Crops	• Broad acre crops are likely to benefit from higher CO_2 concentrations, but these benefits may be negated by higher temperatures and reduced rainfall. • With higher temperatures and reduced water supplies, increased evaporation rates, cotton production is likely to decline. • Increased CO_2 concentrations is expected to have positive effects on wheat production (C_3 plants), (see also Table 3.1) however, reduced water supplies, higher rates of salinity and an increased frequency of droughts would put enormous stress upon crop yields. • Though increasing CO_2 concentrations increases wheat grain yield but this could result in a decrease in wheat grain protein content (due to effects of heat stress or heat waves) leading to deterioration in bread or pasta quality (doubling CO_2 could reduce grain nitrogen content by 4-14%).
Australia	Horticulture and viticulture	• A decrease in frost frequency due to warming would reduce the risk of damage of those fruits that are sensitive to frost. • Temperate fruits (apples, stone fruits) requiring winter chilling (for normal bud burst and fruits setting) would be at risk of lower yields and reduced fruit quality. • Decreased frosts may lead to a prolonged dormancy of temperate fruits. • With decreased rainfall and increased temperature, viticulture is expected to be significantly affected (e.g. earlier ripening and reduction in quality) and in extreme rainfall events, this could greatly impact grape yields. • In cooler climates, warming (e.g. Tasmania), may allow to grow new varieties of grapes.
Australia	Livestock	• With higher temperatures and overall decreased water supplies, beef- cattle and sheep are likely to experience increased incidences of stress related deaths (physiological effects of heat stress include increased susceptibility to parasites; reduced food intake, weight loss, decreased reproduction rates; reduction in milk yields). • Dairy cattle's are also expected to experience an increase in stress related deaths.
Australia	Pastures	• A small increase in temperature and reduced rainfall - pasture growth could be reduced by 10-50% across different parts of Australia. • A large warming and reduced rainfall - average growth of pasture across the continent could be reduced between 30% and 70% by 2030. • A large warming and increased rainfall would cause moderate gain in pasture production by 2030. • High temperatures and low rainfall may reduce pastures production and constraint animal production. • In higher rainfall areas pastures are likely to benefit from higher CO_2 concentrations.
New Zealand	Crops, kiwi fruit	• C_3 crops (wheat) may increase due to $2xCO_2$ response, whereas C_4 crops (maize) may have a reduction in growth • Predicted warmer temperature would affect kiwi fruit growth and quality.

3.4: Adaptation measures

There are many adaptation options available (Table 3.11) to counteract projected climatic impacts on agriculture. These include use of heat/drought tolerant and salt tolerant crops and breeding livestock for greater tolerance to heat stress. The other measures that may be considered are better farm management practices, diversification of crop varieties, shifting cropping seasons and improved farm technologies. Adaptation measures would significantly improve GDP (Table 3.12).

Table 3.11: ***Adaptation measures in agriculture*** *[Cruz et al. 2007].*

Sectors	Adaptation measures
Agriculture	• Use of more heat/drought tolerant crop varieties in areas under water stress. • Use of more disease and pest tolerant crop varieties. • Use of salt tolerant crop varieties. • Introduce higher yielding, earlier maturing crop varieties in cold regions. • Alerted application of nutrients/fertiliser. • Altered application of insecticides/pesticides. • Altering the timing or location of cropping activities. • Change planting date to effectively use the prolonged growing season and irrigation. • Altering amounts and timing of irrigation and other water management. • Develop adaptive management strategy at farm levels.
Livestock	• Breeding livestock for greater tolerance and productivity. • Increase stocks of forages for unfavourable time periods. • Improve pasture and grazing management including improved grasslands and pastures. • Improve management of stocking rates and rotation of pastures. • Increase the quantity of forages used to graze animals. • Plant native grassland species. • Increase plant coverage per hectare.
Agricultural bio-technologies	• Development and distribution of more drought, disease, pest and salt tolerant crops. • Develop improved processing and conservation technologies in livestock production. • Improve crossbreeds of high productivity animals.
Agricultural infrastructure	• Improve pasture water supply. • Improve irrigation system and their efficiency. • Improve use/store of rain and snow water. • Improve information exchange system or new technologies at national as well as regional and international levels. • Improve sea defence and flood management. • Improve timely weather forecasting.

Table 3.12: ***Global impacts on agriculture (Gross Domestic Product, GDP) for a 2.5°C increase in global temperature in per cent changes from reference projection of GDP*** *[Tol 2002; FAO 2007].*

Regions	Without adaptation	With adaptation	Mean
Latin America	-0.8 ±0.6	0.6±0.7	-0.1
South and South East Asia	-0.7±0.3	0.6±0.3	0.0
Middle-East	-0.4 ±0.4	0.6±0.5	0.1
Africa	-0.2 ±0.2	0.5±0.3	0.1
OECD-P	-0.2 ±1.6	0.8±1.6	0.3
OECD-A	-0.3±1.3	1.0±1.3	0.4
OECD-E	0.6 ±1.0	2.1±1.1	1.3
Eastern Europe and Former Soviet Union	0.9±1.2	2.7±1.1	1.8
Centrally Planned Asia	1.7±1.0	3.1±1.0	2.4

3.5: Conclusion

Crop cultivation as well as livestock production could be significantly affected by climate change. With regard to crop cultivation, variations in climatic parameters such as temperature, precipitation, drought, and CO_2 levels are projected to have notable consequences on crops. Livestock losses are expected in many countries resulting from heat waves and heat stress; climate change may also increase susceptibility of livestock to parasites; reduce food intake, weight loss, decrease reproduction rates and reduction in milk yields.

The climate change may increase agriculture production in colder regions and decrease production in warmer regions (low latitude developing countries). The likely impacts of climate change on agriculture would vary from country to country due to its geographical locations. However, different adaptive measures suited to the specific regional areas would reduce such impacts. The adaptive measures may include but not limited to, are breeding of plants that would be able to adapt to higher ambient CO_2 concentrations and increased temperature and reduced rainfall conditions; development of crop varieties that are adaptive to greater tolerance to stresses, drought, heat and flood. Currently, the International organisations such as CGIAR, IRRI are investigating to develop drought-tolerant maize, wheat and beans; drought-escaping rice (rice that tolerate harsh growing conditions) or water-proof rice (plants that can survive completely submersed conditions for up to two weeks). It would be essential to adopt new techniques and management approaches to compensate for the negative impacts of climate changes. Water conservation technologies, integrated water resources management, use of adaptive crop varieties, improved seeds including hybrids, efficient use of fertilizer, irrigation and crop diversification may reduce the likely impacts of climate changes. Selection of naturally hardy food crops (barley, cassava, millet and sorghum) which are widely grown in warmer climate or breeding of heat and drought resistant crop and livestock varieties that may be better adapted to new climatic and atmospheric conditions is required. An awareness campaign about the need to follow the diversification of crops to cope with new situations may also be required.

3.6: References

Akita, S. and Moss, D. N. 1973. Photosynthetic responses to CO_2 and light by maize and what leaves adjusted for constant stomatal apertures. *Crop Science* 13: 234-237.

Cruz, R.V., Harasawa, H., Lal, M., Wu, S., Anokhin, Y., Punsalmaa, B., Honda, Y., Jafari, M., Li, C. and Ninh, N. H. 2007. Asia. *Climate change 2007 Impacts, adaptation and vulnerability. Contribution of Working Group II to the Fourth Assessment Report of the Intergovernmental Panel on Climate Change.* Parry, M.L., Canziani, O.F., Palutikof, J. P., van der Linden, P. J. and Hanson, C. E. (eds.). Cambridge University Press, Cambridge, UK. pp. 469-506.

Easterling, W. E., Aggarwal, P.K., Batima, P., Brander, K. M., Erda, L., Howden, S. M., Kirilenko, A., Morton, J., Soussana, J. F., Schmidhuber, J. and Tubiello, F. N. 2007: Food, fibre and forest products. Chapter 5, pp. 273-313. In: *Climate change 2007: Impacts, adaptation and vulnerability.* Contribution of Working Group II to the Fourth Assessment Report of the Intergovernmental Panel on Climate Change. Parry, M.L., Canziani, O.F., Palutikof, J. P., van der Linden, P. J. and Hanson, C. E. (eds.). Cambridge University Press, Cambridge, UK. (Can also be accessed via http://www.ipcc.ch/pdf/assessment-report/ar4/wg2/ar4-wg2-chapter5.pdf).

FAO 2007. *Adaptation to climate change in agriculture, forestry and fisheries : perspective, framework and priorities.* FAO Interdepartmental Working Group on Climate Change, Food and Agriculture Organisation of the United Nations. 00153, Rome, Italy. 24 pp.

GoV 2007 http://www/greenhouse.vic.gov.au/greenhouse/wcmn302.nsf/LinkView/57A8049B13F0c6E (climate change impacts, state Government of Victoria) accessed on 3.12.07

Hatfield, J., Boote, K., Fay, P., Hahn, L., Izaurralde, C., Kimball, B. A., Mader, T., Morgan, J., Ort, D, Polly, W., Thompson, A. and Wolfe, D. 2008. Agriculture. In : *The effects of climate change on agriculture, land resources, water resources, and biodiversity in the United States.* A report by the U.S. Climate Science Program and the subcommittee on Global Change Research. Washington, DC., USA. 362 pp. http:/www.climatescience.gov/Library/sap/sap4-3/final-report/default.htm

Hennessy, K., Fitzharris, B., Bates, B. C., Harvey, N., Howden, S. M., Hughes, L., Salinger, J. and Warrick, R. 2007. Australia and New Zealand. *Climate change* 2007. Chapter 11. pp. 507-540. *Impacts, adaptation and vulnerability.* Contribution of Working Group II to the Fourth Assessment Report of the IPCC. Parry, M.L., Canziani, O.F., Palutikof, J. P., van der Linden, P. J. and Hanson, C. E. (eds.). Cambridge University Press, Cambridge, UK. (Can also be accessed via http://www.ipcc.ch/pdf/assessment-report/ar4/wg2/ar4-wg2-chapter5.pdf).

IPCC 2007. Climate change 2007: Impacts, adaptation and vulnerability . Working group II contribution to the Intergovernmental Panel on Climate Change Fourth Assessment Report. Summary for policy makers. (http://www.ipcc.ch).

Nicholls, R. J., Wong, P. P., Burkett, V. R., Codignotto, J. O., Hay, J. E., McLean, R. F., Ragoonaden, S and Woodroffe, C.D.2007. Coastal systems and low-lying areas. *Climate change 2007. Adaptation and Vulnerability. Contribution of Working Group II to the Fourth Assessment Report to the Intergovernmental Panel on Climate Change.* Parry, M. L., Canziani, O. F., Palutikof, J. P., van der Linden, P. J. and Hanson, C. E. (eds.). Cambridge University Press, Cambridge, UK. pp. 173-210. http://www.ipcc.ch/pdf/assessment-report/ar4/wg2/ar4-wg2-chapter6.pdf.

Parry, M. L. 1990. "*Climate Change and World Agriculture*", Earthscan Publications. London. http://www.ciesin.columbia.edu/docs/004-038a.html

Pittock, B. (editor). 2003. *Climate change: An Australian guide to the science and potential impacts.* Commonwealth of Australia. Australian greenhouse Office. 228 pp. plus index (11 pp).

Rosenzweig, C. and Hillel, D. 1995. Potential impacts of climate change on agriculture and food supply. *Consequence* 1(2) : http://www.gcrio.org/consequence/summer95/agriculture.html.

Slingo, J.M., Challinor, A.J., Hoskins, B.J. and Wheeler, T.R. 2005. Introduction: food crops in a changing climate. *Philosophical Transaction of the Royal Society B.* 360:1983-1989. doi:10.1098/rstb.2005.1755.

Smith, P., Martino, D., Cai, Z., Gwary, D., Jazen, H., Kumar, P., McCarl, B., Ogle, S., O'Mara, F., Rice, C., Scholes, B., and Sirotenko, O. 2007. Agriculture, pp. 497-540. In Climate change 2007: Mitigation. Contribution of Working Group III to the Fourth Assessment Report of the IPCC on Climate Change. Metz, B., Davidson, O. R., Bosch, P. R., Dave, R. and Meyer, L. A. (eds.). Cambridge University Press, Cambridge, United Kingdom.

Tol, R.S. J. 2002. New estimates of the damage costs of climate change, part I: Benchmark estimates, *Environmental and Resource Economics,* 21 (1): 47-73.

Tubiello, F.N., Soussana, J-F. and Howden, S.M. 2007. Crop and pasture response to climate change. *PNAS* 104 (50):19686-19690. www.pnas.org/cgi/doi/10.1073/pnas.0701728104.

Website

http://www.cana.net.au/water/farmers/impacts.html (Impacts on Australian agriculture and livestock)

http:/www.gcrio.org/Consequnences/summer95/agriculture.html (potential impacts of climate change on agriculture and food supply by Cynthia Rosenzeweig and Daniel Hillel) updated 11 November 2004.

http://www.cs.ntu.edu.au/homepages/jmitroy/sid101/uncc/fs101.html (The impact of climate change on agriculture, climate change fact sheet 101, Information Unit on climate change (IUCC), UNEP, Switzerland. accessed on 11 November 2007).

http://www.greenhouse.gov.au/impacts/agriculture.html (Australia's Agriculture - Impacts of Climate Change).

http://10.10.2.119:8080/00123f24cda4/comfort-file/5d15_3b635476_a76b_11dc_9935_00123f24cda4/j9271e.pdf (adaptation to climate change in agriculture, forestry and fisheries: perspective, framework and priorities, FAO, 2007).

ftp://ftp.fao.org/docrep/fao/005/y4252e/y4252e13.pdf (Climate change and agriculture: physical and human dimensions- chapter 13).

http://www.fao.org/docrep/W5183E/w5183e00.htm

□□□

4 Climate Change Impacts on Fisheries and Aquatic Ecosystems

Key facts

- Fish is an important source of animal protein for billions of people (estimate 2.6 billion people). About 200 million people and their dependants worldwide, mostly in developing countries, live by fishing and aquaculture.
- Fish living in temperate and Polar Regions could be benefited from climate changes since an increase in water temperature would extend the fish growing season. Moreover, rising temperature could reduce over-wintering stress normally experienced by temperate fishes.
- Rise of temperature may cause a decrease in dissolved oxygen (DO) or hypoxic conditions (as temperature and DO are inversely related) and can impair fish reproduction and may cause endocrine disruption in fish (increase incidence of malformations and may alter sex differentiation and sex determination in fish).
- Bioavailability and toxicity of common pollutants such as pesticides and heavy metals may be enhanced with increasing temperatures. Therefore, exposure of toxicants at elevated temperatures may cause a decline in recreational and commercial fisheries.
- Climate changes would make the ocean more acidic (due to higher CO_2 levels), adversely affecting many organisms that use calcium carbonate for their skeletons and shells, such as corals, molluscs and some phytoplankton.
- Rise of sea surface temperature may cause the following effects on marine fisheries: distribution of marine organisms towards north or south poles; enhance shallow water fish yields; a change in match-mismatch hypothesis relating to fish recruitment.
- A comparison of the different levels of temperature rise reveals the following effects on marine, freshwater and terrestrial ecosystems :*Marine ecosystems-* extinction or loss of corals and polar animals; loss of habitat of marine birds; loss of bivalves, pteropod and mammals, and reductions in krill populations. *Freshwater ecosystems* - decline of freshwater fish population, loss of fish habitat and loss of wetlands; *Terrestrial ecosystems* - loss or extinction of plant species, butterflies, reptiles, frogs, birds and mammals.
- Climate change may bring in more frequent storm surges thereby could damage coastal fish ponds causing loss of aquaculture stocks and coastal infrastructure. It will introduce disease or predators into aquaculture facilities.
- Though sea level rise would expand areas suitable for brackish water aquaculture, however, it will cause loss of land for freshwater aquaculture, loss of freshwater fisheries and reduce freshwater available for inland aquaculture.
- 33 countries are highly vulnerable to the effects of global warming on fisheries, of which 21 in Africa, 9 in Asia and 3 in South America.
- Adaptation measures that can be taken in anticipation of future effects of climate change on aquaculture and fisheries include breeding of fish tolerant to high water temperature, development of fisheries management capabilities to cope with impacts of climate change (long range fishing crafts, installation of onboard processing equipment), alternative means of transport when coastal roads are flooded and construction of appropriate infrastructures for storm forecasting, and signalling systems and safe refuge for dealing with possible rising of sea level and increased storms.

Contents

List of Tables

List of Figures

List of Boxes

Acronyms and Glossary

Anoxic: Condition referring to absence of oxygen.

Anthropogenic: Human influenced causes.

Aragonite: A form of calcium carbonates and named for Aragon, Spain, where it was first found in 1790. It is transparent to translucent and can range in colour from honey-colored to pale reds, blues and greens to clear or white. Composes the calcium carbonate skeletons of marine organisms including reef corals, pteropods (e.g. marine snails), bivalves (e.g. oysters) and cephalopods (e.g. squids). Aragonite is more sensitive to ocean acidification than calcite.

Bioclimatic: A climate that is influenced by biological organisms.

Calcification: Process in which tissue becomes hardened as a result of calcium deposits.

Coral bleaching: Is the loss of colour from corals under stressful environmental conditions.

DO: Dissolved oxygen.

Ectotherms: Cold blooded organisms such as fish, whose body temperature is mostly determined by the surrounding environment.

ENSO events: El Niño Southern Oscillation (ENSO), is a general term used to describe both warm (El Niño) and cool (La Niña) Ocean atmosphere events in the tropical Pacific. ENSO is associated with floods, droughts and other disturbances in a range of locations around the world.

Epilimnion: Upper strata of water.

Euryhaline organisms: Organisms those are able to adapt to a wide range of water salinities (e.g. Tilapia, salmon, trout, and molly). For example, euryhaline fish molly (*Poecilia* sp.) can live in fresh, brackish or salt water. Euryhaline organisms are commonly found in habitats such as estuaries and tidal pools where the salinity changes regularly. However, some organisms are euryhaline as their life cycle involves migration between fresh- and marine environments, as in the case with salmon, eels and Indian shad.

Eutrophication: A naturally occurring change that takes place after a water body receives inputs of nutrients, mostly nitrates and phosphates from erosion and run-off from surrounding lands; this process can be accelerated by human activities.

Filter feeder: An organism that feeds by passing water through its body and filtering out the edible plant and animal matter. Examples include molluscs such as clams and oysters and many species of fish and birds.

Hypolimnion: Lower strata of water.

Hypoxic: Condition referring to dissolved oxygen concentration < 2.8 mg/l or low.

Homeostasis: Maintenance of the body's internal environment.

Hydrology: The study of water occurrence, distribution, movement and balances in ecosystems; the seasonal patterns of a river's flow.

LC_{50}: Lethal concentration at which 50% test animal dies.

Lentic: Related to still waters such as ponds, lakes or swamps.

Lotic: Related to fast-moving water, such as in most streams and rivers.

Maquis: Is a type of high ground covered in thick vegetation (shrubland).

Omega-3- polyunsaturated fatty acids : EPA (eicosapentaenoic acid, 20:5 (n-3)) and DHA (Docosahexaenoic acid, 22:6 (n-3)).

Phenology: The timing of events in an animal's life, such as when it lay eggs, migrates, or shows peak in abundance. As this event may be sensitive to climate, phenology studies provide evidence of climate impacts on biology.

Pteropod: Sea butterflies, also known as Thecosomata or flapping snails, are a taxonomic suborder of small pelagic swimming sea snails. These are holoplanktonic opisthobranch gastropod molluscs in the order Opisthobranchia. These creatures, which are about the size of a lentil, are eaten by various marine species, including a wide variety of fishes that are, in turn, consumed by penguins and polar bears. As the ocean becomes more acidic and warmer, these creatures will not be able to survive. By 2050 they may not be able to make a shell anymore. If we lose these organisms, the impact on the food chain will be catastrophic.

Savanna: A flat grassland of tropical or subtropical regions.

Stenohaline organisms: Organisms tolerating only a narrow range of salinities; opposite to euryhaline. Most freshwater organisms are stenohaline and will die in seawater and similarly most marine organisms are stenohaline, and can not live in freshwater.

Stratification: The upper water column typically comprises several layers differing in temperature and or salinity.

4.1: Introduction

Fish is an important source of animal protein for billions of people (estimate 2.6 billion people) (Allison *et al.* 2009). It is also an important source of essential vitamins and fatty acids (e.g. omega- 3 poly-unsaturated fatty acids). About 200 million people and their dependants worldwide, mostly in developing countries, live by fishing and aquaculture. Fish provides an important source of cash income for many rural poor. It provides support for local economies as well as a source of foreign exchange (see Table 4.1).

Table 4.1: ***World fisheries and aquaculture production in some selected countries where fish plays an important role in supplying animal protein (only countries supplying >25% of animal protein is included)*** *[original from FAO 2004 cited in Earth Trends data tables, http://earthtrends.wri.org/pdf_library/data_table/watcoal_2005.pdf].*

Continent/ Country	Inland and Marine Fisheries Production (thousand metric tons) (three years average) Capture		Aquaculture		Trade in Fish and Fisheries Products (million $US) (three years average)		Number of Fishers 2000	Fish Protein as a Percent of Animal Protein Supply 2002
	1990-92	2000-2002	1990-92	2000-2002	Imports	Exports		
Asia								
Bangladesh	684.2	1,058.8	210.1	718.8	6.2	328.3	1,320,480	52
Cambodia	106.3	372.9	7.2	14.3	3.2	27.9	73,425	57
China	7,449.7	16,690.0	7,206.8	26,132.7	1,927.0	4,029.1	12,233,128	19
India	2,867.6	3,799.4	1,212.6	2084.6	23.1	1,351.8	5,958,744	14
Indonesia	2,704.3	4,300.8	522.6	855.6	88.2	1,536.6	5,118.571	57
Japan	8,598.8	4,715.7	808.7	797.7	14,204.2	786.3	260,200	45
Korea, Dem Peoples Rep	2,321.9	1,828.6	364.9	294.9	1,619.9	1,195.9	176,928	40
Malaysia	966.3	1,270.6	65.8	158.4	335.9	359.6	100,666	38
Maynamar/Burma	731.6	1,183.1	14.0	113.8	1.4	210.4	610,000	46
Philippines	1,875.4	1,961.2	391.8	423.9	89.0	396.4	990,872	39
Sri Lanka	185.9	290.9	5.5	9.3	73.2	106.3	146,188	51
Thailand	2,664.2	2,950.3	338.7	702.4	947.7	4,027.6	354,495	40
Vietnam	826.1	1,483.0	164.4	515.9	44.9	1,764.2	29	
Europe								
Iceland	1,375.8	2,031.0	2.7	3.9	65.2	1,309.5	6,100	29
Lithuania	330.3	127.0	4.5	1.9	78.5	57.4	4,700	27
Norway	2,015.3	2,710.0	147.5	518.6	627.9	3,488.7	23,552	26
United Kingdom	788.0	726.0	55.9	167.3	2,249.4	1,305.9	17,847	10
Africa								
Angola	121.3	250.6	-	-	17.5	22.4	30,364	34
Cameroon	70.7	114.4	0.1	0.2	23.7	0.5	24,500	34
Congo	44.4	43.3	0.2	0.2	19.2	2.2	10,500	43
Congo, Dem Rep	171.7	214.6	0.7	3.6	33.5	0.4	108,400	43
Gabon	22.0	43.7	0.0	0.2	12.4	13.5	8,258	33
Gambia	21.5	36.4	0.0	0.0	0.7	2.8	2,000	61
Ghana	393.9	423.6	0.4	5.7	100.4	74.8	230,000	43
Guinea	49.5	100.2	0.0	0.0	6.6	2.0	10,707	43
Liberia	8.3	11.5	0.0	0.0	2.1	0.1	5,143	26
Malawi	68.9	41.6	0.2	0.6	0.4	0.2	42,922	26
Nigeria	287.5	458.2	13.3	26.9	197.6	17.6	481,264	29
Senegal	334,9	393,7	0.0	0.1	1.0	245.5	55,547	44
Sierra Leone	63.6	77.6	0.0	0.0	4.1	13.7	17,990	61
Tanzania, United Rep	357.1	331.1	0.4	0.4	0.4	107.4	92,529	27
Togo	13.0	22.1	0.1	0.4	10.9	6.3	14,120	40
South America								
Guyana	39.6	50.1	0.1	0.6	2.4	55.9	6,571	38
Peru	7,0897.7	9,137.2	5.9	8.2	20.9	1,1361.1	66,361	25
Oceania								
Australia	221.8	193.1	14.4	35.3	529.5	933.5	13,800	7
Fiji	29.1	43.6	0.0	1.7	21.5	38.1	8,985	30
Solomon Islands	49.7	28.8	0.0	0.0	0.5	15.0	11.000	76

Fisheries ecosystems and fishing based livelihoods are subjected to a range of climate-related variability, from extreme weather events, floods, droughts, through changes in aquatic ecosystem structure and productivity to changing patterns and abundance of fish stocks. Human induced climate change is likely to cause major shifts in ocean system productivity and surface freshwater availability. The impact of long term trends in climate change is less understood in fisheries. However, global ocean circulation patterns, sea level rise, changes in ocean salinity and pH and rise in inland water and sea surface temperature and extreme events all of which would affect the biological properties and distribution of species. Moreover, climate change due to enhanced greenhouse gases could affect global fisheries production through changing of water quality and hydrology of water bodies (UNEP 1994).

4.2: Effects on freshwater ecosystems

(based on Handisyde et al. 2006; Ficke et al. 2007; Muir and Allison 2007; Noyes et al. 2009; Wu 2009].

4.2.1: Rising inland temperature and fish biology

The potential effects of climate change i.e. increasing temperature can alter the physiological functions of fish such as thermal tolerance, growth, metabolism, food consumption, reproduction and homeostasis. Fish living in temperate and polar-regions could be more beneficial since a larger temperature changes are expected in the higher latitudes (as compared to tropical areas where a small temperature increase predicted). An increase in water temperature would extend the fish growing season in temperate areas. Moreover, rising temperature could reduce over-wintering stress normally experienced by temperate fishes. Thus, longer growing season and lower winter stress may enhance productivity of temperate fisheries. Global warming may result in the northward expansion of warm water and cool water (Table 4.2) species in North America, Europe and Asia and the southward expansion of warm water species in Australia and South America. Tropical species are expected to move pole wards.

Changing of global temperatures may increase susceptibility of fish to diseases since bacterial disease, whirling disease in aquaculture systems often peak at high temperature (see Table 4.3 for positive and negative effects of climate change on fisheries and aquaculture). Furthermore, fish migrating to pole wards from warmer regions may serve as hosts or vectors for parasites and diseases in the new environment. As temperature tolerance ranges are species-specific to fish, therefore, a change in thermal temperature regimes may cause increase or decrease of fish abundance from an ecosystem.

Though, an increase in temperature seems beneficial for temperate fish, however, rise of temperature would cause a decrease in dissolved oxygen supply (oxygen solubility in water is inversely related with temperature) (Figure 4.1). Rising temperature would also increase in the biological oxygen demand. In general, dissolved oxygen (DO) concentrations of 5 mg/l or more are suitable for most aquatic organisms and concentration below 2.8 mg/l is considered hypoxic. Decreased levels of dissolved oxygen due to rise of temperature may affect fish's growth, reproductive success and survival (Table 4.3). For example, Wu *et al.* (2003) and Wu (2009), reported that hypoxia (low levels of dissolved oxygen) can impair fish reproduction, alter reproductive behaviours, reduce fertilisation and egg hatching success and can cause endocrine disruption in fish (e.g. increase incidence of malformations in fish and may alter sex differentiation and sex determination). Therefore, hypoxic conditions could be a threat to the sustainability of natural fish populations (chapter 11 discusses the effects of various endocrine disrupting chemicals on fish).

Table 4.2: *Temperature tolerances of some common cold water, cool water, warm water and tropical fish species* [Ficke et al. 2007].

Temperature guild	Species	*ILLT-ULLT (° C)	Optimal range (° C)
Cold water	Arctic char (*Salvelinus alpinus*)	0-19.7	6-15
	Atlantic salmon (*Salmo salar*)	-0.5-25	13-17
Cool water	Walleye (*Stizostedion vitreum*)	0-30	20-23
	Striped bass (*Morone saxatilis*)	2-32	13-24
Warm water	European eel (*Anguilla anguilla*)	0-39	22-23
	Channel catfish (*Ictalurus punctatus*)	0-40	20-25
Warm water/tropical	Common carp (*Cyprinus carpio*)	0-35.7	26.7-29.4
Tropical	Redbelly tilapia (*Tilapia zillii*)	7-42	28.8-31.4
	Guinean tilapia (*Tilapia guineensis*)	14-34	18-32

*ILLT is the lower incipient lethal temperature below which a fish cannot survive for an indefinite time; ULLT is the upper incipient lethal temperature above which a fish cannot survive.

It is known that the toxicity of common pollutants such as organophosphates and heavy metals to fish increases at higher temperatures. The uptake of pollutants by fish also increases at higher temperatures. The higher toxicant uptake may reduce fish's ability to reproduce or fish accumulating high level of toxicants could be unfit for human consumption. A recent review by Noyes *et al.* (2009) found that bioavailability and toxicity of pesticides (atrazine, dieldrin, endosulfan) to fish (e.g. rainbow trout and other aquatic biota) enhanced with increasing temperatures (see Table 4.4). Therefore, exposure of toxicants at elevated temperatures may cause a decline in recreational and commercial fisheries [see chapter 8 and 9 for the toxicity values (LC_{50}) of pesticides and heavy metals to fish].

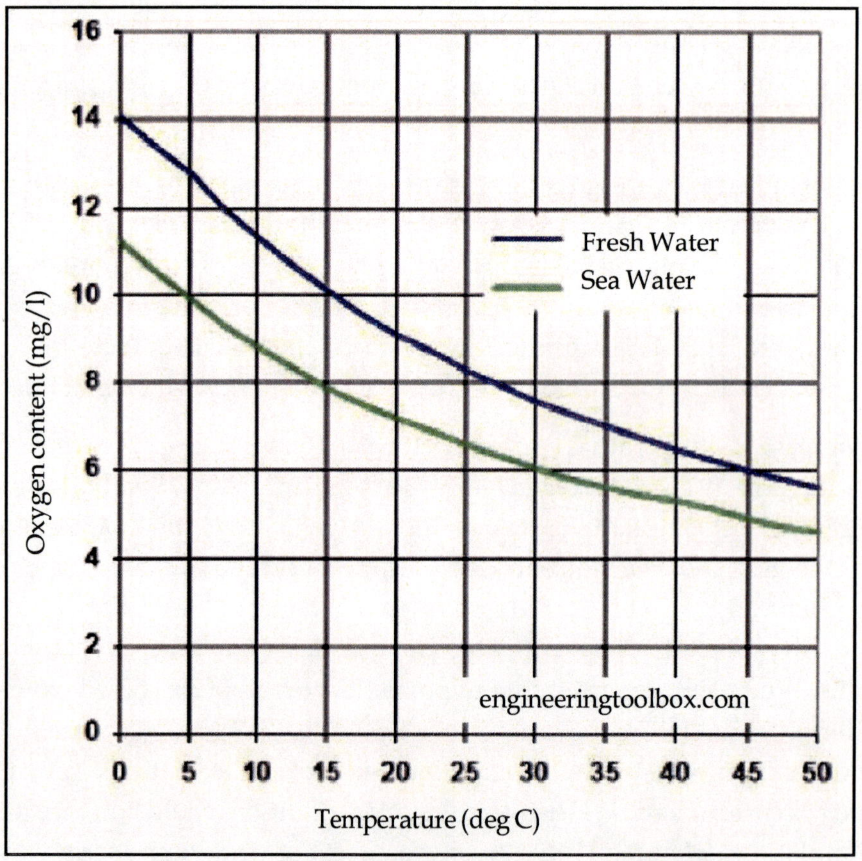

Figure 4.1: *Oxygen solubility vs. various temperatures* [*http://www.engineeringtoolbox.com/oxygen-solubility-water-d_841.html*].

Table 4.3: ***Some selected examples of negative and positive effects of climate change on freshwater and marine aquaculture and fisheries*** [*FAO 2005-08; Handisyde et al. 2006; Muir and Allison 2007*].

Parameters	Positive effects	Negative effects
Higher inland water temperatures	- improved feed conversion ratio and growth rates - enhanced primary productivity	- increased stratification and reduced mixing of water in lakes. - reduced water quality, especially dissolved oxygen (Figure 4.1). - change in range and an abundance of pathogens, predators, and invasive species
Changes in precipitation and water availability (drought, floods)	- intense precipitation means more water available for aquaculture - floods/ high rivers flows is stimulus to natural breeding of some native fish	**Drought** - less water availability for aquaculture. - changes in fish migration and recruitment patterns. - poor water quality causing more disease. - increased competition for water resources. - changes in lake and river levels affecting abundance. distribution and composition of fish stocks.
Increase in frequency and / or intensity of storms	- not clearly understood	**Floods** - changes in freshwater and brackish salinity. - introduction of disease or predators into aquaculture facilities causing loss of aquaculture stock and damage to aquaculture facilities and fishing gear. - higher risk to fishers and increased insurance costs.
Changes in sea surface temperature	- enhanced primary productivity benefiting filter feeders - longer growing season - lower natural mortality in winter	- frequent harmful algal blooms and less dissolved oxygen. - increased incidence of disease, parasites, pests, fouling organisms, nuisance species and or/ predators.
Rising sea level	- salt water intrusion into inland and availability of more area for brackish water aquaculture	- loss of land for freshwater aquaculture. - loss of freshwater fisheries. - reduced freshwater availability for aquaculture - loss of coastal ecosystems (e.g. mangrove forests).

Table 4.4: ***Climate change (temperature) induced toxicological effects of contaminants on wildlife*** [*compiled by Noyes et al. 2009*]. POP = persistent organic pollutants

Climate change induced effect	Relationships/interactions
Altered uptake and elimination	↑ temperature = ↑ uptake of toxicants. ↑ temperature = ↑ elimination. ↑ temperature = remobilization of bio-accumulated POPs.
Increased toxicity	-↑ temperature = ↑ toxicity. -↑ temperature = ↑ metabolism and potentially altered metabolic profiles. - toxicant exposure may limit capacity of species and populations to acclimate to altered temperatures. - pollutant-exposed ectotherms and species at the edge of their physiological tolerance range may be specially sensitive to temperature increases.

4.2.2: Climate change and lotic system

Fishes living in lotic systems (streams and rivers) have adapted to a specific set of hydrological conditions. For example, seasonal flood or river flows determines the success of some freshwater fisheries. However, climate change may alter the hydrological regimes and may impact on fish recruitment or fish migration or fish abundance (see Table 4.3 for negative effects of climate change on fisheries). In higher latitudes (temperate rivers), expected increase in precipitation and decrease in snow packs may result in an earlier smaller flood pulse. This could affect important fisheries as low flows may block migration of important fish living in the system. Further, the tropical river systems may experience an increase of evapotranspiration or faster water loss (climate models predict about 10-25% increases in evapotranspiration at each 2-3^{0}C increase in ambient temperature). Increased evapotranspiration could reduce the volume and quality of available fish habitat during the dry period. As a result tropical areas may experience low water flows of riverine system. The low flows in rivers will dispose fish in a more crowded and stressful conditions in small refuge pools. The spawning habitat thus also is reduced during low flows. Many climate models predicted an increase in hydrological variability or extreme events in weather (large floods or droughts or prolonged dry spells). The large floods can injure larval and juvenile fish and may displace adults whereas prolonged dry spells can cause a decline in fisheries through reduced abundance and altered species composition.

4.2.3: Climate change and lentic system

Climate change can profoundly affect primary production and the trophic state of inland water. It is projected that stratification in lakes may be increased or prolonged due to rise of temperature. Increased temperatures will affect the patterns of thermal stratification and overturn that governed the productivity of deep lakes (see Figure 4.2). The lakes are likely to stratify earlier in the year and overturn later and in case of productive lakes the duration and extent of low dissolve oxygen levels in the bottom layers may be greater. Toxicants such as ammonia and hydrogen sulphide can build up during extended periods of low dissolved oxygen. This is likely to mean poorer water quality when the lake overturns. The other effects of climate change on lake ecosystems are summarised in Table 4.5.

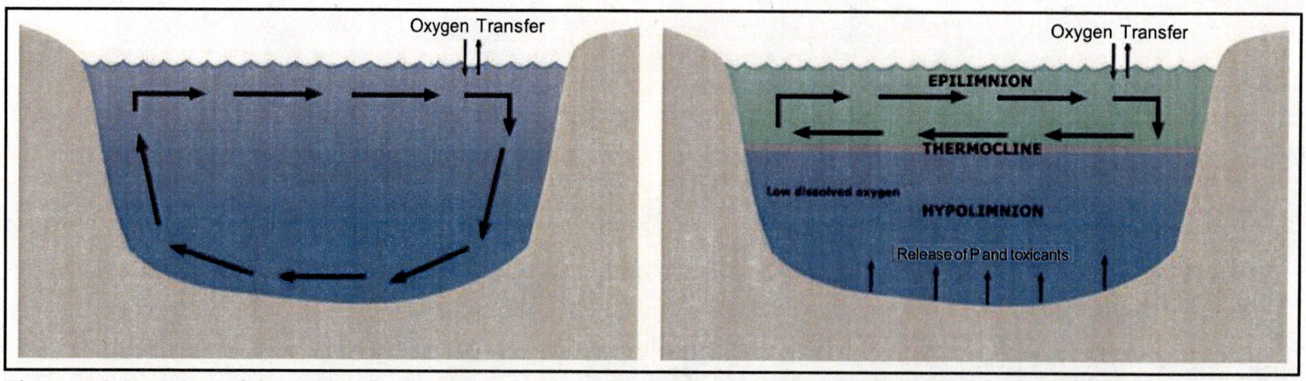

Figure 4.2: ***Patterns of thermal stratification in lakes. Left: a well – mixed lake with oxygen transfer occurring throughout the water column.*** *Right: stratification into layer with hypolimnion (lower layer) having low dissolved oxygen. Climate change will increase or prolonged stratification in lakes [EPA, Victoria, 2008].*

Nevertheless, the predicted increases in temperature can increase the productivity of a body of water by increasing algal growth, bacterial metabolism and nutrient cycling rates. Growth of algae and other micro-organisms could be beneficial for filter feeding fishes. Contrary to the above, rise of temperature which may stimulate macrophyte growth can reduce the amount of fish habitat. In addition, rise of temperature may accelerate algal blooms in water bodies and when algae die-off fish kills can occur due to depletion of dissolved oxygen. Climate change, such as extreme storms could cause sudden and complete overturns of lake anoxic hypolimnia, this may cause frequent fish kills in the future.

Table 4.5: *Projected impacts of climate change drivers of ecological change in lake ecosystems* [EPA Victoria 2008].

Climate change driver of ecological change	Potential impacts on Victorian lakes	Lake types at most risk
Increased temperature	• Loss of temperature sensitive species • Lower dissolved oxygen • Increased distribution of pets species • Changes in productivity and biogeochemical cycles in Lakes • Increased period of thermal stratification with potential for nutrient and toxicant release from sediments	• Upland lake • All lakes • All lakes • All lakes • Deep lakes
Increase evaporation rate	• Lakes may dry up and experience longer times between filling • Increased salinity leading to major ecosystems change	• Shallow lakes • Shallow lakes
Decreased average rainfall	• Lakes may dry up and experience longer times between filling	• Lakes associated with river systems (flood plains and terminal lakes)
Increased frequency and intensity of droughts	• Increased salinity, leading to major ecosystem changes • Lakes may dry up and experience longer times between filling	• Shallow lakes • Shallow lakes
Increased frequency and intensity of bushfires	• Increased organic matter loads	• Lakes in forested catchments
Sea-level rises	• Increased salinity, leading to major ecosystems changes	• Coastal lakes
Reduced groundwater levels	• Lakes may dry up and experience longer times between filling	• Groundwater fed lakes
Increased ultraviolet light	• Direct damage to aquatic life from UV light	• All lakes
Reduced snow fall	• Changes in quantity and timing of freshwater inputs	• Upland lakes
Increased carbon dioxide levels	• Increased productivity of plants in catchment, potentially leading to greater organic matter inputs • Increased lake acidity	• All lakes • All lakes

4.3: Effects on marine ecosystems

Marine fisheries are an important global food supply source. Pressure on marine fish population comes from many sources including global warming/ or climate change. Climate change is one of the most important factors that can affect marine fisheries, of which, four factors appear to contribute or have major impacts in future marine fisheries production:

- Acidification of ocean waters due to rise of CO_2 concentrations.
- Ocean warming due to rise of sea surface temperatures.
- Rise in sea levels.
- Extreme weather events.

4.3.1: *Rising CO_2 concentrations*

Global climate models projected more than a doubling of global atmospheric CO_2 by the end of 2100 compared with pre-industrial levels (280 ppm) (see Figure 4.3a). Such rise in CO_2 will have some dramatic consequences for physical and chemical stressors (e.g. a decline in water pH) of marine systems (Figure 4.3b). Increases in CO_2 levels will make the ocean more acidic (see Box 4.1), adversely affecting many organisms that use calcium carbonate for their skeletons and shells, such as corals, molluscs and some phytoplankton. These affects are due to reduced calcification and increased bleaching *[http:www.climatescience.gov/library/sap/sap4-3/final-report/default.htm/]*. Coral bleaching has been occurring world wide for a decade or so, which is a sign of acidification of seawater. Studies conducted suggested that 60% of coral reefs, as a result, could be lost by 2030. Acidification can also affect the respiratory systems of aquatic animals and other animals having calcareous skeletons or shells. On the other hand, rising CO_2 concentrations may enhance productivity of mangrove and seagrasses. Table 4.6 highlighted the important pathways via which climate change (pH, temperature, sea level rise and extreme events) may affect the future fisheries.

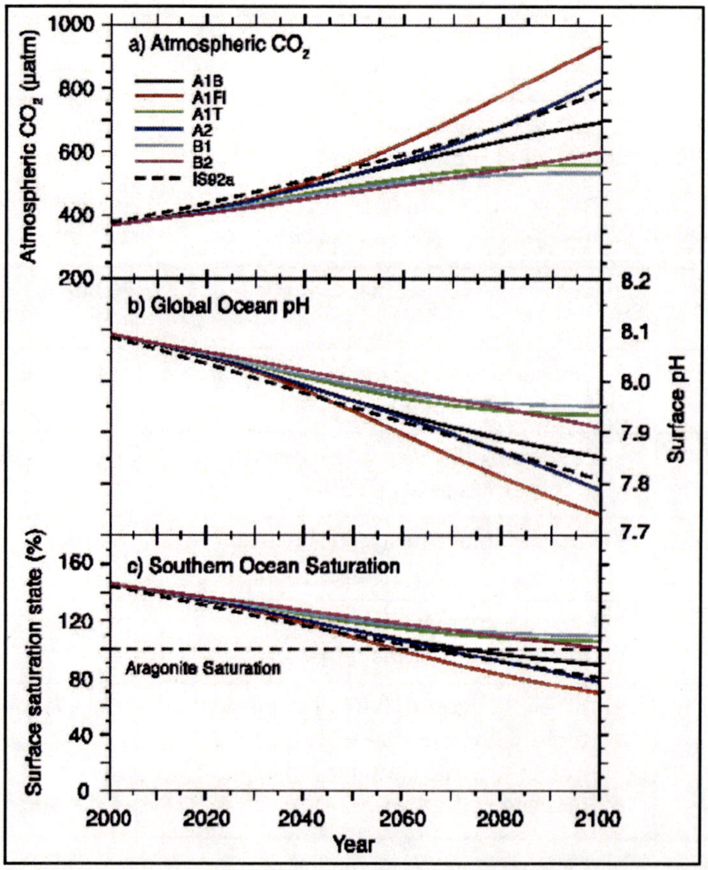

Figure 4.3: ***Changes in global average CO_2 (based on various scenarios), projected global average surface water pH and average saturation state (with respect to aragonite) in the southern ocean.*** *[For further information see Meehl et al. 2007].*

Box 4.1: ***The fate of CO_2 in ocean waters*** *[Roessig et al. 2005; Hobday et al. 2006].*

As CO_2 enters the ocean it undergoes the following reactions: $CO_2 + H_2O \leftrightarrow H_2CO_3 \leftrightarrow H^+ + HCO_3 \leftrightarrow 2H + CO^{2-}_3$ *Increasing CO_2 concentrations in the surface ocean via anthropogenic CO_2 uptake will have two effects. First, it will decrease the surface ocean carbonate ion concentration (CO^{2-}_3) and decrease the calcium carbonate saturation state. Secondly, when CO_2 dissolves in water, it forms a weak acid (H_2CO_3) that dissociate into bicarbonate generating hydrogen ions (H^+), making the water more acidic (lowering water pH). The pH is an important water quality indicator because fish and other organisms are sensitive to pH. Ocean surface has already decreased by 0.1 pH units in colder waters and almost 0.09 pH units in warmer waters. If atmospheric CO_2 concentrations continue to increase, another 0.3 pH unit decrease of oceanic surface waters may occur.*

Table 4.6: ***Examples of climate impact pathways on fisheries*** *[http://10.10.2.119:8080/00123f24cda4/comfort-file/30cb_131f7d84_0097_11dd_90f4_00123f24cda4/a1115e00.pdf; Allison et al. 2005].*

Type of changes	Climate variable	Impacts	Potential outcome for fisheries
Physical environment *(effects of pH, Temp.°C, Sea level)*	Changes in pH through increased CO_2 and acidification	- effects on calcareous animals; e.g. molluscs, crustaceans, corals, echinoderms and some phytoplankton	- Potential declines in production for calcareous marine resources
	Warming upper layers of the ocean	- warm water species replacing cold water species - plankton species moving to higher latitudes - timing of phytoplankton blooms changing - changing zooplankton composition	- Shifts in distribution of plankton, invertebrates, fishes, birds towards the north or south poles, reduced species diversity in tropical waters. - Potential mismatch between prey (plankton) and predators (fish populations) and declines in production and biodiversity.
	Sea level rise	- loss of coastal breeding and nursery habitats e.g: mangroves, coral reefs	- Reduced production of coastal and related fisheries.
Fish stocks *(effects of Temp.°C, Ocean current)*	Higher water temperatures	- changes in sex ratios - altered time of spawning - altered time of migrations - altered time of peak abundance	- Possible impacts on timing and levels of productivity across marine and freshwater systems.
	Changes in ocean current	- increased invasive species, diseases and algal blooms - affects fish recruitment success	- Reduced production of target species in marine and freshwater. - Abundance of juvenile fish affected and therefore production loss in marine and freshwater.
Ecosystems *(effects of droughts, ENSO)*	Reduced water flows and increased droughts	- changes in lake water levels - changes in dry water flows in rivers	- Reduced lake productivity. - Reduced river productivity.
	Increased frequency of ENSO events	- changes in timing and latitude of upwelling - coral bleaching and die-off	- Changes in pelagic fisheries distribution - Reduced coral –reef fisheries productivity.
Coastal infrastructure and fishing operations	Sea level rise	- coastal profile changes, loss of harbours, homes - increased exposure of coastal areas to storm damage	- Costs of adaptation make fishing less profitable, risk of storm damage increases cost of insurance and / or rebuilding, coastal households' vulnerability increased.
	Increased frequency of storms	- more days at sea lost to bad weather, risks of accidents increased - aquaculture installations (coastal ponds, sea cages) more likely to be damaged or destroyed	- Increased risk of both fishing and coastal fish farming, making these less viable livelihood options for the poor; reduce profitability of larger scale enterprises, insurance premium rise.
Inland fishing operations and livelihoods	Changing levels of precipitation	- where rainfall decreases, reduced opportunities for fishing and aquaculture as part of rural livelihood systems	- Reduced diversity of rural livelihoods; greater risk to agriculture; greater reliance on non-farm income.
	More drought or floods	- damage to productive assets (fish ponds, weirs, rice fields) and homes	-Increased vulnerability of riparian and floodplain households and communities.
	Less predictable rain/dry season	- decreased ability to plan livelihood activities-e.g. farming, fishing seasonality	- Increased vulnerability of riparian and floodplain households and communities.

4.3.2: Rising sea surface temperatures

In addition to pH, higher sea temperature is also believed to be a major cause of coral bleaching and damage to reef ecosystems around the world. Coral reef ecosystems are habitat for many fish species (estimated to be 4,000 species of fish, 700 species of coral and thousands of other plants and animals) and it is projected that with a rise of as little as 1.0°C in water temperature would cause stress to coral reefs and mass bleaching (Islam 2008). Rise of sea surface temperature may cause the following additional effects:

- Shifts in distribution of marine organisms towards north or south poles (Table 4.6, Box 4.2),
- Enhance shallow water fish yields (section 4.3.2.1 and Figure 4.4),
- Decline of fish recruitment and fisheries (Figure 4.5).

Box 4.2: ***Examples of rising temperature effects on species distribution and water quality*** *[based on F and A Europe 2007].*

Changing sea temperature and current flows will likely bring shifts in the distribution of marine fish stocks. A decline in North Sea and Baltic Sea cod stocks has been found related with shifts in the geographical range of plankton (Calanus finmarchius) from North Sea to the Arctic Ocean in search of colder waters on which cod larvae feed. Its (plankton) biomass in the North Sea has dropped by 70% since the 1960s. The plankton is the most visible expression of climate change. It has also been observed that fish (e.g. sardines, anchovies), molluscs and crustaceans are moving further north in search of colder waters, since these organisms requires a specific water temperature range which is no longer available in their natural habitat (the natural habitat either become too warm or these species are following movement of their food organisms such as vegetation, plankton, or other marine organism on which they feed, which are also migrating north). Species that habitually live along the African coasts are moving northwards, such as black codling (Physiculus dalwigki), rockling (Giadropsarus granti) and snake eel (Pisodonophis semicinctus) which are now spotted in Galician waters. Despite the facts cited above, the movement also create some new fishing zones.

For the last 30 years or so eutrophication has become more intense because of higher water temperature and eroded earth due to sea level rise, resulting in an increasing number of green, red and brown tides and excessive growth of plankton on European coasts. This have caused depletion of oxygen and release of toxic substances into environment and creating problem for coastal aquaculture.

4.3.2.1: Rising temperature effects on the growth of shallow- and deep-water marine fish

Marine fish growth can be enhanced by climate change, a study conducted by Thresher *et al.* (2007) found that global rise of temperature has enhanced the growth of shallow water marine fish (redfish, jackass morwong and banded morwong) living less than 250 m depth of south west Pacific east of Tasmania, Australia during the last 100 years. The study found that the growth rates of shallow water juvenile morwong were 28.5% faster in the 1990s compared to 1950s (see Figure 4.4). On the other hand, growth rates of deepest dwelling species living at depths of around 1000 m (orange roughy, and black oreo, smooth oreo, spiky oreo and warty oreo) have been decreased significantly over the time.

The study suggests that water temperature has been a primary factor in juvenile fish growth and the decrease in growth rates of the deep-water species could slow or may be reversed with increase of global warming. The study was based on fish otoliths or ear bones and examined 555 specimens ranging from 2 to 128 years, with birth years from 1861 to 1993.

4.3.2.2: Potential change of global temperatures on fish recruitment

The climate change may alter the match-mismatch hypothesis of Cushing (1990). The hypothesis is based on understanding, that larval fish survival is dependent on the exact timing of plankton availability when the resources of the fish larvae in the yolk-sac are depleted and thereby affecting marine fish recruitment. However, it is now suggested by Cury *et al.* 2008, that global warming can effect marine fish recruitment in three ways:

- Change in the mean relative timing of prey (algae) (Figure 4.5a),
- Change in the level of prey abundance (Figure 4.5b) or
- Change in the amplitude of year-to-year variations in prey timing in regions where interannular variability in temperature is expected to increase (Figure 4.5c).

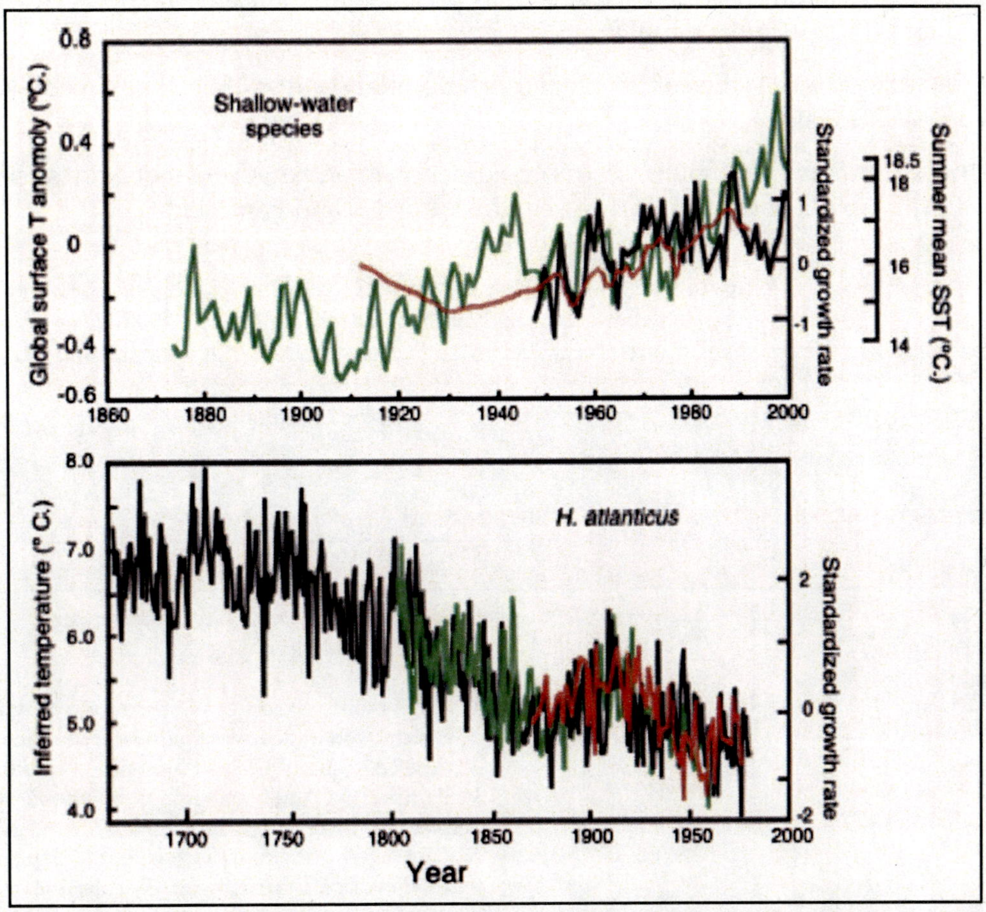

Figure 4.4: *Comparison of long-term temperature records with growth rates for pooled shallow-water species (redfish, jackass morwong and banded morwong) (upper) and deep water fish Orange roughy, H. atlanticus (lower) [Thresher et al. 2007]. Global surface temperature anomaly (green line), Maria island sea surface temperatures record (black line) and juvenile growth rates (red line) [see Thresher et al. 2007 for details and further information].*

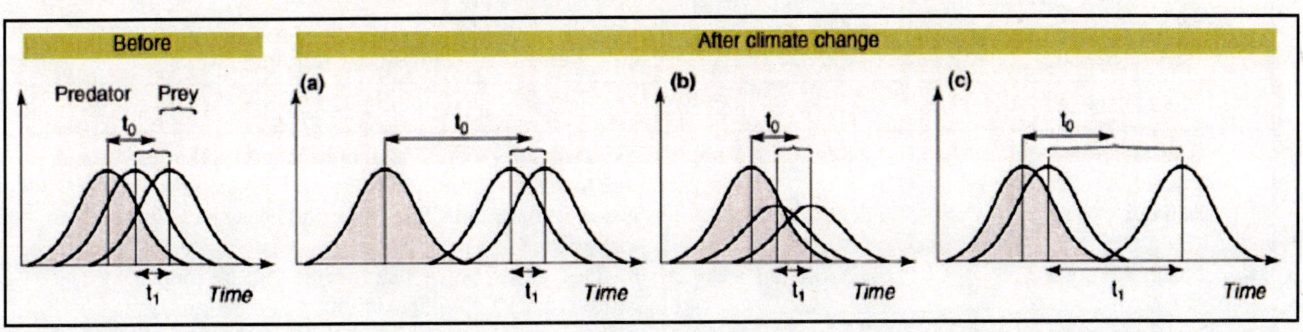

Figure 4.5: *The match-mismatch hypothesis (Cushing 1990) and climate change related three possible effects on marine fish recruitment [Cury et al. 2008; there are now three possible effects of climate change on fish recruitment (as shown in figure 4.5a; 4.5b and 4.5c; .Note: grey curve = seasonal timing of spawning which is fixed, and white curves= primary production which varies from year to year; t_0 is the time lag between the predator and prey populations (degree of time mismatch) and t_1 is the inter-annular variability in the timing of the prey population [see Cury et al. 2008 for further information].*

4.3.2.4: *Effects of temperature rise on different aquatic ecosystems*

There could be a wide range of effects for different levels of temperature rise on ecosystems. These effects ranged from loss and extinction of species and species habitat. Table 4.7 summarised projected effects on aquatic ecosystems in comparison to terrestrial ecosystems, which shows that climate change could have significant effects in all ecosystems. The main effects are as follows:

- *Marine ecosystems*: extinction or loss of corals; and polar animals (bears, walrus, seals); loss of habitat of marine birds; loss of bivalves, pteropod and mammals, reductions in krill populations which may affect penguin populations.
- *Freshwater ecosystems*: Decline of freshwater fish population, loss of fish habitat and loss of wetlands (Kakadu, Australia).
- *Terrestrial ecosystems*: possibilities of loss or extinction or endangerment of plant species, butterflies, reptiles, frogs, birds and mammals in several countries and continent.

Table 4.7: ***Projected impacts of climate change on marine, freshwater and terrestrial ecosystems for different levels of global mean annual temperature rise*** *(all data and information compiled by Fischlin et al. 2007). [The table include projected impacts on terrestrial ecosystems for comparison with marine and freshwater systems [for all references and further information see Fischlin et al. 2007].*

Region	Mean temperature (⁰C) change (above pre-industrial climate)	Impacts to unique or widespread ecosystems or population systems
I: MARINE		
Australia	1.7 (range 1-2.3)	All coral reef bleached (Great Barrier Reef)
Caribbean, Indian Ocean, Australia	0.6	Increased coral bleaching
Indian Ocean	2.5	Extinction of coral reef ecosystems (overgrown by algae)
Ocean basins	2.9 (range1.4-4.1)	70% reduction in deep-sea cold-water aragonitic corals
Antarctica, Arctic	0.6	Marine ecosystems affected by continued reductions in krill impacting Adelie penguin populations; Arctic ecosystems increasingly damaged
Arctic	2.8 (range 2.5-3)	62% loss of Arctic summer ice extent;high possibilities of extinction of polar bears, walrus, seals; Arctic ecosystem stressed
Arctic	2.9	50% loss of existing tundra, 5% eventual gain; 5-57% of Arctic shorebirds breeding area may be lost; 5-56% of breeding area of geese species could be lost
Southern Oceans	2.3 (1.5-2.7)	Loss of Antarctic bivalves and limpets
Southern Oceans	2.9 (range 1.6-4.1)	Loss of aragonitic pteropods
USA	2.8 (range 2.1-3.1)	Loss of 9-62% of mammals from Great basin montane
USA (Delaware)	3 (range 1.9-3.5)	20% loss of coastal migratory bird habitat
II: FRESHWATER		
Australia (wetland)	2.8 (range 1.2-4.5)	Extensive loss/conversion of habitat in Kakadu wetland due to sea-level rise and salt-water intrusion
Malawi	2.3 (2-2.5)	Fish populating declines, wetland ecosystems dry and disappear
North America	1.3 (range 1.1-1.6)	8% loss of freshwater fish habitat; 15% loss in Rocky Mountains; 9% loss of salmon
North America	2.3 (range 1.6-2.3)	24% loss of freshwater fish habitat; 40% loss in Rocky mountains, 27% loss of salmon
USA	2.8 (range 1.9-3.8)	38-54% loss of waterfowl habitat in Prairie wetland (Northern Great Plain)

Contd...

III: TERRESTRIAL		
Australia	1.6 (range 1.1-2.1)	Suitable climates for 25% eucalyptus exceeded
Australia	3.6 (range 2.6-4.3)	Bioclimatic limits of50% of eucalyptus exceeded
Australia	>4	Bioclimatic limits of 73% of eucalyptus exceeded
Australia	>4	Eventual total extinction of all endemic species of Queensland rainforest
Brazil	1.7 (range 1.2-2.6)	38-45% of the plants in the Cerrado committed b extinction
China	2.9	Substantial loss of boreal forest
Europe	2.3	Changes in ecosystem composition
Europe	2.6	4-21% of plants may be extinct
Europe	2.1 (1.4-2.6)	Bioclimatic envelopes of 2-10% plants exceeded, leading to endangerment or extinction; mean species turnover of 48% (range 17-75%); mean species loss of 27% (range 1-68%)
Europe	2.2	Possibilities of 3-16% plant extinction
New Zealand	4 9 (3-5.1)	Likely extinction of 200-300 species (32-63%)
South Africa, Namibia	2.1	41-51% loss in endemic plant species richness
South Africa	2.3 (range 1.5-2.7)	Extinction (100% potential range loss) of 10% endemics; 51-65% loss of Fynbos; succulent Karoo reduced by 80%; threatening 2,800 plant species with extinction; 5 park loss >40% of plant species
Africa	2.2 (range 1.7-3.2)	8-12% of medium/large mammals in 141 national parks critically endangered or extinct; 22-25% endangered
Africa	3.6 (range 2.6-3.7)	30-40% of 277 mammals in 141 parks critically endangered/extinct; 15-20% endangered
Australia	2.1	13-23% butterflies may be extinct
Australia	2.9	21-36% butterflies may be extinct
Australia	3.3 (range 2.3-3.9)	Substantial loss of alpine zone, and its associated flora and fauna (e.g. alpine ski lily and mountain pygmy possum)
Australia	1.9 (range 1.6-2.4)	Projected biodiversity loss in Queensland: 7-14% reptiles; 8-18% frogs; 7-10% birds; and 10-15% mammals
Australia	>4	Projected biodiversity loss in Queensland: 38-67% reptiles; 48-80% frogs; 49-72% birds as 85-90% of suitable habitat lost
Australia	>4	57 endemic frogs/mammal species eventually extinct, 8 endangered
Australia	<1.9 (range <1.6-2.4)	Range loss begins for golden bowerbird; 40-60% range loss for golden bowerbird
Australia	3.1 (2.5-4)	High risk of extinction of golden bowebird as habitat reduced by 90%
Australia, Costa Rica and Spain	0.6	Amphibians extinctions/ extinction risks on mountains due to climate change induced disease outbreaks
Europe	3.3	4-38% of birds committed to extinction
Europe	3.1 (range 1.8-4.2)	Risk of extinction of alpine species
Europe	3.9	4-24% plants critically endangered/extinct; mean species turnover of 63% (spatial range 22-90%); mean species loss of 42% (spatial range 2.5-86%)
Mediterranean	2.6 (range 1.6-3.5)	Forest fire weather index > 45;Increased fire frequency converts forest and maquis into scrub leadng to more pest outbreaks
Mexico	1.7 (1.3-3)	Projected biodiversity extinction: 2-18% mammals; 2-8% birds; 1-1% butterflies
Mexico	3.5 (range 2.3-4.1)	Loss of temperate forest wintering habitat of monarch butterfly
South Africa	2.3 (range 2.3-4)	Projected biodiversity extinction: 24-59%; mammals; 28-40% birds; 13-70% butterflies; 18-80% other invertebrates; 21-45% other reptiles; 66% animal species may lost from Kruger National Park

Contd...

South America, Globe	2.5 (2-3)	Major loss of Amazon rainforest with large losses of biodiversity
USA	5.2	62-100% loss of bird habitat at 4 major coastal sites
USA	2.5	20-70% loss of coastal bird habitat at 4 sites
USA	3.3 (range 2.8-3.8)	Risk of extinction of Hawaiian birds (honeycreepers) as suitable habitat reduced by 62-89%
USA	3.6 (range 3-3.9)	Lose of 30-57% neotropical migratory bird species richness
IV: GLOBAL		
Corals	3.1 (range 2.3-3.7)	Extinction of coral reef ecosystems (overgrown by algæ)
Corals	3.3 (range 2-4.5)	Reduced growth in warm-water aragonitic corals by 20%-60%; 5% decrease in global phytoplankton productivity
Ecosystems	3.7	Few ecosystems can adapt
Endemic species	3.5 (range 2-5.5)	Predicted extinction of 15-40% endemic species in global biodiversity hotspots
Fish	2.4	63 of 165 rivers studied loss of >10% of their fish species
Forest	1.6	Bioclimatic envelopes eventually exceeded, leading to 10% transformation of global ecosystems; loss of 47% wooded tundra, 23% cool coniferforest; 21% scrubland; 15% grassland/steppe; 14% savanna; 13% tundra and 12% temperate deciduous forest
Forest	2.7	Bioclimatic envelopes exceeded leading to 16% transformation of global ecosystems; loss of 58% wooded tundra, 31% cool conifer forest; 25% scrubland; 20% grassland/steppe; 19% savanna; 21% tundra and 21% temperate deciduous forest
Forest	3.7	Bioclimatic envelopes eventually exceeded, leading to eventual transformation of 22% of global ecosystems; loss of 68% wooded tundra, 44% cool coniferforest; 34% scrubland; 28% grassland/steppe; 27% savanna; 38% tundra and 26% temperate deciduous forest
Plants	>2.5	Sink service of terrestrial biosphere saturates and begins turning into a net carbon source
Reserves	3.7	50% of all nature reserves cannot fulfil conservation objectives
Species	2.9 (range 2.6-3.3)	21-52% species committed extinction
Species	1.6 (range 1.2-2)	9-31% (mean 18%) of species to extinct
Species	2.1 (range 2.1-2.3)	15-37% (mean 24%) of species to extinct
Wetlands	3.4	6-22% loss of coastal wetlands; large loss of migratory bird habitat in particular in USA, Baltic and Mediterranean

4.3.3: Rising sea levels

The rising sea level would most likely damage or destroy many coastal ecosystems including mangroves and salt mashes. These ecosystems are essential habitat for wild fish stocks and a source of natural seed for aquaculture. Climate change may bring in more frequent storm surges thereby could damage coastal fish ponds and coastal infrastructures. Higher sea levels can lead to intrusion of saline water into lowland and deltaic regions causing a destruction of agriculture and freshwater aquaculture facilities and contamination of groundwater. On the other hand, sea level rise would expand areas suitable for brackish water aquaculture (see Tables 4.3 and 4.6). A rise in sea levels would increase the depth of water above coral reefs, resulting lower light penetration to support photosynthetic algae living within coral (e.g. *Zooxanthellae).* If the water depth increases faster than the corals can grow, they could effectively destroy habitats for fish upon which many artisanal fisheries are dependent.

4.3.4: Extreme events

Climate change would result more extreme events such as cyclones, storm surges and inland flooding. This would have impacts on fisheries, aquaculture through loss of fish stocks; and damage to facilities and infrastructures (see Table 4.3 and 4.6)

4.4: Vulnerable countries to fisheries due to climate change

A recent study conducted by World FishCentre (Allison *et al.* 2009) found that 33 countries (out of 132 countries investigated) are highly vulnerable to the effects of global warming on fisheries (see Figure 4.6). These countries produce 20 percent of world's fish exports and are dependent on fish for protein (about 27% dietary protein). Two thirds of countries that are vulnerable are located in tropical Africa where fisheries are important to the poor. A total of 22 of the 33 countries in the most vulnerable quartile are classified as Least Developed Countries. The 33 vulnerable countries include 21 in Africa, 9 in Asia and 3 in South America as follows (least developed countries are listed in italic):

Africa : *Angola, Burundi, DR Congo,* Cóte d'Ivorie, *Gambia,* Ghana, *Guinea, Guinea-Bissau, Malawi, Mauritania,* Morocco, *Mozambique, Niger,* Nigeria, *Senegal, Mali, Sierra Leone, Tanzania, Uganda, Zambia,* Zimbabwe

Asia : Algeria, *Bangladesh, Cambodia,* Pakistan, Russian Federation, Turkey, *Yemen,* Ukraine, Vietnam,

South America : Colombia, Peru, Venezuela

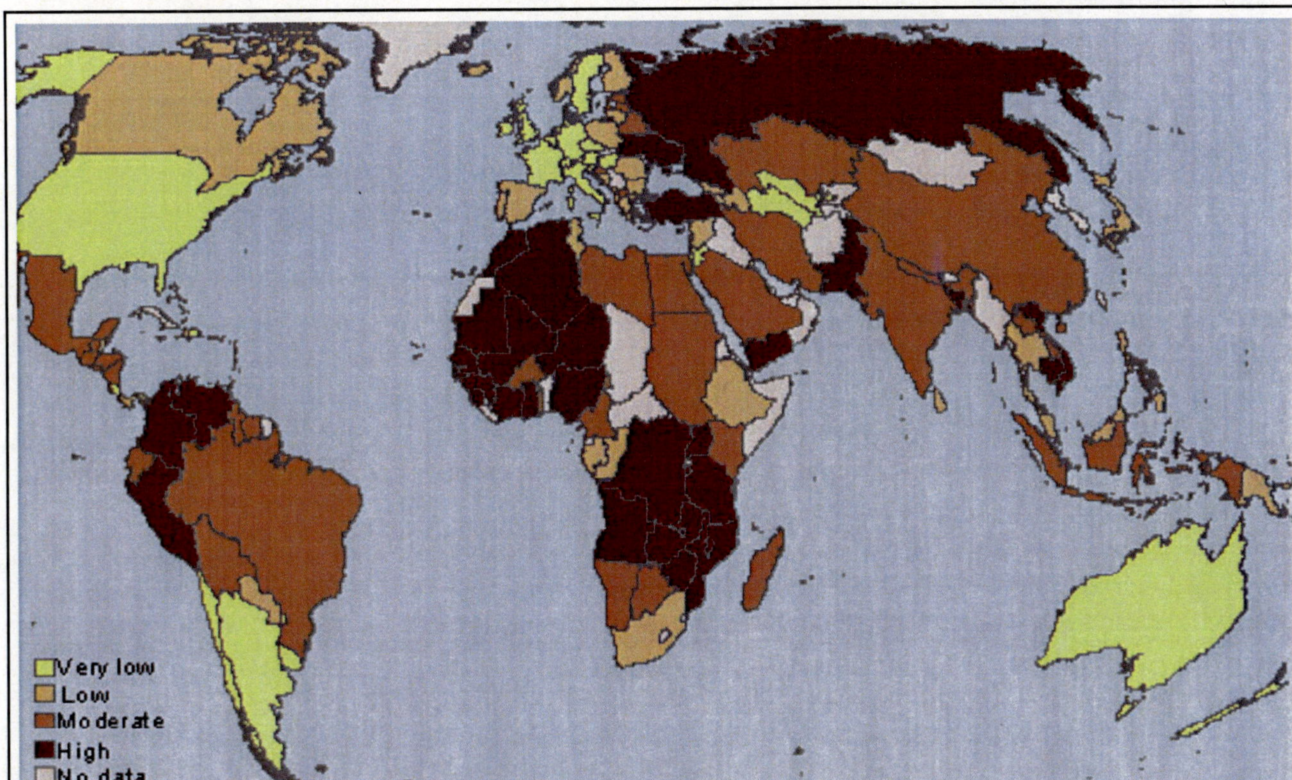

Figure 4.6: ***Vulnerability of national economies due to potential climate change impacts on fisheries under IPCC scenario B2 (local development, lower emissions).*** *Dark brown = upper quartile (highest index value); yellow = lowest quartile; grey = no data available [Allison et al. 2009].*

4.5: Case studies - Asia-Pacific and Australasia Region

4.5.1: Asia

[based on Cruz et al. 2007].

The Asia-Pacific region is the worlds' largest producer of fish, from both aquaculture and capture fishery sectors. The climate models predicted a reduction of primary production in the tropical oceans due to changes in oceanic circulation in a warmer atmosphere. The following changes have been projected due to global warming:

- Large scale changes of skipjack tuna habitat under warming scenario in the equatorial Pacific (note: tuna catch in East Asia and South-East Asia is nearly one fourth of the world's total).
- The migration routes and migration patterns of principal marine fishery (ribbon fish, small and large yellow croakers) in China could be greatly affected by global climate change.
- A measurable decline in fish larval abundance in the coastal waters due to increased frequency of El Niño which would cause general decline in fishery prediction in coastal waters of East, South and South-East Asia.

However, moderate warming would improve the conditions for cod and herring. Higher temperatures and reduced ice cover could increase productivity of fish-prey and extend habitats in the Arctic region. The timing and amount of precipitation could affect the migration of fish species from the river to the floodplains for spawning, dispersal and growth. According to IPCC (2001), future changes in the ocean currents, sea level rise, sea surface temperature, salinity, wind speed and direction, strength of upwelling, the mixing layer of thickness and predator response to climate change will substantially alter fish breeding habitats and food supply for fish and ultimately the abundance of fish populations in Asian waters (see Table 4.8).

Table 4.8: ***Projected climate change impacts on Asian fishery*** *[Cruz et al. 2007].*

Fishery	Area	Projected change	Effects
Arctic Fishery	Arctic	Improve productivity due to rising temperature	+ve
Cod and herring		Moderate warming would enhance the fishery	+ve
Larval abundance	Coastal waters of East, South and South -East Asia	Declines in coastal fishery due to increased El Niño	-ve
Marine fishery	China	Change of migration routes	-ve
Shrimp fishery	North Asia	Decrease of fishery due to rise of temperature	-ve
Tuna	East Asia and South Asia	Changes in habitat	-ve

4.5.2: Australia

[based on Hennessy et al. 2007].

The gross annual value of Australian fisheries is US$ 1.7 billion, of which 68% is wild-catch and 32% aquaculture. In New Zealand, the combined value of fisheries is US$ 0.8 billion, of which 80% is from the commercial catch. A comprehensive study was carried out on the possible effects of climate change on Australian marine organisms including phytoplankton, zooplankton, corals, pelagic, demersal and benthic fishes, mangroves, seagrasses, kelp, turtle and sea birds. It predicted that (see Table 4.9), the distribution, and abundance of Australian marine organisms could be affected due to climate change including southwards migration of some species to avoid higher temperature and a decline of population with calcareous shells. Some benthic and demersal fish population may either decline or move south ward (see Table 4.9).

Table 4.9: ***Expected climate change impacts on Australian marine life including fisheries*** *[Hobday et al. 2006].*

Groups	Projected climate change impacts
Phytoplankton	• species with calcareous shells such as coccolithophores and foraminifera may decline in abundance • earlier timing of the peak in production of some phytoplankton
Zooplankton	• large southward movements of tropical and temperate species as oceans waters warm; • species with calcareous shells, such as echinoderms, crustaceans and molluscs may not be able to maintain shell integrity due to acidification of ocean • increased incidence of jellyfish swarms with warming oceans

Contd...

Seagrasses	• possible increase in seagrass biomass due to rise of CO_2
	• shifts in species distribution southwards and alteration of species composition of seagrass beds as temperature warms up
	• disappearance of UV intolerant species from shallow waters as UV radiation levels rise
Mangroves	• possible increase in mangroves productivity as atmospheric CO_2 levels rises
	• sea level rise will destroy mangroves habitat
Kelp	• ranges will shift southwards
Rocky reefs	• species range will shift southwards as temperatures warms up
Coral reefs	• acidification and warming will cause calcification problems and coral bleaching
Coldwater corals	• ocean acidification will dissolve reefs
Soft sediment fauna	• acidification will erode the shells of molluscs, corals, benthic foraminifera and other groups with carbonate shells
	• warming will shift the ranges and distributions of many species of soft sediment fauna differentially leading to new assemblages of organisms
Benthic and demersal fishes	• southward movements of species along the east and west coast
	• some population may decline where ranges are bounded to the south
	• UV radiation may adversely affect larval fish development and survival
	• changes in temperatures, current patterns and primary and secondary production may affect larval fish health and transport thereby influencing recruitment potential
Pelagic fishes	• pelagic fish such as tuna moves south with warming
	• UV radiation may adversely affect larval fish development and survival
	• changes in temperatures, current patterns and primary and secondary production may affect larval fish health and transport thereby influencing recruitment potential
Turtles	• skewing of sex ratios toward females as temperature rises
	• climate-induced alteration of food supplies will impact turtle populations
	• climate - related increases in wave energy and storm events may erode nesting beaches and reduce egg survival
Sea birds	• warming will expand or shift distributions southward

4.6: Adaptation measures

A wide range of adaptation measures can be taken in anticipation of future effects of climate change on aquaculture and fisheries (Tables 4.10). These include breeding of fish tolerant to high water temperature, development of fisheries management capabilities to cope with impacts of climate change (e.g. planning infrastructure adaptations in response to shifting populations and species, for example, longer range fishing crafts, installation of onboard processing equipment), alternative means of transport when coastal roads are flooded and construction of appropriate infrastructures for storm forecasting and signalling systems and safe refuge for dealing with possible rising of sea level and increased storminess. FAO (2005-08). A number of adaptation strategies suitable for aquaculture and fisheries are summarised in Table 4.10.

Table 4.10: *Potential adaptation measures in aquaculture and fisheries to cope with climate change impacts* [FAO 2005-08].

Climate change elements	Adaptive measures
Aquaculture	
Warming	- selective breeding and genetic improvements for higher temperature tolerance
	- establish regular monitoring and emergency procedures to cope up eutrophication and upwelling
	- set up bio-security measures and monitoring to reduce health risks of aquacultured species
	- increase and improve harmful algal blooms monitoring and early warning systems and change water abstraction points where feasible
Sea level rise and other circulation changes	- shift stenohaline species upstream; introduce marine or euryhaline species in old facilities
	- provide alternative livelihoods through aquaculture (where loss of agriculture land is projected)
Water stress and drought conditions	- encourage non-consumptive water use in aquaculture and encourage development of mariculture where feasible
	- use different/faster growing fish species
	- increase efficacy of water sharing with primary users (e.g. rice-fish aquaculture)
Extreme weather events	- improve siting and design to minimise damage, loss and mass escapes
	- encourage use of native species to minimise impacts on biodiversity
	- use non-reproducing stocks in farming system
	- encourage uptake of individual/cluster insurance
Fisheries	
Reduced yield	- access higher value markets/shifting targeted species
	- increased effort or fishing power
	- diversify livelihood
	- exit the fishery
Increased variability of yield	- diversify livelihood portfolio
	- design insurance schemes
Changes in distribution of fisheries	- migration of fishing effort/strategies and processing/distribution facilities
Vulnerability of infrastructure and communities to flooding, sea level and surges	- rehabilitate infrastructure, design disaster response
	- set up early warning systems and education
	- add new or improved physical defences
Increased danger of fishing	- set up weather warning systems
	- invest in improved vessel stability/safety/communications
Influx of new fishers	- support existing local management institutions, diversity livelihoods

4.7: Conclusion

Climate change would affect the productivity of fisheries through increased water temperature, decreased dissolved oxygen levels, increased incidence of disease, and increased toxicity of pollutants and alterations of hydrological conditions. Artisanal, recreational and commercial fishers could be affected if there is a decline of fisheries due to climate change. Adaptive measures as suggested in Table 4.10 could be considered to cope with climate change impacts. However, more research are required to monitor the effect of climate change so that appropriate measures and policy can be adapted to reduce risks from climate change on fisheries and aquatic ecosystems.

4.8: References

Allison, E. H., Adger, W. N., Badjeck, M. C., Brown, K., Conway, D., Dulvy, N. K., Halls, A., Perry, A. and Reynolds, J. D. 2005. Effects of climate change on the sustainability of capture and enhancement fisheries important to the poor: analysis of the vulnerability and adaptability of fisher folk living in poverty. *Project No. R4778J. Final Technical Report, Fisheries Management Science Program, MRAG/DFID, London. 164 pp. www.fmsp.org.*

Allison, E. H., Perry, A. L., Badjeck, M-C., Adger, W. N., Brown, K., Conway, D., Halls, A. S., Pilling, G. M., Reynolds, J. D., Andrew, N. L. and Dulvy, N. K. 2009. Vulnerability of national economies to the impacts of climate change on fisheries. *Fish and Fisheries.* Doi:10.1111/j.1467-2979.2008.00310x

Cruz, R. V., Harasawa, H., Lal, M., Wu, S., Anokhin, Y., Punsalmaa, B., Honda, Y., Jafari, M., Li, C. and Huu Ninh, N. 2007. Asia. *Climate change 2007: Impacts, Adaptation and Vulnerability. Contribution of Working Group II to the Fourth Assessment Report to the Intergovernmental Panel on Climate Change.* Parry, M. L., Canziani, O. F., Palutikof, J. P.., van der Linden, P. J. and Hanson, C. E. (eds.). Cambridge University Press, Cambridge, U.K. pp. 469-506 (chapter 10). .http://www.ipcc.ch/pdf/assessment-report/ar4/wg2/ar4-wg2-chapter6.pdf

Cury, P.M., Shin, Y-J., Planque, B., Durant, J. M., Fromentin, J-M., Kramer-Schadt, S., Stenseth, N.C., Travers, M. and Grimm, V. 2008. Ecosystem oceanography for global change in fisheries. *Trends in Ecology and Evolution. 23: 338-346.*

Cushing, D.H. 1990. Plankton production and year class strength in fish populations: an update of the match/mismatch hypothesis. *Advance marine Biology.* 26: 249-293.

Environmental Protection Agency (EPA), Victoria. 2008. How will climate change affect Victorian Lakes? *Environment Report.* Publication 1242, June. Victoria, Australia.

FAO 2004. http://earthtrends.wri.org/pdf_library/data_tables/watcoal_2005.pdf (world resources and fisheries Earth Trend data Tables: Coastal and marine ecosystems, Freshwater resources)

FAO 2005-08. Issues Fact Sheets – Fisheries and global climate change. Text by Everett, J. and Garcia, S. M. In: FAO *Fisheries and Aquaculture Department* [online]. Rome. Updated 27 May 2005. [Cited 11 September, 2008]. http://www.fao.org/fishery/topics/13789.

F and A Europe. 2007. Climate change: What impact on fisheries? *Fisheries and Aquaculture in Europe* (F&A Europe). 35. August 2007. page 4-7.

Ficke, A. D., Myrick, C.A. and Hansen, L. J. 2007. Potential impacts of global climate change on freshwater fisheries. *Reviews in Fish Biology and Fisheries.* 17(4): 581-613.

Fischlin, A., Midgley, G. F., Price, J. T., Leemans, R., Gopal, B., Tyrley, C., Rounsevell, M. D. A., Dube, O. P., Tarazona, J., Velichko, A. A. 2007. Ecosystems, their properties, goods, and services. *Climate change 2007: Impacts, Adaptation and Vulnerability. Contribution of Working Group II to the Fourth Assessment Report to the Intergovernmental Panel on Climate Change.* Parry, M. L., Canziani, O. F., Palutikof, J. P.. van der Linden, P. J. and Hanson, C. E.. (eds.). Cambridge University Press, Cambridge, U.K. pp. 211 – 272 (Chapter 4). http://www.ipcc.ch/pdf/assessment-report/ar4/wg2/ar4-wg2-chapter6.pdf

Handisyde, N. T., Ross, L. G., Badjeck, M-C. and Allinson, E. H. 2006. The effects of climate change on world aquaculture: A global perspective. *www.aquacutlure.stir.ac.uk/GISAL/gis-gruop/climate.php.*

Hobday, A. A., Okey, T. A., Poloczanska, E. S., Kunz, T. J. and Richardson, A. J. (editors) 2006. Impacts of climate change on Australian marine life: Part A. Executive Summary. CSIRO Marine and Atmospheric research. Report to the Australian Greenhouse Office, Canberra, Australia, September 2006. 36p. http://www.greenhouse.gov.au/impacts/publications/pubs/marinelife-partb.pdf

Hennessy, K., Fitzharris, B., Bates, B. C., Harvey, N., Howden, S. M., Hughes, L., Salinger, J., and Warrick, R. 2007. Australia and New Zealand. *Climate change 2007: Impacts, Adaptation and Vulnerability. Contribution of working group II to the Fourth Assessment Report of the Intergovernmental Panel on Climate Change.* Parry, M. L., Canziani, O. F., Palutikof, J. P., van der Linden, P. J. and Hanson, C. E. (eds.). Cambridge University Press, Cambridge, UK, chapter 11: 507-540.

IPCC 2001. Intergovernmental Panel on Climate Change, Working Group II. 2001. *Climate change 2001. Impacts, Adaptation and Vulnerability.* IPCC, Third Assessment report, Cambridge, UK: Cambridge University Press.

Islam, M. S. 2008. Increased CO_2 levels threatens Coral reefs. The Daily Star, Bangladesh. Sub-editorial. 15 February, 2008.

Meehl, G. A., Stocker, T. F., Collins, W. D., Friedlingstein, P., Gaye, A. T., Gregory, J. M., Kitoh, A., Knutti, R., Murphy, J. M., Noda, A., Raper, S. C. B., Watterson, I. G., Weaver, A. J and Zhao, Z. C. 2007. Global climate change projections. Chapter 10 (747-845). In: *Climate change 2007: The physical science basis.* Contribution of working group I to the Fourth Assessment Report of the Intergovernmental Panel on Climate Change. [Solomon, S. D., Qin., Manning, M., Chen., Z., Marquis, M., Averyt, K. B., Tignor, M. and Miller, H. L. (eds). Cambridge University Press, Cambridge, United Kingdom and New York, USA.

Muir, J. and Allison, E. 2007. The threat to fisheries and aquaculture from climate change-A policy brief. Policy brief. World Fish Centre. http://www.worldfishcenter.org/resource_centre/ClimateChange2.pdf (accessed on 8 November 2007).

Noyes, P. M., McElwee, M. K., Miller, H. D., Clark, B. W., Van Tiem, L. A., Walcott, K. C., Erwin, K. N. and Levin, E. D. 2009. The toxicology of climate change:Environmental contaminants in a warming world. *Environmental International* doi:10.1016/j.envint.2009.02.006.

Roessig, J. M., Woodley, C. M., Cech, J. J. and Hansen, L. J. 2005. Effects of global climate change on marine and estuarine fishes and fisheries. *Reviews in Fish Biology & Fisheries.* 14: 251-275.

Thresher, R. E., Koslow, J. A., Morison, A. K. and Smith, D. C. 2007. Depth-meditated reversal of the effects of climate change on long term growth rates of exploited marine fish. *Proceedings of the National Academy of Science.* 104 (18): 7461-7465. www.-pnas.org/cgi/doi/10.1073/pns.0610546104

United Nation's Environment Program (UNEP) 1994. *The impacts of climate on fisheries.* United Nations Environment Program (UNEP). Nairobi. (UNEP Environment Library No 13). 36 pp.

Wu, R. S. S., Zhou, B. S., Randall, D. J., Woo, N. Y. D. and Lam, P. K. S. 2003. Aquatic hypoxia is an endocrine disruptor and impairs fish reproduction. *Environmental Science and Technology* 37:1137-1141.

Wu, R. 2009. Effects of hypoxia on fish reproduction and development. *Fish Physiology* 27 (3): 79-141.

Websites

http://earthtrends.wri.org/pdf_library/data_table/watcon_2005.pdf (Earth Trends data Tables; Coastal and Marine Ecosystems, Freshwater Resources)

www.aquaculture.stir.ac.uk/GISAP/gis-group/climate.php (The effects of climate change on world aquaculture: A global perspective. By N. T. Handisyde, L. G. Ross, L. G, M-C., Badjeck and E. H. Allison. 2006)

http://www.oceansatlas.org.servlet/CDSServlet?status=ND1maWdczEZNzg5LmZpZ2lzv (Fisheries and aquaculture: issues: climate change: fisheries and global climate change)

http://www.fao.org/fi/website/FIRetrieveAction.do?dom=topic&fid=13789 (Fisheries and global climate change)

❏❏❏

5 Climate Change Impacts on Human Health

Key facts

- The climate change is likely to affect millions of people both *directly* (via increased deaths, disease and injury from heatwaves, floods, storms, fires and droughts) and *indirectly* (through changes in the water, air, food quality and quantity and ranges of vectors, such as mosquitoes, water-borne pathogens).
- Recent observed climate change related effects include: (a) an increase in heat related deaths across the globe (Australia, Canada, France, Germany, the Netherlands, Portugal, Switzerland, UK and USA; (b) spread of vector- and rodent-borne diseases such as malaria (Africa, India) and dengue and (c) increase in food poisoning (New Zealand).
- Heatwaves can cause adverse health effects such as heat cramps, fainting, heat exhaustion and heat stroke. Human deaths due to heatwaves are associated with cardiovascular and respiratory diseases and in 2003 in Europe 70,000 deaths were recorded from heatwaves alone.
- Projections : (a) cold temperature is expected to become less frequent, while heat waves will become more frequent; (b) increases in heat related deaths and illness across the globe including Australia, Germany, Portugal, UK, and USA; (c) malaria zone may expand in Australia, Africa (highlands), India and 260-320 million more people will be affected by malaria by 2080; (d) dengue area is likely to increase as climate would be more suitable and by 2085, about 6 billion people will be at risk of contracting dengue fever including Australia and New Zealand; (e) diarrhoeal disease may increase with temperature increase; and (f) cholera outbreaks may rise as a result of more plankton blooms providing nutrients for *Vibrio cholerae*.
- Flooding due to climate change can cause deaths, injuries and increase incidence of infectious diseases and exposure to toxic contaminants posing health risk to humans.
- Health consequences of drought include diseases resulting from malnutrition. Due to shortage of water, contaminated water can be used which increases the risk of diarrhoeal diseases (due to faecal contamination) and water washed diseases.
- In south Asia, excessive monsoon rainfall and high humidity has been the major influence in mosquito breeding and survival. Research studies found that mosquito abundance can be amplified with warming, with an average of over ten-fold increase with every unit increase (0.1^{0}C) in temperature.
- The rise of sea surface temperature could contribute to more phytoplankton blooms followed by zooplankton blooms and consequently more cholera cases. For example, in Bangladesh, the incidence of cholera vibriosis increases as populations of zooplankton increase, especially during the summer monsoon months. Zooplankton (a single copepod) can harbour up to 10,000 vibrios, so ingesting just one or two copepods could cause clinical cholera. A simple method of filtering water through four to eight layers of old sari clothes can reduce the number of cases of cholera in Bangladesh.
- The mosquitoes that transmit dengue breed in urban environments in artificial containers (plant pots) and are less sensitive to rainfall patterns, but higher temperatures associated with El Niño may have an effect on the transmission of the virus.
- The predicted temperatures rise may facilitate transmission of *Salmonella* and *Campylobacter* that cause food poisoning. The skin cancer incidence and sunburn severity may increase due to stratospheric ozone depletion and higher temperatures. In addition, the rising levels of CO_2 will bring an increase in the production of pollen in some plants, producing more allergens and thereby making the situation worse for those people suffering from asthma, emphysema and other respiratory illness. Moreover, rise of sea temperature could accelerate growth of harmful algal blooms which produce toxins. Warmer seas may thus contribute to increased cases of shellfish and reef fish poisoning.
- Public health training, effective surveillance, emergency response systems, sustainable prevention and control programs would reduce the impacts of climate change on human health.

Contents

List of Tables

List of Figures

List of Boxes

Acronyms and Glossary

Aeroallergens: Any of various airborne substances, such as pollen or spores carried by winds that can cause an allergic response.

Arbovirus: A class of viruses transmitted to humans by arthropods such as mosquitoes and ticks. The first two letters of the words arthropod' and borne, make up the 'arbo' that now designates this group of viruses as arthropod-borne.

Biotoxins: Naturally occurring toxic compounds produced by certain organisms.

Boutonneuse fever or : Mediterranean spotted fever (Also called Mediterranean spotted fever, fièvre boutonneuse, or Marseilles fever) is a fever of a Rickettsial infection caused by the bacterium, *Rickettsia conorii* and transmitted by the dog tick, *Rhipicephalus sanguineus*. Boutonneuse fever is endemic in many countries surrounding the Mediterranean Sea.

Cryptosporidiosis: Infection of young farm animals (calves, lambs, fowls or piglets) with protozoa of the genus Cryptosporidium, which may be associated with or contribute to enteric disease.

Encephalitis: An inflammation of the brain usually caused by a virus; symptoms include headache and neck pain and drowsiness and nausea and fever.

El Niño: Warming of sea-surface temperatures in the equatorial Pacific Ocean which influences atmospheric circulation, and consequently rainfall and temperature in specific areas around the world.

ENSO events: El Niño Southern Oscillation (ENSO) is a general term used to describe both warm (El Niño) and cool (La Niña) Ocean atmosphere events in the tropical Pacific. ENSO is associated with floods, droughts and other disturbances in a range of locations around the world. The name El Niño, coined from Spanish for "the little boy", refers to the Christ child, because the phenomenon is usually noticed around Christmas time in the Pacific Ocean off the west coast of South America. La Niña, similarly means "the little girl".

Excedens days: Excedens days are days with an 8 hour average ozone concentration of more than 110 μg/m^3 (the threshold value for the protection of human health).

Extreme events: Events or series of events that occur with greater intensity or frequency than most climate events and cause a disproportionate economic and/or social disruption in the area affected.

Falciparum: A protozoan parasite, one of the species of *Plasmodium* that cause malaria in humans. It is transmitted by *Anopheles* mosquitoes. *P. falciparum* is the most dangerous of these infections as *P. falciparum* malaria has the highest rates of complications and mortality. In addition it accounts for 80% of all human malarial infections and 90% of the deaths. It is more prevalent in sub-Saharan Africa than in other regions of the world.

Flood: A large flow of water and inundation.

Hantavirus pulmonary : syndrome (HPS) Is a deadly disease transmitted by infected rodents through urine, droppings, or saliva. The name *hantavirus* is derived from the Hantaan River, where the Hantaan virus (the etiologic agent of Korean hemorrhagic fever) was first isolated by Dr. Ho-Wang Lee and colleagues. Humans can contract the disease when they breathe in aerosolized virus. HPS was first recognized in 1993 and has since been identified throughout the United States.

Heat waves: Prolonged period of excessively hot weather, heat wave which may be accompanied by high humidity.

Helminthiases: A disease in which a part of the body is infested with worms such as pinworm, roundworm or tapeworm. Typically, the worms reside in the gastrointestinal tract but may also burrow into the liver and other organs.

E. coli: *E. coli* is the abbreviated name of the bacterium of the Family Enterobacteriaceae named *Escherichia* (Genus) *coli* (Species). It is an indicator of water quality.

Emphysema: Is a chronic obstructive pulmonary disease (COPD), formerly termed as chronic obstructive lung disease (COLD). It is often caused by exposure to toxic chemicals, including long-term exposure to tobacco smoke. Emphysema is characterized by loss of elasticity (increased Pulmonary compliance) of the lung tissue from destruction of structures supporting the alveoli and destruction of capillaries feeding the alveoli due to the action of alpha 1 antitrypsin deficiency. Thus the small airways collapse during exhalation, as alveolar collapsibility has increased. This impedes airflow and traps air in the lungs, as with other obstructive lung diseases. Symptoms include shortness of breath on exertion and later at rest, hyperventilation and an expanded chest.

Leishmaniasis: Leishmaniasis is a disease caused by flagellate protozoan parasites that belong to the genus *Leishmania*. This disease is also known as Leichmaniosis, Leishmaniose, leishmaniose, and formerly, Orient Boils, Baghdad Boil, kala azar, black fever, sandfly disease, Dum-Dum fever or espundia. It is transmitted by the bite of certain species of sandfly. There are several different forms of leishmaniasis. The disease was named in 1901 for the Scottish pathologist William Boog Leishman. The most common forms are cutaneous leishmaniasis, which causes skin sores, and visceral leishmaniasis, which affects some of the internal organs of the body (for example, spleen, liver and bone marrow).

Lyme disease or borreliosis : An emerging infectious disease caused by at least three species of bacteria belonging to the genus *Borrelia. Borrelia burgdorferi* is the predominant cause of Lyme disease in the United States, whereas *Borrelia afzelii* and *Borrelia garinii* are implicated in most European cases. Lyme disease is the most common tick-borne disease in the Northern Hemisphere. *Borrelia* is transmitted to humans by the bite of infected hard ticks belonging to several species of the genus *Ixodes*. Early manifestations of infection may include fever, headache, fatigue, depression and a characteristic skin rash called 'erythema migrans'.

Morbidity: Rate of occurrence of disease or other health disorders within a population.

Mortality: Rate of occurrence of death within a population within a specified time period.

NO_x: Is a generic term for mono-nitrogen oxides (NO and NO_2). These oxides are produced during combustion, especially combustion at high temperatures.

Onchocerciasis or river blindness : Is the world's second leading infectious cause of blindness. Or a parasitic disease in the tropical regions of Africa and America. It is caused by *Onchocerca volvulus*, a nematode (filarial worm) that can live for up to 15 years in the human body. It is transmitted to people through the bite of a black fly. The worms spread throughout the body, and when they die, they cause intense itching and a strong immune system response that can destroy nearby tissue, such as the eye. The primary treatment is a drug, ivermectin. For best effect, entire communities are treated at the same time. A single dose may kill first-stage larvae (microfilariae) in infected people and prevent transmission for many months in the remaining population. About 18 million people are currently infected with this parasite; approximately 300,000 have been irreversibly blinded by it.

Oropouche: A virus of the Ortho Bunyavirus genus causing disease in humans in the Caribbean and Central and South America. The pathogen is transmitted by the tiny biting midge *Culicoides paraensis* and causes a self-limiting, acute, dengue-like febrile illness called ORO fever.

Phenology: It is the scientific study of periodic biological phenomena, such as flowering, breeding, and migration in relation to climatic conditions. Because many such phenomena are very sensitive to small variations in climate, especially to temperature, phenological records can be a useful proxy for temperature in historical climatology, especially in the study of climate change and global warming.

Photochemical smog: A mix of photochemical oxidant, air pollutants produced by reaction of sunlight with primary pollutants, especially hydrocarbons.

Red tide: Some marine phytoplankton such as Dinoflagellates (e.g. *Gymnodinium*) produce toxin during their growth and blooms of these dinoflagellates may cause sea water to turn reddish in colour (accumulation of carotenoid pigments). This condition is called a 'red tide'.

Rift Valley Fever (RVF): Is a viral zoonosis (affects primarily domestic livestock, but can be passed to humans) causing fever. It is spread by the bite of infected mosquitoes, typically the *Aedes* or *Culex* genera. RVF outbreaks occur across sub-Saharan Africa, with outbreaks occurring elsewhere infrequently (but sometimes severely) - in Egypt in 1977-78, several million people were infected and thousands died during a violent epidemic. In Kenya in 1998, the virus claimed the lives of over 400 Kenyans. In September 2000 an outbreak was confirmed in Saudi Arabia and Yemen.

Ross River Virus: An arbovirus (a virus mainly transmitted by arthropods) widely distributed throughout Australia. The virus causes epidemic polyarthritis, which consists of arthritic symptom that persist for several months and can be severe and debilitating.

Sandfly: Sandfly (or sand fly) is a colloquial name for any species or genus of flying, biting, blood-sucking Dipteran encountered in sandy areas.

Sari: A kind of fine long plain cloth (4.5 x 18 feet printed or coloured or white) wore by Asian (Bangladesh, India and Pakistan) women.

Schistosomiasis: Also known as bilharzia, bilharziosis or snail fever, is a parasitic disease caused by several species of fluke of the genus *Schistosoma*. Schistosomiasis is known as bilharzia or bilharziosis in many countries, after Theodor Bilharz, who first described the cause of urinary schistosomiasis in 1851. It is most commonly found in Asia, Africa, and South America, especially in areas with water that is contaminated with freshwater snails, which may carry the parasite. The disease affects many people in developing countries, particularly children who may acquire the disease by swimming or playing in infected water. Although it has a low mortality rate, schistosomiasis often is a chronic illness that can damage internal organs and, in children, impair growth and cognitive development. Schistosomiasis is the second-most socio-economically devastating disease after malaria.

Sentinel site: Specific health facility, usually a general/family practice, which undertakes to maintain surveillance and report certain specific predetermined events such as cases of certain infectious diseases.

Stratosphere: From the Latin word "stratus" meaning a spreading out. The stratosphere contains the ozone layer, the part of the Earth's atmosphere which contains relatively high concentrations of ozone. "Relatively high" means a few parts per million—much higher than the concentrations in the lower atmosphere but still small compared to the main components of the atmosphere. It is mainly located in the lower portion of the stratosphere from approximately 15 to 35 km (50,000 – 115,000 ft) above Earth's surface, though the thickness varies seasonally and geographically.

Temperate: The area or region between the tropic of Cancer and the Arctic circle or between the tropic of Capricorn and the Antarctic.

Tick-borne disease: Diseases or illnesses transmitted by ticks. Tick-borne illnesses are caused by infection with a variety of pathogens, including rickettsia and other types of bacteria, viruses, and protozoa. Because ticks can harbor more than one disease-causing agent, patients can be infected with more than one pathogen at the same time, compounding the difficulty in diagnosis and treatment. Major tick-borne diseases include: Lyme disease, Tick-borne meningoencephalitis, Rocky Mountain Spotted Fever.

Tularemia: Also known as "rabbit fever", "deer-fly fever", "Ohara fever" and "Francis disease", is a serious infectious disease caused by the bacterium *Francisella tularensis*. A small, gram-negative, non-motile coccobacillus, the bacterium has several sub-species with varying degrees of virulence. The most important of those is *F. tularensis tularensis* (Type A), which is found in lagomorphs in North America and is highly virulent for humans and domestic rabbits. *F. tularensis palaearctica* (Type B) occurs mainly in aquatic rodents (beavers, muskrats) in North America and in hares and small rodents in northern Eurasia. It is less virulent for humans and rabbits. The primary vectors are ticks and deer flies, but the disease can also be spread through other arthropods. The disease is named after Tulare County, California.

West Nile virus: Or WNV is a virus of the family *Flaviviridae*; part of the Japanese encephalitis (JE) antigenic complex of viruses, it is found in both tropical and temperate regions. It mainly infects birds, but is known to infect humans, horses, dogs, cats, bats, chipmunks, skunks, squirrels and domestic rabbits. The main route of human infection is through the bite of an infected mosquito.

WHO: World Health Organisation of the United Nation.

Vector: An organism (as an insect) that transmits a pathogen.

Vibrio: Any comma shaped bacterium. Generally, vibrios are gram-negative, motile and aerobic.

Vivax: *Plasmodium vivax* is a protozoal parasite and a human pathogen. The most frequent and widely distributed cause of recurring malaria. *P. vivax* is one of four species of malarial parasite that commonly infect humans. It is less virulent than *Plasmodium falciparum*, the deadliest of the four and seldom fatal. *P. vivax* is carried by the female *Anopheles* mosquito, since it is the only sex of the species that bites.

VOC: Volatile organic compounds - a group of chemical compounds.

Zoonoses: Zoonosis or zoonose is any infectious disease that is able to be transmitted (by a vector) from other animals, both wild and domestic, to humans or from humans to animals (the latter is sometimes called reverse zoonosis). The word is derived from the Greek words *zòon* (animal) and *nosos* (ill). Many serious diseases fall under this category.

5.1: Introduction

Health includes physical, social and psychological well being. The climate change is likely to affect millions of people both directly and indirectly: **directly** via increased deaths, disease and injury from heatwaves, floods, storms, fires and droughts and **indirectly** through changes in the water, air, food quality and quantity, ecosystems, agriculture, livelihoods and infrastructure, ranges of vectors, such as mosquitoes, water-borne pathogens (Watson *et al.* 2005). The direct and indirect effects of climate change on human health are summarised in Table 5.1.

Table 5.1: ***Climate change related direct and indirect effects on human health*** *[Haines et al. 2000; IPCC 2007].*

Mediating processes	Health outcome
Direct effects	
Exposure to thermal stress (heat waves)	Increased risk of heat related mortality (especially for elderly, chronically sick, very young people).
Changed frequency or intensity of other extreme events (floods, droughts, cyclones)	Increased risk of deaths; injuries, infectious and respiratory and skin diseases; increased risk of food and water shortage; increased risk of malnutrition; psychological disorders; and damage to public health infrastructure.
Indirect effects (disturbances of ecological systems)	
Changed food productivity (crops) through changes in climate and associated pests and diseases	Malnutrition and hunger and consequent impairment of child growth and development.
Sea level rise with population displacement and damage to infrastructure	Increased risk of infectious diseases, psychological disorders; increased risk of deaths and injuries by drowning in floods; human displacement and migration related health effects.
Change in air pollutants (pollens and spores)	Asthma, and allergies; other acute and chronic respiratory disorders and deaths.
Changed local ecology of water borne and food-borne infective agents	Changed incidence of diarrhoeal and other infectious diseases.
Effect on range and activity of vectors and infective parasites	Changes in geographical ranges and incidence of vector-borne diseases.
Social, economic and demographic dislocation through effects on economy, infrastructure, and resource supply	Wide range of public health consequences; mental health and nutritional impairment, infectious diseases, civil conflict.

According to WHO (2008) (<*http://www.who.int/mediacentre/factsheets/fs266/2n/print.html*>), the health effects of rapidly changing climate change could include the following as listed below:

- Increasing frequencies of heatwaves;
- More variable precipitation patterns may increase risks of water-borne diseases;
- Rising temperatures and variable precipitation are likely to decrease the production of staple foods, thereby increasing risks of malnutrition;
- Rising sea levels increase the risk of coastal flooding, and may necessitate population displacement.

The actual health impacts will be influenced by local environmental conditions, socio-economic circumstances and other technological and behavioural adaptations taken to reduce the threats posed to health.

5.2: Recent observations of climate change related health effects

Recent observations of climate change related effects include an increase in heat related deaths across the globe, spread of vector- and rodent-borne diseases such as, malaria, dengue and increase in food poisoning (Table 5.2).

Table 5.2: ***Recent observations of climate change related health effects in some selected countries since 2000*** *[Confalonieri et al. 2007].*

Country	Observations and findings
Australia	Increase in heatwave-related deaths; drowning from floods; diarrhoeal disease in indigenous communities; potential changes in the geographical range of dengue and malaria; likely increase in environmental refugees from Pacific islands.
Bolivia	Intensification of malaria and leishmaniasis transmission.
Bhutan	Loss of life from frequent floods; glacier lake outburst floods; landslides; hunger and malnutrition; spread of vector-borne diseases into higher elevations; loss of water resources; risk of water-borne diseases.
Canada	Increase in heatwave-related deaths; increase in air pollution-related diseases; spread of vector- and rodent- borne diseases; increased problems with contamination of both domestic and imported shell-fish; increase in allergic disorders; impacts on particular populations in northern Canada.
Finland	Small – increase in heat related mortality; changes in phenological phases and increased risk of allergic disorders; small reduction in winter mortality.
Germany	Observed excess deaths from heatwaves; changing ranges in tick-borne encephalitis.
India	Increase in communicable diseases; malaria projected to move higher latitudes and altitudes.
Japan	Increased risk of heat- related emergency visits; Japanese cedar pollen disease patients; food poisoning; sleep disturbances.
The Netherlands	Increase in heat-related mortality; air pollutants; risk of Lyme disease, food poisoning and allergic disorders.
New Zealand	Increases in enteric infections (food poisoning); changes in some allergic conditions; injuries from more intense floods and storms; a small increase in heat related deaths.
Panama	Increase of vector-borne and other infectious diseases; health problems due to high ozone levels in urban areas; Increase in malnutrition.
Portugal	Increase in heat-related deaths and malaria; food and water-borne diseases; West Nile fever, Lyme disease and Mediterranean spotted fever; a reduction in leishmaniasis risk in some areas.
Spain	Increase in heat-related mortality and air pollutants; potential change of ranges of vector- and rodent-bone diseases.
Tajikistan	Increase in heat-related deaths.
Switzerland	Increase of heat-related mortality; changes in zoonoses; increase in cases of tick-borne encephalitis.
United Kingdom	Health impacts of increased flood events; increased risk of heatwave-related mortality; and increased ozone-related exposure.

5.3: Extreme weather events and future projections

[based on Hales et al. 2003; Patz et al. 2003; Confalonieri et al. 2007; Costello et al. 2009].

5.3.1: *Heatwaves*

An extreme heatwave is a prolonged period of excessive day time and night time heat. For example, in Canada, a heatwave is defined as three or more days of 32°C temperatures and beyond. With climate change, cold temperature is expected to become less frequent, while heat waves will become more frequent. Hot days, hot nights and heatwaves have become more frequent in recent times in the world (Table 5.2). Eighteen heat waves were reported from India between 1980 and 1998. In the summer of 2003, severe heat waves affected most of Europe, including France, the UK, Belgium, Germany, the Czech Republic, Italy, Portugal, Spain, Switzerland, and the Netherlands which has attributed in excess of 70,000 deaths. Climate variability model suggests that major cities in Europe and North America will have substantial increases in both frequency and duration of severe heatwaves by 2090, for example, by 2050; the annual summer time mortality in New York is estimated to increase to several folds (see

Table 5.4 for future projections of health related effects due to climate change). Heatwaves can cause adverse health effects including heat cramps (muscular pains and spasms); fainting, heat exhaustion, and heat stroke (Poumadére *et al.* 2005). Heat waves are responsible for a number of human deaths (Table 5.3), most of which are associated with cardiovascular and respiratory diseases.

Table 5.3: ***Human death estimate from heat waves*** *[Confalonieri et al. 2007]*

Place and Date	Number of human death
France, November 2003	14,800
India (Andhra Pradesh) 2003	3000
India (Orissa), 1998-2000	212 0
India (ten states), 1988	1,3000
USA (Chicago), July 1995	514

Human typically display an optimum temperature zone called 'comfort zone' at which the death rate is lowest, outside of which morbidity and mortality increases. In general, people in hotter regions/ towns are more affected by colder temperatures, and people in colder regions/towns are more affected by warmer temperatures. Regions where housing provides poor protection against temperature extremes can have higher human mortality than expected for that location. Epidemiological studies conducted in Europe found a positive correlation between heatwaves and mortality with elderly people (aged over 75 years). These elderly people had diminished their physiological capacity for thermoregulation. Mentally ill people, children and others in thermally stressful occupations are also found to be vulnerable to extreme heat waves.

Table 5.4: ***Projected impacts of climate change relating to heat, ozone related mortality, and malaria, dengue fever and other infectious diseases*** *[from various sources compiled by Confalonieri et al. 2007].*

Area	Health effect	Main projections
I. Heat and cold related mortality		
Australian capital cities	- Heat related mortality in people older than 65 years	- Increase in temperature-attributable death rates from 82/100,000 across all cities under current climate to 246/1000,000 (by 2100).
Germany, Baden Wuertmberg	- Heat and cold related mortality	- About a 20% increase in heat related mortality.
Portugal (Lisbon)	- Heat related mortality	- Increase in heat-related mortality from 5.4. to 6 deaths/100,000 (2003) to 5.8-15.1 deaths/1000,000 (by 2020) and to 7.3-35.9 deaths/1000,000 (by 2050), 19.5-248.4 deaths/1000,000 (by 2080).
UK	- Heat and cold related mortality	- Annual heat related deaths increase from 798 (1990) to 2,793 (by 2050) and to 3,519 (by 2080) - Annual cold related deaths decreases from 80,313 (1990) to 60021 (by 2050) and to 51,243 (by 2080)
USA (four cities in California)	- Annual number of heatwave day - length of heatwave season - Heat-related mortality	- Increase in annual number of days - Number of heat wave day's increases 4 –fold to 6-8 folds (by 2080). - Annual number of heat-related deaths increases from 165 (1990) to 319-1,182 (by 2080).
II. Ozone related health effects		
England and Wales	- Excedens days (ozone particulates, NOx)	- Pollutants such as ozoneconcentrations may increase
USA (50 cities)	- Ozone-related hospitalisations and deaths	- Maximum ozone concentrations increased for all cities - 0.31% increase of cardiovascular disease mortality.
USA (New York metro)	- Ozone-related deaths by country	- 4.4-4.5% increase in ozone related deaths

Contd...

III. Malaria		
Global and regional	- Population at risk in areas where climate conditions are suitable for malaria transmission	- Additional population at risk for >1 month transmission range from >220 million to > 400 million
Africa	- Person-months at risk for stable *falciparum* transmission	- 16-18% increase in person-months of exposure (by 2100) - 5-7% increase in mainly altitudinal distribution and limited latitudinal expansion.
Africa	- Map of climate suitability for stable *falciparum* transmission (minimum 4 months suitable per year)	- Decreased transmission in 2020 in south-east Africa - Localised increases in highland andupland areas, and decreases around Sahel and south central Africa (by 2050s and 2080s).
Africa (Zimbabwe)	- Climate suitability for transmission	- Highlands become more suitable for transmission.
Britain	- Probability of malaria transmission	- Increase in risk of local malaria transmission of 8-15%; - Indigenous malaria may be re-established.
Portugal	- Favourable temperature for disease transmission	- Significant increase in the number of days suitable for survival of malaria vectors if infected vectors are present.
Australia	- Geographical area suitable/ unsuitable for maintenance of vector	- 'Malaria receptive zone' expands southward to include some regional towns by 2050s.
India (all states)	- Climate suitability for *falciparum* and *vivax* malaria transmission	- Geographical range projected to shift away from central regions towards south-western and northern states (by 2050s).
IV: Dengue		
Global	- Population at risk	- Global population at risk due to population growth and climate change is estimated tobe 5-6 billion (by 2085) - Global population at risk with climate change only is estimated to be 3.5 billion (by 2085).
New Zealand	- Currently not present	- Potential risk of dengue outbreaks in some regions under the current climate. - Climate change projected to increase risk of dengue in more regions.
Australia	- Current	- Regions climatically suitable may increase southwards including Sydney.
V: Lyme disease		
Canada	- Geographical range and abundance of Lyme diseases vector, *Ixodes scapularis*	- Northward expansion of approximately 200 km (by 2020s) and 1000 km (by 2080s). - Tick abundance increases 30 to 100% (by 2020s) and 2 to 4 folds (by 2080s).
VI: Tick-borne encephalitis		
Europe	- Geographical range	- Northeast ward expansion from current range under low to high degrees of climate change - Remain in central and eastern Europe under low to medium degrees of climate change.
VII: Diarrhoeal disease		
Global (14 world regions)	- Mortality	- Generally diarrhoeal disease increases with temperature increase
Australia (Aboriginal community-Alice Springs)	- Hospital admissions in children aged under 10	- No significant increase (by 2020). - An annual increase of 5 to 18% (by 2050)
VIII: Food poisoning		
England and Wales	- Notified cases of food poisoning (non-specific)	- for +1, +2 and +3°C temperature increases, absolute increases of approx. 4,000, 9,0000, and 14,000 notified cases of food poisoning

5.3.2: Floods and storms

Floods are low probability and high impact events. Inland floods result from interaction of rainfall, surface run-off, evaporation, catchment size, local topography and climate etc. It is projected that flooding frequency could increase with climate change. Flooding can cause deaths, injuries and increase incidence of infectious diseases and exposure to toxic contaminants (heavy metals, pesticides, oil and other hazardous waste) posing health risk to humans. Flooding and excessive rainfall facilitates entry of human and animal wastes into waterways and drinking water supplies, which are potential for water borne diseases (see Table 5.5). Drowning by storm surge is the major killer causing large number of deaths in the Bay of Bengal region, in particular in Bangladesh.

Water borne diseases, in particular, gastroenteritis, may be caused by *Norovirus*, *Giardia*, *Cryptosporidium*, *Campylobacter*, *E. coli* and *Shigella*. Examples of water borne outbreaks soon after heavy rainfalls are *Cryptosporidiosis* outbreak in Milwaukee, USA in 1993, (1,000 fatalities) and an *E. coli* outbreak in Walkerton, Canada in 2000. An increase of infectious diseases such as cholera, cryptosporidiosis and typhoid fever can occur soon after flood events (reported from India, Bangladesh and Brazil soon after flood events and in case of Bangladesh, cholera outbreaks generally occur during the monsoon season) (Table 5.5).

Extreme rainfall and run-off events may increase the total microbial load in watercourses and drinking-water reservoirs. In the USA, it was found that there is a relationship and association between extreme rainfall events and monthly report of outbreaks of water-borne diseases. In Bangladesh, cholera (*Vibrio cholerae*) is found to be correlated with sea-surface temperatures and seasonal abundance of plankton (see section 5.5.1). Rainfall patterns can influence the transport and distribution of infectious agents while temperature can affect their growth and survival.

Table 5.5: ***Water and food-borne agents: connection to climate*** *[Patz et al. 2003].*

Pathogen group	Pathogenic agent	Food-borne agents	Water-borne agents	Indirect weather effect	Direct weather effect
Viruses	Enteric viruses (hepatitis A virus, Coxsackie B virus	Shell-fish	Groundwater	Storms can increase transport from faecal and waste water sources	Survival increases at reduced temperatures and sunlight (ultraviolet)
Bacteria	Vibrio (*V. vulnificus*, *V. parahaemolyticus*, *V. cholerae*)	Shell-fish	Recreational, wound infections	Enhanced zooplankton blooms	Salinity and temperature associated with growth in marine environment
Cyanobacteria	*Anabaena* spp.,				
Dinoflagellates	*Gymnodinium* spp. *Pseydibutzschia* spp.)				
Protozoa	Enteric protozoa (*Cryptosporidium*, *Cyclospora*)	Fruit and vegetables	Recreational and drinking water	Storms can increase transport from faecal and wastewater sources	Temperature associated with maturation and infectivity of *Cyclospora*

Weather extremes - such as heavy rains, floods, and disasters like Hurricane Katrina that devastated New Orleans, USA in August 2005 - endanger human health as well as destroy property and livelihoods. Approximately 600 000 deaths occurred world wide as a result of weather-related natural disasters in the 1990s, some 95% of which took place in developing countries. According to WHO (Fact sheet no 2666, 2007-climate and health), the previous death tolls or the number of people affected due to floods and storms are listed below:

- Bangladesh (Cyclones during 1963-2009): caused about a million deaths (Box 5.1).
- China (2003): 130 million affected.
- India (Orissa, October 1999): 10,000 died; 10-15 million affected.
- Mozambique (2000/2001): 1813 died.
- Venezuela (December 1999): 30,000 died.

Box 5.1: ***El Niño and La Niña events and its relationships with cyclones in Bangladesh and human casualties*** *[Chowdhury 2009].*

The Intergovernmental Panel on Climate Change (IPCC 2007) projected that the intensity of tropical cyclone activity is likely to increase during the 21st century. The major cyclones that hit Bangladesh from 1963 to 2009 (see below) are thought to be related to climate change of extreme weather events such as El Niño and La Niña. An analysis of historical records reveals that the occurrences of major cyclones in Bangladesh are on the rise in recent times (1963-2009). For example, the number of severe cyclones was only three during the period of 1795-1896; however, it has increased to 13 during 1897-1947 and 51 during the period 1948-1998. It is known that El Niño and La Niña years have considerably increased during 1981-2009 compared to 1950-1980, which scientists believed are due to the effects of climate change or global warming. These have caused an increase in both frequency and intensity of extreme weather events, such as major cyclones and human fatalities in Bangladesh. However, massive preparedness and evacuation initiatives taken during 2009 have reduced human casualties.

- ***2009*** *(25 May): Tropical cyclone 'Aila' (transition from La Niña to ENSO conditions).*
- ***2009*** *(14-15 April): Cyclonic storm 'Bijli' hit Chittagong and Cox's Bazar area with heavy rains and winds of upto 90 km/h (transition from La Niña to ENSO conditions).*
- ***2007*** *(15 November): Cyclone 'Sidr' hit Bangladesh; winds of 215 km/h; human deaths of 5,000-10,000*
- ***1997*** *(29-30 November): A severe cyclone hit Chittagong; max wind speed of 224 km/h (a strong El Niño was active during that time).*
- ***1991*** *(29-30 April): A severe cyclone hit coastal regions of Bangladesh; maximum wind speed of 225 km/h; human deaths of 150,000 (a moderate El Niño was active).*
- ***1985*** *(24-25 May): A severe cyclone hit Chittagong, Cox's Bazar, Noakhali, and coastal islands with 10-15 feet of tidal surge; human deaths of 11,069 (a moderate La Niña. was active).*
- ***1970*** *(12 November): A cyclone hit the entire coast, max. wind speed of 222 km/h; 10-12 feet tidal surge; human deaths of 500,000-1200,000 (a strong to moderate La Niña was active)*
- ***1963*** *(28-29 May): A severe cyclone storm lashed Chittagong, Noakhali, Cox's Bazar and coastal islands of Sandwip, Kutubdia, Maheshkhali and Hatya; human deaths of 11,520 (a moderate El Niño event was active).*

5.3.3: *Drought*

A drought can be defined as a period of abnormally dry weather which persists long enough to produce a serious hydrological imbalance. There are four general types of drought, all of which can impact on humans in different ways.

- Meteorological (when precipitation is unusually low for a particular region)
- Agricultural (when the amount of moisture in the soil is no longer sufficient for crops)
- Hydrological (when surface water and groundwater supplies are below normal)
- Socio-economic (when lack of water affects the economic capacity of people to survive from non-agricultural production)

Health consequences of drought include diseases resulting from malnutrition. Due to shortage of water, contaminated water can be used which increases the risk of diarrhoeal diseases (due to faecal contamination) and water washed diseases (Table 5.6). The effect of drought may also include deaths, infectious and respiratory diseases. Drought diminishes dietary diversity and reduces overall food consumption by people and can lead to macronutrient deficiencies. During a drought in Gujarat, India in 2000, diets were found to be deficient in energy and several vitamins. Drought and lack of food are associated with an increased risk of mortality in Bangladesh. Drought may trigger population movements from rural to urban areas and it can lead to increases in communicable diseases and poor nutritional status resulting from overcrowding, and lack of safe water, food and shelter. During drought, anxiety and depression could lead to deterioration of farmer's mental health or even suicidal due to diminishing incomes. Dust storms could adversely affect people's respiratory system and it may exacerbate asthma symptoms.

Table 5.6: ***Mechanism by which above-average rainfall (I) and below-average rainfall (II) can affect human health*** *[Hales et al. 2003].*

Event	Type	Description	Potential health impact
I : Impact of above-average rainfall			
Heavy precipitation	Meteorological	'extreme event'	Increased mosquito abundance or decreased (if breeding sites are washed away)
Flood	Hydrological	River/stream over tops its banks	Changes in mosquito abundance contamination of surface water
Flood	Social	Property or crops damaged	Changes in mosquito abundance contamination of water with faecal matter and rat urine (leptospirosis)
Flood	Catastrophic flood/ 'disaster'	Flood leading to > 10 killed or > 200 affected or government call for external assistance	Changes in mosquito abundance contamination of water with faecal matter and rat urine and increased risk of respiratory and diarrhoeal disease, deaths (drowning) injuries health effects associated with population displacement loss of food supply psychosocial impacts
II : Impact of below above-average rainfall			
Drought	Meteorological	Evaporation exceeds water absorption, soil moisture decreases	Changes in vector abundance if vector breeds in dried up river beds
Drought	Agricultural	Drier than normal conditions leading to decreased crop production	Depends on socioeconomic factors
Drought	Social	Reduction in food supply or income, reduction in water supply and quality	Food shortage, illness, malnutrition (increased risk of infection), increased risk of disease associated with lack of water for hygiene
Drought	Food shortage/famine/d rought disaster	Food shortage leading to death > 10 killed or > 200 affected or government call for external assistance	Death (starvation) malnutrition (increases risk of infection), health impacts associated with population displacement

5.3.4: Forest Fire

Forest and bush fire has been linked to El Niño episode. Changes in temperature and precipitation are projected to increase the frequency and severity of fire events. It can cause human injury, burns and damage human health from smoke inhalation. Bush fire can exacerbate asthma and other respiratory and cardiovascular conditions. During bush fire, dioxins are released into atmosphere, which have the potential for the risk of cancer in adult humans (see chapter 10 on the effects of dioxins). In Australia, the frequency of fire is expected to rise by 4-25% by 2020 and 15-70% by 2050. Pollutants from bush fires can affect air quality for thousands of kilometres.

5.3.5: El Niño and climate extremes

The term El Niño is nowadays used to refer to the periods of strong and prolonged warm weather, which influence the climate worldwide. The periods of the warm waters in eastern Pacific (El Niño) and periods of cooler waters (La Niña) are accompanied by changes of air pressure in the east and west Pacific: these are called the Southern Oscillation. The whole cycle is now referred to as El Niño Southern Oscillation (ENSO). The effects of La Niña are generally less pronounced and the trends are opposite to those of El Niño.

El Niño is associated with death and disease, most of which result from weather-related disasters such as floods and droughts (Figure 5.1 and Table 5.6). The regions where El Niño has a strong effect on climate are those with the least resources such as southern Africa, parts of South America, South-east Asia. The El Niño cycle is associated with increased risks of some of the diseases transmitted by mosquitoes, such as malaria, dengue and Rift Valley fever (see section 5.4 vector-borne diseases). Malaria transmission is particularly sensitive to weather conditions. In dry climates, heavy rainfall can create puddles that provide good breeding conditions for mosquitoes. In very humid climates, droughts may turn rivers into strings of pools, preferred breeding sites of other types of mosquitoes.

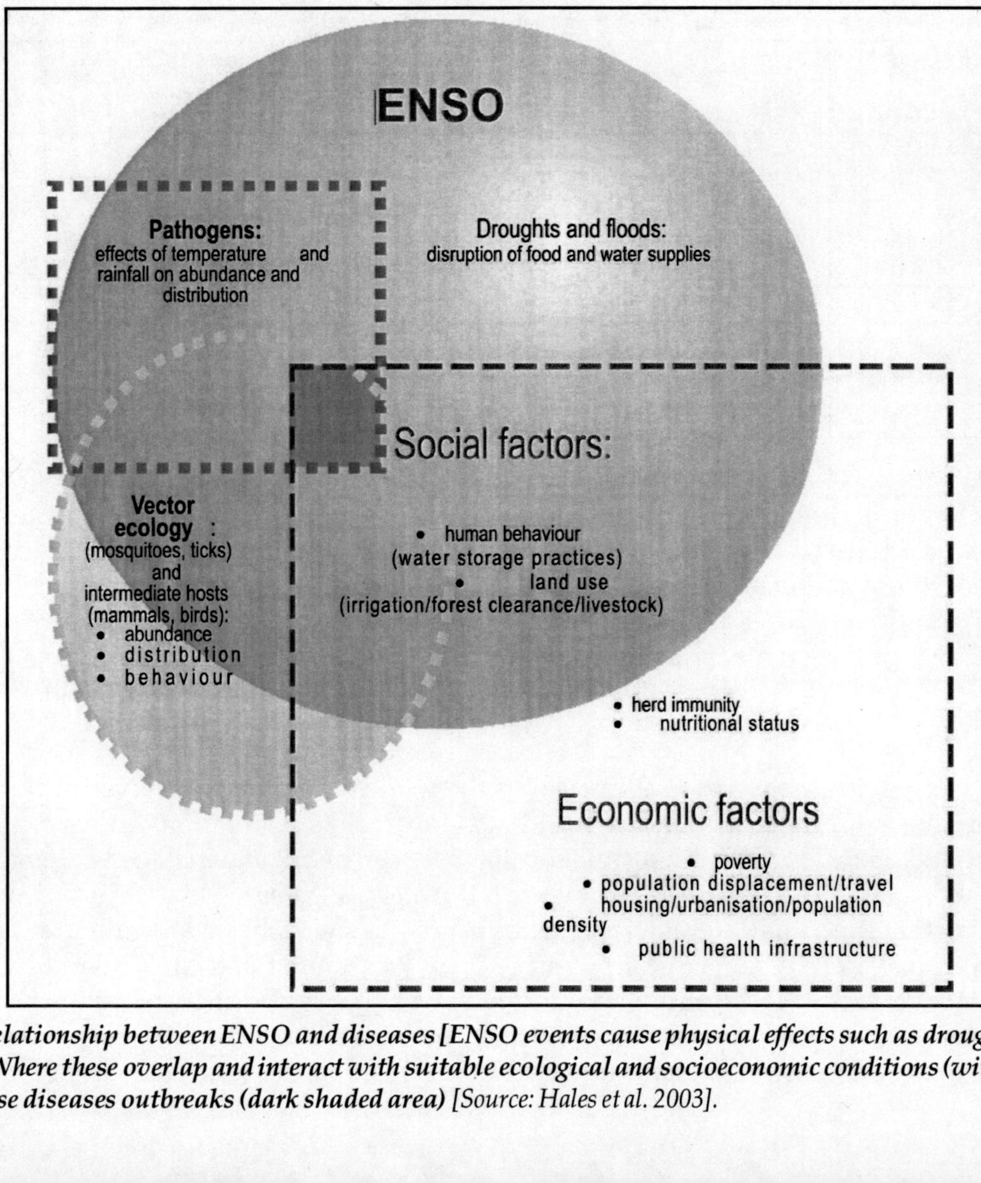

Figure 5.1: ***Relationship between ENSO and diseases [ENSO events cause physical effects such as droughts and floods (blue circle). Where these overlap and interact with suitable ecological and socioeconomic conditions (with dotted lines) that may cause diseases outbreaks (dark shaded area)*** *[Source: Hales et al. 2003].*

5.4: Vector-borne disease

[based on Wilson 2001; McMicheal 2003b, 2006; Costello et al. 2009; http://en.wikipedia.org/wiki/Dengue].

Vector-borne disease (VBD) involves the transmission of infectious agents (viruses, bacteria and parasites) by blood sucking arthropods such as mosquitoes, ticks, fleas and sandflies. The transmission of many infectious diseases is affected by climatic factors, for example, reproduction and survival rates of different vectors are strongly affected by temperature, rainfall patterns and extreme weather events (e.g. floods). Other factors are moisture, surface water, humidity, wind, soil moisture, and changes in and forest distribution. Table 5.7 provides some examples of how diverse climate changes affect the occurrence of infectious diseases. In general, increased warmth and moisture would enhance the transmission of VBDs. Climate change could shift in the distribution of tick vectors in Europe and North America.

Table 5.7: ***Diverse environmental changes and occurrence of infectious diseases*** *[Cited in WHO 2003a, original from Wilson 2001].*

Environmental changes	Example of diseases	Pathway of effect
Dams, canals, irrigation	Schistosomiasis Malaria Helminthiases	Snail host habitat, human contact Breeding sites for mosquitoes Larval contact due to moist soil
Agricultural intensification	Malaria Venezuelan haemorrhagic fever	Crop insecticides and vector resistance Rodent abundance, contact
Urbanization, urban crowding	Cholera Dengue	Water contamination Water-collecting trash, *Aedes aegypti* mosquito breeding sites
Deforestation and new habitation	Cutaneous leishmaniasis Malaria Oropouche Visceral leishmaniasis	Proximity, Sandfly vectors Breeding sites and vectors, immigration of susceptible people Contact, breeding vectors Contact with Sandfly vectors
Reforestation	Lyme disease	Tick hosts, outdoor exposure
Ocean warming	Red tide	Toxic algal blooms, eating of shell-fish
Elevated precipitation	Rift valley fever Hantavirus pulmonary syndrome	Pools for mosquito breeding Rodent food, habitat abundance

Climate change will affect the potential incidence, seasonal transmission and geographic range of various vector-borne diseases, such as malaria, dengue fever and yellow fever (mosquito borne), leishmaniasis (sand flies), Lyme disease (ticks) and onchocerciasis (spread by black flies). VBDs are a major cause of illness and death in tropical countries, where millions of people are at risk from VBDs. An estimated number of people at risk and infected by VBDs are given in Table 5.8. Works of McMicheal (2003b and 2006) are referred here as further reading.

5.4 1: Malaria

Malaria *(Plasmodium vivax* and *P. falciparum)* is the world's most important vector borne-disease. It is affecting millions of people around the world (40% of world's population) and is reported from 92 countries (see Table 5.8). There is a strong relationship between ENSO and malaria epidemics in southern Asia and South America (see sections 5.3.5, Figure 5.1). *Anopheles farauti sensu lato* is the most important Australian vector. Precipitation and temperature are the two important climatic factors in the transmission of malaria. In south Asia, excessive monsoon rainfall and high humidity has been the major influence in mosquito breeding and survival. Malaria is also increasing in the east African highlands in association with local warming. Further, a positive link was found with malaria epidemics and El Niño event - the risk of malaria reported to increases in the year after an El Niño event (see Figure 5.2).

Table 5.8: ***People at risk from Vector-borne disease*** *[http://www.grida.no/climate/vital39.htm].*

Disease	Vector	Population at risk (million)	No. of people currently infected or new cases /year	Present distribution	Likelihood of altered distribution
Malaria	Mosquito	2,400	300-500 million	Tropics and subtropics	Highly likely
Schistosomiasis	Water snail	600	200 million	Tropics and subtropics	Very likely
Lymphatic Filariasis	Mosquito	1094	117 million	Tropics and subtropics	Likely
African Trypanosomiasis (Sleeping sickness)	Tsetse fly	55	250,000 to 300,000 cases per year	Tropical Africa	Likely
Dracunculiasis (Guinea worm)	Crustacean (copepod)	100	1000,000 per year	South Asia, Arabian Peninsula, Central-West Africa	Unknown
Leishmaniasis	Phlebotomine sand fly	350	12 million infected, 500,000 new cases per year	Asia, Southern Europe, Africa, Americas	Likely
Onchocerciasis (River blindness)	Black fly	123	17.5 million	Africa, Latin America	Very likely
American Trypanosomiasis (Chagas disease)	Triatomine bug	100	18 million	Central and South America	Likely
Dengue	Mosquito	1,800	10-30 million per year	All tropical countries	Very likely
Yellow fever	Mosquito	50	More than 5000 cases per year	Tropical South America, Africa	Likely

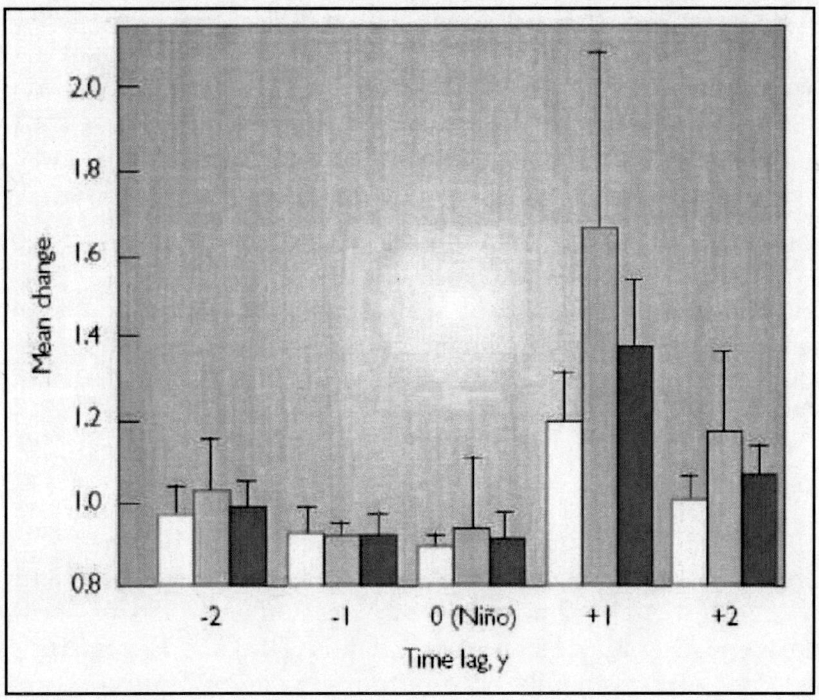

Figure 5.2: ***Relationship between malaria cases and El Niño in Venezuela [Average relative change in malaria incidence two years before (-1 and -2 years), during (El Niño year-year 0) and two years after El Niño (+1, +2). Open bars = 1910-1935 malaria deaths; grey bars = 1975-1995 malaria deaths; black bars = average of both open and grey bars*** *[referred in Patz et al. 2003; original from Boumaet and Dye 1997].*

Overall, the climate change predicted an increase in malaria transmission in endemic malaria areas as well as at higher latitudes within malaria areas. Research studies conducted found that mosquito abundance can be amplified with warming, with an over ten-fold increase with every unit increase (0.1^{0}C) in temperature (Costello *et al.* 2009). In addition, malaria zone may expand to warm high altitudes, in places once was free of the disease and it is projected that additional 260-320 million people will be affected by malaria by 2080 as a consequence of new transmission zones (Costello *et al.* 2009). Recently, there has been growing recognition of the links between El Niño and diseases. In some highland regions higher temperatures possibly linked to El Niño may increase malaria transmission. This has been shown to occur in higher latitude parts of Asia, such as northern Pakistan. At the beginning of this century, periodic epidemics of malaria flared up in Punjab region (north-east Pakistan and north-west India) after excessive monsoon rainfall. In Venezuela and Colombia, malaria cases increase by more than one third following dry conditions associated with El Niño. In Southern Africa, countries have recently experienced malaria epidemics following unusual rainfall.

5.4 2: Dengue

Dengue is the most important viral tropical disease transmitted by *Stegoyma* (previously called *Aedes*) *aegypti* mosquitoes occurring in tropical and sub-tropical region worldwide. Dengue is one of the most important vector-borne viral diseases. The geographical spread of dengue includes northern Australia and northern Argentina, and the entire Singapore, Malaysia, Taiwan, Thailand, Vietnam, Indonesia, Honduras, Costa Rica, Philippines, Pakistan, India, Sri Lanka, Bangladesh, Mexico, Suriname, Puerto Rico, Jamaica, Bolivia, Brazil, Guyana, Venezuela, Barbados, Trinidad and Samoa. Dengue is transmitted to humans by the *S. (Aedes) aegypti* or more rarely the S. (Aedes) *albopictus* mosquito, which feeds/bites during the day.

The abundance and distribution of dengue mosquitoes are related to temperature, moisture and humidity hence are expected to be affected by climate change. Though, high rain fall or high temperature can lead to an increase in transmission of dengue, however, household's water storages could provide favourable breeding grounds of the vector. The mosquitoes that transmit dengue breed in urban environments in artificial containers (plant pots) and are less sensitive to rainfall patterns, but higher temperatures associated with El Niño may have an effect on the transmission of the virus. Preliminary studies have shown a link between ENSO and incidence of dengue in countries where ENSO has a strong effect on the weather (e.g. some Pacific Island nations and Indonesia) (Hales *et al.* 2002). In 1998, many countries in Asia suffered an unusually high level of dengue and dengue haemorrhagic fever, some of which may be attributable to El Niño-related extreme weather.

Climate change is likely to increase the number of regions affected by arbovirus, such as Australia and New Zealand. Approximately one-third of the world's population lives in regions where the climate is suitable for dengue transmission. By 2080, about 6 billion people may be at risk of contracting dengue fever as a consequence of climate change. Dengue haemorrhagic fever and dengue shock syndrome are life-threatening. Dengue mosquito is a morning/evening biter whereas malaria mosquito does not breed in urban environments and is a night biter.

5.4.3: Rodent–borne diseases

Rodents act as reservoirs for a number of diseases either as intermediate hosts or as hosts for arthropod vectors such as ticks. Certain rodent's diseases are associated with flooding including leptospirosis, tularaemia and viral haemorrhagic diseases. Other diseases associated with rodents and ticks include plague, Lyme disease, tick-borne encephalitis (TBE) and Hantavirus pulmonary syndrome (HPS). Infection by Hantaviruses mainly occurs from inhalation of airborne particles from rodents excreta. The 1998 increases of Hantavirus in the USA were linked with El Niño, when there were wet and warm winters (these conditions can increase food sources such as insects and nuts for rodents and promote breeding). Lyme, TBE and Hantavirus infections are all projected to increase as a result of global climate change (Costello *et al.* 2009).

5.5: Other climate change related health effects

[based on ICDDR, B. 2002; Colwell et al. 2003; McMicheal et al. 2003a; WHO 2003a, 2003 b; Watson et al. 2005; Confalonieri et al. 2007; Costello et al. 2009].

5.5.1: *Algal blooms*

Predicted climate change (high temperature and low environmental flows) could cause more frequent harmful algal blooms in waterways (e.g. blue-green algae/ cyanobacteria in freshwater). Blue green algae have the potentials to produce powerful biotoxins. The biotoxins are health hazard to human through drinking water and recreational water use and can affect fish and livestock that may be used for human food. The blue-green algae can produce three types of toxins as listed below affecting human health (see chapter 13 for more information on freshwater biotoxins).

- *Endotoxins*: cause skin rashes, eye irritation and gastroenteritis
- *Neurotoxins*: cause nerves damage and affect mussels
- *Hepatotoxins*: can damage the liver and may also increase the risk of some types of cancer in the long term

With the rise of ocean temperature due to global warming and more intense El Niños, cholera outbreaks might increase as a result of more plankton blooms (Colwell 1996) providing nutrients for *Vibro cholera (*see Box 5.2*).*

Box 5.2: *Relationship between cholera in Bangladesh and climate change* *[ICDDR, B. 2002; Colwell et al. 2003].*

Cholera (Vibrio cholerae) is caused by bacteria that live on microscopic animal called zooplankton. In 2001, WHO reported 184,000 cases of cholera in 58 countries, with more than 2,700 deaths. V. cholerae is believed to respond to climatic phenomenon like global warming. Global warming will increase sea-surface temperatures and the sea surface heights, which drives seawater up into the estuary. The rise of sea surface temperature could contribute to more phytoplankton blooms followed by zooplankton blooms and consequently more cholera cases. Research carried out at the University of Maryland has found an association between V. cholerae and copepod zooplankton. The pathogen of V. cholerae is attached to copepod surfaces, egg cases and the oral region. In Bangladesh, the incidence of cholera vibriosis increases as populations of zooplankton increase, especially during the summer monsoon months (Figure 5.3). Zooplankton (a single copepod) can harbour up to 10,000 vibrios, so ingesting just one or two copepods could cause clinical cholera. A simple method of filtering water through four to eight layers of old sari clothes can reduce the number of cases of cholera in Bangladesh. Old sari clothes can remove 99% of the vibriosis when water was filtered through four layers. The threads of old sari clothes are frequently frayed and thus produced a smaller mesh size compared to new sari.

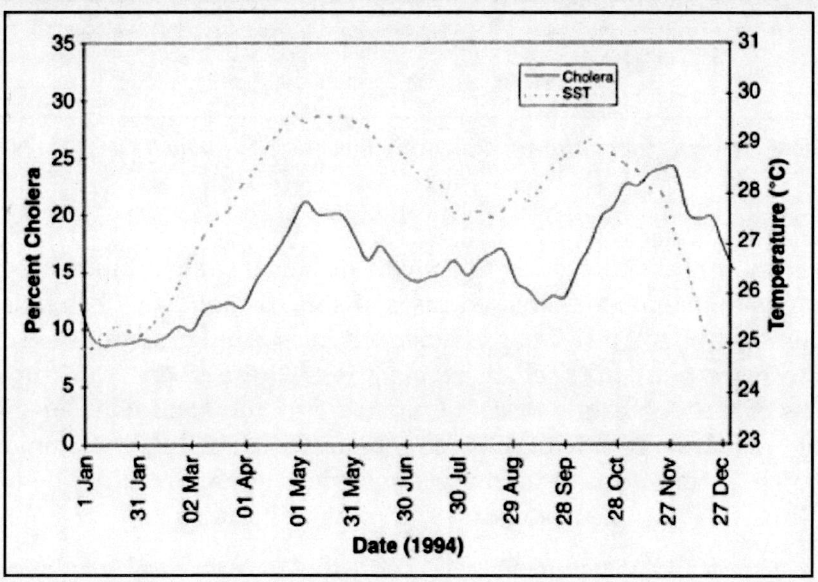

Figure 5.3: ***Relationship between sea-surface temperature (SST) and cholera case data in Bangladesh from January to December 1994*** *[Colwell 1996].*

5.5.2: *Water contamination*

Water borne diseases are caused by pathogens (viruses, bacteria and protozoa). Climate change could cause an increase in the incidence of water and food borne illness since most of the viruses, bacteria and protozoa thrive in warm water and weather. Furthermore, increased rainfall events due to climate change can lead to these pathogens being deposited in aquifers, wells and drinking water. Microbial contamination in freshwater and marine waters can cause eye, ear, nose, and skin, respiratory, gastrointestinal and other infections.

5.5.3: *Food contamination/poisoning*

Food poisoning is associated with warm weather. Higher temperatures generally favour the multiplication of harmful bacteria, such as *Salmonella,* so it would be expected that higher temperatures would be associated with greater risk of illness. For example, over the last 30 years the number of notification of salmonella infections (food poisoning) in New Zealand has been found to be during the summer months with the rise of temperature (Figure 5.4). The predicted temperatures rise may facilitate transmission of *Salmonella* and *Campylobacter* that cause food poisoning. Flies and other pests which may affect food safety, are also supposed to increase in warmer temperatures.

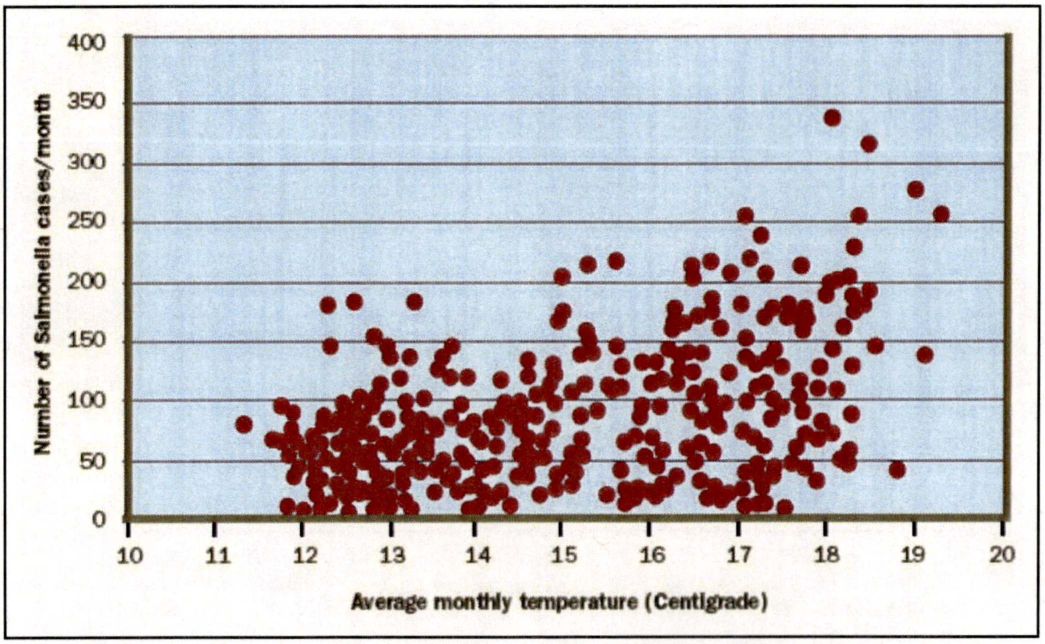

Figure 5.4: ***Relationship between temperature and monthly reports of Salmonella cases in New Zealand*** *(1965-2000) [WHO 2003a or2003 b].*

Rise of sea temperature could accelerate growth of harmful algal blooms which produce toxins. Warmer seas may thus contribute to increased cases of shell-fish and reef fish poisoning (ciguatera-see chapter 13 for the ciguatera effects on humans). Toxic red tides could proliferate as seawater temperatures increase, leading to more fish and shell-fish poisoning. These toxins can find there way through recreational contacts or through eating of contaminated fish and shell-fish causing illness as paralytic shell fish poisoning, diarrheic shell-fish poisoning, neurotoxic shell-fish poisoning and ciguatera fish poisoning (see Chapter 13 for effects of marine biotoxins on human health). The toxins involved evoke a variety of gastro-intestinal and neurological symptoms in humans.

Vibrio parahaemolyticus and *V. vulnificus* are responsible for non-viral infections related to shell-fish consumption in the USA, Japan and South-East Asia. A large outbreak in 2004 (Alaskan coastal waters) due to the consumption of contaminated oysters by *V. parahaemolyticus* was found to be linked with a typically higher temperatures. Similarly, during 1981-1994, high numbers of *Vibrio* bacteria illness from shell-fish poisoning was found to be related with summer high temperatures. Therefore projected global warming may increase shell-fish poisoning.

Mercury is the most toxic heavy metal. Increasing sea temperature due to climate change is projected to increase the rate of methylation of mercury and mercury uptake in fish. Eating fish contaminated with mercury can disrupt the nervous system and damage brain functions; birth defects and miscarriages (for detail see Chapter 8 on bio-accumulation of mercury by fish). Mycotoxins in food could be a problem with increasing temperature and humidity. For example, aflatoxin produced by fungi on grains and nuts are toxic to the liver and carcinogenic.

5.5.4: *Air pollution*

Climate change could increase air pollution levels. It is suggested that the formation of smog will enhance the increase of temperature since smog precursors such as NOx and VOCs react faster at higher temperatures and create more ground level ozone. In addition, climate change will affect natural sources of air pollutant emissions (e.g. forest fires) since increase in drought may influence the occurrence of forest fires. As a result large amount of CO, CO_2, NO_3, SO_2, and VOCs and dioxins may be released into the environment.

5.5.5: *Aeroallergens*

Climate change could also change the distribution and types of airborne allergens, such as pollen. Studies conducted indicate that rising levels of CO_2 will bring an increase in the production of pollen in some plants, producing more allergens and thereby making the situation worst for those people suffering from asthma, emphysema and other respiratory illness.

It is projected that climate change can introduce new aeroallergens into an area through introduction of new invasive plant species with highly allergenic pollen (for example ragweed, *Ambrosia artemisiifolia*), which is a risk for human health. Furthermore, projected increase of CO_2 and rise of temperatures could further increase ragweed pollen production. An increase of aeroallergens is likely to increase rates of asthma and hay fever. About 300 million people suffer from asthma, and 255,000 people died of the disease in 2005. Warmer temperature and increased levels of CO_2 in the atmosphere can promote the growth of some plants and fungi that causes allergy when pollens or spores are inhaled.

5.5.6: *Ultraviolet radiation (UVR)*

Solar radiation (radiation emitted by sun, also referred to as short-wave radiation) has been linked as a cause of skin cancer (melanoma and other types) in fair-skinned humans. High intensity UVR also damages the eye's outer tissues (ocular equivalent of sunburn). The cortical and subcapsular cataracts are associated with UVR exposure. Furthermore, UVR exposure can cause both local and whole-body immunosuppression. It is projected that skin cancer incidence and sunburn severity will increase due to stratospheric ozone depletion and higher temperature. The possible health impacts of ozone depletion are given in Table 5.9.

5.5.7: *Food shortage*

Climate variability such as rising temperature and changing patterns of rainfall would decrease crop production/ yields in many developing countries, in particular countries which depend primarily on rain-fed agriculture and where food security is already a problem. According to Battisti and Naylor (2009) half of worlds population could face severe food shortages by the end the century due to fall of rice and maize harvests by 20% to 40% as a result of increased temperatures during the growing seasons in tropical and subtropical regions. This situation would aggravate the burden of under-nutrition and malnutrition and predispose people to much infectious diseases such as malaria, diarrhoea and respiratory illnesses. WHO (2008) estimated that under-nutrition causes 3.5 million deaths each year in developing countries. Unless adaptive measures are taken, climate change is projected to approximately double the percentage of population at risk of hunger and associated health effects by the 2050s.

Table 5.9: ***Possible effects of solar radiation on human health*** *[McMichael et al. 2003a].*

Effects on skin	Effects on eye	Effects on immunity and Infection	Other effects (beneficial)
Malignant melanoma; non-malanocytic skin cancer; sunburn; chronic sun damage; photodermatosis	Acute photo keratitis and photoconjunctivitis; climatic droplet keratopathy; pterygium; cancer of the cornea and conjunctiva; lens opacity (cataract) - cortical, posterior subcapsular; uveal melanoma and muscular degeneration	Suppression of cell mediated immunity; increased susceptibility to infection; impairment of prophylactic immunization; activation of latent virus infection	Cutaneous vitamin D production (prevention of rickets, osteomalacia and osteoporosis, possible benefits for hypertension, ischaemic heart disease and tuberculosis, possible decreased risk of schizophrenia, breast cancer, prostate cancer and possible prevention of Type I, (usually insulin dependent) diabetes; Non-Hodgkins lymphoma; altered general well-being (sleep/wake cycles, seasonal affective disorder and mood).

5.5.8: Water shortage

Climate change is projected to increase the scarcity of freshwater (e.g. Australia, Asia and Africa - see Chapter 2). For example, more variable rainfall patterns are likely to compromise the supply of fresh water. Water scarcity already affects four out of every 10 people globally. Furthermore, reduced river flows and increased water temperature will lead to declining water quality as the dilution of the contaminants is reduced, less oxygen is dissolved in water and microbiological activity increases. Lack of water and poor water quality can compromise hygiene and health. This increases the risk of diarrhoea, which kills approximately 1.8 million people every year, as well as trachoma (an eye infection that can lead to blindness) and other illnesses. Moreover, water scarcity encourages people to transport water long distances and store supplies in their homes. This can increase the risk of household water contamination, causing illnesses.

5.5.9: Environmental refugees

Climate change will cause sea level rise and flooding of coastal lands which will lead to salination or contamination of freshwater and agricultural lands. If it continues, this, along with extreme weather, may cause displacement of large populations by forced migration, leading to overcrowding elsewhere and rises in the level of infectious diseases. For example, in Bangladesh, nearly 16-20% of the land area would be lost from a 1 m rise in sea level, affecting over 10% of the population. While in Egypt, the corresponding figures are 10-13%. (see also chapter 2, section 2.6 for impacts of sea level rise).

5.6: Case studies-Asia-Pacific and Australasia Region

[based on Cruz et al. 2007; Hennessy et al. 2007].

5.6.1: Asia

Climate change poses substantial risks to human health in Asia since it can increase a number of diseases in Asia including dengue, vector- and water-borne diseases (see Table 5.8). Climate change related diarrhoea and malnutrition are the largest cause of mortality and morbidity in South-east Asian countries including Bangladesh, Bhutan, India, Maldives, Myanmar (Burma) and Nepal in 2000. The relative risks of these conditions are expected to increase in the future (Table 5.10). It is further projected that rise of temperatures along the coast lines of South and South-east Asia would support higher phytoplankton blooms. The phytoplankton blooms are excellent habitats for survival and spread of infectious bacterial diseases such as, cholera (see Box 5.2). In addition, water-borne diseases caused by *Giardia, Salmonella* and *Cryptosporidium* could also become common with the contamination of drinking water. Higher precipitation followed by frequent floods and sea-level rise in the future will degrade the surface water quality (see section 5.3.2) and can lead to more water-borne infectious diseases (e.g. gastrointestinal, cardiovascular and dermatosis).

Table 5.10: ***Projected climate change effects on human health in Asia*** *[Cruz et al. 2007.]*

Mortality and morbidity	Projections
Air quality	- Air quality in industrialised areas in China (Chongqing), and Indonesia (Jakarta) could deteriorate significantly in the future causing heat stress and smog induced cardiovascular and respiratory illness in the region.
Cedar Pollen disease	- In Japan, the cedar pollen disease may increase with the rise of summer temperatures.
Dengue	- The risk of dengue fever will be larger in India and China.
Heat stress	- Mortality and morbidity due to heat stress will be very high in India and China and Indonesia.
Vector borne diseases (VBD)	- Natural habitats of VBD may increase; a warmer and more humid climate would favour the propagation and invasiveness of infectious insect vector.
Water borne diseases	- Cholera and diarrhoeal diseases could become common in the future.

5.6.2: *Australia and New Zealand*

Extreme heat conditions are a regular phenomenon in Australia in recent times, for example, during 2004-2007, many people collapsed from heat exhaustion (temperature of more than 42° C recorded in Melbourne in December 2007, 48.5°C in Western Sydney in February 2004). In Australia, the annual heat related deaths are estimated to be more than 1,000 in people aged over 65 years, which is predicted to increase by 2050 (see Table 5.11). It is further projected that Australian temperate cities are likely to experience higher heat-related deaths (see Table 5.11) than tropical cities. In New Zealand the winter peak deaths is projected to decline due to rise of temperature.

Table 5.11: ***Projected climate change effects on human health in Australia and New Zealand*** *[Hennessy et al. 2007].*

Mortality and Morbidity	Projections
Australia-Heat-related deaths	- Heat-related deaths may rise in Australian main cities (Adelaide, Melbourne, Perth, Sydney and Brisbane) from current 1,115 to 2,300-2,500 (by 2030s); 4,300-6,300 (by 2050s).
New Zealand-Heat-related deaths	- Heat-related deaths may rise in New Zealand main cities (Christchurch, Auckland) from current 14 (over 65 years) to 28, 51 and 88 deaths for warming of 1, 2 and 3°C.
Australian-Ross River disease (RRv)	- Rising temperature and changing rainfall patterns are likely to have significant impacts on the transmission of RRv disease in temperate and semi-arid Australia.
Australia-Dengue	- Dengue virus is likely to increase in tropical and temperate Australia affecting additional population of 0.1-0.3 million in 2020s and 0.6 to 1.4 million by 2050s.
New Zealand-dengue	- Parts of the North Island could become suitable for breeding of the major dengue vector.
Australia-mental health of rural people	- Increase of mental health risks in Australian rural farmers/rural communities (due to drought).
Australia- Bush fire	- Bush fire is projected to increase which can exacerbate asthma and other respiratory and cardiovascular conditions.

Ross river virus (RRv) or Barmah Forest virus (BFv) is the most common mosquito disease in Australia. The virus can cause severe morbidity such as epidemic polyarthritis (arthritic symptoms that persists for several months) and is a significant public health issue in Australia (51,761 cases between, 1991-2002). Currently, there is no treatment for the disease. Studies conducted have found a positive relationship between climate variables and the breeding and survival of RRv vector. Outbreaks of Australian Encephalitis (Murray Valley Encephalitis - MVE), another mosquito-transmitted disease, occur in temperate south-east Australia after heavy rainfall and flooding associated with La Niña events. Dengue (*S. aegypti*) is a substantial threat in Australia, the climate of the far north already supports dengue virus, where dengue outbreaks have increased over the past decade. Dengue has been projected to increase in the future (Table 5.11). Warmer temperatures and increased rainfall variability are likely to increase the intensity and frequency of food-borne and water-borne diseases in both Australia and New Zealand.

5.7: Adaptation measures

Efficient implementation of adaptation strategies can significantly reduce adverse health impacts (morbidity and mortality) caused by climate change. Adaptive action plans are categorized into primary, secondary and tertiary preventions. *Primary prevention* is an intervention implemented before the disease or injury can occur, examples are supplying of bed/mosquito nets to prevent exposure to malaria and early warning systems about extreme heat health warnings. All of these primary measures will help to avoid or reduce risks. *Secondary prevention* involves intervention implemented after disease has begun but before it is symptomatic (e.g. early detection or screening), and subsequent treatment that averts full progression to disease. *Tertiary prevention* is to minimise the adverse effects of an already present disease or injury through better treatment, for example, of heat stroke and improved diagnostic of vector-borne diseases. The other measures may include the followings:

- Heat-health action plans
- Adjusting clothing and activity levels, increasing fluid intake (during hot days)
- Weather advisories to alert public about dangerous heat conditions
- Urban tree plantation to moderate temperature increases
- Emergency medicals services
- Improving climate sensitive disease monitoring
- Cost effective surveillance and emergency response systems
- Safe water and improved sanitation
- Improvement of primary care, disease control, sanitation and disaster preparedness
- Research on barriers and adaptive capacity to protect human health
- Appropriate public health training programs

5.8: Conclusion

The climate change will increase the rates of some diseases (e.g. cardiovascular diseases in the tropical regions, diarrhoeal diseases in low income regions) and malnutrition. In addition, coastal flooding is projected to cause large scale mortality and human displacement. The risk of malaria and dengue is projected to expand to new areas. Infectious diseases those transmitted by insect vectors or water are sensitive to climatic conditions. Currently, there are few data available on the impacts of climate change on human health from less developed countries. Monitoring would be required to gather data relevant to climate change and human health impacts as diseases like malaria, dengue and diarrhoeal diseases are likely to increase in poorer countries. Public health training, effective surveillance and emergency response systems, and sustainable prevention and control programs would reduce the impacts of climate change on human health. Adaptation to climate change (i.e. house with thermally efficient housing or air conditioning, improved health care) can reduce heatwaves related morbidity and morality risks.

5.9: References

Battisti, D.S. and Naylor, R.I. 2009. Historical warnings of future food insecurity with unprecedented season heat. *Science* 323: 607-610.

Boumaet, M. J. and Dye, C. 1997. Cycles of malaria associated with El Niño in Venezuela. *Journal of the American Medical Association* 278(21): 1772-1774.

Chowdhury, M. R. 2009. Cyclone Aila and climate change. El Niño/La Niña intensifying the threat? Sub-editorial, The Daily Star, 13 June 2009, Dhaka, Bangladesh.

Colwell, R. R., Huq, A., Islam, S., Aziz, M. A., Yunus, M., Khan, N. H., Mahmud, A., Sack, R. B., Nair, G. B., Chakraborty, J., Sack, D. A. and Russek-Cohen, E. 2003. Reduction of cholera in Bangladesh villages by simple filtration. *Proceedings of the National Academy of Sciences* 100 (3): 1051-1055.

Colwell, R.R. 1996. Global climate and infectious disease: the cholera paradigm. *Science* 274: 2025-2031.

Confalonieri, U., Menne, B., Akhtar, R., Ebi, K.L., Hauengue, M., Kovats, Revich, B. and Woodward, A. 2007. Human health. *Climate change 2007: Impacts, Adaptation and Vulnerability. Contribution of working group II to the Fourth Assessment Report of the Intergovernmental Panel on Climate Change.* M. L. Parry, O. F. Canziani, J. P. Palutikof, P. J. van der Linden and C. E. Hanson, (eds.). Cambridge University Press, Cambridge, UK, Chapter 8: 391-431.

Costello, A., Abbas, M., Allen, A., Ball, S., Bell, S., Bellamy, R., Friel, S, Grace, N., Johnson, A. 2009. Managing the health effects of climate change. *The Lancet Commission* 373 : 1693-1730. www.thelancet.com.

Cruz, R. V., Harasawa, H., Lal, M., Wu, S., Anokhin, Y., Punsalmaa, B., Honda, Y., Jafari, M., Li, C. and Huu Ninh, N. 2007. Asia. *Climate change 2007: Impacts, Adaptation and Vulnerability. Contribution of Working Group II to the Fourth Assessment Report to the Intergovernmental Panel on Climate Change.* M. L. Parry., O. F. Canziani., J. P. Palutikof., P. J. van der Linden and C. E. Hanson (eds.). Cambridge University Press, Cambridge, U.K. 469-506 pp. (Chapter 10). http://www.ipcc.ch/pdf/assessment-report/ar4/wg2/ar4-wg2-chapter6.pdf

Haines, A., McMichael, A. J. and Epstein, P. R. 2000. Environment and health: 2 Global climate change and health. *Canadian Medical Association Journal.* 163(6): 729-734.

Hales, S., de Wet, N., Maindonald, J. and Woodward, A. 2002. Potential effect of population and climate changes on global distribution of dengue fever : an empirical model. *Lancet* 360: 830-834.

Hales, S., Edwards, S. J. and Kovats, R. S. 2003. Chapter 5. *Impacts of health of climate extremes.* 79- 102. Chapter 5. *In*: McMicheal, A. J., Campbell-Lendrum, D. H., Corvalán, C. F., Ebi, K. L., Githeko, A. K., Scheraga, J. D. and Woodward, A. (eds.). Climate change and human health: Risks and Responses. World Health Organisation, Geneva.

Hennessy, K., Fitzharris, B., Bates, B. C., Harvey, N., Howden, S. M., Hughes, L., Salinger, J., and Warrick, R. 2007. Australia and New Zealand. *Climate change 2007: Impacts, Adaptation and Vulnerability. Contribution of working group II to the Fourth Assessment Report of the Intergovernmental Panel on Climate Change.* Parry, M. L., Canziani, O. F., Palutikof, J. P., van der Linden, P. J.and Hanson, C. E. (eds.). Cambridge University Press, Cambridge, UK, chapter 11: pp. 507-540.

International centre for Diarrhoeal Disease Research, Bangladesh (ICDDR,B) 2002. 1. *Filtering out cholera*: Bangladesh. International Centre for Diarrhoeal Disease Research, Bangladesh.

Intergovernmental Panel on Climate Change (IPCC) 2007. Climate change 2007a: *Impacts, adaptation and vulnerability.* Working group II contribution to the Intergovernmental Panel on Climate Change Fourth assessment report. Summary for policy makers. (http://www.ipcc.ch).

McMicheal, A. J., Lucas, R., Ponsonby, A-L. and Edwards 2003a. *Stratospheric ozone depletion, ultraviolet radiation and health.* Chapter 8. pp. 159-180. *In*: Climate change and human health: Risks and Responses. Editors. McMicheal, A. J., Campbell-Lendrum, D. H., Corvalán, C. F., Ebi, K. L., Githeko, A. K., Scheraga, J. D. and Woodward, A. World Health Organisation (WHO), Geneva.

McMicheal, A. J., Campbell-Lendrum, D. H., Corvalán, C. F., Ebi, K. L., Githeko, A. K., Scheraga, J. D. and Woodward, A. 2003b. Climate change and human health: Risks and Responses. World Health Organisation, Geneva.

McMicheal, A. J., Woodruff, R. E. and Hales, S. 2006. Climate change and human health: present and future risks. *Lancet* 367: 859-869.

Patz, J. A., Githeko, A. K., McCarty, J. P., Hussein, S. and Confalonieri, U. 2003. *Climate change and infectious diseases.* Chapter 6. pp. 103-132. *In*: McMicheal, A. J., Campbell-Lendrum, D. H., Corvalán, C. F., Ebi, K. L., Githeko, A. K., Scheraga, J. D. and Woodward, A. (eds.). *Climate change and human health: Risks and Responses.* World Health Organisation, Geneva.

Poumadére, M., Mays, C., Sophie, Le Mer, and Blong, R. 2005. The 2003 heat wave in France: Dangerous climate change here and now. *Risk Analysis.* 25(6): 1483-1494.

Watson, R. T., Patz, J., Gubler, D. J., Parson, E. A. and Vincent, J. H. 2005. Environmental health implications of global climate change. *Journal of Environmental Monitoring* 7: 834-843.

World Health Organisation (WHO) 2003a. *Health impacts of climate change*: http://www.who.int/globalchange/climate/en/ccSCREEN.pdf

World Health Organisation (WHO) 2003b. Climate change and human health-risks and responses. World Health Organisation. A.J. McMichael, D.H. Campbell-Lendrum. C.F. Corvalan, K.L. Ebi, A.K. Githeko, K.J.D. Scheraga and A. Woodward (eds.) 329 p.

World Health Organisation (WHO) 2008. *Protecting health from climate change*-World health day 2008. World Health Organisation, Switzerland. http://www/who.int/world-health-day/en/ 25p.

Wilson, M. L. 2001. Ecology and infectious disease, In: *Ecosystem change and public health: A Global Perspective*, J. L. Aron and J. A.Patz (eds.). Johns Hopkins University Press: Baltimore. pp. 283-324.

Websites

(http://www.grida.no/climate/vital39.htm) vital climatic graphics: potential impacts of climate change. United National Environment Program (UNEP)

http://www/who.int/mediacentre/factsheets/fs266/en/print.html (Climate and health)

http://www/who.int/mediacentre/factsheets/fs192/en/print.html (El Niño and its health impact)

(http://www.who.int/features/factfiles/climate_change/en/index.html

http://www.searo.who.int/LinkFiles/Regional_Health_Forum_Volume_12_No_1_Climate_change_and_its_impact.pdf (climate change and its impact on health of Bangladesh-Atiq Rahman)

http://en.wikipedia.org/wiki/Dengue

□□□

Part-2 : Chemical Impacts

6 Chemicals in the Environment

Key facts

- The use of chemicals has many benefits, however, there are potential risks that some of these chemicals may cause harm to organisms and human health if they enter into the environment.
- Harmful chemicals are those chemicals which are persistent, bio-accumulative, as well as toxic (PBT).
- Environmental or ecological risk assessment (ERA) consists of hazard identification, effect assessment, exposure assessment, risk characterization, risk classification, risk-benefit analysis, risk reduction and monitoring.
- The five types of environmental monitoring are: chemical monitoring, bio-accumulation monitoring, biological effect monitoring, health monitoring and ecosystem monitoring.
- An integrated monitoring program is a coordinated monitoring activity comprising of both chemical and biological measurements in a variety of environmental compartments. There are four different monitoring levels in an integrated ERA, these are: suborganismals (biomarkers), organisms (bioassays), populations (bio-indicators) and ecosystems (ecological indicators).
- A biomarker is defined as a change in a biological response (ranging from molecular through cellular and physiological responses to behavioural changes) which can be related to exposure to or toxic effects of chemicals. Biomarkers act as early signals, since toxicants have an impact at molecular and sub-cellular levels before their effects are observed at the whole-organism and population level.
- A bioassay involves the use of a biota for testing chemical toxicity. Bioassays are experiments that use living organisms to test the toxicity of chemicals. Fish, various aquatic invertebrates, protozoans, duckweeds and algae all are used for bioassays of aquatic samples.
- Sampling of environment for chemicals can be done by grab/spot sampling, bio-monitoring and passive sampling (time-integrated or continuous monitoring). Passive sampling technology has been widely accepted throughout the world as a tool for environmental sampling or chemical's risk assessment as it is more advantageous compared to traditional grab sampling and bio-monitoring.
- Aquatic organisms (fish or sessile bivalve molluscs) have been used as a tool to quantify the hydrophobic or non-polar contamination/persistent organic pollutants (e.g. PCBs, metals, organochlorine pesticides) and to obtain toxicologically relevant information from the aquatic environment. These contaminants are concentrated or accumulated in the fatty tissue of the aquatic organisms via fish gills, skin and food.
- Passive sampling can provide important information on temporal and spatial variability of pollutants, biological/toxicological effects, contaminant bioavailability, as well as early detection of pollutants.
- Quality assurance (QA) and quality control (QC) should be the top priority for any chemical contaminant analysis.

Contents

List of Tables

List of Figures

Acronyms and Glossary

Analytes: The physical and chemical indicators to be determined.

BCF: The bio-concentration factor is the concentration of chemicals in the tissue of an organism.

Bio-accumulative: Chemicals which have a strong tendency to remain in the fatty tissue of organisms which they enter. A substance is classed as bio-accumulative if it accumulates in the fatty tissue relative to water in a ratio (quantity-wise) of 10,000 to 1 or more.

Bioassay: Procedure for estimating the concentration or biological activity of a substance (e.g. vitamin, hormone, plant growth factor, antibiotic and enzyme) by measuring its effect on a living system compared to a standard system.

Bioavailability: The fraction of the bulk amount of the chemical present in soil/sediment and (interstitial) water that can potentially be taken up during the organism's lifetime into the organisms tissues (excluding the digestive tract).

Biomarker: A biomarker is defined as a change in a biological response (ranging from molecular through to cellular and physiological responses to behavioural changes) which can be related to exposure to or toxic effects of chemicals. A biomarker can be any kind of molecule indicating the existence, past or present, of living organisms. In medicine, a biomarker can be a substance that is introduced into an organism as a means to examine organ function or other aspects of health. It can also be a substance whose detection indicates a particular disease state, for example, the presence of an antibody may indicate an infection. More specifically, a biomarker indicates a change in expression or state of a protein that correlates with the risk or progression of a disease, or with the susceptibility of the disease to a given treatment.

Blank solution: A solution that does not contain a detectable amount of the analyte of interest. The blank solution is typically used for calibration purposes. Depending on its purpose, the following blank solutions can be defined: *Calibration blank* (the solution used for creating the zero concentration point of the calibration graph- this solution contains only the diluent used for making the standard solution); *Reagent blank* (a blank solution that contains the reagent used to dissolve the samples such as acids used for digestion - the reading for this solution is typically subtracted from sample readings) and

Method blank (a blank solution that has been handled similarly as a sample, and to which the same reagents have been added that had contact to the same type of vessels and that was treated by a similar procedure).

Chelating agent: Chelation is the binding or complexation of a bi- or multidentate ligand. These ligands, which are often organic compounds, are called chelants, chelators, chelating agents, or sequestering agent.

EC_{50}: Is the concentration of any chemical that affects 50% of the population exposed (usually refers to an effect other than death).

EDC: Endocrine disrupting chemicals.

Effluent: A complex waste material (e.g. liquid industrial discharge or sewage) that may be discharged into the environment.

Hydrophilic: Having affinity for water (polar) (see also passive sampling).

Hydrophobic: Lacking affinity for water and soluble in low polarity organic solvents (non-polar).

In situ: A latin phrase meaning in the place.

LC_{50} : The concentration that is lethal for 50% of the population exposed. Or The amount of a substance required to kill half a given population, a measure of toxicity

LD_{50} : In toxicology, the median lethal dose, LD_{50} (abbreviation for "Lethal Dose, 50%"),

Log K_{ow} : Is the partition coefficient between octanol and water that relates to the solubility of the chemical in fats. The $\log_{10}$ of the octanol/water partition coefficient of a chemical is expressed as the solubility in octanol relative to water. A log K_{ow} of 3 denotes the compound is 1,000 times more soluble in octanol than water.

Lysosomes: Lysosomes are organelles containing digestive enzymes (acid hydrolases). They are found in animal cells, while in plant cells the same roles are performed by the vacuole. They digest excess or worn-out organelles, food particle, and engulfed viruses or bacteria.

Metallothionein: Metallothionein (MT) is a family of cysteine-rich, low molecular weight (MW ranging from 3,500 to 14,000 Da) proteins. MTs have the capacity to bind both physiological (Zn, Cu, Se) and xenobiotic (Cd, Hg, Ag) heavy metals through the thiol group of its cysteine residues, which represents nearly the 30% of its amino acidic residues. MT's function is not clear, but experimental data suggest that MTs may provide protection against metal toxicity, be involved in regulation of physiological metals (Zn and Cu) and provide protection against oxidative stress. MTs are present in a vast range of taxonomic groups, ranging from prokaryotes (such as the cyanobacteria *Synecoccus* spp.), protozoa (the ciliate, *Tetrahymena* genera), plants (such as *Pisum sativum, Triticum durum, Zea mays, Quercus suber*), yeast (such as *Saccharomyces cerevisiae, Candida albicans*), invertebrates (such as the nematode, *Caenorhabditis elegans*, the mollusc, *Mytilus edulis*) and vertebrates (such as the mammalian, *Homo sapiens*).

NATA: The National Association of Testing Authorities, Australia

NOEC: No observed effect concentration is the highest concentration of the chemical at which no effects are identified.

PBT: Persistent, bio-accumulative and toxic.

Partition coefficient: A ratio of the equilibrium concentration of the chemical between a non-polar and polar solvent.

Passive sampling: A device that collects or accumulates pollutants (e.g. pesticides) independently through a diffusion barrier onto a sorbent medium without use of a vacuum source or energy.

Persistent: Chemicals which take an extensive time to decay once they are released into the environment. A substance is classed as persistent if it does not decay to half of its original quantity within two months (if in water) or six months (if in soil or sediment).

QA/ QC: QA (Quality Assurance - defined as the implementation checks on the success of quality control such as through replicate samples) and QC (Quality Control - defined as the implementation procedures to maximise the integrity of monitoring data via avoidance of contamination or cleaning procedures).

Risk: A statistical concept defined as the expected likelihood or probability of undesirable effects resulting from a specified exposure to known or potential environmental concentrations of a material.

Receptor: A plant, animal, community of organisms or ecosystem that is exposed to stressors in the environment.

T½ - Is the half-life of the chemical in water, soil or sediment.

Toxicity: The inherent potential or capacity of a material to cause adverse effects in a living organism.

Toxic: Chemicals that are toxic and cause direct damage to organisms that are exposed to them.

6.1: Introduction

Chemicals are a vital part of our daily life. They provide us with a wide range of benefits, enhancing agricultural and industrial productivity and against the control of diseases. On the other hand, chemicals have the potential to cause considerable health and environmental problems throughout their life cycle, from production to disposal. For example, the bulk of chemicals applied in the field to control pest's plants and animals often end up in the environment. Effluent discharge from industries may contain harmful chemicals such as heavy metals and pharmaceuticals. Chemicals entering into the environment can contaminate air, soil and water. It is therefore essential that chemicals are managed in a safe and secured manner to protect both the environment and human health (Plant *et al.* 2005).

6.2: Harmful chemicals

Harmful chemicals are synthetic chemicals that are persistent or bio-accumulative and toxic (PBT) in nature. According to the UK chemicals stakeholder forum (http://www.defra.gov.au/environment/chemicals/csf/criteria/detailed.htm) a substance is persistent if it does not decay to half of its original quantity within two months (if in water) or six months (if in soil or sediment). Bio-accumulative chemicals are those that have a strong tendency to dissolve in fats and remain in the fatty tissue of organisms. A substance is classed as bio-accumulative if it accumulates in the fatty tissue relative to water in a ratio (quantity-wise) of 10,000 to 1 or more. A chemical is toxic if it causes direct damage to organisms that are exposed to them. A substance can be classed as being of concern if it is lethal to at least 50% of test aquatic organisms of a concentration of 0.1 mg/l or less. Chemicals that have the potential to damage the environment or human health can be identified based on their persistence or bio-accumulation and toxicity as illustrated in Table 6.1.

Table 6.1: ***First tier criteria for chemicals of greatest concern.*** (*http://www.defra.gov.au/environment/chemicals/csf/concern/index.htm*)

Criterion	First tier thresholds
Persistence (P)	t½ water >2 months. t½ soil/sediment >6 months.
Bio-accumulative (B)	Log K_{ow} > 5 (greatest concern) (unless BCF <50,000). Log K_{ow} >4 (high concern) (unless BCF <500).
Toxicity (T)	LC_{50}/EC_{50} < 1mg/l or NOEC <0.11mg/L. Category 1 or 2 carcinogen, mutagen or reprotoxin.
Endocrine Disrupting Effects (EDC)	Evidence of EDC.
Aquatic ecosystems	P = t½ marine water >60 days , fresh water >40 days or t½ marine /sediment >180 days, freshwater sediment >120 days.

For explanation of BCF; Log K_{ow}; NOEC; LC_{50} and EC_{50} see acronyms and glossary

There are certain chemicals that are highly toxic (acute toxicity LC_{50}/EC_{50} <0.1 mg/l; NOEC <0.01 mg/l; LD_{50} /EC_{50} <0.1 mg/kg) but not persistent or bio-accumulative. These chemicals may still be a cause for concern, especially if they are continually released to the environment. Substances which are actually or potentially very bio-accumulative are those substances with a BCF >10,000 or a log K_{ow} >6. Such chemicals may accumulate in fatty tissue as well as in bone or bind to proteins of biota, etc. Substances, if actually found in biota may be of concern due to their bio-accumulation. There are some substances which are persistent as well as toxic (see Table 6.2). Substances which are both bio-accumulative and toxic may be a cause for concern; especially if the substances are released regularly (i.e. the input load is greater than removal by degradation).

Table 6.2: *Examples of chemicals that are either persistent or bio-accumulative or toxic.*

Chemicals	Persistent	Bio-accumulative	Toxic	EDC
Inorganic chemicals • Metalloids (As) (Chapter-7)	-	√	√	-
Inorganic chemicals • Heavy metals (Cu, Cd, Pb, Hg) (Chapter-8)	√	√	√	√
Persistent organic pollutants • Pesticides (Chapter-9)	√	√	√	√
Persistent organic pollutants • Dioxins (PCDD/Fs, PCBs) (Chapter-10)	√	√	√	√
Persistent organic pollutants • Endocrine disrupting chemicals (EDCs) (Chapter-11)	√	√		√
Drugs • Pharmaceuticals (human & veterinary) (Chapter-12)	√	√	√	-
Natural chemicals/toxins • Freshwater and Marine Biotoxins (Chapter-13)	-	√	√	-

6.3: Environmental or ecological risk assessment (ERA) of chemicals

[based on Commonwealth of Australia 2002; Van der Oost et al. 2003].

Risk assessment can be defined as the process of assessing magnitudes and probabilities to adverse effects of human activities or natural catastrophes. The risk assessment process consists of eight steps as briefly described below:

- *Hazard identification* (identification of the adverse effects, which may be caused by chemicals such as toxicity of a chemical).
- *Effect assessment* (estimation of the relationships between dose or level of exposure to chemicals and the incidence and severity of an effect. It is generally to determine a predicted no effect concentration (PNEC) for other species)).
- *Exposure assessment* (estimation of concentration or doses to which human populations or environmental compartments are or may be exposed. For example, for existing chemicals an assessment can be made by measuring concentrations, whereas for new chemicals a predicted environmental concentration (PEC) be estimated).
- *Risk characterization* (integration of the above three steps in order to estimate the incidence and severity of the deleterious effects likely to occur due to actual or predicated exposure to chemicals. For new chemicals, the risk quotient can be determined from PEC/PNEC ratio - see chapter 12, section 12.7 for estimating hazard or risk quotient and environmental risk assessment of pharmaceuticals).
- *Risk classification* (evaluation of risks in order to decide if risk reduction is required. In general risk classification is performed at two levels - upper limit or maximum permissible level (MPL) and the lower limit or the negligible level (NL)).
- *Risk-benefit analysis* (the risks and benefits of a proposed risk reduction action).
- *Risk reduction* (measures to protect human or the environment from the identified risks such as use of acceptable environmental water quality guidelines as described in Chapter 7 through Chapter 13).
- *Monitoring* (observation of chemicals or biological elements. Monitoring is essential in several stages of the ERA process) (see section 6.4 on monitoring methods).

6.4: Monitoring chemicals in aquatic environments

[based on Van der Oost et al. 2003]

Monitoring is a repetitive observation for defined purposes of one or more chemical or biological elements according to a prearranged schedule over time and space, using comparable and standardized methods. The five environmental monitoring methods which may be performed in order to assess risks of contaminants for organisms and to classify the environmental quality of ecosystems are listed below:

- *Chemical monitoring* (CM) - exposure assessment by measuring levels of a selected set of well-known contaminants in abiotic environmental compartments.
- *Bio-accumulation monitoring* (BAM) - exposure assessment by measuring contaminant levels in biota or determining the critical dose at a critical site (bio-accumulation). Persistent hydrophobic chemicals may accumulate in aquatic organisms (e.g. fish, mussels) via the direct uptake from water by gills, skin and indirectly via uptake of suspended particles (ingestion) and consumption of contaminated food (see Chapter 8.2 on bio-accumulation).
- *Biological effect monitoring* (BEM) - exposure and effect assessment by determining the early adverse alternations that are partly or fully reversible (biomarkers).
- *Health monitoring* (HM) - effect assessment by examining the occurrence of irreversible diseases or tissue damage in organisms.
- *Ecosystem monitoring* (EM) - assessment of the integrity of an ecosystem by making an inventory of, for instance, species composition, density and diversity.

A regular, systematic use of living organisms to evaluate changes in environmental or water quality as in BAM, BEM, HM and EM is called *biological monitoring*. The *toxicity monitoring* is to measure the bimolecular and physiological responses of individual organisms towards toxicants in an experimental set up. An *integrated monitoring* program is a coordinated monitoring activity comprising both chemical and biological measurements in a variety of environmental compartments. It is also called *triad* approach, consisting analyses of chemical contamination, assessment of toxicity using bioassays and determination of faunal community structures. There are four different monitoring levels in an integrated ERA as proposed by Van Gastel and Van Brummelen (1996):

a. *Sub-organismal (biomarkers):* A biomarker is defined as a change in a biological response (ranging from molecular through cellular and physiological responses to behavioural changes) which can be related to exposure to or toxic effects of environmental chemicals. Biomarkers act as early signals, since toxicants have an impact at molecular and sub-cellular levels before their effects are observed at the whole-organism and population level. They are sensitive to concentrations below those causing cyto-toxicity. WHO subdivided biomarkers into three classes (see Vasseur and Cossu-Leguille 2003): biomarkers of exposure; biomarkers of effect and biomarkers of susceptibility (see Table 6.3).

b. *Organisms (bioassays):* A bioassay involves the use of a biota to test for chemical toxicity. Here, survival, growth and reproduction of individuals are chosen as endpoints of the laboratory based ecotoxicity tests (Table 6.4). Fish, various aquatic invertebrates, protozoans, duckweeds and algae all are used for bioassays of aquatic samples. Bioassays may be qualitative or quantitative. Qualitative bioassays are used for assessing the physical effects of a substance that may not be quantified, such as abnormal development or deformity. Quantitative bioassays involve estimation of the concentration or potency of a substance by measurement of the biological response that it evokes. Quantitative bioassays are typically analyzed using biostatistics. Bioassays are essential in the development of new drugs and monitoring pollutants in the environment and are now a requirement before a new chemical is registered for use as a pharmaceutical or an aquaculture chemical. Bioassays are usually conducted *in vitro* (or in the laboratory) and standard end points include LC_{50}, EC_{50}, NOEC, LOEC. Bioassays are generally more sensitive than chemical methods and can detect biologically-active chemicals at very low concentrations.

Table 6.3: ***Examples of biomarkers and their applications*** *[Vasseur and Cossu-Leguille 2003; Allan et al. 2006].* PAH (polycyclic aromatic hydrocarbons; PCBs=polychlorinated biphenyls

Biomarkers class	Biomarkers	Pollutants	Comments
Biomarkers of exposure (cover the detection and measurement of an exogenous substance, its metabolites, or the product of an interaction between a xenobiotic agent and target molecules or cells, in a compartment within organism; molecular biomarkers of exposure are mainly composed of proteins)	HSP (heat shock proteins)	Thermal shock, metals/heavy metals	- HSP is activated by thermal shock and other environmental conditions (hypoxia/anoxia, osmotic pressure, presence of oxidizing agents, heavy metals and toxic compounds)
	Cytochrome P450 (CYP)	PAHs, PCBs, dioxin	- family of proteins; the expression of CYP is used to indicate contact or contamination with toxins (PAHs, PCBs; dioxins and pesticides)
	Metallothionein	metals	- exposure to metals and heavy metal pollution can be detected by the induction of metallothionein (MT) synthesis. MTs are involved in essential heavy metal homeostatsis, cellular protective mechanisms after toxic metal exposure and have redox activity and antioxidant function
	Glutathione S transferase	hydrocarbons, PCBs, organochlorines	-
Biomarkers of effect (include measurable biochemical, physiological or other alterations within tissues or body fluids of an organism that can be recognised or associated with an established or possible health impairment of health or diseases; integrity of cellular membranes (peroxidation of lipids), intracellular redox state, or the integrity of DNA molecules can constitute biomarkers of toxic effects, often correlated to concentrations or exposure time to a cytotoxic pollutant)	Lysosomes	stress	- lysosomes are used as biomarkers of environmental effects (stress) due to uptake and accumulation of xenobiotics that can cause measurable changes in the volume, size and number of lysosome present; these biomarkers can be compound-specific (PAH, PCBs or metals) or non-specific (generalised pollution of a water body)
	Antioxidant enzyme	PAHs, PCBs, organochlorine pesticides, organophosphorous compounds, carbamates,	- the decrease in the activity of antioxidant enzymes (superoxide dismutase, glutathione peroxides or glutathione reductase) can be used to study the effect of PAHs, PCBs and organochlorine pesticides on aquatic organisms
	Acetylcholinesterase (AchE)	Endocrine disruptors	- AchE activity is inhibited in the presence of organophosphorous compounds and carbamates
	Vitellogenin	Endocrine disruptors	- induction of vitellogenin in male species (see chapter 11) of test species is a good biomarker of the effect of the presence of hormonal (endocrine disrupting) compounds in water
Biomarkers of susceptibility (it indicates the inherent or acquired ability of an organism to respond to the challenge of exposure to a specific xenobiotic substance and include genetic factors and changes in receptors that alter the susceptibility of an organism to the exposure)	Paraoxonase	organophosphorous	- this type of biomarker has been mostly investigated in the medical field - paraoxonase (PON1) (a liver and plasma enzyme involved in lipids oxidation) is a sensitive biomarker for sensing exposure to organophosphorous
	Aryl human receptors	PAHs	-

Table 6.4: ***Environmental contaminant and bioassays.***

Chemicals	Bioassays
Metalloids (arsenic, selenium)	Algae, Rats
Heavy metals (copper, cadmium, lead, mercury, zinc)	Aquatic :Algae, daphnia, fish tests Terrestrial: soil microflora and soil fauna (e.g. earthworms), plants (seed germination and root elongation) Induction of MT (metallothionein)
Pesticides	Algae, daphnia, fish tests Rats
Dioxins	Chemically activated luciferase gene expression bioassay system (CALUX); enzyme immunoassays (EIA)
Endocrine Disrupting Chemicals (EDC)	Yeast estrogen screen (YES), YAS (yeast androgen screen), Estrogen responsive chemically activated luciferase gene expression bioassay system (ER-CALUX)
Pharmaceuticals	Microbial Inhibition plate test
Freshwater cyanotoxins Marine phycotoxins (ASP, CTX, DSP, NSP, PSP)	Mouse bioassay; ELISA (enzyme linked immunosorbent assay); PPI (protein phosphate inhibition assay)

c. *Populations (bio-indicators):* At this level, effects are measured as changes in the genetic structure, the age structure or the abundance of a population.

d. *Ecosystems (ecological indicators):* At this level, changes in species composition, abundance and diversity may be indicative of the effects of pollution on communities.

6.4.1: *Sampling of chemicals residue in the environment*

Evidence of potential adverse effects or the risk of chemical(s) can be identified by chemical measurements of the chemical concentrations in an environment. Monitoring helps to predict the level of risks (whether the risk is high, moderate or low) to various receptors such as human, wildlife or aquatic biota and whether the measured concentrations are within the recommended guideline threshold values for the protection of human and environmental health. The actual concentrations of hydrophobic and hydrophilic chemicals can be measured by grab sampling or passive sampling, whereas the toxicity of chemicals to biological organisms can be measured by bioassays. Merits of garb sampling, biological monitoring and passive sampling are highlighted in the following sections.

6.4.1.1: *Grab/spot/bottle sampling*

[based on Kot et al. 2000; Vrana et al. 2005; Allan et al. 2006; Söderstrom et al. 2009].

Grab sampling involves a single sample or measurement taken at a specific time or over a short period as is feasible. A grab sample provides a snapshot of the chemical concentration in the water at a particular moment in time. It provides information on water quality only at the instant that the sample is taken and is not representative of average water quality. It requires relatively large number of samples to be taken from any location over the entire duration of sampling, which is time consuming and can be very costly. Moreover, decomposition of the sample during transportation and storage may occur. For some analytes it is necessary to add preservatives (e.g. for biocides and metals) to ensure integrity of the sample during transport and storage. Despite addition of preservative, some sample can be modified by processes such as adsorption to the walls of the sample container, volatilisation, and either chemical and/or microbial degradation (Greenwood *et al.* 2009). Grab/spot/bottle sampling has the following additional limitations:

- Spot sampling fails to detect and account for temporal variation in pollutant concentration
- Do not provide a truly representative picture of the extent of contamination

- Fails to account for bioavailability of pollutants in water (especially for non-polar organics and certain heavy metals)
- Fails to provide any information on the toxicity of water samples since spot sample provide information on total concentration.

6.4.1.2: *Bio-monitors/bioindicators*

[based on Vrana et al. 2005; Greenwood et al. 2009; Söderstrom et al. 2009].

Spot sampling problems can be overcome by using bioindicators. Aquatic organisms (fish or sessile bivalve molluscs) have been used as a tool (Table 6.5) to quantify the hydrophobic or non-polar contamination/persistent organic pollutants (e.g. PCBs, metals, organochlorine pesticides) and to obtain toxicologically relevant information in the aquatic environment. These contaminants are concentrated or accumulated in the fat of the aquatic organisms. Here, organisms are deployed over an extended period of time and changes in the levels of pollutants of interest are measured at the beginning and at the end of the trial. Use of bio-organisms can provide an estimate of the average environmental concentrations of pollutants over the deployment period (up to several months) and bio-accumulation of pollutants can provide a qualitative indication of level of pollutants between sites and between times. However there are some limitations of using bio-indicators to monitor chemical pollutants which are listed below:

- Bio-organisms may not be suitable to monitor polar compounds since polar organic compounds show low bio-accumulation and adsorption in biota
- There are some inherent problems with bio-organisms (e.g. metabolism, clearance of chemicals by organisms, mortality of test organisms, life-cycle stage and growth can affect bio-accumulation of pollutants plus limited distribution and availability of uniform species/same species worldwide)
- It is not possible to expose organisms in harsh environments where concentration of pollutants exceeds toxic levels (e.g. some industrial and domestic discharges)
- Test species may contain measurable levels of some pollutants before deployment, which requires to depurate the organism(s) before use and need large representative control samples at the start of monitoring campaign
- The analysis of biota tissue samples is expensive, time consuming and extraction and analysis of animal tissues are more complex than for passive samplers
- Deployment of fish is very expensive and ethically questionable
- Some pollutants can be eliminated by the test animals

Table 6.5: ***Living organism commonly used as bioindicators*** *[Monirith et al. 2003; Isobe et al. 2007; Leung et al. 2008; Wu et al. 2007; Esteve-Turrillas et al. 2008; Vicente-Martorello et al. 2009]*

Biomonitoring group	Biosampler/bio-organism	Chemical compounds evaluated
Molluscs	Blue mussel (*Mytilus edulis*)	PAHs, PCBs, OCPs, organotin, PBDEs, heavy metals (Cd, Cu, Pb, Zn)
	Blue mussel (*Mytilus trossulus*)	PAHs
	Blue mussel (*Mytilus gallorovincialis*)	PCBs; OCPs (DDT, CHL, HCB)
	Ezo mussel (*Crenomytilus grayamus*)	PCBs; OCPs (DDT, CHL, HCB)
	Green mussel (*Perna viridis*)	PAHs, PCBs, OCPs (DDT, CHL, HCB); EDCs (nonylphenol, octylphenol, Bisphenol-A), heavy metals (Cd, Cr, Cu, Pb, Zn)

Contd....

	Lake mussel (*Andonta piscinalis*) Mussel (*Elliptio complanata*) Pacific oyster (*Crassostrea gigas*)	OCPs, pesticides surfactants PAHs
Fish	Brown trout (*Salmo trutta*) Carp (*Cyrpinus carpio*) Fish (Crucian carp, Eel, Rudd, Trench) Fish (*Sparus aurata; Solea senegalensis*) Pink salmon egg Rainbow trout (*Oncorhynchus mykiss*)	PCBs PAHs, OCPs Musk compounds, PAHs Heavy metals (As, Cd, Cu, Pb, Zn) PAHs PAHs, OCPs
Annelids	Oligochaetes (*Lumbriculus variegatus*)	PAHs, PCBs
Plants	Pine needles	PAHs in air

To overcome the shortcomings of both spot sampling and biomonitoring procedures, passive samplers have been developed to mimic the uptake of pollutants by living organisms as discussed in great detail in section 6.4.1.3.

6.4.1.3: Passive sampling
[based on Vrana et al. 2005; Allan et al. 2006; Wu et al. 2007; Seethapathy et al. 2008; Greenwood et al. 2009; Söderstrom et al. 2009]

Passive sampling can be defined as the sampling technique based on free flow of analyte molecules from the sampled medium to a receiving phase in a sampling device as a result of a difference between the chemical potentials of the analyte in the two media (i.e. uptake of a chemical from the environment in passive sampler occurs by diffusion). The net flow of the analyte molecules from one medium to the other continues until equilibrium is established in the system or until the sampling period is stopped (Figure 6.1). The accumulation in passive sampler can reflect either the equilibrium concentration or the time-weighted average (TWA) concentration over the deployment period (days to months). In passive sampling, a reference (or receiving phase) is exposed to the water phase and all passive sampling devices absorb/adsorb pollutants from water. Most passive samplers measure concentrations of freely dissolved analytes (both bio-available and toxic fractions) and provide useful data on the variability of contaminant concentrations or temporal changes in toxicity. However, in order to predict TWA water concentrations of contaminants from levels accumulated in passive samplers, extensive calibration studies are necessary to characterize the uptake of chemicals into a passive sampler. Moreover, uptake of chemicals can be influenced by environmental variables such as temperature, flow rate, turbulence and bio-fouling of the sampler surfacer. Despite the above, there are several advantages in using passive samplers which are listed below:

- useful tool for monitoring of both polar and non-polar pollutants
- easy to deploy and retrieve
- cost effective alternative to biomonitoring
- provide reliable and robust information required to support registration applications (continuous monitoring)
- can be stored for indefinite period without change in the concentration
- based on diffusion (no supply of power or energy is required - auto-sampling)
- show low variability due to the standard configuration of the synthetic samplers
- only one device is necessary at a given sampling location for the duration of sampling
- ability to sample large volumes of water
- reduced effort required for deployment and sample processing compared to other commonly used methods
- can be used to monitor chemicals with a broad range of physico-chemical properties (Figure 6.2)
- standard tool for all over the world.

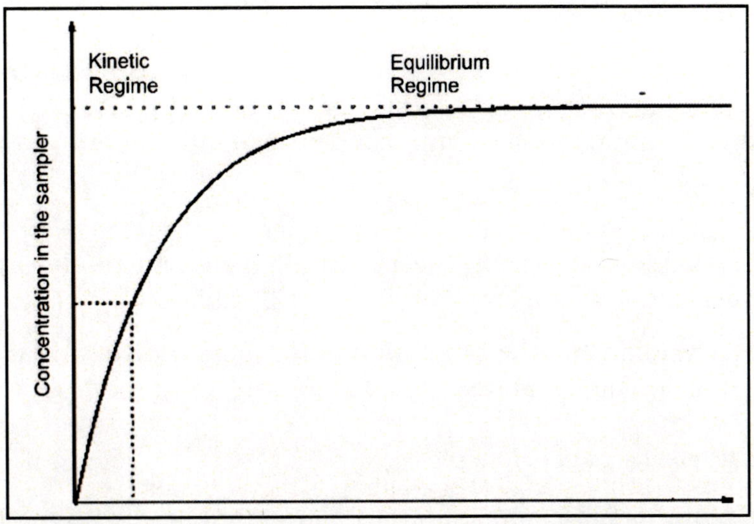

Figure 6.1: ***Analyte mass uptake profile of passive sampling devices. Passive sampling devices operate in two main regimes (kinetic and equilibrium)*** *[Vrana et al. 2005].*

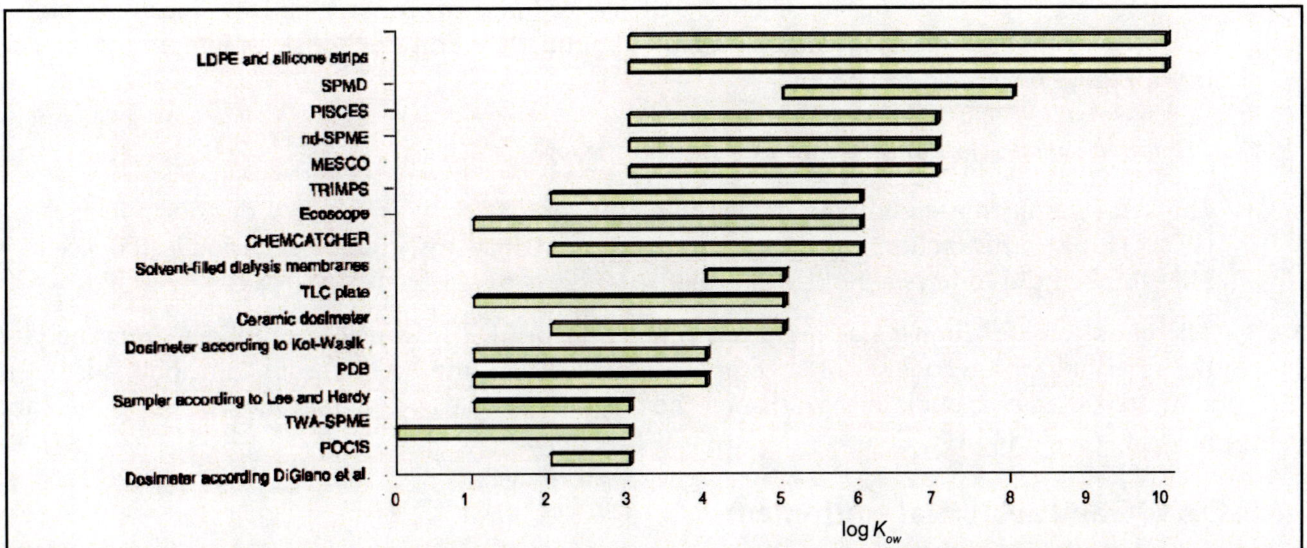

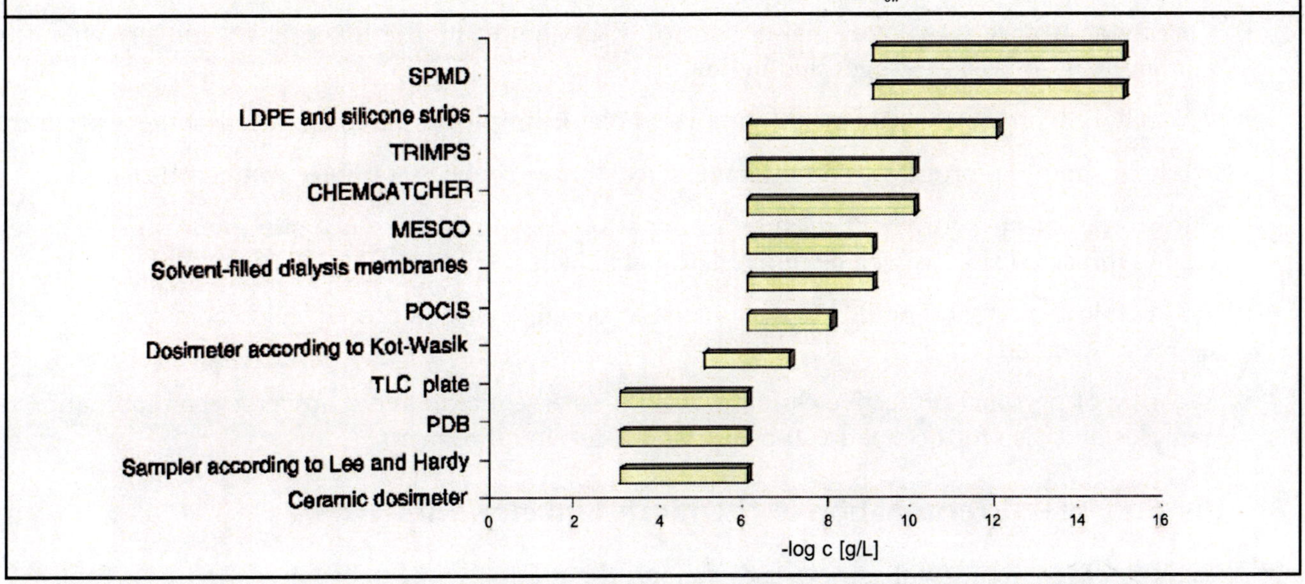

Figure 6.2. ***Typical hydrophobicity range of organic compounds sampled by selected passive sampling devices mentioned in Table 6.6 (characterised by the value of octanol/water partition coefficient, log*** K_{ow}***) (top Figure) and typical detectable concentrations of organic compounds by selected passive sampling devices (bottom Figure)*** *[Vrana et al. 2005].*

6.4.1.3.1: Passives samplers for metals

- **Diffusive gradients in thin films (DGT)** sampler where a thin hydrogel layer forms the diffusion limiting membrane and this overlay a chelating agent receiving phase (Figure 6.3). The DGT has been used to monitor cadmium, chromium, copper, lead and zinc in a wide range of aquatic environments (Figure 6.3. Table 6.6.).
- **Artificial mussels (AM)**: This device consists of non-permeable Perspex tubing, in which 200 mg Chelex-100 is suspended in seawater/freshwater inside the tubing. Both ends of the plastic tube are capped by a layer of polyacrylamide gel (Figure 6.3. Table 6.6).
- **Chemcatcher® metal version** (receiving phase with Empore™ disk) and the diffusion-limiting membrane of cellulose acetate membrane (Figure 6.3. Table 6.6).

6.4.1.3.2: Passives samplers for polar organic pollutants ($logK_{ow}$<4)

- **The chemcatcher® (polar organic version) and polar organic integrative sampler (POCIS)** are used to monitor TWA concentrations of polar organic pollutants ($logK_{ow}$<4) such as polar herbicides, pharmaceuticals and personal care products. In both samplers the diffusion-limiting membrane is a polyethersulphone sheet with water-filled micropores and the receiving phases comprise a range of adsorbent materials, either bound in an Empore™ disk, or in a free particulate from (Figure 6.3. Table 6.6).

6.4.1.3.3: Passives samplers for non-polar organic pollutants (3<$logK_{ow}$<8)

- The **semi-permeable membrane device (SPMD)** consists of a lay-flat low density polyethylene (LDPE) tube (diffusion limiting layer) containing small quantity (upto 1 ml of triolein-the receiving phase). The lipid mimics a biological organism (Figure 6.3. Table 6.6).
- **The chemcatcher® (non-polar organic version)** has a LDPE diffusion-limiting membrane, and the receiving phase consists of an Empore™ disk containing a C_{18} chromatographic adsorbent saturated with *n*-octanol. In this sampler the receiving phase and membranes are supported and held by a watertight PTEF body (Figure 6.3. Table 6.6).

6.5: QA/QC and analytical equipment

- Quality assurance (QA) and quality control (QC) should be the top priority for any chemical contaminant analysis as suggested below:
- All analytical protocol should include analysing blank samples as part of quality assurance program.
- All field sampling programs should have at least three replicate samples from each risk site.
- Sites to be selected for chemical monitoring should include reference site (non-impacted site), highly impacted sites as well as minimal impacted sites.
- All chemical analysis should be done as soon as possible, or be stored in cold room or in a freezer as recommended.
- Wearing of personal protective clothing or 'PPE' is essential to avoid any cross contamination of samples and also for occupational health and safety requirements.

6.6: Instrumental determination of chemicals and elements

Analysis of environmental sample uses state of the art analytical instrumentation (see Table 6.7) and such analysis should be done in any ISO 17025 accredited laboratory for the analyte in question for accuracy of results.

Table 6.6: ***Different types of passive sampler device used for monitoring of inorganic and organic (non-polar and polar) compounds*** [*based on Vrana et al. 2005; Wu et al. 2007; Hyne and Aistrope 2008; Söderstrom et al. 2009*]. BTEX = benzene, toluene, ethyl-benzene, and xylene; PAH = Polycyclic aromatic hydrocarbons; SVOC = semi-volatile organic compound, 99^{tc} = Technetium-99; VOC = Voltile Organic Compounds; MTBE = Methye tertiary butyl either

Passive sampler	Analytes	Construction	Deployment period	Comments
A. Passive sampling devices for inorganic contaminants				
AM (artificial mussel)	Cadmium, chromium, copper, lead, mercury and zinc	The device consists of non-permeable Perspex tubing (length: 6 cm; diameter 2.5 cm), in which 200 mg Chelex-100 is suspended in seawater/freshwater inside the tubing. Both ends of the plastic tube are capped by a layer of polyacrylamide gel.	14 days-1 month	- uptake in AM are not affected by salinity and temperature - currently used as Global AM watch program in eight countries including Australia
Chemcatcher	Cadmium, copper, lead, nickel, and zinc	Comprises an immobilised chelating acceptor resin on a PTEF base and a cellulose acetate membrane filter acting as a thin diffusion layer	14 days -1 month	- selectivity of the sampler can be adjusted using appropriate combination of membrane and Empore disks and calibration data available for many chemicals - require acid extraction
DGT (diffusion gradient in thin films)	55 metallic elements including common heavy metals, phosphorous sulphide, and ^{99}TC	Two layers of acrylamide gel mounted in a holder device, one containing an acceptor phase, the other acting as a thin diffusion layer.	1 week	- versatile well documented -complicated preparation device - require acid extraction
PIMS (passive integrative mercury sampler)	Neutral mercury species	LDPE lay-flat tubing	Weeks-months	- membrane characteristics may be altered for control of sampling rates - direct analysis of the receiving phase
PLM (permeation liquid membrane)	Copper, lead	Microporous hydrophobic support separating test solution from receiving solution	Hours	- selectivity of the sampler can be adjusted using appropriate combination of carrier media and receiving phase - require solvent extraction
SLM (supported liquid membrane)	Doubly charged cations	A strip solution with strong complexing agent is separated from the test solution by a macroporous hydrophobic membrane.	Days	- versatile, selectivity of the sampler can be adjusted - direct analysis, can be coupled on-line for real time monitoring
SLMD (stabilized liquid membrane device)	Divalent metal ions	LDPE lay-flat tubing containing an acidic solution with high affinity for the target elements	Days-weeks	- early development stage - require acid extraction

Contd...

B. Passive sampling devices for organic contaminants				
Ceramic dosimeter (ceramic dosimeter and toximeter)	PAHs, BTEX; chlorinated hydrocarbons	Ceramic tube (5 x 1 cm) filled with a sdid-phase sorbent material, closed with PTFE lids	Upto 1 year	- no need for extensive laboratory calibrations; suitable for long term monitoring and sorbent material of the 'toximeter' variant can be tested in contact bioassays. - low sensitivity - require solvent extraction or thermal desorption
Chemcatcher (universal passive sampler using Empore disk)	Polar and non-polar organics (pharmaceuticals and pesticides)	Made of inert plastic (PTEF) containing a disk of solid receiving phase bound in a porus polymer, and a disk of diffusion-modulating membrane	14 days – one month	- selectivity of the sampler can be adjusted using appropriate combination of membrane and Empore disks and calibration data available for many chemicals - require solvent extraction
CIDS (cellulose sampling device)	Polar herbicides (atrazine, diuron, metolachlor, molinate)	Dialysis membrane tubing containing solvent mixture of 1-dodecanol and 2,2,4-trimethylpentane	22 days	
Dosimeter	BTEX and atrazine	Activated carbon receiving phasein a perforated acrylic housing	Upto 2 months	- require solvent extraction
Ecoscope (solvent filled dialysis membranes and chelating sorbent discs)	Non-polar organics	A plastic housing containing a chelating sorbent disc for sampling metals and dialysismembrane filled with solvents		
Gaiasafe	Metals, anions, organic compounds	Paper or fabric strips impregnated with a solution of binding agent	2 days-2 months	- require solvent extraction
Gore-sorber	BTEX, MTBE, PAHs, VOCs, SVOCs	Various sorbent materials filled in a carrier hose made of Gore-Tex	14 days	- require thermal desorption
LDPE (Low density polyethylene) and silicone strips	Hydrophobic organic compounds	Low density polyethylene or silicone strips	1 months	- simple construction, inexpensive, simple sample processing and calibration data available for many analytes - smaller sampling capacity than SPMDs - require solvent extraction
MESCO (membrane-enclosed sorptive coating)	PAHs, PCBs, organochlorine pesticides	PDMS-coated stir bar used in SBSEor a PDMS rod enclosed in a membrane made of regenerated cellulose or low-density polyethylene	2 weeks	- both non-polar and polar analytes accumulated in a sampler equipped with a cellulose membrane - miniaturised sampler, non depletive matrix extraction,solventless sample processing. - low membrane stability of the sampler variant with cellulose dialysis membrane - require thermal desorption

Contd...

nd-SPME (negligible depletion-solid phase micro extraction)	Hydrophobic chemicals (PAHs, PCBs, petroleum hydrocarbons, organochlorine pesticides, aniline, phenols)	A fibre coated with a liquid (polymer), a solid (sorbent), or a combination of both	Hours	- negligible depletion extraction and a cheap disposable device - low sensitivity - require thermal desorption in GC inlet
Passive (sampler according to Lee and Hardy 1998)	Chlorobenzens, nitrobenzenes and nitrotoluenes	Silicone polycarbonate permeation membrane and an adsorbent receiving phase	Upto 1 day	- require solvent extraction
PDB (passive diffusion bag samplers)	Polar organic compounds, VOCs, metals, trace metals	Dialysis membrane or a low-density polyethylene bag filled with distilled water	2 weeks	- relatively expensive and sample recovery is rapid - not suitable for sampling of semi-volatile compounds - conventional analysis of the receiving phase
PISCES (passive *in situ* concentration-extraction sampler)	PCB	Hexane in a polyethylene membrane	2 weeks	- volume reduction of the receiving phase
POCIS (polar organic chemical integrative sampler)	Polar herbicides and pharmaceuticals (drugs, antibiotics, personal-care products) with log K_{ow} <3; Estrogenic substances (hormones)	Solid sorbent receiving phase material enclosed in a polyethersulphone membrane	Upto 2 months	- high sensitivity; sampler can be adjusted using appropriate sorbent materials; membrane has low susceptibility to biofouling, and calibration data available for many chemicals - require solvent extraction
Porous (sampler according to De Jonge and Rothenberg 2005).	Wide range of contaminants	A water permeable porous sampler that acts as semi-infinitive adsorptive sink	1 month	- traces integrated in the sampler store information of water volume that passed the sampler during deployment - require solvent extraction
Stainless steel housing (sampler according to Kot-Wasik 2004)	Herbicides, phenols, triazines	A stainless steel housing, containing organic solvent in chamber separated from water by a membrane	1 month	- a sample of the receiving phase solvent can be taken without affecting the integrity of the sampler - low sensitivity and receiving phase solvent may diffuse out of the sampler during field deployment - analysis of a sub-sample of a solvent taken and analysed without further clean-up steps

Contd...

Solvent filled dialysis membranes	Hydrophobic organic compounds	Non-polar solvent immiscible with water filled in a cellulose dialysis membrane	1 month	- non prone to biofouling - low sensitivity; solvent diffuses out of the sampler during deployment - volume reduction of the receiving phase
SPATT (solid phase adsorption toxin tracking)	Polar phytotoxins	Porous synthetic resin filled polyester fabric sachets	1 week	- require solvent extraction
SPMD (semi-permeable membrane devices)	Hydrophobic semi-volatile organic compounds	Flat tube of LDPE filled with triolein	1 month	- widely use and commercially available; -calibration data available for many analyte classes; high sensitivity - susceptible to biofouling; complicated sample clean up - dialysis in organic solvents, size exclusion chromatography
TLC plate (thin-layer chromatography plate)	Organophosphates		1 month	- require solvent extraction -good sensitivity because of large surface area
TRIMPS (trimethylpentane containing passive sampler)	Pesticides (endosulfan, chlorpyrifos)	2,2,4-trimehtylpentane filled in a low density polyethylene membrane	1 month	- simple sample clean up - receiving phase solvent diffuses out of the sampler during deployment - direct analysis of the receiving phase solvent
TWA-SPME (solid-phase microextraction applied for determination of TWA concentrations)	BTEX	A fibre coated with a liquid (polymer), a solid (sorbent), or a combination of both	A few minutes	- no need for extensive laboratory calibrations; sampling rate can be estimated using empirical mass-transfer models - short-term sampling only; fibre susceptible to damage or fouling mass transfer models - thermal desorption in GC inlet

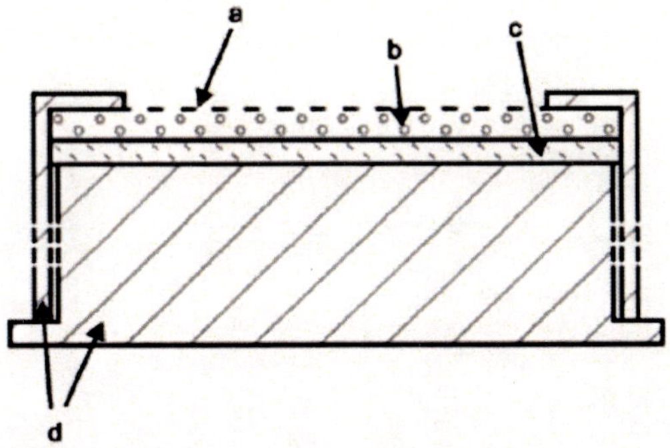

Diffusive gradients in thin films (DGT)
(a) protective membrane-prefilter cellulose nitrate; (b) diffusion membrane-hydrogel; (c) receiving phase-gel containing chelating agent; (d) sampler body-plastic

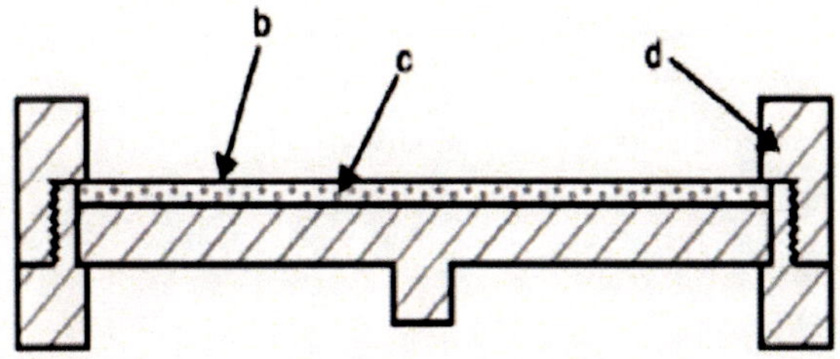

Chemcatcher®
(b) diffusion membrane-low density polyethylene or cellulose acetate; (c) receiving phase-Empore™ disk (C_{18} poly(styrenedivinylbenzene)
copolymer or chelating resin; receiving phase-gel containing
chelating agent; (d) sampler body- polytetrafluoroethylene (PTFE)
also called TEFLON

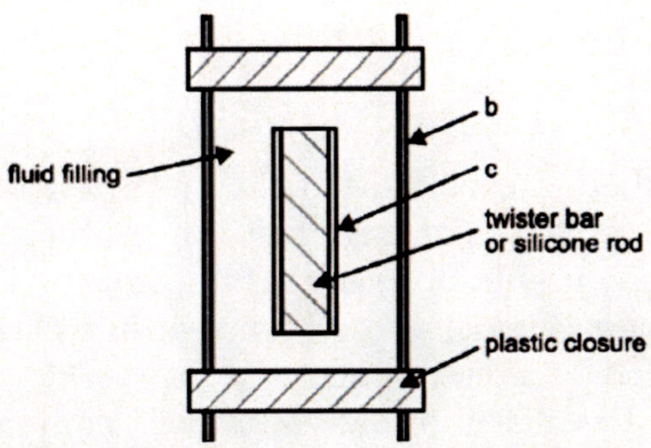

Membrane-enclosed sorptive coating (MESCO)
(b) diffusion membrane-dialysis membrane or low density polyethylene; (c) receiving phase-polydimethylsiloxane or silicone

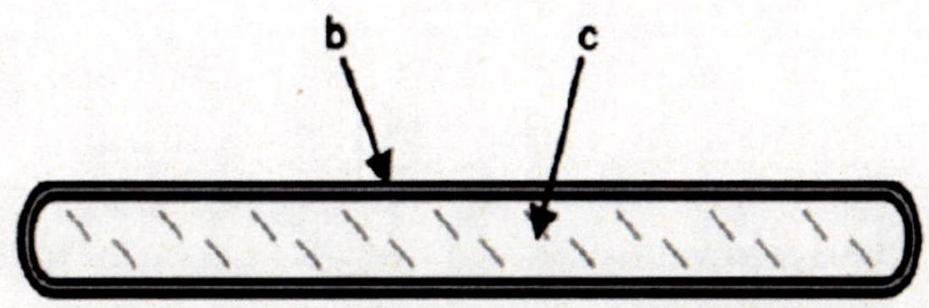

Semi-permeable membrane device (SPMD)
(b) diffusion membrane-low density polyethylene;
(c) receiving phase-triolein;

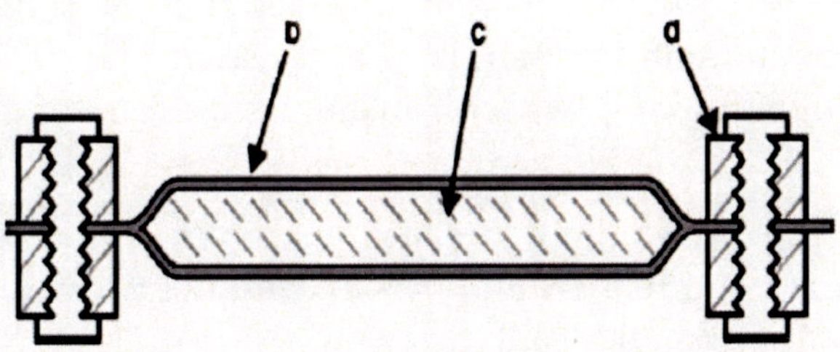

Polar organic integrative sampler (POCIS)
(b) diffusion membrane-polyethersulfone; (c) receiving phase-chromatographic phase (oasis); (d) sampler body-stainless steel flanges

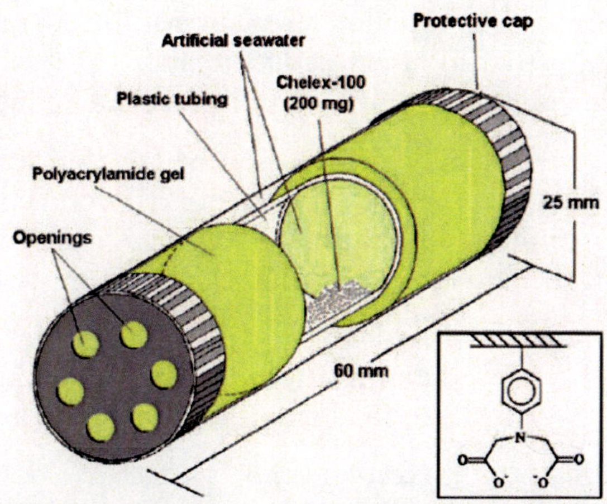

Artificial mussels
Protective membrane-non-permeable Perspex tubing (length: 6 cm; diameter 2.5 cm), receiving phase -200 mg Chelex-100;
diffusion membrane- polyacrylamide gel.

Figure 6.3 : ***Schematic diagrams of the main designs of passive sampler showing the components*** *[from Wu et al. 2007; Greenwood et al. 2009].*

Table 6.7: *Environmental contaminant analysis and bioassays.*

Chemicals	Analytical equipment[1]
Metalloids (arsenic, selenium)	Hydride-FAAS, EAAS, HydrideICP/MS
Heavy metals (copper, cadmium, lead, mercury, zinc)	EAAS, ICP-MS,
Pesticides	GC-PFPD, GC-ECD, GC-NPD, GC-MS
Dioxins	High resolution GC-MS
Endocrine Disrupting Chemicals (EDC)	GC-MS/MS, HPLC, ELISA, LC-MS/MS
Pharmaceuticals	LC-MS/MS
Freshwater cyanotoxins	HPLC
Marine phycotoxins (ASP, CTX, DSP, NSP, PSP)	Various LC-MS/MS

[1]EAAS = Electrothermal atomic absorption spectrometry; ECD = electron capture detector, FAAS = flame atomic absorption spectrometry; ELISA = Enzyme linked immuno-sorbent assay; GC-MS = Gas chromatography-mass spectrometry; HPLC = High pressure liquid chromatography; ICP = Inductively coupled plasma, LC-MS/MS = Liquid chromatography-mass spectrometry (LC-MS, or alternatively HPLC-MS); PFPD = pulsed flame photometric detector, NPD = nitrogen phosphorus detector.

6.7: Conclusion

The use of chemicals has many benefits, however, there are potential risks that some of these chemicals may cause harm to environmental organisms and human health if they enter into the environment. A regular monitoring program to assess the concentrations of chemicals in food, water and environment would therefore be vital to assess and manage any risks posed from harmful chemicals. In recent time, passive sampling technology has been widely accepted throughout the world as a tool for environmental sampling or chemicals risk assessment as it is more advantageous compared to traditional grab sampling. Passive sampling can provide important information on temporal and spatial variability of pollutants, biological/toxicological effects, account for contaminant bioavailability, as well as early detection of pollutants. In addition, passive samplers have the potential to replace the use of living organisms in assessing bioavailability. The other merits of passive samplers are lower cost, greater repeatability, smaller variability and greater acceptability on ethical grounds. About 75% of the organic micropollutants listed in the water quality criteria of the EU and US and the EU Water Framework Directive and the recommendations of the Convention for the protection of the marine Environment of the North-East Atlantic (OSPAR) can be monitored using passive sampling devices.

Human activities resulting from agricultural, industrial and mining activities continually release chemicals that may change our environment through contamination of air, land and water. The effects of some of these harmful chemicals are described in the following chapters with an emphasis on chemicals in the aquatic environment.

Chapter 7: ***Arsenic*** (world arsenic contamination; sources; speciation; Arsenic species in water, soil and water; toxicity to biota, plants and human; guideline standards in drinking water and in food and case study with arsenic problem in Bangladesh).

Chapter 8: ***Heavy metals*** (sources of Cd, Cu, Pb, Hg and Zn; speciation; bio-accumulation of metals; effects of heavy metals on aquatic organisms; national and international water quality guidelines; case studies with heavy metal inputs in Asia-Pacific and Australian environment).

Chapter 9: ***Pesticides*** (classification of pesticides; history of pesticide use; global and regional pesticide use; physio-chemical properties; toxicity and ecotoxicity; pesticide residues; effects of pesticides; national and international water quality guidelines and case studies with persistent organochlorine pesticides in Asia-Pacific and Australia regions).

Chapter 10: ***Dioxins, furans and PCBs*** (dioxin compounds; sources of dioxins; physical and chemical characteristics; effects of dioxins and risk to human health; tolerable daily intake (TDI) of dioxins in food; and case studies with dioxins related compounds in Asia-Pacific and Australia).

Chapter 11: ***Endocrine disrupting chemicals or EDCs*** (human endocrine systems; categories of EDCs; sources of EDCs; physico-chemical properties; environmental and biological impacts of EDCs. EDCs drinking water quality guidelines and case studies EDCs in Asian seafood, EDCs in USA and Australian environment).

Chapter 12: ***Pharmaceuticals (human and veterinary)*** (pharmaceutical industry; classification of pharmaceuticals; usage and sources; physico-chemical properties; environmental risk assessment of pharmaceuticals; pharmaceuticals in the environment; environmental and biological impacts of pharmaceuticals residues; management of pharmaceuticals and environmental water quality guidelines).

Chapter 13: ***Biotoxins*** (**13.1:** ***freshwater biotoxins***: conditions favouring cyanobacterial blooms; toxin producing cyanobacteria; biota toxicity of cyanobacteria; cyanobacteria alert levels and human health guidelines; **13.2:** ***marine biotoxins***: causes of marine algal blooms, types of phycotoxins; amnesic shellfish poisoning, diarrhetic shellfish poisoning, neurotoxic shellfish poisoning, paralytic shellfish poisoning, ciguatera shellfish poisoning and regulatory limits).

6.8: References

Allan, I. J., Vrana, B., Greenwood, R., Mills, G .A., Roig, B. and Gonzalez, C. 2006. Review-A "toolbox" for biological and chemical monitoring requirements for the European Unions water framework directive. *Talanta* 69: 302-322.

Commonwealth of Australia. 2002. *Environmental Health Risk Assessment Guidelines for assessing human health risks from environmental hazards.* page 4. Department of Health and Ageing and Health Council, Canberra. www.health.gov.au/publhealth/strateg/envhlth/risk/

De Jonge, H. and Rothenberg, G. 2005. New device and method for flux-proportional sampling of mobile solutes in soil and groundwater. *Environmental Science and Technology* 39 (1) : 274-282.

Esteve-Turrillas, F. A., Yusà, V., Pastor, A. and de la Guardia, M. 2008. Review-new perspectives in the use of semi-permeable membrane devices as passive samplers. *Talanta* 74: 443-457.

Greenwood, R., Mills, G.A. and Vrana, B. 2009. Review- potential applications of passive sampling for monitoring non-polar industrial pollutants in the aqueous environment in support of REACH. *Journal of Chromatography A.* 1216: 631-639.

Hyne, R. V. and Aistrope, M. 2008. Calibration and field application of a solvent-based cellulose membrane passive sampling device for the monitoring of polar herbicides. *Chemosphere* 7: 611-620.

Isobe, T., Takada, H., Kanai, M., Tsutsumi, S., Isobe, K. O., Boonyatumanond, R. and Zakaria, M. P. 2007. Distribution of polycyclic aromatic hydrocarbons (PAHs) and phenolic endocrine disrupting chemicals in South and Southeast Asian mussels. *Environmental Monitoring and Assessment* 135: 423-440.

Kot, A., Zabiegala, B. and Namieœnik, J. 2000. Passive sampling for long-term monitoring of organic pollutants in water. *Trends in Analytical Chemistry* 19 (7): 446-459.

Kot-Wasik, A. 2004. A new passive sampler as an alternative tool for monitoring of water pollutants. *Chem. Analy.* 49. 691-705

Lee, H. L. and Hardy, J. K. 1998. Passive sampling of monocyclic aromatic priority pollutants in water. *International Journal of Environmental and Analytical Chemistry* 72 (2): 83-97.

Leung, K. M. Y., Furness, R. W., Svavarsson, J., Lau, T. C. and Wu, R. S. S. 2008. Field validation, in Scotland and Iceland, of the artificial mussel for monitoring trace metals in temperate seas. *Marine Pollution Bulletin* 57: 790-800.

Monirith, I., Ueno, D., Takahashi, S., Nakata, H., Sudaryanto, A., Subramanian, A., Karuppiah, S., Ismail, A., Muchtar, M., Zheng, J., Richardson, B. J., Prudente, M., Hue, N. D., Tana, T. S., Tkalin, A. V., and Tanabe, S. 2003. Asia-

Pacific mussel watch: monitoring of persistent organochlorine compounds in coastal waters of Asian countries. *Marine Pollution Bulletin* 46: 281-300.

Plant, J. A., Korre, A., Reeder, S., Smith, B. and Voulvoulis, N. 2005. Chemicals in the environment: implications for global sustainability. *Applied Earth Science (Trans. Inst. Min. Metall. B)*. 114 (June): B65-B97.

Seethapathy, S., Górecki, T. and Li, X. 2008. Review. Passive sampling in environmental analysis. *Journal of Chromatography A* 1184 : 234-253.

Söderstrom, H., Lindberg, R. H. and Fick, J. 2009. Strategies for monitoring the emerging polar organic contaminants in water with emphasis on integrative passive sampling. *Journal of Chromatography A* 1216: 623-630.

Van der Oost, R., Beyer, J. and Vermeulen, N. P. E. 2003. Fish bioaccumulation and biomarkers in environmental risk assessment: a review. *Environmental Toxicology and Pharmacology* 13: 57-149.

Van Gestel, C. A. M. and Van Brummelen, T. C. 1996. Incorporation of the biomarker concept in ecotoxicology calls for a redefinition of the terms. *Ecotoxicology* 5: 217-225.

Vrana, B., Mills, G. A., Allan, I. J., Dominiak, E., Svensson, K., Knutsson, J., Morrison, G. and Greenwood, R. 2005. Passive sampling techniques for monitoring pollutants in water. *Trends in Analytical Chemistry* 24 (10): 845-868.

Vasseur, P. and Cossu-Leguille, C. 2003. Biomarkers and community indices as complementary tools for environmental safety. *Environment International* 28(8):711-717.

Vicente-Martorell, J. J., Galindo-Riaòo, M. D., García-Vargas, M. and Granado-Castro, M. D. 2009. Bioavailability of heavy metals monitoring water, sediments and fish species from a polluted estuary. *Journal ofhazardous materials* 162: 823-836.

Wu, R. S. S., Lau, T. C., Fung, W. K. M., Ko, P. H. and Leung, K. M. Y. 2007. An artificial mussel for monitoring heavy metals in a marine environments. *Environmental pollution*, 145, 104-110.

Webssites

http://www.nicnas.gov.au/publications/guideance_notes/risk_assessment pdf.pdf (risk assessment of chemicals: assessment of exposure from all sources) accessed 27 January 2008.

http://www.defra.uk/environment/chemicals/csf/annrpt0001/annexb.htm (chemicals stakeholder forum: annual report 2001-2001 (accessed 27 January 2008)

http://www.defra.gov.au/environment/chemicals/csf/criteria/detailed.htm

http://www.defra.gov.au/environment/chemicals/csf/concern/index.htm

□□□

7 Chemical Impacts - Arsenic

Key facts

- Arsenic (As) is a metalloid, with properties intermediate between those of a metal and a non-metal. Arsenic can exit in the environment and biological systems in various inorganic and organic forms (species). These are arsenite - As^{III}, arsenate - As^{V}, monomethyl arsonic acid (MMAA), dimethyl arsinic acid (DMAA), arsenocholine, arsenobetaine and arsenosugars.
- Arsenicals have a wide range of toxicity. In general, inorganic arsenicals are more toxic than oragnoarsenicals to biota and trivalent arsenic (arsenite) is more acutely toxic than pentavalent (arsenate) form in the following orders: Arsenous acid (As^{III}) $\rightarrow$ arsenic acid (As^{V}) $\rightarrow$ MMAA $\rightarrow$DMAA $\rightarrow$ arsenocholine $\rightarrow$ arsenobetaine.
- Under oxidizing and aerobic conditions, the predominant form of arsenic (As) in soil and water is arsenate (As^{V}), whereas under anaerobic and reducing/or waterlogged conditions, arsenite (As^{III}) predominates.
- Inorganic As^{III} and As^{V} are the major As species in groundwater, soils, terrestrial plants, whereas the main species found in marine animals is Arsenobetaine (AsB), which is considered to be non-toxic.
- The LD_{50} values of different arsenicals show that arsenite (As^{III}) is most toxic (LD_{50} range 15-42 mg/kg) and arsenobetaine is least toxic (LD_{50} > 10,000 mg/kg).
- Millions (60-100 million globally including 50 million alone in Bangladesh) of people worldwide are at risk from arsenic due to natural, mining and geothermal occurrences. Drinking water in Bangladesh, China, India (West Bengal), Vietnam and Argentina is severely contaminated with arsenic. Mining related arsenic problems have been reported from Thailand, Ghana, Greece and Austria, parts of USA, South Africa, and Zimbabwe. Arsenic associated with geothermal waters including hot springs have been reported from parts of the USA, Japan, Chile, Argentina, France, and New Zealand.
- Contaminated soils and agrochemicals are major source of arsenic in plants, whereas contaminated groundwater used for irrigation water could also be a source of arsenic in plants. Some plants such as the ferns, *Pteris vittata* and *Pityrogramma calomelanos* are capable of accumulating high concentrations of arsenic; therefore ferns could be ideal ecological indicators of arsenic contamination.
- Fish, shell-fish and seaweed products are the main source of arsenic (As) to humans depending on food habit and geological location.
- Rice and vegetables grown in contaminated soil and contaminated irrigation water can contain high inorganic As. Several research studies conducted in Bangladesh have found high concentrations of arsenic in vegetables and rice in areas where concentrations of arsenic in soil and water are also high.
- Arsenic has caused the following effects on various organisms: inhibition of growth in micro-algae and macro-algae; retardation/inhibition of growth in soil micro-organisms; mortality in juvenile copepods; significant reduction in populations of zooplankton; growth of fathead minnow and flag fish significantly reduced and effects on crop yield or growth in crops/plants.
- Among the plants, beans are classified as having no tolerance to As, while carrot and tomatoes are regarded as very tolerant. Highly sensitive As tolerant crops such as rice, bean, and oats can suffer from phytotoxicity at a soil arsenic concentration of 20 mg/kg, whereas for medium to low As tolerant crops radish and wheat this is 100 mg/kg.
- Arsenic is not essential for humans but one of the few substances that can cause cancer to humans from consumption of elevated level of arsenic (via drinking water or food).

- In some Asian countries (Bangladesh, Vietnam, Cambodia, Laos), the acceptable maximum level of arsenic in drinking water are much higher (50 μg/l) than WHO recommended 10 μg/l.
- Chronic exposure to arsenic has been linked to carcinogenic effects in both humans and animals. These include cancer of the skin and various internal organs (lung, bladder, liver and kidney). Non- cancer end points include hyper-pigmentation, hypo-pigmentation and keratosis of skin, peripheral vascular disease (black foot disease), cardiovascular disease, hypertension and neurotoxicity. There are some claims that chronic exposure of arsenic may also cause diabetes development and prostate cancer.
- The most important remedial action is prevention of further exposure of arsenic by providing safe drinking water and facilitating to obtain low-arsenic groundwater through accessing safe shallow groundwater or deeper aquifers (greater than 200 m in case of Bangladesh); rain water harvesting; pond-sand-filtration; household chemical treatment and piped water supply from safe or treated sources.

Contents

List of Tables

List of Figures

List of Boxes

Acronyms and Glossary

Alloy: A substance composed of two or more metals.

Alluvial: Soil material (clay, silt, sand or gravel) deposited by running water.

Arsenate, As^{IV}: An arsenate (compound) is a compound that contains the arsenate ion: AsO_4^{3-}. Arsenate is much like phosphate. In acid conditions we have arsenic acid, H_3AsO_4; in weakly acid conditions we have the dihydrogen arsenate ion, $H_2AsO_4^-$; in weakly basic conditions we have hydrogen arsenate ion $HAsO_4^{2-}$; and finally, in basic conditions, the arsenate ion AsO_4^{3-}.

Arseniasis, arsenicosis: Chronic arsenic poisoning or effect of arsenic poisoning, usually over a long period such as from five to twenty years.

Arsenic: Arsenic is a chemical element that has the symbol As and atomic number 33. Its Atomic Mass is 74.92. Its Ionic charge is (3^-). This is a notoriously poisonous metalloid that has many allotropic forms: yellow (molecular non-metallic) and several black and grey forms (metalloids) are a few to mention that are seen.

Arsenite, As^{III}: The salts of the arsenous acid H_3AsO_3.

Arsenobetaine (AB): Organoarsenical $(CH_3)_3As{+}CH_2COOH$, a major arsenic containing species in marine invertebrates and fish.

Arsenocholine (AC): Organoarsenical $(CH_3)^3As{+}CH_2CH_2OH$, a major arsenic containing species in marine invertebrates and fish.

Arsenopyrite: Is an iron arsenic sulfide (FeAsS). It is a hard (5.5-6) metallic, opaque, steel grey to silver white mineral with a relatively high specific gravity of 6.1. When dissolved in nitric acid, it releases elemental sulfur. When arsenopyrite is heated, it becomes magnetic and gives off toxic fumes. With 46% arsenic content, arsenopyrite, along with orpiment, is a principal ore of arsenic. When deposits of arsenopyrite become exposed to the atmosphere, usually due to mining, the mineral will slowly oxidize, converting the arsenic into oxides that are more soluble in water, leading to acid mine drainage.

Arsenosugars: Major arsenic compounds in algae but are reported to be present in significant concentrations in other marine organisms such as bivalves. Arsenoribofuranosides are organoarsenicals containing a pentose moiety as part of their molecular structure which explains why they are commonly referred to arsenosugars.

Artesian well: Bored well. Of water, rising to the surface under internal hydrostatic pressure; "an artesian well"; "artesian pressure

As: Arsenic.

Black foot disease: Blackfoot disease (BFD) is a severe form of peripheral vascular disease (PVD), in which the blood vessels in the lower limbs are severely damaged, resulting eventually in progressive gangrene. It has been observed in certain areas along the south-western coast of Taiwan where the average concentration of dissolved arsenic in the wells is 671 ± 149 μg/l with an average As^{III}/As^{V} ratio of 2.6. The first symptoms of the disease are the spotted discoloration of the skin on the feet that turns from brown to black as the disease develops until amputation of the affected extremities, the final resort to save the BFD victims, becomes necessary.

Body burden: The total amount of a substance in the body. Some substances build up in the body because they are stored in fat or bone or because they leave the body very slowly.

Cladocerans: Cladocerans are small crustaceans commonly found in most freshwater habitats, including lakes, ponds, streams and rivers.

Colloidal soil: Clay, mud or silt has finely layered structure and splits easily.

Delta: Triangular shaped piece of land made by deposits of mud and sand at the mouth of a river.

DMAA: Dimethylarsinic acid.

Eh: Redox potential, represented by Eh.

Geothermal: Relating to the Earth's internal heat; commonly applied to springs or vents discharging hot water or steam.

Holocene: The most recent geological epoch of the 'quaternary period', comprising roughly the past 10,000 years since the end of 'Pleistocene up to the present'.

Hyperaccumulator: Any plant that can accumulate large quantities of trace elements from its environment, and is often used in phytoremediation.

IARC: The International Agency for Research on Cancer is an intergovernmental agency forming a part of the World Health Organisation of the United Nations.

Lesion: An abnormal structural change in an organ or part due to injury or disease.

Metalloid: Elements that are neither metals nor non-metals such as arsenic. Such substances tend to have the physical properties of metals, in that they have a shiny appearance and a high melting and boiling point but the chemical properties of non-metals (do not react with acids).

Micaceous: The name of a group of hydrous aluminosilicate minerals characterized by highly perfect cleavage, so that they readily separate into very thin leaves, more or less elastic. Or A soil or rock type which contains an appreciable amount of the mineral mica.

MMAA: Monomethylarsonic acid.

ng/l: Nanogram per litre.

Ores: A mineral containing a constituent for which it is mined and worked.

Pleistocene: The earlier epoch of the 'quaternary period'. It extended from the end of the Pliocene, about 2 million years ago, to the beginning of the Holocene, about 10,000 years ago.

Redox: A reaction involving both reduction and oxidation.

Shell: A rock that is formed by the consolidation of calcium.

UNICEF: United Nations Children's Emergency Fund.

7.1: Introduction

Arsenic (As) is a metalloid, with properties intermediate between those of a metal and a non-metal. As is a natural element (atomic number 33), found throughout the earths crust at average concentrations of 2-5 mg/kg and is primarily associated with igneous and sedentary rocks.

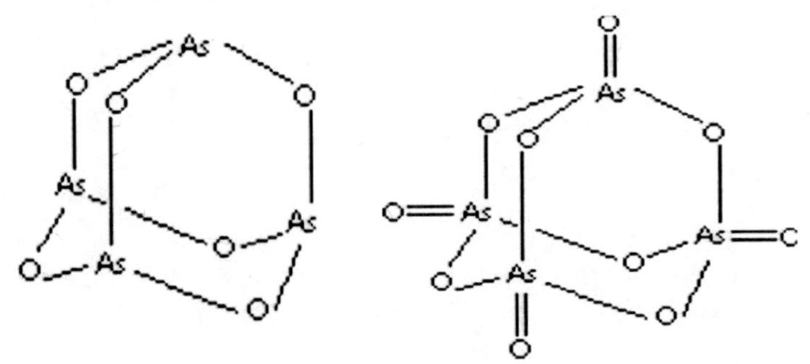

Arsenic trioxide (As^{III})   Arsenic pentoxide (As^{V})

Figure 7.1: ***Structure of arsenic.***

Arsenic is present in soils and it is also reported in many parts of the world in drinking water obtained from wells drilled into arsenic-rich ground strata. It occurs both in inorganic and organic form, of which inorganic arsenic is generally more toxic. Arsenic is found in two valencies, trivalent arsenite (As^{3+}) and pentavalent arsenate (As^{5+}) (see Figure 7.1). The arsenite is more soluble, mobile and is absorbed faster than the arsenate whereas arsenate is less soluble. Among the inorganic As, arsenite (As^{III}) is 60 times more toxic than arsenate (As^{V}). Inorganic As^{III} and As^{V} are the major As species in groundwater; other species of As are monomethylarsonic acid (MMAA) and dimethyl-arsinic acid (DMAA); methylated As species are also present in natural water in some particular circumstances. This toxic metal is widespread in groundwater.

As is commonly found in association with coal, gold, nickel, cobalt, antimony and iron ores or as arsenides or sulphides in copper, lead and zinc sulphide ores. Arsenicals are used commercially and industrially. Human activities (mining and manufacturing industries) can increase the amount of arsenic in the environment. Industrial effluents also contribute arsenic to water in some areas (see Table 7.1, Figure 7.1).

Table 7.1: ***Anthropogenic sources of Arsenic*** *[Commonwealth of Australia 2005; Naidu et al. 2006].*

Sources	Examples
Agriculture (pesticides and herbicides)	Because of the toxicity, As compounds were widely used as pesticides and herbicides which resulted in substantial accumulations in the soil (currently phased out in many countries). Lead arsenates was used in horticulture, arsenic trioxide were used in termite control as pesticide. Monosodium methyl arsenate, disodium methyl arsenate and sodium arsenate were extensively used in agriculture as herbicides. Old orchards can have high As concentrations due to accumulations of As in the soil. Arsenic trioxide (0.2% As_2O_3) was widely used in Australia to control the cattle tick (*Boophilus microplus*).
Animal farms	Organic As species such as arsanilic acid and carbarsone are used as a growth promoter (in feed) for poultry and pigs, manure used from these farms could be a source of As (note: poultry litter found to contain 15-25 mg/kg total As).
Atmospheric deposition	Emissions of As into the atmosphere can come from smelting and refining of non-ferrous metals, iron, and steel production and coal burning. Other sources are wood burning, agricultural chemicals, glass and cement manufacture and waste incineration. These activities can lead atmospheric As emissions that are deposited onto the land.

Contd...

Coal	The coal combustion process produces fly and bottom ashes which may contain substantial concentrations of As. Deposition of flying ash onto land could increase soil As level. Natural coal may contain an average of 1.05- 3.4.2 mg/kg of total As. As level in coal from Czechoslovakia varies around 900-1,500 mg/kg. In Guizhou province, China some coals often contains As level as high as 100-9,600 mg/kg.
Glass making	Arsenic trioxide and pentoxide are used to produce clear glass.
Irrigation	Waste effluents of irrigation can lead to an accumulation of As in soils. Use of As contaminated groundwater (e.g. Bangladesh and West Bengal, India) resulted in gradual accumulation of As in top soils.
Laboratory procedures	Arsenic trichloride is used in organo-arsenic chemistry and in chemical analysis procedures.
Mining and smelting	Arsenic is a natural constituent of many alloys, including those of copper, lead, zinc, silver and germanium. Soils can be contaminated with As during mining and smelting processes. It is often removed as an impurity during smelting of copper, lead and zinc. As in streams, shallow wells and bore holes in gold mining areas of Ghana ranges from <2 to 175 μg/l; and As in Topolnitza River (copper smelter area), Bulgaria was reported to vary form 750 to 1,500 μg/l.
Tannery (Hide preservation)	Arsenic trioxide and sodium arsenate are used as hide preservatives. Tannery wastes are therefore contaminated with As.
Wastes (Urban and industrial)	Sewage sludge (bio-solids) can contain As concentrations of 2-50 mg/kg of dry sludge. Waste from pulp and paper industry may contain small amounts of As.
Wood preservatives	Copper chrome arsenate is used to impregnate or preserve wood fibres in Australia, New Zealand, USA; timber treatment sites could be a good source of As into the environment and soil by way of spills, leaks and run-off from treated wood; soil As concentrations in some timber treatment sites is reported to be around 500-10,000 mg/kg.

7.2: Sources of arsenic contamination

7.2.1: Natural source

Natural contamination of groundwater aquifers by arsenic is reported from various parts of the world (over 70 countries) (Figure 7.2). The most noteworthy occurrences are from Asia and America, where millions of people are at risk due to drinking of arsenic contaminated water.

- *Asia*: Bangladesh, West Bengal (India), inner Mongolia (China), Taiwan, Vietnam.
- *America*: Chile, Argentina, Mexico and USA (south-west).

Arsenic contamination generally occurs with fluctuating water tables and flooding cycles in particular in acid sulphide/sulfate soils or where iron and /or manganese-enriched layers, or saline-layer aquifers occur (see Table 7.2).

7.2.2: Mining and geothermal

Mining related arsenic problems have been reported from Thailand (associated with tin mining), Ghana (associated with gold mining), Greece (Lavrion region associated with lead and silver mining), Austria, parts of USA (gold mining, Pb-Zn-Ag mining), South Africa, and Zimbabwe (Figure 7.2). Arsenic associated with geothermal waters including hot springs have been reported from parts of the USA, Japan (Kyushu: 500-4,600 μg/l), Chile, Argentina, France, New Zealand (3,800 μg/l) (see Figure 7.2).

Table 7.2: ***Areas of arsenic contamination, population at risk, concentration range, water quality guideline values of drinking water around the world*** *[adapted from Ng 2005; Naidu et al. 2006; Sharma and Sohn 2009 and http:/www.bgs.uk/arsenic].*

Country	Population at risk (thousands)	Sources of As exposure	As concentration (μg/l)	Guideline for As in drinking water (μg/l)	Discovery year	Remarks
ASIA						
Bangladesh	50,000	Drinking water: artisan wells	<1-4,700	50	1980s	50 out of 64 districts were identified as having groundwater concentration of As exceeding 50 μg/l (see section 7.6, case study with Bangladesh); As problems in groundwater related to alluvial and deltaic aquifers; affected aquifer (of Holocene age) are shallow (<100-150 m deep) comprised a mixed sequence of micaceous sands, silts, clays deposited by the Rivers, deeper groundwater's (>100-150 m depth) have generally low As concentrations (<10 μg/l) (Pleistocene age); chemical characteristics of high arsenic ground-waters are high iron (>0.2 mg/l), manganese (>0.5 mg/l), bicarbonate (>500 mg/l), and often phosphorus (>0.5 mg/l); As III is the dominant species found in groundwater of Bangladesh.
China (Guizhou)	20	Burning of coal	100-10,000 mg/kg	8 mg/kg	1950s	-
China (inner Mongolia)	600	Drinking water: bores	1-2,400	50	1990s	As concentrations in deep artesian groundwater's ranged between >40 and 750 μg/l; shallow ground-waters (non-artesian) As concentrations ranged between >10 and 60 μg/l.
China (Xinijiang)	100	Drinking water: artisan wells	1-8,000	50	1980s	As concentrations in deep artesian groundwater's ranged between >40 and 750 μg/l; shallow ground-waters (non-artesian) As concentrations ranged from >10 to 60 μg/l.
China, whole country	5630	Drinking water	>50	50	1980s	As IV is the dominant species found in groundwater in China.
Cambodia	-	Groundwater	1-1340		2004-2006	Mekong River flood plain.
India (West Bengal)	1000	Drinking water: artisan wells	<10-1,000	50	1980s	As problems in groundwater related to alluvial and deltaic aquifers; affected aquifers (of Holocene age) are shallow (< 100-150 m deep); deeper groundwater's (>100-150 m depth) have generally low As concentrations (<10 μg/l); chemical characteristics of high arsenic ground-waters are high iron (>0.2 mg/l), manganese (>0.5 mg/l), bicarbonate (>500 mg/l), often phosphorus (>0.5 mg/l), low chloride (<60 mg/l), sulphate (<1 mg/l), nitrate and fluoride (< 1 mg/l), and pH (~ >7).
India (Kolkata)	-	Near pesticide production plant	<50-23,080	-	1990-1997	-
Japan (Fukuoka)	-	Natural origin	0.001-0.293	-	1994	-
Nepal	550	Drinking water	8-2,620	50	1980s	90% people in Terai regions rely on groundwater for drinking water (47% of Nepal's population); out of 15,000 tube wells analysed, 23% contained As above 10 μg/l and 5% above 50 μg/l; about 0.5 million people in Terai region are at risks of As poisoning.

Contd...

Taiwan	200	Artisan wells	10-1,820	10	1950s	High As concentrations were found in deep artesian well water abstracted from sediments include fine sands, mud and black shale; groundwaters from shallow wells have low As concentrations.
Thailand	1	Water contaminated by tin mining waste	1-5,000	50	1980s	Worst case of As poisoning due to mining is Ron Pibool, Southern Thailand located in the famous Tin belt.
Thailand	-	Shallow (alluvial) groundwater, mining	1.25-5114	-	1994	Nakhon Si Thammarat province.
Vietnam	Millions	Drinking water	1-3,050	10	2001	About 150,000 tube wells are used for extraction of drinking water from the Mekong and Red Rivers deltas; deltas have high concentrations of iron, magnesium and ammonium; the first reported As contamination from Hanoi was in 2001; private small-scale tube wells contained average As 159 μg/l with highly affected rural areas where the groundwater As was 430 μg/l.
AMERICAS						
Argentina	2000	Drinking water	100-1,000	50	1981	The Chaco-Pampean Plain is one of the largest regions of high As in groundwater's (1 million km^2); other areas with high As includes Cordoba, La Pampa, Santa Fe, Buenos Aires; the groundwater often have high salinity; metal oxides in the sediments (Fe, Mn oxides and hydroxides) all thought to be the main source of dissolved As.
Chile	437	Drinking water	900-1,040	50	1962	High As concentrations reported from arid region (Northern Chile-Antofagasta, Calama, Tocopilla); the aquifers are composed of volcanic sediments and high salinity.
Bolivia	25	Drinking water	Not reported	50	1971	-
Mexico	400	Drinking water	10-4,100	50	1983	High As contaminated areas include arid Lagunera region (north of Central Mexico) and State of Sonora; high pH (6.3-8.9) and fluoride (2-305 μg/l) concentrations identified from the affected areas.
Peru	250	Drinking water	500	50	1983	-
USA (Western)	-	Drinking water	10-48,000	10	1988	Worst affected areas include Nevada, California, Arizona , Wisconsin, Minnesota, South Dakota, Oklahoma, where As concentrations exceeding >10 μg/l; the groundwater's in those areas have high pH, (>8) and P (>4 mg/l), low dissolved oxygen concentrations and high concentration of Fe, Mn and C.
EUROPE						
Hungary	220	Drinking water (deep groundwater)	10-176	10	1974	As >50 μg/l in alluvial sediments in Southern Hungary and neighbouring Romania; groundwater's in the Great Hungarian Plain (80-560 m depth) with high As have also high concentration of Fe, NH_4 and high concentration of humic acid.
Romania	36	Drinking water : bores	10-176	10	2001	-
Finland		Well waters; natural origin	19-980		1993-94	South-west Finland

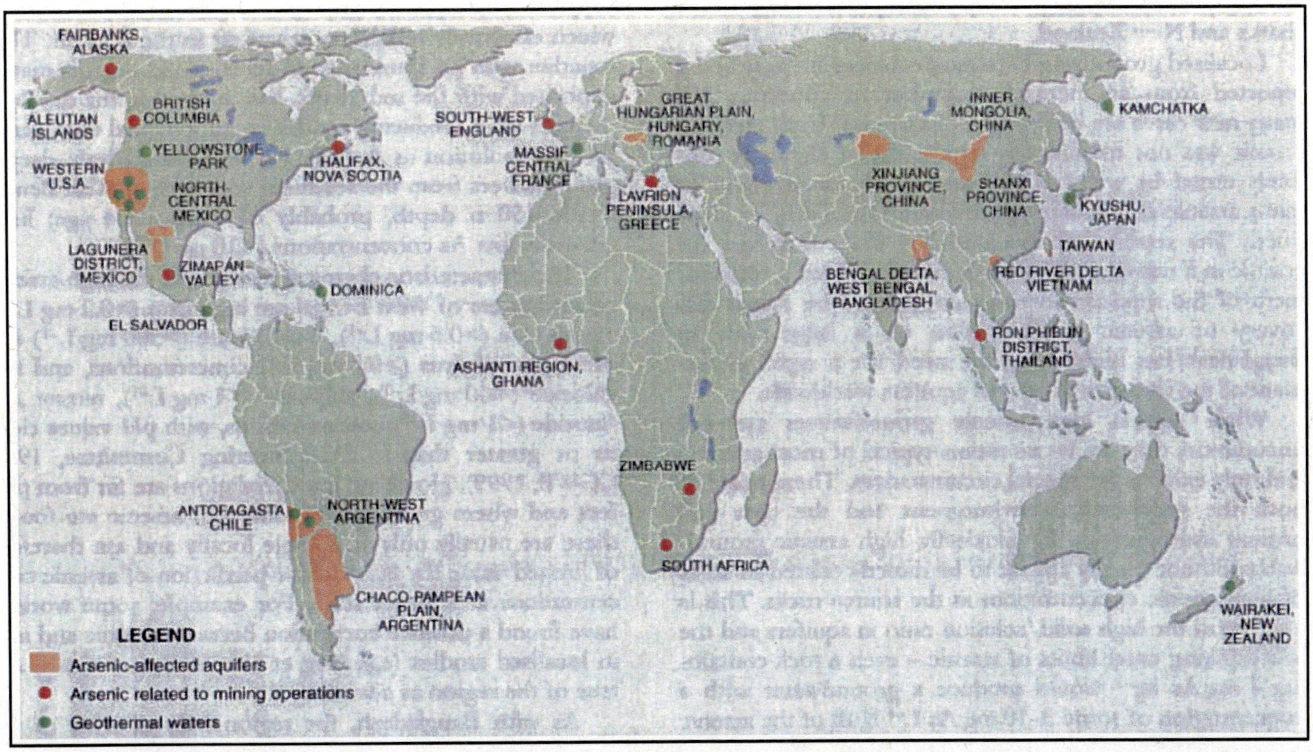

Figure 7.2: ***The documented cases of groundwater contamination of arsenic due to natural, mining and geothermal occurrences in the world*** *[http://www.bgs.ac.uk/arsenic/Bangladesh/reports/chapworldReview.pdf- Arsenic in groundwater's across the world, page 11, BGS and DPHE 2001)].*

7.2.3: Environmental levels of arsenic

Arsenic is found in all environmental media. The total arsenic concentrations in air of remote and rural areas are generally less compared to urban and industrial areas (Table 7.3). The typical range of arsenic in open seawater is 1-2 μg/l and the levels in rivers and lake are below 10 μg/l (Ng 2005). Though the levels of arsenic in groundwater could be less (1-2 μg/l), however elevated levels of arsenic can be found in areas with volcanic and sulphide mineral deposits. Sediments receiving inputs from copper smelters, gold mines, pesticide and other industrial plants can have high levels of arsenic. Most natural soils contain low levels of arsenic, but industrial wastes (mine) and pesticide applications can increase the concentrations. Higher arsenic levels in biota collected from mine waste sites, arsenic-treated areas, near smelter and mining areas have been reported.

7.3: Arsenic speciation in the environment

[based on IPCS 2001; Ng 2005; Sharma and Sohn 2009].

Arsenic can enter into the environment and biological system in various inorganic and organic forms (species) (Table 7.4). These are arsenous acid (As^{III}), arsenic acid (As^{V}), monomethyl arsonic acid (MMAA), dimethyl arsinic acid (DMAA), arsenocholine, arsenobetaine and arsenosugars. Arsenobetaine and arsenocholine are collectively called as 'fish arsenic'. Speciation of inorganic As in soil is largely controlled by reduction and oxidation processes (redox). Under oxidizing and aerobic conditions, the predominant form of arsenic in soil and water is arsenate (As^{V}), whereas under anaerobic and reducing/ or waterlogged (<200 mV) conditions, arsenite (As^{III}) predominates. The rate of conversion is dependent on the Eh and pH of the soil as well as other physical, chemical and biological factors. In short, at moderate or high Eh, inorganic arsenic occurs as a series of pentavalent arsenate oxyanions (As^{V}) such as $H_3AO_4^-$, H_2AO_4, $HAsO_4^{2-}$ and AsO_4^{3-}. However, under most reducing (acid and mildly alkaline conditions (low Eh values) arsenite (As^{III}) predominates. Microbial activity can influence As speciation via various mechanisms such as redox reactions with Fe and As and via (de)methylation of As species.

Table 7.3: *Arsenic levels in the environment.*

Environment	Levels	Reference
Air - remote and rural areas	Background levels: 0.2-1.5 ng/m^3. Mean range: 0.02-4 ng/m^3 (total arsenic).	IPCS 2001
Air-industrial	Background levels: <50 ng/m^3 Mean range: >1000 ng/m^3 (total arsenic).	IPCS 2001
Air-Urban	Background levels: 0.5-3 ng/m^3. Mean range: 3-200 ng/m^3 (total arsenic).	IPCS 2001
Groundwater	1-2 µg/l (average). 3 mg/l (areas with volcanic rock and sulphide minerals).	IPCS 2001 IPCS 2001
Natural levels in soils	1-40 mg/kg (mean 5 mg/kg). Contaminated soils (contaminated with mine, smelter waste) up to 26 530 mg/kg).	IPCS 2001 IPCS 2001
Open ocean	1-2 µg/l. Polluted with mine waste: 100 µg/l.	IPCS 2001; Ng 2005
Sediments (dry weight)	Uncontaminated: 5-15 mg/kg. Contaminated sediments: much higher levels (up to 2- 23,200 mg/kg).	IPCS 2001 IPCS 2001
Surface Freshwater (rivers and lakes)	Generally <10 µg/l Varies 1-100 µg/l.	IPCS 2001 Francesconi 2007
Marine organisms	<1 to >100 mg/kg (predominantly organic arsenic).	IPCS 2001
Terrestrial and freshwater biota	Background levels: < 1 mg/kg (fresh weight).	IPCS 2001

Table 7.4: *Examples of environmental arsenicals [IPCS 2001; Ng 2005; Sharma and Sohn 2009].*

Type of arsenic	Name	Formula	Comments
Inorganic (trivalent)	Arsenous acid	$As(OH)_3$	Some species identified in water
	Arsenenous acid	$HAsO_2$	-
	Arsenic trioxide	As_2O_3 or As_4O_6	-
	Arsenic chloride	As_2Cl_3	-
	Arsenic sulfide	As_2S_3	-
Inorganic (pentavalent)	Arsenic pentoxide	As_2O_5	-
	Arsenic acid	$AsO(OH)_3$	-Some species identified in water
	Arsenenic acid	$HAsO_3$	-
	Arsenates, salts of ortho-arsenic acid	$H_2AsO^-_4$, $HAsO^{2-}_4$	-
Organic	Methylarsine	CH_3AsH_2	-
	Dimethylarsine	$(CH_3)_2AsH$	-
	Trimethylarsine	$(CH_3)_3As$	-
	Monomethylarsonic acid	$CH_3AsO(OH)_2$,MMA^V	-Some species identified in water
	Monomethylarsenous acid	$CH_3As(OH)_2$, MMA^{III}	-Some species identified in water
	Dimethylarsinic acid	$(CH_3)_2AsO(OH)$, DMA^V	-Some species identified in water
	Dimethylarsenous acid	$(CH_3)_2AsOH$, DMA^{III}	-
	Trimethylarsinic oxide	$(CH_3)_3AsO$, TMAO	-
	Tetramethylarsonium ion	$(CH_3)_4As^+$, TMA	-
	Arsenobetaine (fish arsenic)	$(CH_3)_3As^+CH_2COO^-$, AB	-
	Arsenocholine (fish arsenic)	$(CH_3)As^+CH_2$ CH_2OH, AC	-
	Dimethylarsinoylribosides	1-11	-
	Triaklylarsonioribosides	12,13	-
	Dimethylarsonoulribitol sulfate	14	-
	Glycerophosphorarsenochloine	15	-
	Glycerophosphatidylarsenochloine	16	-

Speciation determines how arsenic compounds interact with their environment. For example, the behaviour (mobility) of arsenite (As^{III}) and arsenate (As^{IV}) in soil differs considerably depending on soil pH and soil type. At a lower pH (e.g. pH 5.8), mobility of arsenate is slightly more than arsenite. When pH changes from acidic to neutral or basic level, mobility of both arsenite and arsenate increases with increasing pH. In strongly adsorbing soils, transport rate and speciation are influenced by organic carbon content and microbial population. Both arsenite and arsenate are transported at a slower rate in strongly adsorbing soils than in sandy soils.

7.4: Environmental and biological impacts of arsenic

[based on 2001; Ng 2005; Commonwealth of Australia 2005; Burgess and Ahmed 2006; Naidu et al. 2006; Sharma et al. 2008].

7.4.1: Water

Inorganic As^{III} and As^{V} are the major As species in groundwater followed by MMAA and DMAA; As^{V} could be 90% of the total As in oxygenated natural water. Weathering and leaching of As-rich geological formations and mining wastes could result in elevated concentrations of As in natural waters.

7.4.2: Soil

As^{III} and As^{V} are the major As species in most soils. However, the distribution of As species in soils depends on the type and amounts of sorbing components of the soil, pH and the redox potential. The main source of As in soils is the parent rock, from which the soil is derived. The form and behaviour of As vary greatly between flooded soils, such as paddy fields, and non-flooded soils. According to Heikens (2006), the most important As species are arsenate (As^{V}) under non- flooded conditions and arsenite (As^{III}) under flooded conditions. As^{III} has a higher solubility than As^{V}, resulting in a higher mobility of As in flooded soils.

7.4.3: Terrestrial plants

The dominant As species in plants are As^{III} and As^{V} whereas methylated As (MMMA and DMAA) are also common. The concentrations of arsenic in terrestrial organisms are generally low, and inorganic species are dominant. On the contrary, mushrooms and earthworms have been reported to have many organic arsenic compounds found in marine and freshwater conditions. Contaminated soils and agrochemicals are major source of arsenic in plants, whereas contaminated groundwater used for irrigation water could also be a source of arsenic in plants. Terrestrial plants can accumulate arsenic by root uptake from the soil or by adsorption of air borne arsenic deposited on the leaves. Some plants such as the ferns, *Pteris vittata* and *Pityrogramma calomelanos* are capable of accumulating high concentrations of toxic As^{III} (see Table 7.5), therefore ferns could be ideal ecological indicators of arsenic contamination. These plants (*Pteris vittata* and *P. calomelanos*) are called hyper-accumulating plants.

7.4.4: Marine biota

Marine animals contain various As species and the concentrations of As in marine animals are higher than in the surrounding water. The main species found in marine animals is Arsenobetaine (AsB) (which is considered to be non-toxic) (see Box 7.1). Bivalves contain both AsB and arsenosugars (AsS). The major species found in marine algae are organoarsenicals such as AsS. Therefore, when human consume seafoods they are usually obtaining most of the arsenic in organic form. Inorganic arsenic is present in seafoods and marine organisms but in trace quantities.

Table 7.5: ***Arsenic accumulation in some selected ferns*** *[information compiled by Chang et al. 2009].*

Strain/Ferns	Total arsenic concentration in ferns accumulated	Comments
Deparia lobatorcrenata As-5	9.02 mg/kg (As-contaminated soil)	nonhyperaccumulator
Deparia lobatorcrenata As-8	11.15 mg/kg (As-contaminated tailing)	-
Deparia petersenii As-4	2.69 mg/kg (As-contaminated soil)	nonhyperaccumulator
Deparia petersenii As-6	12.10 mg/kg (As-contaminated tailing)	-
Pityrogramma calomelanos	88-8000 μg As/g (As-contaminated soil)	**hyperaccumulator**
Pteridim aquilinium As-1	Not detected (unpolluted soil)	nonhyperaccumulator
Pteridim aquilinium As-2	17.5 mg/kg (As-contaminated soil)	-
Pteridim aquilinium	$3x10^{-3}$ mg/kg (As-contaminated mine)	-
Pteridim aquilinium	1.83 mg/kg (As-contaminated substrate)	-
Pteris tremula	100 mg/kg (As-contaminated soil)	nonhyperaccumulator
Pteris vittata	69-28,522 μg/g (As-contaminated soil)	**hyperaccumulator**
Pteris vittata	2500-22,6300 mg/kg (As-contaminated soil)	-

Box 7.1: *Toxic arsenic vs mercury in seafood* *[Francesconi 2007].*

In terrestrial foods, both arsenic and mercury occur at low concentrations as inorganic species. In seafood, such as fish muscle (fillets), mercury occurs at high concentrations as the highly toxic methyl mercury (MeHg); whereas arsenic, although it also occurs at high concentrations, is present mainly as the harmless aresobetaine. Inorganic arsenic is a well know toxic substance. It constitutes the majority of arsenic in most terrestrial foods, but the concentrations are generally very low (<0.05 mg/kg). Organic arsenics are found at high concentrations in seafood as the harmless arsobetaine (>50 mg/kg wet mass).

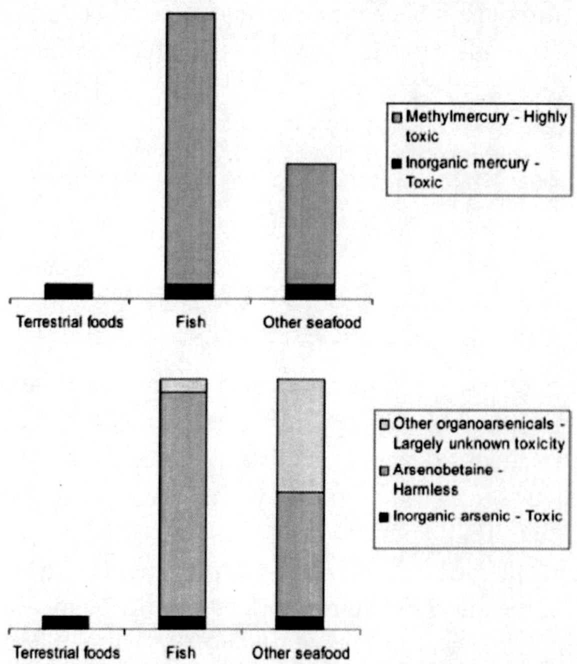

7.4.5: Food

Fish, shell-fish and seaweed products are the main source of As to humans depending on food habit and geological location. Raw rice, fruits and vegetables can contain high inorganic As^{III}, whereas beef, chicken, pork, eggs and butter can contain a small fraction of inorganic As but a higher total As. Rice and vegetables (tomatoes, radishes, turnips) grown in contaminated soil and contaminated irrigation water can contain high inorganic As (see section 7.6). Typical arsenic concentrations in various European foods are given in Table 7.6. (see also Table 7.14 for guideline values of arsenic for water and food).

Table 7.6: ***Typical arsenic concentrations (mg/kg wet weight) in various foods in Europe*** † *[Francesconi 2007].*

Product	Typical levels (mg/kg)
Vegetables	0.01-0.02
Fruits	0.006-0.01
Cereals	0.01-0.02
*Meats (various farm animals)	0.01-0.02
Offal (including liver and kidney - various farm animals)	0.01-0.02
Crustaceans (muscle)	1-20
Bivalve molluscs	1-20
**Fisheries products	1-20
Fish muscle	1-20

† *No maximum permissible concentrations set for arsenic in Europe*
*Farm animals include cattle, pigs, sheep and poultry.
**Fisheries products include molluscs, crustaceans and most fish.

7.4.6: Toxicity of arsenic to biota, plants and human

Arsenic is toxic to both animals and plants. The toxicity of arsenic in the environment is affected by temperature, pH, Eh, organic content, phosphate concentration, and adsorption to solid matrices, presence of other substances and toxicants, duration of exposure and the arsenic species present. Arsenicals have a wide range of toxicity. In general, inorganic arsenicals are more toxic than organo-arsenicals to biota and trivalent arsenic (arsenite) is more acutely toxic than pentavalent (arsenate) form in the following orders:

Arsenous acid (As^{III}) → arsenic acid (As^{V}) → MMAA →DMAA → arsenocholine → arsenobetaine

Table 7.7: ***LD_{50} of different arsenic species*** *[http://www.epa.gov/fishadvisories/forum/2004/presentations/wednesday/abernathy.pdf].*

Arsenic compound	LD_{50} (mg/kg)
Inorganic arsenite [As(III)]	15-42
Inorganic arsenate [As(V)]	20-200
MMA-monomethylarsonic acid	700-1,800
DMA-dimethyarsinic acid	1,200 -2,600
AsC-arsenocholine (marine)	6,500
AsB-arsenobetaine (marine)	≥10,000

The LD_{50} values of different arsenicals are given in Table 7.7, which shows that arsenite (As^{III}) is most toxic (LD_{50} range 15-42 mg/kg) and arsenobetaine is least toxic (LD_{50} > 10,000 mg/kg). LC_{50}/EC_{50} values of various arsenic species (As^{III}, As^{V}, MMAA) to aquatic and terrestrial organisms are shown in Table 7.8.

7.4.6.1: Effects of arsenic to biota

Arsenic has caused the following effects on various organisms as listed below:

- Inhibition of growth in micro-algae and macro-algae.
- Retardation/inhibition of growth in soil micro-organisms.
- Survival and reproduction of invertebrates- *Collembola* (springtails).

- Wilting, stress, crop yield, and growth in crops/plants.
- Accumulation of arsenic in earthworms (Table 7.9)
- Mortality in juvenile copepods.
- Significant reduction in populations of zooplankton (rotifers, copepods and cladocerans).
- Growth of fathead minnow and flag fish.

Table 7.8: ***Laboratory based toxicity data of various arsenic species of aquatic and terrestrial organisms*** *[based on various sources and compiled in WHO (2001); http://www.inchem.org/documents/ehc/ehc/ehc22.htm].*

Biota	Value range
Micro-organisms-Algae (14 days EC_{50}) (As/l)	48 μg to 202 mg (As^{V}).
Terrestrial organisms (7 D LC_{50})	Fruit fly: 40.5 mg/kg (As^{III}). Fruit fly: 59.3 mg/kg (As^{V}).
28-D EC_{50}	*Collembola* survival and reproduction: 3 mg/kg (As^{III}). *Collembola* survival and reproduction: 119 mg/kg (As^{III}).
LD_{50} mg/kg	Birds: 47.6 – 386 mg As/kg (As^{III}). Birds: >5000 mg/kg (MMAA).
Aquatic invertebrates (48 h LC/EC_{50})	0.68 to 73.5 mg/l (As^{III}). 3.6 to 49.6 mg/l (As^{V}).
Aquatic vertebrates (96 h LC/EC_{50})	Freshwater fish: 10.8-91 mg/ (As^{III}). Freshwater fish: 4.8- >360 mg/l (As^{V}). Freshwater fish: 1.9-1412 mg/l (MMAA). Marine fish: 12.7-28.5 mg/l (As^{III}). Marine fish: 21.4-157 mg/l (As^{V}).

Table 7.9: ***Concentrations of total arsenic in earthworm tissue*** *[data compiled by Langdon et al. 2003; Schultz and Joutti 2007]*

Species	Soil concentration (mg/kg)	Tissue concentration (mg/kg)	pH	Bio-concentration factor
Aporrectodea rosea (rosy-tip worm)	79.1	40	5.7	0.5
Eisenia fetida (red worms/tiger worms)	23-87	418-902	-	10.30-18.10
Lumbricus rubellus (red wriggler or red marsh worms)	79.1-8930	41.4-620	5.1-7.5	0.069-0.52
Lumbricidae	5.0-79.7	3.2-17.9	5.6-7.7	0.10-0.64
Lumbricus terrestris (night crawler)	73.9	50	-	0.67

Bio-concentration factor = earthworm tissue concentration/soil concentration

Lab based toxicity experiments conducted reveals that arsenite is generally more toxic than arsenate to biota. For example, the 96-h LC_{50} for freshwater fish ranges from 10.8 to 91 mg/l for trivalent arsenic (As^{III}) and from 4.8 to >360 mg/l for pentavalent arsenic (see Table 7.8).

7.4.6.2: Bio-accumulation of arsenic in fish

Fish can bio-accumulate arsenic in different organs including muscle tissues (the part eaten by human) (Table 7.10). The higher accumulation of arsenic in fish muscles therefore can be deleterious to human health.

Table 7.10: *Bio-accumulation of arsenic in different parts of fish species from various countries.*

Country	Muscles (μg/g)	Gills (μg/g)	Liver (μg/g)	Intestine (μg/g)	Comments	Reference
Bangladesh	0.097-1.32				-As in foodstuffs on sale in the UK imported from Bangladesh	Ali Ramalli *et al.* 2005
Canada	0.57-1.15		0.42-2.25	1.48-8.92	-Five freshwater species, Back Bay, near Yellowknife	Rosemond *et al.* 2008
Croatia	0.038-0.309	0.005-0.255	0.005-0.227		-Five fish species of River Neretva	Has-Schon *et al.* 2006
Pakistan	2.12-15.2	1.04-10.4		1.01-11.2	-Fish from highly arsenic contaminated Manchar Lake; -High levels of As (μg/g) found in the muscle tissues of *Labeo rohita* (7.6), *L. calbasu* (9.3), *Mystus gulio* (8.6), *M. seenghara* (12.1), *Catla catla* (15.2). -These species are common fish food in Bangladesh, India, Nepal and Pakistan.	Shah *et al.* 2008
Spain	0.396-12.58	-	-	-	-Twenty five fish species	Delgado-Andrade *et al.* 2003
Thailand	0.05-0.806	-	-	-	-Arsenic contaminated water	Jankong *et al.* 2007

7.4.6.3: Phytotoxicity of arsenic

The phytotoxicity of As varies with the plant species, soil As levels, soil characteristics, types of compound and temperature. As^{III} is more toxic than As^{V} to plants and plants can absorb As via the roots and or/foliage. Plant As levels correlate with total soil-As concentrations. In general, As toxicity to plants is highest in soils with high water-soluble (bio-available) As content such as coarse textured soils with little colloidal material and low ion exchange capacity. Among the plants, beans are classified as having no tolerance to As while carrot and tomatoes are regarded as very tolerant (see Table 7.11). Highly sensitive As tolerant crops such as rice, bean, oats can suffer from phytotoxicity at a soil arsenic concentration of 20 mg/kg, whereas for medium to low As tolerant crops radish and wheat this is about 100 mg/kg (Abedin *et al.* 2002). Natural levels of As in vegetation rarely exceed 1-2 mg/kg (dry weight) and elevated levels of As in vegetations in general result from environmental contamination or highly contaminated sites such as smelting operations, and mining sites. Soil type and soil texture are the most important variable for toxicity of inorganic As to plants, for example, inorganic As is five times more toxic in a sandy soil than in a clayey soil.

Table 7.11: *Sensitivity of various plants to arsenic [cited in Horswell and Speir 2006 original from Sheppard 1992].*

Sensitivity	Crop
High	Bean, soybean, rice, spinach, green beans, other legumes, onion, cucumber, alfalfa.
Medium	Apple, cherry, corn, cotton, potato, radish, strawberry, blueberry.
Low	Asparagus, tomatoes, carrot, wheat, oats, corn, cabbage, potatoes, peanuts, barley, pine.

The health based food standards for As is 2 mg/kg in Australia (NRS 2006). Studies conducted found that As accumulation in edible plant potions (fruit and grain) is low and do not exceed the guideline standards. The highest quantities of As residues are found in plant roots and moderate As accumulation has been found in the vegetative parts such as leaves and stem of plants and the smallest quantities in fruits and seeds. Because of highest As accumulation or concentrations in roots, humans can be exposed to As from eating of root vegetables such as carrot, beet, potato or arum. The effects of elevated levels of soil As on plants include a reduction of crop yield or growth of plants, inhibition of seed germination and seedling establishment. According to a recent review by Brammer and Ravenscroft (2009), the effects of As toxicity on rice include the following symptoms:

- delayed seedling emergence
- reduced plant growth
- yellowing and wilting of leaves
- brown necrotic spots on older leaves
- reduced grain yields (e.g. rice yields in conventional paddy fields in Bangladesh decreased from 8.92 to 2.99 t/ha with an increase of soil As levels from 26.3 to 575 mg/kg).

Table 7.12: ***Classification of metals and/or their compounds as carcinogen by International Agency for Research on Cancer, IARC (IARC 2006)*** *[http://monographs.iarc.fr/ENG/Classification/index.php].*

Substances	IARC carcinogen group*	Comments
Arsenic and its compounds	1	Carcinogenic to humans.
Cadmium and its compound	1	Carcinogenic to humans.
Chromium metal	3	Not classifiable as to carcinogenicity in humans.
Chromium (VI) compounds	1	Carcinogenic to humans.
Chromium(III) compounds	3	Not classifiable as to carcinogenicity in humans.
Lead metal	Ø	Not classified
Lead compounds	2A	Probably carcinogenic to humans.
Mercury and its compounds	2B	Possibly carcinogenic to humans.
Nickel metal	2B	Possibly carcinogenic to humans.
Nickel compounds	1	Carcinogenic to humans.

*Category 4 (not in the Table) probably not carcinogenic to humans.

7.4.6.4: Arsenic toxicity and effects on human health

7.4.6.4.1: Arsenic exposure routes

Arsenic is a well-documented human carcinogen and IARC (2006) classified it as class 1 carcinogen group among metals (Table 7.12). Humans and animals are exposed to many different forms of inorganic and organic arsenic species (arsenicals) in food, water, and other media.

7.4.6.4.2: Metabolism of arsenic in mammals

Humans and other mammals can be environmentally exposed to two common arsenic species such as inorganic arsenic prevalent in drinking water as arsenite (As^{III}) or arsenate (As^{V}). As^{III} and As^{V} is metabolised in the liver through successive oxidative methylation and reduction steps to its trivalent and pentavalent mono- and di-methylated metabolites (see Figure 7.3). Controlled experimental results indicate that both As^{III} and As^{V} are extensively methylated in humans, with dimethylarsinic acid ($DMAA^{V}$) being the principal methylated metabolite excreted in human urine.

7.4.6.4.3: Effects on human

Arsenic is toxic, bio-accumulative and carcinogenic. Arsenic is not essential for humans and has been classified as group/class 1 carcinogenic (Table 7.12).

The exposure of humans to arsenic contaminated water and foods can lead to some physical changes on the skin such as the appearance of small black or white marks (*melanosis*), then thickening of the skin on the palms and the feet (*keratosis*), followed by skin lesions and eventually skin cancer. The development of internal cancer in humans may take 10 years to develop and is often the result of long term exposure to arsenic. The long term ingestion or exposure (10-15 years) of arsenic can lead to a disease called 'Arseniasis, arsenicosis, and arsenicism'. The diseases/syndromes of arsenic poisoning in humans is given in Table 7.13.

Chronic exposure to arsenic has been linked to carcinogenic effects in both humans and animals. These include cancer of the various skin and various internal organs (lung, bladder, liver and kidney). Non-cancer endpoints include hyper-pigmentation, hypo-pigmentation, keratosis of skin, peripheral vascular disease (black foot disease), cardiovascular disease, hypertension and neurotoxicity. There are some claims that chronic exposure of arsenic may also cause diabetes development and prostate cancer (Naidu et al. 2006; Brammer and Ravenscroft 2009).

The arsenic in drinking water may cause effects in utero and child development, child health and adult disease. It may further cause spontaneous abortion, still births, reduced birth weight, and infant mortality. There is some evidence which supports that arsenic in drinking water causes increased risks of coronary artery disease, including increased mortality. Smith and Steinmaus (2009) reviewed information relating to the health consequences of early life exposure to different concentrations of arsenic in drinking water for different countries including Bangladesh, China, Chile, Hungary, India, Thailand and the USA.

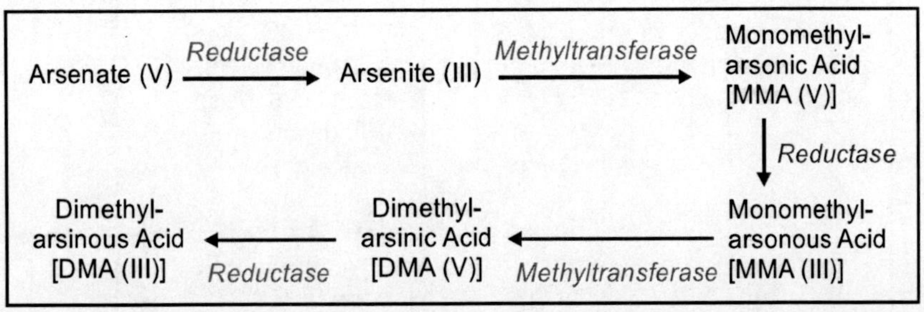

Figure 7.3: ***Proposed mammalian arsenic metabolism*** *[from Aposhian and Aposhian 2006 referred in Beyersmann and Hartwig 2008, page 497].*

Table 7.13: ***Symptoms of arsenic poisoning in humans*** *[www.sos-arsenic.net/english/toxic_effect/index.html]*

Symptom	Human body part
Darkening of the skin (diffuse melanosis).	- Palm of the hand
Spotted pigmentation (spotted melanosis).	- Chest, back or limbs
Leucomelanosis (white and black spots side by side).	-
Buccal mucus membrane melanosis (diffuse, patchy or spotted mela nosis).	- Tongue, gums, lips.
Keratosis (late feature of arsenical dermatosis).	- Palm of the hand and the sole of the foot (moderate severe toxicity). - Rough dry skin, palpable nodules (spotted keratosis) in dorsum of hands, feet, and legs (severe cases).
Severe cases	- Conjunctival congestion and nonpitting swelling (solid edema) of the feet; liver enlargement (hepatomegaly); spleen enlargement (speno-megaly), fluid on the abdomen (ascitis) .
Advanced cases.	- Squamous cell carcinoma; basal cell carcinoma; bowen disease; carcinoma affecting the lung, uterus, bladder, genitourinary tract.

7.5: Arsenic environmental and food safety guideline values

Arsenic is not essential for humans but one of the few substances that can cause cancer to humans from consumption of elevated level of arsenic (drinking or taking food). In some Asian countries (Bangladesh, Vietnam, Cambodia, Laos, the acceptable maximum level of arsenic in drinking water are much higher (50 μg/l) than WHO recommendations of 10 μg/l and Australian standard of 7 μg/l. Until 2005, there were no arsenic guideline values for food except in Australia (see Table 7.14). In October 2005, the ministry of Health of China adopted a new food safety standard on As in foods (GB 2762-2005), which covered a variety of food products and this standard can be used as a guideline for other countries (see Table 7.14).

Table 7.14: *Guideline values for arsenic in drinking water, food and soil in various countries.*

Water/food/soil	Australia, (1, 2)	Bangladesh and other Asian countries (3)	Chinese food safety standard for inorganic arsenic (mg/kg) (4)	WHO (5)	Europe (6, 7)
Drinking water MRL for Fisheries products	-7 μg/l -2 mg/kg (crustacean, fish); -1 mg/kg (molluscs) (NRS, 2006)	- 50 μg/l -	- - 0.10 (fish); - 0.50(shellfish).	- 10 μg/l -	-10 μg/l -MPC has not been set for food.
MRL for Food	-1-2 mg/kg (FSANZ, 2000; Abedin et al. 2002)	-	- 0.15 (rice); 0.10 (flour); - 0.20 other cereals; - 0.05 (vegetables); - 0.10 (beans/pulses); - 1.50 (algae); - 0.05 (fruit); - 0.05 (poultry); 0.05 (egg); -0.25 (milk power); - 0.05 (fresh milk)	-	-MPC has not been set for food.
Soil in residential area	-100 mg/kg - Health investigation level)	-	-	-	- 20 mg/kg with plant uptake. - 20 mg/kg without plant uptake. - 20 mg/kg allotments. - 500 mg/kg in commercial/ industrial.
Aquatic ecosystems protection trigger values (95% protection- μg/l)[1]	- Total arsenic: 50 μg/l. - As^{III}: 95% protection in freshwater 24 μg/l. - As^{V}: 95% protection in freshwater-13μg/l. - Marine water: Insufficient data.	-	-	-	-
Recreation[2] Irrigation[1]	50 μg/l 100 mg/l LTV 2000 mg/l STV	- -	- -	- -	- -
Livestock[2] Aquaculture/fish[1]	500 mg/l Freshwater: <50 μg/l Marine water: <30 μg/l	- -	- -	- -	- -

MPC = maximum permissible concentrations; MRL = maximum residue levels; FW = freshwater; MW = marine water; LTV = long term trigger value; STV = short term trigger value

1. ANZECC/ARMCANZ 2000; 2. ANZECC 1992; 3. Ng 2005; 4. Heikens 2006; 5. Environment Agency, UK. 2002; 6. WHO 2001; 7. Francesconi 2007.

7.6: Case studies- Arsenic contamination in Bangladesh

7.6.1: Arsenic in aquifers

Arsenic contamination of groundwater in Bangladesh is regarded as the worst mass poisoning in human history since more than 50 million people are at risk from groundwater contamination. In 1970, the UNICEF installed shallow tube wells across Bangladesh to eradicate the surface water borne diseases that affected much of the population. Since, then the rural population in Bangladesh relies on hand-pumped shallow tube wells (20-75 m deep) for drinking water (total tube wells estimated to be 8.6 million). By 1995, it was confirmed that most of the tube wells installed by UNICEF were contaminated with arsenic at levels above the world health organisation's safety limit of 10 µg/l or of Bangladesh 50 µg/l (see Figure 7.4 and Table 7.15). Arsenic concentration was found to be greatest in the range of 0-3,000 µg/l in tube wells of less than 75 m deep and is relatively free from As in tube-wells deeper than 150 m (see Figure 7.4 and 7.5).

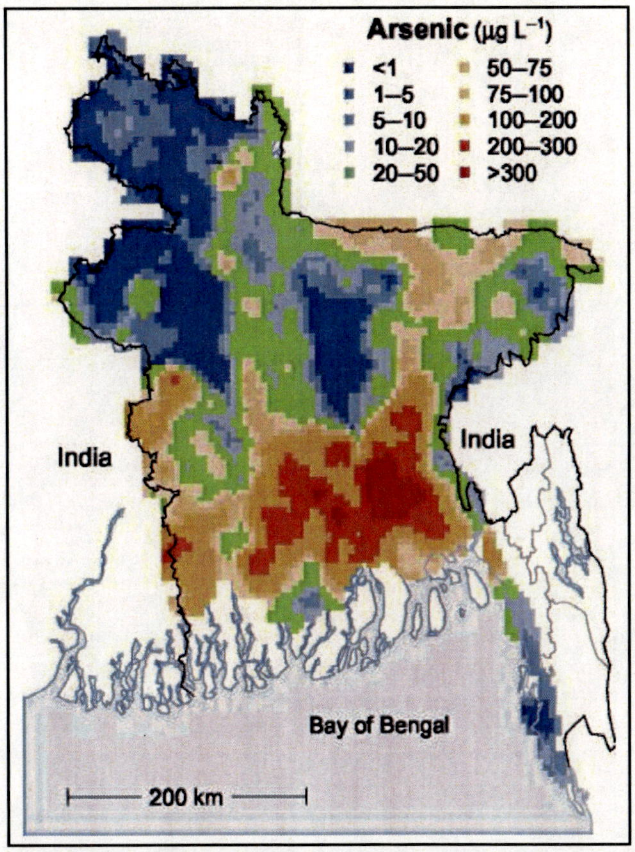

Figure 7.4: ***Map showing the estimated arsenic concentration in shallow tube-wells (<150 m deep) in all 64 districts of Bangladesh.*** *[Kinniburg and Smedley 2001 (BGS and DPHE, 2001), volm 1. summary, page 3 ; htttp://www.bgs.ac.uk/arsenic/bangladesh/reports/chapstatistics.pdf].*

According to the British Geological Survey (BGS), a combination of three factors is responsible for groundwater As problem in Bangladesh and these are listed below (see also Box 7.2):

- natural presence of As in the aquifer sediments.
- release of As from the sediments to the groundwater.
- flushing of As during natural groundwater circulation.

Several research studies conducted have found that the worst As affected aquifers are in alluvial deposits beneath the recent floodplains (older sediments are not significantly affected by As). Among the floodplains (the Ganges, Tista, Meghna and Brahmmaputra), the aquifers that lie beneath the Meghna floodplains of southeast Bangladesh have been most affected (see Figure 7.4, red and brown areas).

This is confirmed by high levels of As detected in aquifers of southern districts. On the contrary, north of Bangladesh has been least contaminated (see Figure 7.4, blue and green areas).

Table 7.15: ***Mean arsenic concentrations in some selected districts of Bangladesh found in DPHE/BGS national hydro chemical survey (μg/l)*** *[BGS and DPHE 2001; highlighted districts have exceeded Bangladesh guideline for As in drinking water of 50 μg/l].*

District	As (μg/l)	District	As (μg/l)
Bagerhat	**156**	Lalmonirhat	1
Bandarban	-	**Madaripur**	**191**
Barisal	**92**	Magura	28
Barguna	1	Mymensingh	12
Bhola	10	**Munshiganj**	**189**
Bogra	18	Naogaon	6
Brahmanbaria	**101**	**Narail**	**88**
Chandpur	**366**	Narayanganj	48
Chittagong	32	Narshinghdi	41
Chuadanga	**79**	Natore	1
Comilla	**142**	Nawabganj	6
Cox's Bazar	3	Netrokona	40
Dhaka	41	Nilphamari	2
Dinajpur	3	**Noakhali**	**162**
Faridpur	**140**	Pabna	32
Feni	**54**	Panchagam	3
Gaibandha	22	Patuakhali	3
Gazipur	4	Pirojpur	30
Gopalganj	**187**	Rajbari	49
Habiganj	22	Rajshahi	7
Jaipurhat	1	Rangamati	-
Jamalpur	14	Rangpur	8
Jessore	**70**	**Satkhira**	**133**
Jhalakati	23	**Shariatpur**	**151**
Jhenaidah	46	Sherpur	22
Khagrachari	-	Sirajganj	31
Kishorganj	**51**	Sunamganj	48
Khulna	35	Sylhet	22
Kurigram	22	Tangail	20
Kushtia	**104**	Thakurgaon	1
Lakshmipur	**179**		

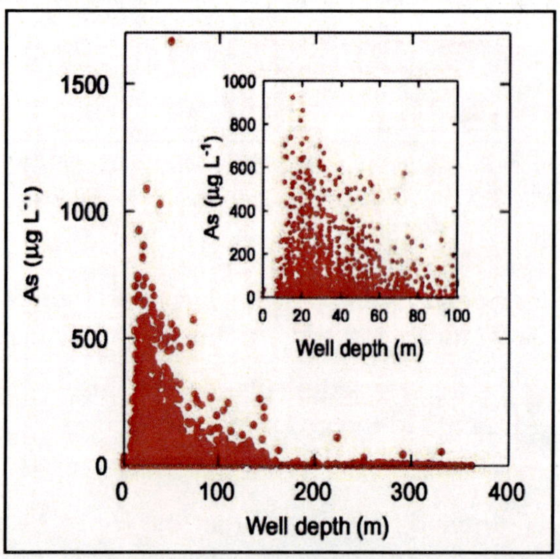

Figure 7.5: ***Arsenic concentration of groundwater (tube-wells) in Bangladesh. The highest As concentrations and the greatest range are found in groundwater pumped from tube-wells shallower than 75 m. Tube-wells deeper than 150 m are relatively free of As*** *[Burgess and Ahmed 2006; original data from Kinniburg and Smedley 2001].*

Box 7.2: ***Theories regarding mobilisation of Arsenic into groundwater in Bangladesh*** *[Commonwealth of Australia 2005].*

1. *Pyrite oxidation or man made causes:* *As is released from the sulphide minerals (arseno-pyrite) in the shallow aquifer due to oxidation. Due to extraction of groundwater, air and water containing significant dissolved oxygen is drawn from underground where it reacts with arsenopyrite. This oxidation reaction releases arsenic into the groundwater. Thus the increased use of shallow tube wells may be implicated as a cause of arsenic problem in Bangladesh.*

2. *Oxy-hydroxide reduction or natural causes:* *As is derived by desorption from ferric hydroxide minerals under reducing conditions. Ferric hydroxide minerals are present as coatings in the groundwater aquifer sediments. The As contaminated sediments along with abundant organic matter are deposited into the alluvial plains of the Bengal Basin. In anaerobic groundwater, the organic matter consumes the oxygen and conditions changes from initial oxidizing to a reducing stage. In the reducing stage the adsorbed As is released into water. This theory implies that the release of arsenic into the groundwater has not been triggered by human activities. The arsenic contaminated ground-waters in Bangladesh is high in iron, manganese, bicarbonate (see Table 7.2), therefore the oxy-hydroxide theory may be applicable for mobilisation of As.*

7.6.2: Arsenic in soils and in the food chain

Arsenic concentration in uncontaminated soils in some Bangladesh districts ranged between 0.10 and 2.75 mg/kg. In contrast, in areas where irrigation is carried out with contaminated groundwater, the soil As level is reported to be upto 81 mg/kg (Huq *et al.* 2006b). The elevated level of As in soil has resulted in elevated concentrations of As in food since crops receiving arsenic contaminated irrigated water can uptake As during the phyto-extraction process and bio-accumulate in different degrees in different parts of plants (e.g. roots, stems, and grains).

Table 7.16: ***Arsenic in common plants of Bangladesh from uncontaminated and contaminated areas*** *[Huq et al. 2006a and 2006b].*

Common name (local name)	Botanic name	Uncontaminated areas (mg/kg)	Contaminated areas (mg/kg)*
Arum (kochu)	*Colocassia antiqourum*	0.077-0.387	0.13-153.2
Aubergine	*Solanum melongena*	0.23	2.3
Bean (shim)	*Dolicos lablab*	0.092	0.3-1.16
Bitter gourd (karola)	*Momordicum charantia*	1.56	2.12
Chilli (morich)	*Capsicum spp.*	0.41	1.52
Green papaya (pepe)	*Carica papaya*	0.212-0.46	0.04-2.22
Indian spinach (pui shak)	*Brasilia alba*	0.102-0.146	0.07-1.00
Long bean (barboti)	*Vicia laba*	0.3	0.37-2.83
Potato (alu)	*Solanum tuberosum*	0.62	0.71-2.43

*Recommended Australian guideline threshold levels of As in food is 1 mg/kg.

Several research studies have found high concentrations of arsenic in vegetables and rice in areas where concentrations of arsenic in soil and water are also high for example, the greatest concentration of arsenic was found in leafy vegetables (in particular in arum or *kochu;* upto 153.2 mg/kg) followed by other vegetables (Table 7.16). Higher concentrations of arsenic were also reported in rice plants (*boro* rice) in the following orders: rice roots → rice stem → rice leaf → rice grain → rice husk (Table 7.17). It is evident that, the food chain could be a significant pathway of As ingestion by the people of Bangladesh (see Figure 7.6). Figure 7.6 is a simplified version of rice and vegetables contamination with arsenic in Bangladesh and its exposure routes to humans and livestock.

Table 7.17: ***Distribution of arsenic (mg/kg) in rice plant, grain, leaf, stems and root*** *[Huq et al. 2006a].*

Plant part	Minimum-Maximum*	Average
Root	9.06-73.4	29.1
Stem	0.35-25	4.29
Leaf	0.06-3.32	1.21
Grain	0.02-2.81	0.29
Husk	0.02-0.15	0.07

**Recommended guideline threshold levels in food as As is 1 mg/kg*

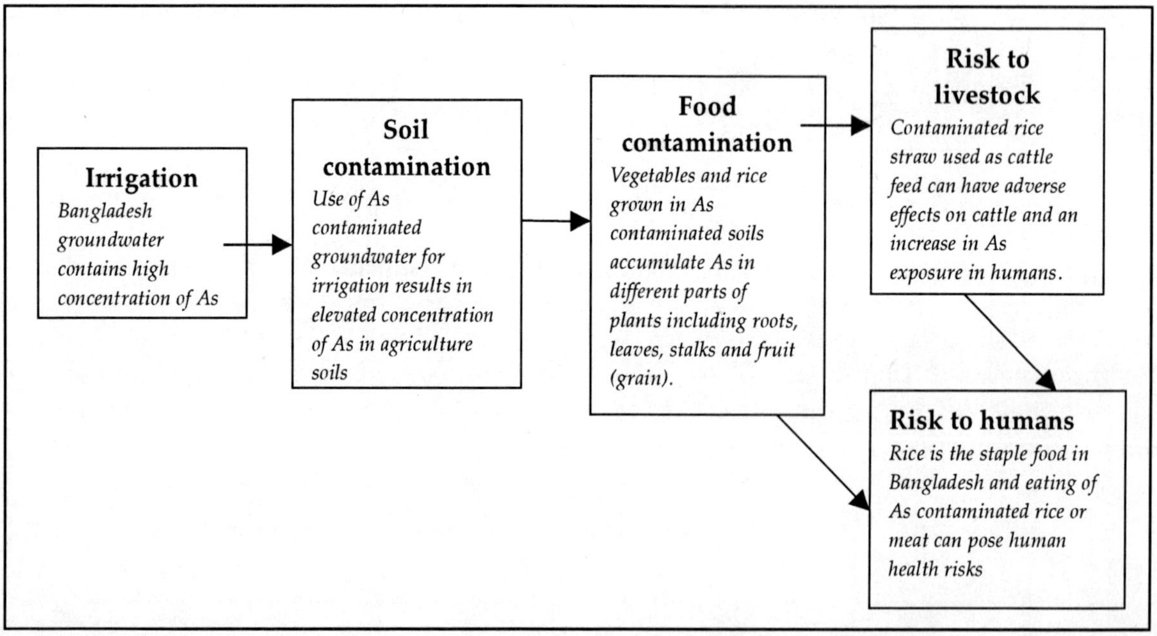

Figure 7.6: ***A simplified version of rice and vegetables contamination with arsenic in Bangladesh and its exposure routes to human and livestock.***

Rice contamination data (Table 7.18) shows that inorganic arsenic comprised about 77% of the total As present in rice grown in Bangladesh compared to 37% in rice from the USA (see Table 7.18). Rice samples analysed data further reveals that raw rice and brown rice had higher total As and higher percentages of inorganic As compared to white and parboiled rice, which may suggest that parboiling and polishing removes As from the rice and that the As is mainly present in the outer husk and bran layer.

Table 7.18: ***Arsenic speciation data for rice*** *[based on various sources and compiled by Heikens 2006].*

Country	Rice type	Total Arsenic (μg/kg, range)	Inorganic arsenic (μg/kg, range)	Percent of inorganic arsenic
Bangladesh	Aman and Boro	0.03-0.31	0.01-0.22	34-94 (mean 77.16%)
China	Unknown	0.180-0.22	0.07-0.164	37-91 (mean 64%)
India	Various	0.03-0.08	0.02-0.05	36-67 (mean 51.5%)
Italy	Various	0.19-0.22	0.061-0.14	53-65 (mean 59%)
Spain	Paella rice and white	0.149-0.17	0.062-0.126	48-85 (mean 66.5%)
Thailand	Various	0.11-0.20	0.06-0.10	44-74 (mean 59%)
USA	White rice, Brown rice and other rice	0.160-0.40	<0.021- 0.271	11-62 (mean 37.4%)

7.7: Arsenic remedial measures – prevention and control

Drinking water : The most important remedial action is prevention of further exposure of arsenic by providing safe drinking water and facilitating to obtain low-arsenic groundwater through accessing safe shallow groundwater or deeper aquifers (greater than 200 m in case of Bangladesh); rain water harvesting; pond-sand-filtration; household chemical treatment; and piped water supply from safe or treated sources (Table 7.19).

Other measures could be the following :

- to help people to distinguish between high-arsenic and low-arsenic water sources, for example, by painting the hand-pumps shallow tube wells with red and green to discriminate between contaminated vs. non contaminated shallow tube wells;
- innovation of new types of treatment technologies to treat arsenic contaminated water;
- effective health education.

Table 7.19: ***Alternative arsenic safe drinking water for Bangladesh*** *[BGS and DPHE 2001; Milton et al. 2006].*

Option	Advantages	Disadvantages	Cost
Rainwater harvesting	- Bangladesh receives heavy rainfall during the rainy season, the water quality of which is reported to be very good. - The average rainfall of Bangladesh is 2200-2,800 mm - Rain water seems to be adequate for cooking and drinking of all populations of As affected areas	-May experience water shortage during the dry season. - Storage tank need to be kept clean. - Water quality of rain water in storage tank should be checked for chemicals and microbiological contaminant levels from time to time.	US$ 150 (storage tank of 3200 L); US$ 10 (500 L earthen tank).
Dug well	- Dug well is usually arsenic free. -Arsenic concentrations in dug well are reported to be low (around 10 μg/l or less) (Bangladesh guideline for arsenic is 50 μg/l). - Dug well can produce water for 20- 30+ families. - Well protected dug well is generally microbiologically free	- Only suitable for certain areas. - Can be installed only during the dry season (prior to monsoon).	US$ 300-500 (installation cost).
Deep tube well	-Deep tube well in Bangladesh is relatively free from arsenic (> 150 m depth). - Arsenic concentrations are usually less than 1 μg/l.	-Deep aquifer does not exit all over the country.	US$ 750 (installation cost).
Pond sand filter	-Popular source of drinking water in coastal areas. - Effectively treats surface water. - Can remove pathogens to a safe or tolerable limit.	- Ponds needs to be kept free from cattle bathing, washing, or fish culture and using chemical fertilizers.	US$ 400-5000 (installation cost)
Three kolshi (pitcher) filter/Sono three kolshi filter	- The Filter uses sand, iron fillings, charcoal and brick chips, which are locally available. - Arsenic removal efficiency is high (produces 40 l/12 hrs). - Can be manufactured at the community level.	- Filter media requires regular cleaning to prevent bacteriological contamination and clogging.	US$6. US$ 1.01 for pitcher (kolshi) replacement, iron fillings and coarse sand.

Soil contamination : According to Brammer (2009), the possible mitigation of arsenic contamination of irrigated paddy soils in South and South-East Asia may include the following : water treatment (e.g. filtration/chemical treatment, coprecipitation with iron); alternative irrigation supply (e.g. use of deep aquifer, river/lake); agronomic methods (e.g. growing of dry land crops such as maize, wheat, millet since dry land crops use much less water than rice therefore less As is added to soils, growing crops on raised beds, or use of arsenic tolerant varieties or growing of rain fed crops); soil amendment (addition of ferric iron, or phosphorous); soil rehabilitation (e.g. growing hyperaccumulator plants, or removal of top soil).

7.8: Conclusion

Arsenic contamination of drinking water is a worldwide problem and some Asian countries are severely affected including Bangladesh, West Bengal of India, China and Vietnam. Millions of people are now at health risk from contamination of drinking water and food with arsenic from groundwater. In addition, there are some potential human health risks related to livestock and fisheries as these can be exposed to arsenic via drinking water, pond water and feeds. According to Ng *et al.* (2003), 60-100 million people globally may be at risk of exposure to excessive levels of arsenic.

Long term use of arsenic contaminated water could result in arsenic accumulation in soils and an increase in arsenic concentrations in soils and crops thereby may result in reduction of soil quality and crop yields. Prolonged irrigation with arsenic contaminated water has already caused rice and vegetables contamination with arsenic in Bangladesh.

A number of measures that may be taken to reduce the risks of arsenic exposure to humans would be :

- Use of rain water (e.g. Bangladesh is one of the wettest regions of the world, with two metres of rainfall a year. Currently rainwater is lost to the ocean or evaporates during the dry season).
- Extraction of drinking and irrigation water from deep tube wells (tube wells >150 m deep found virtually As free in Bangladesh, the main zone of As release in Bangladesh is in the relatively shallow zone and installation of deeper wells therefore may be an option to obtain As free water).
- Use of simple and inexpensive water filtering device such as 'sono filters' to remove As from water collected from shallow tube wells (see http://www.physics.harvard.edu/~wilson/arsenic/remediation/SONO/As%20filtration%20pictures.pdf).
- Promotion of cropping patterns that require less irrigation water.
- Use of arsenic tolerant crop varieties (Brammer 2009)
- Breeding of rice plants that are tolerant to arsenic and have a limited arsenic uptake.
- Use of hyper-accumulating plants such as Chinese brake fern, *Pteris vittata* which are reported to be extremely efficient in extracting arsenic from soils.
- Clean up As contaminated top soils.
- Cooking of rice in excess water or removal of tuber/vegetable skins (Brammer 2009)

7.9: References

Abedin, M. J., Cotter-Howells, J. and Meharg, A. A. 2002. Arsenic uptake and accumulation in rice (*Oryza sativa* L.) irrigated with contaminated water. *Plant Soil*, 240: 311-319.

Ali Ramalli, S. W, Haris, P. I., Harrington, C. F. and Ayub, M. 2005. A survey of arsenic in foodstuffs on sale in the United Kingdom and imported from Bangladesh. *Science of the Total Environment*, 37: 23-30.

Australian and New Zealand Environment and Conservation Council (ANZECC) 1992. Australian Water Quality Guidelines for Fresh and Marine Water Quality. Chapter 1-4.

Australian and New Zealand Environment and Conservation Council (ANZECC) and Agriculture and Resource Management Council of Australia and New Zealand (ARMCANZ) 2000. Australian and New Zealand Guidelines for Fresh and Marine Water Quality, Volume 1. The Guidelines (Chapters 1-7).

Aposhian, H. V. and Aposhian, M. M. 2006. Arsenic toxicology: five questions. *Chem. Res.Toxicol.* 19: 1-15.

Beyersmann, D. and Hartwig, A. 2008. Carcinogenic metal compounds: recent insight into molecular and cellular mechanisms. *Arac. Toxicol.* 82: 493-512.

Brammer, H. 2009. Mitigation of arsenic contamination in irrigated paddy soils in south and south-east Asia. *Environment International* 35 : 856-863.

Brammer, H and Ravenscroft, P. 2009. Review article - Arsenic in groundwater: A threat to sustainable agriculture in south and south-east Asia. *Environment International* 35 : 647-654.

British Geological Survey (BGS) and Directorate of Public Health and Engineering (DPHE) 2001. Arsenic contamination of groundwater of Bangladesh. Volume 2: Final report. D. G. Kinniburgh and P. L. Smedley (eds.). *British Geological Survey Report WC/00/19*. British Geological Survey, Keyworth.

Burgess, W. and Ahmed, K.M. 2006. Arsenic in aquifers of the Bengal basin. Chapter 2 , 31-44. In : Naidu, R., Smith, E, Owens, G., Bhattacharya, P. and Nadebaum, P. (eds.). *Managing arsenic in the environment. From soil to human health*. CSIRO, Victoria, Australia.

Chang, J-S., Yoon, I-H. and Kim, K-W. 2009. Short communication-Heavy metal and arsenic accumulating fern species as potential ecological indicators in As-contaminated abandoned mines. *Ecological Indicators* doi:10.10.16/ j.ecoind.2009.03.011.

Commonwealth of Australia. 1989. *Arsenic and its compounds*. National Occupational Health and Safety Commission. 26 pp.

Commonwealth of Australia. 2005. *Managing arsenic in water supplies*. Interim AusAID guide and operating procedures. Commonwealth of Australia. 22p. http://www.dcita.gov.au/cca

Delgado-Andrade, C., Navarro, M., Lopez, H. and Lopez, M. C. 2003. Determination of total arsenic levels by hydride generation atomic absorption spectrometry in foods from south-east Spain: Estimation of daily dietary intake. *Food Additives and Contaminants* 20: 923-932.

Environment Agency, UK 2002. Soil guideline values for arsenic contamination. Department of Environment, Food and Rural Affairs. The Environment Agency. Bristol, 14 pp. UK. www.environment-agency.gov.uk

FSANZ 2000. Food Standards Australia and New Zealand. http://www.foodstandards.gov.au/newsroom/ publications/annualreport/anzfaannualreport2001254.cfm

Francesconi, K. A. 2007. Toxic metal species and food regulations-making a healthy choice. *The Royal Society of Chemistry-Analyst*. 132: 17-20.

Has-Schon, E., Bogut, I. and Strelec, I. 2006. Heavy metal profile in five fish species included in human diet, domiciled in the end flow of River Neretva (Croatia). *Archives of Environmental Contamination and Toxicology* 50: 545-551.

Heikens, A. 2006. Arsenic contamination of irrigation water, soil and crops in Bangladesh: Risk implications for sustainable agriculture and food safety in Asia. RAP/Publication 2006/20. FAO, Bangkok, Thailand. <http:// www.fao.org/docrep/009/ag105e/ag105e00.htm>

Horswell, J. and Speir, T. 2006. Arsenic phytotoxicity. P. 183-207. In: Naidu *et al.* (eds.). 2006. *Managing arsenic in the environment. From soil to human health*. CSIRO, Victoria, Australia.

Huq, S. M. I., Correl, R. and Naidu, R. 2006a. *Arsenic accumulation in food sources in Bangladesh*. Chapter 15: 283-293 pp. In : Managing arsenic in the environment. From soil to human health. Edited by *Naidu et al.* CSIRO, Victoria, Australia.

Huq, S. M. I., Joarder, J. C., Parvin, S., Correll, R. and Naidu, R. 2006b. Arsenic contamination in food-chain: Transfer of Arsenic into food materials through groundwater irrigation. *J. Health Population and Nutrition*. September 24(3): 305-316. [on line http://www.bioline.org.br/request?hn06039].

International Agency for Research on Cancer (IARC) 2006. Inorganic (IARC). IARC monographs on the evaluation of carcinogenic risks to humans. http://monographs.iarc.fr/ENG/Classification/index.php.

International Program on Chemical Safety (IPCS) 2001. *Arsenic and arsenic compounds*. Geneva. World Health Organisation. International Program on Chemical Safety -Environmental Health Criteria 224. Published under joint sponsorships of UNEP, ILO, WHO, Geneva. http://www.inchem.org/documents/ehc/ehc/ehc224.htm.

Jankong, P., Chalhoub, C., Kienzi, N. B., Goessler, W., Francescon, K. A., and Visoottviseth, P. 2007. Arsenic accumulation and speciation in freshwater fish living in arsenic-contaminated waters. *Environmental Chemistry* 4: 11-17.

Kinniburg, D. G. and Smedley, P. L. 2001. Arsenic contamination of groundwater Bangladesh (4 vols). British Geological Survey, Keyworth.

Langdon, C.J., Piearce, T.G., Mehrag, A.A. and Semple, K.T. 2003. Review- Interactions between earthworms and arsenic in the soil environment : a review. *Environmental Pollution* 124 : 361-373.

Milton, A. H., Smith, W., Dear, K, Cadwell, B, Sim, M. and Ng, J. 2006. *Arsenic mitigation*. Chapter 19. page 355-362. In: Naidu, R., Smith, E, Owens, G., Bhattacharya, P and Nadebaum, P (eds.) 2006. *Managing arsenic in the environment. From soil to human health*. CSIRO, Victoria, Australia.

Naidu, R., Smith, E, Owens, G., Bhattacharya, P. and Nadebaum, P. (eds.) 2006. *Managing arsenic in the environment. From soil to human health*. CSIRO, Victoria, Australia.

Ng, J. C. 2005. Environmental contamination of arsenic and its toxicological impacts on humans. *Environmental. Chemistry* 2: 146-160.

Ng, J.C., Wang, J. and Sharim, A. 2003. A global health problem caused by arsenic from natural sources. *Chemosphere* 52:1353-1359.

NRS (National Residue Survey). 2006. National residue Survey. Annual report 2005-2006. Australian Government Department of Agriculture, Fisheries and Forestry, Meat and Livestock Australia. Canberra, ACT 2601. 186p.

Rosemond, S., Xie, Q. and Liber, K. 2008. Arsenic concentration and speciation in five freshwater fish species from Back Bay near Yellowknief, NT, Canada. *Environmental Monitoring Assessment*. Doi: 10.1007/s 10661-007-0112.

Schultz, E. and Joutti, A. 2007. Arsenic ecotoxicity in soils. Risk assessment and risk management procedure for arsenic in the temperate region. Finish Environment Institute. 53p.

Shah, A. Q., Kazi, T. G., Arain, M. B., Jamali, M. K., Afridi, H. I., Jalbani, N., Baig, J. A and Kandhro, G. A. 2008. Accumulation of arsenic in different freshwater fish species-potential contribution to high arsenic intakes. *Food Chemistry* 112: 520-524.

Sharma, V.K. and Sohn, M. 2009. Review article -Aquatic arsenic: toxicity, speciation, transformation and remediation. *Environment International* doi:10:1016/j.envi.2009.01.005.

Sheppard, S.C. 1992. Summary of phytotoxic levels of soil arsenic. *Water, Air, Soil Pollution*. 64: 539-550.

Smith, A.H. and Steinamaus, C.M. 2009. Health effects of arsenic and chromium in drinking water : recent human findings. *Annual Review of Public Health* 30:9.1-9.16 doi:10.1146/annualreview.pubhealth.031308.100143.

World Health Organization of the United Nations (WHO) 2001. Arsenic in drinking water. Fact sheet 210. http://www.who.int/mediacentre/factsheets/fs210/en/print.htm.

Websites

http://monographs.iarc.fr/ENG/Classification/index.php (International Agency for Research on Cancer)

http://www.bgs.ac.uk/arsenic/bangladesh/reports/Chapstatistics.pdf

http://www.bgs.ac.uk/arsenic/bangladesh/reports/ChapWorldReview.pdf

http://www.ccaresearch.org/tag10/year4.ppt#1 (year 4 research-fate of CCA-treated wood)

http://www.epa.gov/fishadvisories/forum/2004/presentations/wednesday/abernathy.pdf (bioaccumulation of arsenic in fish and toxicity of As species)

http://www.publish.csiro.au/samples/ManagingArsenicSample.pdf

http://www.unicef.org. Arsenic mitigation in Bangladesh

www.sos-Arsenic.net/English/toxic_effect/asinfood.html. (Arsenic in the food chain)

ftp://ftp.fao.org/docrep/fao/009/ag105e/ag105e00.pdf (arsenic contamination of irrigation water; soil crops in Bangladesh: risk implications for sustainable agriculture and food safety in Asia)

http://www.inchem.org/documents/ehc/ehc/ehc224.htm

http://www.greenfacts.org/en/arsenic/arsenic-greenfacts-level2.pdf

http://www.who.int/mediacentre/factsheets/fs210/en/

□□□

8 Chemical Impacts-Heavy Metals

Key Facts

- The heavy metals (HM) or trace metals in an aquatic environment can come from natural (e.g. geological minerals, wind blown silicate dust, volcanic emissions, sea spray and combustion) and anthropogenic sources (e.g. mining, industry and sewage treatment discharges as well as electronic waste and agriculture fertilisers). In aquatic systems, the heavy metals of greatest concerns are cadmium, copper, lead, mercury and zinc.
- Metals may exist in a number of physico-chemical states in aqueous systems such as particulate matter and dissolved forms. Metal speciation in freshwater is strongly influenced by hardness, alkalinity, pH, natural dissolved organic matter (DOM) and redox potential.
- Water hardness is quantitatively related to metal toxicity in freshwater (i.e. toxicity of heavy metals decrease with increasing hardness and alkalinity; on the other hand, toxicity increases with decreasing salinity, dissolved organic matter and chelators). The uptake and toxicity of heavy metals of most metals in marine waters is low as compared to freshwater because metals are easily complexed with chloride ions.
- The toxicity of HMs is related to the dissolved ionic form of the metal rather than total concentration of the metal. Each different physico-chemical forms of an element has a different toxicity, for example, if the element (e.g. Cu) is absorbed on a colloidal particles it is expected to be of little or no effect on aquatic organisms, however, if the copper is available as free Cu (II) ion, few aquatic organisms may survive.
- Metals in aquatic ecosystems are mainly deposited in sediments, which act as a sink and also as a source of heavy metals. Most metals are insoluble in water, soluble in fats and resistant to biological and chemical degradation - properties that result in their accumulation in the tissues of organisms and their biological amplification in food chain and food webs.
- Aquatic flora and fauna such as algae, invertebrates and fish can accumulate HMs many times as higher than the background level. There are three main mechanisms by which bio-accumulation of metals in aquatic organisms can occur: (a) via respiration (gills or skin surface); (b) by adsorption on to the body surface or (c) from ingestion of food. Larger fish such as marlin, shark, sword fish and tuna can accumulate very high concentrations of metals such as mercury.
- Higher levels of cadmium, copper, lead, mercury can affect survival and growth of aquatic organisms including gill damage. The toxic effect of HMs may cause fish population decline as a result of increased susceptibility to diseases or mortality or reduced fecundity.
- A recent study conducted has found a variation in bio-accumulation of heavy metals by different ecophysiological earthworm groups (epigeic, endogeic and anecic) signifying to consider different ecological categories of earthworms when bio-monitoring of pollutants in terrestrial environment.
- Cultivation of crops for human or animal consumption on contaminated soil can potentially lead to uptake and accumulation of trace metals in edible plant parts with a resulting risk to human and animal health.
- Exposure of humans to heavy metals can cause damage to the kidney and nervous system. Lead and cadmium are considered as potential carcinogen. Organomercurials are readily absorbed into fish and can make its way up the food chain through fish. The major toxic effects of methyl mercury (MeHg) are on the nervous system. In addition, the surviving victims and their offspring displayed mental retardation, cerebral palsy, muteness, which was observed in children exposed while in the foetal stage.
- A comparison of global road side dusts data reveals that heavy metal pollutants level in most developed and developing countries have exceeded the recommended soil quality guideline values.
- Biosorbents or biological materials such as algae, bacteria, yeasts and fungi can be used because they are cheaper and more effective alternatives for the removal of metallic elements such as heavy metals from aqueous solution.

Contents

List of Tables

List of Figures

List of Box

Acronyms and Glossary

Acute toxicity: A rapid adverse effect (e.g. death) of exposure to a substance in a living organism.

Anecic: A habitat classification term. These earthworms live in deep vertical burrows, but feed at or near the soil surface, especially at night.

Biosorbents: It can be defined as the removal of metal or metalloid species, compounds and particulates from solution by biological material.

Bio-accumulation: A process by which chemical substances are accumulated by aquatic organisms from water, either directly or through consumption of food containing the chemicals.

Bio-concentration: A process by which there is a net accumulation of a chemical directly from water into aquatic organisms.

Bio-concentration factor (BCF) : A unitless value describing the degree to which a chemical can be concentrated in the tissues of an organism in the aquatic environment.

Cerebral palsy: Cerebral refers to the cerebrum, which is the affected area of the brain. Palsy refers to disorder of movement. Cerebral palsy describes a group of permanent disorders of the development of movement and posture, causing activity limitation that are attributed to nonprogressive disturbances that occurred in the developing foetal or infant brain.

Chelator: The type of co-ordination compound in which a central metal ion is attached by co-valent links to two or more non-metal atoms in the same molecule, called ligands

Chelation: Is the binding or complexation of a bi- or multidentate ligand. These ligands, which are often organic compounds, are called chelants, chelators, chelating agents, or sequestering agent. The ligand forms a chelate complex with the substrate. The term is reserved for complexes in which the metal ion is bound to two or more atoms of the chelating agent.

Colloid: A colloid is a type of mechanical mixture where one substance is dispersed evenly throughout another. Because of this dispersal, some colloids have the appearance of solutions. A colloidal system consists of two separate phases: a dispersed phase (or internal phase) and a continuous phase (or dispersion medium). A colloidal system may be solid, liquid, or gaseous.

Cytosol: The cytosol or intracellular fluid (or cytoplasmic matrix) is the liquid found inside cells. In eukaryotes this liquid is separated by cell membranes from the contents of the organelles suspended in the cytosol, such as the mitochondrial matrix inside the mitochondrion. The entire contents of a eukaryotic cell, minus the contents of the cell nucleus, are referred to as the cytoplasm. In prokaryotes, most of the chemical reactions of metabolism take place in the cytosol, while a few takes place in membranes or in the periplasmic space. In eukaryotes, while many metabolic pathways still occur in the cytosol, others are contained within organelles.

Diffusion: Diffusion is the movement of particles from a high concentration to a low concentrations.

DOM: Dissolved organic matter.

EDTA: EDTA is a widely used abbreviation for the chemical compound ethylenediaminetetraacetic acid. EDTA refers to the chelating agent with the formula $(HO_2CCH_2)_2NCH_2CH_2N(CH_2CO_2H)_2$. This amino acid is widely used to sequester di- and trivalent metal ions (for example Ca^{2+} and Mg^{2+}). EDTA binds to metals via four carboxylate and two amine groups. EDTA forms especially strong complexes with Mn (II), Cu (II), Fe (III), Pb (II) and Co (III).

EC_{50}: Is the concentration of chemical that affects 50% of the population exposed (usually refers to an effect other than death).

Endogeic: These earthworms burrow continuously to form a network of channels - some vertical and some horizontal.

Etiology: The cause of a disease. Aetiology is the preferred spelling in some countries, including the UK, whereas 'etiology' without an 'a' has taken over in the US. The word comes from the Greek 'aitia', cause + 'logos', discourse.

Fulvic acids: High molecular weight organic acids resulting from plant decay.

Gum disease: Illnesses that inflame and, if untreated, can destroy the gum tissues and bones that support and anchor teeth.

Half-life: Time required to reduce by one-half the concentration of a material in a medium (e.g. soil or water) or organism (e.g. fish tissue) by transport, degradation, transformation or depuration.

Hallucination: Hallucination, is a perception in the absence of a stimulus. Hallucinations can occur in any sensory modality – visual, auditory, olfactory, gustatory, tactile, proprioceptive, equilibrioceptive, nociceptive and thermoceptive. A mild form of hallucination is known as a disturbance and can occur in any of the senses above. These may be things like seeing movement in peripheral vision, or hearing faint noises and voices.

Hardness: Hardness is a measure of the concentration of calcium and magnesium ions in water and is frequently expressed as mg/l calcium carbonate.

Hepatopancreas: The hepatopancreas is an organ of the digestive tract of arthropods, gastropods and fish. It provides the functions which in mammals are provided separately by the liver and pancreas.

Humic: Material obtained from organic matter of soils, produced by decomposition of plant or animal matter.

Humic Acids: Humic acids are complex, highly condensed organic matter, with a high molecular weight which results in a deep, rich dark brown colour to these products.

LC_{50}: The concentration that is lethal for 50% of the population exposed.

Lethal: Causing death by direct action, in case of aquatic organisms, it is the cessation of all visible signs of biological activity.

Ligand: A molecule, ion or atom that is attached to the central atom of a co-ordination, a chelate or other complex. May also be called as a complexing agent.

Lipophilicity: High affinity for lipids (fats).

Methylation: The introduction of methyl groups (CH_3) into organic and inorganic compounds.

Methyl mercury: The most common form is CH_3Hg^+, although $(CH_3)Hg$ also occurs. Both are extremely potent toxicants and can lead to secondary poisoning through bio-magnification.

mg/kg: Milligram per kilogram

Muteness: The condition of being unable or unwilling to speak.

NHMRC: National Health and Medical Research Council.

Osmoregulation: The regulation of the movement of water by osmosis into and out of cells; the maintenance of water balance within the body.

Performance indicators: These are the indicators used to assess the risk that a particular issue will occur (they are used in the guideline packages to compare against the trigger values). They are generally median (or mean) concentrations in the ambient water and may be stressor and/or condition indicators.

pH: Value that represent the acidity or alkalinity of an aqueous solution. It is defined as the negative logarithm of the hydrogen ion concentration of the solution.

µg/l: Micrograms per litre

ng/l: Nanograms per litre

Redox potential: An expression of the oxidizing or reducing power of a solution relative to a reference potential. Oxidation = gain of oxygen or loss of electrons and reduction = loss of oxygen (gain of electrons).

Sequestering Is a procedure of isolating different types of physical processes or different particle species by separating them geometrically in additional dimensions of space.

Speciation: The intimate chemical environment of the indicator, that is the compound or ion of which it forms a part.

STPs: Sewerage treatment processes.

Sub-lethal: Involving a stimulus below the level that causes death.

Threshold concentration: A concentration above which some effect (or response) will be produced and below which it will not.

Toxicity: The inherent potential or capacity of a material to cause adverse effects in a living organism.

Trace metals: Trace metals are metals in extremely small quantities, almost at the molecular level, that reside in or are present in animal and plant cells and tissue. They are a necessary part of good nutrition, although they can be toxic if ingested in excess quantities. Trace metals include iron, magnesium, zinc, copper, chromium, nickel, cobalt, vanadium, arsenic, molybdenum, and selenium. Trace metals are sometimes referred to as trace elements, though the latter is a broader category. Trace metals and heavy metals, these two terms are now regarded as synonymous to each other and used in this book.

Trigger values (TV): These are the concentrations (or loads) of the key performance indicators measured for the ecosystem, below which there exists low risk that adverse biological (ecological) effects will occur. They indicate a risk of impact if exceeded and should 'trigger' some action, and either further ecosystems specific investigations or implementation of management/ remedial actions are needed.

Uptake: A process by which materials are absorbed and incorporated into a living organism.

8.1: Introduction

Heavy metals or trace metals are elements with an atomic weight greater than 20 and characterised by similar atomic electron configurations in the outer orbitales. Metals can exist in waters and organisms in a wide variety of chemical forms and in combination with other materials. In normal waters, metals can exist as free metal ions (e.g. Cd, Cu), inorganic complexes, organic complexes and compounds (e.g. methyl mercury CH_3Hg^++ CH_3Hg), and associated with colloidal or particulate matter. The concentration and forms of metals in water can be influenced by pH, Eh (redox potential), water run-off, and hardness etc. In aquatic systems, the heavy metals of greatest concerns are cadmium, copper, lead, mercury and zinc (Table 8.1). These elements are toxic to organisms above specified threshold concentrations but may also be essential (e.g. copper and zinc) for metabolism at lower concentrations.

8.2: Sources of heavy metals

The heavy metals located in an aquatic environment come from natural and anthropogenic sources (Table 8.1). Such sources include geological minerals, wind blown silicate dust, volcanic emissions, sea spray and combustion. Human activities such as mining, industry and sewage treatment discharges as well as electronic waste (computers, printers, photocopy machines, TV sets, mobile phones and toys) and agriculture are some of examples of anthropogenic sources contributing to the elevated levels of heavy metals.

Table 8.1: *Sources of some toxic heavy metals in aqueous systems due to human activities*

Metals	Atomic number	Anthropogenic source
Cadmium (Cd)	48	Phosphate fertilisers; sewage effluents; zinc mine tailings; battery manufacture; industry (electroplating; nickel plating; smelting); corrosion of galvanised pipes; discharge from refineries; natural deposits.
Copper (Cu)	29	Fungicides, algicides, copper mining, antifouling paints, coal and wood combustion; iron and steel production; corrosion of household plumbing systems; natural deposits.
Lead (Pb)	82	Batteries exhaust emissions; lead ore mine wastes, lead smelters, sewage discharges, stormwater run-off; agriculture run-off from fields fertilised with sewage sludge, lead pipes, batteries, lead petrol; corrosion of household plumbing systems; erosion of natural deposits.
Mercury (Hg)	80	Industry, agriculture, caustic soda, pulp and paper production, paint manufacture, fossil fuel combustion; run-off from landfills and croplands; natural deposits.
Zinc (Zn)	30	Metal treatment plants; chemical plants, foundries; mining, urban run-off, waste incineration.

8.3: Physico-chemical properties of metals

8.3.1: Speciation of metals

Metals may exist in a number of physico-chemical states in aqueous systems. These include particulate matter and dissolved forms such as, simple inorganic species, organic complexes and elements adsorbed on a variety of colloidal particles (see Table 8.2). Under normal environmental conditions, majority of Cu, Hg and Pb are associated with suspended particulate (60-80%), while Zn and Cd are likely to remain predominantly in solution (30-40% as particulates).

According to Snodgrass (1980), the following factors affect the solid-phase speciation of metals:

- Direct adsorption onto inorganic clays.
- Adsorption/complexation with organic particulates; detritus or bacteria.

- Association with humic and fulvic acids.
- Ionic exchange with clay colloids.
- Precipitation.

Table 8.2: ***Possible forms of trace element*** *[Florence et al. 1992].*

Physico-chemical forms	Possible examples	Approximate diameter (nm)
Simple ionic species	ZN $(H_2O)_6{}^{2+}$	0.8
Inorganic complexes	$CdCl^+$, $PbCO_3$	1
Differing valency states	Cr (III), Cr(VI)	1
Lipid soluble complexes	CH_3HgCl	1
Organometallic species	$CH_3As(OH)_2$	1
Weak complexes	Cu fluvate	2
Adsorbed on colloidal particles	Cu^{2+}/humic acid/ Fe_2O_3	10-500
Particulate	Metals adsorbed into or contained within clay particles	>450

8.3.2: Factors relating to speciation of metals

Metal speciation in freshwater (salinity = 2.5 ppt) is strongly influenced by a variety of physico-chemical parameters (Table 8.3) including hardness (Ca and Mg concentration), alkalinity, pH, natural dissolved organic matter (DOM) and redox potential. Water hardness is reported to be the only physico-chemical variable that has been quantitatively related to metal toxicity in freshwater. The natural DOM, redox potential and salinity are relevant to estuarine and marine systems.

8.3.3: Bioavailability of metals

The bioavailability (ability of a metal to bind or traverse the cell membrane) of a metal is generally dependent on he physico-chemical form, or speciation of a metal. Furthermore, the bio-availability of metal [Cd, Cr (III), Cu, Ni, Pb, Zn] in freshwater decreases with increasing hardness, with Ca and or Mg competing with metals for surface binding sites on cell membranes (Figure 8.1). A decrease in pH may increase the bio-availability of metals to freshwater organisms, resulting in metal desorption from colloidal and particulate matter, and dissociate some inorganic and organic metal complexes. For example, an increase in proton (H^+) concentration competitively inhibits the binding of the free metal ion (e.g. Cd) at the cell membrane surface and hence, reduces metal (Cd) bio-availability.

Natural DOM (humic and fulvic acids) may reduce metal bio-availability in surface waters by binding metals in metal-organic complexes. Copper generally forms strong complexes with natural DOM and are able to form strong bonds with ligands such as fulvic or humic acids. This is due to the borderline, hard/soft acid nature of the Cu^{2+} ion, which makes it as a versatile electron acceptor (Cu-DOM form strong complexes ranging from 84 to 99% of total Cu in freshwater, whereas Cd-DOM forms weak complexes). Synthetic organic ligands (found in agricultural or industrial effluents) such as oxines or xanthates may complex metals to form lipid-soluble metal complexes and may diffuse directly through the hydrophobic cell membrane (see section 8.3.3 for metal uptake process in biota) transporting both the ligand and metal into the cell. However, such type of lipid soluble complexes represents less than 1% of the total metal concentration in natural waters.

It is generally established that metal bio-availability decreases with increasing salinity due to formation of metal-chloride complexes, which are considered to have a relatively low bio-availability. Further information about bio-availability of metals can be sourced from Markich *et al.* (2001).

Table 8.3: ***Metal speciation and factor for controlling it*** *[Bubb and Lester 1991; Markich et al. 2001].*

Metals	Complexes	Predominant species in water	Factors controlling speciation
Cadmium	Form a wide range of soluble complexes such as with carbonate, sulphate, chloride, hydroxide and with synthetic chelating agents (EDTA and NTA), of which cadmium carbonate and Cd^{2+} are the dominant inorganic forms in river water (pH dependent) 70-90% of Cd in freshwater stream can be present as the free hydrated ion or other labile complexes. Unlike mercury does not form organometallic complexes.	Cd^{2+}	pH*, inorganic complexation, organic complexation.
Copper	90% of Cu in average river water may be complexed by dissolved humic matter.	Cu^{2+} +, (most common); copper hydroxide and carbonate complexes (at higher pH).	pH*, inorganic complexation, organic complexation.
Lead	Main Pb complexes are carbonates (pH 7-9) and lead hydroxides (in potable and good quality river water).	Pb^{2+}, lead hydroxides (pH~6); Lead hydroxide and carbonate species (at higher pH).	pH*, inorganic complexation, organic complexation.
Mercury	In oxygenated surface water, mercuric ion is (Hg^{2+}) predominant. In the absence of organics, mercury exits as hydroxy species in river water [$Hg(OH)O_2$, ($HgOCl_2$) and in seawater as $HgCl_2$	Hg^0, Hg^{2} (mercuryII), Organo-mercurial (mono and dimethyl mercury).	CH_3Hg content, inorganic complexation, organic complexation.
Zinc	In aerobic water, a significant fraction of zinc exists as simple ionic forms such as carbonate, and hydroxyl complexes, hydrated and stable organic compounds. Also adsorbed onto organic and inorganic colloids (e.g. Humics) in particular at higher pH (above pH 7).	Zn^{2+} (predominant form); $ZnOH^+$, Zn $(OH)_2$ (zinc hydroxides at higher pH).	pH*, inorganic complexation, organic complexation.

*Important in freshwaters only

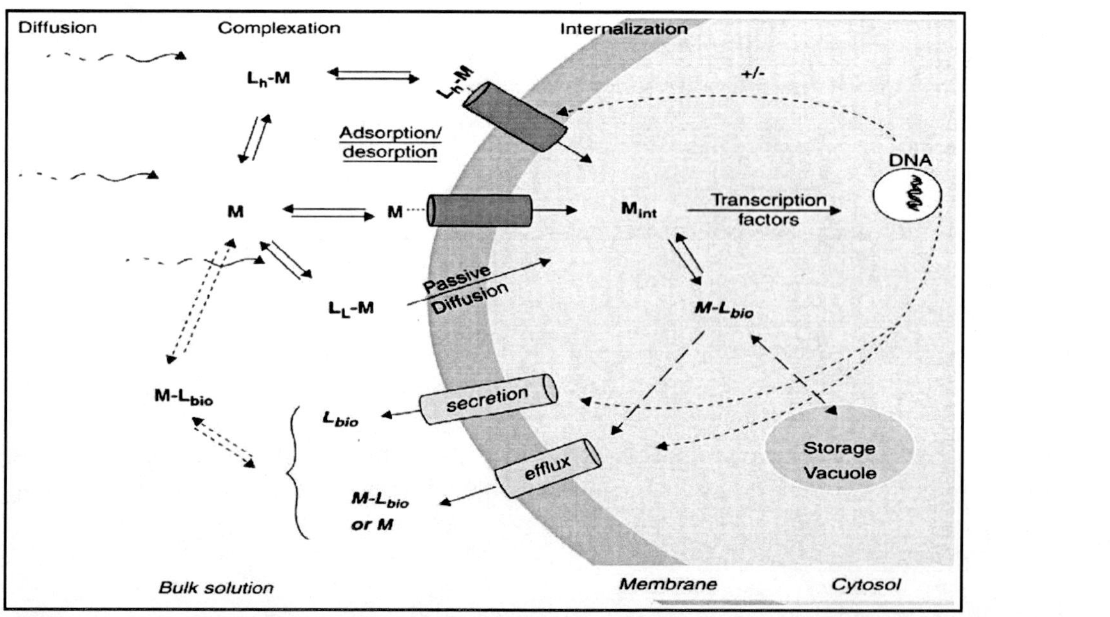

Figure 8.1: ***Conceptual model of some of the important physico-chemical processes leading to and following the uptake of a trace metal by aquatic micro-organisms*** *[Worms et al. 2006; L = ligand that complex the metal (M); subscripts; L = lipophilic; h = hydrophilic; bio = biological; int = internalized].*

8.3.4: Bio-uptake process of metals

The bio-accumulation of trace metals in natural waters is related to their ability to cross biological barriers, such as plasma membrane. According to Worms *et al.* (2006), the bio-uptake of metals includes the following steps:

- *DIFFUSSION*: Trace metals and their complexes must first diffuse from the external medium to the surface of the organism (mass transport).
- *COMPLEXATION*: Metals may dissociate and reassociate (complexation/dissociation) during diffusion to the biological surface.
- *INTERNALIZATION*: Metals must react with a sensitive site on the biological membrane (adsorption/desorption) followed by biological transport (internalization).

The key process that control trace metal uptake is shown in Figure 8.1. Heavy metals ions, such as copper (Cu^{2+}) and cadmium (Cd^{2+}) are readily adsorbed on the cell surface of algae, or by the gill mucous of fish. The adsorbed metals slowly diffuse through the cell membrane to the interior cell, where it may participate in injurious reactions. In general, if the metal complex is lipid soluble, the much more rapid process of direct diffusion can take place. During this diffusion, the ligand is also transported into the cytosol. The fraction of total metal in solution that can be transported across a membrane surface is equivalent to the bio-available or toxic fraction. This in turn is dependent on the relative affinity of the metal for solution ligands and the receptor molecule or the solution-membrane partition coefficient for a lipid-soluble complex (see Figure 8.1).

8.4: Environmental and biological impacts of heavy metals

8.4.1: Bio-accumulation of metals in aquatic organisms [see also Ramamoorthy and Moore 1984]

Most metals are insoluble in water, soluble in fats and resistant to biological and chemical degradation - properties that result in their accumulation in the tissues of organisms and their biological amplification in food chain and food webs. Aquatic flora and fauna such as algae, invertebrates and fish can accumulate heavy metals to several orders of magnitude above background levels in the aquatic environment. When the rate of excretion or loss is slower than the rate of absorption, metal accumulation in organisms will occur in body tissues. Heavy metals can accumulate in algae and macrophytes, in particular, phytoplankton can store metals over a long period of time. Thus zooplankton grazing on phytoplankton can accumulate heavy metals if phytoplankton were enriched. In this way heavy metals can enrich a food chain or a food web. The notable examples of metal accumulation are methyl mercury in fish (Box 8.1), arsenic in lobsters and cadmium, copper, lead, zinc and mercury in oysters and mussels. The other inorganic and organic fish contaminants are also listed in Box 8.1.

Box 8.1: ***Fish contaminants, Minamata disease*** *[Bubb and Lester 1991; http://en.wikipedia.org/wiki/Minamata_disease].*

Fish contaminants

Several organic and inorganic compounds can find their way into fish and seafood. These aspects are covered in this book in the following chapters as folllows :

- *Inorganic chemicals - arsenic (Chapter 7); cadmium, lead, mercury, copper, zinc (see Chapter 8);*
- *Organic compounds - pesticides/insecticides (chlorinated hydrocarbons) (see Chapter 9); Dioxins; polychlorinated biphenyls (PCBs) (see Chapter 11)*
- *Drug residues – hormones (see Chapter 10); antibiotics (see Chapter 12)*
- *Biotoxins – (see Chapter 13)*

Minamata disease

Sometimes referred to as Chrisso-Minamata disease, is a neurological syndrome caused by severe mercury poisoning. The Minamata incident happened in 1950s when inorganic and methylated-mercury were discharged into Minamata Bay by the Chrisso-Corporation chemical factory during acetaldehyde production, which uses mercuric sulphate as a catalyst. Methylation in the sediments with subsequent bio-accumulation and transfer through the food chain caused methyl-mercury accumulation in fish and shellfish, which were subsequently consumed by the local Japanese. This resulted in 21 deaths and 784 officially designated cases of Minamata disease. As of March 2001, 2,265 victims had been officially recognised (1,784 of whom had died) and over 10,000 had received financial compensation from Chrisso. Symptoms of mercury poisoning include ataxia, numbness in the hands and feet, general muscle weakness, narrowing of the field of vision and damage to hearing and speech. In extreme cases insanity, paralysis, coma and death follows within weeks of the onset of symptoms. A congenital form of the disease can also affect foetuses in the womb.

Table 8.4: ***Heavy metal distribution in different tissues of brown tiger prawn (Penaeus esculentus) (mg/kg on a wet-weight basis) caught from Torres Strait*** *[Evans-Illidge 1997].*

Metal	Head	Tail	Shell	Ovary
As MPC = 1.0	18.2	14.06	7.9	20.4
Cd MPC = 0.2	5.8	0.16	0.06	0.24
Cu MPC = 10	54.8	3.9	9.88	3.8
Hg MPC= 0.5	0.02	0.04	0.03	0.01
Pb MPC = 1.5	0.05	0.05	0.06	0.24
Zn MPC = 150	56.5	14.5	4.74	61.9

MPC = maximum permissible concentration based on NHMRC (1986).

There are three main mechanisms by which bio-accumulation of metals in aquatic organisms can occur:

- via respiration (gills and skin surface.)
- by adsorption onto the body surface and,
- from ingestion of foods (solids and dissolved matter).

However, the dominant pathway of accumulation appears to be via food.

It is well known that metal accumulation in aquatic organisms does not occur uniformly across tissues. This may be due to the complex forms of the metals and due to certain organs and tissues performing specific roles in metal uptake, retention and excretion. For example, in crustaceans, such as in prawns, accumulation of metals is highest in the head (due to high metal storage in hepatopancreas), whereas tail flesh contains the lowest concentrations of all metals (except arsenic and mercury) (see Table 8.4). For burrowing and detritus-feeding organisms such as eels, worms and bivalves the majority pathway appears to be through ingestion of sediment. For pelagic fish, exposure of metals would be limited since these fish have minimal contact with bottom sediments.

Molluscs and bivalves (e.g. mussels, oysters) when living in heavily polluted environments can accumulate very high concentrations of heavy metals such as Cd, Pb and Zn. Because of their ability to concentrate heavy metals, bivalves have been used widely to monitor heavy metal levels in the aquatic environment. In fish, most of the metals are accumulated in the kidney and liver and to a lesser extent in the muscles (the part eaten by human). However, methyl-mercury is reported to accumulate in fish tissues at higher concentrations, which may be related with high bio-concentration factors of methyl-mercury for fish (see Table 8.5 for bio-concentration factor in aquatic organisms). Large fish, known for their longevity, such as shark, sword fish, marlin and tuna can accumulate very high concentrations of mercury (see Table 8.6). Methyl-mercury has a half-life of approximately 02 years in fish, which allows larger and older fish (predatory species) to contain higher levels of methyl-mercury. Therefore, consumption of fish contaminated with methyl-mercury is the main source of exposure of humans to methyl-mercury. The notable 'Minamata disease' is a classical example of bio-accumulation and bio-magnification of mercury (Box 8.1).

Table 8.5: ***Bio-concentration factors (BCF) of toxic heavy metals to aquatic organisms*** *[note that highest BCF is with methyl-mercury].*

Metals	Aquatic species	Bio-concentration factors	Source
Cadmium	Mussel	10,000-20,000	Cossa (1988)
Copper	Phytoplankton, zooplankton, macrophytes, macro-invertebrates, fish	100-26,000	Spear and Pierce (1979)
Lead	Invertebrates and fish	499-1,700	USEPA (1985)
Methyl-mercury	Fish	1000,000-100,000,000	ANZECC (2000b)
Zinc	Freshwater animals	50-1,130	USEPA (1987)

Table 8.6: ***Mercury levels in different oily and non-oily fishes (Hg in mg/kg, fresh weight)*** *[Maycock and Benford 2007] [Note: highest Hg levels are in sword fish; marlin and shark]*

Fish	Minimum	Maximum	Median	Mean
Oily fish				
Orange roughy	0.53	0.65	0.60	0.60
Salmon	0.03	0.08	0.05	0.05
Sword fish	0.15	2.7	1.4	1.4
Trout	0.01	0.1	0.05	0.06
Tuna-fresh	0.1	1.5	0.31	0.40
Non-oily fish				
Halibut	0.04	0.62	0.29	0.29
Hoki	0.08	0.31	0.19	0.19
Marlin	0.41	2.2	0.88	1.1
Monk fish	0.1	0.30	0.20	0.20
Pollack	0.007	0.02	0.01	0.01
Sea bass	0.03	0.09	0.07	0.07
Sea bream	0.05	0.06	0.05	0.05
Shark	1.0	2.2	1.4	1.5
Tuna-canned	0.03	0.71	0.15	0.19

Recommended Australian standard for mercury and methyl-mercury in fish = 1 mg/kg) (NRS 2006; In 2003, the Codex Joint Expert Committee on Food Additives (JECFA), administered by FAO and the World Health Organization (WHO), revised the guideline for mercury in fish to 1.6 micrograms of methyl mercury intake per day per kilogram of body weight, nearly half the original guideline of 3.3 micrograms (http://www.fao.org/fishery/topic/14815).

8.4.2: Effects of heavy metals on aquatic organisms

8.4.2.1: Toxicity of heavy metals

Some heavy metals are essential to aquatic life and some are non-essential. Metals such as copper (Cu) and zinc (Zn) are known to be essential, whereas Cd, Hg and Pb are regarded as highly toxic to aquatic life. However, an abundance of both essential and non-essential heavy metal supplies could be hazardous and highly toxic to living organisms.

The toxicity of heavy metals is related to the dissolved ionic form of the metal rather than total concentration of the metal. A number of factors influence the exhibited toxicity, these are chemical forms of the metal (inorganic and organic complexed forms; see section 8.3.1), presence of other metals, salts, environmental factors (temperature, salinity, hardness, dissolved oxygen content) and the condition of organisms (stage of life history, size acclimatisation, etc.). For example, the uptake and toxicity of heavy metals in freshwater organisms has been shown to decrease with increasing water hardness and alkalinity (see Markich *et al.* 2002a and 2002b). On the other hand, toxicity increases with decreasing salinity, dissolved organic matter and chelators. There are a number of tests that can be carried out to assess the toxicity of heavy metals in aquatic systems. These are given in Table 8.7.

Table 8.7: ***Tests to evaluate adverse effects of metals on aquatic ecosystems*** *[Connell 1997].*

1. **Physico-chemical properties**
   - Partitioning coefficient (K_{ow} value)
   - Aqueous solubility
2. **Degradation and accumulation**
   - Biodegradability
   - Fish bio-accumulation
3. **Effects on biotic communities**
   - Growth inhibition test of algae
   - Acute toxicity test of zooplankton
   - Acute toxicity test of fish

8.4.2.2: Factors relating to toxicity of metals

Each different physico-chemical forms of an element has a different toxicity, for example, if the element (e.g. Cu) is absorbed on a colloidal particle, it is expected to be of little or no effect on aquatic organisms, however, if the copper is available as free Cu (II) ion, few aquatic organisms may survive. Toxicity of metals is a function of metal speciation and is controlled by a wide range of physico-chemical factors as listed below (Bubb and Lester 1991):

- pH can increase or decrease toxicity to aquatic organisms (e.g. Hg(II) and toxicity in fish has been shown to decrease with increased pH).
- temperature generally enhances toxicity of metals (increase of water temperature can increase the rate of metabolic rates resulting in enhanced uptake and thereby enhance the toxicity of metals).
- high suspended solid loadings have been shown to reduce metal availability to organisms and to geochemical cycling.
- presence of sediments can reduce toxicity to aquatic organisms (e.g. mercury toxicity and accumulation by fish and other organisms was effectively reduced by sediments).
- high levels of alkalinity and hardness tend to decrease toxicity through complexation and competition with Ca and Mn ions and organic ligands.
- free (hydrated) metal ion is the most toxic and bio-available form.
- some inorganic complexes such as Cu^{2+} and $CuOH^{+}$ were reported to be toxic to fish in solutions with low organic content, whereas copper carbonate complexes appeared to be non-toxic.
- lipid soluble metal complexes are particularly toxic forms of heavy metals because they can diffuse rapidly through a bio-membrane and carry both metal and ligand into the cell (e.g. Copper xanthalates, 8-hydroxyquinolinate and alkyl mercury compounds).

Connell (1993) proposed that the toxicity of heavy metals to aquatic organisms is in the following orders:

	Silver	Tin
	Copper	Aluminium
Mercury >	Zinc >	Nickel
	Cadmium	Iron
	Lead	Chromium
Most toxic ←		→ Least toxic

8.4.2.3: Effect of toxic heavy metals on aquatic organisms

8.4.2.3.1: Cadmium (Cd)

Cadmium is highly toxic and non-essential to plants and animals. The toxicity of cadmium can damage the nervous and reproductive systems and can reduce the ability of aquatic organisms to osmoregulate. Furthermore, higher levels of cadmium can have an affect on the survival, growth, and tissue structure of aquatic species. The toxicity of cadmium is dependent on the hardness of water (see Figure 8.2). For instance, the uptake and toxicity of cadmium in freshwater organisms decreased with increasing water hardness and alkalinity (i.e. toxicity is reduced with increasing levels of calcium and magnesium in the water). The recommended guidelines for growing freshwater and marine organisms is also based on hardness (see Table 8.8). Moreover, cadmium toxicity increases with decreasing salinity. At higher water temperatures, cadmium levels increase and fish survival decreases under low dissolved oxygen conditions. Dissolved organic matter, chelators and zinc in water reduce the toxicity of cadmium to fish. Toxicity values (LC_{50}) of Cd, Cu, Pb, Hg and Zn to marine organisms are summarized in Table 8.9. Cadmium toxicity data of some selected Australian freshwater biota are shown in Table 8.10.

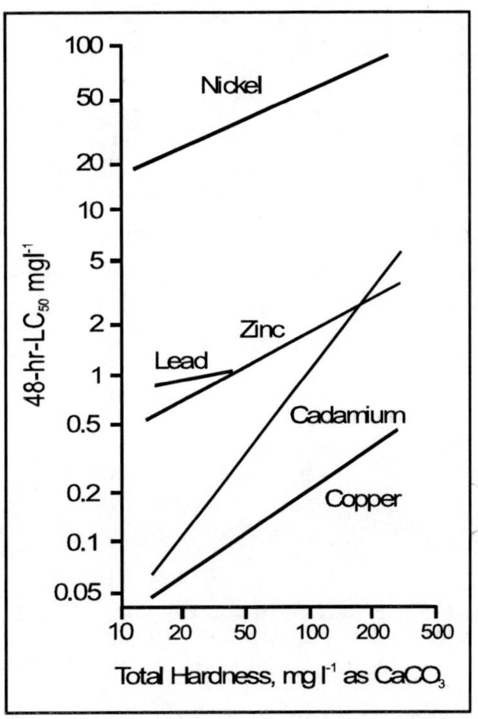

Figure 8.2: ***Correlation between total hardness of water and 48 h LC_{50} for rainbow trout to cadmium, copper, lead, zinc and other metals** [Brown 1968].*

Table 8.8: ***Recommended water quality guidelines for cadmium (aquatic foods production)*** *[ANZECC 2002c; page 9.4-33].*

Group	Guideline (mg/l)	Comments (Hardness as $CaCO_3$/l)
Recommended guidelines for Cd	<0.0002-0.0018 Freshwater	0.0002 at hardness of 0-60 mg/l 0.0008 at hardness of 60-120 mg/l 0.0013 at hardness of 120-180 mg/l 0.0018 at hardness of >180 mg/l
	<0.005-0.0005 salt water	Saltwater 0.005 at hardness of >100 mg/l 0.0005 at hardness of <100 mg/l

Key points for Cd toxicity (ANZECC 2000b)

- Cd toxicity is hardness dependent.
- Cd is adsorbed strongly by suspended material.
- Cd complexes with chloride and results in reduced toxicity at higher salinity.
- Cd has a variable tendency to bio-accumulate but bio-concentration can be significant for bivalves.

Table 8.9: ***Lethal toxicity of metals to some selected marine organisms LC_{50} (mg/l) at 96 h*** *[Connell 2005].*

Class of organisms	Cadmium	Copper	Lead	Mercury	Zinc
Crustaceans	0.015-45	0.17-100	-	0.05-0.5	0.4-50
Fish	25-55	2.5-3.2	188	0.8-0.23	60
Molluscs	2.2-35	0.14-2.3	-	0.058-32	10-50
Polychaetes	2.5-121	0.16-0.5	7.2-20	0.02-0.09	1.8-55

8.4.2.3.2: Copper (Cu)

Free cupric copper ions (Cu^{2+}) are the most toxic form of copper. Copper is usually more toxic to freshwater fish and invertebrates than any other heavy metal except mercury (Figure 8.2). In freshwater, particularly in acidic soft waters with low complexation capacity, copper may be highly toxic. However, Cu is less toxic to marine fish due to the high complexing capacity of salt water. Dissolved organic matter (e.g. humic and fluvic acids) reduces uptake and toxicity of copper in freshwater organisms. Zinc exacerbates toxicity of copper. Copper can cause gill damage, reduced growth and kidney damage of aquatic organisms. Like cadmium, copper toxicity decreases with increasing water hardness and alkalinity and increases as salinity decreases. Similarly, the water quality guideline is dependent on hardness (see Table 8.11). Toxic effects of copper occur when the rate of uptake exceeds the rates of physiological or bio-chemical detoxification and excretion.

Key points for Cu toxicity (ANZECC 2000b)

- Cu toxicity is hardness dependent, i.e. toxicity decreases with increasing hardness and alkalinity.
- Dissolved organic matters in freshwaters are sufficient to remove copper toxicity but often not in soft waters.
- Cu is adsorbed strongly by suspended material.
- Cu toxicity in algae, invertebrates and fish generally increases as salinity decreases.
- Cu can bio-accumulate in aquatic organisms, but being an essential element, it is regulated by the organisms.

Table 8.10: ***Metal toxicity data of some selected Australian freshwater biota (algae, crustaceans, molluscs and fish)*** *[all data compiled by Markich et al. 2002a, Markich et al. 2002b; Khan and Nugegoda 2007]. NR = Not recorded*

Metal	Species	Test end point	Duration of exposure (h)	Temperature (° C)	pH	Conductivity (µS/cm)	Hardness (mg $CaCO_3$/l)	Alkalinity (mg $CaCO_3$/l)	Organic carbon (mg/l)	Water concentration (µg/l)
Cadmium	Duckweed (*Lemna dispema*) (adult)	Food production	7d	22±2	5.0	NR	NR	NR	NR	590 (EC_{50})
Cadmium	Duckweed (*Lemna dispema*) (adult)	Food production	7d	22±2	5.5	NR	NR	NR	NR	260 (EC_{50})
Cadmium	Mollusca Mussel, *Hyridella depressa* (adult)	Valve movement	48	20±1	6.5±0.1	220	38	31	4.2	126 (EC_{50})
Cadmium	Mollusca Mussel, *Hyridella depressa* (adult)	Valve movement	48	20±1	7.0±0.1	220	38	31	4.2	130 (EC_{50})
Cadmium	Crustacea - Shrimp, *Paratya australiensis* (adult)	Survival	96	20±1	6.0	NR	29	12	NR	15 (LC EC_{50})
Cadmium	Crustacea - Shrimp, *Paratya australiensis* (adult)	Survival	96	20±1	7.8	NR	NR			180 (LC_{50})
Cadmium	Crustacea - *Cherax destructor* (yabby)	Survival	96	25						494 (LC_{50})
Cadmium	Fish -Jollytail, *Galaxias maculatua* (adult)	Survival	96	15±1	7.2	NR	25	48	NR	3800 (LC_{50})
Cadmium	Fish -Jollytail, *Galaxias maculatus* (adult)	Survival	96	25±1	7.2	NR	25	48	NR	670 (LC_{50})
Copper	Green alga, *Chlorella sp.*	Population growth	72	27±1	5.7±0.5	23	3.9	3.3	NR	35 (EC_{50})
Copper	Green alga, *Chlorella sp.*	Population growth	72	27±1	6.5±0.5	23	3.9	3.9	NR	1.5 (EC_{50})
Copper	Mollusca –Mussel, *Velesunio ambiguus (adult)*	Valve movement	48	20±1	7±0.1	220	38	31	<0.2	22.8 (EC_{50})
Copper	Mollusca -Mussel, *Velesunio angasi* (adult)	Valve movement	48	20±1	7.5±0.1	220	38	31	11.2	698 (EC_{50})

Contd...

Copper	Crustacea -Crayfish, *Cherax destructor* (69 days old)	Survival	94	25±1	7.5±0.1	150	54	27	NR	1400 (LC_{50})
Copper	Crustacea - Crayfish, *Cherax destructor* (90 days old)	Survival	96	20±1	7.2	NR	5.1	19	NR	480 (LC_{50})
Copper	Crustacea - *Cherax destructor* (yabby)	Survival	96							379 (LC_{50})
Copper	Fish -Chequered rainbow fish *Melanotaenia splendida inomata* (adult)	Survival	96	25±1	6	64	8	4	NR	720 (LC_{50})
Copper	Fish -Chequered rainbow fish *Melanotaenia splendida inomata* (adult)	Survival	96	25±1	7.5	150	54	4	NR	460 (LC_{50})
Lead	Duckweed, *Lemna disperma*	Plant damage	28 days	24±1	6.7	NR	NR	NR	NR	16300 (EC_{50})
Lead	Crustacea-Water flea, *Daphnia carinata*	Survival	48	18±1	6.9±0.3	75	19	NR	NR	170 (LC_{50}); 230 (LC_{50})
Lead	Crustacea -Shrimp, *Paratya australiensis*	Survival	48	18±1	6.9±0.3	75	19	NR	NR	3700 (LC_{50})
Lead	Fish -Rainbow trout, *Oncorhynchus mykiss*	Survival	96	18±1	6.9±0.3	75	19	NR	NR	530 (LC_{50})
Lead	Fish -Chequered rainbow fish *Melanotenia splendida inomata*	Survival	96	25±1	6.0±0.4	64	8	4	11	3200 (LC_{50})
Lead	Fish -Jollytail, *Galaxias maculatus*	Survival	96	18±1	6.9±0.3	75	19	NR	NR	4900 (LC_{50}); 2600 (LC_{50})
Mercury	Duckweed, *Lemna disperma*	Plant damage	28 days	24±1	6.7	NR	NR	NR	NR	1000 (EC_{50})
Mercury	Crustacea -Water flea, *Daphnia carinata*	Survival	48	18±1	6.9±0.3	75	19	NR	NR	2.9 (LC_{50})

Contd...

Mercury	Crustacea -Shrimp, *Paratya australiensis*	Survival	48	18±1	6.9±0.3	75	19	NR	NR	71 (LC_{50})
Mercury	Fish -Rainbow trout, *Oncorhynchus mykiss*	Survival	96	18±1	6.9±0.3	75	19	NR	NR	85 (LC_{50})
Mercury	Fish -Tilapia, *Oreochromis niloticus*	Survival	96	29	7.3±0.2	425	94	245	NR	20 (NOEC) 38 (LC_{50})
Mercury	Fish -Jollytail, *Galaxias maculatus*	Survival	96	18±1	6.9±0.3	75	19	NR	NR	184 (LC_{50}) 17 (LC_{50})
Zinc	Duckweed, *Lemna disperma*	Frond production	7 d	22±2	5.0	NR	NR	NR	NR	14500 (EC_{50})
Zinc	Duckweed, *Lemna disperma*	Frond production	7 d	22±2	5.5	NR	NR	NR	NR	9290 (EC_{50})
Zinc	Crustacea -Water flea, *Daphnia carinata*	Survival	48	18±1	6.9±0.1	75	19	NR	NR	220 (LC_{50}) 240 (LC_{50})
Zinc	Crustacea -Shrimp, *Paratya australiensis*	Survival	48 96	15±1	7.5±0.1	150	54	27	NR	16000 (LC_{50}) 6900 (LC_{50})
Zinc	Fish -Rainbow trout, *Oncorhynchus mykiss*	Survival	48 96	15±1	7.5±0.1	150	54	27	NR	380 (LC_{50}) 340 (LC_{50})
Zinc	Fish -Crimson spotted rainbow fish *Melanotaema fluviatilis*	Survival	96	25±0.5	7.2±0.2	92	20	NR	NR	270 (LC_{50})
Zinc	Fish -Jollytail, *Galaxias maculatu*	Survival	96	15±1	7.5±0.1	150	54	27	NR	9600 (LC_{50})

Table 8.11: *Recommended water quality guidelines for copper (aquatic foods production) [ANZECC 2000c; page 9.4-37]*

Group	Guideline mg/l	Comments (Hardness as $CaCO_3$/l)
Recommended guidelines for Cu	<0.005 Freshwater and saltwater	0.006 at hardness < 100 mg/l 0.03 at hardness >100 mg/l

8.4.2.3.3: Lead (Pb)

High lead levels of Pb can cause damage to gill epithelium and affected fish may die from suffocation. Pb is less toxic than copper, cadmium, zinc and mercury (Figure 8.2). The uptake and toxicity of lead in freshwater organisms generally decreases with increasing water hardness and alkalinity (Figure 8.2, Table 8.12). Both acute and chronic toxicity of lead in several species of freshwater animals was found to be greater in soft water than in hard water. Lead is strongly complexed by dissolved organic matter in natural waters and this complexation is likely to reduce lead toxicity.

Table 8.12: *Recommended water quality guidelines for lead (aquatic foods production) [ANZECC 2000c; page 9.4-41].*

Group	Guideline mg/l	Comments (Hardness as $CaCO_3$/l)
Recommended guidelines	<0.001-0.007 Freshwater and saltwater	<0.001 at hardness of 0-60 mg/l <0.002 at hardness of 60-120 mg/l <0.004 at hardness of 120-180 mg/l <0.007 at hardness of >180 mg/l

Key points for Pb toxicity (ANZECC 2000b):

- Pb toxicity is inversely hardness dependent, i.e. toxicity decrease with increasing hardness and alkalinity.
- Pb is strongly complexed by dissolved organic matter in most natural waters.
- Pb is adsorbed strongly by suspended clay, humic substances and other suspended materials.
- Pb can bio-accumulates in aquatic organisms but is generally not available at sufficient concentrations to cause significant problems.
- Pb speciation in seawater is dominated by chloride complexing. Increasing salinity reduces Pb toxicity.

8.4.2.3.4: Mercury (Hg)

Inorganic mercury has low toxicity to aquatic organisms whereas organic mercury complexes (e.g. methyl mercury) are more toxic and tend to bio-accumulate. Organo mercury compounds are 4 to 31 times more toxic than inorganic mercury (II). Mercury compounds can damage vital tissues and organs including gill, liver, kidney, brain and skin in fish and may reduce reproductive success. In fish, the pre-hatch stage is most susceptible to intoxication. High phosphate and organic carbon can reduce mercury toxicity in freshwater algae. An increase in hardness of fresh water would reduce the toxicity of mercury compounds, whereas in marine environments toxicity usually increases with decreasing salinity. Moreover, the acute toxicity of mercury compounds is inversely related to the dissolved oxygen content of water. The water quality guidelines for Hg for aquatic food production is given in Table 8.13

Table 8.13: ***Recommended water quality guidelines for mercury (aquatic food production)*** *[ANZECC 2000c; page 9.4-43].*

Group	Guideline mg/l	Comments
Recommended guidelines	<0.001 Freshwater and saltwater	In general, for freshwater fish 0.001 mg/l and for marine crustacean <0.0025 mg/l suggested.

Key points for Hg toxicity (ANZECC 2000b):

- Hg occurs in the environment as mercury (II) and organomercurial compounds of which organomercurial compounds are particularly toxic (see Chapter 7, Box 7.1 for a comparison between inorganic/organic arsenic vs. inorganic/organic mercury with reference to toxicity).
- Hg toxicity is hardness dependent.
- Hg is strongly adsorbed by particles and is more often associated with sediments.
- Hg has a strong affinity for chlorine and sulphur-containing ligands, particularly sulphide.
- Toxicity of inorganic mercury in marine environment usually increases with decreasing salinity.

8.4.2.3.5: Zinc (Zn)

Zinc poisoning is similar to copper (i.e. gill damage, reduced growth and kidney damage). It is more toxic in soft water, although fish such as rainbow trout are especially sensitive to zinc toxicity; however the resistance increases with age. Zinc toxicity is hardness dependent, that is, toxicity decreases with increasing hardness (Figure 8.2) and alkalinity. Its uptake and toxicity generally decreases as salinity increases. The water quality guidelines for Zn for aquatic food production are given in Table 8.14.

Table 8.14: ***Recommended water quality guidelines for zinc (aquatic foods production)*** *[ANZECC 2000c; page 9.4-51].*

Group	Guideline mg/l	Comments
Recommended guidelines	<0.005 freshwater and saltwater	USEPA (1993) <0.11 for freshwater <0.086 for saltwater

Key points for Zn toxicity (ANZECC 2000b)

- Like Cu, Zn is an essential element and is required by many aquatic organisms.
- Zn toxicity is both hardness and alkalinity dependent, toxicity decrease with increasing hardness and alkalinity.
- Zn uptake and toxicity generally decreases as salinity increases.
- Levels of dissolved organic matter in freshwaters are sufficient to remove zinc toxicity but often not in soft waters.
- Zn forms complexes with dissolved organic matter (which is pH dependent).

8.4.3: Heavy metals accumulation in terrestrial biota - earthworms

Earthworms constitute upto 80% of the total biomass of the soil fauna and are known to be an efficient accumulator of heavy metals from soils. It is a key indicator for ecosystem health and many studies have been performed on the response of earthworms to metals. Metals have been shown to cause mortality, reduced fertility, cocoon production, cocoon viability and growth. The ecotoxicity data on earthworm can be sourced from the following web sites <http://

contamsites.landcareresearch.co.nz/literature_ecotox_data.htm and http://contamsites.landcarereserrch.co.nz_new_zealand_ecotoxicity_data.htm>. A recent study conducted by Ernst *et al.* (2008) has found a variation in bio-accumulation of heavy metals by different ecophysiological earthworm groups (epigeic, endogeic and anecic). For example, tissue Pb concentrations were found higher with anectic (*Lumbricus terrestris, Aporrectodea longa*) and epigeic earthworms (*Lumbricus rubellus*) which may be related with Pb concentrations in soils and leaf litter. In contrast, highest Cd and Hg concentrations were found in the endogeic species *Octolasion cyaneum* that feed on soil efficiently. Lowest tissue Cd and Hg concentrations were found in epigeic *L. rubellus* which occurs mostly in the top layer of the soil, thereby may have accumulated low concentrations of HM. The study signifies to consider different ecological categories of earthworms when bio-monitoring of pollutants in terrestrial environment (soils).

8.4.4: Heavy metals impact on human health

Cultivation of crops for human or animal consumption on contaminated soil can potentially lead to uptake and accumulation of trace metals in edible plant parts with a possible risk to human and animal health. For example, the food chain (soil-plant-human) pathway is recognised as one of the major pathways for human exposure to metals. It is known that serious and systematic health problems can develop as a result of excessive dietary metals such as cadmium and lead in the human body (Table 8.15). According to Mills *et al.* (2004), the consumption of vegetables and crops is the main source of cadmium for humans. In several places around the world (e.g. China, Mexico, and Pakistan) cropland is irrigated with wastewater, therefore there is risk of transfer of cadmium from crops to humans in these countries via wastewater irrigation. For example, rice has the ability to accumulate soil cadmium in grains (Peralta-Videa *et al.* 2009). Ma (1996) reported that the main way of exposure of humans and mammals to lead is via the food chain. High accumulation of heavy metals (Cd, Pb) in wheat grains grown in the soil dressed with sewage sludge has been reported from Pakistan (Jamali *et al.* 2009). Studies conducted in Tarragona, Spain found high lead content in edible vegetables (onions and parsley) (Capdevila *et al.* 2003).

Mercury contamination through organomercurial is more harmful than mercury itself (see section 8.4.2.3.4 and Table 8.6). Organomercurial are readily absorbed into fish, therefore organic mercury in the form of methyl and ethyl mercury makes its way up the food chain through fish (Peralta-Videa *et al.* 2009). The major toxic effects of methyl mercury (MeHg) are on the nervous system, the documented cases of mercury poisoning in Japan resulted many deaths (see Box 8.1). Additionally, the surviving victims and their offspring displayed mental retardation, cerebral palsy, muteness, which was observed in children exposed while in the foetal stage.

Though zinc and copper are essential elements, however, their excessive concentration in food and feed plants are of great concern because of their toxicity to humans and animals. Lead and cadmium are considered as potential carcinogen and are associated with etiology of a number of diseases (Table 8.15), especially cardiovascular, kidney, nervous system, blood as well as bone diseases. (Zhuang *et al.* 2008). It is reported that the soil and vegetables polluted with lead and cadmium in Copsa Mica and Baia Mere (Romania), significantly contributed to decreased human life expectancy within the affected areas, reducing the average age at death by 9-10 years (Lăcătusu *et al.* 1996). High concentration of heavy metals in the soil, fruit and vegetables caused the high prevalence of upper gastrointestinal cancer rates in the Van region in Turkey (Türkdoðan *et al.* 2002).

Table 8.15: ***Heavy metal's effect on human health*** *[source WHO (1996); NHMRC (2004); Peralta-Videa et al. (2009); http://www.lentech.com/periodic-chart-elements].*

Heavy metals	Potential human health effects (from high exposure of metals)	WHO guideline values	Australian drinking water quality (health value) (NHMRC 2004)
Cadmium (Cd)	Kidney damage, bone fracture, reproductive failure and infertility; damage to central nervous and immune system; possible DNA damage or lung cancer development; acts as estrogens in breast cancer; alterations in uterus, ovaries and oviduct; excessive risk of cardiovascular mortality; endocrine disruption	0.003 mg/l	0.002 mg/l
Copper (Cu)	Long term exposure to copper may cause irritation of nose, mouth and eyes, headaches, stomach-aches, dizziness, vomiting, and diarrhoea; high uptake of copper may cause liver and kidney damage.	2 mg/l	2 mg/l
Lead (Pb)	A rise in blood pressure; anaemia; kidney damage; miscarriage and subtle abortions; brain damage, declined fertility of men through sperm damage; damages central nervous system, excretory system, circulatory system, and cardiovascular system	0.01 mg/l	0.01 mg/l
Mercury (Hg)	Disruption of the nervous system; damage to brain functions; DNA damage and chromosomal damage; allergic reactions, resulting in skin rashes, tiredness and headaches; sperm damage, birth defects and miscarriages; kidney damage; gum disease, delusions and hallucinations.	0.001 mg/l	0.001 mg/l
Zinc (Zn)	Though essential for human health but too much zinc can still cause some health problems such as stomach cramps, skin irritations, vomiting, nausea and anaemia; damage the pancreas, respiratory disorders.	-	< 3 mg/l

8.5: Heavy metal pollution and water and food quality guidelines

Heavy metals can pollute water bodies through natural deposits, waste discharged from mining, industrial and agricultural activities (see Table 8.1). Any water contaminated with heavy metals could be unsuitable for human drinking, livestock drinking, irrigation, aquatic ecosystems protection, recreation and aquaculture. Therefore, monitoring of heavy metal pollutants in the aquatic environment is essential to safeguard the health of biota and humans.

An aquatic ecosystem is considered to be polluted with heavy metals if metal levels are several times higher than background (see Table 8.16). To protect an ecosystem from heavy metal pollution ANZECC (2000a) suggested some guideline trigger values (Table 8.17). The trigger values are the concentrations of the key performance indicators measured for the ecosystem below which there exists a low risk to the ecosystem. Trigger levels are listed at 90% (majority disturbed systems), 95% (moderately disturbed systems) and 99% (undisturbed systems) species protection.

The metals in the aquatic system are mainly deposited in sediment. Therefore, sediments act as a major source and as a sink of heavy metals. This represents a source of bio-available metals to benthic biota, leading potentially to metal bio-accumulation in the aquatic food chain. Many studies have found direct relationships between metal levels in sediments and the biota. Therefore, an ecosystem that has a high conservation and ecological value and thus needs protection should have the metal levels in sediments not exceeding the guideline values (see Table 8.18).

Table 8.16: ***Concentration of heavy metals in unpolluted and polluted aquatic environment.*** *[Note: Table 8.16 should be read in conjunction with Table 8.17 - sediment quality guidelines].*

Metals	Sediments (dry weight)	Freshwater	Ocean Water	Reference
Cadmium	• Unpolluted marine: 0.01 mg/kg • Polluted freshwater: >3000 mg/kg • Polluted marine: 50 mg/kg	• Natural: 10-100 ng/l • Polluted: 450-17,000 ng/l	• Unpolluted: 0.02-0.12µg/l	Bruland *et al.* (1979); Ramamoorthy and Moore (1984); Bryan and Langton (1992).
Copper	• Unpolluted marine and freshwater: <20 mg/kg • Polluted freshwater and marine: 100-2000 mg/kg	• Uncontaminated: 0.5-1.0 µg/l • Polluted with mines: 50-2000 µg/l	• Unpolluted: <0.1µg/l • Polluted: >15µg/l	Tyler and Buckney (1973); Bruland and Franks (1983); Ramamoorthy and Moore (1984).
Lead	• Natural freshwater: 20 mg/kg • Unpolluted marine: <4 mg/kg • Polluted freshwater: 295-3700 • Polluted marine: 200-11,400 mg/kg	• Uncontaminated: <3 µg/l • Polluted: 20-89µg/l	• Unpolluted : 0.0014-0.014µg/l	Flegal and Patterson (1983); Forstner and Wittman (1983); Ramamoorthy and Moore (1984).
Mercury	• Unpolluted freshwater: 0.08-0.26 mg/kg • Unpolluted marine: 0.03 mg/kg • Polluted freshwater: upto 7.65 mg/kg • polluted marine: 0.6-630 mg/kg	• Unpolluted: 0.02-0.1 µg/l • Background levels in lakes-methyl mercury: 0.05 ng/l • Polluted lakes- methyl mercury: 0.5-2 ng/l • Polluted: upto 0.5µg/l	• Unpolluted: <0.001-0.004 µg/l • Polluted: upto 1.09 µg/l	Olafsson (1983); ANZECC (2000b).
Zinc	• Uncontaminated freshwater: <50 mg/kg • Natural marine: 6 mg/kg • Polluted freshwater: 81-3800 mg/kg • Polluted marine water from industry and mines: 60-20,000	• Unpolluted: 0.5-15 µg/l • Polluted: 100-3000 µg/l	• <1 µg/l (unpolluted)	Bruland *et al.* (1979); Ramamoorthy and Moore (1984).

Table 8.17: ***Water quality guidelines trigger values (TV, µg/l) for selected heavy metals*** *[ANZECC 2000a[1]; ANZECC 1992[2] and NHMRC 2004[3]; FW = freshwater; SW = saltwater; values in blue shade are the TV applied to typical slightly and moderately disturbed systems; ID = insufficient data to derive a TV; LTV = long term TV; STV = short term TV].*

Heavy metal	Drinking water (health value)[3] (µg/l)	Raw waters for drinking town supply[2] (µg/l)	Aquatic ecosystems protection trigger values (99% or 95% or 90% protection µg/l)[1] FW = freshwater; MW = marine water	Recreation[2] (µg/l)	Irrigation[2] (µg/l)	Livestock[2] (µg/l)	Aquaculture[1] (µg/l)	Potential sources of contamination
Cadmium (Cd) (varies with hardness)	2	5	FW: 99% -0.06; **95% -0.2**; 90% -0.4 MW: **99% -0.7**; 95% -5.5; 90%-14	5	10-LTV 50-STV	10	<0.2-1.8 FW[1] <0.5-5 SW[1]	Naturally occurring (secondary) Industry (primary) Phosphate fertilisers
Chromium (Cr)	50	50	FW: Insufficient data MW: 99% -0.7; 95% -5.5; 90%-14	50	100-LTV 50-STV	1,000-	<20 FW; <20 SW	Naturally occurring (primary) Industry (primary)
Copper (Cu) (varies with hardness)	2,000		FW: 99% -1.0; **95% -1.4**; 90%-1.8 MW: 99% -0.3; **95% -1.3**; 90%-3	1,000	200-LTV 5000-STV	400 (Sheep); 1,000 (cattle) 5,000 (pigs); 5,000 (poultry)	<5 FW[1]; <5 SW[1]	Naturally occurring (secondary) Industry (secondary) Human settlements (secondary) Agricultural activities (use of fungicides)
Lead (Pb) (varies with hardness)	10	50	FW: 99% -1.0; **95% -3.4**; 90%-5.6 MW: 99% -2.2; **95% -4.4**; 90%-6.6	50	2000-LTV 5000-STV	100	<1-7 FW <1-7 SW	Naturally occurring (secondary) Industry (secondary) Human settlements (secondary)
Mercury (Hg) (inorganic)	1	1	FW: **99% -0.06**; 95% -0.6; 90%-1.9 MW: **99% -0.1**; 95% -0.4; 90%-0.7	1	2-LTV 2-STV	2	<1 FW <1 SW	Naturally occurring (secondary) Industry (primary)
Mercury (Hg) (methyl)			FW: 99% -ID; 95% -ID; 90%-ID MW: 99% -ID; 95% -ID; 90%-ID					
Zinc (Zn)	ID but aesthetic value is 3,000		FW: 99% -2.4; **95% -8.0**; 90%-15 MW: 99% -7; **95% -15**; 90%-23	5,000	2,000-LTV 5,000-STV	20,000	<5 FW <5 SW	Naturally occurring (primary) Industry (primary)

Trigger values (TV) for aquatic ecosystems protection:

- These are the concentrations (loads) of the key performance indicators measured for the ecosystem below which there exists a low risk and adverse biological (ecological) effects will occur. It indicates a risk of impact if exceeded and should 'trigger' some action such as, some ecosystem specific investigations/implementation of management /remedial actions;
- The trigger values protection levels signifies the percentage of species to be protected;
- In most cases 95% protection level trigger values should apply to ecosystems (bold highlighted) that could be classified as slightly-moderately disturbed (highlighted);
- An ecosystem which is very sensitive and requires high conservation, the trigger value indicated at 99% protection level should be used.

Table 8.18: ***Recommended sediment quality guidelines (mg/kg dry weight)*** *[ANZECC 2000a].*

Contaminant	ISQG-low trigger value	ISQG-high trigger value
Cadmium	1.5	10
Copper	65	270
Lead	50	220
Mercury	0.15	1
Zinc	200	410

ISQG = interim sediment quality guideline

As mentioned in section 8.4.4, the food chain pathway is one of the major pathways for human exposure to metals. The maximum permissible concentrations for lead, cadmium and mercury in food are given in Table 8.19.

Table 8.19 : ***Maximum permissible concentrations (MPCs) for food set by the European Communities for lead cadmium and mercury (mg/kg wet weight)*** *[Fancesconi 2007].*

Food types	Lead	Cadmium	Mercury
Horticulture			
Vegetables	0.1-0.3	0.05,0.1 or 0.2	-
Onion –Australian standard	0.1	-	-
Wheat grain	0.2	0.1	-
Fruits	0.2	0.05	-
Meat			
Meats (various farm animals)	0.1	0.1	-
Offal (including liver and kidney various farm animals)	0.5	0.5 or 1.0	-
Seafood			
Crustaceans (muscle)	0.5	0.5	0.5
Bivalve molluscs	1.5	1.0	0.5
Fisheries products (including molluscs, crustaceans, and most fish species)	-	-	0.5
Fish muscle (particular species)	0.2 or 0.4	0.005, 0.1 or 0.3	1.0
Fish-yellowfin tuna	0.1	No limit	1

8.6: Case studies - Heavy metal monitoring in world's terrestrial and aquatic environment

8.6.1: Global trends of heavy metal pollutants concentration in road dust

[based on Wang et al. 1998; Ahmed et al. 2007; Faiz et al. 2009].

Road dusts can contain high levels of heavy metals originating from atmospheric aerosols or as a result of different human activities such as vehicles on road (e.g. vehicles exhaust particles, lubricating oil residues, tire wear particles), weathered street surface particles, industries (e.g. building construction and renovation, corrosion of galvanised metal structures, heating systems). The most common heavy metals released from vehicles on road comprised Cd, Cu, Pb, Ni, and Zn, of which many of them become bound to the surface of soil and road side dust. For example, Pb and Ni are dominant source from exhaust emissions; whereas Cu and Zn are due to brake wear (break wear could contribute 90% of the road traffic Cu emissions). Intake of dust (via dust ingestion, dermal contact or breathing) containing high concentration of heavy metals can be deleterious to human health including children.

Street dusts are also a potential source of pollution in terrestrial and aquatic environments. A comparison of global road side dusts data reveals that heavy metal pollutants level in most developed and developing countries (Table 8.20) have exceeded the recommended soil quality guideline values and shows the following trends:

- **Cadmium** concentration ranged between 0.3 and 72 mg/kg with highest level reported from Bahrain and Jordan (Amman). In general, Cd concentration in most places (31 places out of 35) was below the recommended guideline values of 3 mg/kg (see Table 8.20).
- **Copper** concentration ranged between 11.3 and 1075 mg/kg with highest level (Cu >200 mg/kg) reported from France (Paris), USA (Ohio), UK (London) and China (Hong Kong); 22 places had exceeded the recommended soil quality guideline values of Cu (50 mg/kg) (see Table 8.20).
- **Lead** concentration ranged between 11.2 and 1,927 mg/kg with very high level (Pb >300 mg/kg) reported from 12 places including Spain (1,927 mg/kg), France (1,450 g/kg), Jordan (210-1,131 mg/kg), Bahrain (697.2 mg/kg), USA (649.7 mg/kg), Italy (544 mg/kg), India (536 mg/kg), Turkey (105-556 mg/kg). 24 places had exceeded the recommended soil quality guideline values of Pb (100 mg/kg) (see Table 8.20).
- **Nickel** concentration ranged between 4.2 and 177 with very high level (Ni >100 mg/kg) reported from USA (Hawaii), UK (Coventry) and Bahrain. Only 8 places had exceeded the recommended soil quality guideline values of Ni (50 mg/kg) (Table 8.20).
- **Zinc** concentration ranged between 13.1 and 3,840 mg/kg with very high level (Zn >500 mg/kg) reported from 8 places including China/Hong Kong (3,840 mg/kg), France (840 mg/l), China (Shanghai) (733.8 mg/kg), China (Baoji) (715.1 mg/kg), China (Guangzhou) (586 mg/kg), UK (Manchester) (653 mg/kg), UK (Birmingham) (534 mg/kg); 17 places had exceeded the recommended soil quality guideline values of Pb (300 mg/kg) (see Table 8.20).
- The global comparison showed that Bangladesh (Dhaka), Jordan (Aqaba, Karak), Turkey (Tokat) had the least road dust pollution (Table 8.20).

8.6.2: Water quality with respect to heavy metals in the 'Water Tower of Asia' Tibetan Plateau [based on Huang et al. 2008].

The water used by 85% of the Asian population originates in Tibetan Plateau (1.22 million km^2). The Plateau is called 'Water Tower of Asia' because it is the source of the seven greatest rivers in Asia, the water resources of which serves one third of the world's population (see Table 8.21). The water quality of the Plateau is considered to be pristine being located in remote parts of the world, however, impact of two of the largest developing nations (China and India which are situated on both sides of the Plateau) can cause long range transport and deposit of air pollutants, such as heavy metals on the Plateau that might impact the water quality. An investigation was carried out during spring 2006 across 42 sampling sites of the four major Tibetan rivers (Salween, Mekong, Yangtze and Yarlung Tsangpo) for various elements including cadmium, copper, lead, mercury and zinc,. The results of this study found negligible amounts of Cd, Cu, Hg, Zn in rivers, though elevated concentration of Pb was found in a tributary to the Salween River (Table 8.21). Future mining activities in the region may increase heavy metals input into the rivers since China's largest copper deposit belt (6.5 million tonnes of reserves) has been discovered recently in the area located between the Mekong River and the Yangtze River watersheds.

Table 8.20: ***Average Global trend in heavy metal concentrations (mg/kg) in road dusts. The data were compared with soil guideline values*** *[based on Wang et al. 1998; Ahmed et al. 2007; Faiz et al. 2009]*

[Note: Highlighted box indicate places where soil guideline value have exceeded; soil guideline values and background soil data (Bangladesh, Pakistan and Asian paddy soils) are also included at the bottom of the table for comparison].

Country (City) & Publication year	Cadmium (Cd)	Copper (Cu)	Lead (Pb)	Nickel (Ni)	Zinc (Zn)
Angola (Luanda) 2005	1.15	41.78	351.3	10	316.6
Bahrain 1993	72	-	697.2	126	151.8
Bangladesh (Dhaka) 2007	-	47	75	25	152
Canada (Ottawa) 2000	0.37	65.84	39.05	15.2	112.5
China (Baoji), 2009	-	123.2	408.4	48.8	715.1
China (Beijing), 2007	1.67	42	126	72	167
China (Guangzhou) 2006	2.41	176	240	23	586
China (Hong Kong) 1996	-	296	652	-	2305
China (Hong Kong) 2001	2.18	24.8	93.4	-	168
China (Hong Kong) 2003	-	110	120	28.6	3840
China (Shanghai) 2008	1.23	196.8	294.9	83.98	733.8
China (Urumqi) 2003	1.17	94.54	53.53	43.28	294.47
China (Xi'an) 2006	-	94.98	230.5	-	421.5
France (Paris) 2001	1.7	1075	1450	25	840
India (Calcutta), 1999	3.12	44	536	42	159
Italy (Palermo) 2003	1.1	98	544	14	207
Jordan (Amman) 2007	3.1-11.2	66.5-350	210-1131	43-88	166-410
Jordan (Aqba-Shuna) 2004	5	-	79	40	79
Jordan (Aqaba) 2007	1.9-2.9	21-56	93-212	51-115	103-160
Jordon (Karak) 2004	-	11.3	11.2	4.2	13.1
Korea (Taejon) 1998	-	47-57	52-60	-	172-214
Mexico (Hermosillo) 2007	4.25	26.34	36.15	4.70	387.98
Norway (Oslo) 1997	1.4	123	180	41	412
Pakistan (Islamabad) 2009	5	52	104	23	116
Spain (Madrid) 1997	-	188	1927	44	476
Sweden 2006	0.32	79	53	12	220
Turkey (Istanbul) 2003	1.5-2.3	49-234	105-556	30-33	447-594
Turkey (Kayseri) 2006	2.53	36.9	74.8	44.9	112
Turkey (Sivas) 2003	2.6	84	197	68	206
Turkey (Tokat) 2003	1.63	38	53.63	54	294.9
UK (Coventry) 2003	0.9	226.4	47.1	129.7	385.7
UK (London) 1995	-	300	897	NA	1,866
UK (Manchester) 2003	-	113	265	-	653
USA (Cincinnati/Ohio) 1998	-	253	649.7	-	-
USA (Honolulu/Hawaii) 2000	-	167	106	177	434
Soil guideline values					
Acceptable values in common soil (Sezgin *et al.* 2003)	3	50	100	50	300
Swedish soil guideline values (mg/kg dw) (OIS 1998)	-	40	50	50	150
Background soil data					
Asian paddy soils (average)	-	33	-	22	66
Bangladesh soils (average)	-	27	23.50	22	68
Pakistan soils (Islamabad) (average)	5.95	25	62.5	32	91.05

Table 8.21: *Water quality (heavy metals levels range) in the Water Tower of Asia-Tibetan Plateau* **(mg/l)** *[Huang et al. 2008].*

Rivers	Cadmium (Cd)	Copper (Cu)	Lead (Pb)	Mercury (Hg)	Zinc (Zn)	Comments
Salween River	<0.002-<0.002	<0.01-1.23	<0.10-781	<0.001-<0.001	0.54-18.2	Northern Tibetan Plateau draining through Thailand and Myanmar; 325,00 km^2 total catchment; 2nd longest river in S.E. Asia; 3200 km long and annual run-off of 70 billion m^3
Mekong River	<0.002-<0.002	0.86-2.20	2.52-15.2	<0.001-<0.001	0.83-3.56	Headwaters of the Tibetan Plateau and draining through China's Yunnan province, Myanmar, Thailand, Cambodia, Laos and Vietnam; 810,00 km^2 total catchment; 12th longest river in the world, the sevenths longest in Asia, 4,500 km long and annual run-off of 74 billion m^3
Yangtze River	<0.02-0.06	<0.01-14.6	<0.10-27.6	<0.001-<0.001	0.68-4.83	The river starts from the Geladandong glaciers (6,621 m above sea level) in the eastern part of the Tibetan Plateau and drains south-eastwards, then eastwards and ends up in the East China Sea. One third of China's population lives within Yangtze River basin, which covers nearly 20% of China's land surface; 1.8 million km^2 total catchment; longest river in the Asia and third longest in the world after the Nile and the Amazon; 6,300 km long and annual run-off of 975 billion m^3
Yarlung Tsangpo River (Brahmmaputra or Big river)	<0.02-<0.002	<0.01-12.6	<0.10-130	<0.001-<0.001	<0.02-29.2	The river starts from the Geladandong glaciers (6621 m above sea level) in the eastern part of the Tibetan Plateau and drains south-eastwards, then eastwards and ends up in the East China Sea. One third of China's population lives with in Yangtze River basin, which covers nearly 20% of Chinas land surface; 630,000 km^2 total catchment; longest river in the Asia and third longest in the world after the Nile and the Amazon; 1,200 km long and annual runoff of 165 billion m^3
CNEQS Guideline values (WHO)	1 3	10 2000	10 10	0.00005 0.001	50 n	

CNEQS = Chinese national environmental quality standards; WHO = World Health Organisation guideline for drinking water quality; n = no health based guidelines value is proposed.

8.6.3: *Heavy metal inputs in Australian waters* [*see also NRS 2006*]

Urban drainage, mining and industrial discharges (smelters, power stations, paper mills, port facilities, sewage treatment works, oil refineries and other chemical and industrial manufacturing plants) contribute a significant quantity of heavy metals to the waters of Australian Bays and estuaries such as, Port Phillip Bay (Victoria) and the Sydney estuary (NSW). Port Phillip Bay receives significant amount of heavy metals originating from rivers, creeks, drains and sewage treatment Plant (Table 8.22). The Yarra River in Melbourne is the greatest single contributor of heavy metal pollution into the bay, which may possibly be the cause of relative high concentrations of cadmium and mercury in fish and shell-fish found in Port Phillip Bay (Fabris *et al.* 1999). On the other hand, copper, lead and zinc are among the main pollutants of urban run-off experienced in the Sydney Estuary.

Table 8.22: *Estimated annual load (t) of heavy metal pollutants into Port Phillip Bay, Melbourne, Australia* [*Harris et al. 1996*].

Metal	Yarra River	Rivers, Creeks, Drains (excluding Yarra river)	Western treatment Plant	Ground water
As	1.4	0.9	-	0.07-0.09
Cd	0.17	0.14	0.02	0.2-0.3
Cu	6.6	8.0	2.7	0.5-0.7
Hg	0.04	0.04	0.02	-
Pb	19	11	1.1	1.0-1.4
Zn	63	58	5.8	1.8-2.4

Mercury contamination of sediments and biota is widespread in Victorian streams and lakes, in particular, where gold mining activities have occurred. High concentrations of mercury have been reported in sediments and biota of the Gippsland's Lakes (concentrations were higher in sediments than any reported values of the world). It is believed that during times of flood, sediments containing mercury are scoured from the riverbeds and swept downstream to the lakes. Gold mining activities in the Upper Goulburn River in Victoria caused high mercury accumulation in trout and red-fin fishes and in stream sediments. Investigations conducted during early 1980s, reported that mercury concentrations in fish from Port Phillip Bay (Victoria) and Townsville's (Queensland) coastal waters exceeded NHMRC guideline value of 0.5 μg/g (wet-weight). Elevated concentrations of other metals have in the past detected in fish, oysters and seagrasses nearby smelters, refineries and other heavy industries. It is also known that the Derwent River in Tasmania has been heavily polluted by mercury, cadmium; lead and zinc, which originated from the earlier operation of a Zinc refinery plant (see Table 8.23).

A compiled dissolved heavy metal monitoring data (1978 -1992) from estuaries of New South Wales, Victoria, South Australia, Tasmania and Queensland reveals that metal concentrations in waters were below the recommended ANZECC guideline values with the exception of Port Pirie (SA) (see site 8 of Table 8.23), where the effluents from the nearby lead-zinc smelter may have caused higher zinc levels. In addition, total metal concentrations was found to be higher at sites received high particulate load such as Derwent River, Tasmania and Ross River, Queensland (see site 10 and 11 of Table 8.23).

Table 8.23: ***Dissolved[1] heavy metal concentrations (μg/l) in some Australian estuarine waters (data compiled by Batley, 1995 based on published information between 1978-1991*** [*see Bately 1995*].

Estuary	Cadmium (Cd)	Copper (Cu)	Lead (Pb)	Mercury (Hg)	Zinc (Zn)
ANZECC (2000) guidelines for marine water	*0.7*	*1.3*	*4*	*0.1*	*15*
1.Victoria - Central Port Phillip Bay	<0.05	0.6	<0.8	<0.002	<2
2. Victoria - Corio Bay	0.2	1.1	<0.8	<0.002	<2

Contd...

3. New South Wales - Port Hacking Estuary	0.2	0.5	0.4	-	-
4. New South Wales - North Lake Macquarie	1.9	1.5	1.6	-	5.2
5. New South Wales - South Lake Macquarie	0.2	1.2	0.1	-	1.0
6. New South Wales - Lake Munmorah	0.1	1.5	0.2	0.02	2.7
7. South Australia - Port Augusta	0.37	0.45	0.54	-	<10
8. South Australia – Port Pirie (offshore)	0.32	0.25	5.1	-	47
9. Tasmania – Macquarie Harbour	0.03	7	-		
10. Tasmania – Derwent River	0.05 (0.51)[2]	1.2 (7.3)[2]	0.23 (4.6)[2]	0.034 (3.83)[2]	3.4 (40)
11. Queensland – Ross River	0.07 (0.20)[2]	0.24 (1.21)[2]	0.11 (0.59)[2]	-	-

[1]filtered through a 0.45 μm membrane filter; [2]numbers in parenthesis are total metals.

Data from Australian estuarine sediments are given in Table 8.24. It shows that heavy metal concentrations are within the ANZECC sediment quality guidelines with the exception of some harbour and industrial sites (site 10,11, 12 and 13 of Table 8.24, i.e. Sydney Harbour, Port Kembla and Lake Macquarie North).

Table 8.24: ***Trace metal concentrations (μg/g) in some Australian estuarine sediments)*** *[data compiled by Batley (1995) based on published information between 1978-1992].*

Estuary and depth	Cadmium (Cd)	Copper (Cu)	Lead (Pb)	Zinc (Zn)
ANZECC (2000) guidelines for sediments quality: low TV	*1.5*	*65*	*50*	*200*
ANZECC (2000) guidelines for sediments quality: high TV	*10*	*270*	*220*	*410*
1. Victoria - Corio Bay offshore (surface)	0.1–13	2-50	2-210	4-440
2. Victoria - Corio Bay (surface)	0.2-9	4-35	14-100	14-166
3. Victoria - Port Phillip Bay near shore (surface)	0.8	1.5	8	21
4. Victoria - Port Phillip Bay offshore (surface)	2	8	22	40
5. Victoria – Port Phillip Bay near Werribee Treatment Complex (surface)	<5	<5-<75	<20-140	9-300
6. New South Wales - Lake Munnmorah - 5 cm	-	70	40	150
7. New South Wales - Lake Munnmorah - 60 cm	-	50	20	50
8. New South Wales - Tuggerah Lakes – 5 cm	-	20	40	110
9. New South Wales - Tuggerah Lakes – 70 cm	-	30	30	80
10. New South Wales - Lake Macquarie North – 55 cm	160	170	1200	2400
11. New South Wales - Lake Macquarie South	4	20	68	150
12. New South Wales – Blackwattle Bay, Sydney Harbour – 10 cm	3	180	520	1150
13. New South Wales - Port Kembla Harbour -10 cm	2	113	113	380
14. New South Wales - Quibray Bay, Botany Bay -10 cm	0.5	3	10	25
15. Queensland - Cleveland Bay -10 cm	-	15-70	<5-53	24-460
16. Queensland - Saunders Bay -surface	-	1-3	<0.5-5	6-18

TV = trigger value

8.7: Conclusion

Metal pollution can harm aquatic organisms through lethal and sub-lethal effects and can reduce or eliminate species from an ecosystem through increased susceptibility to fish disease, mortality and decreased fecundity. Aquatic flora and fauna such as invertebrates (oyster, mussels) and fish can bio-accumulate (thousands to million times) heavy metals thus human and other predators at the top of the food chain can be at risk from eating heavy metal contaminated sea foods. Irrigation water may transport dissolved heavy metals to agricultural fields and they accumulate in soils. The heavy metal transfer from soils to plants is a key pathway of human exposure to metal contamination. Human exposure to heavy metals may cause kidney damage, reproductive failure and infertility, damage to nervous and immune systems. Therefore, monitoring of heavy metal pollutants in the aquatic environment is essential to safeguard the health of biota and humans.

Heavy metals can adversely affect aquatic organisms and the ecosystem in general via the following ways as summarised below:

- Inhibition of growth and reproduction.
- Inhibition of settlement of marine invertebrate larvae.
- Interference with reproduction, metabolism and behaviour.
- Decrease in population numbers.
- Changes in behaviour.
- Reduced fecundity.
- Spinal fracture.
- Inhibition of normal functioning of gill such as ion regulating mechanisms of gills.
- Death.

In recent years, application of biotechnology in controlling and removing metal pollution has received much attention. Biosorbents such as bacteria, fungi and algae (see Table 8.25 for biosorption capacity of bacteria, fungi, and algae) can be used to remove heavy metals from aquatic ecosystems because they are cheaper and more effective alternatives for the removal of metallic elements such as heavy metals from aqueous solution. These biosorbents posses' metal-sequestering property and can decrease the concentration of heavy metal ions from ppm to ppb level (Wang and Chen 2009).

Table 8.25: ***Biosoprtion of metals by different species of bacteria, fungi (mg/g) and algae (nmol/g) tested*** *[see Wang and Chen 2009 for further details]*

Bio-group	Metal ions	Species tested	Biosorption capacity (range)
Bacteria	Cd	*Streptomyces pimprina; S. rimosus*	30.4-64.9
	Cd	*Aeromonas caviae*	155.3
	Cd	*Enterobacter* spp.	46.2
	Cd	*Pseudomonas putida; P. aeruginosa; Pseudomonas* spp.	8-278
	Cd	*Staphylococcus xylosus*	250
	Cu	*Bacillus* spp.; *B. subtilis*	16.3-20.8
	Cu	*Pseudomonas stutzeri; P. aeruginosa; P. cepacia; P. putida*	22.9-96.9
	Cu	*Enterobacter* spp.	32.5
	Cu	*Micrococcus luteus*	33.5
	Cu	*Sphaerotilus natans*	5.4-60

Contd...

	Cu	*Streptomyces coelicolor*	66.7
	Cu	*Thiobacillus ferrooxidans*	39.8
	Pb	*Bacillus* spp.; *B. firmus*	92.3-467
	Pb	*Pseudomonas aeruginosa; P. putida*	0.7-270.4
	Pb	*Streptomyces rimosus*	30-135
	Pb	*Corynebacterium glutamicum*	567.7
	Pb	*Enterobacter* spp.	50.9
	Zn	*Bacillus firmus*	418
	Zn	*Pseudomonas putida*	6.9-17
	Zn	*Streptomyces rimosus*	30-80
	Zn	*Thiobacillus ferrooxidans*	82.6-172.4
	Zn	*Streptoverticillium cinnamoneum*	21.3
Fungi	Cd	*Penicillium spinulosum; Penicillium* spp.; *P. digitatum; P. notatum; P. canescens; P.chrysogenum*	0.4-210.2
	Cd	*Aspergillus niger; A. terreus; A. oryzae*	1.31-3.43
	Cu	*Penicillium italicum; P. griseofulvum; P.chrysogenum; Penicillium spp.; P. notatum; P. spinulosum; P. simplicissimum*	0.4-207.68
	Cu	*Aspergillus niger; Aspergillus carbonarius; A. flavus; A. fumigatus; A. awamori; A. oryzae; A. terreus*	0.75-224
	Hg	*Penicillium canescens; P. purpurogenum*	54.8-70.4
	Hg^{2+}; CH_3Hg^+	*Aspergillus niger*	
	Pb	*Penicillum digitatum; P. canescens; P.chrysogenum; P. purpurogenum; P. simplicissimum; Penicillium spp.*	5.5-298.01
	Pb	*Aspergillus niger*	2.25-93
	Zn	*Penicillium spinulosum; Penicillium italicum; P. notatum; P.chrysogenum*	0.2-25.5
	Zn	*Aspergillus niger; A. nidulans; A. oryzae*	
Algae	Cd	*Codium fragil (G)e; Chaetomorphia linum (G); Chlorella vulgaris (G); Coralline officinalis (R); Fucus vesiculosus (B); Gracilaria edulis (R); Gracilaria salicornia (R); Padina* spp.*(B);Padina tetrastomatica (B); Porphira columbina (B); Sargassum* spp. *(B); Sargassum baccularia (B); S. natans (B); S.siliquosum; Ascophyllium nodosum (B)*	0.0827-1.913
	Cu	*Chlorella vulgaris (G); Chlorella miniata (G); Scenedesmus obliquus (G); Sargasum* spp (B).	0.254-1.08
	Pb	*Galaxaura marginate (R); Gracilaria corticata (R); Padina gymnospora (B); Padina tetrastomatica (B); Polysiphonia violacea (R); Cladophora glomerate (G); Ascophyllium nodosum (G); Chondrus crispus (R); Sargassum fluitans (B); S. hystrix (B); S. natans (B); S. vulgare; Ulva lactuca (G); Codium taylori (G); Chlorella vulgaris (G); Fucus vesiculosus (B)*	0.121-2.896

B=brown alga; G=green alga; R-red alga

8.8: References

Ahmed, F., Bibi, M. H. and Ishiga, H. 2007. Environmental assessment of Dhaka city (Bangladesh) based on trace metal contents in road dusts. *Environmental Geology* 51: 975-985.

Australian and New Zealand Environment and Conservation Council (ANZECC) 1992. Australian Water Quality Guidelines for Fresh and Marine Water Quality. Chapter 1-4.

Australian and New Zealand Environment and Conservation Council (ANZECC) and Agriculture and Resource Management Council of Australia and New Zealand (ARMCANZ), 2000a. Australian and New Zealand Guidelines for Fresh and Marine Water Quality, Volume 1, The Guidelines (Chapters 1-7). http://www.mincos.gov.au/publications/Australian_and_New_Zealand_guidelines_for_fresh and_marine_water_quality.

Australian and New Zealand Environment and Conservation Council (ANZECC) and Agriculture and Resource Management Council of Australia and New Zealand (ARMCANZ), 2000b. *Australian and New Zealand Guidelines for Fresh and Marine Water Quality*, Volume 2, Aquatic ecosystems - rationale and background information. National Water Quality Management Strategy. (Chapter 8). http://www.mincos.gov.au/publications/australian_and_new_zealand_guidelines_for_fresh_and_marine_water_quality/volume_2.

Australian and New Zealand Environment and Conservation Council (ANZECC) and Agriculture and Resource Management Council of Australia and New Zealand (ARMCANZ), 2000c. Australian and New Zealand Guidelines for Fresh and Marine Water Quality, Volume 3, Primary Industries - Rationale and Background Information. http://www.mincos.gov.au/publications/australian_and_new_zealand_guidelines_for_fresh_and_marine_ water_quality/volume_3.

Bately, G. E. 1995. Heavy metals and tributyltin in Australian coastal and estuarine waters. The State of the Marine Environment Report for Australia: Pollution - Technical Annex 2. Dept. of the Environment, Water, Heritage and the Arts. Ocean Rescue 2000 Program.

http://www.environment.gov.au/coasts/publications/somer/annex2/batley.html.

Brown, V. M. 1968. The calculation of the acute toxicity of mixtures of poisons to rainbow trout. *Water Research*. 2: 723-733.

Bruland, K. W., Franks, R. D., Knauer, G. A., and Martin, J. H. 1979. Sampling and analytical methods for the determination of copper, cadmium, zinc and nickel at the nanogram per litre level in seawater. *Anal. Chim. Acta*. 105: 233-245.

Bruland, K. W. and Franks, R. P. 1983. *Mn, Ni, Cu, Zn and Cd in the Western North Atlantic*. In Trace Metals in Sea Water. Ed. C. S. Wong, E. Boyle, K. W. Bruland, J. D. Burton and E. D. Goldberg. Plenum Press. New York and London, pp. 395-414.

Bryan, G. W. and Langston, W. J. 1992. Bioavailability, accumulation and effects of heavy metals in sediments with special reference to United Kingdom estuaries: a review. *Environmental Pollution*. 76: 89-131.

Bubb, J. M. and Lester, J. N. 1991. The impact of heavy metals on low land rivers and the implications for man and the environment. *The Science of the Total Environment* 100: 207-233.

Capdevila, F., Nadal, M., Schuhmacher, M. and Domingo, J. L. 2003. Intake of lead and cadmium from edible vegetables cultivated in Tarragona province, Spain. *Trace Element Electr*. 20: 256-261.

Connell, D. W. 1993. *Water pollution - Causes and effects in Australia and New Zealand*. Third Edition. University of Queensland Press, St Lucia, Queensland, Australia.

Connell, D.W. 1997. *Basic Concepts of Environmental Chemistry*. Lewis Publishers, Boca Raton, Florida.

Connell, D. W. 2005. *Basic concepts of environmental chemistry*. 2nd edition. Taylor and Francis Group, LLC. CRC Press. Boca Raton, copper, cadmium, zinc and nickel at the nanogram per litre level in seawater. *Anal. Chim. Acta*. 105: 233-245.

Cossa, D. 1988. Cadmium in *Mytilus* spp.; Worldwide survey and relationship between seawater and mussel content. *Marine Environmental Research* 26: 265-284.

Ernst, G., Zimmermann, S., Christie, P. and Frey, B. 2008. Mercury, cadmium and lead concentrations in different ecophysiological groups of earthworms in forest soils. *Environmental Pollution* 156 : 1304-1313.

Evans-Illidge, E. 1997. *Heavy metals in commercial prawns and crayfish species in the Torres Strait*. Commonwealth Coastal Action Program. Great Barrier Reef Marine Park Authority; 111 pp.

Fabris, G. J., Monahan and Batley, G. E. 1999. Heavy metals in waters and sediment of Port Phillip Bay. Marine and Freshwater Research 50: 503-513.

Faiz, Y., Tufail, M., Javed, M.T., Chaudhary, M.M. and Siddique, N. 2009. Road dust pollution of Cd, Cu, Ni, Pb and Zn along Islamabad Expressway, Pakistan. *Microchemical Journal*. Doi:101016:/j.microc.2009.03.009.

Flegal, A. R and Patterson, C. C. 1983. Vertical concentration profiles of lead in the central pacific at 15° N and 20° S. Earth Planet Science Letter 64: 19-32.

Florence, T. M., Morrison, G. M. and Stauber, J. L. 1992. Determination of trace element speciation and the role of speciation in aquatic toxicity. *The Science of the Total Environment* 125: 1-13. Florida, USA. 462 pp.

Förstner, U. and Wittmann, G. T. W. 1983. Metal pollution in the aquatic environment. Springer-Verlag, Berlin.

Francesconi, K. A. 2007. Toxic metal species and food regulations-making a healthy choice. *The Royal Society of Chemistry-Analyst*. 132 : 17-20.

Harris, G., Batley, G., Fox, D., Hall, D., Jernakoff, P., Molloy, R., Murray, A., Newell, B., Parslow, J., Skyring, G., Walker, S. 1996. *Port Phillip Bay Environmental Study*. Final Report, CSIRO: Canberra, Australia.

Huang, X., Sillanpää, M., Duo, B. and Gjessing, E.T. 2008. Water quality in the Tibetan Plateau : Metal contents of four selected rivers. *Environmental Pollution* 156: 270-277.

Jamali, M.K., Kazi, T.G., Arain, M.B., Afridi, H.I., Jalbani, N., Kandhro, G.A., Shah, A.Q. and Baig, J.A. 2009. Heavy metal accumulation in different varieties of wheat (*Triticum aestivum* L.) grown in soil amended with domestic sewage sludge. *Journal of Hazardous Materials* 164: 1386-1391.

Khan, S. and Nugegoda, D. 2007. Sensitivity of juvenile crayfish *Cherex destructor* (Decapoda :Parastacidae) to trace metals. *Ecotoxicology and Environmental Safety* 68: 463-469.

Lãcãtusu, R., Rãutã, C., Cârstea and Ghelase, I. 1996. Soil-plant-man relationships in heavy metal polluted areas in Romania. *Applied Geochemistry* 11: 105-107.

Ma, W-C. 1996. Lead in mammals. In: Byer, W.N, Heniz, G., Redmon-Norwwod, A.W (eds.). *Environmental contaminants in wildlife.* SETAC special publication series. Boca Raton, Florida: CRC Lewis:1996. p.281-296.

Markich, S. J., Brown, P. L., Batley, G. E., Apte, S. C. and Stauber, J. L. 2001. Incorporating metal speciation and bio-availability into water quality guidelines for protecting aquatic ecosystems. *Australasian Journal of Ecotoxicology* 7: 109-122.

Markich, S. J., Warne, M. S. T. J., Westbury, A-M. and Roberts, C. J. 2002a. A compilation of data on the toxicity of chemicals to species in Australasia. *Australasian Journal of Ecotoxicology*. 8 (1): 1-72.

Markich, S. J., Warne, M. S. T. J., Westbury, A-M. and Roberts, C. J. 2002b. A compilation of data on the toxicity of chemicals to species in Australasia. *Australasian Journal of Ecotoxicology*. 8 (1), part 2: 73-138.

Maycock, B. J. and Benford, D. J. 2007. Risk assessment of dietary exposure to methylmercury in fish in the UK. *Human and Experimental Toxicology* 26: 185-190.

Mills, P. Ramsey, M.H. and John, E.A. 2004. Heterogeneity of cadmium concentration in soil as a source of uncertainty in plant uptake and its implications for human health risk assessment. *Science of the Total Environment* 326:49-53.

National Health and Medical Research Council (NHMRC) 1986. *National Food Standard A12*: Metals and Contaminants in Food. National Health and Medical Research Council, Canberra.

National Health and Medical Research Council (NHMRC) 2004. Australian drinking water guidelines 6. National Health and Medical Research Council (NHMRC), EH 19.

National Residue Survey (NRS) 2006. National Residue Survey Annual Report 2006-2007. Australian Government Department of Agriculture, Fisheries and Forestry, Canberra. 218 pp.

OIS. 1998. Swiss ordinance relating to impacts on the soil : 1st July 1998. SR 814.12. Switzerland.

Olfasson, J. 1983. Mercury concentrations in the North Atlantic in relation to cadmium, aluminium, and oceanographic parameters. In: C. S. Wong, E. Boyle and K. W. (eds.). *Trace Metals in Sea Water*.

Peralta-Videa, J.R., Lopez, M.L., Narayan, M., Saupe, G. and Gardea-Torresdey, J. 2009. Review-The biochemistry of environmental heavy metal uptake by plants: implications for the food chain. *The International Journal of Biochemistry and Cell Biology* doi:10.1016/j.biocel.2009.03.005.

Ramamoorthy, S. and Moore, J. W. 1984. *Heavy metals in natural waters.* Applied Monitoring and impact assessment. Springer-Verlag, New York Inc., U.S.A.

Sezgin,N., Ozcan, H.K., Demir, G., Nemlioglu,S. and Bayat. C. 2003. Determination of heavy metal concentrations in street dust in Istanbul E-5 highway. *Environment International* 29: 979-985.

Sondgrass, W.J. 1980.*Distribution and behaviour of nickel in the aquatic environment.* In : J.O. Nriagu (ed.). Nickel in the Environment. John Wiley and Sons, New York, Chapter 9.pp 203-274.

Spear, P. A. and Pierce, R. C. 1979. Copper in the aquatic environment: Chemistry, distribution and toxicology. *National Research Council of Canada.* NRCC #17589. Ottawa, Canada.

Türkdoðan, M.K., Kilicel, F., Kara, K., Tuncer, I. and Uygan, I. 2002. Heavy metals in soil, vegetables, and fruits in the endemic upper gastrointestinal cancer region of Turkey. *Environmental Toxicology and Pharmacology* 13: 175-179.

Tyler, P. A. and Buckney, R. T. 1973. Pollution of a Tasmanian river by mine effluents. I. Chemical evidence. Internationale Revue der Gesamten *Hydrobiologie* 58: 873-883.

USEPA 1985. *Ambient water quality criteria for lead-1984.* EPA-440/5-84-027. United States Environment Protection Agency, Washington DC.

USEPA 1987. *Ambient water quality criteria for zinc-1987.* EPA-440/5-87-003. Criteria and standard division, US Environment Protection Agency, Washington, DC.

USEPA 1993 Methods for measuring the acute toxicity of effluents and receiving waters to freshwater and marine

organisms. EPA/600/4-90/027F. August 1993.

Wang, J. and Chang, C. 2009. Research review paper-Biosorbents for heavy metals removal and their future. *Biotechnology Advances* 27: 195-226.

Wang, W. H., Wong, M. H., Leharne, S. and Fisher, B. 1998. Fractionation and biotoxicity of heavy metals in urban dusts collected from Hong Kong and London. *Environmental Geochemistry and Health* 20: 185-198.

WHO. 1996. Guidelines for drinking water quality, 2nd edition. Vol. 2 *Health criteria and other supporting information,* (pp. 940-949) and Addendum to Vol. 2. 1998 (pp. 281-283), Geneva, World Health Organisation. http://www.who.int/water_sanitation_health/GDWQ/summary_tables/sumtab.htm.

Worms, I., Simon, D. F., Hassler, C. S. and Wilkinson, K. J. 2006. Bioavailability of trace metals to aquatic organisms: importance of chemical, biological and physical processes on biouptake. *Biochimie.* 88: 1721-1731.

Zhuang, P., McBride, M.B., Xia, H., Li, Ningyu, L. and Li, Z. 2009. Health risk from heavy metals via consumption of food crops in the vicinity of Dabaoshan mine, South China. *Science of Total Environment* 407: 1551-1561.

Website

http://h2osparc.wq.ncsu.edu/info/hmetals.html

http://www.lennteh.com/Periodic-chart-elements/Hg/Cd/Zn/Pb-en.htm

http://www.nhmrc.gov.au/publications/synopses/files/adwg_11_06_fact_sheets.pdf.

□□□

9 Chemical Impacts - Pesticides

Key facts

- Pesticides are chemicals that are broadly used to protect crops, livestock and other animals and plants from pests and diseases. They are classified according to the type of pest to be controlled (algicides, herbicides, fungicides and insecticides).
- Commercial production of DDT began in 1943. At that time, DDT was considered to be a wonderful invention. It was cheap to produce, very toxic to insects and much less toxic to mammals and its usage resulted in a sharp decline of mosquito borne diseases. In 1962, it was discovered that DDT was causing egg-shell thinning in bird eggs and thus was leading to the near extinction of bird species such as peregrine falcons and bald eagles.
- Pesticides use has greatly helped to increase and improve the world food production. Use of pesticides also increases farmer's profits by preventing crop losses due to pest attacks and weed infestation.
- Pesticides have been grouped by their mode of action to allow pesticides resistance to be managed. Use of pesticides with the same modes of action contributes to the development of resistance problems by killing the susceptible pests and leaving only those with resistance to the class of pesticides that work through similar mechanisms. Therefore, careful selection of pesticides from different mode of action groups is essential to achieve effective chemical control of weeds and insects. Grouping of pesticides define mode of actions and attributes generalised structure and properties.
- World pesticides production has exceeded 2 billion kilograms in 2000, of which herbicides account for the largest proportion of global pesticide use. Overall, North America, Western Europe and Asia are the major users of pesticides. China and India are the largest producers and users of pesticides in Asia.
- A number of banned or restricted pesticides are still being used in some Asian countries, such as DDT, HCH, methamidophos, monocrotophos etc.
- Among the pesticides, carbamates, organophosphates, synthetic pyrethroids are more toxic to aquatic invertebrates and fish than herbicides, but their toxicity varies from one species to another. Synthetic pyrethroids generally low toxic to mammals and birds but are highly toxic to fish and invertebrates.
- Pesticides applied in the fields to control pest plants and animal can contaminate air, land and water and may cause adverse effects on non-target organisms including human, livestock, wildlife and environment (e.g. fish kills; reproductive failure in birds and illness in humans and possible endocrine disruption and killing of pollinators such as bees).
- Pesticide residues can enter into human body via food chains or drinking water. The tendency of a pesticide to accumulate in the tissues of living organisms is generally related with log K_{ow} of pesticides (log K_{ow} values of 3 to 7). Pesticide chemicals, as residue, if consumed by humans in excess than the recommended maximum residue limits (MRL) can be hazardous.
- Water contaminated with pesticides can be unfit for human consumption, irrigation, food processing, livestock drinking, recreational use, aquatic ecosystems protection and aquaculture.
- Long term exposure of pesticides to humans may cause cancer and other tumours; brain and nervous system damages; birth defects; miscarriages, infertility, male sterility and other reproductive problems. It may cause damage to liver, kidneys, lungs and other body organs.
- Residue surveys (1999-2007) conducted in Asia (as part of global POPs contamination monitoring) found that DDTs in human breast milk from China, Vietnam, Cambodia and Malaysia were relatively higher than those from other Asian developing countries and developed countries.

- As hydrophobic pesticide residues can be transferred via food chains to humans, an annual residue monitoring program of agri-produce (e.g. crops, fish, vegetable and fruit) would help to protect the health of people and production of high quality food for domestic and export markets. Monitoring of residues further helps to monitor the use of currently registered pesticides and whether pesticides are used according to good agriculture (aquaculture and mariculture as well) practices. Furthermore, monitoring of surface and groundwater for high risk pesticide residue and comparing those results with guideline reference threshold values would help to define environmental impacts.

Contents

List of Tables

List of Figures

List of Boxes

Acronyms and Glossary

Acaricides: Pesticides that kill mites.

Acetylcholinesterase inhibitor or anti-cholinesterase: A chemical that inhibits the cholinesterase enzyme from breaking down acetylcholine, so increasing both the level and duration of action of the neurotransmitter acetylcholine.

Agent orange : 2,4,5-T was widely used as a defoliant by the USA in the Vietnam war in the 1960s, and combined with 2,4-D, this spray was named Agent Orange. A defoliant is any chemical sprayed or dusted on plants to cause its leaves to fall off. Defoliants differ from herbicides in that the former seeks mainly to strip leaves from plants, and the latter is used to destroy or inhibit the growth of plants. More than 21,000,000 US gallons (79,000,000 L) of Agent Orange were sprayed across South Vietnam. 4.8 million Vietnamese people were exposed to Agent Orange, resulting in 400,000 deaths and disabilities, and 500,000 children born with birth defects.

Anuran: Frogs and toads. Frogs are amphibians in the order Anura (meaning "tail-less", from Greek *an-*, without + *oura*, tail). Frogs and toads are generally tailless, with long hind legs modified for hopping. From taxonomic perspective, all members of the order Anura are frogs, but only members of the family Bufonidae are considered "true toads". The use of the term "frog" in common name usually refers to species that are aquatic or semi-aquatic with smooth and/or moist skins, and the term "toad" generally refers to species that tend to be terrestrial with dry, warty skin. An exception is the fire-bellied toad (*Bombina bombina*): while its skin is slightly warty, it prefers a watery habitat.

ATPases: Are a class of enzymes that catalyze the decomposition of adenosine triphosphate (ATP) into adenosine diphosphate (ADP) and a free phosphate ion.

Cholinesterase inhibition or acetylcholinesterase: An enzyme that hydrolyses the neurotransmitter 'acetylcholine' to choline and acetate. Organophosphorous insecticides act by inhibiting the action of cholinesterase.

Concentration: The amount of active ingredient or pesticide equivalent in a quantity of diluent.

Contaminant: Any biological, chemical, physical and radiological substance or matter that has an adverse response (effect) on air, water, soil or living things.

Cytochromes: Any of a class of iron-containing proteins important in cell respiration as catalysts of oxidation-reduction reactions.

EC_{50} or Median effective concentration: The concentration of material in water that is estimated to be effective in producing some lethal response in 50% of the test organisms.

Enzymes: Enzymes are biomolecules that catalyze (*i.e.*, increase the rates of) chemical reactions. Almost all enzymes are proteins.

GABA: Gamma aminobutyric acid: an amino acid, that is found in the central nervous system and acts as an inhibitory neurotransmitter.

Half-life: The amount of time it takes for half of a substance or entity to undergo some specified process. For example, the half-life of a radioactive substance is the amount of time it takes for half of its atoms to decay, and the half-life of a drug is the amount of time it takes before half of the active elements are either eliminated or broken down by the body. Depending on the particular substance involved, a half-life may be significantly shorter or longer than half of the substance's full period of activity. Residual chemicals tend to have longer half-lives.

Hydrophilic: Having affinity for water (polar).

Hydrophobic: Lacking affinity for water (non-polar).

Incineration: The act of burning something completely; reducing it to ashes.

Isomerism: The relation of two or more chemical compounds, radicals, or ions that contain the same numbers of atoms of the same elements but differ in structural arrangement and properties.

K_{oc}: Soil Organic Carbon-Water Partitioning Coefficient (K_{oc}). K_{oc} values are useful in predicting the mobility of organic soil contaminants; higher K_{oc} values correlate to less mobile organic chemicals while lower K_{oc} values correlate to more mobile organic chemicals.

LC_{50} or median lethal concentration: The concentration of material in water that is estimated to be lethal to 50% of the test organisms. The LC_{50} is usually expressed as a time-dependent value, e.g. 24 hour or 96 hour; LC_{50}, the concentration estimated to be lethal to 50% of the test organisms after 24 or 96 hours of exposure.

LD_{50} or median lethal dose: The concentration of material in water that is estimated to be lethal to 50% of the test organisms. Appropriate to use with test animals such as rats, mice and dogs. It is rarely applicable to aquatic organisms because it indicates the quantity of a material introduced directly into the body by injection or ingestion rather than the concentration of the material in water in which aquatic organisms are exposed during toxicity tests.

Limit of reporting: The limit of reporting (LOR) is the minimum concentration of a residue used for reporting, when analysed in biota or environmental sample.

Liphophilicity: High affinity for lipids (fats).

Log K_{ow}: Logarithm of the octanol/water partitioning coefficient e.g. for a Kow =10,000 for a compound.

Maximum level (ML): A maximum level is defined as the maximum tolerable concentration of a contaminant (e.g. pesticide or metal or natural toxicant) in or on a food, agricultural commodity or animal feed.

Maximum residue limit (MRL): The maximum residue limit is the maximum concentration of a residue that is legally permitted or recognised as acceptable in or on a food, agriculture commodity or animal feed. It results from the officially authorised safe use of an agriculture or veterinary chemical.

NOEC: No observed effect concentration.

NOEL: No observed effect level.

Organophosphate/ organophosphorous: Organophosphates and Organophosphorous are used in this book as synonymous.

Octanol/water partitioning coefficient (K_{ow}) : It quantifies the liphophilicity of a substance and is assumed to be an index of the ability to pass through biological membranes and to bio-accumulate in living organisms. It is utilized as a measure of the affinity of a pesticide for the biota. Bio-accumulating pesticides have a log K_{ow} >3.

Phloem: The plant tissue that transports dissolved nutrients from the leaves to the other parts of the plant.

pK_a: An acid dissociation constant; K_a, (also known as acidity constant, or acid-ionization constant), is a quantitative measure of the strength of an acid in solution. It is the equilibrium constant for a chemical reaction known as dissociation in the context of acid-base reactions.

Primipara: A woman who has given birth to her first child.

Resistance : A heritable change in the sensitivity of a pest population that is reflected in the repeated failure of a product to achieve the expected level of control when used according to the label recommendation for that pest species

Tadpoles: Tadpoles are the aquatic larvae form of adult frogs. Tadpoles are aquatic, free swimming larvae with distinctive head and tail regions. As tadpoles develop towards adulthood they develop hind legs. Soon afterwards, their forelimbs emerge and their tail begins to atrophy and begins to look like a small stub. Lastly, the tail is fully reabsorbed and the metamorphosis is complete.

Teratogenic: Of or relating to substances or agents that can interfere with normal embryonic development.

USDA: United States Department of Agriculture

USFWS: US Fish and Wildlife Service

Xylem: A vascular tissue in land plants primarily responsible for the distribution of water and minerals taken up by the roots.

9.1: Introduction

Pesticides are chemicals that are broadly used to protect crops, livestock and other animals and plants from insect-pests and diseases. Any substance or mixture of substances intended for preventing, destroying, repelling, or mitigating any pest or weed is a pesticide. These agriculture chemicals include insecticide; herbicide and fungicide (see Figure 9.1 for chemical structure of some common pesticides).

Insecticide	***Endosulfan*** 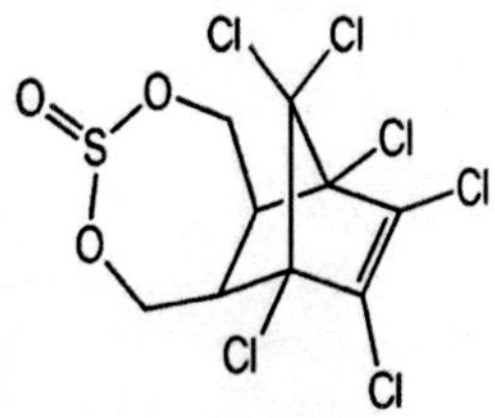 MF: $C_9H_6Cl_6O_3S$ MWt: 406.925 logK_{ow}: 4.74-4.79	***Azinphos- methyl*** 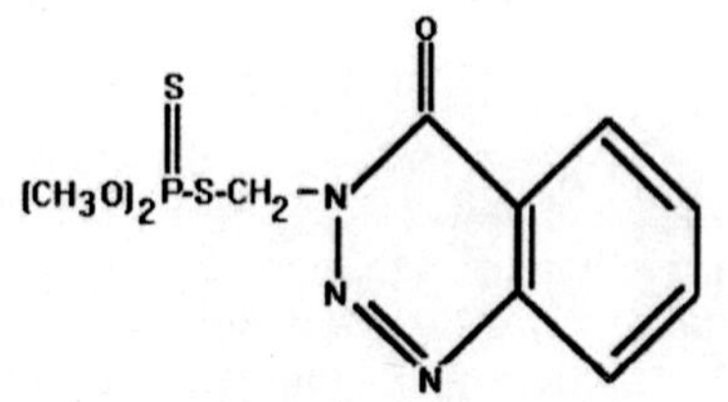 MF: $C_{10}H_{12}N_3O_3PS_2$ MWt: 317.324 logK_{ow}: 2.544	***Chlorpyrifos*** 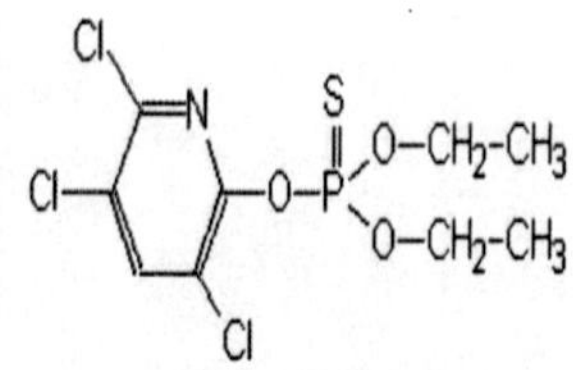 MF: $C_9H_{11}C_{13}NO_3PS$ MWt: 350.586 Log K_{ow}: 4.77
Herbicide	***Atrazine*** 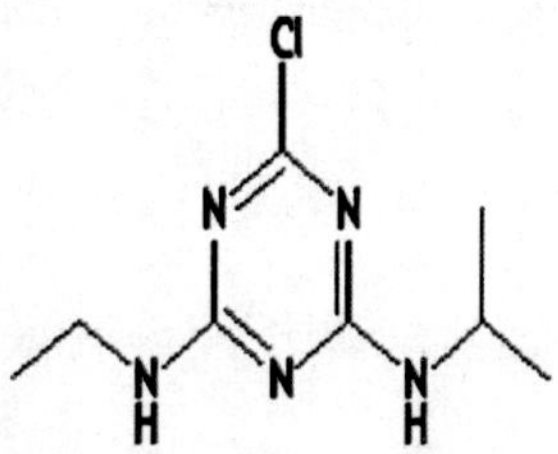 MF: $C_8H_{14}ClN_5$ MWt: 215.683 logK_{ow}: 2.632	***Molinate*** 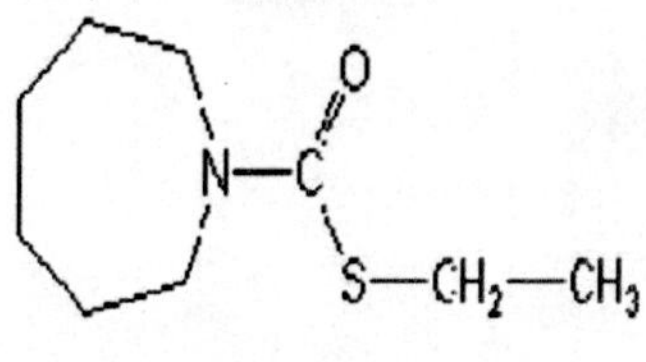 MF: $C_9H_{17}NOS$ MWt: 187.302 logK_{ow}: 2.88	***Glyphosate*** 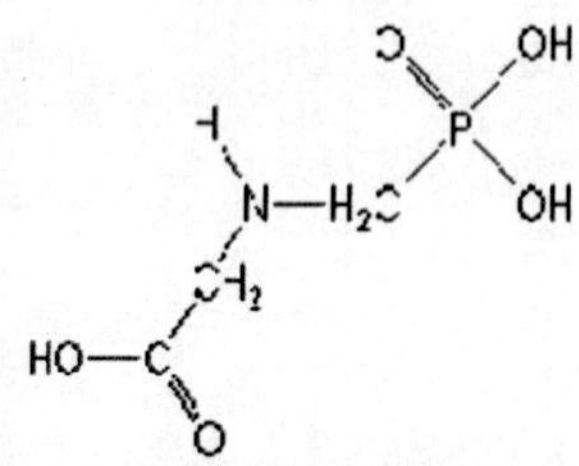 MF: $C_3H_8NO_5P$ MWt: 169.073 logK_{ow}: -3.2
Fungicide	***Copper hydroxide*** $H—O^- \quad Cu^{2+} \quad H—O^-$ MF: CuH_2O_2 MWt: 97.6	***Mancozeb*** 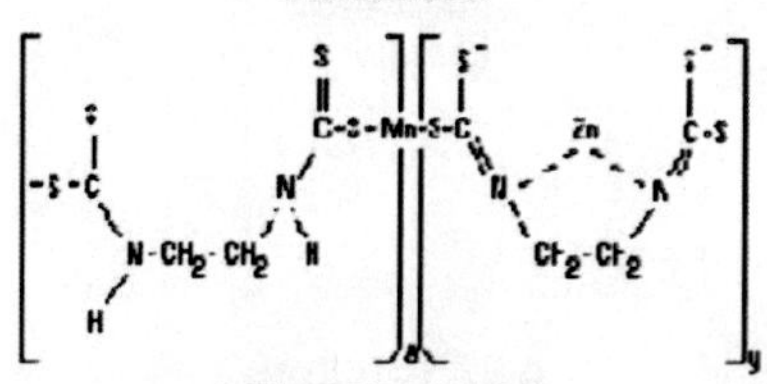 MF: $C_4H_6MnN_2S_4$ MWt: 265.302 logK_{ow}: 1.656	***Thiram*** 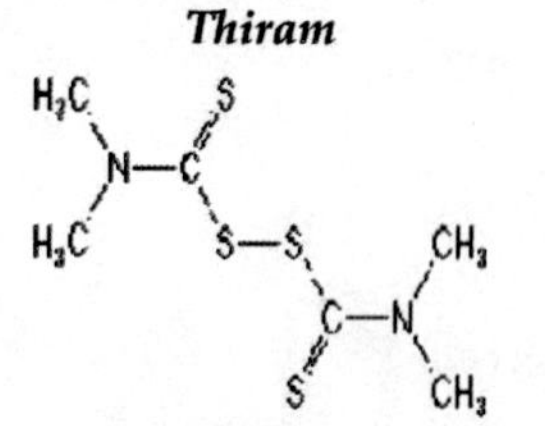 MF: $C_6H_{12}N_2S_4$ MWt: 240.433 logK_{ow}: 1.73

Figure 9.1: ***Chemical structure of some common pesticides*** *[(Structures are based on eMolecules - http://www.emolecules.com/].*

9.2: Classification of pesticides

Pesticides are classified according to the type of pest to be controlled (algicides, herbicides, fungicides and insecticides) (Table 9.1). They are also classified based on their chemicals classes (Table 9.2) and mode of action. Biological or biopesticides include micro-organisms (bacteria, fungi, viruses and protozoa) that are effective in controlling target pests. These agents usually do not have toxic effects on animals and humans and do not leave toxic or persistent chemical residues in the environment (Table 9.3). Broad-spectrum pesticides are those that kill an array of species, while narrow-spectrum, or selective pesticides only kill a small group of species.

Table 9.1: *Major classification of pesticides.*

Pesticides	Purpose of use
Algicides or Algaecides	Control of algae
Avicides	Control of birds
Bactericides	Control of bacteria
Fungicides	Control of fungi and oomycetes
Herbicides	Control of weeds
Insecticides	Control of insects including killing of ovicides (eggs), larvicides (larvae) or adulticides (adult insects)
Miticides or Acaricides	Control of mites
Molluscicides	Control of slugs and snails
Nematicides	Control of nematodes
Rodenticides	Control of rodents
Virucides	Control of viruses

Pesticides are also classified into systemic and non-systemic pesticides. *Systemic pesticides* penetrate plant tissues and are translocated to other parts of plant via diffusion through the cuticle of leaves or through the outer layer of roots (soil applications) into the epidermal cells, into either the xylem or phloem. *Non-systemic or contact pesticides* have little or no ability to penetrate plant tissues (leaves or roots) and translocate from site of contact to distal parts of the plant.

Table 9.2: ***Pesticides, their chemical classes and important chemical properties*** *[based on Roberts and Huson 1999; Roberts et al. 1998; Tomlin 2006; http://www.chem.unep.ch/ddt/documents/Ritter_en_ChemProperties.pdf; PAN-http://www.pesticideinfo.org/Detail_Chemical.jsp?Rec_Id=PC34514; EXTOXNET-http://extoxnet.orst.edu/].*

Chemical class symbols: A = acaricide; Al = algicide; B = bird repellent; Bi = biological insecticide; Bc = bactericide; F = Fungicide; H = herbicide; I = insecticide; IP = insect pheromone; Ix = ixodicide; M = molluscicide; N = nematicide; P = plant growth regulator; R = rodenticide; S = soil sterilant; Rr = rodent repellent; Wp = wound protectant; other symbols: d = days; h = hour; mo = month; wk = week; y = year.

Active ingredient (usage)	Chemical class	Half-life in soil (DT_{50})	Water solubility (mg/l)	K_{oc} (l/kg)	log K_{ow}	pKa
2,4-D (H)	pheoxycarboxylic acid	<7 d	45,000	60	2.58-2.83	2.73
2,4-DB (H)	pheoxycarboxylic acid	<7 d	46	440	3.9	4.8
2,4,5-T (H)[note1]	trichlorophenoxyacetic acid	2- 10 wk	Slight; 280	-	-	-
8-hydroxy quinoline sulphate (F, Bc)	bactericide	24-36 h	300	-	-	-
Abamectin (I,A,N)	macrocyclic lactone	35 d	0.007-0.01	4000	4.4	-
Acephate (I)	organophosphate	2-7 d	790	2	- 0.89	-
Acetamiprid (I)	neonicotinoid	0.8-5.4 d	4,250	71.1-267	0.8	0.7
Acifluorfen (H)	diphenyl ether	108 -200 d	60-62	131-1,955	1.19	3.86
Acrolein (H)	ungrouped	7.5-10 h	208 g/kg		1.08	-
Aldicarb (I, A,N)	carbamate	2-12 d	4.93 g/l	21-68	1.15	-
Aldrin (I)	cyclodiene organochlorine	see dieldrin	almost insoluble	see dieldrin	see dieldrin	-
Allethrin (I)	pyrethroid	-	in soluble	-	4.96	-
Alpha-Cypermethrin (I)	pyrethroid	see cypermethrin	0.00067	see cypermethrin	6.94	4.1
Ametryne (H)	triazine	51 d	200	100-930	2.63	2.56

Contd...

Aminopyralid (H)	ungrouped	25	2,480	10.8	0.201	4.2
Amitraz (A,I)	amidine	<1 d	<0.1	1,000-2,000	5.5	4.2
Amitrole (H)	triazole	<5-56 d	>1384 g/l	93	- 0.97	4.82
Asulam (H)	carbamate	7-14 d	4,000	40	-	1.7
Atrazine (H)	triazine	16-77 d	33	39-173	2.5	1.7
Azamethiphos (I)	organophosphate	6 h	1,100	-	1.05	3.6
Azimsulfuron (H)	sulfonylurea	-	72.3-6,536 (pH dependent)	-	4.43 (pH 5)	-
Azinphos-methyl (I)	organophosphate	several weeks	28	1,000	2.96	-
Azoxystrobin (F)	strobilurin	14 d	6	500	2.5	-
Bacillus thuringiensis	bacterium	-	insoluble	-	-	-
Benalaxyl (F)	acylalanine	77 d	28.6	2,728-7,173	3.54	8.8
Bendiocarb (I)	carbamate	0.5-10 d	280	28-40	1.72	5.2 or non-ionised
Benomyl (F)	Benzimidazole	19 h	1.9-3.6 (pH dependent)	1900	1.37	-
Bensulfuron (H)	sulfonylurea	88.5 d	2.1 -3100 (pH dependent)	-	2.18	-
Bensulide (H)	phosphorodithioate	1-6 mo	25	-	4.2	-
Bensultap (I)		3-35 d	0.448		2.28	3.3
Bentazone (H)	benzothiadiazinone	12 d	570	13.3-176	0.46 (pH 7)	
Benzofenap (H)	pyrazole	38 d	0.13		4.69	
Beta-cyfluthrin (I)	pyrethroid	30	1.9-2.9 µg/l	100,000	5.9	-
Bifenazate (A)		=5 d	2.06	1,778	3.4	-
Bifenthrin (A, I)	pyrethroid	65-125 d	<1 µg/l	131,000-302,000	>6	-
Bioallethrin (I)	pyrethroid	48 -1447 h	4.6	55	4.68	-
Bioresmethrin (I)	pyrethroid	strongly adsorbed	<0.3	-	>4.7	-
Bitertanol (F)	traizole	195 d	2.7	2,040	4.1	
Boscalid (F)	carboxamide	28 - 200 d	4.6	808.5	2.96	9.27
Bromacil (H)	uracil	5 mo	700-1,287	32	1.88	
Bromoxynil (H)	hydroxybenzonitrile	<1 d	89	10,000	1.04	4.4
Bupirimate (F)	aminopyrimidinol	35-90 d	13.06	200	3.9	-
Buprofezin (A,I)	ungrouped	80 - 104 d	0.386	-	4.8	
Butachlor (H)	chloroacetamide	6-10 wk	20	-	-	-
Butafenacil (H)	pyrimidindione	1-2 d	10	365	3.2	4.36
Butroxydim (H)	cyclohexanedione oxime	150	6.9	6-1,270	1.9	-
Cadusafos (I,N)	organophosphate	11-55 d	248	144-351	3.9	-
Captan (F)	phthalimide	1 d	3.3	200	2.8	-
Captafol (F)	Phthalimide	-	1.4	-	3.8	-
Carbaryl (I,P)	carbamate	7-28 d	120	100-600	1.85	4.2
Carbendazim (F)	benzimidazole	8-32 d	29	200-250	1.38	-
Carbetamide (H)	carbamate	1 mo	3,500	-	-1.58	-
Carbofuran (I,N)	carbamate	30-60 d	320-351	22	1.52	-
Carbosulfan (I)	carbamate	3-30 d	3	-	5.4	<0.5
Carboxin (F)	carboxamide	<1 d	147	71	2.3	-
Carfentrazone-ethyl (H)	traizolinone	2.5-4 d	12 µg/ml	15-35	3.36	-

Contd...

Cartap (I)	2-dimethylaminoproponate -1,3 dithiol	3 d	200 g/l	-	-	-
Chlordane (I)	cyclodiene organochlorine	1 yr	0.1	-	6	-
Chlorfenapyr (A,I)	arylpyrrole	1370	insoluble	>10,000 ml/g	4.83	
Chlorfenvinphos (A, I)	organophosphate	71-270 d (pH deps.)	7.3-121	1,500	3.85-4.22	-
Chlorfluazuron (I)	benzoylurea	6 weeks - few months	0.012		5.9	-
Chloropicrin (I,N)	fumigant	4 d	1.62-2.27 g/l	15,556 -13,5976		-
Chlorothalonil (F)	chloronitrile	0.3-28 d	0.81	1,600-14,000	2.92	
Chlorpropham(H,P)	carbamate	35-65 d	89		3.79	
Chlorpyrifos (I)	organophosphate	33-56 d	1.4	1,250-12,600	4.7	-
Chlorsulfuron (H)	sulfonylurea	1.5-33 d	590-31,800 (diff. pH)	40	-0.99	-
Chlorthal-dimethyl (H)	benzenedicarboxylic acid	100 d	0.5	5,000	4.28	
Clethodim (H)	cyclohexanedione oxime	1-3 d	soluble	-		-
Clodinafop-propargyl	aryloxyphenoxypropionate	< 2 h	4	1,404	3.9	
Clofentezine (A)	mite growth inhibitor	16.8-132 d	2.5 µg/l	240	4.1	
Clomazone (H)	isoxazolidinone	30-135 d	1,100	356	2.5	2.0
Clopyralid (H)	pyridinecarboxylic acid	45	118-143 g/l	4.64	-1.81	~2.5
Copper hydroxide (F-Bc)	inorganic	1000	5.06×10^{-4} g/l	5000	0.44	
Copper octanoate (F, Bc, Al)	inorganic	-	-	-	-	-
Copper oxychloride (F, Bc)	inorganic	100	1.19×10^{-3} g/l	1,000	-	-
Copper sulfate (Al, F)	inorganic	-	148-335 g/l	-	-	-
Cryolite (I)	selective feeding blocker	-	250	-	-	-
Cyanazine (H)	triazine	14 d	171	190	2.1	0.63
Cyfluthrin (I)	pyrethroid	185		100,000	5.9-6	-
Cyhalofop-butyl (H)	aryloxyphenoxypropionate	2-10 h	0.44-0.46	5,247	3.31	3.80
Cypermethrin (I)	pyrethroid	60 d	0.004	26,492-144,652	6.6	-
Cyproconazole (F)	triazole	90 d	93	377	3.1	-
Cyprodinil (F)	anilinopyrimidine	20-60 d	13-20	1,470	3.9	4.44
Cyromazine (I)	triazine	-	13,000	-	-0.06	5.22
Dalapon or 2,2-DPA (H)	halogenated alkanoic acid	7.5	629	1	-	1.74-1.84
DDT (I)	organochlorine	2.0-15.6 yr	0.012-0.0055	243,000	4.89-6.914	-
Deltamethrin (I)	pyrethroid	23	<0.2 µg/l	8,380,000	4.6	-
Diafenthiuron (A,I)		<1 h-1.4 d	0.06		5.76	
Diazinon (A, I)	organophosphate	11-21 d	60	332	3.3	2.6
Dicamba (H)	benzoic acid (auxin)	<14 d	6,600	2	-0.55	1.97
Dichlobenil (H)	benzonitrile	1- 6 mo	21	0.06	2.7	
Dichlorprop (H)	aryloxyalkanoic acids		350	50	5.91	3.0
Dichlorvos (A, I)	organophosphate	<1	18,000	10	1.9	

Contd...

Diclofop (H)	aryloxyphenoxypropionate		450	100	2.8	3.43
Diclofop-methyl (H)	aryloxyphenoxypropionate	1-57 d	0.8	14,000-24,400	4.58	-
Dicofol (A)	ungrouped	80	0.8	5,917-8,383	4.3	-
Dieldrin (I)	chlorinated hydrocarbon	0.5-3 yr	0.17	7,413-12,027	4.32	-
Difenoconazole (F)	triazole	50-150 d	15	3,759	4.4	1.1
Diflubenzuron (I)	benzoylurea	< 7d	0.08		3.89	-
Diflufenican (H)	pyridinecarboxamide	85.6-282 d	<0.05	1,978	4.9	-
Dimethoate (A, I)	organophosphate	2-4.1 d	23,300	16.25-51.88	0.704	-
Dimethomorph (F)	cinnamic acid amide	30	41.8-81.1 (diff. pH)	2,000	2.63	-
Dinocap (F, A)	dinitrophenyl crotonate	3.6-30 d	0.151	2,889-310,200	4.54	-
Diquat-dibromide (H)	bipyridylium	<1 wk	700 g/l	1,000,000	-4.6	-
Disulfoton (A, I)	organophosphate	30	25	600	3.95	
Dithianon (F)	quinone	na	0.14		3.2	-
Diuron (H)	urea	90-180 d	37.4	400	2.85	-
Dodine (F)	guanidine	na	630	na	1.65	-
DSMA (methylarsonic acid, H)	organoarsenic	55 d	279 g/kg	250-2,850	<0	-
Emamectin benzoate (I)	macrocyclic lactone	rapid degradation	0.024 g/l	1,284	5	-
Endosulfan (I,A)	cyclodiene organochlorine	30-70 d	0.32-0.33	3,000-20,000	4.74-4.79	
Endothal (H, Al, P)	ungrouped	8.5 d	100 g/kg	na	-2.09	3.4-6.7
Endrin (I)	cyclodiene organochlorine	> 10 yrs	insoluble	-	5.6	-
EPN (I, A) (*0*-ethyl-*0*-4-nitrophenyl phenylphosphonothioate)	Organophosphate	< 15 d	0.92	-	>5.02	-
Epoxyconazole (F)	triazole	2-3 mo	$6.6x10^{-4}$ g/ 100 ml	957-2,647	3.33	
EPTC (H)	thiocarbamate	4-6 wk	375	11,500	3.2	
Esfenvalerate (I)	pyrethroid	287 d	0.002	5,300	6.22	
Ethion (A, I)	organophosphate	90 d	2		4.28	
Ethofumesate (H)	bezofuran	84-407 d	50	203	2.7	
Etoxazole (A)	mite growth inhibitor	7-25 d	0.225		6.9	
Etridiazole (F)	aromatic hydrocarbon	3-9.5 d	117	272-396	3.37	2.77
Fenamiphos (N)	organophosphate	several weeks	400	low mobility	3.3	-
Fenarimol (F)	pyrimidine	14-130 d	13.7	500-992	3.69	-
Fenbutain oxide (A)	oraganotin acaricide		0.0152	low mobility	5.2	-
Fenhexamid (F)	hydoxyanilide	=1 d	20	853	3.51	-
Fenitrothion (I)	organophosphate	4-18 d	14	2,000	3.43	-
Fenoxaprop-P-ethyl (H)	aryloxyphenoxypropionate	1-10 d	0.7	11,200	4.58	-
Fenoxycarb (I)	carbamate	few to 31 d	7.9	100	4.07	-
Fenpyroximate (A)	ungrouped	26.3-49.7 d	1.5x10-2 mg/l		5.01	-
Fenthion (I)	organophosphate	1.5	4.2	1,500	4.84	-
Fenvalerate (A, I, Ix)	pyrethroid	75-80 d	0.01	5,300	5.01	-

Contd...

Fipronil (I)	phenylpyrazole	readily degraded	1.9	427-1,278	4	-
Flamprop-M (H)	arylaminopropionic acid		12	1,000	3.69	3.7
Florasulam (H)	triazolopyrimidine	7-31 d	0.121 g/l	2-69 (av. 22)	- 1.22	4.54
Fluazifop-butyl (H)	arylphenoxypropionate	< 3 wk	1	5,700	4.5	-3.1
Fluazinam (F)	2.6-dinitroaniline	26.5 d	0.135	1,705-2,316	4.03	7.34
Fludioxonil (F)	phenylpyrrole	10-25 d	1.8	1,614	4.12	<0-14.1
Flumethrin (I)	pyrethroid	-	-	-	-	
Flumetsulam (H)	triazolopyrimidine	1-2 mo	49	5-182	- 0.68	4.6
Flumiclorac -pentyl (H)	N-phenylphthalimide	0.48-4.4 d	0.189	-	4.99	-
Flumioxazin (H)	N-phenylphthalimide	15-27 d	1.79	1,412	2.55	-
Fluometuron (H)	urea	10 - 100 d	110	31-117	2.38	-
Flupropanate (H)	halogenated alkanoic acid	immobile	3.9 kg/l	-	<-1.9	-
Fluquinconazole (F)	triazole	50-300 d	1	>875	3.24	-
fluroxypyr (H)	aryloxyalkanoic acids	-	91	51-81	- 0.2	2.94
Flusilazole (F)	triazole	95	45	1,511	3.74	2.5
Flutolanil (F)	carboxamide	190-320 d	8.01	-	3.17	-
Flutriafol (F)	triazole	-	130	-	2.3	-
Folpet (F)	phthalimide	4.3 d	0.8	304-1164	3.11	-
Fosetyl-aluminium (F)	phosphonate	20 m-40 h	111.3	-	- 2.1	-
Furalaxyl-M (F)	acylalanine	31-65 d	251.5	-	2.98	-
Gamma-cyhalothrin (I)	pyrethroid	32-51 d	2.1x10-6 mg/l	86,602 (av.)	0.695	-
Glufosinate-ammonium (H)	phosphinic acid	7-20 d	1370 g/l	10-1,230	<0.1	-
Glyphosate	glycine derivative	1-130 d	10.5 g/l	24,000	<-3.2	2.34
Guazatine (F)	guanidine	immobile	>3 kg/l		- 1.2	
Halosulfuron -Methyl (H)	sulfonylurea	55	0.015-1.65	75	- 0.02	3.44
HCH (I)	cyclodiene organochlorine	-	8.35-8.53	-	3.5	-
Heptachlor (I)	cyclodiene organochlorine	0.75-2 yr	0.18	3,020	5.27	-
Haloxyfop (H)	aryloxyphenoxypropionate	1-20 d	0.58-1.91	128	4.33	2.9
Hexachlorobenzene (F)	chlorinated hydrocarbon	-	0.006	-	5.86	-
Hexaconazole (F)	triazole	rapid degradation	0.017 g/l	1,040	3.9	-
Hexaflumuron (I)	benzoylurea	100-280	0.027	5,338-70,977	5.64	-
Hexazinone (H)	1,2,4-triazinone	1-6 mo	29.8 g/l	54	1.2	2.2
Hexythiazox (A)	mite growth inhibitor	8 d	0.5	6,200	2.53	-
Hydramethylnon (I)		7 -22 d	0.005-0.007	-	2.31	-
Imazalil (F)	imidazole	100 d	0.0951 g/l	2,080-8,150	3.82	-
Imazamox (H)	imidazolinone	5-41 d	116->628	2-374	- 1.03	2.3-10.8
Imazapic (H)	imidazolinone	31- 410 d	2,150	-	0.393	2.0-11.1
Imazapyr (H)	imidazolinone	24-143 d	9.74-11.3 g/l	100	0.11	1.9
Imazethapyr (H)	imidazolinone	90	1,400	10	1.04-1.49	-

Contd...

Imidacloprid (I)	neonicotinoid	-	610		0.57	-
Imiprothrin (I)	pyrethroid	-	93.5	-	2.9	-
Indoxacarb (I)	oxadiazine	3-186 d	0.2	3,300-9,600	4.65	-
Iodosulfuron-methyl-sodium (H)	sulfonylurea	1-5 d	0.16-65 g/l	0.8-152	1.07	3.22
Ioxynil (H)	hydroxybenzonitrile	10 d	50	200	2.2	4.1
Iprodione (F)	dicarboximide	20-80 d	13	202-543	3	-
Isoxaben (H)	benzamide	3-4 mo	1.42	-	3.94	-
Isoxaflutole (H)	isoxazole	1.3 d	6.2	112	2.34	-
Kresoxim-methyl (F)	oximinoacetate	< 3 d	2	219-372	3.4	-
Lambda-cyhalothrin (I)	pyrethroid	6-40 d	0.005	330,000	7	>9
Lindane (I) (also known as gamma-hexachlorocyclohexane, (ã-HCH) or benzene hexachloride (BHC)	organochlorine	1-2 yrs	7.3 mg/l	3529	3.8-3.99	-
Linuron (H)	urea	2-5 mo	63.8	500-600	3	
Lufenuron (I,A)	benzoylurea	13-20 d	<0.006	38	5.12	>8
Malathion (A, I)	organophosphate	7 d	145	1,800	2.75	
Mancozeb (F)	dithiocarbamate	< 1 d	6.2	998	0.26	10.3
MCPA (H)	pheoxycarboxylic acid	< 7 d	293.9-320.1	10	2.75	3.73
Mecoprop (H)	aryloxyalkanoic acids		880	13	1,004	3.78
Mesosulfuron -methyl (H)	sulfonylurea	8-68 d	0.483	92	1.39	4.35
Metalaxyl-M (F)	acylalanine	21 d	26,000	45	1.71	
Metalaxyl (F)	acylalanine	22-48 d	8,400	70	1.75	<0
Methamidophos (A, I)	organophosphate	<2 d	> 200 g/l	-	- 0.8	-
Methidathion (A, I)	organophosphate	3-18 d	200	-	2.2	-
Methiocarb (A, Br, I, M)	cabamate	rapid degradation	27	-	3.08	-
Methomyl (A, I)	carbamate	4-8 d	57.9 g/l	72	0.093	-
Methoprene (I)	juvenile hormone mimic	10 d	0.515	-	>6	-
Methoxyfenozide (I)	diacylhydrazine	23-268 d	3.3	-	3.7	-
Methyl bromide (A, I, N, R, S)	fumigant	-	17.5 g/l	-	-	-
Metiram (F)	dithiocarbamate	2.7 d	insoluble	500,000	0.3	-
Metolachlor (H)	chloroacetamide	20 d	488	121-309	2.9	-
Metosulam	triazolopyrimidine	25 d	100-5,600 (pH dependent)	<500	0.978	4.8
Methoxychlor (1)	organochlorine	120 d	0.1	-	4.83-5.08	-
Metribuzin (H)	triazinone	1-2 mo	1.05 g/l	60	1.6	
Metsulfuron -methyl (H)	sulfonylurea	52 d	0.548-213 g/l (pH dependent)	35	0.018	3.8
Mevinphos (A, I)	organophosphate	3	Miscible	117	0.127	-
Milbemectin (A,I)	macrocyclic lactone	16-33 d	0.88	-	5.3	-
Mirex (I)	organochlorine	10 yrs	not soluble ; 0.07 mg/l	$2.4X10^{7}$	5.28	-

Contd...

Molinate (H)	thiocarbamate	8-25 d	900-990	190	2.88	-
Monocrotophos (I)	organophosphate	1-5 d	very soluble	1	-0.22	-
Monolinuron (H)	urea	45-60 d	735	250-500	2.2	-
Myclobutanil (F)	triazole	66	115-132	224-920	2.94	-
Napropamide (H)	alkalamide (acetamide)	230-670 d	7.4	600	3.3	-
Naptalam (H)	phthalimide	3-8 wk	200	2,152	0.104 (pH 5)	4.6
Nicotine (I)	ungrouped	rapid degradation	miscible	-	0.93	3.1
Norflurazon (H)	pyridazinone (PDS)	6-9 mo	34	218-635	2.45	-
Omethoate(A, I)	organophosphate	7 d	miscible	50	- 0.74	-
Oryzalin (H)	dinitroaniline	1.2 month	2.6	700-1,100	3.73	9.4
Oxadiazon (H)	oxadiazole	3-6 mo	1	1,400	4.91	-
Oxadixyl (F)	oxazolidinone	2-3 mo	3.4 g/kg	-	0.65-0.8	-
Oxamyl (A, I, N)	carbamate	7 d	280 g/l	25	- 0.44	-
Oxycarboxin (F)	carboxamide	2.5- 8 wk	1.4 g/l	95	0.772	-
Oxyfluorfen (H)	diphenyl ether	5-55 d	0.116	2891-32381	4.47	-
Chinomethionat /Oxythioquinox (F, A)	ungrouped	1-3 d	1	45-90 d	3.78	-
Paclobutrazol (P)	triazole	0.5-1 yr	26	-	3.2	-
Paraquat-dichloride (H)	bipyridylium	1000 d	620 g/l	10,000	- 4.5	-
Parathion-methyl (I)	organophosphate	14	55	5,000	3	-
PCP (pentachlorophenol) (I, F, H)	Ungrouped	Very persistent	80	-	5.1	4.71
Penconazole (F)	triazole	133-343 d	73	2,500	3.72	1.51
Pencycuron (F)	phenylurea		0.3		4.68	
Pendimethalin (H)	dinitroaniline	3-4 mo	0.33	5,000	5.2	
Pentanochlor (H)	anilides		8	540	3.6	-
Permethrin (I)	pyrethroid	< 38 d	6x10-3	100,000	6.1	-
Phenmedipham (H)	phenyl carbamate	25 d	4.7	2,400	3.59	-
Phenthoate (I,A)	organophosphate	£1 d	10	-	3.69	-
Phorate (A, I, M)	organophosphate	7-10 d	50	543	3.92	-
Phosmet (A, I)	organophosphate	rapid degradation	25	820	2.95	-
Phosphamidon (I, A)	Organophosphate	-	Miscible with water, acetone and other common organic solvents	-	0.79	-
Phosphine (I, R)	ungrouped		26 cm^3/100 ml	-	-	-
Phosphonic acid (F)	phosphonate	16 w	425 g/100 g	-	-	-
Picloram (H)	pyridinecarboxylic acid	30-90 d	0.056 g/100 ml	16	1.9	2.3
Picolinafen (H)	pyridinecarboxamide	23-31 d	$3.9x10^{-5}$ g/l	15,000-31,800	5.37	
Pinoxaden (H)	phenylpyrazoline	<1 d	200		3.2	
Pirimicarb (I)	carbamate	7-234 d	3 g/l	23	1.7	4.44
Pirimiphos-methyl (A, I)	organophosphate	3.5-25 d	9.7-11	-	4.2	4.3

Contd...

Prallethrin (I)	pyrethroid	-	8	-	4.49	-
Pretilachlor (H)	chloroacetamide	30 d	50	-	4.08	-
Prochloraz (F)	imidazole	-	34.4	-	4.12	-
Procymidone (F)	dicarboximide	4-12 wk	4.5	-	3.14	-
Profenofos (A, I)	organophosphate	1 wk	28	2,000	4.44	-
Profoxydim (H)	cyclohexanedione oxime	3-13 d	5.31		3.9	5.91
Prometryn (H)	triazine	14-158 d	33	400	3.1	4.1
Propachlor (H)	chloroacetamide	4 d	580	80	1.4-2.3	
Propamocarb (F)	carbamate	< 30 d	>900 g/l		0.84	9.5
Propanil (H)	anilide	1	130	239-800	3.3	-
Propaquizafop (H)	aryloxyphenoxypropionate	15	0.63 g/m^3	10,000	4.78	-2.3
Propargite (A)	ungrouped	40-67 d	0.215	4,000		>12
Propazine (H)	triazine	8-100 d	5	65-268	3.01	1.7
Propiconazole (F)	triazole	29-70 d	100	950	3.72	1.09
Propineb (F)	dithiocarbamate	7	0.01 g/l	500	- 0.26	-
Propoxur (I)	carbamate	high mobility	1.75 g/l	-	1.56	-
Propyzamide (H)	benzamide	30 d	15	800	3.3	-
Prothiofos (I)	organophosphate	1-2 mo	0.07		5.67	-
Pymetrozine (I)	selective feeding blocker	2-69 d	0.29 g/l	246-7875	- 0.18	4.06
Pyraclostrobin (F)	phenylpyrazole	1-7 d	0.082	1949	3.49	-
Pyrazophos (F)	phosphorothiolate	10-21 d	4.2	1,332-2,670	3.8	-
Pyrethrins/pyrethrum (A, I)	ungrouped	-	0.2-9	-	4.3-5.9	-
Pyridaben (A,I)	ungrouped	<21 d	0.012	-	6.37	-
Pyrimethanil (F)	anilinopyrimidine	7-54 d	0.121 g/l	-	2.84	3.52
Pyriproxyfen (I)	juvenile hormone mimic	-	-	-	5.37	
Pyrithobac-Na (H)	pyrimidinyloxybenzoic	60 d	264-728	-	0.6	2.34
Quinalphos (I, A)	organophosphate	21 d	17.8	-	4.44	-
Quinclorac (H)	quinolinecarboxylic acid	mobile	0.065	-	- 0.74	4.34
Quinoxyfen (F)	quinoline	> 1 yr	0.047-0.128	15,415-75,900	4.66	3.56
Quintozene (F)	nitroaniline	4-10 mo	0.1	6,030	5.1	-
Quizalofop (H)	arylphenoxypropionic acids	< 1 d	0.3	477	4.28	3.1
Rimsulfuron (H)	sulfonylurea	10-20 d	<10	40	0.288	4
Rotenone (A,I)	ungrouped	3	0.142 μg/l	4000	4.16	-
Sethoxydim (H)	cyclohexanedione oxime	<1 d	25-4,700	100	4.51	-
Siduron (H)	urea		18		3.8	-
Simazine (H)	triazine	27-102 d	6.2	103-277	2.1	1.62
Spinosad (I)	spinosyn	9.4-17.3 d	89-235		2.8-5.2	-
Spiroxamine (F)	spiroketalamine	35-64 d		659-6,417	1.28-2.79	6.9
Sulfometuron (H)	sulfonylurea	11-47 d	17.6-1,627	33	- 0.77	3.51
Sulfosulfuron (H)	sulfonylurea	11-47 d	1,627	33	- 0.77	3.51
Sulfur (F,A)	inorganic	100	insoluble	30	-	-

Contd...

Tau-fluvalinate (I, A)	pyrethroid	12-92 d	1.03 ppb	>110,000	4.26	-
Tebuconazole (F)	triazole	610	36	1,121	3.7	-
Tebufenozide (I)	diacylhydrazine	36.5	0.83	622.5	4.25	-
Tebufenpyrad (A)	ungrouped	2-22 d	2.61	1,894-9,310	4.93	-
Tebuthiuron (H)	urea	-	2.5 g/l	-	1.82	-
Temephos (I)	organophosphate	12 d	0.03	-	4.91	-
Tepraloxydim (H)	cyclohexanedione oxime	5.2-14 d	0.4 g/l	-	0.2	4.58
Terbacil (H)	uracil	120	710	-	1.91	9.5
Terbufos (I, N)	organophosphate	9-27 d	4.5	-	2.77	
Terbutryn (H)	triazine	14-50 d	22	2,000	3.65	4.3
Tetramethrin (I)	pyrethroid		1.83	-	4.6	
Thiabendazole (F)	benzimidazole	29 h	0.03	110	2.39	4.73
Thiacloprid (I)	neonicotinoid	7-21 d	185	615	1.26	-
Thiamethoxam (I)	neonicotinoid	51 d	4.1 g/l	-	- 0.13	-
Thifensulfuron-M (H)	sulfonylurea	1-7 d	2,240	45	0.02	4.0
Thiobencarb (H)	thiocarbamate	2-3 wk	16.7	3170	4.23	-
Thiodicarb (I, M)	oxime carbamate	3-8 d	0.2219	350	1.62	-
Thiophanate-M (F, Wp)	thiophanate	3-4 wk	insoluble	1,830	1.5	7.28
Thiram (F)	dimethyldithiocarbamate	0.5 d	18	762	1.73	-
Tolclofos-methyl (F)	chlorophenyl/nitroaniline	44 d	1.1	-	4.56	-
Tolylfluanid (F)	sulfamide	2-11 d	-	0.9	3.9	-
Toxaphene (I)[note2]	chlorinated camphene, camphechlor	1-14 yrs	Negligible; 3	$2.1x10^5$	3.33-6.44	-
Tralkoxydim (H)	cyclohexanedione oxime	2-5 d	6-6.7	30-300	2.1	4.3
Transfluthrin (I)	pyrethroid		5.7x10-5 g/l		5.46	-
Triadimefon (F)	triazole	12	64	300	3.11	-
Triadimenol (F)	triazole	110-375 d	33	640	3.08-3.28	-
Triallate (H)	thiocarbamate	8-11 wk	4	2,400	4.6	-
Tribenuron-M (H)	sulfonylurea	3.5-5.1 d	0.05-18.3	-	0.78	4.7
Trichlorfon (I)	organophosphate	-	120 g/l	20	0.43	-
Triclopyr (H)	aryloxyalkanoic acids	-	408	59	0.42-0.45	2.68
Tridemorph (F)	morpholine	13-130 d	1.1	2,500-10,000	4.2	6.5
Trifloxystrobin (F)	oximinoacetate	4.2-9.5	0.61	1,642-3,745	4.5	
Trifloxysulfuron-Na (H)	sulfonylurea	49-78 d	25,700	29-584	- 0.43	4.76
Triflumuron (I)	benzoylurea	-	0.025	-	4.91	-
Trifluralin (H)	dinitroaniline	25-201 d	0.184-0.221	4,400-40,000	4.83	-
Triflusulfuron-M (H)	sulfonylurea	3 d	1-11,000	-	0.96	4.4
Triforine (F)	piperazine	3 wk	9	-	2.2	10.6
Triticonazole (F)	triazole	105-268 d	9.3	418	3.92	-
Zineb (F)	dithiocarbamate	16	10	1,000	=1.3	-
Ziram (F, B, Rr)	dimethyldithiocarbamate	42 h	1.58-18.3	400	1.23	-

Note 1 : Agent Orange, a defoliant used by the U.S. in the Vietnam War, was equal parts of 2,4,5-T and 2,4-D.
Note 2 : Toxaphene is a mixture of more than 175-200 components produced by chlorination of camphene.

Table 9.3: *Categories and examples of biopesticides.*

Product	Definition	Examples
Microbial pesticides	Micro-organisms that operate as the active ingredient.	Bacteria, fungi, viruses, virus coat proteins.
Plant-pesticides	Substances that plants produce from genetic material that has been added to them.	*Bacillus thuringiensis* pesticidal protein, potato leaf roll virus (PLRV) resistance gene produced in potato plants.
Biochemical pesticides	Naturally occurring substances that control pests by non-toxic mechanisms.	Pheromones, floral attractants and plant volatiles, natural insect-growth regulators, plant-growth regulators and herbicides, repellents.

9.3: History of Pesticide Use

People have been using pesticides for thousands of years to control various types of pests such as insects, weeds, bacteria, rodents and other biological organisms. The earliest pesticides were inorganic substances as follows:

- *Arsenic* (wood preservatives).
- *Ash* (various purposes).
- *Chromium* (wood preservatives).
- *Copper* (algicide, wood preservatives).
- *Lead* (insecticide).
- *Mercury* (wood preservatives).
- *Sulphur* (insecticide).

Some of the above inorganic pesticides are still used today, for example copper oxychloride is widely used as a fungicide and is sanctioned for use in 'organic' crops.

The Sumerians used sulfur to control insects and mites 5,000 years ago, whereas the Chinese used mercury and arsenic compounds to control body lice and other pests. The Greeks and Romans used oil, ash, sulfur, and other materials to protect themselves, livestock's and crops from various pests. People in various cultures have used smoke, salt, spices and insect-repelling plants to preserve food and keep pests away. By the 15th century, toxic chemicals such as arsenic, mercury and lead were being applied to crops to kill pests. In the 17th century, nicotine sulfate was extracted from tobacco leaves for use as an insecticide. The 19th century saw the introduction of two more natural pesticides, pyrethrum which is derived from chrysanthemums (a kind of flowering plant) and rotenone which is derived from the roots of tropical vegetables (*Derris* spp.). In 1939, it was discovered that dichlorodiphenyltrichloroethane (DDT) was a very effective insecticide. It quickly became the most widely-used pesticide in the world. In the 1940s manufacturers began to produce large amounts of synthetic pesticides and their use became widespread.

Commercial production of DDT began in 1943. At that time, DDT was considered to be a wonderful invention. It was cheap to produce, very toxic to insects, and much less toxic to mammals and its usage resulted in a sharp decline of mosquito borne diseases. DDT and other organochlorine insecticides were used for many years to control mosquitoes and as a broad-spectrum insecticide against insect pests that damaged food and crops. Unfortunately, scientists learned later that many organochlorine insecticides were persistent in the environment and were bio-accumulating in birds, humans, and other animals. In 1962, Rachel Carson wrote the book *Silent Spring,* in which she reported that DDT was causing egg-shell thinning in bird eggs and thus was leading to the near extinction of bird species such as peregrine falcons and bald eagles. Today most of the organochlorine pesticides have been banned in the United States by the Environment Protection Agency (EPA) because of the tendency of these compounds to persist in the environment and bio-accumulate in animals. Even though many of these substances are effective pesticides, the use of some of these materials has been banned or restricted because of health and environmental concerns. Lead and arsenic are no longer used as insecticides, the use of mercury as a fungicide has been restricted and the U.S. Environmental Protection Agency (EPA) is phasing out the use of arsenic as a wood preservative.

As a group, the organophosphate and carbamate pesticides are probably the most widely used insecticides, although many are being restricted by the EPA because of their toxicity. It is considered that 1940s and 1950s have been the start of the "pesticide era." Pesticides use has increased 50-fold since 1950. Though most of pesticides usages are in developed countries, but use in developing countries is increasing.

Pyrethroid pesticides were developed as synthetic versions of the naturally occurring pesticide pyrethrin, which is found in chrysanthemums. Most pyrethroids are safer than the organochlorines, organophosphates, and carbamates, although some synthetic pyrethroids are toxic to the nervous system. Pyrethroids have been modified to increase their stability in the environment and many different pyrethroids are in use today.

9.4: Merits and problems of pesticide usage

Pesticides use has greatly helped to increase and improve world food production. These are mainly used to increase farmer's profits by preventing crop losses due to pest attacks and weed infestation. In the USA alone farmers get an estimated four-fold return on money they spend on pesticides. Application of pesticides further helps to protect human health by killing mosquitoes (which cause diseases such as Nile virus and malaria) and preventing moulds that grows in human food. Application of pesticides reduces the risk of environmental damages through eradicating invasive weeds in parks and wilderness, and controlling algae in ponds and lakes (which interfere in swimming, fishing and produces unpleasant odour) [see Box 9.1 for usage of some common pesticides].

Box 9.1: ***Information on usage of some selected pesticides*** *[based on Extoxnet (http://extoxnet.orst.edu/tibs/movement.htm) and Tomlin (2006)].*

Atrazine (H): *Pre- and post emergence control of annual broad-leaved weeds and annual grasses in maize, sorghum, sugarcane, pineapples, grassland, macadamia nuts, conifers, and industrial weed control.*

Azinphos-methyl (OP): *It is used to control of sucking and chewing insects of the orders Coleopteran, Diptera, Homoptera, Hemiptera and Lepidoptera of fruit trees, including citrus, vines, strawberries, nuts, vegetables, potatoes, cereals, maize, cotton, ornamentals, beets, soya beans, tobacco, rice, coffee, sugarcane and forestry and other crops.*

Chlorpyrifos (OP): *It is used to control Coleopteran, Diptera, Homoptera, and Lepidoptera in soil or on foliage in over 100 crops, including pome fruit, stone fruit, citrus fruit, nut crops, strawberries, figs, bananas, vines, vegetables, potatoes, beet, tobacco, soybeans, sunflowers, sweet potatoes, peanuts, rice, cotton, alfalfa, cereals, maize, sorghum, asparagus, glasshouse and outdoor ornamentals, turf and in forestry. Also used for control of household pests (Blatellidae, Muscidae, and Isopetera), mosquitoes (larvae and adults), and pests of domestic animals. Also used for termite control under building foundations.*

Copper sulfate (F,A): *Used for control of most species of algae in ponds, lakes, potable water, fish hatcheries, rice fields, streams, ditches, swimming pools, etc. Used as a fungicide to control bacterial and fungal diseases of fruit, vegetable, nut and field crops.*

DDT (OC) (Dichlorodiphenyltrichloroethane): *It is used mainly to control mosquito-borne malaria. DDT has been banned for general use in the USA since 1972 and in Australia since 1987. DDT was commonly used as an insecticide in cotton and tobacco farming and still in use in Asia, Africa and Central America for control of vectors of malaria, yellow fever and sleeping sickness.*

Endosulfan (OC): *Technical grade endosulfan comprises two stereoisomers, (α) alpha-endosulfan and (β) beta-endosulfan present in the ratio of 64-67% α and 29-32% β. It is used to control sucking, chewing and boring insects and mites to a wide range of crops including citrus, vines, olives, potatoes, ornamentals, cucurbits, cotton, tea, coffee, rice, cereals, maize, sorghum, oil seed crops, hops, hazels, and sugarcane, tobacco, alfalfa, mushrooms, glass house crops and forestry.*

***Glyphosate* (H):** *Herbicide, used to control annual and perennial plants including grasses, broad-leaved weeds, pre-harvest, in cereals, peas, beans, oilseed rape, flax and mustard; control of annual and perennial grasses, and broad-leaved weeds in stubble and post-planting/pre-emergence of many crops; as a direct spray in vines and olives in pasture, orchards, forestry and industrial weed control; and also as an aquatic herbicide.*
***Molinate* (H):** *It is used to control germinating broad-leaved and grass weeds (in particular Echinochola spp., in rice) and post-emergence in other type of rice culture.*
***Pyrethrins* (I, A):** *It is used to control a wide range of insects, and mites on fruit, vegetables, field crops, ornamentals, glasshouse crops, and house plants. Also used as personal insect repellents and for human head lice control.*
***Simazine* (H):** *It is used to control most germinating annual grasses and broad-leaved weeds in pome and stone fruits, bush and cane fruits, citrus fruits, vines, strawberries, nuts, olives, pineapples, field beans, French beans, peas, maize, sweet corn, asparagus, hops, alfalfa, lupins, oilseed rape, artichokes, sugarcane, cocoa, coffee, rubber, oil palms, tea, turf and ornamentals.*
***2,4-dichlorophenoxyacetic acid* (H):** *Used for post-emergence control of annul and perennial broad-leaved weeds in cereals, maize, sorghum, grassland, established turf, grass seed crops, orchards (pome and stone fruit), cranberries, asparagus, sugarcane, rice, forestry, and on non-crop land; also used in controlling broad-leaved aquatic weeds.*

A = algicide, F = fungicide, H = herbicide, I = insecticide, OC = organochlorine, OP = organophosphorus.

One estimate from USA indicates that crop yield losses could be 24–57% (depending on the crop species), if no pesticides or alternative crop protection measures were used. Although agriculture accounts for two-thirds of all expenditures on pesticides, non-agricultural uses (house holds, home gardens) are also significant.

Despite above benefits, there is some concern about the adverse effects of pesticides on non-target organisms including humans, fish and wildlife. For example, non-target pesticide poisoning could cause fish kills; reproductive failure in birds and illness in humans and possible endocrine disruption. Systemic insecticides which poison pollen and nectar in the flowers may kill needed pollinators such as bees. It has been estimated that less than 0.1% of the pesticide applied to crops actually reaches the target pest whereas the rest enters the environment via contamination of air, soil and water. A number of pesticides can persist in the environment for a long period, are bio-accumulative and toxic.

9.5: Global pesticides usage

World pesticide use has exceeded 5 billion pounds (2 billion kilograms) in 2000, of which herbicides usage accounted for the largest portion (44%) followed by insecticides and fungicides. World total pesticide expenditures totalled more than US$ 32.5 billion in 2000 (Table 9.4). Overall, North America, Western Europe and Asia are the major users of pesticides (Table 9.4). According to *http://www.farmingsolutions.org/facts/factcontent_det.asp?cnt=6&id=40*, the world agro-chemical market by crop comprised as follows: fruits and vegetables - 26%; cereals - 15%; maize - 12%; rice -10%; soybeans - 9.4%; cotton - 8.6%; sugar beet - 2.8%; oilseed rape - 1.6% and other crops - 14%.

9.6: Regional pesticide usage - Australia and Asia

According to the Australian Academy of Technological Sciences and Engineering (Radcliffe 2002; www.atse.org.au), the principal form of pesticides used in Australia are herbicides, followed by insecticides, fungicides and growth promotants (see Table 9.5). On the otherhand, insecticides are used extensively in Asian countries such as Bangladesh, China and India. In addition, a number of banned or restricted pesticides are still being used in Asia, these include DDT, HCH, methamidophos and monocrotophos (see Table 9.5).

Table 9.4: ***Global pesticide consumption during 1983-2000 (US$ million) and Global pesticide use and expenditure in 2000.*** *[http://www.farmingsolutions.rg/facts/factscontent_det.asp?cnt=&id=4;http://www.greenpeace.org/china/en/campaigns/food-and-agrucltu re/pestcides/pestic; Frederick M. Fishel@http://eds.ifas.ufledu/P1180*].*

Global pesticide consumption (1983-2000)				
Region	**1983**	**1993**	**1998**	**2000***
North America	3,991 (19.5%)	7,377 (28.8%)	8,980 (26.3%)	29.1%
Latin America	1,258 (6.1%)	2,307 (8.4%)	3,000 (8.8%)	na
South America	na	na	na	14.9%
Western Europe	5,847 (28.5%)	7,173 (26.1%)	9,000 (26.3%)	19.8%
Eastern Europe	2,898(14.1%)	2,571(9.3%)	3,190 (9.3%)	2.7%
Africa/Mideast	942(4.6%)	1,258(4.5%)	1,610 (4.7%)	na
Asia/Oceania	5,572 (27.1%)	6,814 (24.8%)	8,370 (24.5%)	27.5%
Rest of world	na	na	na	6%
World total	***20,507***	***27,500***	***34,150***	***na***
Global pesticide use				
Pesticides category	**Millions pounds (a.i)**	**%**	**US$ million**	**%**
Herbicides	1,944	36	14,319	44
Insecticides	1,355	25	9,102	28
Fungicides	516	10	6,384	19
Other	1,536	29	2,964	9
Total	***5,046***	***100***	***32,769***	***100***

na = not available; a.i = active ingredient;
*Frederick M. Fishel, Pesticide Information Office; Florida Cooperative Extension Service, Institute of Food and Agricultural Sciences, University of Florida, Gainesville, FL 32611.

Table 9.5: ***Annual pesticide consumption in Australia and some selected Asian countries*** *[Radcliffe 2000; Huang and Thiemann 2002; FAO 2005; Hong-Yan et al. 2007; Poolpak et al. 2008; Tariq et al. 2009; Abhilash and Singh 2009; and http://www.pan.-uk/pestnews/Issue/Pn34/pn34p18.htm; http://chinaenvironmentandhealth.org/article.asp?id=48; http://econ.worldbank.org/WBSITE/EXTERNAL/EXTDEC/EXTRESEARCH/EXTPROGRAMS/EXTIE/NIPREXT/0,,contentMDK:21007293~menuPK:3201823~pagePK:64168182~piPK:64168060~theSitePK:1909289,00.html; http://www.agnet.org/library/ac/1998e/; http://www.nihs.go.jp/GINC/meeting/7th/7profile/thai-rep.pdf; http://www.pan-uk.org/pestnews/Issue/pn29/pn29p11.htm]*

Country	Tonnes (active ingredient)	Tonnes consumed	Value in US$' 000	Remarks
Australia	na	38,500	na	- Herbicides are the largest group used (comprising glyphosate 15,000 tonnes; atrazine 3,000 tonnes; simazine 3,000 tonnes; 2,4-D and MCPA 1,000 tonnes); others - insecticides (13,000 tonnes including parathion methyl, endosulfan, metham sodium) and fungicides (3,000 tonnes including mancozeb and captan); and plant growth regulators (500 tonnes) (data of 2000). - The Australian Pesticides and Veterinary Medicines Authority (APVMA) is responsible for the assessment and registration of pesticides and veterinary medicines in Australia.

Contd...

Bangladesh	3,635	22,100	75,000	- Pesticides use in Bangladesh has been more than doubled since 1992 due to expansion of agricultural activities (from 7,350 tonnes in 1992 to 22,116 tonnes in 2004). - About 70% of pesticides are used in rice/paddy production and the rest 30% in tea, sugarcane, potato, mango, banana and vegetables. - Main pesticides being used are butachlor, carbofuran, cartap, chlorpyrifos, cypermethrin, diazinon, sulphur, mancozeb, phenthoate, pretilachlor, propiconazole, and quinalphos. - Insecticides is the main pesticide used (64.8%), followed by fungicides (19.3%), herbicides (15.6%) and other (acaricide and rodenticide -0.3%). - The following pesticides are banned for use in agriculture in Bangladesh: BHC, DDT, dichlorvos, dieldrin, heptachlor, methamidophos, methylbromide, monocrotophos, phosphamidon, 2,and 4, 5-T. - The pesticide industry in Bangladesh comprised of private and foreign companies.
Cambodia	42	198	226	- Cambodia does not produce or export pesticides but imports predominantly from Viet Nam, Thailand, China, Malaysia, France, Singapore and Taiwan etc. - In 2002, the pesticides used comprised insecticides (73.3%), rodenticides (15%), herbicides (8.1%) and fungicides (3.5%).
China	258,000	1,000,000	5,670,000	- There are about 2,600 pesticide producers (including those producing public health insecticides) in China, with a production capacity of 850,000 tonnes, of over 600 varieties, 800,000 tonnes of formulation and over 20,000 formulated products, and becoming the largest consumer and the second largest producer country in the world. - In 2005, China produced 1,039,000 tones of pesticides and exported 428,000 tones. - Around 470 million hectares are treated with pesticides each year in vegetables (23.8%), rice (18.1%) and wheat (10.4%) farming which account for half of all pesticide usage. - Among the pesticides usage, insecticides account for 74%, herbicides 13%, fungicides 10% and other 3% during 1996.
DPR Korea	3,000	12,000	60,000	- Information lacking.
India	41,020	164,080	820,400	- India still manufactures, uses and exports some of toxic chlorinated pesticides including DDT, HCH and PCP. - Pesticides commonly used in India are: organophosphates (chlorpyrifos, dimethoate, fenthion, and monocrotophos**, phorate***, profenofos**, quinalphos); organochlorines (DDT, endosulfan, and lindane); carbamates (carbaryl, carbendazim, carbofuran**) and pyrethroid (cypermethrin**).

Contd...

				- Three commonly used pesticides (lindane (g-HCH), DDT and malathion)) accounts for 70% of the total usage. - Insecticides account 80% of the total pesticide used, followed by herbicides and fungicides. - The higher insecticides use in India apparently related to warmer, humid and tropical climate that provides a favourable breeding environment for insects with shorter life cycle and higher hatching rate). - The major uses of pesticides are in Andhra Pradesh, Karnataka, Maharashtra, Gujarat and Punjab. - Pesticides are used mainly to grow cotton, paddy and cole crops. - Application rates of pesticide is only 0.5 kg/ha (compared to 6.6 kg/ha in Korea and 12.0 kg/ha in Japan). - India alone produces 90,000 tonnes of pesticides (the largest industry in Asia after China) and twelfth largest in the world.
Indonesia	na	na	na	- Total active ingredients of insecticides, herbicides and fungicides registered in Indonesia are 94, 70 and 61 respectively. - Rodenticides, nematicides, bactericides, are also used but in small quantity. - A number of pesticides have been banned including 2,3,5-T, aldrin, aldicarb, chlordane, DDT, dieldrin, endpin, EPN, HCH, heptachlor, lindane, methoxychlor, parathion-methyl, PCP, toxaphene etc.
Laos	10	40	200	- The use of chemicals in plant protection is of relatively low importance in Lao PDR. - The country does not produce any active ingredients, nor does it formulate any pesticides locally. All pesticides are imported and used mainly on dry-season irrigated rice, vegetables and marketable high-value crops - Pesticides use comprised insecticides (19 kinds), fungicides (10 kinds), herbicides (5 kinds), and control of snails (3 kinds) and nematode (2 kinds). - Herbicides and fungicides are rarely used. - 26 kinds of pesticides are prohibited for usage.
Malaysia	51,065	204,260	85,020	- In the year 2004, 51,065.51 metric tonnes of pesticides was imported. - Endosulfan being phased out and nine pesticides such as aldrin, dieldrin, DDT, endrin, chlordane, hexachlorobenzene, mirex, toxaphene and heptachlor are no longer used in Malaysia. - Pesticides such as folpet, PCP, captafol, parathion-methyl, azinphos-ethyl and benomyl have been deregistered.
Myanmar (Burma)	758	3,030	15,095	- There are three pesticide formulation plants in Myanmar. Two are neem (*Azadirachta indica*)

Contd...

				formulation plants in Paleik and Pakokku another one is pilot pesticide formulation plant in Hmawbi established by UNIDO Aid in 1990. - Pesticide consumption is very low compared to many neighbouring countries and 90% pesticides are used in agriculture. - Annual consumption of pesticides is about 350-450 tonnes (most of them imported). - 77% pesticides are used on rice, groundnuts, cotton and vegetables, of which 90% are insecticides. - Fungicides are mostly used on oil seeds, cereals, sugarcane, horticultural crops and legumes. - Most commonly used pesticides are aldrin, carbofuran, cartap hydrochloride, cypermethrin, diazinon, deltamethrin, fenitrothion, malathion, monophenthorate, phosphamidon, phosphine.
Nepal	145	580	2,100	- Most pesticides used are imported from India, some from China and other countries on the basis of registration. - Nepal has limited use of pesticide per capita and or / per ha. - In comparison to other countries in the Asia Pacific Region, the use of chemical pesticides is one of the lowest. - Use of pesticides is mainly for farming of vegetables, fruits, tea, rice and cotton. - According to the latest estimate the annual import of pesticides is about 177 tons (a.i.) and current annual use is more than 142 tons (a.i.) with 48.3% as insecticides, 46.2% fungicide, 4.4% Herbicide, 1.1% others. - About 319 types of pesticides by trade name (Insecticides-213, Fungicides-71, Herbicides-23, Rodenticides-8, Acaricides-2 and others-2) have been registered for use under Pesticides Act and Rules. - Highly persistent types of pesticides such as phosphamidon and organo-mercury fungicides are banned in Nepal.
Pakistan	32,500	129,589	172,300	- More than 108 types of insecticides, 30 types of fungicides, 39 types of herbicides, 5 types of acaricides, and 6 different types of rodenticides are used in Pakistan (data of 2002). - Insecticides (organophosphates) account for 60% of the total use. - 80% of pesticides are used on cotton production and the reminder is used for paddy, sugarcane, maize, fruits, vegetables and tobacco crops. - Use of pesticides has increased by 1,169% in the last 20 years. - All pesticides are imported as either technical grade or ready formulations.

Contd...

				- 25 pesticides banned in Pakistan including captafol, dalapon, DDT, dieldrin, endrin, heptachlor, methyl parathion, toxaphene, methamidophos, monocrotophos.
Philippines	7,934	31,735	158,675	- The fertiliser and pesticide authority (FPA) is the national authority for registration. - Pesticides use comprised insecticides, fungicides, herbicides, molluscicides, nematicides, rodenticides, fumigants and termiticides and others (surfactants, emulsifiers, wood preservatives). - Organotin and endosulfan have been severely restricted from use in lowland rice or near water bodies and monocrotophos is restricted for use on bean fly control only. - Main pesticide used are as follows: rice (46.2%), banana (15.3%), vegetables (15.1%), mango (9.3%) and pineapple (7.0%).
Republic of Korea	26,610	100,000	842,638	- The rural development administration (RDA) is responsible for pesticide regulation and control. - Extremely toxic pesticides are used in forestry (e.g. aldicarb G is only allowed for use in forests and aluminium phosphide only on stored grain). - Most highly toxic pesticides are prohibited for use in paddy rice.
Sri Lanka	1,696	6,329	49,000	- Pesticides are imported as ready-to-use products in handy packages, bulk formulations or technical materials for local formulations. - Over 1,000 brand pesticide products are registered (approximately 55 insecticide active ingredients, 32 fungicide active ingredients and 30 weedicide active ingredients) which are marketed and/or handled by more than 120 private sector organizations. - In 2000, pesticide use (active ingredient) comprised insecticides (22.2%), herbicides (52%), and fungicides (25.7%).
Thailand	49,108	132,509	253,537	- Most of pesticides are imported including endosulfan (OC-insecticide), carbendazim (fungicide), copper oxychloride (fungicide/bactericide), methamidophos OP-insecticide), methyl- parathion (OP-insecticide), monocrotophos (OP-insecticide), sulphur (fungicide/acaricide). - Insecticides were the most imported pesticides, followed by herbicides and plant growth regulators. - Organochlorines such as DDT, HCH, lindane, dieldrin and aldrin may still be in use for agricultural and public health purposes (though most of the above pesticides are banned, perhaps enter via illegal ways).
Vietnam	24,473	50,000	159,000	- About 99% of pesticides quantities used in Vietnam are imported.

Contd...

				- Before 1992, there were only 77 active ingredients with 96 product names belonging to 25 foreign firms that have been permitted for use in Vietnam. By 2005, there have been more than 400 active ingredients with more than 1,000 product names which are supplied by more than 100 foreign and domestic companies. - In 1991 insecticide made up 83.3%, fungicide 9.5%, herbicide 4.1%, others 3.1%, respectively, however in 2004 that ratio was 38.0%, 28.0 percent, 30% and 1.8%, respectively. - Banned or restricted pesticides such as DDT, methamidophos, monocrotophos and parathion methyl may still be in use. - Application rates in Vietnam are around 1.0 kg/ha (a.i.), (compared to 11.8 kg/ha in Korea and 19.4 kg/ ha in Japan).

*na = data not available; *** = extremely hazardous;** = highly hazardous according to WHO classification (WHO 2004); DDT= p,p'-dichlorodiphenyltrichloroethane; HCH = hexachlorocyclohexane; OC = Organochlorines, OP = Organophosphorus; PCP= pentachlorophenol Cole crops: Crops such as cabbage, broccoli, Brussels sprouts, and other crucifers; a.i = active ingredients.*

Neem[a] = Neem is a tree (*Azadirachta indica*) of the mahogany family, Meliaceae. It (neem) is native to India, Myanmar, Bangladesh, Sri Lanka and Pakistan; growing in tropical and semi-tropical regions and is a source of environment-friendly biopesticides. Products made from neem have proven to be of medicinal properties, being antihelmintic, antifungal, antidiabetic, antibacterial, antiviral, anti-fertility, and sedative (see http://en.wikipedia.org/wiki/Neem).

9.7: Physico-chemical properties of pesticides

9.7.1: Bio-accumulation properties and K_{ow}

Bio-accumulation is an increase in the concentration of a chemical in biological organisms over time, compared to the chemical's concentration in the environment. The pesticides with high bio-accumulation potential have a modest diversity of chemical structure and have very stable bonds and molecular structure show a high degree of symmetry (see Figure 9.2 for the molecular structure of DDT compared to azinphos methyl). In most cases, the bonds present are C-C aliphatic, C-C aromatic; C-H, C-Cl or other halogens. These groups of compounds are relatively non-polar, having low water solubility but have high fat or lipid solubility. They are called 'lipophilic compounds' or 'hydrophobic compounds' and are most likely to bio-accumulate in biota.

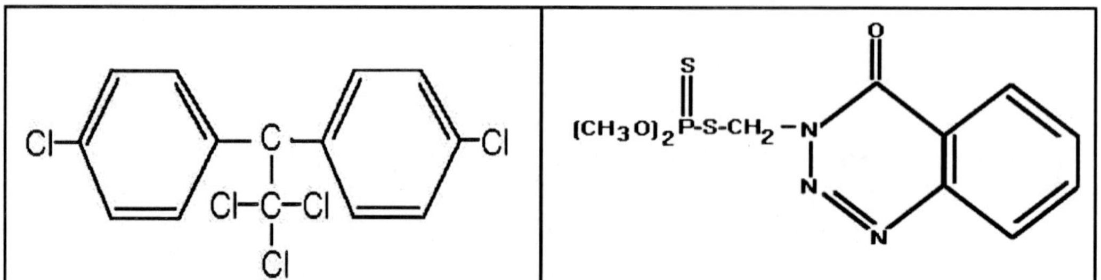

DDT *Azinphos-methyl*

Figure 9.2: ***Structure of organochlorine (DDT) with simple bonds (C-C, C-Cl) and organophosporus pesticides (azinphos methyl) with many functional groups.***

A good indication of the potential of organic compounds to bio-accumulate is given by the octanol-water partitioning coefficient or expressed as log K_{ow}. The K_{ow} is the ratio of the concentration of a chemical in *n*-octanol (a surrogate for animal lipid) to the concentration in water both at equilibrium and at a constant temperature. In general, bio-accumulation of organic compounds increase with increasing liophilicity. However, there is an upper limit to log K_{ow} above which bio-accumulation will not occur. Most of the potentially bio-accumulating compounds have log K_{ow} values between 3 and 7 (see Table 9.2 for log K_{ow} values for different pesticides). Chemicals with log K_{ow} values below 3 are not considered to be bio-accumulative. The chlorinated hydrocarbons and polyaromatic hydrocarbons exhibit the highest capacity for bio-accumulation. For example, DDT, endosulfan, chlordane have a high potential to bio-accumulate in biota (log K_{ow} >4). However, bio-accumulation tends to decrease beyond log K_{ow} 6 due to increasing molecular size and decreasing solubility in fat.

Pesticides which have more limited capacity for bio-accumulation are structurally quite diverse (some herbicides and organophosphours insecticides). These compounds contain many functional groups which are susceptible to attack in their structure by biotic and abiotic degradation processes (see Figure 9.2). These groups generally contain lesser number of very stable bonds compared to chlorinated and polyaromatic hydrocarbon groups. The relatively low lipophilicity leads to a lack of bio-accumulation capacity in this group showed by lower log K_{ow} values (log K_{ow} of some herbicides: acrolein (1.08); amitrole (-0.97); glyphosate (<-3.2); 2,4-D (2.58); simazine (2.1)) (see also Table 9.2).

9.7.2: Persistence properties and half-life

Insecticides such as organochlorine pesticides or chlorinated hydrocarbons (e.g. DDT, heptachlor and endosulfan) have limited range of bonds which are relatively resistant to attack by abiotic or biotic agents in the environment. This group of pesticides tend to have low polarity and low solubility in water. This property means that environmental degradation of oragnochlorine pesticides is restricted. As a result most compounds in this group persist in soils with long half-lives (0.04-15.6 years). Table 9.2 reflects the half-life (DT_{50}) values of a range of pesticides.

Alternatively, many herbicides are a more chemically reactive group with greater water solubility than the organochlorine pesticides (see Table 9.2 for water solubility and half-lives for a range of pesticides). They are susceptible to hydrolysis when they come in contact with water. The half-lives of herbicides are considerably less in soils than the oragnochlorine pesticides. It means that herbicides are less persistent in soils.

Persistence is measured as the time it takes for half of the initial amount of a pesticide to breakdown. Thus, if a pesticide's half-life is 30 days, half will be left after 30 days, one-quarter after 60 days, and one-eighth after 90 days and so on. Pesticides which have a short half-life and are also very soluble in water can move rapidly through certain soils. As the pesticide moves away from the surface, it moves away from the degrading agents such as sunlight and bacteria and as it gets deeper into the soil, it degrades more slowly and thus has a chance to end up in groundwater (leaching). In general, a pesticide which has high solubility in water is more easily carried-off with rainwater as run-off. Water soluble pesticides can have adverse effects on fish and other organisms. If the pesticide is very insoluble in water, it is likely to stick to soil or sediments or biota (see Figure 9.3). The greater the half-life period of a pesticide, the greater is its persistence in the environment. Persistent chemicals tend to have a half-life in soil of greater than 2-3 weeks (see Table 9.2 and 9.6).

Table 9.6: ***Classification of pesticides according to their mobility and persistence in soil*** *[Shaw and Chadwick 1998]*.

Mobility class	K_{oc} (cm^3 g^{-1}) of organic carbon	Persistence class	t½ (days)
Non-mobile	>4000	Non-persistent	<5
Slightly mobile	4000-500	Slightly persistent	5-21
Moderately mobile	499-75	Moderately mobile	22-60
Very mobile	<15	Very persistent	>60

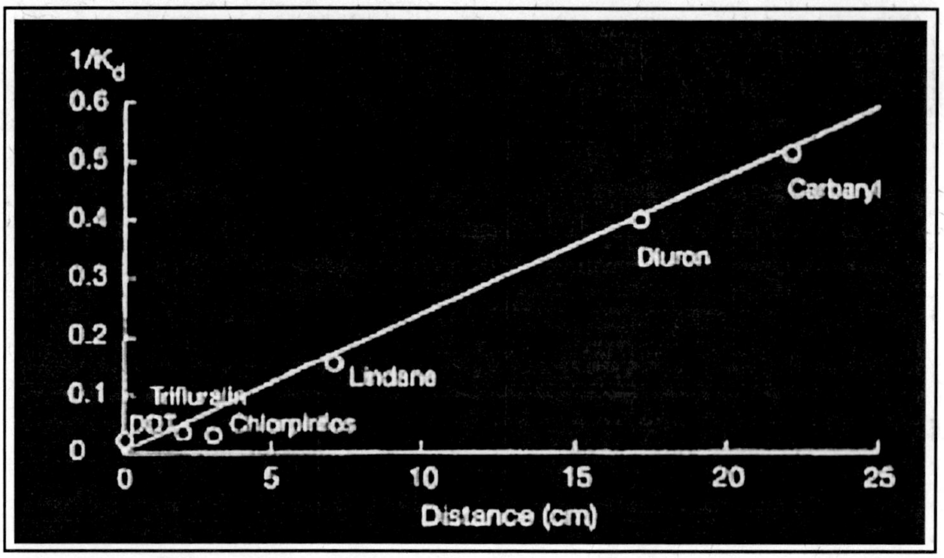

Figure 9.3: ***The relation between leaching distance of several pesticides and their co-efficient of distribution (K_d) in the soil.*** *[Navarro et al. 2007; Note: The hydrophobic products, such as DDT remain on the soil surface longer, while other, more water-soluble products, such as carbaryl or diuron, is more liable to leach].*

9.7.3: Mobility and K_{oc}

Mobile chemicals are those which tend to leach from soil into water. The prediction of mobility of a chemical in an ecosystem can be done by determining its binding capacity to organic material (e.g. humus). This is termed as the organic carbon binding constant (K_{oc}). K_{oc} is the ratio of the mass of a chemical that is adsorbed in the soil per unit mass of organic carbon in the soil per the equilibrium chemical concentration in solution. Higher K_{oc} values correlate to less mobile organic chemicals, while lower values correlate to more mobile organic chemicals (Table 9.6). Mobile chemicals have K_{oc} values of less than 500 (see Table 9.6). Table 9.2 reflects the K_{oc} values of a range of herbicides, insecticides and fungicides.

9.7.4: Mode of action of pesticides

[Based on CropLife Australia 2008; http://feql.wsu.edu/entom490/Lect4MOASpray.pdf and http://plant-disease.ippc.orst.edu/articles.cfm?article_id=12; http://www.irac-online/document/MoA%20classification-aug08-v6.i.pdf; http://www.entmclasses.umd.edu/peap/leaflets/pil43.pdf]

The toxicity of pesticides is manifested by interactions with macromolecules like enzymes or cytochromes to cause inhibition of function. For example, some fungicides and insecticides function by inhibiting the function of cytochromes important for electron transfers. While in some cases normal function of macromolecule is mimicked thereby enhancing its effect to the detriment of the pest (e.g. mimicking of plant growth hormones like auxin by 2,4-D; dicamba, triclopyr; clopyralid). According to the mode of action, pesticides can be classified as contact and systemic herbicides or insecticides (see Table 9.7).

Some herbicides (simazine and oryzalin) are applied directly to the soil before the emergence of weeds, whereas some herbicides are sprayed directly on germinating weed (e.g. glyphosate, which is a true phloem mobile systemic, whereas paraquat is a non-systemic contact herbicide). Some herbicides are broad spectrum and can kill both monocots (grasses) and dicots (broadleaf plants). Glyphosate inhibits aromatic amino acid synthesis and paraquat causes extensive oxidation of cell macromolecules. Oryzalin inhibits root elongation by inhibiting cell division and elongation. Simazine is used as a pre-emergent herbicide to control broadleaf plants. It inhibits photosynthesis. 2,4-D is used as a post-emergent herbicide to control broadleaf plants. It mimics the action of the auxin hormone.

Table 9.7: ***Classification of pesticides based on their mode of action*** *[Broadley et al. 1990].*

Herbicides	Insecticides	Fungicides
Contact herbicides • Kill only the plant tissues contacted via a rapid collapse of affected tissue. • Used for control of weeds. • Examples are diquat, paraquat, dinoseb and bromoxynil.	*Contact insecticides* • Kill insect pests by direct contact. • Pesticide enters the insects' body by absorption through the cuticle with pesticides droplets and dust. • Examples are many insecticides including organochlorines, organophosphorous, carbamates, synthetic pyrethroids.	*Protectant fungicides* • These chemicals are applied before infection occurs and inhibit spore germination by lethal or sub-lethal effects on fungal hyphae. • Examples are copper sulphate plus lime, copper oxychloride, dithiocarbamates.
Systemic herbicides • These chemicals move throughout plant tissues via the vascular system. • Effective against both annual and perennial weeds. • Examples are glyphosate, 2,4-D MCPB.	*Systemic insecticides* • Chemicals are absorbed by the roots or leaves and translocated systemically to other plant tissues. • Effective against aphids which feed on plant sap. • Examples are aldicarb, carbofuran, and disulfoton.	*Systemic fungicides* • Movement within plants occur via translocation of chemicals across leaf lamina from one side to the other.
Residual herbicides • These chemicals are applied to the soil, to be taken up by the germinating weeds. • Examples are trifluralin, atrazine, hexazinone and bromacil.	*Fumigant insecticides* • These chemicals enter the insects' body through the spiracles when in a gaseous state. • Used mainly to control pests of stored grain. • Examples are methyl bromide and phosphine.	*Eradicant fungicides* • Taken up by the plant and attack fungus once it has entered the plant. • Examples are dodine (used against apple scab) and sulphur (against powdery mildews).

Insecticides tend to have limited mobility within the plant. Many insecticides are neurotoxins and kill insects rapidly. Carbamates and organophosphates inhibit the neuromodulatory enzyme acetylcholinesterase resulting in excessive nerve signal transmission in the central nervous system and ultimately kill by respiratory failure. Examples are: carbamates (carbaryl, methomyl) and organophosphates (dimethoate, chlorpyrifos and malathion). Imidacloprid and acetamiprid are neurotoxic and function as mimics of the neurotransmitter acetylcholine by binding to its receptor at nerve endings. Avermectins are also neurotoxins and used as acaricides (that kill mites). This blocks the action of the inhibitory neurotransmitter GABA (gamma aminobutric acid). Propargite used as acaricide, which adversely affects cellular energy production by inhibiting the enzyme-ATPase. These compounds mimic the action of the insect moulting hormone ecdysone and cause premature moult.

Non- systemic fungicides act as protectants and are applied to prevent establishment of disease organisms. Inorganic sulfur soaks up electrons that would be used for energy and copper compounds bind to amino acid in enzymes and inhibit activity. Captan (chlorothioalkyl heterocyclics) probably affects multiple biochemical targets; iprodione (dicarboximides) disrupts mitochondrial membranes affecting energy production. Trifloxystrobin and Kresoxim-methyl (strobilurins) inhibit electron transfers and thus inhibits energy production in the electron transport chain of the mitochondria. Fenhexamid (hydroxyanilide) inhibits sterol biosynthesis in fungal disease organism.

Pesticides (herbicide, insecticides, fungicides) have been grouped by their mode of action (i.e. how the pesticide works) to allow pesticides resistance to be managed (see Table 9.8). Use of pesticides with the same mode of action contributes to the development of resistance problems by killing the susceptible pests and leaving only those with resistance to the class of pesticides that works through similar mechanisms. Development of pest resistance can be avoided or delayed by rotating pest control chemicals that work through different modes of action. It is therefore suggested that careful selection of pesticides from different mode of action groups are essential to achieve effective chemical control of weeds and insects. For example, if a weed becomes resistant to an herbicide, it is often becomes resistant to other herbicides in the same mode of action group. Examples of herbicide resistance weeds in Australia are given in Box 9.2.

Table 9.8: ***Pesticides' group based on mode of action - herbicides, insecticides and fungicides*** *[www.apvma.gov.au; CropLife Australia 2008]*

A: HERBICIDES GROUP - BASED ON MODE OF ACTION	
Group A - Mode of action: inhibitors of acetyl coA carboxylase (inhibitors of fat synthesis/ACCase inhibitors); Resistance risk - high	
Chemical family	**Active ingredient**
Aryloxy-pheoxypropionates	*Clodinafop; Cyhalopop; Diclofop; Fenoxaprop; Fluazifop; Haloxyfop; Propaquizafop;Quizalofop*
Cyclohexanediones	*Butroxydim; Clethodim; Sethoxydim; Tepraloxydim; Tralkoxydim*
Phenylpyrazoles	*Pinoxaden*
Group B - Mode of action: Inhibitors of acetolactate synthase (ALS inhibitors); Resistance risk - high	
Imidazolinoes	*Imazamox; Imazapic; Imazapyr; Imazethapyr*
Pyrimidinylthiobenzoates	*Pyrithobac-Na*
Sulfonylureas	*Azimsulfuron; Bensulfuron; Chlorsulfuron; Halosulfuron; Iodosulfuron; Mesosulfuron; Metsulfuron; Rimsulfuron; Sulfometuron; Sulfosulfuron; Thifensulfuron; Triasulfuron; Tribenuron; Trifloxysulfuron*
Triazolopyrimidines (Sulfonamides)	*Florasulam; Flumetsulam; Metosulam*
Group C - Inhibitors of photosynthesis at photosystem II (PS II inhibitors); Resistance risk – moderate	
Amides	*Propanil*
Benzothiadiazinones	*Bentazone*
Nitriles/Hydroxybenzonitriles	*Bromoxynil; Ioxynil*
Phenylcarbamates	*Phenmedipham*
Pyridazinones	*Chloridazon*
Triazines	*Ametryn; Atrazine; Cyanazine; Prometryn; Propazine; Simazine; Terbutryn*
Triazinones	*Hexazinone; Metribuzin*
Uracils	*Bromacil; Terbacil*
Ureas	*Diuron; Fluometuron; Linuron; Methabenzthiazuron; Siduron; Tebuthiuron*
Group D – Inhibitors of microtubule assembly; Resistance risk - moderate	
Benzamides	*Propyzamide*
Benzenedicarboxylic acids	*Chlorthal*
Dinitroanilines	*Oryzalin; Pendimethalin; Trifluralin*
Group E - inhibitors of mitosis/microtubule organisation; Resistance risk - moderate	
Carbamates	*Carbetamide; Chlorpropham*

Contd...

Group F - Inhibitors of carotenoids biosynthesis at the phytoene desaturase step (PDS inhibitors); Resistance risk - moderate	
Nicotinanilides/ pyridienecarbozxamide	*Diflufenican*
Picolinamides/ pyridienecarbozxamide	*Picolinafen*
Pyridazinoes	*Norflurazon*
Group G – Inhibitors of protoporphyrinogen oxidase (PPOs); Resistance risk - moderate	
Diphenyl ethers	*Acifluorfen; Oxyfluorfen*
N-phenylphthalimides	*Flumiclorac; Flumioxazin*
Oxadiazoles	*Oxadiazon*
Triazolinones	*Carfentrazone*
Pyrimidindiones	*Butafenacil*
Group H - Bleachers: inhibitors of 4-hydroxyphenyl-pyruvate dioxygenase; Resistance risk - moderate	
Isoxazoles	*Isoxaflutole*
Pyrazole	*Benzofenap*
Group I – Disruptors of plant cell growth; Resistance risk - moderate	
Isoxazoles	*Isoxaflutole*
Pyrazole	*Benzofenap*
Group I – Disruptors of plant cell growth; Resistance risk - moderate	
Benzoic acids	*Dicamba*
Phenoxycarboxylic acids	*2,4-D; 2,4-DB; MCPA; MCPB; Mecoprop*
Pyridine carboxylic acids	*Aminopyralid; Clopyralid; Picloram; Triclopyr*
Quinoline carboxylic acids	*Quinclorac*
Group J – Inhibitors of fat synthesis (Not ACCase inhibitors); Resistance risk - moderate	
Chlorocarbonic acids/ halogenated alkanoic acid	*Dalapon or 2,2-DPA; Flupropanate*
Benzofurans	*Ethofumesate*
Phosphorodithioates	*Bensulide*
Thiodicarbamates	*EPTC; Molinate; Thiobencarb; Tri allate; Vernolate*
Group K - Inhibitors of cell division/inhibitors of very long chain fatty acids (VLCFA inhibitors); Resistance risk - moderate	
Acetamides/ alkanamide	*Napropamide*
Chloroacetamides	*Metolachlor; Propachlor*
Group L – Inhibitors of photosynthesis at photosystem; Resistance risk - moderate	
Bipyridyls/Bipyridylium	*Diquat; Paraquat*
Group M – Inhibitors of ESPS synthase; Resistance risk - moderate	
Glycines	*Glyphosate*
Group N - Inhibitors of glutamine synthase; Resistance risk - moderate	
Phosphinic acids	*Glufosinate*
Group O – Inhibitors of cell wall (cellulose) synthesis; Resistance risk - moderate	
Benzonitriles	*Dichlobenil*
Benzamides	*Isoxaben*

Contd...

Group P – Inhibitors of auxin transport; Resistance risk - moderate	
Phthalamates	*Naptalam*
Group Q – Bleachers: Inhibitors of cartooned biosynthesis unknown target; Resistance risk - moderate	
Triazoles	*Amitrole*
Isoxazolidinones	*Clomazone*
Group R – Inhibitors of dihydropteroate synthase (DHP inhibitors); Resistance risk - moderate	
Carbamates	*Asulam*
Group Z – Herbicides with unknown and probably diverse sites of action; Resistance risk - moderate	
Arylaminopropionic acids	*Flamprop*
Dicarboxilic acids	*Endothal*
Organoarsenicals	*DSMA, MSMA*
B : INSECTICIDES GROUP - BASED ON MODE OF ACTION	
Group 1A and 1B – Acetyl choline esterase inhibitors	
Carbamates	*Aldicarb; Bendiocarb; Carbaryl; Carbofuran; Carbosulfan; Methiocarb; Methomyl; Oxamyl; Pirimicarb; Propoxur; Thiodicarb*
Organophosphorous	*Acephate; Azamethiphos; Azinphos-methyl; Cadusafos; Chlorfenvinphos; Chlorpyrifos; Chlorpyrifos methyl; Diazinon; Dichlorvos; Dimethoate; Disulfoton; Ethion; Fenamiphos; Fenitrothion; Fenthion; Malathion / Maldis; Methamidophos; Methidathion; Mevinphos; Omethoate; Parathion-methyl; Phorate; Phosmet; Pirimiphos-methyl; Profenofos* *Prothiofos; Tempephos; Terbufos; Trichlorfon*
Group 2A, 2B, 2C – GABA-gated chloride channel antagonists	
Cyclodiene organochlorins	*Endosulfan*
Polychlorocycloalthanes	*Dicofol; Tetradifon*
Fipronil	*Fipronil*
Group 2D – Neuronal inhibitor (unknown mode of action)	
Bifenazate	*Bifenazate*
Group 3A - Mode of action: sodium channel modulators	
Pyrethroids	*Allethrin; Alpha-Cypermethrin; Beta-cyfluthrin; Bifenthrin; Bioallethrin; Bioresmethrin; Cyfluthrin; Cypermethrin; Cyphenothrin; Deltamethrin; Esbiothrin; Esfenvalerate; Fenvalerate* *Flumethrin; Gamma-cyhalothrin; Imiprothrin; Lambda-cyhalothrin; Permethrin; Prallethrin;* *Pyrethrins; Tau-fluvalinate; Tetramethrin; Transfluthrin*
Group 4A, 4B, 4C - Mode of action: nicotinic acetylcholine receptor agonists/antagonists	
Neonicotinoids	*Acetamiprid; Imidacloprid ; Thiacloprid; Thiamethoxam*
Nicotine	*Nicotine*
Dimethylaminopropane -1, 3-dithiol Bensultap	*Cartap; Bensultap*
Group 5A - Mode of action: nicotinic acetylcholine receptor agonists (allosteric) (not group 4)	
Spinosyns	*Spinosad*
Group 6A, 6B - Mode of action: chloride channel activators	
Avermectins	*Abamectin/Avermectins; Emamectin benzoate*
Milbemectins	*Milbemectin/Milbemycin*

Contd...

Group 7A, 7B, 7C - Mode of action: juvenile hormone mimics	
Juvenile hormone analogues Fenoxycarb Pyriproxyfen	*Methoprene* *Fenoxycarb* *Pyriproxyfen*
Group 8A, 8B - Mode of action: compounds of unknown or non specific mode of action (fumigants)	
Methyl bromide Phosphine generating compounds	*Chloropicrin; Methyl bromide* *Aluminium phosphide; Magnesium phosphide;Phosphine*
Group 9A, 9B, 10A, 10B - Mode of action: compounds of unknown or non specific mode of action (selective feeding blockers)	
Pymetrozine/pyridine Cryolite Clofentezine Hexythiazox Etoxazole	*Pymetrozine* *No registered actives* *Clofentezine* *Hexythiazox* *Etoxazole*
Group 11A, 11B, 11C, 11D, 11E - Mode of action: microbial disruptors of insect midgut membranes (includes transgenic B.t. crops)	
Biological insecticide / *Bacillus thuringiensis*	*Bacillus thuringiensis tenebrionis* *Bacillus thuringiensis israelensis* *Bacillus thuringiensis kurstaki* *Bacillus thuringiensis aizawai* *Bacillus thuringiensis sphaericus* *Bacillus thuringiensis tolworthi*
Group 12A, 12B - Mode of action: inhibitors of oxidative phosphorylation, disruptors of ATP formation (inhibitors of ATP synthase)	
Organotin miticides/acaricide Diafenthiuron	*Fenbutatin oxide* *Diafenthiuron*
Group 13A - Mode of action: uncoupler of oxidative phosphorylation via disruption of H proton gradient	
Pyrazole/arylpyrrole	*Chlorfenapyr*
Group 14A - Mode of action: inhibition of magnesium stimulated ATPase	
Propargite	*Propargite*
Group 15A - Mode of action: inhibitors of chitin biosynthesis, type o Lepidopteran	
Benzoylureas	*Chlorfluazuron; Diflubenzuron; Hexaflumuron; Lufenuron; Triflumuron*
Group 16A - Mode of action: ecdysone agonists/moulting disruptors	
Diacylhydrazines	*Methoxyfenozide; Tebufenozide*
Group 17A - Mode of action: inhibitors of chitin biosynthesis type 1 homopteran	
Buprofezin	*Buprofezin*
Group 18A - Mode of action: moulting disruptor, Dipteran	
Cyromazine	*Cyromazine*
Group 19A - Mode of action: octopaminergic agonists	
Amitraz	*Amitraz*

Contd...

Group 20A - Mode of action: mitochondrial complex III electron transport inhibitors (coupling site II)	
Hydramethylnon	*Hydramethylnon*
Group 21A - Mode of action: mitochondrial complex I electron transport inhibitors	
METI acaricides Rotenone *Rotenone*	*Fenpyroximate; Pyridaben; Tebufenpyrad*
Group 22A - Mode of action: voltage dependent sodium channel blocker	
Indoxacarb	*Indoxacarb*
C : FUNGICIDES GROUP - BASED ON MODE OF ACTION	
Group A - Mode of action: inhibition of beta-tubulin biosynthesis; high risk	
Benzimidazole	*Carbendazim; Thiabendazole; Thiophanate-methyl*
Group B - Mode of action: inhibition of NADH cytochrome c reductase in lipids ; high risk	
Dicarboximide	*Iprodione; Procymidone*
Group C - Mode of action: inhibition of C-14 demethylation in sterol biosynthesis; medium risk	
Imidazole	*Imazalil; Prochloraz*
Piperazine	*Triforine*
Pyrimidine	*Fenarimol*
Triazole	*Bitertanol; Cyproconazole; Cyproconazole (+ iodocarb); Diclobutrazol; Difenoconazole; Epoxyconazole; Fluquinconazole; Flusilazole; Flutriafol; Hexaconazole; Myclobutanil; Paclobutrazol; Penconazole; Propiconazole; Propiconazole (+ cyproconazole); Tebuconazole; Triadimefon; Triadimenol; Triticonazole*
Group D - Mode of action: inhibition of RNA polymerase, high risk	
Phenylamide-Acylalanine	*Benalaxyl-M; Furalaxyl; Metalaxyl (+mancozeb); Metalaxyl-m; Metalaxyl-m (+copper oxychloride); Metaxyl-m (+mancozeb)*
Phenylamide -Oxazolidinone	*Oxadixyl; Oxadixyl (+mancozeb); Oxadixyl (+propineb)*
Group E - Mode of action: inhibition of ^8 and ^7 isomerase and ^14 reductase in sterol biosynthesis; low medium risk	
Morpholine	*Tridemorph*
Spiroketalamine	*Spiroxamine*
Group F - Mode of action :n/a	
Phosphorothiolate-organophosphorous	*Pyrazophos*
Group G - Mode of action: inhibition of succinic acid oxidation; medium risk	
Carboxamide/Anilide	*Boscalid; Carboxin; Flutolanil; Oxycarboxin*
Group H - Mode of action?	
Hydroxypyrimidine/pyrimidinol	*Bupirimate*
Group I : Mode of action: inhibition of Methionine biosynthesis; medium risk	
Anilinopyrimidine	*Cyprodinil; Cyprodinil (+fludioxinil); Pyrimethanil; Pyrimethanil (+chlorothalonil)*
Group K : Mode of action: inhibition of Mitochondrial synthesis in cytochrome bc1; high risk	
Strobilurn	*Azoxystrobin; Azoxystrobin (+cyproconazole); Kresoxim-methyl; Pyraclostrobin; Trifloxystrobin*

Contd...

Group L - Mode of action: inhibition of Multisite contact	
Phenylpyrroles	*Fludioxonil; Cyprodinil (+fludioxinil)*
Group M: Mode of action: n/a	
Quinoline	*Quinoxyfen*
Group Y - Mode of action: multisite activity	
Carbamate	*Iodocarb; Propamocarb*
Phosphonate	*Fosetyl-al; Phosphorous acid/phosphonic acid*
Inorganic	*Chlorine dioxide; Copper cuprous oxide; Copper hydroxide; Copper octanoate; Copper oxychloride; Copper sulfate; Hydrogen peroxide + peroxyacetic acid; Iodine; Mercury; Sodium metabisulphite; Sulphur*
Dithiocarbamate	*Mancozeb; Metiram; Propineb; Thiram; Zineb; Zineb (+copper oxychloride); Ziram*
Phthalimide	*Chlorothalonil*
Chlorophenyl	*Quintozene*
Quinone	*Dithianon*
Hydroxyquinoline	*8-hydroxy quinoline sulphate*
Pyradinamine	*Fluazinam*
Cyclic imide	*Captan*
Gunaidine	*Dodine; Guazatine*
Group X - Mode of action: unspecified	
Cinnamic acid derivative	*Dimethomorph*
Sulfamide	*Tolylfluanid*
Dinitrophenyl	*Dinocap*
Organophosphorous	*Tolclofos-methyl*
Thiadiazole	*Etridiazole*
Quinoxaline	*Oxythioquinox/chinomethionate; Pencycuron*

n/a =not available

Box 9.2: ***List of herbicide resistance weeds in Australia*** *[CropLife Australia 2008; www.croplifeaustralia.org.au-list of herbicide weeds in Australia]. http://www.croplifeaustralia.org.au/files/resistancemanagemen/herbicides/2008%20Herbicide%20Resistant%20Weeds%20List%20and%20Integrated%20Weed%20Management%20Strategies.pdf*

Note: Resistance has been confirmed at least one mode of action group in Australia as listed in Table 9.8

Grass weeds and Groups	**Broadleaf weeds**
Annual rye grass (*Lolium rigidium*) (Groups -A, B, C, D, Q, M)	Wild radish (*Raphanus raphanistrum*) (Groups - B, C, F, I)
Wild oat (*Avena* spp.) (Groups -AB, Z)	Indian hedge mustard (*Sisymbrium orientale*) (Groups - B, I)
Northern barley grass (*Hordeum glaucum*) (Groups - A,B,L)	African turnip weed (*Sisymbrium montevidensis)* (Groups - B)
Brome grass (*Bromus rigidius*) (Groups - A, B)	Arrowhead (*Sagittaria montevidensis*) (Groups – B)
Large crab grass (*Digitaria sanguinalis*) (Groups - A, B)	Black bindweed (*Fallopia convolvulus*) (Groups – B)
Barley grass (*Hordeum leporinum*) (Groups - A, L)	Calomba daisy (*Pentzia suffrutocosa*) (Groups – B)
Brome grass (*Bromus diandrus*) (Groups – A)	Charlock (*Sinapsis arvensis*) (Groups – B)
Paradoxa grass (*Phalaris paradoxa)* (Groups - A)	Common sowthistle (*Sonchus oleraceus*) (Groups – B)

Contd...

Barnayrd grass (*Echinochola crus-galli)* (Groups- C, C, M, M)	Dirty dora (*Cyperus difformis*) (Groups – B)
Liverseed grass (*Urochloa panicoides)* (Groups - C, C, M, M)	Iceplant (*Mesembryanthemum crystallinum*) (Groups – B)
Awnless barnyard grass (*Nassella trichotoma*) (Groups - C, C, M, M)	Lincoln weed (*Diplotaxis tenuifolia*) (Groups – B)
Giant Parramatta grass (*Sporobolus fertilis*) (Groups – J)	Patersons curse (*Echinum plantagineum*) (Groups – B)
Serrated tussock (*Nassella trichotoma*) (Groups – J)	Prickly lettuce (*Lactuca serriola*) (Groups – B)
Silver grass (*Vulpia bromoides*) (Groups - C, L)	Starfruit (*Damasonium minus*) (Groups – B)
	Turnip weed (*Rapistrum rugosum*) (Groups – B)
	Wild turnip (*Brassica tournefortii*) (Groups – B)
	Stinging nettle (*Utrica urens*) (Groups – C)
	Dense-flowered fumitory (*Fumaria densiflora*) (Groups – D)
	Capeweed (*Arctotheca calendula*) (Groups – L)
	Square weed (*Mitracarpus hirtus*) (Groups – L)

9.8: Environmental and biological impacts of pesticides

9.8.1: Toxicity and eco-toxicity of some common pesticides

All pesticides are toxic to some forms of life. There is a standard toxicity classification schemes for rating the relative toxicity of pesticides to different groups of organisms (mammals, birds and aquatic organisms). The classification helps in interpreting toxicity data and pesticide risk assessment as given in Table 9.9. Different classes of pesticides often show general patterns for toxicities. Organophosphorous and carbamate pesticides are usually more toxic to invertebrates than fish (see Table 9.10), but their toxicity varies from one species to another. Synthetic pyrethroids generally low toxic to mammals and birds but are highly toxic to fish and invertebrates. Many herbicides have low toxicity to fish and invertebrates. Table 9.10 provides a comprehensive mammalian (rat) and ecotoxicology values (birds, fish, daphnia, algae and bees) of various herbicides, insecticides and fungicides respectively.

Table 9.9: ***Toxicity classification of pesticides*** *[Kamrin 1997].*

Mammals	Birds	Aquatic organisms
High toxicity -I Oral LD_{50}: 0-50 mg/kg Dermal LD_{50}: 0-200 mg/kg Inhalation LC_{50}: 0-0.2mg/l; skin and eye irritation "severe".	***Very high toxicity*** Oral LD_{50}: <10 mg/kg Dietary LC_{50}: <50 ppm	***Very high toxicity*** Acute LC_{50}: <0.1 mg/l; Chronic LC_{50} <0.01 mg/l or <10 μg/l)
Moderate toxicity-II Oral LD_{50}: >50-500 mg/kg Dermal LD_{50}: >200-2000 mg/kg Inhalation LC_{50}:>-0.2-2.0 mg/l; skin and eye irritation "moderate".	***High toxicity*** Oral LD_{50}: 10-50 mg/kg Dietary LC_{50}: 50-500 ppm	***Very high toxicity*** Acute LC_{50}: 0.1-1 mg/l; Chronic LC_{50} <0.01-0.1 mg/l
Slight toxicity-III Oral LD_{50}: >500-5000 mg/kg Dermal LD_{50}: >2000-20000 mg/kg Inhalation $LC5_{50}$:>-2-20 mg/l; skin and eye irritation "slight".	***Moderate toxicity*** Oral LD_{50}: >50-500 mg/kg Dietary LC_{50}: >500 -1000 ppm	***Moderate toxicity*** Acute LC_{50}: >1-10 mg/l; Chronic LC_{50} >0.1-1 mg/l
Non-toxic-IV Oral LD_{50}: >5000 mg/kg Dermal $LD5_{50}$: >20000 mg/kg Inhalation LC_{50}:>20 mg/l; skin and eye irritation "no irritation".	***Slight toxicity*** Oral LD_{50}: >500-5000 mg/kg Dietary LC_{50}: >1000-5000 ppm	***Slight-toxicity*** Acute LC_{50}: > 10-100 mg/l; Chronic LC_{50} >1-10 mg/l

LD_{50} or LC_{50} is the median lethal dose (LD_{50}) or median lethal concentration (LC_{50}), which is the dose or concentration of the test substance that kills 50% of test organisms within a given time period.

Table 9.10: ***Herbicides, insecticides and fungicides toxicity and ecotoxicity values for mammals, birds, fish, daphnia, algae and bees*** *[Roberts et al. 1998; Tomlin 2006; PAN- http://www.panna.org/; EXTOXNET-http://extoxnet.orst.edu/].*

*LD_{50} rats= oral (male or female); LD_{50} birds = mallard ducks or bob white quail; * indicate other bird species*

LC_{50} fish=96 h for rainbow trout (indicate other than Rainbow trout); SL-solubility limit of a.i; NT=non toxic*

LC_{50} daphnia= 48h; NE= no effect; NT=non toxic

EC_{50} algae= 72 h for Selenastrum capricornutum or Scenedesmus panonicus (indicate other algal species); T=toxic to algae)*

LD_{50} bees= HT=highly toxic; NT=non toxic; T=toxic

C=carbamate; OC=organochlorine; Ne=neonicotinod; OP=organophosphate; P=pyrethroid

HERBICIDES						
Active ingredient	**LD_{50} Rats (mg/kg) oral**	**LD_{50} Birds (mg/kg) oral**	**LC_{50} Fish (mg/l) 96 h**	**LC_{50} Daphnia (mg/l) 48 h/72 h**	**EC_{50} Algae (mg/l) 72 h/120 h**	**LD_{50} Bees (μg/bee) contact/oral**
Azimsulfuron	>5000	>2250	154	>1000	0.012	>25
2,4-D	639-764	>5620	>100	235	33.2	NT; 104.5
2,4-DB	370-700	-	4	-	-	NT
2,4,5-T	500	310 (chicken)	0.98-3.26	5	-	-
Acifluorfen	1370-1540	325	17	77	>260	-
Acrolein	29	19-20.2	0.15	22	0.050	-
Ametryn	1160	>5620	5	28	0.0036	>100
Aminopyralid	>5000	>2250	>100	>100	27-30*	>120
Amitrole	>10000	>2150	>1000	>18	2.3-5.1	NT;>150
Asulam	>4000	>4000	>5000	-	-	NT
Atrazine	1869-3090	940-2000	4.5-11	6.9	0.043	>97-100
Bensulfuron	>5000	>2510	>66	>130	0.02	>100
Bensulide	270-360	1386	1.1	0.58	-	1.6
Bentazone	>1000	1140	>100	125	47.3*	NT; 100
Benzofenap	>15000	-	>10	>10	-	-
Bromacil	1300-2000	2250	36	121	6.8 ppb	NT
Bromoxynil	81-177	217	29.2*	12.5	0.65-44	150
Butachlor	2000	>4640	0.52	2.4	0.0018-0.0027	>90
Butafenacil	>5000	2250	3.9	>8.6	0.0025	>25-100
Butroxydim	1635-3476	1221->2000	>6.9	>3.7	0.71	>200
Carbetamide	>2000	>2000	>100	36.5	-	NT
Carfentrazone	5143	>1000	1.6-43*	9.8	0.012	>35
Chloridazon	2140-3830	>2000	32-46	132	1.9*	>200
Chlorpropham	4200	>2000	7.5	4	3.3	NT; 466
Chlorsulfuron	5545	5000	>250	>112	0.05*	>100
Chlorthal - dimethyl	>10000	>2250	>4.7	>4.6	-	ST
Clethodim	1360-1630	>2000	67	>120	57.8*	100
Clodinafop	1392-2271	>1455	0.39	>74	25	>100

Contd...

Clomazone	2077	>2510	19	5.2	2.1*	-
Clopyralid	2675-3738	1465->2000	103.5	225	6.9-7.3	NT; >100
Cyanazine	182-334	>2000	10-16*	42-106	<0.1*	100-190; NT; >99
Cyhalofop	> 5000	> 5620	>0.49	-	>1	>100
Dalapon or 2,2-DPA or 2,2-Dichloropropionic acid	7570-9330	5660	>100	-	100	NT
Dicamba	1707	>2000	135	110	41->250	NT;>100
Dichlobenil	>2000	683	5-13*	6.2	2	>11
Diclofop -methyl	>481-693	>1100->1600	0.23	0.23	0.53	NT
Diflufenican	>5000	>2150	108.8	0.24	0.00045*	NT
Diquat	408	155	39	0.0022	0.019	22
Diuron	>2000	1140	14.7	1.4	0.022	NT
DSMA (methylarsonic acid)	2833	425	>167	77.5	7.6	68
Endothal	38-54	344-5000	49	92	>4.8	NT
EPTC (S-ethyl dipropylthiocarbamate)	>2000	20000	19	14	-	11
Ethofumesate	>5000	3552-8743	11.9-20.2	13.5-22.0	3.9*	>50
Fenoxaprop	3150-4000	>2000	0.46	0.56	0.51	>300
Flamprop	>4000	>4640	3.19	3	6.8*	NT;>100
Florasulam	>6000	1046	>96	>292	0.00894	>100
Fluazifop	>3030	>17000	1.37	>316	-	LT
Flumetsulam	>5000	>2250	>379*	NT	4.9 (5d)	>100
Flumiclorac	>5 000	>2250	1.1	>38	-	>196
Flumioxazin	>5000	-	2.3	-	-	-
Fluometuron	>6000	2974	30	10	0.16*	>155
Flupropanate	9600-11900	6750-11000	>10000	-	-	-
Glufosinate -ammonium	1620-2000	>5000	710	560-1000	=1000	NH; >100
Glyphosate	>5000	>3851	86	780	485	>100
Halosulfuron	8866	>2250	>131	>107	0.0053	>100
Haloxyfop	337	>2150	1.18	4.64	-	100
Hexazinone	1100	2258	320	442	0.007	>60; NT
Imazamox	>5000	>1846	122	122 NOEC	>0.037*	>25
Imazapic	>5000	>2150	>100	>100	>51.7	>100
Imazapyr	>5000	>2150	>100	>100	71	>100
Imazethapyr	>5000	>2150	340	>1000	50 NOEL	>100
Iodosulfuron	2678	0.2	>100	>100	0.152*	>150
Ioxynil	114-178	62*	0.043	7.7	0.15*	3.27
Isoxaben	>10000	>2000	>1.1	>1.3	>1.4	>100
Isoxaflutole	>5000	>2150	1.7	1.5	0.016	>100

Contd...

Linuron	1500-5000	940	3.15	0.75	0.015*	>1600
MCPA ((4-(4-chloro-2-methylphenoxy)acetic acid)	962-1470	377	50-560	>190	>392	>200
MCPB (4-(4-chloro-2-methylphenoxy)butanoic acid)	1780	282-5000	4.3	55	0.42	>25
Mecoprop	930-1166	>486-1000	150-240	420	532	10-100
Mesosulfuron -methyl	>5000	>2000	100	>1000	0.21*	13
Methabenzthiazuron	>5000	-	15.9	30.6	-	NT
Metolachlor	2672	>2510	1.2	11.2	0.006	85
Metosulam	>2000	>2000	SL	SL	0.075	100
Metribuzin	322-510	164-680	74.6	49.6	0.021	35
Metsulfuron	>5000	>2510	>150	>120	0.157**	>50
Molinate	369-450	>5000	13	-	-	-
Napropamide	>5000	>2200	9.4	24	4.5*	100
Naptalam	1770	>4640	76.1	118.5	-	NT
Norflurazon	>5000	>2510	8.1	>15	0.0176	NT
Oryzalin	>10000	>500	3.26	1.4	0.051	25
Oxadiazon	>5000	>1000	1.2	>2.4	0.006-3*	>400
Oxyfluorfen	>5000	>2150	0.41	15	-	NT
Paraquat	129-157	175-199	26	6.1	0.10-0.28	36-150
Pendimethalin	>5000	1421	0.14	0.28	-	>101.2
Phenmedipham	>8000	>2100	1.4-3.0	3.8	0.13*	NT; 23
Picloram	>5000	6000	5.5	34.4	36.9	>100
Picolinafen	>5000	>2250	0.68	0.45	0.18	>200
Pretilalchlor	60999	>10000 (non toxic)	0.9	13	0.002	93
Profoxydim	>2000	>2000	13-18	18.1	33*	>200
Prometryn	>2000	>5000	5.5	12.66	0.023	NT
Propachlor	>550-1800	91	0.17	7.8	0.015	>197
Propanil	>2500	375	8-11*	4.8	-	-
Propaquizafop	>5000	>2000->2198	1.2	>2	>2.1	>20
Propyzamide	5620-8350	8770	>4.7	>5.6	-	>100
Pyrithobac-Na	3200-3300	>2250	>1000	>1100	0.107	>25
Quinclorac	2680	>2000	>100	113	NT	NT
Quizalofpo	1480-1670	>2000	10.7	2.1	>3.2	>50
Rimsulfuron	>5000	>2250	>390	>360	125 g/L	>100
Sethoxydim	2676-3200	>5000	38	1.5	-	NT
Siduron	>7500	>10000	14	18	0.25	-
Simazine	>7000	>10000	17.5	-	-	NT;>99
Sulfometuron	>5000	>5000	>12.5	>12.5	-	>100

Contd...

Sulfosulfuron	>5000	>2250	>95	>96	0.221	>30
Tebuthiuron	528-620	>500	144	297	0.05	>100
Tepraloxydim	5000	>2000	>100	>100	76	>200
Terbacil	934	>56000*	46.2	68	-	NT
Terbutryn	500-10000	>2000	>100	>100	0.042	>100-225
Terbutryn	2500	5000	1.1	2.66	0.013	
Thifensulfuron	>5000	>2510	>410	970	15.7*	>12.5
Thiobencarb	1033-1130	7800->10000	1.1	1.1	0.017	>100
Tralkoxydim	934-1258	>3020	>7.2	>175	7.6*	>0.1
Tri allate	1100	>2251	1.2	0.43	0.12	NT
Triasulfuron	>5000	>2150	>100	>100	0.35	>100
Tribenuron	>5000	>2250	>738	894	>0.0208*	>100
Triclopyr	577-692	1698	117	133	45	NT; >100
Trifloxysulfuron	>5000	2000-2250	103	>108	0.0065	>25
Trifluralin	>5000	>2000	0.088	0.245	12.2	>100
Triflusulfuron	>5000	2250	760	960	0.62*	1000
Vernolate	1500-1550	12000	4.6	-	-	NT

INSECTICIDES

Active ingredient	LD_{50} Rats (mg/kg) oral	LD_{50} Birds (mg/kg) oral	LC_{50} Fish (mg/l) 96 h	LC_{50} Daphnia (mg/l) 48 h/72 h	EC_{50} Algae (mg/l) 72 h/120 h	LD_{50} Bees (µg/bee) contact/oral
Acephate (OP)	1030-1447	350	>1000	67.2	>980	1.2
Abamectin/ Avermectins	10	84.6	0.0032	0.34 ppb	>100*	T
Acetamiprid	146-217	180	>100*	>200	>98.3	14.5
Aldicarb (C)	0.93	71	>0.56	0.18	1.4	HT; 0.285
Allethrin (P)	900-2150	5620	0.134*	-	-	-
Alpha-Cypermethrin (P)	57	>2025	0.0028	0.0001	>0.1*	T;0.059
Aluminium phosphide/ phosphine	8.7	37.5	$9.7x10^{-3}$	0.2	-	-
Amitraz	>1600	788	0.74	0.035	>12	50
Azamethiphos (OP)	1180	30.2	0.115-0.2	0.00067	-	T; <0.1
Azinphos-methyl (OP)	9	32	0.02	0.0011	7.15	T
Bacillus thuringiensis israelensis -(Bi)	>5000	3077	>370	>25	-	-
Bacillus thuringiensis tenebrionis -(Bi)	-	1961	>100	>12	-	>100
Bacillus thuringiensis aizwai -(Bi)	>5050	>1714	>100	-	275	16.5
Bacillus thuringiensis kurstaki -(Bi)	5050	-	143.5	-	-	NT
Bacillus thuringiensis sphaericus -(Bi)	-	-	-	-	-	-

Contd...

Bacillus thuringiensis tolworthi -(Bi)	-	-	-	-	-	-
Bendiocarb (C)	40-156	3.1-19	1.55	0.33	-	T; 0.1
Bensultap	1105-1120	311	0.76	40	-	LT; 25.9
Beta-cyfluthrin (P)	380	>2000	89 ng/L	0.0003	>0.01	<0.1
Bifenazate	>5000	1142	>0.76	0.5	0.9	>100
Bifenthrin (P)	54.5	1800	0.00015	0.00015	0.00016	0.1-0.015
Bioallethrin (P)	709-1042	2030	0.0094*	0.0356	-	-
Bioresmethrin (P)	7070-8000	>10000*	0.00062	0.0008	-	HT; 2ng
Buprofezin	2198	>2000	>1.4	0.42	>2.1*	NH
Cadusafos(OP)	37.1	16-230	0.13	0.0016	5.3*	-
Carbaryl (C)	264-500	>2179	1.3	0.006	1.1	T; 1
Carbofuran (C)	8	2.5-5	22-29	0.0386	-	T
Carbosulfan (C)	185-250	10	0.042	0.0015	20	T; 1.046
Cartap	325-345	-	1	-	-	MT
Chlorfenapyr	441-1152	10	0.0074	0.00611	132 ppb	0.2
Chlorfenvinphos (OP)	10	16-107*	<0.32*	0.0003	1.6	0.55
Chlorfluazuron	>8500	>2510	=1071*	0.00091	0.39*	>100
Chloropicrin (Fu)	250	-	0.168*	0.91	-	-
Chlorpyrifos (OP)	135-165	490	0.007	0.0017	>0.4	360 ng
Chlorpyrifos methyl (OP)	>3000	>1590	0.41	0.016	0.57	0.38,HT
Clofentezine	>5200	>3000	>0.015	0.00145	NT	>252.6
Cryolite	>5000	>2000	-	-	-	-
Cyfluthrin (P)	500	>2000	0.00047	0.00016	>10	T
Cypermethrin (P)	250-4150	>10000	0.69	0.00015	-	HT
Cyromazine	3387	1785	>100	>9.1	124	NT
DDT	113-118	2240	7	0.0011	-	5
Deltamethrin (P)	135->5000	>4640	0.00091	0.0035	>9.1	79 ng/bee
Diafenthiuron	2068	>1500	0.0038	0.00015	>50	1.5
Diazinon (OP)	1250	2.7	2.6-3.2	0.00096	1	HT
Dichlorvos (OP)	50	24	0.2	0.00019	52.8	0.29
Dicofol	578-595	3010	0.12	0.14	0.075	NT; 140
Dieldrin	38-46	50	0.0037-5.2*	11.9-5370*	-	-
Diflubenzuron	>4640	>4640	>0.2	0.0071	>0.3	NH;>100
Dimethoate (OP)	387	10.5-42	24.2	2	90.4-282	0.12-0.15
Disulfoton (OP)	12-Feb	39	3	0.013	>4.7*	T
Emamectin benzoate ()	76-89	46	0.174	0.00099	-	T
Endosulfan (OC)	70	204-245	0.002*	0.075	>0.56*	NT

Contd...

Endrin	17.8-7.5	-	0.0003-3.4*	0.00004-0.333*	-	-
EPN	34-36	220	0.21	0.0071	-	-
Esbiothrin (P)	-	-	-	-	-	-
Esfenvalerate (P)	75-88	381	0.00026	0.00024	-	0.017
Ethion (OP)	208	128	0.52-0.72*	0.0056	-	T
Etoxazole	>5000	>2000	>40	>40	1	>200
Fenamiphos (OP)	6	0.7-1.6	0.0721	0.0019	11	-
Fenbutain oxide	3000-4400	5065	0.27	0.05-0.08	>0.005	>200
Fenitrothion (OP)	1700-1720	23.6-1190	1.3	0.0086	1.3	T
Fenoxycarb	>100000	>7000	1.6	0.4	1.1	>1
Fenpyroximate	245-480	>2000	0.0055	0.00326	9.98*	NH
Fenthion (OP)	250	7.2	0.83	0.0057	1.79	0.16
Fenvalerate (P)	451	>1600-9932	0.0036	-	-	0.23
Fipronil	95-97	11.3	0.248	0.19	0.068	HT
Flumethrin (P)	0.004	-	-	-	-	-
Gamma-cyhalothrin (P)	>50	>2000	110 ng/L	66.7 ng/L	-	0.005
HCH (OC)	88-270	120-130	0.022-0.028	1.6-2.6	0.78*	0.011
Heptachlor	100-220	2080	0.0074-0.02	-	-	-
Hexaflumuron	>5000	>2000	>0.5	0.00011	>3.2	>100
Hexythiazox	>5000	>2510	>300	1.2	-	NT
Hydramethylnon	1131	>2510	0.16	1.14	-	30
Imidacloprid	450	31-152	211	85	>100	Harmful
Imiprothrin (P)	900-1800	>5620	0.038	0.051	-	-
Indoxacarb	268-1732	98	0.65	0.6	-	23.33
Isofenphos (OP)	20	8.7	3.3	0.0039	6.8	T
Lambda-cyhalothrin (P)	56-79	>3950	0.00036	0.00036	1	39 ng/bee
Lindane (OC)	60-250	>2000	1.7-90 mg/l	460 mg/l (96 h)	-	HT
Lufenuron	>2000	>2000	>73	T;1100	-	1600
Magnesium phosphide	11.2	as above	as above	as above	-	-
Malathion /Maldison (OP)	1375-2800	3500	0.1*	0.001	13*	T;0.71
Methamidophos (OP)	13-15.6	10	25	0.27	>178	T
Methidathion (OP)	25-54	23.6-28	0.01	-	22	ST
Methiocarb (C)	33-47	7.1-9.4	0.436-4.7	0.019	1.15	NT
Methomyl (C)	30-34	24.2	2.49	0.017	>100	T; 00.28
Methoprene	>10000	>4640*	760 ppb	360 ppb	1.33	-
Methoxychlor (OC)	5000-6000	2000	0.02-0.065	0.1	-	NT
Methoxyfenozide	1800	425	-	77.5	-	68
Methyl bromide	3.03	73	3.9	2.6	-	NH
Mevinphos (OP)	3-12	4.63	0.017	-	-	T;0.027

Contd...

Milbemectin/Milbemycin	456-762	968-1005	0.0045	>100	=7.3	0.46
Mirex	235-312	-	0.2-30	0.001	-	-
Monocrotophos	18-20	4.8	7	0.00024	-	HT
Nicotine	50-60	T	4	0.24	-	T
Omethoate (OP)	25	79.7	9.1	0.022	167.5	T
Oxamyl (C)	2.5-3.1	3.16-3.83	4.2	0.319	3.3*	HT; 0.038
Parathion-methyl (OP)	19-30	1044	2.7	0.0073	3	T
PCP (Pentachlorophenol)	210	5139*	0.052-0.17	-	-	-
Permethrin (P)	430-4000	>9860	0.0054	0.0006	-	0.098
Phrnthoate						
Phorate (OP)	1.6-3.7	0.62	0.013	-	-	T;10
Phosmet (OP)	113-160	507	0.23	0.0085	-	1
Phosphamidon	17.9-30	3.8	7.8	0.01-0.22	highly toxic	HT
Pirimicarb (C)	107-142	28.5	79	0.017	140	NT; 4
Pirimiphos-methyl (OP)	1180-1414	40-140	0.64	0.00008	1	0.22
Prallethrin (P)	460-640	1171	0.012	0.0062	-	-
Profenofos (OP)	358	70-200	0.08	0.001	1.16	T; 0.102
Propargite	2800	>4640	0.043	0.014	>1.08	>100
Propoxur (C)	50	2828	3.7-13.6	0.15	-	HT
Prothiofos (OP)	1390-1569	100-200	0.5-1	0.014	2.3	NH
Pymetrozine	5820	>2000	>100	87	47.1	>117
Pyrethrins (P)	1030-2370	>5620	0.0052	0.012	-	22 ng/bee
Pyridaben	820-1350	>2250	1.1-3.1	0.00059	-	0.55
Pyriproxyfen	>5000	>2000	>0.325	0.4	0.064	-
Quinalphos (OP)	71	37	0.005	0.00066	-	HT
Rotenone	132-1500	-	1.9	-	-	NT
Spinosad	3783-5000	>2000	30	14	>105.5	0.0029
Tau-fluvalinate (P)	261-281	>2510	0.0027	0.001	>2.2	6.7
Tebufenozide	>5000	>2150	5.7	3.8	>0.64	>234
Tebufenpyrad	595-997	>2000	0.03	0.046	0.54*	LT
Tempephos (OP)	4204->10000	1200	9.6	-	-	HT; 1.55
Terbufos (OP)	1.6	15	0.01	-	-	NT;4.1
Tetramethrin (P)	>5000	>2250	0.01	-	-	-
Thiacloprid (Ne)	396-836	2716	30.5	=85	97	17.32
Thiamethoxam (Ne)	>1563	576-1552	>100	>100	>100	0.024
Thiodicarb (C)	66	2023	>3.3	0.027	-	MT
Toxaphone	50	-	0.01	0.0034-1.06	-	-
Transfluthrin (P)	>5000	>2000*	0.0007	0.0017	>0.1	-
Trichlorfon (OP)	250	-	0.7	0.00096	>10	LT
Triflumuron	>5000	561	>320	0.225	>25	T

Contd...

FUNGICIDES						
Active ingredient	LD_{50} Rats (mg/kg) oral	LD_{50} Birds (mg/kg) oral	LC_{50} Fish (mg/l) 96 h	LC_{50} Daphnia (mg/l) 48 h/72 h	EC_{50} Algae (mg/l) 72 h/120 h	LD_{50} Bees (μg/bee) contact/oral
Azoxystrobin	>5000	>2000	0.47	0.08	0.12	>25
8-hydroxy quinoline sulphate	500	NT	NT	-	-	NT
Benomyl	>5000	>10000	0.27	0.64	2.1-3.1	> 50
Bitertanol	>5000	776-2000	2.14	1.8-7	6.52	104.4-200
Boscalid	>5000	>2000	2.7	5.33	3.75*	166
Bupirimate	>4000	>5200	1.4	7.3	-	0.2
Captan	9000	>5000	0.034	10-Jul	-	788
Captafol	5000-6200	>101700	0.5	3.34	-	NT
Carboxin	2864	3302	2.3	>57	0.48*	>181
Chlorothalonil	>5000	>4640	0.047	0.07	0.21	>63
Copper hydroxide	489-1280	223	10	0.0422	22.5*	49
Copper octanoate	>2000	-	T	-	-	NT
Copper oxychloride	950-1862	167.3	0.217	0.29	56.3	18.1
Copper sulfate	-	600-1000*	VT	2.3	-	T
Dimethomorph	3900	>2000	6.2	>10.6	29.2	>102
Dinocap	990-1212	>2150	0.013	0.0042	>105	LT; 29
Dithianon	300	290-430	0.1*; T	2.4	12*	>100
Dodine	>1000	788-1142	0.53-0.6*	0.13	0.0051	>200
Epoxyconazole	>5000	>2000	2.2-4.6	8.7	2.3*	>100
Etridiazole	1028	560-1640	2.4	3.1	0.072	-
Fenhexamid	>5000	>2000	1.34	>18.8	8.81	>200
Fluazinam	>5000	1782-4190	0.036	0.22	0.16	>100
Fludioxonil	>5000	>20000	0.5	1.1	0.93	>329
Fludioxonil	>5000	>20000	0.5	1.1	0.93	>329
Fludioxonil	>5000	>20000	0.5	1.1	0.93	>329
Fluquinconazole	112	>2000	1.9	>5	0.014	-
Flusilazole	674-1100	>1590	1.2	3.4	-	NT
Flutolanil	>10000	>2000	5.4	> 6.8	1.85	>208.7
Flutriafol	1140-1480	>5000	61	78	-	LT
Folpet	>9000	>2000	Toxic	>1.46	>10	NT
Fosetyl-al	>7080	>8000	>122	>100	21.9	21.9
Furalaxyl	500-2000	-	-	-	-	-
Guazatine	360	82	1.41	0.15	-	>200
Hexaconazole	2189-6000	>4000	3.4	2.9	-	>100
Hydrogen peroxide	-	-	-	-	-	-

Contd...

Iodine	-	-	-	-	-	-
Iodocarb	-	-	-	-	-	-
Iodocarb	-	-	-	-	-	-
Kresoxim-methyl	>5000	>2150	190 ppb	0.186	0.063*	14
Mancozeb	>5000	>5500	1	3.8	0.044	>209
Mercury	-	-	T	-	-	-
Metiram	>5000	>2150	0.33	0.11	0.063-0.3*	>80
Oxythioquinox/ chinomethionate	1095-2541	196	0.131	0.12	0.14	NT;>100
Paclobutrazol	1300-2000	>7900	27.8	33.2	-	>0.002
Penconazole	2125	1590-2424	1.7-4.3	11-Jul	0.83	NT;>5
Pencycuron	>5000	>2000	>690	0.27	0.56	>100
Phosphorous acid/ phosphonic acid	-	-	>100	ST	-	NT
Prochloraz	1023	662-1954	1.5	4.3	0.1	141
Propamocarb	2000-2900	>1842	>99	106	>85	>100
Propiconazole	1517	2510-2825	4.3-5.3	10.2	0.02	NT;>100
Propineb	>5000	>5000	0.4	4.7	2.7	NT; >70
Pyraclostrobin	>5000	>2000	0.006	0.016	>0.843*	>73.1
Pyrazophos	151-778	118-480	0.48-1.14	0.00036	65.5	0.25
Pyrimethanil	4150-5971	>2000	10.6	2.9	1.2	>100
Quinoxyfen	>5000	>2250	0.27	0.08	0.058	>100
Quinoxyfen	>5000	>2250	0.27	0.08	0.058	>100
Quintozene	>5000	2000	0.55	0.77	-	>100
Sodium metabisulphite	-	-	-	-	-	-
Spiroxamine	500-595	565	18.5	6.1	0.012	>100
Sulphur	>5000	>5000	NT	>665	>232*	NT
Tebuconazole	1700-4000	2912-4438	4.4	4.2	3.8	83
Thiram	2600	>2800	0.128	0.21	0.065	73.5
Tolclofos-methyl	5000	>5000	0.72	-	-	-
Tolylfluanid	>5000	>2000	0.045	0.69	>1	>197
Triadimefon	1000	>4000	17.4	11.3	1.71	>100
Triadimenol	700	>2000	21.3	51	3.7	NT
Tridemorph	480	1388	3.4	1.3	0.28*	>200
Trifloxystrobin	>5000	>2000	0.015	0.016	0.0053	>200
Trifloxystrobin	>5000	>2000	0.015	0.016	0.0053	>200
Triforine	>16000	>5000	>1000	117	380	NH
Triticonazole	>2000	>2000	3.6	9	>1	>100
Zineb	>5200	-	6-8*	-	-	NT
Ziram	2068	97	1.9	0.048	-	NT; >100

9.8.2: Pesticide residues

Residue is the remains of a small amount of chemical either the chemical itself or as breakdown products that remain in or on a product. The pesticides found in or on foods or in water are called residues. Residues may enter into the environment via surface or ground water as a result of spray drift, vaporisation, surface run-off, unlawful acts, spills and drainage discharges, and through leaching or soil dusts (see Figure 9.4).

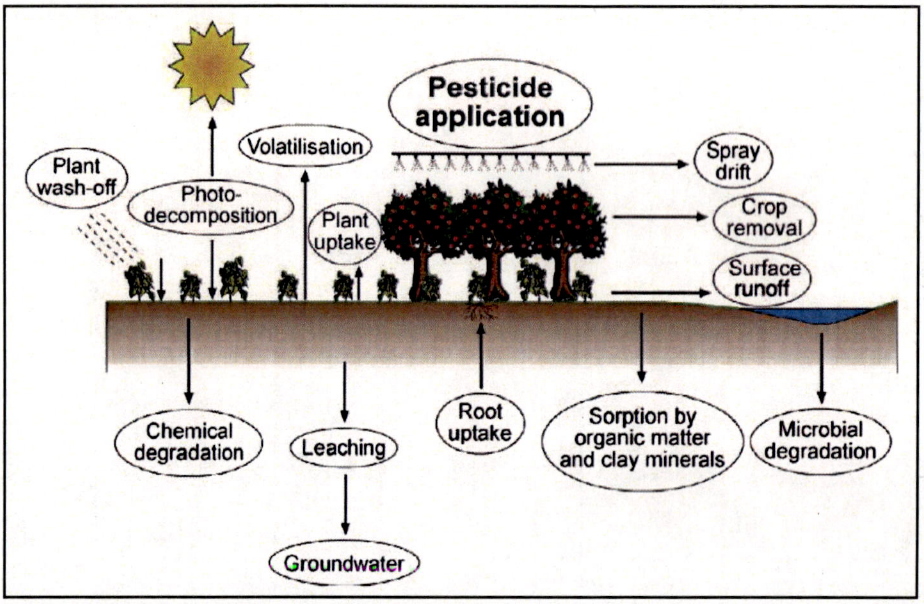

Figure 9.4: ***Mechanisms by which agrochemicals (pesticides) can be transported from the site of application to off-site environments.***

Residues in food and agricultural produce may result from the application of pesticides (e.g. fungicides used to control fungal diseases in plants and plant products; insecticides and acaricides used to control insect and mite pest and herbicides used to control weeds in crop), accidental contamination or intentional treatment of animals or feed with pesticides or from environmental contamination or unintended exposure of plants and animals to pesticides that are no longer registered in Australia and other countries, such as organochlorines - dieldrin, DDT but remain in the soil for long periods (Poolpak *et al.* 2008).

Humans can be exposed to pesticides residues via contaminated water and food or non-target drift or application. Most importantly, pesticide residues may enter into the human body through food chains. For example, livestock can be exposed to pesticides through intake of grain, fodder; forage as well as feeding of processed by-products and from ingestion of contaminated soil (see Box 9.3). These pesticides can then be accumulated in the animals because of the lipophilic (fat loving) nature of some of pesticides that may cause them to be incorporated and accumulated in the tissues (e.g. fat, milk) of living organisms. However, the tendency of a pesticide to accumulate in fat and milk of animals is generally related with the physico-chemical properties of the pesticide such as log K_{ow}. Therefore, the risk of chemicals residues to human are in the risk of residues remaining on the food consumed (such as meat, milk and milk products).

Box 9.3: ***Farmers' seek compensation over chemical residue in meat (June 1997).***
[http://www.fluorideaction.org/pesticides/chlorfluazuron.ici.suit.97.htm; Ref: Food Safety and Hygiene. A bulletin for the Australian Food Industry June 1997]

Four hundred and eighty cattle producers from northern New South Wales and western Queensland, Australia have launched a class action to seek millions of dollars compensation in the Federal Court from chemical company ICI and the NSW and Queensland Government. The case involves contamination of cattle and meat by the chemical chlorafluazuron (CFZ) which was widely used in aerial spraying to control insects in cotton crops between 1989 and 1994. The cattle owners fed their animals with cotton trash and cotton seed remnants during the prolonged drought early in this decade.

The detection of CFZ residues in meat being processed at a plant in northern NSW in October 1994 resulted in the immediate quarantine of about 3,000 beef farms and the rejection of beef exports by Japan and the United States. Some properties where CFZ contaminated feed was never used could still have been affected by spray drift following the aerial application of CFZ it was claimed in the Federal Court. Legal Counsel for the cattle producers said that CFZ was not registered for use in any other country at that time. Hence the finding of any CFZ residue in exported meat would lead to rejection. The plaintiffs are alleging common law negligence by ICI in failing to carry out appropriate testing to establish an appropriate withholding period and to provide adequate warning of risks associated with use of the chemical. The plaintiffs further allege negligence on the part of the NSW Department of Agriculture and the Queensland Department of Primary Industries in allowing the use of CFZ. Counsel claimed that both State Departments were represented on the Technical Committee for Agricultural Chemicals (TCAC) and on the body that replaced it, the Australian Agricultural and Veterinary Chemicals Committee. The TCAC had cleared CFZ for aerial application to cotton at the same time the State Departments were recommending the use of cotton trash for feed.

Pesticide chemicals as residue if consumed by humans more than the recommended maximum residue limits (MRL) can be hazardous (see section 9.8.3.6 for effects of pesticides on human health). It is therefore advisable that a national pesticide residue monitoring program be conducted on a regular basis to assess pesticide risk. The purpose of such monitoring is as follows:

- To estimate of the occurrence (frequency) of pesticide residue in agriculture product or in water,
- To confirm that residues are below the guideline levels for beneficial water usage,
- To confirm that residues in agriculture produce are below the set limit of MRL and
- To take up corrective measures or action.

The survey monitoring wing will inform industries, regulators and marketers of the incidence of pesticide risk for residues and of suggest for adherence to good agriculture practices (GAP) by all concerned.

9.8.3: Effects of pesticides

Pesticides are toxic chemicals and if handled and used incorrectly can lead to adverse effects on the environment, non-target organisms, livestock and humans. The harmful or unwanted effects that may happen from application or use of pesticides on non-target organisms are:

- Fish kills due to accidental spills or uncontrolled washing of pesticides associated with treatment tanks and equipment.
- Alteration of wildlife due to erroneous use of pesticides.
- Damage caused by hydrophobic pesticides residues in the food and fatty tissues of organisms.
- Contamination of groundwater by water soluble or hydrophilic pesticides.
- Alteration of the environment due to phytosanitary treatment such as, the elimination of aquatic weeds (may cause deoxygenation of the water resulting to a threat to fish life) or the terrestrial weeds (may severely affect the wildlife).

9.8.3.1: Ecological effects

The use of pesticides can have unintended effects on the environment since 98% of sprayed insecticides and 95% of herbicides reach a destination other than target species. This includes non-target species, volatilisation, air, water, bottom sediments and food. Pesticides can contaminate air, land and water via spray drift, run-off from fields, leaching as well as during manufacturing process and during application. Pesticides applied to crops can volatilize and may be blown by winds into nearby areas (Figure 9.4), potentially posing a threat to wildlife.

9.8.3.2: *Effects on water quality*

In the USA, organic chemicals including pesticides were found to pollute most streams in a study conducted by the US Geological Survey in 1999-2000. Detected pesticides were carbaryl, *cis*-chlordane, chlorpyrifos, diazinon, dieldrin, lindane, and methyl parathion (Kolpin *et al.* 2002). In another study, the US Environment Protection Agency (EPA) reported that normal agriculture use has lead to the presence of 46 pesticides in groundwater and 76 in surface water bodies, of which atrazine was the most frequently detected. In Australia (New South Wales), atrazine and endosulfan were most often found in surface waters during 1993-1995 (see Table 9.11). A recent study conducted in Northern Victoria, Australia by Goulburn-Murray Water (Rose and Kibria 2007) found that the most frequently detected pesticides in irrigation channels and rivers were endosulfan, atrazine and copper and less frequently detected pesticides were chlorpyrifos, parathion methyl and azinphos methyl. Water contaminated with pesticides can be unfit for human consumption, irrigation, food processing, livestock drinking, recreation use, aquatic ecosystems protection and aquaculture (see Table 9.12 for guidelines of pesticides for various beneficial water usages including aquatic ecosystems protection and aquaculture).

Table 9.11: ***Pesticides detected in surface and groundwater in Australia and in the USA during 1990-2006.*** *[information based on Kolpin et al. 2002; Wightwick and Allinson 2009].*

Environmental matrices	Australia	USA
Surface waters	***Herbicides:*** atrazine; chlorthal-dimethyl; cyanazine; diuron; flumeutron; linuron; metolachlor; metribuzin; molinate; prometryn; propazine; propyzamide; simazine. ***Insecticides:*** chlorpyrifos; chlordane; DDT; endosulfan; lindane; malathion; parathion methyl; profenofos ***Fungicides :*** chlorothalonil	***Herbicides:*** acetochlor; alachlor; atrazine; bentazone; bromacil; cyanazine; DCPA (chorathal-dimethyl/propanil); diuron; EPTC (S-ethyl dipropylthiocarbamate); enthalfluralin; metolachlor; molinate; napropamide; norflurazon; pebulate; prometon; propachlor; simazine; tebuthiuron; thiobencarb; trifluralin; 2-4-D. ***Insecticides:*** azinphos-methyl; carbaryl; carbofuran; *cis*-chlordane chlorpyrifos; DDE; diazinon; dieldrin; endosulfan; fonofos (o-ethyl S-phenyl ethyl phosphonodithioate); lindane, malathion; methyl parathion; propargite ***Fungicides:*** chlorothalonil
Groundwater	***Herbicides:*** amitrole; atrazine; chlorpropham; diuron; hezaninone; linuron; metribuzin; metolachlor; simazine; terbutryn; 2,4,5-T ***Insecticides:*** aldrin; BHC; chlorpyrifos; chloradane; DDT; dieldrin; endrin; glyphosate; HCB; heptachlor ***Fungicides:*** vinclozolin	***Herbicides:*** atrazine; bentazone; broamacil; diuron; metolachlor; molinate; norflurazon; prometon; simazine; thiobencarb; ***Insecticides:*** carbofuran;

9.8.3.3: *Effects on plants*

The presence of pesticides in soil can hinder nitrogen fixation in higher plants. Pesticides can kill bees, thereby may cause a decline in pollinator.

9.8.3.4: *Effects on birds*

Animals may be poisoned by pesticides residue that remain on food after spraying, for example, when wild animals enter fields that were sprayed with pesticides. The organochlorine pesticides are toxic to a wide range of biota. A reduction in the reproductive success of carnivorous birds such as, peregrine falcon has been related to DDE (a metabolite of DDT). In addition, the accumulation of DDT has also been linked to the thinning of egg shells causing them to break during incubation and dramatically reducing the population of predatory birds, such as, falcons which is at the top of the food chain. In the farmland of Britain (1979-1999), birds' population have declined due to loss of plant and invertebrate species on which the birds feed for pesticides use (see Table. 9.13 for the toxicity values of various pesticides on birds). It has been estimated that over 67 million birds are killed by pesticides each year in the USA (see Table 9.10 for LD_{50} values for birds).

Table 9.12: ***Water quality guidelines for safe levels of pesticides for aquaculture/fish farming and aquatic ecosystems protection*** *[information sourced from ANZECC (2000a and ANZECC 2000b)].*

TV = trigger values; FW = freshwater; SW = saltwater; MW = marine water; H = Herbicide; C = carbamate; OC = organochlorine; OP = Organophosphorous; P = pyrethroid

Chemical	Safe level (µg/l) for aquaculture	Safe level (µg/l) for aquatic ecosystems protection, TV for 99% or 95% aquatic species protection	Chemical	Safe level (µg/l) for aquaculture	Safe level (µg/l) for aquatic ecosystems protection, TV for 99% or 95% aquatic species protection
2,4-D (H)	<0.004 - FW	280–FW (95%)	Esfenvalerate (P)	-	0.001 – FW (95%)
2,4,5-T (H)		36 – FW (95%)	Fenitrothion (OP)	<0.2 - FW	0.2 –FW (95%)
2,4-dichlorophenol	<4.0 - FW	-	Fenvalerate (P)	0.083 -FW	-
Aldrin (OC)	<0.01 - FW	-	Glyphosate (H)		370 – FW (95%)
Amitrole (H)	300 - FW	-	Gunthion	0.01 -FW	-
Atrazine (H)	<0.34 - FW	13 – FW (95%)	Hexachlorobenzole (F)	0.00001 -FW	-
Azinphos-methyl (OP)	<0.01 - FW	0.01 – FW (99%)	Heptachlor (OC)	<0.005 -FW	FW (99%)
Carbaryl (C)	0.02 - FW	-	Lindane (OC)	0.01 -FW 0.004 -SW	0.01-0.2 – FW (95%)
Carbamate	<0.1-FW	-	Malathion (OP)	<0.1 - FW	0.05 – FW (95%)
Carbofuran (C)		0.06 – FW (95%	Methomyl (C)		3.5 –FW (95%
Chlordane (OC)	0.01-FW 0.04 -SW	0.03 - FW (99%)	Methoxychlor (OC)	<0.03 - FW	-
Chlorpyrifos (OP)	<0.001-FW	0.01– FW (95%) 0.009 – MW (99%)	Mexacarbate (I)	0.1 - FW	-
Chlorothalonil (F)	<0.0082-FW		Mirex (OC)	<0.001 - FW	-
Cyanazine (H)	0.0035-FW	-	Molinate (H)		3.4 – FW (95%)
Cypermethrin (I)	0.00147-FW	-	Paraquat (H)	<0.01 - SW	
DDT (including DDD and DDE) (OC)	<0.0015-FW	0.006 - FW (99%)	Parathion (OP)	0.04 - FW	0.004 –FW (95%)
Diazine	<1.0-10.0 - FW	-	Pentachloraphenate	<0.1 - FW	-
Diazinon (OP)	0.002 - FW	0.01 – FW (95%)	Pyrethrin (I,A)	<0.001-FW	-
Dicamba (H)	200 - FW	-	Pyrethrum (I)	0.01-FW	-
Diquat (H)		1.4 – FW (95%)	Rotenone (I,A)	<0.003-0.008-SW 10 - FW	-
Dimethoate (OP)		0.15 –FW (95%)	Simazine (H)	10.0-FW	3.2 – FW (95%)
Dieldrin (OC)	<0.005 - FW	-	Tebuthiuron (H)		2.2 – FW (95%)
Dalapon (H)	110 - FW	-	Thiobencarb (H)		2.8 – FW (95%)
Diquat (H)	0.5 - FW	-	Thiram (H)		0.01 – FW (95%)
Diuron (H)	1.5 - FW	-	Trichlorphon	<0.001 - FW	-
Endosulfan (OC)	<0.003 - FW 0.01- SW	0.03 - FW (95%) 0.005 - MW (99%)	Trifluralin (H)		2.6 – FW (95%)
Endrin (OC)	<0.002 -FW	0.01 - FW (99%) 0.004 - MW (99%)	Toxaphene (OC)	<0.002- FW	0.1 for FW

9.8.3.5: Effects on fish and aquatic organisms

Fish and other aquatic biota (invertebrates) can be harmed by water contaminated with pesticides (Table 9.13). For example, application of herbicides can cause fish kills when dead plants rot and use up most of the dissolved oxygen in the water. Copper sulphate (herbicide) is used to kill aquatic plants, however this chemical is very toxic to fish and other aquatic organisms. Repeated exposures of sub-lethal doses of pesticides can cause physiological and behavioural changes in fish therefore may reduce the populations in the wild. The USDA and USFWS estimate that between 6 and 14 million fish are killed by pesticides each year in the USA (Table 9.10 listed LC_{50} values of various pesticides for fish, crustaceans and algae; and Table 9.12 listed water quality guidelines for some selected pesticides for aquaculture and aquatic ecosystems protection).

Table 9.13: ***Relative toxicity to non-target organisms for 100 of the most commonly used pesticides in the U.K*** *[Sotherton and Holland 2003].*
Others = include rodenticides, molluscides, growth regulator, desiccants and wood preservatives.

Type of animal	Herbicide	Insecticide	Fungicide	Others
Birds				
Very toxic	3%	17%	0	17%
Toxic	11%	33%	12%	28%
Harmful	39%	17%	20%	17%
Non-toxic	47%	33%	68%	39%
Total	***100***	***100***	***100***	***100***
Fish				
Very toxic	17%	67%	39%	39%
Toxic	25%	24%	43%	11%
Harmful	17%	6%	18%	17%
Non-toxic	42%	6%	0%	33%
Total	***100***	***100***	***100***	***100***
Invertebrates				
Very toxic	15%	78%	44%	47%
Toxic	29%	17%	32%	12%
Harmful	18%	0%	20%	12%
Non-toxic	38%	6%	4%	29%
Total	***100***	***100***	***100***	***100***

9.8.3.6: Effects on amphibians

Experiments conducted in Canada found that exposing tadpoles with endosulfan at levels found in nearby farming fields may kill tadpoles and cause behavioural and growth abnormalities. Scientists from the USA found that some pesticides used in farming area (Sierra Nevada) may disrupt the nervous system of frogs and may cause a decline in the frogs' populations. The multiple effects of pesticides on frogs include longer period for metamorphosis into adults, deformities in frogs' eye, mouth and limb malformations. According to a recent review (Mann *et al.* 2009), pesticides may cause the following effects on anurans (frogs and toads):

- impaired behaviour and growth (e.g. pyrethroids and fungicide)
- limb deformities (e.g. mixtures of atrazine and carbaryl)
- altered sex ratios (e.g. mancozeb)
- acute and chronic toxicity (e.g. azinphos methyl, glyphosate based products, paraquat, permethrin)

- endocrine disruption (there is some controversy whether atrazine is a potential endocrine disruptors since some research suggested it may adversely affect the reproductive capacity of frogs and toads, whereas other and more recent research did not find links between atrazine exposure and the development of abnormal gonads in *Xenopus laevis, Rana pipienis and Hyla versicolr* (Mann *et al.* 2009).
- delayed metamorphosis, survival and growth (e.g. mixtures of pesticides)

In addition to the above, the interactions between pesticides and natural stressors (abiotic/ biotic factors) may also affect frogs and toads survival and growth (see Table 9.14).

Table 9.14: ***Combined effects of abiotic factors (UV, pH and temperatures) and pesticides on frogs*** *[information and data compiled by Mann et al. 2009].*

GS = gosner stage (stage of tadpoles when the forelimbs emerge)

Abiotic factors	Pesticide	Amphibian species	Effect
UV-B 4 day lab. study	Carbaryl (15 mg/l) *(insecticide)*	Gray tree frog *Hyla versicolor* (tadpoles)	Reduced survival and reduced swimming ability.
UV-B 4 day lab. study	Carbaryl (0.86-1.76 mg/l) *(insecticide)*	African clawed frog, *Xenopus laevis* (tad poles)	Reduced survival and reduced swimming ability.
UV-B Mesocos study	Carbaryl (4.1 mg/l) *(insecticide)*	Dusky gopher frog, *Rana sphenocephala* (2 weeks old)	No effects on survival.
pH (5.5 vs pH 7.5) 96 h lab. study	Triclopyr *(herbicide)*	African clawed frog, *Xenopus laevis* *B. americanus* Green frog, *Rana clamitns* Leopard frog, *Rana pipines* (GS 25)	Reduced survival at low pH.
pH (5.5 vs pH 7.5) 10 day lab. exposures;	Triclopyr *(herbicide)*	Leopard frog, *Rana pipines* (GS 25)	Reduced survival at low pH.
Temperature : 17-27°C 96-h lab. exposures	Carbaryl (10.3-30 mg/l)	Green frog, *Rana clamitns*	Reduced survival at higher temperatures.
Variable temperatures : (21±7.5°C vs constant temperatures (20±7.5°C) 96 h lab. study	Endosulfan (0.8 µg/l) *(insecticide-organochlorine)*	*Blue mountains tree frog, Litoria citropa* (GS 25)	Reduced survival and susceptibility to predation.
Temperatures : During embryonic development 14°C vs 20°C)	Endosulfan (0.3-1.3 µg/l) *(insecticide-organochlorine)*	*Brown - stripped frog, Limnodynastes peronii* (GS 25)	Reduced growth.
Temperatures : During embryonic development 14°C vs 20°C)	Endosulfan (0.3-1.3 µg/l) *(insecticide-organochlorine)*	*Brown - stripped frog, Limnodynastes peronii* (GS 25)	Increased susceptibility to predation if the tadpoles were reared at low embryonic temperature.

9.8.3.7: Effects on human health

Humans can be exposed to pesticides via oral ingestion, dermal absorption and inhalation. The immediate health effects from pesticide exposure includes irritation of nose, throat and skin causing burning, stinging and itching as well as rashes and blisters. Workers who handle pesticides may experience nausea, abdominal pain, headaches, vomiting, dizziness, and diarrhoea as well as skin and eye problems. In many cases, symptoms of pesticide poisoning are similar of colds or the flu. The long term health effects include cancer and other tumours; brain and nervous system damage; birth defects;

miscarriages, infertility, male sterility and other reproductive problems. It may cause damage to liver, kidneys, lungs and other body organs (Agnihorti 1999; USEPA 2006; Abhilash and Singh 2009; see Table 9.15). Investigations revealed a positive relation between exposure of pesticides and different types of cancers including kidney, pancreatic, prostate (atrazine related) and stomach cancer. However, the chronic effects may not appear even after years of exposure. Some pesticides act as endocrine disruptors (e.g. atrazine, simazine, endosulfan, DDT) and have been shown to cause serious harm to animals, including cancer, sterility and developmental problems (see Chapter 11 on Chemical Impacts - EDCs). Other pesticides related effects are disruption of endocrine system, and suppression of immune systems. In Australia, during 1989-1992, at least two fatalities were related to pesticides poisoning (a rate of one fatality every two years). The World Health Organisation and UN Environment Program estimates that about 3 million workers (in agriculture sector) in the developing world experiences severe poisoning from pesticides in each year of which about 18,000 would ultimately die (WHO/UNEP 1990).

Table 9.15: ***USEPA national primary drinking water standards for selected pesticides*** *[USEPA 2006]. [http://www.ehso.com/ehso.php?URL=http%3A%2F%2Fwww.epa.gov/safewater/dwh/c-soc/24-d.html]*

Contaminant	*MCL (mg/l)	Potential health effects from exposure above the MCL	Common sources
Atrazine (H)	0.003	Cardiovascular system or reproductive problems.	Run-off from herbicide used on row crops.
Carbofuran (I, N)	0.04	Problems with blood, nervous or reproductive system.	Leaching of soil fumigant used on rice and alfalfa.
Chlordane (OC)	0.002	Liver or nervous system problems; increased risk of cancer	Residue banned termiticides
Copper (F)	1.3	Gastrointestinal distress (short term exposure); liver or kidney damage (long term exposure).	Corrosion of household plumbing systems; erosion of natural deposits; pesticides.
2,4-D (H)	0.07	Kidney, liver, or adrenal gland problems.	Run-off from herbicide use.
Dalapon (H)	0.2	Minor kidney changes.	Run-off from herbicide use.
Diquat (H)	0.002	Cataracts.	Run-off from herbicide use.
Endothal (H)	0.1	Stomach and intestinal problems.	Run-off from herbicide use.
Endrin (OC)	0.002	Liver problems.	Residue of banned insecticide.
Glyphosate (H)	0.7	Kidney problems; reproductive difficulties	Runoff from herbicide use
Heptachlor (OC)	0.0004	Liver damage; increased risk of cancer.	Residue of banned termiticides.
Hexachlorobenzene (F)		Liver or kidney problems; reproductive difficulties; increased risk of cancer.	Discharge from metal refineries and agricultural chemical factories.
Lindane	0.0002	Liver or kidney problems.	Run-off/leaching from insecticide used on cattle, lumber, gardens.
Methoxychlor (OC)	0.04	Reproductive difficulties.	Run-off/leaching from insecticide used on fruits, vegetables, alfalfa.
Oxamyl (I,A,N)	0.2	Slight nervous system effects.	Run-off/leaching from insecticide used on apples, potatoes, and tomatoes.
Pentachlorophenol (I,F,H)	0.001	Liver or kidney problems; increased risk of cancer.	Discharge from wood preserving factories.
Picloram (H)	0.5	Liver problems.	Herbicide run-off.
Simazine (H)	0.004	Problems with blood.	Herbicide run-off.
Toxaphene (OC)	0.003	Kidney, liver or thyroid problems; increased risk of cancer.	Run-off/leaching from insecticide used on cotton and cattle.
2,4,5-TP (silvex) (H)	0.05	Liver problems.	Residue of banned herbicides.

MCL = Maximum contaminant level in drinking water below which there is no known or expected risk to health.

9.9: Case studies - persistent organochlorine pesticides residue in Asia-Pacific and Australia region

Organochlorine pesticides (OCP) are 'persistent organic pollutants' or POPs, that remain in the environment for long period of time, bio-accumulate in the fatty tissues of the living organisms through the food chain and are toxic to humans and wildlife therefore, pose a risk through causing adverse effects to human health and the environment. Many OCP causes reproductive impairments, immunological and teratogenic dysfunctions in humans (USEPA 2006).

Persistent OC pesticides include dichlorodiphenyltrichloroethane (DDTs), chlordane (CHL), and hexachlorobenze (HCB). In 1995, the Governing Council of the United Nations Environment Programme (UNEP) called for global action to be taken on persistent organic pollutants (POPs) including nine pesticides. These are Aldrin, CHL, DDT, Dieldrin, Endrin, Heptachlor, HCB, Mirex, and Toxaphene. OC insecticides such as DDT and hexachlorocyclohexane (HCH) are still in use in some developing countries (see Table 9.5) and the former Soviet Union. Many developing Asian nations banned or restricted the production and usage of nine of the above OC pesticides during 1970s and 1980s however these chemicals are still being used in some developing nations for agricultural and public health purposes. It is well known that DDT is regularly used in tropical latitudes in African and Asian nations to control mosquito-borne diseases such as malaria.

9.9.1: Persistent organochlorine pesticides residue in the Asia-Pacific Region.

9.9.1.1: DDTs in mussels from Asia-Pacific countries

[based on Monirith et al. 2003]

As part of global persistent organic pollutants monitoring in the Asia-Pacific region (Global Mussel Watch Program), DDTs were examined in mussels (green mussel - *Perna viridis,* blue mussel - *Mytilus edulis* and *M. gallorovincialis)* and ezo mussel - *Crenomytilus grayamus*) from 10 countries during the period 1994-1999 (see Table 9.16). The results of these monitoring found high concentrations of DDTs in mussels collected from China (mean 16,000 ng/g lipid), Hong Kong (mean 7,700 ng/g lipid) and Vietnam (mean 4,440 ng/g lipid). The results indicate that residues of DDTs are being released to the environment in China, Hong Kong and Vietnam and may be related to recent usage of DDTs in those countries. On the other hand, lower levels of DDTs in mussels were observed from the Philippines (mean 21.0 ng/g lipid), Cambodia (mean 23.0 ng/g), Indonesia (mean 70.0 ng/g lipid) and Malaysia (mean 90.0 ng/g lipid) and this may indicate less usage of DDTs in those countries (see Figure 9.5 and Table 9.16 for comparison).

9.9.1.2: DDTs in human breast milk from the Asia-Pacific countries

[based on Tanabe and Kunisue 2007].

Residue surveys conducted in Asia (as part of global POPs contamination monitoring) found that DDTs in human breast milk from China, Vietnam, Cambodia and Malaysia were relatively higher than those from other Asian and developed countries (see Table 9.17 and Figure 9.5) as follows:

- China (3,600 ng/g lipid) (maximum)
- Vietnam (2,300 ng/g lipid) (maximum)
- Cambodia (1,800 ng/g lipid) (maximum),
- Malaysia (1,600 ng/g lipid) (maximum)

In China, DDTs were predominant in environmental samples collected from coastal areas and animal feed, which may indicate that DDTs are still being used in China though officially banned. Furthermore, in Vietnam higher levels of DDTs in sediments from populated locations was also reported indicating recent application of DDTs for public health purposes, such as malaria control. Over 20,000 tonnes of DDTs are currently used annually in Vietnam and current results of DDTs in human breast milk may indicate recent usage of DDTs in Vietnam. DDTs is used to control parasites in fish cage aquaculture in Cambodia and higher levels of DDTs also detected in fishes in Vietnam. High levels of DDTs from Malaysia may indicate DDTs illegally use in some Asian countries. In contrast, DDTs level in human breast milk in developed countries (Australia, Germany, Japan, Sweden, UK, and USA) is much lower (170-470 ng/g lipid) compared to Asian developing countries. Overall, DDTs in breast milk had gradually decreased in China and Vietnam.

Table 9.16: ***Concentrations of organochlorine compounds (DDTs)*** *[ng/g, lipid weight] in mussel samples from Asian countries [Monirith et al. 2003].*
Note: Shaded box indicate detection of higher concentrations from those countries

Country Number of sites monitored and year of sampling	Site description	DDTs
Cambodia (7 locations); 1998	6 in fishing village and one site in international port.	23 (mean) 16-48 (range)
China (10 locations); 1999	Industry, aquaculture, ferry, marine traffic, high population and small harbour.	16,000 (mean) 830-54,000 (range)
Hong Kong (10 locations); 1998, 1999	Agriculture, aquaculture, harbour, industry, ferry, shipping traffic, high populations, power station.	7700 (mean) 640-61000 (range)
India (17 locations); 1998	Urban, harbour, aquaculture, fishing harbour, industry.	380 (mean) 29-3000 (range)
Indonesia (9 locations); 1998	Fishing port, industry, port, aquaculture, and fishing.	70 (mean) 6.5-160 (range)
Japan (8 locations, all Tokyo Bay); 1994	Tokyo Bay (all 8 sites).	270 (mean) 70-790 (range)
Korea (10 sites); 1998	Aquaculture, oyster farms, small harbour, Island, shipyard, repairing shipyard.	150 (mean) 14-350 (range)
Malaysia (13 sites); 1998	Agriculture and aquaculture, urban, recreational beach, industry, port.	90 (mean) 16-270 (range)
Philippines (7 locations); 1998	Aquaculture sites.	21 (mean) 3.3-3.8 (range)
Russia (2 locations); 1999	-	630 (mean) 520-730 (range)
Singapore (1 location); 1999	-	110
Vietnam (9 locations); 1997	floating habitats, fishing, aquaculture, remote area, urban, shipping traffic, fishing village.	4440 (mean) 220-34000 (range)

Table 9.17: ***DDTs in human breast milk (ng/g lipid) collected from general public in Asian countries*** *[data compiled by Tanabe and Kunisue 2007; Devanathan et al. 2009; Haraguchi et al. 2009].*
Note: Data from developed countries included for comparison.

Country	DDTs	Remarks
ASIAN DEVELOPING COUNTRIES		
Cambodia (1999-2000)	1,800	
Cambodia (2000)	1,500	49 samples
China (Beijing, 1998)	2,000	60 samples
China (Guanghou, 2000)	3,600	54 samples
China (Dalian, 2002)	2,100	20 samples
China (Shenyang, 2002)	870	20 samples
China (Beijing, 2007)	1,300	25 samples (all 25 primipara)
Czech Republic (1996)	1,050	17 samples
Hong Kong (1999)	2,870	132 samples

Contd...

India (2000)	430	
India (New Delhi, 2005-2006)	1,500	21 samples
India (Mumbai, 2005-2006)	450	26 samples
India (Kolkata, 204-2005)	1,100	17 samples
India (Chennai, 2002-2003)	1,200	12 samples
Indonesia (Jakarta, 2001)	640	16 samples
Indonesia (Purwakarta, 2002)	1,300	18 samples
Indonesia (2001-2003)	1,000	55 samples
Korea (Seoul, 2007)	180	29 samples (including 20 primipara)
Malaysia (2003)	1,600	17 samples
Philippines (2000)	190	10 samples
Taiwan (Taichung, 2000-2001)	330	36 samples (including 30 primipara)
Vietnam (Hanoi, 2000)	2,100	42 samples
Vietnam (north, 2000-2001)	2,200	48 samples
Vietnam (Hochiminh, 2001)	2,300	44 samples
Vietnam (Hanoi, 2007)	1,200	20 samples (all 20 primipara)
DEVELOPED COUNTRIES		
Australia (Melbourne, 2002-2003)	480	39 samples
Canada (1996)	470	12 samples
Germany (1995-1997)	240	246 samples
Japan (1999)	430	
Japan (Osaka, Fukuoka, 1998)	290	49 samples
Japan (Fukuoka, 2001-2004)	340	(38 primipara)
Japan (Takayama (Gifu), 2007))	97	20 samples (including 6 primipara)
Japan (Sendai (Miyagi), 2007))	260	20 samples (including 14 primipara)
Japan (Kyoto, 2007-2008)	160	20 samples (including 11 primipara)
Russia (Arkhangelsk, 1996-1997)	1,400	(40 primiparae)
Russia (Buryatia, 2003-2004)	660	35 samples
Spain (1991)	610	
Sweden (1997)	170	20 samples
UK (1997-1998)	470	168 samples
UK (Lancaster/London, 2001-2003)	150	54 samples
USA (Massachusetts, 2004)	64	38 samples

9.9.2: Results of annual Australian pesticide residue monitoring of food

There is an annual pesticide residue survey program in Australia to monitor residue levels in food including meat, fisheries, plants and horticulture products. From recent survey of Australia's agriproducts undertaken under National Residue Survey (NRS) and Food Standards Australia and New Zealand (FSANZ) show a high rate of compliance with food standards. For example, the National Residue Survey (NRS 2007) conducted during 2006-2007 found that more than 97% agricultural produce contains agrochemicals pesticides which are below the MRLs or recognised as acceptable in or on a food. The survey results suggest that Australian produce are clean and green and are of high quality (see Table 9.18 for comparisons).

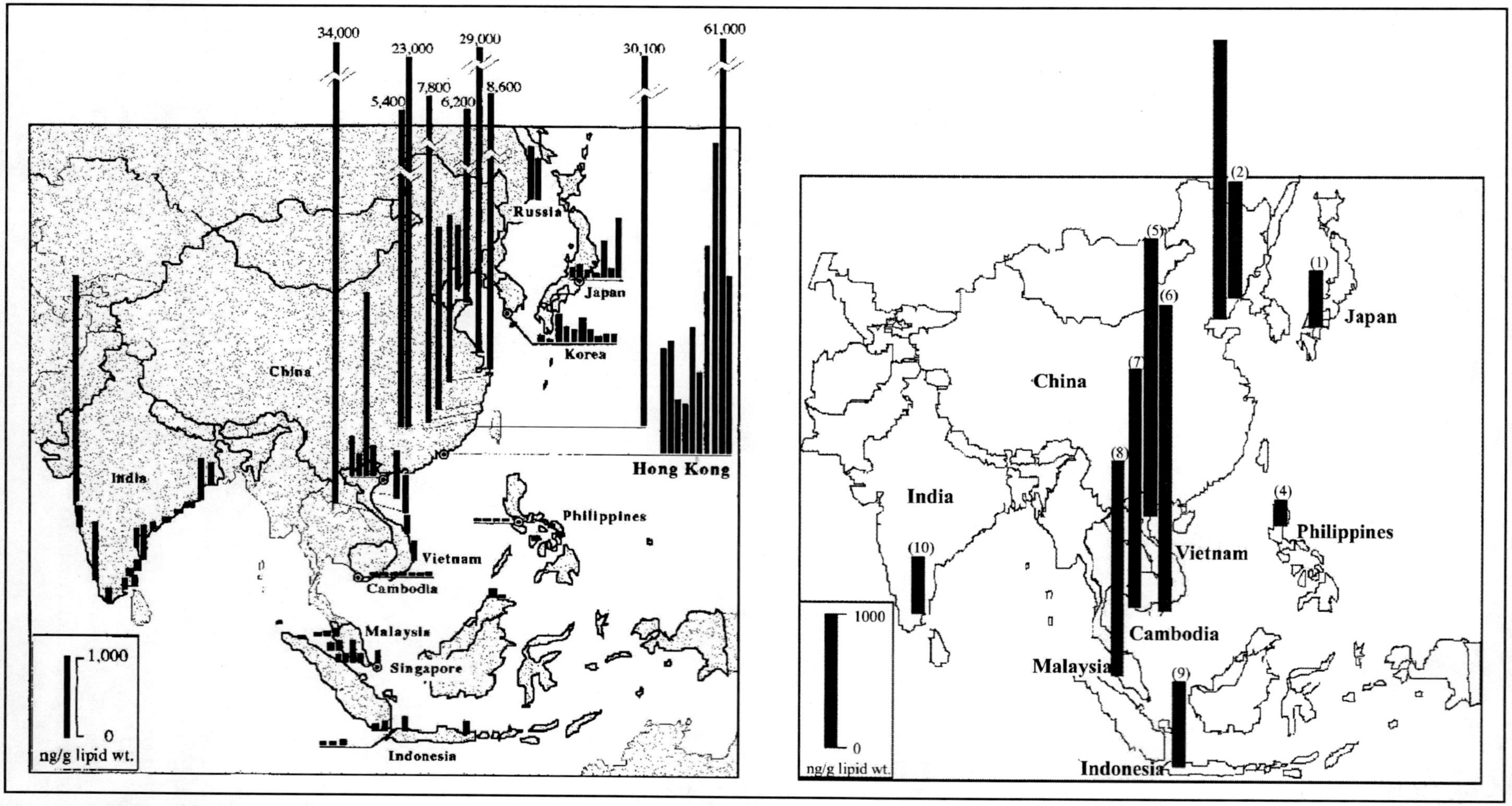

Figure 9.5: ***Concentrations of DDTs in mussels collected from coastal waters of Asia (left figure) and human breast milk collected from general public in Asian countries (right figure).***

Left Figure *sourced from Monirith et al. (2003, page 289);*

Right Figure *sourced from Tanabe and Kunisue (2007, page 403). site explanation: (1) Ehme, Japan; (2) Shenyang, China; (3) Dalian, China; (4) Quenzon, Philippines; (5) Hanoi, Vietnam; (6) Hochiminh, Vietnam; (7) Phnom Pen, Cambodia; (8) Penang, Malaysia; (9) Jakarta; Indonesia; (10) Palaverkadu, India].*

Table 9.18: *Summary results of Australian National Residue Survey on agriculture produce* **2006-2007** [NRS 2007].

Agriculture products	Residue of pesticides generally monitored	Selected agri produces and their maximum residue limits (MRL) or Australian Standard of selected pesticides (mg/kg)	Agriculture chemical residues that exceeded the Australian Standard (MRL)	Samples complaint with relevant standards (overall %)
Animal products Meat (camel, cattle, deer, goat, horse, kangaroo, pig, poultry, ostrich, sheep, wild boar).	***Orgnochlorines*** (aldrin and dieldrin, chlordane, DDT, endosulfan, HCB, HCH, heptachlor, lindane, methoxychlor). ***Organophosphates*** (chlorfenvinphos, chlorpyrifos, chlorpyrifos -methyl, coumaphos, diazinon, ethion, famphur, fenitrothion, fenthion, malathion, phosmet, temephos). ***Synthetic pyrethroids*** (bifenthrin, bioresmethrin, cyfluthrin, cyhalothrin, cypermethrin, deltamethrin, fenvalerate, flumethrin, permethrin).	**Agri. produce : Cattle, goat (fat)** ***MRL : Organochlorines***: dieldrin and Aldrin - 0.2; DDT – 5; endosulfan 0 0.2; heptachlor - 0.2; lindane (Y-HCH) – 2; ***Oragnophosphates***: chlorpyrifos - 0.5; diazinon - 0.7; fenthion – 1; phospmet : 0.05-1; temephos - 5. ***MRL : Synthetic pyrethroids***: bifenthrin – 2; cyfluthrin - 0.5; cypermethrin - 0.5; fenvalerate – 1; permethrin - 1.	2 cattle samples out of 5290 2 goat samples out of 251	99.96% 99.20%
Honey	No pesticides residue monitored; only antibiotics, metals.	-	-	-
Egg	No pesticides residue monitored; only antibiotics and anticoccidials.	-	-	-
Fisheries (molluscs - abalone, scallop; crustaceans - rock lobster, prawn; finfish - blue grenadier, saddletail sanpper, swordfish, yellowfin tuna, whitting)	***Orgnochlorines*** (aldrin and dieldrin, chlordane, DDT, HCB, HCH, heptachlor, mirex).	**Agri. produce : Fish** (Blue grenadier, saddletail snapper, swordfish, yellowfin tuna, whiting) ***MRL : Organochlorines***: aldrin and dieldrin - 0.1; chlordane - 0.05; DDT – 1 HCB -0.1;Heptachlor 0.05.	0 out of 25 fish samples	100%
Plant products Grains (barley, canola, chickpea, faba bean, field pea, lupin, oat, sorghum, triticale, wheat [grain, bran, flour, durum, semolina]	***Orgnochlorines*** (aldrin and dieldrin, chlordane, DDT, endosulfan, HCB, HCH, heptachlor, lindane, methoxychlor, mirex). ***Organophosphates*** (azamethiphos, chlorfenvinphos, chlorpyrifos, chlorpyrifos-methyl, coumaphos, diazinon, dichlorvos, dimethoate, ethorprofos, fenitrothion, malathion, methacrifos, phosmet, primiphos-methyl, profenofos, terbufos). ***Synthetic pyrethroids*** (bifenthrin, bioresmethrin, cyfluthrin, cyhalothrin, cypermethrin, deltamethrin, fenvalerate, permethrin, phenothrin); ***Acaricides*** (indoxacarb); ***Benzoyl ureas*** (diflubenzuron, triflumuron); ***Carbamates*** (carbaryl); **Fumigants** (phosphine), ***Fungicides*** (captafol, flutriafol, hexaconazole, iprodione, penconazole, propiconazole, tebuconazole, triadimefon).	**1.Agri. produce : Canola, chick pea** (whole) ***MRL : Organochlorines***: endosulfan – 1; Lindane (Y-HCH) - 0.05. ***MRL : Oragnophosphates***: chlorpyrifos - 0.05; dimethoate - 0.01 ***MRL : Synthetic pyrethroids***: bifenthrin – 0.02; Cyfluthrin : 0.5- 0.05; cypermethrin : 0.01-0.2; fenvalerate : 0.1-0.5. ***MRL : Fungicides***: iprodione : 0.02. **2. Agri. Produce : Wheat** (whole grain) ***MRL :Organochlorines***: dieldrin and aldrin - 0.02; DDT - 0.1; endosulfan - 0.2; heptachlor - 0.02; lindane (Y-HCH) - 0.5 ***MRL : Organophosphates:*** chlorpyrifos- 0.1; dimethoate - 0.05; fenitrothion – 10; malathion – 8; phosmet - 0.05; terbufos - 0.01. ***MRL : Synthetic pyrethroids:*** bifenthrin – 2; cyfluthrin – 2; cyhalothrin – 0.05; cypermethrin -0.2; deltamethrin -2; fenvalerate -2; permethrin -2;, phenothrin -2. ***MRL : Fungicides***: flutriafol - 0.02; propiconazole - 0.05; tebuconazole - 0.2; triadimefon - 0.5.	0 out of 185 canola samples; 0 out of 8 chick pea samples 0 out of 2054 wheat grain samples	100% 100% 100%
Horticulture (apple, blueberry, macadamia nut, onion, pear)	***Orgnochlorines*** (aldrin and dieldrin, dicofol, endosulfan, heptachlor, lindane). ***Organophosphates*** (acephate, azinphos-methyl, chlorpyrifos, diazinon, dimethoate, fenamiphos, fenitrothion, fenthion, malathion, methidathion, parathion, parathion-methyl, prothiofos, phorate, trichlorfon). ***Synthetic pyrethroids*** (bifenthrin, cyfluthrin, cypermethrin, deltamethrin, permethrin); ***Acaricides*** (chlorfenapyr, fenpyroximate, indoxacarb, propargite, propargite, tebufenozide, tebufenpyrad); ***Benzimadazoles*** (thiabendazole); ***Carbamates*** (carbaryl, fenoxycarb, pirimicarb); ***Fungicides*** (benalaxyl, captan, captafol,carbendazim, difenoconazole, dimethomorph, dithiocarbamates, fenarimol, flutriafol, hexaconazole,imazalil, iprodione,metaxyl, kresoxsim-methyl, penconazole, procymidone, propiconazole, tebuconazole, triadimefon, trifloxystrobin); ***Herbicides*** (diquat, glyphosate, glufosinate, ioxynl, methbenzthiazuron, oxyfluofen, paraquat, pendimethalin).	**1. Apple, pear** (whole) ***Organochlorines***: dieldrin and aldrin - 0.05; DDT – 1; endosulfan – 2; lindane (Y-HCH) - 2. ***Organophosphates***: azinphos-methyl – 2; chlorpyrifos - 0.5; dimethoate – 5; fenitrothion : 0.5; fenthion – 2; malathion 2; parathion-methyl - 0.5; trichlorfon 0.1. ***Synthetic pyrethroids***: bifenthrin - 0.05; cypermethrin 1. ***Fungicides***: captan – 10; carbendazim – 5; dithiocarbamates – 3; imazalil – 5; iprodione – 3; trifloxystrobin - 0.3. **2. Onion** (whole) ***Organochlorines***: dieldrin and aldrin - 0.1; DDT – 1; endosulfan - 0.2; lindane - 2. ***Organophosphates***: chlorpyrifos - 0.01; dimethoate – 2; malathion – 2; phorate - 0.5. ***Synthetic pyrethroids***: cypermethrin - 0.01. ***Herbicides***: ioxynl - 0.02; methbenzthiazuron - 0.05; oxyfluofen - 0.05; pendimethalin - 0.05.	9 out of 455 apple samples 1 out of 91 pear samples 0 out of 98 onion samples	98.02% 98.90% 100%

9.10: Conclusion

Intensive agriculture can contribute pesticides input into surface waters (rivers, creeks) via spray drift, drainage discharge, run-off and volatilisation. Pesticide residues in irrigation tail water, water discharged from irrigated crops and water from subsurface drains if not properly managed could be a major source of pesticides in the environment. Many surface waters near the intensive agriculture areas may be contaminated with a variety of pesticides. Presence of mixtures of pesticides in the environment could result in additive toxic effects on various aquatic organisms. The current Australian ANZECC (2000a and 2000b) water quality guidelines to protect aquatic biodiversity are derived from examining the effects of single pesticide exposure and are mostly based on ecotoxicity data derived from North America or Europe. Therefore, there is a need that each country generates local and regional toxicity data based on native species.

As hydrophobic pesticide residues can be transferred via food chains to humans, a residue monitoring program of agri-produce (e.g. crops, fish, vegetable and fruit) would help to protect the health of people and production of high quality food for domestic and export markets. Monitoring of residues further helps to monitor the use of currently registered pesticides and whether pesticides are used according to good agriculture practices. Furthermore, monitoring of surface and groundwater for high risk pesticide residue and comparing those results with guideline reference threshold values would help to define environmental impacts (FAO 2005). Pesticides such as 2,4-D; atrazine, DDT, endosulfan, simazine have been identified/suspected as endocrine disruptors (EDC), therefore measures are required to reduce risks from such chemicals (see Chapter 11 on Chemical Impacts - EDCs).

9.11: References

Abhilash, P. C. and Singh, N. 2009. Review Pesticide use and application: An Indian scenario. Journal of Hazardous Materials 165: 1-12.

Agnihorti, N. P. 1999. Pesticide safety and monitoring. All India Coordinate Research Project on Pesticides Residues. Indian Council of Agricultural Research, New Delhi, India.

Australian and New Zealand Environment and Conservation Council (ANZECC) and Agriculture and Resource Management Council of Australia and New Zealand (ARMCANZ) 2000a. Australian and New Zealand Guidelines for Fresh and Marine Water Quality, Volume 1, The Guidelines Chapters 1-7). http://www.mincos.gov.au/publications/australian_and_new_zealand_guidelines_for_fresh and_marine_water_quality

Australian and New Zealand Environment and Conservation Council (ANZECC) and Agriculture and Resource Management Council of Australia and New Zealand (ARMCANZ) 2000b. Australian and New Zealand Guidelines for Fresh and Marine Water Quality, Volume 3, Primary Industries Rationale and Background Information.

http://www.mincos.gov.au/publications/australian_and_ new_zealand_ guidelines_for_fresh_ and_marine_ water_quality/volume_3

Broadley, R., Collinge, M. and Middleton, K. 1990. Pesticide application manual. 2nd edition. Queensland Department of Primary Industries, Brisbane. Queensland Government. 154 pp.

CropLife Australia 2008. Fungicide, herbicide and insecticide mode of action http://www.croplifeaustralia.org.au/ files/resistancemanagemen/fungicides/Fungicide%20Activity%20Group%20Table%20-%20150908.pdf; http:/ /www.croplifeaustralia.org.au/files/resistancemanagemen/herbicides/2008%20Herbicide%20Mode%20of %20Action%20Table.pdf;http://www.croplifeaustralia.org.au/files/resistancemanagemen/insecticides/ 2008%20Insecticide%20Mode%20of%20Action%20Table.pdf

Devanathan, G., Subramanian, A., Someya, M., Sudaryanto, A., Isobe, T., Takahashi, S., Chakraborty, P. and Tanabe, S. 2009. Persistent organochlorines in human breast milk from major metropolitan cities in India. Environmental Pollution 157: 14 8-154.

FAO (Food and Agriculture Organization of the United Nations) 2005. Proceedings Asia regional Workshop on the implementation, monitoring and observance. International Code of Conduct on the distribution and use of pesticides. Food and Agriculture Organisation of the United Nations, Office for Asia and the Pacific, RAP publication 2005/29. Bangkok, Thailand.

Haraguchi, K., Koizumi, A., Inoue, K., Harada, K. H., Hitomi, T., Minata, M., Tanabe, M., Kato, Y., Nishimura, E., Yamamoto, Y., and Watanabe, T. 2009. Levels and regional trends of persistent organochlorines and polybrominated diphenyl ethers in Asian breast milk demonstrate POPs signatures unique to individual countries. Environmental Pollution doi:10.10.16/j.envint.2009.06.003

Hong-Yan, Z., R., Rau-Tai, G., Yuan-Fang, H, Xiao-Hong, J. I. A. and Shu-Ren, J. 2007. Spatial variability of organochlorine pesticides (DDT and HCHs) in surface soils from alluvial region of Beijing, China. Journal of Environmental Sciences 19: 194-199.

Huang, D.Q. and Thiemann, W. 2002. Contamination by selected chlorinated pesticides in surface waters in Hanoi, Vietnam. Chemosphere 47: 357-367.

Kamrin, M. A. 1997. Pesticide profile: Toxicity, environmental impact, and fate. CRC, Lewis Publishers, Boca Raton, Florida.

Kolpin, D. W., Furlong, E. T., Meyer, M. T., Thurman, E. M., Zaugg, S. D., Barber, L. B. and Buxton, H. T. 2002. Pharmaceuticals, hormones, and other organic wastewater contaminants in US streams, 1999-2000: A national reconnaissance. Environmental Science and Technology 36: 1202-1211.

Mann, R. M., Hyne, R . V., Choung, C. B. and Wilson, S. P. 2009. Amphibians and agriculture chemicals: Review of the risks in a complex environment. Environmental Pollution. doi: 10.1016/j.envpol.2009.05.015.

Monirith, I., Ueno, D., Takahashi, S., Nakata, H., Sudaryanto, A., Subramanian, A., Karuppiah, S., Ismail, A., Muchtar, M., Zheng, J., Richardson, B. J., Prudente, M., Hue, N. D., Tana, T. S., Tkalin, A. V., and Tanabe, S. 2003. Asia-Pacific mussel watch: monitoring of persistent organochlorine compounds in coastal waters of Asian countries. Marine Pollution Bulletin 46: 281-300.

Navarro, S., Vela, N. and Navarro, G. 2007. Review: An overview on the environmental behaviour of pesticide residues in soils. Spanish Journal of Agriculture Research 5(3): 357-375.

NRS (National Residue Survey) 2007. National residue Survey Annual report 2006-200., Australian Government Department of Agriculture, Fisheries, Forestry, Canberra. 218p. http://wwwdaff.gov.au/nrs

Poolpak, T., Pokethitiyook, P., Kruatrchue, M., Arjarasirikoon, U. and Thanwaniwat, N. 2008. Residue analysis of organochlorine pesticides in the Mae Klong River of central Thailand. Journal of Hazardous Materials 156: 230-239.

Radcliffe, J. C. 2002. Pesticide use in Australia. A review undertaken by the Australian Academy of Technological Sciences and Engineering. Australian Academy of Technological Sciences and Engineering. ANL Publishing, Parkville, Victoria. 174 pp. www.atse.org.au

Roberts, T. R., Huston, D. H., Lee, P. W., Nicholls, P. H. and Plimmer, J. R. 1998. Metabolic pathways of agrochemicals. Part 1; Herbicides nod plant growth regulators. The Royal Society of Chemistry. 849 pp.

Roberts, T. R., and Hutson, D (editors). 1999. Metabolic pathways of agrochemicals. Part 2: insecticides and fungicides. The Royal Society of Chemistry. 1475 pp.

Rose, G. and Kibria, G. 2007. Pesticide and heavy metal residues in Goulburn-Murray Irrigation Water 2004-2006. Australasia Journal of Ecotoxicology 13: 65-79.

Sarmah, A.K., Midler, K. and Ahmad, R. 2004. Fate and behaviour of pesticides in the agroecosystem—a review with a New Zealand perspective. *Australian Journal of Soil Research* 42 : 125-154

Shaw, I. C. and Chadwick, J. 1998. Principles of Environmental Toxicology. Taylor and Francis. UK. 124p.

Sotherton, N. and Holland, J. 2003. Indirect effects of pesticides on farmland wildlife. pp. 1173-95.In: Handbook of Ecotoxicology. 2nd edition. Edited by D.J. Hoffman, B.A. Rattner, G.A.Burton Jr. and J. Cairns Jr. Lewis Publishers, Boca Raton, Florida

Tanabe, S. and Kunisue, T. 2007. Persistent organic pollutants in human breast milk from Asian countries (review). Environmental Pollution. 146: 400-413.

Tariq, M.L., Afzal, S., Hussain, I. and Sultana, N. 2007. Review- pesticides exposure in Pakistan. Environmental International 33: 1107-1122.

Tomlin, C. D. S. 2006. A world compendium The Pesticide Manual. 14th edition. Editor: C. D. S. Tomlin. British Crop Production Council. 7 Omni Business Centre, Omega Park, Alton, Hampshire, GU34 2QD, London. 1349 pp.

USEPA (United States Environment Protection Agency) 2006. List of drinking water contaminants and MCLS. http://www.epa.gov/safewater/mcl.html.

WHO (World Health Organization) /UNEP (United Nations Environment Program). 1990. Working group. Public health impact of pesticides used in agriculture. Geneva: World Health Organisation.

WHO (World Health Organization) 2004. WHO recommended classification of pesticides by hazard and guidelines to classification. World Health Organisation/ International Programme on Chemical Safety, Geneva.

Wightwick, A and Allinson, G. 2007. Pesticide residues in Victorian waterways : A review. Australasia Journal of Ecotoxicology 13 : 91-112.

Websites

EXTOXNET: http://extoxnet.orst.edu/tibs/movement.htm

http://en.wikipedia.org./wiki/Pesticide (accessed on January 2008)

http://en.wikipedia.org/wiki/Pesticide (Pesticide. Retrieved on January 2008)

http://pmep.cce.cornell.edu/profiles/extoxnet/TIB/movement.html) (Extoxnet,)

http://www.emolecules.com

http://www.epa.gov/pesticides/factsheets/riskassess.htm. (USEPA)

http://www.ocfa.on.ca/local/files/communications/current%20issues/pesticides/final%20paper%2023APR2004.pdf (The Ontario College of Family Physicians 2004. pesticides literature review-Toronto, Ontario,Canada.

http://www.pesticide-residue.co.uk

http:/www.pesticidereform.org/article.php?list=type&type=38 (Californians for pesticide reform-pesticides and human health)

http://eds.ifas.ufledu/P1180 (pesticide use trends in the US: Global comparison) accessed on 19 October 2008)

http://www.chem.unep.ch/ddt/documents/Ritter_en_ChemProperties.pdf (UNEP)

http://plant-disease.ippc.orst.edu/articles.cfm?article_id=12 (Fungicide theory and mode of action, Oregon State University, USA);

http://feql.wsu.edu/entom490/Lect4MOASpray.pdf) (ENTOM 490 grape pest management, specific mode of action, pesticide label

http://www.wilsoncenter.org/topics/docs/pesticides_feb28.pdf

http://www.pan-uk.org/pestnews/Issue/pn29/pn29p11.htm

http://www.fao.org/docrep/008/af340e/af340e00.HTM#Contents

□□□

10 Chemical Impacts - Dioxins, Furans and PCBs

Key facts

- Dioxins are the most dangerous and 'super toxic' chemicals so far known on the earth and three closely related families are collectively known as dioxins. These are Polychlorinated dibenzo-*p*-dioxin or dioxins (PCDDs), Polychlorinated dibenzofurans or furans (PCDFs) and polychlorinated biphenyls (PCBs).
- 2, 3, 7, 8-TCDD (or called TCDD) is the most potent of all the congeners. The toxicity equivalency factor or TEF of TCDD has been assigned as 1, while all other congeners assigned in order of toxicity ranking magnitude relative to TCDD.
- There are 75 compounds/congeners in the dioxin family (PCDD), 135 compounds in the furan family (PCDF) and 209 possible compounds in the PCBs based on the possible number of chlorine atoms.
- The toxicity of dioxins and dioxin-like compounds vary substantially among the different PCDDs, PCDFs and PCBs. It is generally accepted that only 29 out of the 419 congeners of dibenzo-*p*-dioxins, dibenzofurans and polychlorinated biphenyls are toxic to humans and other organisms. The TCDD is most toxic and registered by the WHO as carcinogenic for humans.
- Dioxins are ubiquitous all over the world including the Arctic and Antarctic due to long range aerial transport. It is released into the environment as a result of combustion activities including power generation, waste incineration (e.g. municipal, medical waste), metal smelting, chlorine bleaching of pulp and paper mill and manufacture of some chemicals (e.g. pesticides, herbicides). They are also produced from natural processes such as bushfires/forest fires, compost burning, wood stoves and volcanoes.
- Dioxins have been detected in almost all environmental matrices. **Air**: range <1-15,000 fg TEQ/m^3); **Soils**: range 0.1-98,000 pg TEQ/g; **Sediments**: range 0.03-101,000 pg TEQ/g lipid; **Vegetation**: 0.3-1.9 pg TEQ/g; **Biota**: Kangaroo (0.001-25 pg TEQ/g lipid); Birds of prey (maximum 3,900 pg TEQ/g lipid); Bird-spotted owlet (9.2-270 pg/g fat); fish (range 0.029-329 pg/g fat); **Food-animal origin (PCDD/Fs)**: dairy products (range 0.0023-2.8 pg TEQ/g fat); pork (range 0.253-1.8 pg/g fat); chickens (range 0.390-2.18 pg/g fat); chicken eggs (0.248-2.24 pg/g fat); **Food-plant origin (PCDD/Fs):** vegetables (range 0.0086-0.44 pg/g); cereals (range 0.016-0.94 pg/g); rice (range 0.0039-0.05 pg/g); **Human women milk (PCDD/Fs):** from 43 countries, global range 0.15-48.7 pg WHO-TEQ/g fat; **Human women milk (PCBs):** reported from 24 countries, global range 20-1,800 ng/g lipid
- Dietary intake is the main route for human exposure to dioxins. About 95% of human intake of dioxins is through food such as fish, dairy products and meat.
- In 1990, the WHO suggested a TDI of 10.0 pg TEQ/kg of bw to protect the human health from adverse dioxins' effect. However, in 1998 such TDI has been further reduced to 1.0-4.0 pg TEQ/kg of bw.
- Dioxins are highly toxic to fish, birds and mammals but significantly less toxic or non-toxic to amphibians, invertebrates and plants. Lack of sensitivity of dioxins to plants and invertebrates is believed to be related with non-existent of Aryl hydrocarbon receptor (Ah receptors) in those organisms.
- Dioxins (PCDD/Fs and PCBs) have been detected in human (women) breast milk from several countries (e.g. Australia, Belgium, Brazil, Bulgaria, Cambodia, Canada, China, Croatia, Czech Republic, Egypt, Finland, Fiji, France, Germany, Greece, Hong Kong, Hungary, India, Indonesia, Ireland, Italy, Japan, Kazakhstan, Korea, Laos, Latvia, Luxembourg, Malaysia, New Zealand, the Netherlands, Norway, the Philippines, Portugal, Republic of Uzbekistan, Romania, Russia, Slovakia, South Africa, Spain, Sweden, Taiwan, UK, Ukraine, USA and Vietnam as a consequence breast feeding infants could have daily exposure to dioxins from early age.

- The dumping sites and e-waste (wastes of electrical and electronic equipment) recycling sites in Asia are potential sources of dioxins and can be correlated to significantly higher concentrations of dioxin related compounds found in breast milk and agricultural produces in India and China.
- Dioxin levels in Australian agricultural commodities are low as compared to Asia, Europe and North America. Low level of industrial activities and lower average population may have contributed lower dioxins emissions in Australia
- Long term exposure to dioxins can cause cancers of the human blood system; hence TCDD is regarded as class-1 carcinogen.

Contents

List of Tables

List of Figures

List of Boxes

Acronyms and Glossary

Ah: Ah or Aryl hydrocarbon receptor is a cell protein that initiates many of the effects of dioxin-like chemicals.

Bio-concentration: The process by which there is a net accumulation of xenobiotic chemicals directly from the abiotic environment (e.g. water) into organisms, resulting from simultaneous uptake (e.g. via gill, or epithelial tissue) and elimination (e.g. via gill, faeces, skin) and metabolic transformation.

Bio-concentration factor (BCF) : The concentration ratio of the chemical in the organism relative to water.

Biomarker: A biomarker is defined as a change in a biological response (ranging from molecular to cellular and physiological responses to behavioural changes) which can be related to exposure to or toxic effects of chemicals.

Chloracne: Chloracne is an acne-like eruption of blackheads, cysts and pustules associated with over-exposure to certain halogenated aromatic compounds, such as chlorinated dioxins and dibenzofurans. The lesions are most frequently found on the cheeks, behind the ears, in the armpits and groin region.

Congener: Closely related chemicals derived from the same parent compound or a member of the same taxonomic group. Congener is a term in chemistry that refers to one of many variants or configurations of a common chemical structure; for example, polychlorinated biphenyls (PCBs) occur in 209 different forms, or congeners. Each congener has two of chlorine atoms located at specific sites on the PCB molecule.

Coplanar PCBs: Since large number of hydrogen atom is present on the biphenyl nucleus, many different chlorinated compounds (termed "congeners" - isomers of different homologous series) are possible. As many as 209 congeners of the PCBs could be theoretically produced. Based on their structure, PCBs are grouped into: Coplanar – those compounds having chlorinated substituents in both para positions and any/all meta positions. There are 20 coplanar PCBs of which three are very toxic and non-ortho chlorinated depending on the number of Cl atoms on the biphenyl rings; other group of PCBs is divided into mono-, di-, tri-, tetra-, penta-, hexa-, hepta-, octa-, nona- and deca-chlorobiphenyls.

CYP1A1: Cellular protein involved in the metabolism of polycyclic aromatic hydrocarbons. Cytochrome P450, family 1, sub-family A; polypeptide 1 is a protein which in humans is encoded by the *CYP1A1* gene. The protein is a member of the cytochrome P450 superfamily of enzymes.

Cytochrome: Cytochromes are, in general, membrane-bound hemoproteins that contain heme groups and carry out electron transport. The heme group is a highly-conjugated ring system (which means its electrons are very mobile) surrounding a metal ion, which readily interconvert between the oxidation states. For many cytochromes, the metal ion present is that of *iron*, which interconvert between Fe^{2+} (reduced) and Fe^{3+} (oxidized) states (electron-transfer processes) or between Fe^{2+} (reduced) and Fe^{3+} (formal, oxidized) states (oxidative processes). Cytochromes are, thus, capable of performing oxidation and reduction. Because the cytochromes (as well as other complexes) are held within membranes in an organized way, the redox reactions are carried out in the proper sequence for maximum efficiency.

Cytochrome P450: Abbreviated as CYP, P450, infrequently as CYP450 is a very large and diverse superfamily of hemoproteins found in all domains of life. Cytochrome P450 uses a plethora of both exogenous and endogenous compounds as substrates in enzymatic reactions. Usually they form part of multicomponent electron transfer chains, called P450-containing systems. The expression of the CYP1A1 and CYP1B1 genes are regulated by the aryl hydrocarbon receptor, a ligand activated transcription factor.

Dimer: Chemical compound formed by the union of two identical molecules. A dimer is a chemical or biological entity consisting of two subunits called monomers, which are held together by either intramolecular forces (covalent bonds) or weaker intermolecular forces.

Dioxin: The group of persistent chlorinated chemical compounds, the polychlorinated dibenzodioxins (PCDDs or dioxins) including the closely related polychlorinated dibenzofurans (PCDFs or furans) and dioxin-like polychlorinated biphenyls (PCBs), which have certain similar structures and properties, and have similar biological characteristics including toxicity.

Endpoint: An observable or measurable biological event used as an indicator of the effect of a chemical on a biological system (cell, organ and organism).

EROD: 7-ethoxyresorufin-o-deethylase.

Femtogram (fg): $1x10^{-15}$ (0.000 000 000 000 001g).

Half-life: Is the time it takes for half of the amount of chemical to be removed from the environment.

Henry's Law of constant: The tendency of a material to volatise from aqueous solution to air. Sometimes measured, more usually calculated as the ratio of vapour pressure (in Pascals) x molecular weight / solubility (mg/l).

Heterodimer: The term homodimer is used when the two molecules are identical (e.g. A-A) and heterodimer when they are not (e.g. A-B). In biochemistry and molecular biology, dimers of macromolecules like proteins and nucleic acids are often observed. The dimerization of identical subunits is called homodimerization; the dimerization of different subunits or unrelated monomers is called heterodimerization.

Isomers: Any molecules that have the same number and type of atoms but have different structures.

***In vitro*:** In cells in test tubes, or in lab condition.

***In vivo*:** In whole organisms or in nature

LD_{50} or LC_{50}: The lethal dose or lethal concentration of a chemical that is estimated to be fatal to 50% of those organisms exposed under the stated test conditions.

Lipids: Lipids include fats and oils.

LOAEL: Lowest-observed adverse effects level.

Long-Range Transport: Is the potential for a chemical to travel from its point of release to a remote region, typically hundreds of kilometres away (distant).

Microgram (μg): $1x10^{-6}$ (0.000 001g).

Milligrams (mg): $1x10^{-3}$ (0.001g).

Nanogram (ng): $1x10^{-9}$ (0.000 000 001g).

NATO CCMS: North American Treaty Organization Committee on the Challenges to Modern Society.

NOAEL: No observed-adverse-effects level.

PCBs: Polychlorinated biphenyls.

PCDDs: Polychlorinated dibenzodioxins (dioxins).

PCDFs: Polychlorinated dibenzofurans (furans).

PCP: Phencyclidine, phencyclidine hydrochloride, PCP, angel dust (a drug used as an anaesthetic by veterinarians).

Persistence: Is the ability of a chemical to stay unchanged in the environment for a long time.

Picogram (pg): $1x10^{-12}$ (0.000 000 000 001 g) one-trillionth of a gram.

POPs: Persistent Organic Pollutants.

Primipara: A woman who has given birth to her first child.

TCDD: 2, 3, 7, 8-tetrachlorodibenzo-p-dioxin, the most toxic dioxin.

TEF: Toxic equivalency factor; This factor indicates the degree of toxicity compared to 2, 3, 7, 8-TCDD, which is given a reference value of 1. To calculate the total TCDD toxic equivalent (TEQ) of a dioxin mixture, the amounts of each toxic compound are multiplied with their Toxic Equivalency Factor (TEF) and then added together. TEQ = Σ[PCDDi x TEFi] + Σ[PCDFi x TEFi] + Σ[PCBi x TEFi].

TEQ: Toxic equivalents - it allows the toxicity of a complex mixture to be estimated and expressed as a single number.

UNEP: United Nations Environment Program.

WHO: World Health Organisation.

10.1: Introduction

Dioxins are referred to a group of chlorinated organic chemicals that are classified as persistent organic pollutants (POPs) (*see Box 10.1: Stockholm convention on POPs*). These compounds share similar chemical structures, properties (high boiling and melting points, limited solubility in water and low vapour pressure) and biological characteristics including toxicity. They are extremely persistent and can bio-accumulate in the body fat of wildlife, other animals and humans and tend to remain unchanged for long periods. Dioxins are the most dangerous toxic chemical so far known on the earth ('super toxic') and considered as a serious threat to public health. There are three closely related dioxin families as listed below which are referred to as 'dioxins' including furans and PCBs:

- Polychlorinated dibenzo-*p*-dioxins (PCDDs or dioxins) ($C_{12}H_4Cl_4O_2$).
- Polychlorinated dibenzofurans (PCDFs of furans) ($C_{12}H_4Cl_4O$).
- 'Dioxin – like' polychlorinated biphenyls (PCBs).

All the above three groups have varying degrees of toxicity in comparison to the most toxic 2, 3, 7, 8- tetrachlorodibenzo-*p*-dioxin (TCDD). 2, 3, 7, 8 tetrachlorodibenzo-*p*-dioxin is the most toxic dioxin to mammals, birds and fish, which has been the focus of many studies and has received most attention. 2, 3, 7, 8 represent that the four chlorine atoms are attached at position 2, 3, 7, 8 on the dioxin molecule (see Figure 10.1).

Box 10.1: *Stockholm Convention on persistent organic pollutants (POPs).*

Stockholm convention is an international legally binding agreement on persistent organic pollutants or POPs. In 1995, the Governing Council of the United Nations Environment Programme (UNEP) called for global action to be taken on persistent organic pollutants (POPs). POPs are "chemical substances that persist in the environment for long, bio-accumulate in the fatty tissues of the living organisms through the food chain and are toxic to humans and wildlife", therefore, poses risk through causing adverse effects to human health and the environment". POPs circulate globally and can cause damage wherever they travel. The Stockholm Convention on POPs aims to protect human health and the environment by banning the production and use of some of the most toxic chemicals known to human kind. This list includes 12 organochlorine chemicals or 12 worst offenders (dirty dozens) that are listed as POPs under the convention:

- *Dioxins and furans (polychlorinated dibenzo-p-dioxins or PCDDs and polychlorinated dibenzofurans or PCDFs).*
- *PCBs (polychlorinated biphenyls).*
- *Nine pesticides (Aldrin, chlordane, DDT, dieldrin, endrin, heptachlor, mirex, hexachlorobenzene and toxaphene).*

The convention entered into force on May 17th 2004 with ratification by an initial 128 parties and 151 signatories. Its mandate is to protect human health and environment from POPs. In implementing the Convention, Governments are supposed to take measures to eliminate or reduce the release of POPs into the environment. Co-signatories agree to outlaw nine of the twelve chemicals, limit the use of DDT to malaria control only and curtail inadvertent production of dioxins and furans. However, the treaty specifically permits the public health use of DDT for the control of mosquitoes (the malaria vector). Several other substances that are being considered for inclusion in the Convention are: hexabromobiphenyl, octaBDE, pentaBDE, pentachlorobenzene, short-chained chlorinated paraffins, lindane, α- and β-hexachlorocyclohexane, dicofol, endosulfan, chlordecone and PFOS. Further information and details on the Stockholm Convention can be found on the web site (http://en.wikipedia.org/wiki/Stockholm_Convention_on_Persistent_Organic_Pollutants).

10.2: Dioxin compounds

A dioxin consists of two benzene rings connected by two oxygen atoms, whereas a furan consists of two benzene rings connected by only one oxygen atom. Dioxins and furans have a very similar chemical structure. Polychlorinated biphenyls (PCBs) are a class of organic compounds with 1 to 10 chlorine atoms attached to biphenyl which is a molecule composed of two benzene rings each containing six carbon atoms (see Figure 10.1).

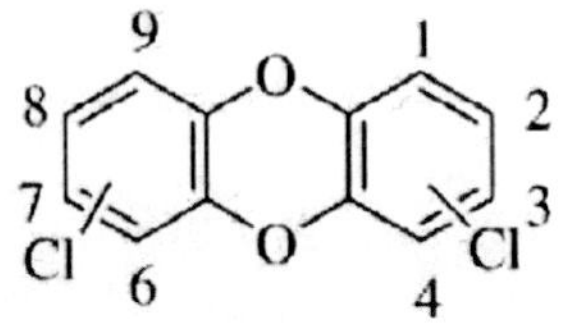

PCDDs

Dioxins have two oxygen atoms joining the two benzene rings (e.g. TCDD (2,3,7,8-tetrachlorodibenzo-p-dioxin).

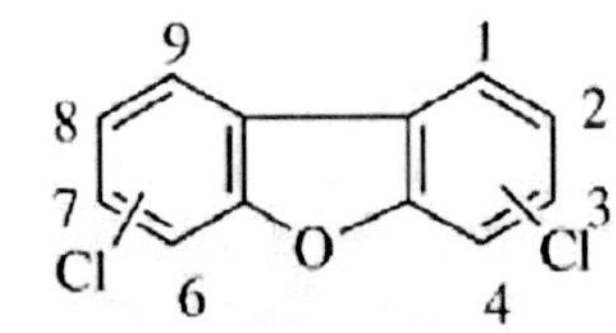

PCDFs

Furans have one oxygen atom joining the two benzene rings (e.g. TCDF (2, 3, 7, 8-tetrachloro-dibenzofuran).

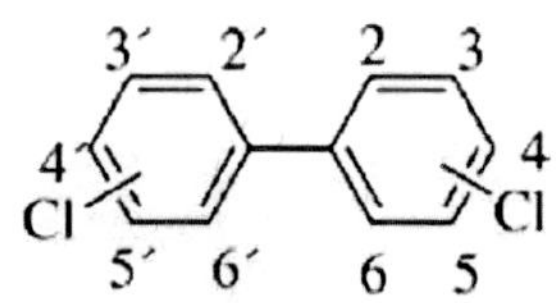

PCBs

PCBs consist of a biphenyl (two benzene rings with a carbon to carbon bond between carbon 1 on one ring and carbon 1′ on the second ring) with a varying number of chlorines.

Figure 10.1: ***Structure of dioxin, furan and PCB*** *[Kulkarni et al. 2008].*

There are 75 compounds/congeners in the dioxin family (PCDD), 135 compounds in the furan family (PCDF) and 209 possible compounds in the PCBs based on the possible number of chlorine atoms (Table 10.1). The number of isomers of PCDDs and PCDFs for a certain number of chlorine atoms is given in Table 10.2.

Table 10.1: ***Dioxin families and their isomers.***

Dioxin families	Congeners	Toxic congeners
Polychlorinated-dibenzo-p-dioxins or PCDDs	75	7
Polychlorinated dibenzo furans or PCDFs	135	10
polychlorinated biphenyls -PCBs	209	12
Total	**419**	**29**

The term "dioxin" is commonly used to describe a family of chemical compounds containing 75 dioxins and 135 furans. However, the only isomers presenting chlorine in the 2, 3, 7, 8 positions have been reported to be toxic. The 75 dioxin compounds differ among themselves only by the location and number of chlorine atoms attached to the molecule.

Table 10.2: ***Number of PCDD and PCDF isomers*** *[Rappe 1996].*

Number of chlorine atoms	Number of PCDD isomers	Number of PCDF isomers
Monochloro	2	4
Dichloro	10	16
Trichloro	14	28
Tetrachloro	22	38
Pentachloro	14	28
Hexachloro	10	16
Heptachloro	2	4
Octachloro	1	1
Total	**75**	**135**

10.3: Sources of dioxins

Dioxins are ubiquitous all over the world including the Arctic and Antarctic due to long range aerial transport. They are environmental contaminants that are found worldwide in a variety of environmental matrices (air, soil, sediments, vegetables, rice, fish, mussels, birds, seals, whales, dolphins, cattle, cows-milk, chicken, eggs, human adipose tissue and human women breast milk). Dioxins are released into the environment as a result of combustion activities including power generation, waste incineration (e.g. municipal, medical waste), metal smelting, chlorine bleaching of pulp and paper mill and the manufacture of some chemicals (e.g. pesticides, herbicides) (see Figure 10.2). They are also produced from natural processes such as bushfires/forest fires, compost burning, wood stoves and volcanoes. In the USA, forest fires were a major and a natural source of dioxins in 2004 (see Table 10.3). Electrical and electronic equipment recycling and disassembling (WEEE or E-waste) sites (see section 10.8.1) and municipal waste dumping sites could also be a potential source of PCDD/Fs, and PCBs.

PCBs are commonly known as chlorobiphenyls and produced by direct chlorination of biphenyl. There are no known natural sources of PCBs. They were first synthesized in the United States in 1929. PCBs are either oily liquids or solids that are colourless to light yellow and were used as coolants and lubricants in transformers, capacitors and in other electrical equipment, surface coatings, printing inks, carbonless duplication paper, waxes, etc. During the time of their manufacturing no effective mechanisms of their disposal were identified and therefore, even now they are found widely distributed in the environment, as they do not easily break down. While industrial use of PCBs has been sharply restricted, significant quantities of PCBs are still being released into the environment from waste dumps and failure of old electrical equipments. Most importantly, although the use of PCBs have been banned all over the world it is being formed as a by-product during the manufacture of various types of polymeric products such as adhesive, plastic, polythene, pesticides, rubber and related polymeric compounds. Besides, they are also being formed during incomplete incineration as well as due to burning of waste materials in the dust bins (see also *http://www.nio.org/WaterFront/PCB/PCBs.jsp#refer*).

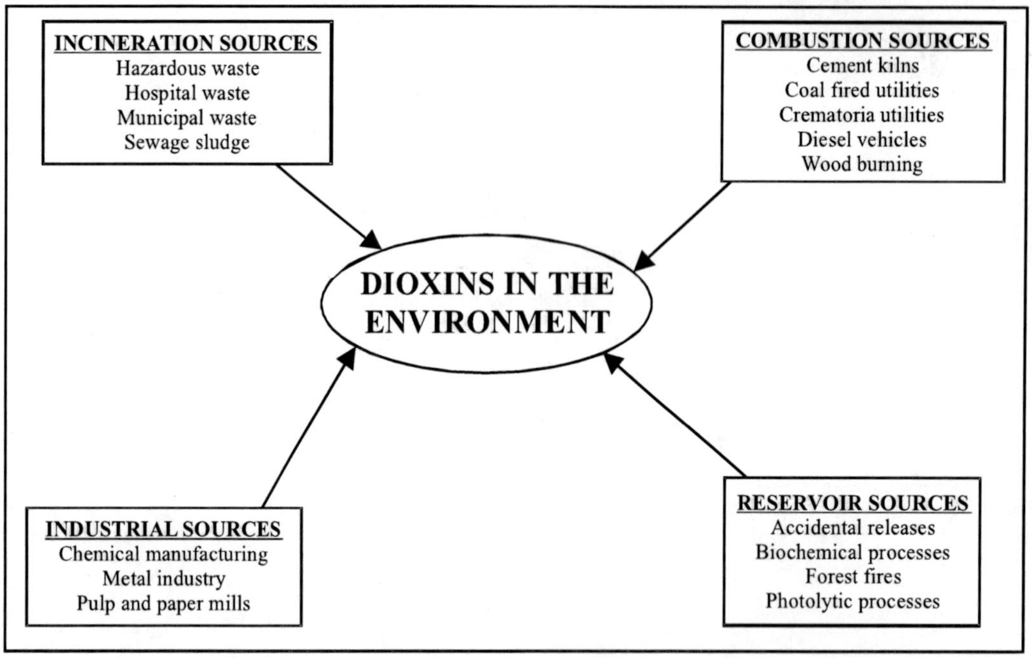

Figure 10.2: ***Dioxin sources and release in the environment*** *[adapted from Kulkarni et al. 2008].*

Table 10.3: ***Dioxin sources in the USA during 2004*** *[http://www.dioxinfacts.org/dioxin_health/dioxin_tissues/threshold.html]*

Sources	Percent	Quantity (g-TEQ)
Natural forest fires	54%	1288.44
Backyard trash burning	26%	620.36
Industrial (waste incineration-2%; energy generation - 4%; metallurgical processes - 3%; chemical manufacturing - 1%; land sludge application - 4%; other high temperature sources - 1%)	15%	357.90
Societal (residential wood combustion, gasoline and diesel use and crematoria	5%	119.3
Total	**100%**	**2386**

10.4: Physico-chemical characteristics of dioxins

10.4.1: Environmental fate

Dioxins are persistent, bio-accumulative and toxic chemicals. They are semi-volatile and can be transported in either gaseous form or attached to dust and ash particles. Both fine particulate and gaseous forms of dioxins are more amenable to long-range atmospheric transport by wind. Gravitational settling and washout in rain are the main pathways via which dioxins enter the terrestrial and aquatic environments. Dioxins deposited in aquatic environments will be adsorbed strongly on particulate organic carbon suspended in the water column and will rapidly redistribute by settling out to bottom sediments. Whereas dioxins deposited on the ground surface in terrestrial environments adsorb to organic matter in the soil and suspended in pore water. Dioxins adsorbed in soils will subsequently enter the aquatic environment via run-off and soil erosion. Dioxins are not expected to leach from the soil to groundwater due to their low water solubility and strong adsorption to particulate organic matter (see Table 10.4 for physico-chemical characteristics of dioxins).

10.4.2: Persistence

Dioxins have a relatively long atmospheric half-life (see Table 10.4), which enables them to be transported for long distances in the atmosphere. In soils and sediment, dioxins bind strongly to organic matter and persist for long periods. In general, the higher chlorinated congeners (e.g. PCDD/Fs, PCBs) are very resistant to chemical and biological degradation. The half-life of TCDD in humans has been estimated to be 7-11 years with individual variation (Srogi 2008). According to Geyer *et al.* (2002), the half-life of different polychlorinated dibenzo-*p*-dioxins (PCDDs) in adult humans varies between 7.78-170 years as cited below:

- 2,3,7,8-Tetrachlorodibenzo-*p*-dioxin or TCDD (range 5.8-9.7 years)
- 1,2,3,7,8-Pentachlorodibenzo-*p*-dioxin or PeCDD (range 12.6-15.7 years)
- 1,2,3,4,7,8-Hexachlorodibenzo-*p*-dioxin or HxCDD (range 8.4-45 years)
- 1,2,3,4,6,7,8-Heptachlorodibenzo-*p*-dioxin or HpCDD (range 25-102 years)
- 1,2,3,4,6,7,8,9-Octachlorodibenzo-*p*-dioxin or OCDD (range 850-132 years)

Table 10.4: ***Physical and chemical characteristics of dioxins*** *[Adapted from Gatehouse 2004; Srogi 2008 and http://www.environment.gov.au/settlements/publications/chemicals/dioxins/report-11/index].*

Characteristics	PCDD or dioxins	PCDF or Furans	PCB
Octanol/water partition coefficient (log K_{ow})	6-9	6-9	4.46-8.18
Water solubility (µg/l)	0.019	0.069	2.7-590
Henrys law of constant (atm m^3/mol)	1.6×10^{-5} to 1.0×10^{-4}	1.48×10^{-5}	5.2×10^{-4} to 2.0×10^{-3}
Half-life or persistence in soil	9-15 years	7 years	
Half-life or persistence in subsurface soil	25-100 years		
Half-life or persistence in human	7-11 years		2-8 years
Bio-accumulation factor (BAF) or Bio-concentration factor (BCF)	130,000	61,000	60,000-270,000

10.4.3: Bio-accumulation

Dioxins have low solubility in water and high solubility in lipids as represented by its high log K_{ow} values (Table 10.4). Because of their high affinity to lipids, dioxins can accumulate in the bodies of aquatic and terrestrial organisms including humans (e.g. adipose tissue) when they are exposed to these chemicals in water, soil, sediment, or in their diet. Dietary intake is the primary mechanism governing the levels of dioxins and dioxin-like chemicals in biota (see section 10.5.1 and 10.5.2).

10.4.4: Mode of action

Dioxins-like PCDD, PCDFs and PCBs share a common toxic mechanism, which involves binding to a cellular protein, known as the aryl hydrocarbon (AhR) receptor, where they form complexes that translocate to the nucleus and initiate changes in gene expression (Figure 10.3). AhR is high-affinity binding protein, present in low numbers per cell. Binding to this receptor is believed to be the initial step leading to biochemical, cellular and tissue level changes occurring in organisms exposed to dioxins. For example, dioxins can induce a broad spectrum of biological responses, including induction of gene expression of cytochrome P450, CYP1A1, and CYP1A12, disruption of normal hormone signalling pathways, reproductive and developmental defects (Kulkarni *et al.* 2008). The translocation of the ligand-binding subunit into the nucleus requires interaction with a protein called "arnt" (aryl hydrocarbon receptor nuclear translocating protein). Figure 10.4 provide a schematic model of the mode of action of dioxin in cell (see also Box 10.2).

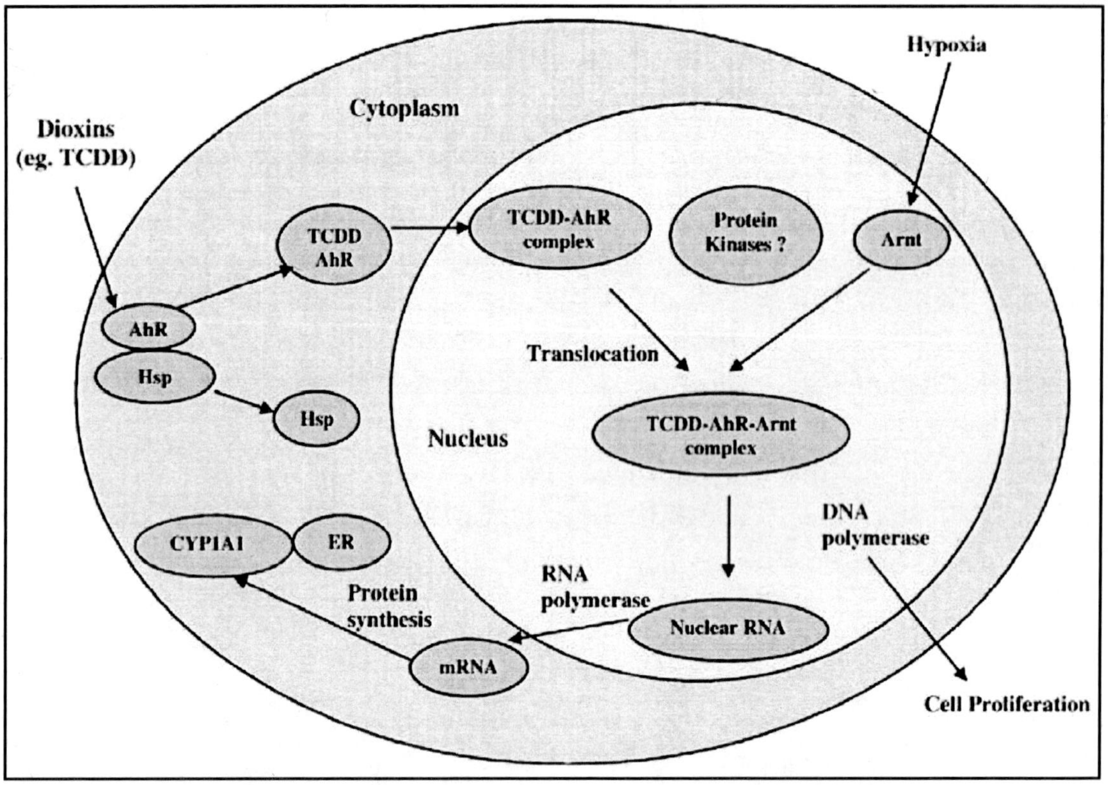

Figure 10.3: *A schematic model of the action of dioxins in cell [Mandal 2005].*

10.4.5: Toxicity and toxicity equivalent of dioxins

The toxicity of dioxins and dioxin-like compounds varies substantially among the different PCDDs, PCDFs and PCBs. It is generally accepted that only 29 out of the 419 congeners of dibenzo-p-dioxins, dibenzofurans and polychlorinated biphenyls are toxic to humans and other organisms (see Table 10.5). The 2, 3, 7, 8-tetrachlorodibenzo-*p*-dioxin (TCDD) is most toxic and registered by the WHO as carcinogenic for humans.

Dioxins exist in the environment as complex mixtures rather than as single chemical compounds, with the potency of each individual dioxin-like congener in a mixture differing considerably. As a result, the concept of Toxicity Equivalent Factors (TEF) or weighing factors for dioxins has been developed (WHO 1998), whereby the toxicity of a mixture is expressed in terms of a single value or TCDD. Toxicity Equivalent Concentration (TEQ) reflects the combined toxicity of all dioxins in the mixture. The TCDD TEQ (or in short TEQ) in an environmental sample (e.g. sediment, soil, organism) is calculated by analytically determining the levels of each congener in the sample, multiplying the concentration of each congener by a congener specific TEF and then adding all the products to give a TEQ. The TEQ can be obtained by the following equation:

$$TEQ = \Sigma[PCDD_i \times TEFi] + \Sigma[PCDF_i \times TEFi] + \Sigma[PCB_i \times TEFi]$$ (WHO 1998; Reiner *et al.* 2006).

TEF describes the potency of each dioxin-like congener relative to 2, 3, 7, 8-TCDD (or called TCDD), (*the most potent of all the congeners)*. The TEF of TCDD has been assigned as 1, with all other congeners assigned an order of magnitude of toxicity ranking relative to TCDD (see Table 10.5). More recently it has been discovered that certain polychlorinated biphenyls are similar in nature to dioxin and appear to pose similar risks to both humans and animals. This close toxicological similarity of coplanar polychlorinated biphenyls (which also bind to Ah receptor) to dioxins has led the extension of the TEF to some of the coplanar PCB compounds called dioxin-like PCBs. The WHO system of TEFs includes TEF values for 12 non-ortho and mono-ortho PCBs that have demonstrated dioxin–like toxicity (Table 10.5). Moreover, the WHO has developed TEF for mammals, birds and fish owing to differences in the sensitivity to certain congeners between the classes of organisms (Table 10.5).

Table 10.5: ***Dioxin, furans and PCB toxic equivalent factors (TEFs) for mammals, birds and fish***
[Van den Berg et al. 1998 and 2006; http://www.paceanalytical.com/services/analytical-services/specialty-analytical-services/dioxins-furans/TEF.pdf].

Isomers/congener	Compound name	Mammals	Birds	Fish
Chlorinated dibenzo-p-dioxins				
2, 3, 7, 8-TetraCDD	Tetrachlorodibenzo-*p*-dioxin (TCDD)	1	1	1
1, 2, 3, 7, 8-PentaCDD	Pentachlorodibenzo-*p*-dioxin (PeCDD)	1	1	1
1, 2, 3, 4, 7, 8-HexaCDD	Hexachlorodibenzo-*p*-dioxin (HxCDD)	0.1	0.05	0.5
1, 2, 3, 6, 7, 8-HexaCDD	Hexachlorodibenzo-*p*-dioxin (HxCDD)	0.1	0.01	0.01
1, 2, 3, 7, 8, 9-HexaCDD	Hexachlorodibenzo-*p*-dioxin (HxCDD)	0.1	0.1	0.01
1, 2, 3, 4, 6, 7, 8-HeptaCDD	Heptachlorodibenzo-*p*-dioxin (HpCDD)	0.01	<0.001	0.001
1, 2, 3, 4, 6, 7, 8, 9-OctaCDD	Octachlorodibenzo-*p*-dioxin (OCDD)	0.0003	0.0001	0.0001
Chlorinated dibenzofurans				
2, 3, 7, 8-TetraCDF	Tetrachlorodibenzo-furan (TCDF)	0.1	0.1	0.05
1, 2, 3, 7, 8-PentaCDF	Pentachlorodibenzo-furan (PeCDF)	0.03	0.1	0.05
2, 3, 4, 7, 8-PentaCDF	Pentachlorodibenzo-furan (PeCDF)	0.3	1	0.5
1, 2, 3, 4, 7, 8-HexaCDF	Hexachlorodibenzo-furan (HxCDF)	0.1	0.1	0.1
1, 2, 3, 6, 7, 8-HexaCDF	Hexachlorodibenzo-furan (HxCDF)	0.1	0.1	0.1
1, 2, 3, 7, 8, 9-HexaCDF	Hexachlorodibenzo-furan (HxCDF)	0.1	0.1	0.1
2, 3, 4, 6, 7, 8-HexaCDF	Hexachlorodibenzo-furan (HxCDF)	0.1	0.1	0.1
1, 2, 3, 4, 6, 7, 8-HeptaCDF	Heptachlorodibenzo-furan (HpCDF)	0.01	0.01	0.01
1, 2, 3, 4, 7, 8, 9-HeptaCDF	Heptachlorodibenzo-furan (HpCDF)	0.01	0.01	0.01
1, 2, 3, 4, 6, 7, 8, 9-OctaCDF	Octachlorodibenzo-furan (OCDF)	0.0003	0.001	<0.0001
Non-ortho-substituted PCBs				
3, 3' ,4, 4'-tetrachlorobihenyl	Tetrachlorobiphenyl (PCB 77)	0.0001	0.05	0.0001
3, 4, 4', 5-tetrachlorobihenyl	Tetrachlorobiphenyl (PCB 81)	0.0003	0.1	0.0005
3, 3', 4, 4', 5'-pentachlorobihenyl	Pentachlorobiphenyl (PCB 126)	0.1	0.1	0.005
3, 3', 4, 4', 5, 5'-hexachlorobihenyl	Hexachlorobiphenyl (PCB 169)	0.03	0.001	0.00005
Mono-ortho-substituted PCBs				
2, 3, 3' 4, 4'-pentachlorobiphenyl	Pentachlorobiphenyl (PCB 105)	0.00003	0.0001	<0.000005
2, 3, 4, 4', 5-pentachlorobiphenyl	Pentachlorobiphenyl (PCB 114)	0.00003	0.0001	<0.000005
2, 3', 4, 4', 5-pentachlorobiphenyl	Pentachlorobiphenyl (PCB 118)	0.00003	0.00001	<0.000005
2', 3, 4, 4', 5-pentachlorobiphenyl	Pentachlorobiphenyl (PCB 123)	0.00003	0.00001	<0.000005
2, 3, 3', 4, 4', 5-hexachlorobiphenyl	Hexachlorobiphenyl (PCB 156)	0.00003	0.0001	<0.000005
2, 3, 3', 4, 4', 5, 5'-hexachlorobiphenyl	Hexachlorobiphenyl (PCB 157)	0.00003	0.0001	<0.000005
2, 3', 4, 4', 5, 5'-hexachlorobiphenyl	Hexachlorobiphenyl (PCB 167)	0.00003	0.0001	<0.000005
2, 3, 3', 4, 4', 5, 5'-heptachlorobiphenyl	Heptachlorobiphenyl (PCB 189)	0.00003	0.0001	<0.000005

10.5: Environmental and biological impacts of dioxins

10.5.1: Occurrence of PCCD/Fs and PCBs

Dioxins have been detected in almost all environmental matrices such as air, soil, sediments, vegetation, biota (kangaroo, platypus, birds, fish, shellfish, marine mammals), food (dairy products, beef, pork, chicken, duck, eggs, vegetables, cereals, rice), human women milk, and human blood/ serum from Austria, Australia, Belgium, Brazil, Bulgaria, Cambodia, Canada, China, Croatia, Czech Republic, Egypt, Finland, Fiji, France, Germany, Greece, Hong Kong, Hungary, India, Indonesia, Ireland, Israel, Italy, Japan, Kazakhstan, Korea, Laos, Latvia, Luxembourg, Malaysia, the Netherlands, New Zealand, Norway, the Philippines, Portugal, Republic of Uzbekistan, Russia, Slovakia, South Africa, Spain, Sweden, Taiwan, UK, Ukraine, USA and Vietnam (see Table 10.6). The following are the summary of PCDD/Fs and PCBs detected in different environmental matrices:

- *Air*: range <1-15,000 fg TEQ/m^3) (contaminated sites had highest PCDD/Fs and rural area had the lowest PCDD/Fs in the air).
- *Soils*: range 0.1-98,000 pg TEQ/g (contaminated/urban soils had the highest PCDD/Fs and rural/ remote areas soils had the lowest PCDD/Fs).
- *Sediments*: range 0.03-101,000 pg TEQ/g lipid (highly contaminated estuarine/marine sediments had the highest PCDD/Fs).
- *Vegetation*: 0.3-1.9 pg TEQ/g (spruce and pine needles).
- *Biota*: Kangaroo, Australia (0.001-25 pg TEQ/g lipid); Birds of prey, Australia (maximum 3,900 pg TEQ/g lipid); Bird-spotted owlet, India (9.2-270 pg/g fat); fish (range 0.029-329 pg/g fat; highest PCDD/Fs level in herring muscle from Finland coast); marine mammals (whale 0.71-13 pg TEQ/g; polar bears 2-23 pg TEQ/g).
- *Food-animal origin (PCDD/Fs)*: dairy products (range 0.0023-2.8 pg TEQ/g fat); pork (range 0.253-1.8 pg/g fat); chickens (range 0.390-2.18 pg/g fat); chicken eggs (0.248-2.24 pg/g fat).
- *Food-plant origin (PCDD/Fs):* vegetables (range 0.0086-0.44 pg/g); cereals (range 0.016-0.94 pg/ g); rice (range 0.0039-0.05 pg/g).
- *Human women milk (PCDD/Fs only):* reported from 43 countries; global range 0.15-48.7 pg WHO-TEQ/g fat; higher level reported from China (48.7 pg/g in 2001), Finland (39 pg/g in 1992-1994), Germany (34 pg/g in 1986-1991), the Netherlands (47 pg/g in 1993), Norway (36 pg/g in 1992-1993) and Russia (42.26 pg/g in 1997); recent global trends show a decline in PCDD/Fs levels in human women milk.
- *Human women milk (PCBs only):* reported from 24 countries; global range 20-1,800 ng/g lipid; higher level reported from Europe (1,800 ng/g from Faroe Island in 2001), Germany (550 ng/g in 1995-97), Russia (520 ng/g in 1996), Kazakhstan (380 ng/g).

10.5.2: Dioxins exposure routes to human

Dioxins travel through air and deposit on water or land. In water, dioxins initially bind in small particles or plankton, whereas on land, it deposit on plants or bind to the soil (see Figure 10.4). Dioxins can be found in meat and dairy produce, poultry, fish and on unwashed fruit and vegetables. Fish can accumulate dioxins through exposure to water; whereas other terrestrial animals can be exposed to dioxins in the air. These animals can bio-accumulate dioxins in their fatty tissues. Dioxins in the air can also land on fruits and vegetables but washing can eliminate these.

Table 10.6: ***Selected global environmental concentrations of TEQ (PCDD/Fs) and PCBs***
[data compiled by Rappe 1996; Commonwealth of Australia 2004a; Reiner et al. 2006; Birch et al. 2007; LaKind 2007; Tanabe and Kunisue 2007; Mai et al. 2007; Srogi 2008; Li et al. 2009; Malarvannan et al. 2009; Schuhmacher et al. 2009; Tsang et al. 2009; Wang et al. 2009; Zhao et al. 2009a].
nd = not detected; dw = dry weight; fw = fresh weight; ww = wet weight

Region/Country	Matrix	Concentration range (PCDD/Fs)	Remarks
I: Environment-air			
Asia	Air	7-1486 fg TEQ/m³	
Australia	Air	14-17 fg TEQ/ m³	
Europe	Air	<1 fg TEQ/m³	Rural area
Europe	Air	15,000 fg TEQ/m³	Contaminated area
South America	Air	3-394 fg TEQ/m³	
Sweden	Air	3.3-6.1 fg TEQ/m³	Ambient air
I: Environment-aquatic			
Japan	Freshwater	0.012-48 pg TEQ/l	2116 sites
Japan	Seawater	nd-3.9 pg TEQ/m³	33 sites
Korea	Freshwater	0.001-1.061 pg TEQ/l	
I: Environment-soil			
Australia	Soil	0.05-23 pg TEQ/g dw	86 locations; dioxins in soils from urban and industrial locations were substantially higher than levels in agricultural and remote locations (see also section 10.8)
Austria	Soil	1.6-14 pg/g	Pasture/rural
Belgium	Soil	2.1-2.7	Pasture/rural
Germany	Soil	1-5 pg TEQ/g	Rural
Germany	Soil	0.1-30 pg/g	Pasture/rural
Germany	Soil	10-30 pg TEQ/g	Urban
Germany	Soil	30-300 pg/g	Industrial areas
Germany	Soil	30,000 pg TEQ/g	Contaminated
Germany	soil	100-80,000 pg/g	Close to major sources
Greece	Soil	2-45 pg/g	
Japan	Soil	220 pg TEQ/g	Agricultural and rural
Japan	Soil	1200 pg TEQ/g	Industrial and urban
Ireland	Soil	0.8-13	Pasture/rural
Italy (Seveso)	Soil	0.91-16 pg TEQ/g	
Italy	Soil	0.1-43 pg/g	Pasture/soil
Luxemburg	Soil	1.4 pg/g	Pasture/rural
Netherlands	Soil	2.2-16 pg TEQ/g	Rural
Netherlands	Soil	2.2-17 pg/g	Pasture/soil
Netherlands	Soil	98,000 pg TEQ/g	Contaminated
New Zealand	Soil	0.17-1.99 pg TEQ/g	Rural
New Zealand	Soil	260-6,670 pg TEQ/g	Urban
Spain	Soil	0.1-8.4	Pasture/rural
UK	Soil	<1-20 pg TEQ/g	Rural
UK	Soil	1,585 pg TEQ/g	Contaminated
Vietnam	Soil	9.9-184 pg/g	Contaminated
I: Environment-sediments			
Australia	Estuarine sediments	32-4352 pg WHO-TEQ/g dw (mean 712 pg/g)	Port Jackson
Australia	Estuarine sediments	667.8-4352.5 pg WHO-TEQ/g dw	Homebush Bay (Industrial site)

Contd...

Australia	sediments	Range 0.002-520 pg TEQ/g dw	Urban/industrial areas had significantly greater levels of dioxins than samples from remote and agricultural locations; Parramatta River (100-520 pg/TEQ/g)
Europe	Marine sediments	101,000 pg TEQ/g	Gulf of Finland; Highly contaminated
Finland	Estuarine sediments	0.7-100 pg WHO-TEQ/g dw	
Germany	Estuarine sediment (River Elbe)	7-150 pg WHO-TEQ/g	
Hong Kong harbour	Estuarine sediment	4-33 pg WHO-TEQ/g	
Italy	Estuarine sediments	nd-570	
Italy	Sediments	0.03 pg TEQ/g	Po River
Japan	Sediments	16.1-50.7 pg TEQ/g	
Japan	Marine sediments	0.012-49.3 pg TEQ/g	
Netherlands	Estuarine sediments	352-1849 pg WHO-TEQ/g dw	St Laurens harbour
New Zealand	Sediments	0.081-2.71 pg TEQ/g	
North America	Sediments	91 pg TEQ/g	Lake Ontario; average;
North Sea	Marine sediments	5.5-17.2 pg TEQ/g	
Norway	Estuarine sediments		
Russia	Sediments	0.03 pg TEQ/g	Lake Baikal
South Africa	Estuarine sediment	0.2-22 pg TEQ/g	
Spain	Marine sediments	0.4-8 pg TEQ/g	Catalan coast
UK	Estuarine sediments	14-24 pg WHO-TEQ/g dw	Umber Estuary
I: Environment-vegetation			
Austria	Vegetation	0.3-1.9 pg TEQ/g	Spruce needles
Germany	Vegetation	0.53-1.64 pg TEQ/g	Pine needles
II: Biota-Terrestrial animals			
Australia	Kangaroo	0.001-25 pg TEQ/g lipid	
Australia	Platypus & echidnas	9.3-60 pg TEQ/g lipid	
Australia	Birds of prey	3900 pg TEQ/g lipid (maximum)	
Canada	Caribou	0.7-6.4 pg TEQ/g lipid	
Japan	Sika deer	3.2-330 pg TEQ/g lipid	
India	Bird	19-24 pg/g fat	Eagle
India	Bird	150-200 pg/g fat	Osprey (hawk)
India	Bird	9.2-270 pg/g fat	Spotted owlet
India	Bird	1300-2700 pg/g fat	Spotted owlet liver
India	Bird	620-1000 pg furans/g fat	Spotted owlet liver
II: Biota-aquatic Food/ fish, shellfish			
Adriatic coast	Fish	0.23-329 pg TEQ/g lipid	Mackerel>Red mullet>Anchovy
Australia	Fish	0.053-0.49 pg TEQ/g ww	23 samples
Australia	Bivalves	0.043-0.2 pg/TEQ/g ww	18 samples
Belgium	Fish	4.5 pg/g fat	2000-2001
China (Pearl River Delta)	Fish	0.26 pg/g fat	2005
Finish coast	Fish	165-329 pg TEQ/g lipid	Hering muscle
USA	Fish	<1-63 pg TEQ/g	Dioxin PEQ; Great Lakes
USA	Fish	<1-182 pg TEQ/g	PCB TEQ; Great Lakes
Hong Kong	Fish	0.407-1.219 pg/g fat	2005; Means of freshwater (big head carp, grass carp, mandarin fish) and marine fish (big-eye, golden thread, snub nose pompano) (pg EROD-TEQ/g fat)

Contd...

Japan	Food	0.03 pg/g fresh weight	2001-2002
Spain	Food	0.01 pg/g fresh weight	2002; 2 samples
Taiwan	Food	0.0086 pg/WHO TEQ/g fw	dark green vegetables
Taiwan	Food	0.0256 pg/WHO TEQ/g fw	light green vegetables
III: Food-plant origin (cereals)			
Belgium	Food	0.9 pg/g fw	2002-2006; 4 samples
Greece	Food	0.94 pg/g fw	2002; 3 samples
Hong Kong	Food	0.07 pg/g fw	2005; 5 samples
Japan	Food	0.0027 pg/g fw	2001-2002;
Spain	Food	0.04 pg/g fw	2002; 2 samples
Taiwan	Food	0.01666 pg WHO-TEQ/g fw	
UK	Food	0.26 pg/g fw	2001
III: Food-plant origin (Rice)			
Hong Kong	Food	0.05 pg/g fresh weight	2005; 5 samples (different brands of rice)
Japan	Food	0.0039 pg/g fresh weight	2001-2002
IV: Human women milk-(PCDD/Fs only)			
America	Women milk	10.2 pg WHO-TEQ/g fat	1995-1996; 5 samples; include summation of PCDD/Fs and coplanar PCBs
Australia	Women milk	9 pg TEQ/g lipid (mean)	2002-2003; 173 samples
Australia	Women milk	5.9 WHO TEQs* (mean)	2002-2003; 157 donors and 17 samples; 12 regions; all milk samples from primipara
Belgium (Leige)	Women milk	29.4 pg TEQ/g fat	2000-2001; Wallonia
Brazil	Women milk	8.1-9.7 pg TEQ/g fat	1992; Rio de Janeiro
Brazil	Women milk	2.73-5.34 TEQS*	2001; various places; 10 donors
Bulgaria	Women milk	5.08-7.11 TEQ/g fat	
Cambodia	Women milk	7.8 pg TEQ/g fat	1999-2000
Cambodia	Women milk	5.6 TEQs*	2000; dumping site
Canada	Women milk	24 pg TEQ/g fat	1990
Canada	Women milk	15.0 pg TEQ/g fat	1986-1987
China	Women milk	0.15-17 TEQS*	2000-2003; several places including urban, rural
China	Women milk	2.6 pg TEQ/g fat	General population
China	Women milk	5.4 pg TEQ/g fat	Exposed to pentachlorophenol
China	Women milk	25.99 pg WHO-TEQ/g fat	2003-2005; 5 samples; Luqiao region; EROD-TEQ/fat (pg/g)
China	Women milk	5.46 pg WHO-TEQ/g fat	2001; 41 samples; Hebei region; TEQ based on summation of PCDD/Fs and total PCBs
China	Women milk	48.7 pg WHO-TEQ/g fat	2001; DR-CALUX
China	Women milk	8.2 pg TEQ/g fat	2002; mainland China
China	Women milk	3.73 pg TEQ/g fat	2007
Croatia	Women milk	6.40 TEQS*	2002; 20 donors
Czech Republic	Women milk	7.44-10.73 pg TEQ/g fat	
Czech Republic	Women milk	12.1 pg TEQ/g fat	1993; Kanado; Uhreskè regions
Czech Republic	Women milk	11.4-23.9 pg WHO-TEQ/g fat	1999-2000; 81 samples
Czech Republic (seven region)	Women milk	17.3 pg TEQ/g fat	1999-2000
Egypt	Women milk	22.33 TEQS*	2002; 90 donors; all milk samples from primipara
Finland	Women milk	39.0 pg TEQ/g fat	1992-1994
Finland	Women milk	12.1 TEQS*	1998; from Helsinki; 11 donors
Finland	Women milk	8.58-.511.8 TEQS*	1998; from Turku area; 3 donors
Fiji	Women milk	3.34 TEQS*	2002; 20 donors
France	Women milk	20.1-23 pg TEQ/g fat	1990; Paris area
France	Women milk	19.56 TEQS*	1998; samples from eight territorial zones; 244 donors; all milk samples from primipara

Contd...

Germany	Women milk	31 pg TEQ/g fat	1990
Germany	Women milk	16.0 pg TEQ/g fat	1995
Germany	Women milk	34 pg TEQ/g fat	1986-1991
Germany	Women milk	11.6 TEQS*	1997; 126 donors; data as I-TEQ
Germany	Women milk	12.9 TEQS*	1998; 69 donors; data as I-TEQ
Germany	Women milk	13.9 TEQS*	2001; 69 donors; North Rhine-Westphalia
Germany (Munich)	Women milk	9.91 pg TEQ/g fat	2005
Germany	Women milk	13.84 pg WHO-TEQ/g fat	2002-2003; 165 samples; Duisburg region
Germany	Women milk	Mean, 7.3 pg WHO-TEQ/g fat	2000-2003; 169 samples; TEQ based on summation of PCDD/Fs and total PCBs
Germany	Women milk	8.2 pg WHO-TEQ/g fat	2005; Bavaria region
Greece	Women milk	7.27-27.27 pg WHO-TEQ/g fat	2003; Athens region
Hong Kong	Women milk	8.25 pg TEQ/g fat	2001-2002
Hong Kong	Women milk	5.88-9.17 pg WHO-TEQ/g fat	2005; 29 samples, TEQ based on summation of PCDD/Fs and total PCBs
Hungary	Women milk	6.79 TEQS*	2002; 30 donors; all milk samples from primipara
India	Women milk	12 pg TEQ/g fat	1985-1994
India	Women milk	21 TEQS*	2000; dumping site; 11 donors; data as I-TEQ values
India	Women milk	8.3 TEQS*	2000; reference site; 8 donors
India	Women milk	6.22 TEQS*	2004; Ahmedabad; 23 donors; all milk samples from primipara
Ireland	Women milk	7.72 TEQS*	2002; 20 donors; all milk samples from primipara
Italy	Women milk	12-15 TEQS*	1999; all milk samples from primipara; Venice area
Italy (Venice, Rome)	Women milk	12.4 pg TEQ/g fat	1998-2001
Italy	Women milk	10.3-12.5 TEQS*	2000; 12 samples from Seveso, Milan and Rural Lombardy
Japan	Women milk	16.5 TEQS*	1999; Osaka city; 48 donors
Japan (all region)	Women milk	13.9 pg TEQ/g fat	All region; 1998-2002
Japan	Women milk	9.4-14 TEQS*	1999-2000; Tokyo region; 35 donors
Japan	Women milk	13.6-20.3 TEQS*	2001-2004; Fukuoka prefecture
Japan	Women milk	7.4 pg TEQ/g fat (mean)	2002-2004; 60 samples; mean; Sapporo region
Kazakhstan	Women milk	7.0-57.2 pg TEQ/g fat	Several locations; 1994
Korea	Women milk	17.62 pg TEQ/g fat	1998; Seoul area
Korea	Women milk	15.13-24.1 TEQS*	1997-1998; Seoul, and urban areas
Korea	Women milk	3.3-7.11 TEQS*	2002; sample from Ansan, Busan, Jeju, Seoul
Laos	Women milk	0.81 TEQS*	2001; 3 donors; Sepone area
Latvia	Women milk	Mean, 8.8 pg WHO-TEQ/g fat	2004; 30 samples
Luxembourg	Women milk	17.7 PEQS*	2000; 22 donors; all milk samples from primipara
Luxembourg	Women milk	14.97 PEQS*	2002; 20 donors; all milk samples from primipara
Malaysia	Women milk	8.6 TEQS*	2003; Penong, Kedah; all milk samples from primipara
New Zealand	Women milk	5.31 TEQS*	1998; rural & urban areas; all milk samples from primipara
New Zealand	Women milk	6.86 TEQS*	2002; 30 donors; all milk samples from primipara;
Netherlands	Women milk	17.09-21.29 pg TEQ/g fat	
Netherlands	Women milk	16.3-17.1 TEQS*	1998; all milk samples from primipara; data given as I-TEQ values
Netherlands	Women milk	47.0 pg TEQ/g fat	1993
Netherlands	Women milk	7.16-7.43 pg TEQ/g fat	

Contd...

Netherlands	Women milk	48.7 pg WHO-TEQ/g fat	2001; 10 samples; DR-CALUX
Netherlands	Women milk	18.27 TEQS*	2002; 30 donors; all milk samples from primipara
Norway	Women milk	36 pg TEQ/g fat	1992-1993
Norway	Women milk	6.0-8.9 TEQS*	2002; 10 donors; all milk samples from primipara
Norway	Women milk	5.5 TEQS*	2005; 27 donors; five counties
Philippines	Women milk	12 pg TEQ/g fat	all milk samples from primipara; median value
Philippines	Women milk	7.5 TEQS*	2000; dumping site; area known to be contaminated
Philippines	Women milk	3.94 TEQS*	2002; 20 donors;
Portugal	Women milk	9.3 pg WHO-TEQ/g fat	1999-2003; mean value; 181 samples
Portugal	Women milk	12.4 TEQS*	2001; 19 donors; Oporto region
Portugal	Women milk	10.8 TEQS*	2002; 52 donors; Lisbon area
Portugal	Women milk	6.5 TEQS*	2004; 58 donors; Madeira island
Repub. Of Uzbekistan	Women Milk	23.9 TEQS*	2001; 41 donors; samples from Karakalpakstan
Romania	Women milk	8.86 TEQS*	2002; 30 donors; all milk samples from primipara
Russia	Women milk	42.26 TEQS*	1997; Chapaevsk region; all milk samples from primipara; area known to be contaminated
Russia	Women milk	16.3-28.5 TEQS*	1998; Irkutsk region; all milk samples from primipara
Russia	Women milk	24 pg TEQ/g fat	1989
Russia	Women milk	10 pg TEQ/g fat	2002
Russia	Women milk	9 pg WHO-TEQ/g fat	2000-2002; 42 samples; mean value; Northern region
Slovak Republic	Women milk	9.07 TEQS*	2002; 40 donors; all milk samples from primipara
South Africa	Women milk	15 pg TEQ/g fat	
Spain	Women milk	13.3 pg TEQ/g fat	1990; Madrid area
Spain	Women milk	14.0 pg TEQ/g fat	1996
Spain	Women milk	11.56-11.9 TEQS*	2002; all milk samples from primipara
Spain	Women milk	7.8 pg TEQ/g fat	2004; 11 samples; Madrid area
Spain	Women milk	7.6 pg WHO-TEQ/g fat	2007; 15 samples; mean value; Tarragona county
Sweden	Women milk	9.0-9.1 TEQS*	1997-1999; Uppsala county; all milk samples from primipara
Sweden	Women milk	8.6-9.58 TEQS*	2000-2002; Uppsala county; all milk samples from primipara
Taiwan	Women milk	14.6 TEQS*	2000; 37 donors; Taichung and Central area
Taiwan	Women milk	7.4-8.53 TEQS*	2001
Taiwan	Women milk	7.62 pg TEQ/g fat	2000-2001
Taiwan	Women milk	7.37 pg TEQ/g fat	2000; 30 samples; only PCDD/Fs
UK	Women milk	11.77 TEQS*	2001; 23 donors
Ukraine	Women milk	8.38-10.16 pg TEQ/g fat	
Ukraine	Women milk	10.04 TEQS*	2002; 30 donors; all milk samples from primipara
USA	Women milk	19.0 pg TEQ/g fat	1980s
USA	Women milk	11.79 (PCDD/Fs + PCBs) pg TEQ/g fat	2002-2003; California
USA	Women milk	7.18 TEQS*	2002; 20 donors; all milk samples from primipara
Vietnam	Women milk	2.1 pg TEQ/g fat	1988; Hanoi area
Vietnam	Women milk	12.1 TEQS*	1999; 16 donors; Aluoi Valley; area known to be contaminated

Contd...

Vietnam Vietnam	Women milk Women milk	5.2-11 pg TEQ/g fat 12 pg TEQ/g fat	1985-1994; South Vietnam 2000
IV: Human women milk-(PCBs only)			
Belgium	Women milk	11.5 pg PCBs/g fat	2000-2001; 20 donors; age 26-38 years; Leige region
Belgium	Women milk	89 ng/g lipid wt	2006; 197 donors
Brazil	Women milk	150 ng PCBs/g fat	1992
Cambodia	Women milk	20 ng PCBs/g fat	1999-2000
Canada	Women milk	250 ng PCBs/g fat	1996
China	Women milk	42 ng PCBs/g fat	2002
China	Women milk	42 ng/g lipid wt	2002; Dalian region
China	Women milk	28 ng/g lipid wt	2002; Shenyang region
China (mainland)	Women milk	1.69 pg PCBs/g fat	2007; 1237 donors; age 18-37 years
Czech Republic (seven regions)	Women milk	57.2 pg PCBs/g fat	1999-2000; 81 donors; age 18-36 years
Europe	Women milk	1800 ng/g lipid wt	1999; Faroe island (Torshavn)
Germany	Women milk	550 ng PCBs/g fat	1995-1997
Germany (Duisburg)	Women milk	13.43 pg PCBs/g fat	2002-2003; 189 donors; age 19-42 years
Germany (Munich)	Women milk	9.88 pg PCBs/g fat	2005; 42 donors;
Hong Kong	Women milk	4.67 pg PCBs/g fat	2007; 316 donors; age 17-42 years
India	Women milk	30 ng PCBs/g fat	2000
Indonesia	Women milk	33 ng PCBs/g fat	2001
Indonesia	Women milk	27 ng/g lipid wt	2001-2003
Italy	Women milk	15.7 pg PCBs/g fat	1998-2001; 39 donors; age 21-40 years; (Venice, Rome)
Japan	Women milk	240 ng PCBs/g fat	1999
Japan	Women milk	8.8 pg PCBs/g fat	1998-2002;839 donors; age 20-39 years; all region
Japan	Women milk	4.9 pg PCBs/g fat	2002-2004; 30 donors; age 21-40 years; Sapporo area
Japan	Women milk	120 ng/g lipid	2001-2004; Fukoka area
Japan	Women milk	10 pg PCBs/g fat	2002-2004; 38 donors; age 21-40 years; Fukoka area
Japan	Women milk		2006; Osaka area
Kazakhstan	Women milk	380 ng PCBs/g fat	1994
Korea	Women milk	6.38 pg PCBs/g fat	1998; 20 donors; age 23-41 years; Seoul
Malaysia	Women milk	80 ng PCBs/g fat	2003
New Zealand	Woman milk	34 ng/g lipid wt	1998; 53 donors
Norway	Women milk	172 ng/g lipid wt	200-2002; 29 donors
Philippines	Women milk	72 ng PCBs/g fat	2000
Philippines	Women milk	70 ng/g lipid ww	2004
Poland	Women milk	153 ng/g lipid wt	2004; Wielkopolska region
Russia	Women milk	520 ng PCBs/g fat	1996
Russia	Women milk	346 ng/g lipid wt	2000; 14 donors; Murmansk region
Russia	Women milk	191 ng/g lipid wt	2002; 23 donors; Arkhangelsk region
Spain	Women milk	3.13 pg PCBs/g fat	2004; 11 donors; Madrid region
Sweden	Women milk	320 ng PCBs/g fat	1997
Taiwan	Women milk	5.19 pg PCBs/g fat	2000-2001; 20 donors; age 25-35 years
UK	Women milk	200 ng/g lipid wt	2001-2003; London
UK	Women milk	110 ng/g lipid wt	2001-2003; Lancaster
Vietnam	Women milk	74 ng PCBs/g fat	2000; Hanoi
Vietnam	Women milk	79 ng/g lipid wt	2001; Hochiminh
V: Human-Blood/serum/plasma (PCDD/Fs only)			
Australia	Blood serum	10.9 pg TEQ/g lipid	Mean value
USA	Blood/serum/plasma	27.5 ng TEQ/l	Recreational fishers; Great Lakes
Israel	Blood/serum/plasma	26.6-32 ng TEQ/g fat	

Contd...

Vietnam (A So)	Blood/serum/plasma	16.6-45.9 pg TEQ/g lipid	
New Zealand	Blood/serum/plasma	12.8 pg TEQ/g serum fat	1834 samples
Japan	Human serum	28.15 pg WHO-TEQ/g fat[a]	1999; 280 samples; TEQ based on summation of PCDD/Fs and total PCBs
Belgium	Human serum	75 pg WHO-TEQ/g fat[a]	1999; 47 pool samples
Belgium	Adipose		
Korea	Women serum	19.89 pg WHO-TEQ/g fat[b]	2001; 12 samples
Germany	Maternal serum	28.36 pg WHO-TEQ/g fat[a]	2000-2003; 169 samples
America	Maternal serum	9.98 pg WHO-TEQ/g fat[b]	1995-1996; 5 samples
Hong Kong	Maternal serum	3.13 pg WHO-TEQ/g fat[c]	2005; 21 samples
America	Cord serum	5.81 pg WHO-TEQ/g fat[b]	1995-1996; 5 samples
China	Cord serum	7.7-16.9 pg WHO/TEQ/g fat[a]	2005; three places (samples from Lin'anPingqiao and Luqiao)
Hong Kong	Cord serum	2.97 pg WHO/TEQ/f fat[c]	2005; 21 samples
V: Human-Adipose tissue, blood/serum/plasma (PCBs only) *[data compiled by Zhao et al. 2009a]*			
Belgium	Adipose	841 ng/g lipid	2000; mean value; 20 samples
Belgium	Adipose	490-763.7 ng/g lipid	2002-2005; mean value range
Belgium	Liver	380 ng/g lipid	2003-2005; 25 samples
China	Serum	65-69 ng/g lipid	2005; mean value range; Guiyu and Haojiang districts (selected PCBs - see Zhao *et al.* 2009a)
China	Kidney	399.4 ng/g lipid	2008; mean value19 samples; Wenling district
China	Liver	455.1 ng/g lipid	2008; mean value; 55 samples; Wenling district
China	Lung	257.9 ng/g lipid	2008; mean value; 7 samples; Wenling district
Denmark	Adipose	1027.4-1924.9 ng/g lipid	1993-1997; 409 samples (selected PCBs -- see Zhao *et al.* 2009a)
Japan	Human milk	63.9 ng/g lipid	2005; 89 samples;
Japan	Serum	37.5 ng/g lipid	2005; 89 samples
Sweden	Liver	459-2085 ng/g lipid	1994; 5 samples; Stockholm district
Sweden	Adipose	561-2343 ng/g lipid	1994; 5 samples; Stockholm district; (selected PCBs-see Zhao et al. 2009a)
Sweden	Adipose	757-1687 ng/g lipid	1996-1999; Stockholm district (selected PCBs-see Zhao *et al.* 2009a)
UK	Serum	170 ng/g lipid	2003; mean value; 154 samples
USA	Serum	1931.4-2264.4 ng/g lipid	1975-1994; different age groups (≤<10 yr-42 yr); Michigan district (Arocolor and Hexabromobiphenyls only)
USA	Serum	35-90 ng/g lipid	1990-2002
USA	Adipose	404 ng/g lipid	2003; 21 samples; California district
USA	Adipose	144 ng/g lipid	2002-2004; 52 samples; New York district

[a] = TEQ calculated based on summation of PCDD/F and total PCB; [b] = TEQ calculated based on summation of PCDD/F and Co-PCB;
[c] = TEQ calculated based on PCDD/F only (Tsang *et al.* 2009); * = Dioxins TEQs (mean, ppt, lipid basis) (WHO-TEF) (LaKind 2007)

People can be exposed primarily from foods contaminated with PCDDs, PCDFs and PCBs as a result of accumulation of these substances in the food chain and in high fat foods. Polychlorinated dioxins and furans may enter our body through breathing, ingestion or absorption through the skin. It is known that about 95% of the human intake of dioxins is through food (see *http://www.dioxinfacts.org/dioxin_health/dioxin_tissues/threshold.html; EC 1991;* see Table 10.7), mainly from animal origin such as fish, dairy and meat, of which fish appears to be the main contributor (see Box 10.2 for comparative risk of dioxin in fish and cardiovascular disease in Norway). On the other hand, children can be exposed to much higher levels of dioxins from breast feeding (see sections 10.8.3 and 10.8.5.4). According to

Wang *et al.* 2009, about 50% of the dietary intake of PCDD/Fs by the US population was from meat and dairy products, whereas marine foods and lipids (including oil and fats) were accounted for 33.7 and 15.3% exposure of dioxins to Tarragona population in Spain. In Taiwan, the consumption of fish was observed to be positively correlated to the level of serum PCDD/Fs. Additionally, accidental chemical exposure could be other routes of exposure to humans. Furthermore, according to Weber *et al.* (2008) PCDD/Fs can reach food and feed directly, some examples of which are listed below:

- In 1998, dairy and meat products were contaminated with high levels of PCDD/Fs originating from citrus pulp made using lime deposits from the chlorine/or organochlorine industry (http://ec.europa/euenvironment/dioxin/pdf001_ubt_final.pdf)
- In 1999, in Belgium, chickens, eggs and other animal products were contaminated with PCDD/Fs via waste PCB oil that was mixed with old fat and used for animal feed.
- In 2002, an animal feed mixture containing choline chloride was discovered to contain high levels of PCDD/Fs, which were later established as originating from PCP via saw mill dust that had been added to the mixture.
- In 2007, contaminated guar gum from India caused a global PCDD/F food scandal. Exported guar gum (widely used as starch for processed food) was contaminated with PCDD/Fs from PCP *(http://www.crl-freiburg.eu/dioxin/Guar%20Gum%20Evaluation.pdf)*.

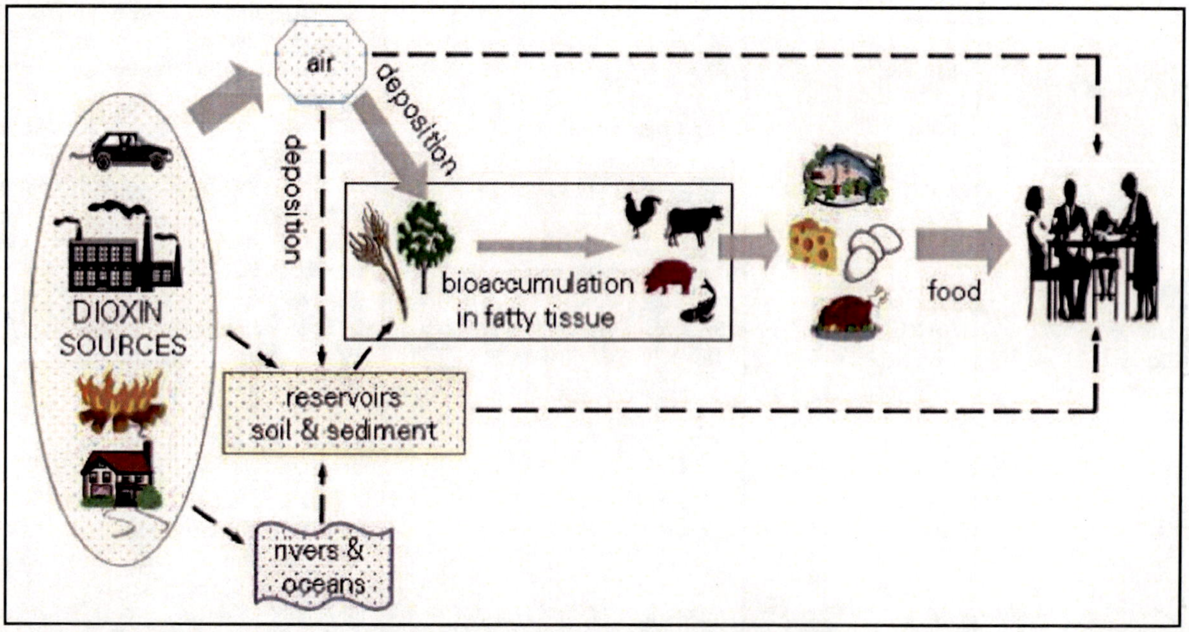

Figure 10.4: ***Pathway for dioxins entering human bodies*** *[Commonwealth of Australia 2004a and 2004b]*

Box 10.2: ***Finnish case study of comparative risk: dioxin in fish, fish consumption and heart disease*** *[source Tuomisto 2005; http://www.dioxinfacts.org/dioxin_health/dioxin_tissues/threshold.html].*

In Finland (according to Dr. J. T. Tuomisto of the Finnish National Public Health Institute), as much as 80 percent of dioxin exposure comes from fish consumption. While reducing fish consumption would reduce dioxin exposure in the population, however, it would also yield an unintended consequence - an increase in the death rate due to cardiovascular disease. In Finland, fish is the main source of omega-3 fatty acids, nutritional components that have been shown to decrease cardiovascular deaths. Tuomisto estimated the net loss of life in Finland if farmed salmon consumption were limited to once a month or less, based on the recommendation put forwarded by the U.S. EPA (2000). He estimated that, with limited farmed salmon consumption, the estimated cancer risk from dioxin is reduced by approximately 50 deaths annually. In contrast, fish consumption is thought to prevent 30,000 cardiac deaths annually. Limited fish consumption would result in losing some, but not all, of this benefit. Tuomisto predicted 7,500 extra deaths per year due from cardiovascular causes if limited fish consumption were practiced in Finland. Comparing 50 extra estimated (cancer) deaths per year to 7,500 extra (cardiovascular) deaths per year should leave no doubt in anyone's mind that fish consumption in Finland should not be limited.

Table 10.7: ***Contribution of foods (%) from dioxins*** *[data compiled by Srogi 2008].*

Country and year study conducted	Dairy	Meat, poultry	Eggs	Fish	Cereal products, vegetable (fats and oils)
Belgium (2001)	30	31	-	39	-
China (2000)	16	35	21	28	-
Finland (1999)	2-16	2-7	2-4	63-94	1-11
France (1999)	33	13	2	26	26
Germany (1998)	39	30	11	11	9
Italy (1996)	26	32	7	35	-
Japan (2000)	2	12	-	71	15
Korea (1999)	1	4	5	39	51
Norway (1997)	22	14	12	46	6
Spain (2000)	27	13	2	30	28
Sweden (1999)	19	15-31	1-2	34-36	14-51
The Netherlands (1999)	24-39	20-21	4-5	2-10	9-40
The United kingdom (1992, 2002)	25-44	18-20	1-4	6	45
USA (1995)	29	30	7	6	28

10.6: Effects of dioxins on biota

There is limited data are available on the toxic effects of dioxins on fish, birds, mammals and wildlife. Laboratory studies showed that dioxins are aryl hydrocarbon receptor (Ah) agonists, which can cause a wide spectrum of adverse effects in many vertebrate species (fish, birds and mammals) with embryos, foetus and new born being especially vulnerable to exposure during gestation and lactation.

Dioxins are most potent toxicants at very low concentrations and can disrupt the endocrine, reproductive, immune and nervous system of the offspring of fish, birds and mammals. Though dioxins are highly toxic to fish, birds and mammals, however, they are slightly toxic or non-toxic to amphibians, invertebrates and plants (Tables 10.8 and Table 10.9). Lack of sensitivity of dioxins to plants and invertebrates is believed to be related with non-existent of Ah receptors in those organisms. Tables 10.9 and 10.10 summarises the toxic concentrations and toxic effects of dioxins to wildlife and other organisms, respectively.

Table 10.8: ***Acute toxicity of TCDD*** *[US EPA 1985; Gatehouse 2004].*

Species	LD_{50} µg/kg
Guineapig	0.6-2.5
Mink	4
Chicken	25-30
Monkey	<70
Rabbit	115-275
Mouse	114-280
Rat	22-320
Dog	>100-<3,000
Frog	>1000
Hamster	1,157-5,051
Earthworm	Low toxicity
Plants	No effect

Table 10.9: *Toxicity of TCDC to a variety of algae/plants, invertebrates and vertebrates* [Landis and Yu 2004; Gatehouse 2004].

Test species	TCDD concentration in water (ng/l)	TCDD concentration in organism (pg/g) (wet weight)	Duration of exposure	Effects
Vertebrates				
Fish, Coho salmon *Oncorhynchus kisutch juvenile* (3.5 g)	0.56		96h/114 d	No toxic effect
Rainbow trout (*Oncorhynvhus mykiss*) eggs	5.60		96 h/56 d	50% mortality
Mink, *Mustela vision* (new born)		1,000	Daily for 12 d	100% mortality after 14 d
Invertebrates				
Annelid, worm *Paranais* sp.	200 (unmeasured)		55 d	No decrease in reproductive success
Mollusc, snail (adult) *Physia* sp.	1,330	502,000	33 d	No toxic effect
Arthropod, Cladoceran (adult) *Daphnia magna*	1,330	1,570,000	33 d	No toxic effect
Algae/plants				
Algae (*Oedogonium cardiacum*)	1,330	2,295,000	33 d	No effects
Vascular plant, Duck weed *Lemna minor*	1,300		33 d	No effects
Vascular plant, Duck weed *Lemna minor*	7,13	30,700	33 d	No effects

Table 10.10: *Effects of dioxins to wildlife and mammals* [Adapted from Gatehouse 2004]. *http://www.environment.gov.au/settlements/publications/chemicals/dioxins/report-11/index]*

Organism	Toxic effects	Remarks
Mammals	Very highly toxic	• Dioxins can cause wide spectrum of adverse toxic effects including lethality, reproductive and developmental effects, endocrine and immune system dysfunction, haemorrhaging, anaemia, wasting syndrome and biochemical effects. • Mink, otter are the most sensitive aquatic-feeding mammals while guineapigs are among the most sensitive terrestrial-feeding mammals. • LD_{50} for mortality in the guineapig is 600-2,000 pg/g; 28d LD_{50} for mortality of female mink is 264 pg/g/bw/day and male mink is 4,200 pg/g/bw/day. • New born mammals are most sensitive to the toxic effects of TCDD than adults (lethal dose is 100 pg TCDD/g/bw).
Birds	Very highly toxic	• Galliformes (domestic chicken, pheasants, quails, turkeys) are most sensitive order of birds to dioxins and eastern bluebird are the least sensitive. • LOAELs (lowest observed adverse effects level) for most sensitive and least sensitive birds causing embryo mortality and developmental effects (lab. studies) are between 10 and 10,0000 pg/g fresh weight TCDD in bird eggs. • Adverse effects in embryos following exposure to dioxins include embryo lethality, developmental abnormalities, deformities of skull, beak, eyes and toes, changes in body and organ weight and edema and low breeding success.
Fish	Very highly toxic	• Very highly toxic to fish, in particular early life stages. • Salmonid fish are most sensitive class of fish, in particular, the lake trout, *Salvelinus namaycush* (LC_{50} for lake trout sac fry is 65 pg TCDD/g and the NOAEL (no observed adverse effect level) is 34 pg TCDD/g egg). • LC_{50} for sac fry for other fishes are in the range of 439 -2,610 pg TCDC/g ww (see Gatehouse 2004). • NOEAL for sac fry mortality for various fish following embryo exposure ranged between 34 pg/g and 1,190 pg/g.

Contd...

		• Fish with longer development times also retain TCDD for longer. • Salt water fish found more tolerant to TCDD exposure than most freshwater fish. • Typical toxic response in early life stages of fish following exposure to dioxins include yolk sac edema, haemorrhaging, arrested development, and head and spinal deformities and in older fish includes lesions, fin necrosis and death.
Reptiles	Unknown	• No data available.
Amphibians	Slightly toxic	• 100-1,000 folds less sensitive to the deleterious effects of TCDD than the early life stages of fishes.
Invertebrates	No toxic effects	• Lack of sensitivity to dioxin is due to non existent of Ah receptor in invertebrates.
Algae and Plants	No toxic effects	• Lack of sensitivity to dioxins is due to non existent of Ah receptor in plants.

ww = wet weight

10.7: Dioxins risk to human health

[based on Srogi 2008; Devanathan et al. 2009; http://www.ejnet.org/dioxin; http://www.deh.gov.au/industry/chemicals/dioxins/index.html]

Short-term exposure of humans to high levels of dioxins may result in skin lesions, such as chloracne and patchy darkening of the skin and altered liver function. Long-term exposure is linked to impairment of the immune system, developing nervous system, endocrine system and reproductive functions. Dioxin exposure has further been linked to birth defects, inability to maintain pregnancy, decreased fertility, reduced sperm counts, endometriosis, diabetes, learning disabilities, lung problems; and lowered testosterone levels (Srogi 2008; *http://www.ejnet.org/dioxin/*).

Chronic exposure of animals to dioxins has resulted in several types of cancer (see WHO 2007). In 1997, a WHO group (International Agency for Research on Cancer, IARC) declared the most potent 2, 3, 7, 8-tetratchlorodibezo-p-dioxin or TCDD as a class-1 carcinogen, meaning a "known human carcinogen".

Breast milk monitoring has been implemented in many countries since breast milk samples offer a convenient and non-invasive means of monitoring humans for the presence of lipophilic compounds hence data provide an insight into the environmental conditions and historical human exposure. As PCDD/Fs and PCBs are lipophilic, they can accumulate in human breast milk. Dioxins (PCDD/Fs and PCBs) have been detected in human (women) breast milk from several countries (e.g. Australia, Belgium, Brazil, Bulgaria, Cambodia, Canada, China, Croatia, Czech Republic, Egypt, Finland, Fiji, France, Germany, Greece, Hong Kong, Hungary, India, Indonesia, Ireland, Italy, Japan, Kazakhstan, Korea, Laos, Latvia, Luxembourg, Malaysia, New Zealand, the Netherlands, Norway, the Philippines, Portugal, Republic of Uzbekistan, Romania, Russia, Slovakia, South Africa, Spain, Sweden, Taiwan, UK, Ukraine, USA and Vietnam) (see sections 10.5, 10.8.2 and 10.8.3), as a consequence breast feeding infants could have daily exposure to dioxins from early age. Other risks of dioxins to biota and human health are highlighted in case studies section (see section 10.8).

10.7.1: Guidelines for dioxins

- The USEPA has set the allowable concentration of 2,3,8-tetraCDD in drinking water from 0.13 to as low as 0.0013 pg/l based on estimated human cancer risks (tumour incidence risk: 0.13 pg/l for 10^{-5} 0.0013 pg/l for 10^{-7}) (USEPA 1984).
- In Australia, the drinking water guideline values for DCCC (2,7-dichlorodibenzo-p-dioxin) and for PCBs (PCB 105, 118 and 167) has been set at 0.0016 ng/l.
- USEPA ecological screening values for total PCDDs and total PCBs are 0.199 and 38.6 pg/g bw respectively (USEPA 2003).

- The Joint FAO/WHO Expert Committee on Food Additives (JECFA) established a provisional tolerable monthly intake of 70 pg of dioxins and dioxins like PCBs (JECFA 2001) and the scientific committee on food of the European commission establishes a 14 pg WHO TEQ/kg bw/day of tolerable intake on a weekly basis (EU SCF 2000).
- Canadian environmental soil quality guideline for dioxins is 4 pg/g.
- The maximum contaminant level (MCL) and the maximum contaminant level goal (MCLG) have been set at 30 pg TEQ/l and 0 pg TEQ/l respectively by the USEPA (USEPA 2001).
- In 1990, the WHO suggested a TDI (Tolerable daily intake) of 10.0 pg TEQ/kg of bw to protect human health from adverse dioxins' effects. However, in 1998 such TDI has been further reduced to 1.0-4.0 pg TEQ/kg of bw. Table 10.11 provides a comparative TDI adopted in different countries.

Table 10.11: *Average daily dioxin intake for adults proposed by different countries and organisations [ww.esdo.org); bw = body weight].*

Countries	Tolerable daily intake (pg-TEQ/kg bw/day)
Canada	10
Japan	1-4
Netherlands	2-4
USA	2
WHO	1-4

10.8: Case studies – Dioxins related compounds in the Asia – Pacific and Australia Region

10.8.1: e-waste disassembling/recycling sites and environment and food contamination by PCDD/Fs and PCBs in China [Sepulveda et al. 2009; Zhao et al. 2009b].

About 3,000 tonnes of waste electrical and electronic equipment (WEEE or e-waste) is produced everyday in the world. This WEEE originates from Australia, Canada, Europe, Japan, Korea and the USA and is exported to Asian countries such as China, India and Pakistan. According to UNEP DEWA/GRID-Europe (2005) nearly 70% of WEEE is exported to China for recycling. e-waste recycling (e.g. old computers, printers, copying machines, television sets and mobile phones) has become a booming industry in southwest of China, in areas such as Luqiao and Guiyu of Zhejiang Province. Furthermore, e-waste has become the fastest growing category of all solid waste found in China due to legal and illegal imports. As a result, a cluster of villages in the south-eastern portion in the Zhejiang province has become a recycling centre for e-waste. These primitive recycling operations involved stripping of metals in open pit acid baths, removing electronic components from circuit boards by heating over a grill and recovering metals by burning cables in (or near) crop land. Due to these operations and through leakage, evaporation, run-off, and leaching many toxic chemicals, such as PCDD/Fs and PCBs can be released to the local environmental matrices (air, sediment and soil) (see Figure 10.5). The following substances can be found or can be released as emissions or outputs during e-waste recycling :

- Leachate from dumping activities.
- Particulate matter (coarse or fine particles) from dismantling activities.
- Fly or bottom ashes from burning activities.
- Flumes from mercury amalgamate 'cooking', desoldering and other burning activities
- Wastewater from dismantling and shredding facilities.
- Effluents from cyanide leaching, other leaching activities or mercury amalgamation (see Figure 10.5).

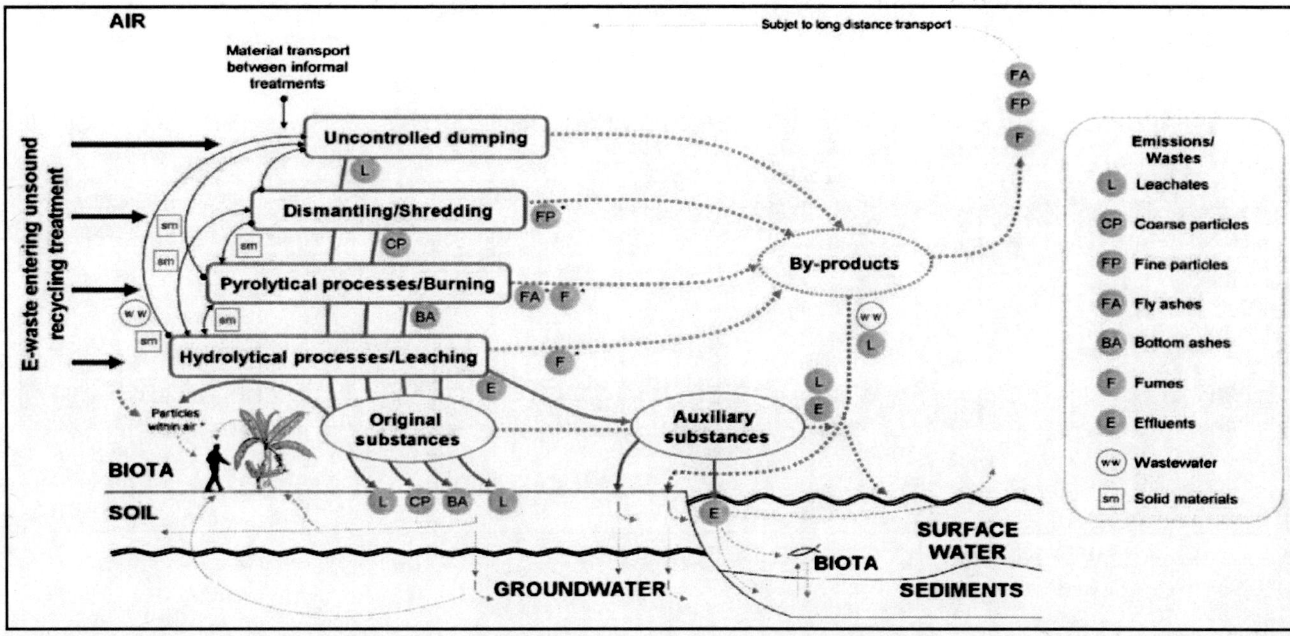

Figure 10.5: ***Principal e-waste/WEEE recycling activities in China and India, types of produced emissions and general environmental pathways*** *[from Sepulveda et al. 2009].*

NOTE : ***Ovals:*** *types of substances contained within emissions;* ***Continuous bold lines:*** *fate of original and auxiliary substances;* ***Dotted bold lines:*** *fate of by-products such as dioxins and furans;* ***Black arrows with a bold dot:*** *material transport fluxes between treatments;* ***Fine dashed arrows:*** *general environmental pathways.* ***Environmental fluxes*** *are driven by processes as* ***atmospheric deposition*** *(dry/ wet),* ***leaching, adsorption–desorption, complexation*** *(by which heavy metal and cyanide secondary products can be formed),* ***uptake, degradation*** *(chemical/biological) and* ***volatilization.*** *In addition, the environmental fate of pollutants depends on the physico-chemical properties of the media.*

Chlorinated dioxins and furans (PCDD/Fs) can be formed during e-waste burning. Once emitted into the atmosphere, PCDD/Fs are dispersed into the environment and because of their semi-volatile and hydrophobic properties; they tend to accumulate in organic rich media. Studies conducted in China have found that, PCDD/Fs concentrations in air around e-waste area of Guiyu is in the range of 64.9 to 2365 pg/M^3, which is the highest documented values of these compounds in the ambient air in the world and are attributed principally to WEEE dismantling activities (Sepulveda *et al.* 2009). Rice field soils and reservoir areas influenced by acid leaching activities had total PCDD concentrations (dioxins) in WEEE recycling sites in Guiyu which were respectively approximately 15,300-10,400 and 2,160 times higher than the USEPA ecological screening values for total PCDDs (see Table 10.12). Similarly, the total PCDF concentrations in soils affected by acid leaching and in rice crop soils in Guiyu were 939 and 17 times higher than the corresponding ecological screening values for furans (Table 10.12).

Table 10.12: ***PCDD/Fs concentrations in soils, rice crops and reservoir of Guiyu (China) and USEPA Ecological screening levels*** *[Sepulveda et al. 2009].*

	Total PCDDs concentration	Total PCDFs concentration
Soils – from an acid leaching area	3,050 pg/g dw	36,250 pg/g dw
Rice crop soils influenced by open burning	2,067 pg/g dw	667 pg/g dw
Reservoir	429 pg/g dw	36.2 pg/g dw
Ecological screening levels (USEPA 2003)	0.199 pg/g dw	38.6 pg/g dw

Zhao *et al.* (2009b) reported elevated levels of PCBs in agricultural produces (vegetables, pork, chicken, rice, fish, hen eggs) grown to close by e-waste disassembling sites in Luqiao and Guiyu of Zhejiang Province of China. The study found the highest dietary intake of PCBs by local residents living around the e-waste disassembly sites through consumption of rice and fish (see Table 10.13). It is assumed that contamination of rice and fish by PCBs may have occurred via dry particulate deposition of PCBs from nearby e-waste recycling areas. The studies of Sepulveda *et al.* (2009) and Zhao *et al.* (2009 a and 2009b) demonstrates that there is risk of growing agriculture produce to areas close to e-waste recycling or disassembly sites as the likelihood of contamination by PCDD/Fs and PCBs is very high.

Table 10.13: ***Total dietary intakes (TDIs in ng/d) of PCBs (ΣPCBs, ΣTEQs) for residents living in the control and the e-waste disassembly site of Zhejiang Province in China*** *[Zhao et al. 2009b].*

	Control site ΣPCBs	E-waste Disassembly site ΣPCBs	Control site ΣTEQs	E-waste Disassembly site ΣTEQs
Chicken	39.9	92.3	3.2	5.2
Drinking water	5.9	6.5	0.3	9.5
Fish	1345.0	2526.3	16.7	45.8
Hen eggs	302.4	449.8	4.0	5.2
Pork	175.2	202.3	8.5	12.8
Pulses	14.2	17.9	0.6	0.5
Rice	2564.8	8599.9	37.7	100
Vegetables	226.2	477.8	5.9	17.6
TDI	4673.7	12372.9	76.9	196.6

Dietary intake of PCB congener = Concentration of PCBs in the food x intake of each food type;
TEQs = toxic equivalence quotients (in pg/d)

10.8.2: PCBs in coastal waters of Asia

As part of global persistent organic pollutants monitoring in the Asia-Pacific region by the Global Mussel Watch Program, PCBs were examined in mussels (green mussel, *Perna virdi;* blue mussel, *Mytilus edulis* and *M. gallorovincialis* and ezo mussel, *Crenomytilus grayamus*) from 10 countries during the period 1994-1999 (see Table 10.14). The results of these monitoring found high concentrations of PCBs in mussels collected from Japan (3,000 ng/g lipid) and Russia (3,200 ng/g lipid). The results suggest that Japanese coastal waters may have been contaminated by PCBs due to continuous release from old transformers and capacitors still in use or in storage. The higher concentrations in Russia may indicate the presence of local PCBs sources since in the former Soviet Union, technical PCBs mixtures have been

Table 10.14: ***Concentrations of PCBs in mussel samples from Asian countries*** *[Monirith et al. 2003].*

Country,number of locations and year of sampling	PCBs concentrations range (ng/g lipid weight)	Comments
Cambodia (7 locations); 1998	35 (mean) <3.8-220 (range)	Highest concentrations at international air port site
China (10 locations); 1999	120 (mean) 20-600 (range)	Highest concentrations at industry and ferry site
Hong Kong (10 locations); 1998, 1999	310 (mean) 40-710 (range)	Highest concentrations at ferry and shipping site
India (17 locations); 1998	220 (mean) 10-2200 (range)	Highest concentrations at industry and harbour site
Indonesia (9 locations); 1998	87 (mean) 5.6-210 (range)	Highest concentrations at fishing port and industry site
Japan (8 locations); 1994	3000 (mean) 510-12,000 (range)	Highest concentrations at Bay site
Korea (10 locations); 1998	170 (mean) 30-340 (range)	Highest concentrations at Harbour site
Malaysia (13 locations); 1998	56 (mean) <4.2-250 (range)	Highest concentrations at port, industry site
Philippines (7 locations);1998	290 (mean) 22-550 (range)	Highest concentrations at Metro Manila site
Russia (2 locations);1999	3200 (mean) 2700-3700	Highest concentrations at Bay site
Singapore (1 location);1999	90	-
Vietnam (9 locations); 1997	160 (mean) 26-450	Highest concentrations at floating habitat site

produced and used as dielectric fluid in the manufacture of power capacitors and transformers. On the other hand, lower concentrations of PCBs were found in mussels from Cambodia, Indonesia, Malaysia, Vietnam, India and China (average range 30-310 ng/g lipid). Despite this, some urban/industrial cities in India (Mumbai, Cochin), the Philippines (Manila) and China (Fuzhou Lian Jian) showed relatively higher PCB concentrations.

10.8.3: PCBs in human breast milk

Elevated levels of PCBs (240 ng/g lipid) were detected from Japanese women's breast milk (Table 10.15) which was comparable to western developed nations' women's breast milk concentrations (250 and 320 ng/g lipid from Canada and Sweden, respectively). PCBs from other Asian countries women's breast milk were much lower (30-80 ng/g lipid) compared to developed Asian country like Japan (see Figure 10.6 and Table 10.15). The higher levels of PCBs were also detected in Japan in Asian mussels watch program (see Table 10.14).

Table 10.15: *PCBs and dioxin in human breast milk (ng/g lipid) collected from general public in Asian countries [Tanabe and Kunisue 2007].*

Country	PCBs	TEQs (PCDD/PCDFs)
Cambodia (1999-2000)	20	7.8
China (2002)	42	8.2
India (2000)	30	12
Indonesia (2001)	33	-
Japan (1999)	240	36
Malaysia (2003)	80	13
Philippines (2000)	72	12
Vietnam (2000)	74	12

10.8.4: Dioxins and dioxins-related compounds (DRCs) in human breast milk in municipal dumping sites in Asia

Human breast milk collected from several Asian countries including China, Cambodia, India, Indonesia, Japan, Malaysia, the Philippines, and Vietnam (1999-2003) detected elevated levels of dioxins related compounds (DRCs) such as PCDDs , PCDFs (Figure 10.7 and Table 10.15). This residue survey indicates that there is considerable dioxin contamination in developing as well as in developed Asian nations. The study found that the critical emission sources of these DRCs, in particular in Asian developing countries were from open dumping sites of municipal wastes. For example, in Asia, a variety of municipal wastes are dumped continuously and burnt under low temperatures possibly resulting uncontrolled emissions of dioxins related compounds.

In another study, Tanabe and Kunisue (2007) also found higher residue levels of DRCs (PCDD/PCDF) in soils from open dumping sites in Asian developing countries than the agricultural and urban (reference sites) soils far from the dumping sites. This finding indicates that the dumping sites in Asia are potential sources of dioxins and can be related to significantly higher concentrations of DRCs in women's breast milk found from residents around the dumping sites, for example from India. In India, people living around the dumping sites consume milk from buffaloes and cows that graze daily on the contaminated garbage or soils in the waste dumping sites, therefore livestock can be exposed to dioxins. In contrast, in Vietnam and Cambodia, though there are also many open dumping sites, but domestic livestock are not reared around the dumping sites and levels of contaminants in human breast milk were not significantly different between dumping and reference sites. The TEQs of DRCs in women's breast milk from the dumping sites of India (mean TEQ of 40.0 pg/g lipid) were close to or higher than those from some developed countries including Japan (TEQs of DRCs in women's breast milk from Japan, Germany and Canada are reported to be 36.0, 34.0 and 24.0 pg/g lipid, respectively).

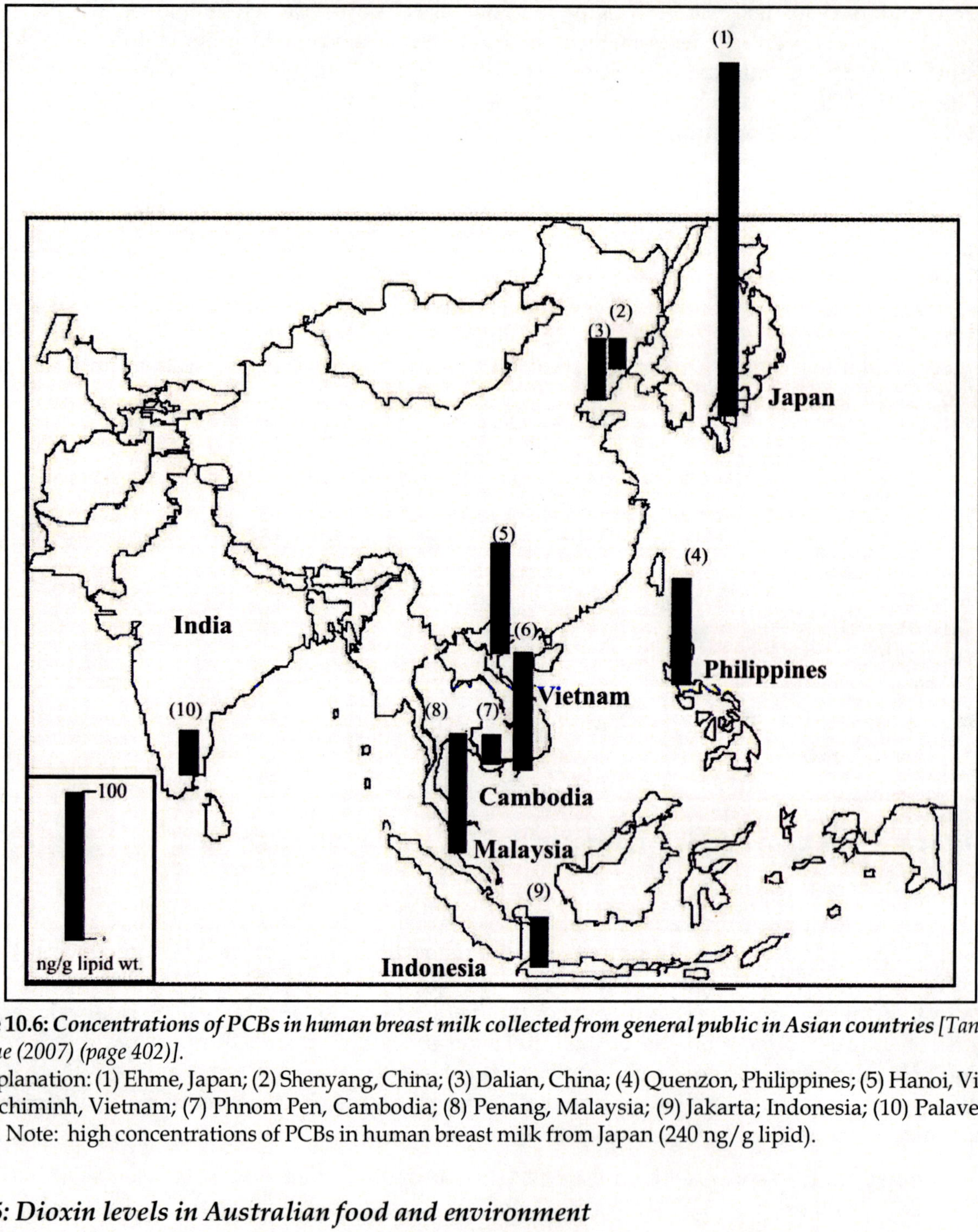

Figure 10.6: ***Concentrations of PCBs in human breast milk collected from general public in Asian countries*** *[Tanabe and Kunisue (2007) (page 402)].*
site explanation: (1) Ehme, Japan; (2) Shenyang, China; (3) Dalian, China; (4) Quenzon, Philippines; (5) Hanoi, Vietnam; (6) Hochiminh, Vietnam; (7) Phnom Pen, Cambodia; (8) Penang, Malaysia; (9) Jakarta; Indonesia; (10) Palaverkadu, India]. Note: high concentrations of PCBs in human breast milk from Japan (240 ng/g lipid).

10.8.5: Dioxin levels in Australian food and environment

The Australian National Dioxins Program carried out a number of investigations during the period 2001-2004, the technical reports of all theses investigations are available on-line [see *http://www.deh.gov.au/industry/chemicals/dioxins/index.html].* The investigation was aimed at assessing the levels of dioxins in Australian environment including the following:

- dioxins emissions from bushfires.
- dioxins emissions from motor vehicles.
- dioxins in air, soils, aquatic environments, fauna, and agriculture commodities.
- dioxins in human population.

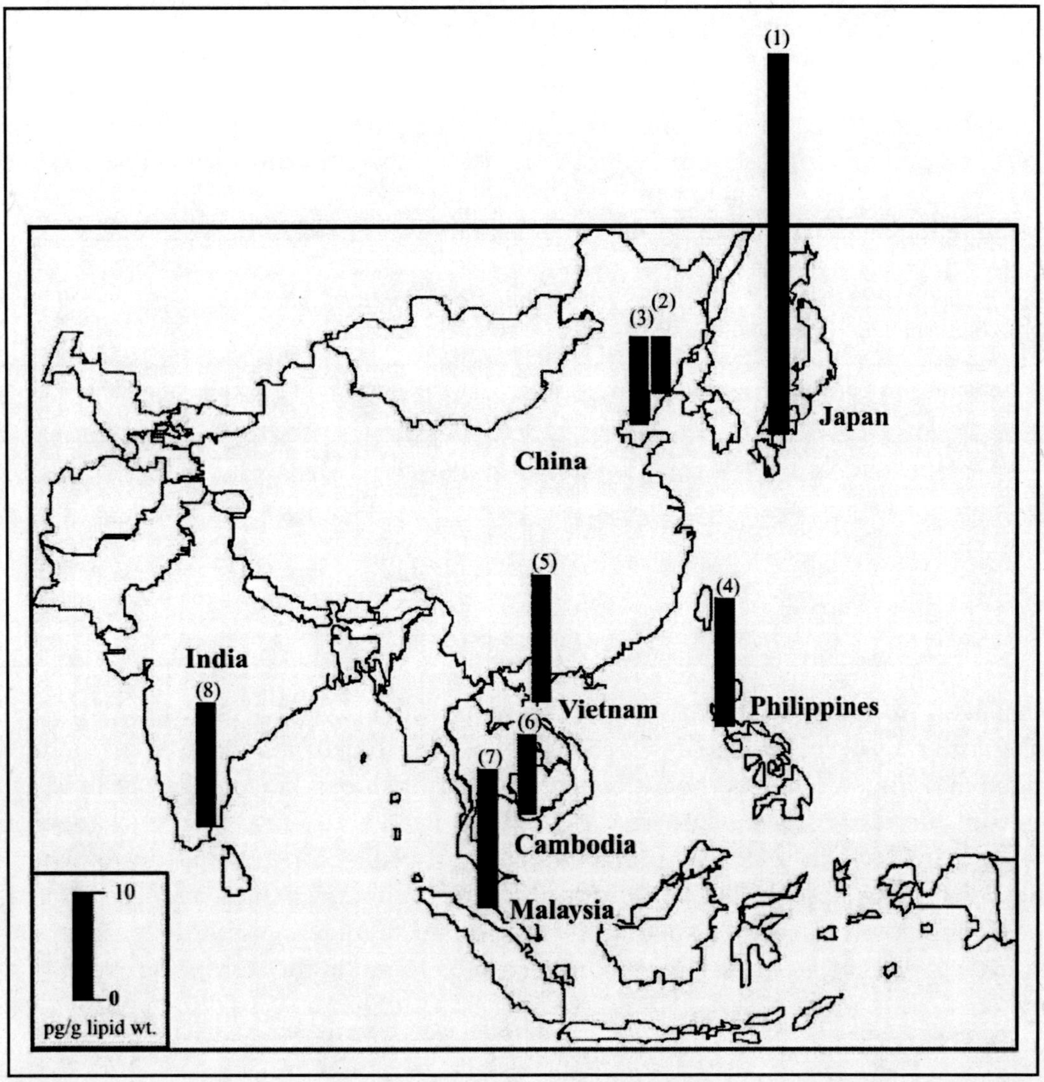

Figure 10.7: ***TEQs (toxic equivalent concentration) of DRCs (dioxins related compounds) (PCDDs, PCDFs, PCDFs and coplanar PCBs) in human breast milk collected from general public in Asian countries*** *[Tanabe and Kunisue (2007) (page 401)]* site explanation: (1) Ehme, Japan; (2) Shenyang, China; (3) Dalian, China; (4) Quenzon, Philippines; (5) Hanoi, Vietnam; (6) Phnom Phen, Cambodia; (7) Penang, Malaysia; (8) Jakarta; Indonesia; (9) Palaverkadu, India]. Note: high concentrations of DRCs in human breast milk from Japan (36 ng/g lipid).

The above investigations revealed that the ambient levels of dioxins in the Australian environment and Australian population are generally low compared to other countries.

10.8.5.1: Dioxins in the Australian environment

The total emissions of dioxins (air, water and land) estimated in Australia are between 160-1,788 g TEQ/year. Uncontrolled combustion including controlled and wildfires, waste burning, accidental fires contributes around 70% of total dioxins emissions to air (1,360 g TEQ/annum), of which burning and wildfires accounts for 20-30%. Ferrous and non-ferrous metal production; production of chemicals and consumer goods such as pulp and paper; fossil fuel power generation; waste disposal, land filling and waste incineration contributes around 23% (420 g TEQ/annum). Domestic wood heaters contribute around 4% of dioxins emissions. Higher levels of dioxins recorded during winter months in Australian urban areas are likely to be due to smoke from these heaters. Motor vehicles account for less than 2% of total dioxin emissions; however, with the banning of leaded petrol in January 2002 the levels have substantially declined.

10.8.5.2: Dioxins in Australian soils, sediments and fauna

Dioxins levels in Australian soils from urban and industrial locations are substantially higher (0.13-23 pg/g) than in agricultural and remote locations (0.03- 5.2 pg/g). However, overall dioxins levels in Australian soils are much lower compared to many industrial countries (see Table 10.16). The dioxins levels in Australian sediment ranged from 0.02 to 520 pg TEQ/g dw. Highest concentrations were found in sediments from urban and industrial environments (100-520 pg TEQ/g dw).

Table 10.16: ***Dioxin levels in soils in Australia and other industrial countries*** *[http://www.environment.gov.au/settlements/publications/chemicals/dioxins/report-11/index]*

Dioxins in Australian soils (pg/g dw) according to land use	Urban and industrial soils compared to other countries (pg TEQ/g)
Industrial: 0.13-11	Australia: 0.13-23
Urban: 0.17-23	Canada: 105
Agricultural: 0.03-4.3	Germany: 115
Remote: 0.041-5.2	Japan: 1200 USA: 35 UK: 90

Dioxins levels in Australian fauna were found highest in birds (ranged from 1.0 pg TEQ/g lipid in the Galah, *Cacatua roseicapilla* to over 3,506.0 pg TEQ/g lipid in a collard sparrowhawk, *Accipiter cirrhocephalus*) and lowest levels in herbivores such as kangaroos (0.27 pg TEQ/g lipid in the Eastern grey kangaroo, *Macropus giganteus* to 25.4 pg TEQ/g lipid in a western grey kangaroo, *Macropus fuliginosus*). Dioxins level in marine mammals ranged between 2.0 pg/g lipid in the Dugong, *Dugong dugong* and 585.0 pg/g lipid in the bottlenose dolphin blubber, *Tursiops aduncus*. Marine mammals living in estuaries close to urban/industrial areas also had relatively higher levels. Detail information about the dioxins levels in Australian environment can be found in the technical report – Ecological risk assessment of dioxins in Australia- *[http://www.environment.gov.au/settlements/publications/chemicals/dioxins/report-11/index]*.

10.8.5.3: Dioxins in Australian agriculture produce

Dioxins emitted to air can deposit on plant, soil and water surfaces, which can then enter the food chain when animals eat leaves, soils or sediments contaminated with dioxins. As mentioned in section 10.5.1, 95% of the human intake of dioxins is through food, mainly from animal origin such as fish, dairy and meat. Dioxins levels in Australian agriculture commodities, including beef, sheep, pigs, milk and fish were found low compared to other countries (well below standards set by the European Union) (see Table 10.17). In particular, dioxins level in Australian fish and dairy milk are much lower compared to Asia and Europe (see Figure 10.8).

Table 10.17: ***European Union (EU) standards for dioxins/furans (upper bound concentrations) (pg TEQ/g fat) compared with Australian agriculture produce results*** *[http://www.deh.gov.au/industry/chemicasl/dioxins/index.html]*

Source	Mean Australian study	Maximum- Australian study	EU standards Maximum pg TEQ/g
Beef	0.557	1.77	3
Fish (cultured salmonids)	0.28	0.350	4
Dairy Milk	0.434	0.749	3
Pig	0.331	0.551	1
Poultry	0.330	0.529	2
Sheep	0.57	2.83	3

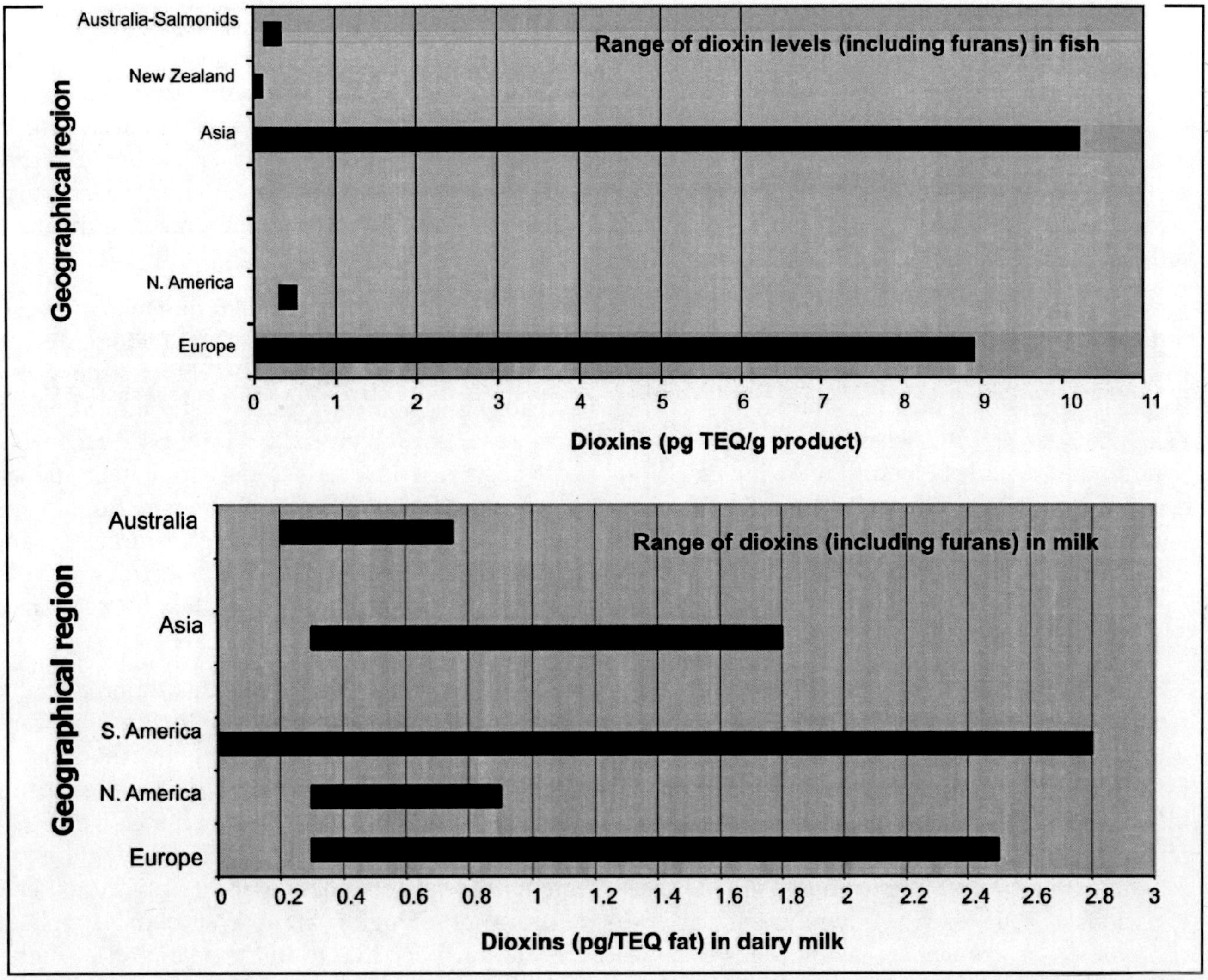

Figure 10.8: *Range of dioxin levels (including furans) in fish (left) and dairy milk (right) in Autralia and other countries [adapted from Commonwealth of Australia 2004b; http://www.deh.gov.au/industry/chemicals/dioxins/index.html page 47 and 50].*

10.8.5.4: Dioxins in Australian women's breast milk

The levels of PCDD/PCDFs and PCBs in the breast milk of Australian women (2002-2003) were found similar across all Australian regions (mean 9.0 pg TEQ/g lipid; median 8.9 pg TEQ/g) and values were much lower than international standards (e.g. 34.0 pg TEQ/g lipid in Germany, 36.0 pg TEQ/g in Norway). The 2002-03 Australian breast milk residue survey result were a factor of two lower than from 1993 survey (the mean and median in 1993 were 16.0 and 16.4 pg TEQ/g lipid, respectively) [see *http://www.deh.gov.au/industry/chemicalsdioxins/factsheet3.html]*.

10.9: Conclusion

Dioxins (PCDDs, PCDFs, and PCBs) exhibit serious health effects when it reaches as little as a few parts *per trillion* in human body fat. The most toxic dioxin, 2, 3, 7, 8-TCDD is considered to be class-1 carcinogen. Several studies suggest that exposure to 2, 3, 7, 8-TCDD could increases the risk of several types of cancer in human and a wide range of deleterious effects including reduced immunity to nervous system, disorders and miscarriages and birth deformity. It is also a powerful hormone disrupting chemical. The non-cancer effects include cardiovascular diseases, diabetes and changes in blood composition.

Though a tolerable daily intake (TDI) of 1.0 to 4.0 pg I-TEQ/kg body weight/day has been established for dioxins by the WHO (Van Leeuwen *et al.* 2000), however, the ultimate goal should be to reduce human intake levels below 1.0 pg I-TEQ/kg body weight/day. Regular monitoring of the levels of dioxins in meat, milk products and fish would help to reduce the risk to humans. Food samples analysed under the Food Standards Australia and New Zealand found that fish, meat and dairy products were the major groups contributing to dioxins exposure to general human population. A recent study conducted by Perelló *et al.* 2009 found that cooking processes are only of limited value as a means of reducing concentrations of PCDD/Fs and PCBs.

The low level of dioxins in Australian agri-food commodities may be attributed due to low level of industrial activity compared with Northern Hemisphere. In addition, lower average population densities in Australia relative to Europe, Asia and America are thought to be the basis of lower exposure to vehicular emissions. Low population in Australia also cause low domestic heating, and motor vehicle usage compared with Northern Hemisphere. Furthermore, action taken by Australian Governments over the last decade helped in reducing emissions of dioxins. Such actions included implementation of cleaner air regulations by state and territories requiring industrial facilities to reduce emissions of dioxins, banning of backyard rubbish burning, phasing out the use of leaded petrol, banning manufacture and use of chemicals known to be source of dioxins (e.g. pentachlorophenol, 2, 4 ,5-T); destroying OCPs and PCBs in destructing facilities, replacing older environment polluting wood heaters by appliances that contribute better air quality.

Since dioxins can pass from the woman's body to the infant through the placenta and breast milk, women planning to have children should avoid eating food containing high levels of dioxins for several years. Breast milk is an ideal medium for assessing exposures to persistent organic pollutants such as PCDD/Fs and PCBs. A periodic assessment of levels of PCDD/Fs in women's milk provides exposure information for human population. Furthermore, the concentrations of PCDD/Fs in women's milk can reflect the levels and trends of contamination in the local environment. The Stockholm Convention will rely on women's milk monitoring to aid in the assessment of the effectiveness of the Treaty in reducing emissions of POPs into the environment. In developed countries, though residue levels of dioxin related compounds in women's breast milk have decreased in recent times due to implementation of highly efficient incinerators and strict regulations on the production and usage of hazardous chemicals. However, improper management may still cause increased release of dioxins related compounds or DRCs in the environment. Despite all, much less is known about the dioxin levels in human specimens such as human serum and cord serum, such information could provide a more complete picture on the body loadings of dioxin plus additional information on the levels of dioxins that might be transferred from a mother (through placenta) to her developing foetus. Continued environmental monitoring and bio-monitoring of PCDD/Fs and PCBs would therefore serve to ensure that exposures don't increase either unexpectedly or unnoticed.

10.10: References

Birch, G. F., Harrington, C., Symons, R. K. and Hunt, J. W. 2007. The sources and distribution of polychlorinated dibenzo-*p*-dioxin and polychlorinated dibenxofurans in sediments of Port Jackson, Australia. *Marine Pollution Bulletin* 54: 295-308.

Commonwealth of Australia. 2004a. National dioxins program. Dioxins in Australia: a summary of studies conducted from 2001 to 2004. Australian Government. Department of the Environment and Heritage. Canberra. 16 pp. http://www.deh.gov.au/indutry/chemicals/dioxins/index.html.

Commonwealth of Australia 2004b. Dioxins in Agricultural commodities in Australia. Department of Environment and Heritage, Canberra. 29 pp. http://www.deh.gov.au/industry/chemicals/dioxins/index.html.

Devanathan, G., Subramanian, A., Someya, M., Sudaryanto, A., Isobe, T., Takahashi, S., Chakraborti, P. and Tanabe, S. 2009. Persistent organochlorines in human breast milk from major metropolitan cities in India. *Environmental Pollution* 157: 148-154.

EC (European Commission DG Environment) 1991. Compilation of the EU dioxin exposure and health. Data task 4- human exposure. http://ec.europa.eu/environment/dioxin/pdf/task4.pdf.

EU SCF (European Union Scientific Committee on Food) 2000. Opinions on the SCF on the risk assessment of dioxins and dioxin-like PCBs in food.

Gatehouse, R. 2004. *Ecological risk assessment of dioxins in Australia-National Dioxins Program*. Technical report no. 11. Australian Government. Department of the Environment and Heritage, Canberra. Commonwealth of Australia. 146 pp. http://www.environment.gov.au/settlements/publications/chemicals/dioxins/report-11/pubs/report-11.pdf

Geyer, H. J., Schramm, K-W., Feicht, E. A., Behechti, A., Steinberg, C., Brüggemann, R., Poiger, H., Henkelmann, B. and Kettrup, A. 2002. Half-lives of tetra-, penta-, hexa-, hepta-, and octachlorodibenzo-*p*-dioxin in rats, monkeys, and humans-a critical review. *Chemosphere* 48: 631-644.

JECFA (Joint FAO/WHO Expert Committee on Food Additives) 2001. Summary and conclusions of the fifty-seventh meeting.

Kulkarni, P. S., Crespo, J.G., and Afonso, A. M. C. 2008. Dioxins sources and current remediation technology- A review. *Environmental International*. 34: 139-153.

Lakind, J.S. 2007. Recent global trends and physiological origins of dioxins and furans in human milk. *Journal of Exposure and Environmental Epidemiology* 17: 510-524.

Landis, W. G. and Yu, M. H. 2004. *Environmental Toxicology, Impacts of chemicals upon ecological systems*. 3rd edition, Lewis Publishers. Boca Raton, Florida. 484 pp.

Li, J., Zhang, L., Wu, Y., Liu, Y., Zhou, P., Wen, S., Liu, J., Zhao, Y. and Li, X. 2009. A national survey of polychlorinated dioxins, furans (PCDD/Fs) and dioxin-like polychlorinated biphenyls (di-PCBs) in human milk in China. *Chemosphere* 75: 1236-1242.

Malarvannan, D., Kuisue, T., Isobe, T., Sudaryanto, A., Takahashi, S., Prudente, M., Subramanian, A. and Tanabe, S. 2009. Organohalogen compounds in human breast milk from mothers living in Payatas and Malate, the Philippines: Levels, accumulation kinetics and infant health risk. *Environmental Pollution* 157: 1924-1932.

Mai, T.A., Doan, T.V., Terradellas, J., de Alencastro, L.F. and Grandjean, D. 2007. Dioxin contamination in soils of Southern Vietnam. *Chemosphere* 67: 1802-1807.

Mandal, P. K. 2005. Dioxin: a review of its environmental effects and its aryl hydrocarbon receptor biology. *Journal Comparative Physiology B. Biochem Syst Environ Physiol*. 175: 221-230.

Monirith, I., Ueno, D., Takahashi, S., Nakata, H., Sudaryanto, A., Subramanian, A., Karuppiah, S., Ismail, A., Muchtar, M., Zheng, J., Richardson, B.J., Prudente, M., Hue, N.D., Tana, T.S., Tkalin, A.V., and Tanabe, S. 2003. Asia-Pacific mussel watch: monitoring of persistent organochlorine compounds in coastal waters of Asian countries. *Marine Pollution Bulletin*. 46: 281-300.

Perelló, G., Martí-Cid, R., Castell, V., Llobet, J.M. and Domingo, J,L. 2009. Influence of various cooking processes on the concentrations of PCDD/PCDFs, PCBs and PCDEs in foods. *Food Control*. doi:10.1016/j.foodcont.2009.05.003.

Rappe, C. 1996. Sources and environmental concentrations of dioxins and related compounds. *Pure and Applied Chemistry*. 68(9): 1781-1789.

Reiner, E. J., Clement, R. E. and Okey, A. B. 2006. Advances in analytical techniques for polychlorinated dibenzo-*p*-dioxins, polychlorinated dibenzofurans and dioxin-like PCBs. *Annal Bioannal Chem* 386: 791-806.

Schuhmacher, M., Kiviranta, H., Ruokojärvi, P., Nadal, M. and Domingo, J. L. 2009. Concentrations of PCDD/Fs, PCBs and PBDEs in breast milk of women from Catalonia Spain: A follow-up study. *Environmental International* 35: 607-613.

Sepulveda, A., Schluep, M., Renaud, F.C., Streicher, M., Kuehr, R., Hagelüken, C. and Gerecke, A.C. 2009. A review of the environmental fate and effects of hazardous substances released from electrical and electronic equipment during recycling: examples from China and India. *Environmental Impact Assessment Review*. doi:10.1016/jeiar.2009.04.001.

Srogi, K. 2008. Levels and congener distributions of PCDDs, PCDFs and dioxin-like PCBs in environmental and human samples: a review. *Environmental Chemistry Letter* 6: 1-28.

Tanabe, S. and Kunisue, T. 2007. Persistent organic pollutants in human breast milk from Asian countries (review). Environmental Pollution. 146: 400-413.

Tsang, H. L., Wu, S. C., Wong, C. K. C., Leung, C. K. M., Tao, S. and Wong, M. H. 2009. Risk assessment of PCDD/Fs levels in human tissues related to major food items based on chemical analysis and micro-EROD assay. *Environmental International* doi:10.1016/j.envint.2009.05.005

Tuomisto, J. 2005. Review-Does mechanistic understanding helps in risk assessment-the example of dioxins. *Toxicology and Applied Pharmacology* 207: S2-S10.

UNEP DEWA/GRID-Europe. 2005. E-waste, the hidden side of IT equipments manufacturing and use. *Environment Alert Bulletin* No. 5. UNEP: Nairobi, Kenya (*http://www.grid.unep.ch/product/publication/download/ew_ewaste.en.pdf*)

USEPA (United States Environmental Protection Agency) 1984. Ambient water quality criteria for 2,3,7,8-TCDD,EPA440/5-84-007.

US EPA (United States Environmental Protection Agency) 1985. Drinking water criteria document for 2, 3, 7, 8-tetrachlorodibenzo-p-dioxin. U.S. Environmental Protection Agency. EPA 600/x-84/94-1.

USEPA (United States Environmental Protection Agency) 2000. Guidance for assessing chemical contaminant data for use in fish advisories. Risk assessment and fish consumption limits, vol 2. Washington, DC, 3rd ed., (available at http://www.epa.gov/ost/fishadvice/volume2/index.html).

USEPA (United States Environmental Protection Agency) 2001. National primary drinking water standards. EPA 816-F-01-007.

USEPA (United States Environmental Protection Agency) 2003. Release of guidance for developing ecological soil screening levels (Eco-SSLs) and Eco-SSls for nine contaminants. USEPA region 5 : Office of solid waste and emergency response: 2003a. 3 pp.

Van den Berg, M., Birnbaum, L., Bosveld, A. T., Brunstorm, B., Cook, P., Feeley, M., Giesy, J. P., Hanberg, A., Hasegawa, R., Kennedy, S. W. 1998. Toxicity equivalency factors (TCFs) for PCBs, PCDDs, PCDFs for humans and wildlife. *Environmental Health Perspectives*, 106 (12): 775-792.

Van den Berg, M., Birnbaum, L., Denison, M., De Vito, M., Farland, W., Feeley, M., Fileder, H., Hakansson, H., Handberg, A., Haws, L., Rose, M., Safe, S., Schrenk, D., Tohyama, C., Tritscher, A., Tuomisto, J., Tysklind, M., Walker, N. and Peterson, E. 2006. The 2005 World Health Organization revaluation of human and mammalian toxic equivalency factors for dioxins and dioxin-like compounds. *Toxicological Sciences* 93(2): 223-241.

Van Leeuwen, F.X., Freeley, M., Schrrenk, D., Larsen, J.C., Farland, W. and Younus, M. 2000. Dioxins: WHO's tolerable daily intake (TDI) revised. *Chemoshere* 40:1095-1101.

Wang, I-C., Wu, Y-L., Lin, L-F. and Chang-Chien, G-P. 2009. Human dietary exposure to polychlorinated dibenzo-*p*-dioxins and polychlorinated dibenzofurans in Taiwan. *Journal of Hazardous Materials* 164: 621-626

Weber, R., Gaus, C., Tysklind, M., Johnston, P., Forter, M., Hollert, H., Heinisch, E., Holoubek, I. et al. 2008. Dioxin- and POP-contaminated sites-contemporary and future relevance and challenges. Overview on background, aims and scope of the series. *Environmental Science and Pollution Research* 15: 363-393.

WHO (World Health Organisation) 1998. Executive summary: assessment of the health risk of dioxins: re-evaluation of the tolerable daily intake (TDI)-WHO consultation. WHO; May 1998, available at http://wwwwho.int/pcs/pubs/dioxin-exec-sum-final.html

WHO. 2007. *Dioxins and their effects on human health.* Fact sheet N°225. November 2007. http://www.who.int/mediacentre/factsheets/fs225/en/index.html

Zhao, G., Wang, Z., Zhou, H. and Zhao, Q. 2009a. Burdens of PBBs, PBDEs, and PCBs in tissues of the cancer patients in the e-waste disassembly sites in Zhejiang, China. *Science of the Total Environment* 407: 4831-4837.

Zhao, G., Zhou, H., Wang, D., Zha, J., Xu, Y., Rao, K., Ma, M., Huang, S. and Wang, Z. 2009b. PBBs, PBDEs, and PCBs in food collected from e-waste disassembly sites and daily intake by local residents. *Science of the Total Environment* 407: 2565-2575.

Websites

http://www.dioxinspin.com/tox_chem_structure1.htm

http://www.ejnet.org/dioxin/ (dioxin home page)

http://environmentalchemistry.com/yogi/chemistry/pcb.html

http://www.dioxinfacts.org/dioxin_health/dioxin_tissues/threshold.html-Dioxin: A Threshold Carcinogen - The Chlorine Chemistry Division of the American Chemistry Council

http://news.bbc.co.uk/2/hi/health/medical_notes/358889.stm

http://effluent-plankton-fish-bird-bird

http://www.ecoinfo.org/env_ind/region/toxindescript/toxindescriptione.cfm Environment Canada- description of some toxic contaminants found in the Pacific and Yukon region. Accessed on 29 Feb 08).

http://www.rsc.org/ebooks/archive/free/BK9780854042258/BK9780854042258-00001.pdf

□□□

11 Chemical Impacts - Endocrine Disrupting Chemicals (EDCs)

Key facts

- Endocrine disrupting chemicals (EDCs) are micropollutants with estrogenic or androgenic activities at very low concentrations and are emerging as a major concern for water quality.
- EDCs include *natural and synthetic hormones* (Estrone or E1, 17β-estradiol or E2, Estriol or E3, 17α-ethynylestradiol or EE2); *alkylphenols* (nonylphenol or NP; octyphenol or OP), *phenols* (bisphenol-A or BPA), *phthalates* (Di-n-butyl phthalate or DBP, Di-2-(ethylhexyl)-phthalate or DEHP); *organotins* (tributyltin or TBT); and *phytoestrogens* (β-sitosterol, coumestrols, genistein).
- Wastewater treatment plants, intensive livestock effluents, industrial wastewaters, landfill, paper and pulp mill effluents and antifouling paints are important sources of EDCs in the environment.
- EDCs are ubiquitous and persistent in the environment and have been detected in sewage treatment effluents, surface waters (rivers, streams), sediments, groundwater, rain water, biota and even in drinking water from Australia, Bangladesh, Belgium, Canada, China, Denmark, Egypt, France, Germany, Greece, Holland, Hong Kong, India, Israel, Italy, Japan, Korea, Malaysia, Norway, Poland, Portugal, Spain, Sweden, Switzerland, Taiwan, The Netherlands, UK, USA, and Vietnam.
- The concentrations detected in surface waters are in the following ranges: E2 (nd-93 ng/l), EE2 (nd-35.6 ng/l), NP (0.007-644 μg/l), BPA (0.016-21 μg/l), phthalates (<0.0005-16.6 μg/l), TBT (500 ng sn/l) and phytoestrogens (nd-143400 ng/l).
- Estrogens such as EE2, and E2 are more potent EDCs and likely to pose greatest risk to the environment and wildlife. EE2 as low as 0.1 ng/l and E2 as low as 1.0 ng/l can cause increased level of vitellogenin (an egg yolk precursor protein) in fish.
- EDCs can bio-accumulate in the fatty tissue of the organisms (algae, fish and molluscs) and increase in concentration as they move up through the food chain. A recent investigation conducted in Italy in 2008 found that almost all of the tested seafood (anchovies, hakes, mackerels, mantin shrimp, mullets, octopus, seabreams, swordfish and tuna) contained the highest concentrations of NP. NP has also been detected in human blood, breast milk, urine and adipose tissue confirming persistent and bio-accumulative nature of NP.
- Studies conducted with animals have found that low doses of BPA have caused a variety of adverse health effects (reduced sperm counts, impaired immune system functioning, increase in prostate tumour proliferation, altered prostate and uterus development, insulin resistance, alteration in brain chemistry, early puberty and behavioural changes).
- Human can be exposed to phthalates via food wrapping materials, use of medical devices or pharmaceutical drugs containing phthalate in their coatings and via carpets in houses. Exposures to high levels of phthalates have been reported to cause miscarriages and pregnancy complications in women. Furthermore, exposure of phthalates could be the leading cause of the reproductive disorders in humans including reduced sperm counts, histological changes in testes and reduced fertility.
- TBT pollution has been related specifically to the development of imposex, i.e., a masculination of prosobranch gastropod females by the development of a vas deferens and / or a penis in addition to their female sex characteristics. Imposex can be induced by TBT at ambient concentration as low as 1.0 ng TBT-Sn/l.
- Most conventional wastewater treatment plants do not have advanced treatment process/technology and therefore these emerging micro pollutants are not removed but are released to the receiving natural waters.

Contents

List of Tables

List of Figures

Acronyms and Glossary

Aerosols: An aerosol is a suspension of fine solid or liquid droplets in a gas.

Androgens: Male sex hormone that is produced in the testes and responsible for typical male sexual characteristics.

Bio-accumulation: A process by which chemical substances are accumulated by aquatic organisms from water, either directly or through consumption of food containing the chemicals.

Bio-concentration: A process by which there is a net accumulation of a chemical directly from water into aquatic organisms.

Body burden: Also known as 'chemical load', it is the amount of harmful chemicals present in a person's body. It is determined by bio-monitoring, which assesses exposure by measuring the chemicals or their metabolites in human specimens such as blood or urine.

Conjugates: Unite chemically so that the product is easily broken apart into the original compounds.

Coumestrol: Coumestrol is a natural organic compound in the class of phytochemicals known as coumestans. It has garnered research interest because of its estrogenic activity and its prevalence in some foods, such as soybeans and herbs such as *Pueraria mirifica*.

DES: Diethylstilbestrol.

e-EDCs: Estrogenic endocrine disrupting chemicals. e-EDCs have the potential to cause an estrogenic response at very low concentrations (parts per billion to parts per trillion). Estrogenic EDCs are a sub-class of EDCs that, when organisms are exposed to them, function as estrogens.

E1: Estrone.

E2: 17β-estradiol.

E3: Estriol

EE2: 17α-ethynylestradiol.

EDC: Endocrine disrupting chemical.

EEC: European Economic Council

Edema: Swelling from excessive accumulation of watery fluid in cells, tissues, or serous cavities.

Estrogens: Are a group of steroid compounds, named for their importance in the estrous cycle, and functioning as the primary female sex hormone. The three major naturally occurring estrogens in women are estradiol, estriol and estrone.

Estrogenic: A female hormone produced in the ovary which is involved in the menstrual cycle and in developing and maintaining female sexual characteristics (e.g. breasts).

Estrone: An estrogenic hormone from the urine of pregnant females.

GA/CA Glucuronic acid is a carboxylic acid. Its structure is similar to that of glucose. However, glucuronic acid's sixth carbon is oxidized to a carboxylic acid.

Gonopodia: A modified anal fin in male mosquito fish (*Gambusia affinis holbrooki*) used to transfer sperm to females.

Gonado somatic index (GSI) : Proportion/ratio (in percentage) of gonad weight in relation to body (somatic) weight.

Hermaphrodite: Any animal possessing both male and female sex organs in the same individual; any plant possessing stamens and carpels in the same flower.

Homeostasis: The ability of a system or living organism to adjust in its internal environment to maintain a stable equilibrium such as the ability of warm-blooded animals to maintain a constant body temperature.

Imposex: The term is applied to some sea snails, marine gastropod molluscs which, under the toxic effects of pollutants, develop sex organs that are in contrast to their actual gender. For example, characteristics such as the development of male sex organs, (penis and the vas deferens) in female gastropods. The only known inducer of imposex is tributyltin (TBT), an anti-fouling agent for boats which affects females of the species *Nucella lapillus* (dog whelk) and more than 200 other marine gastropods. The imposex stages of female dog whelks (*N. lapillus*) and other molluscs (including *N. lima*) are used in the UK and worldwide to monitor levels of tributyltin (TBT).

Intersex: An organism having biological characteristics of both the male and female sexes.

In utero: A term literally meaning "in the uterus". It is used in biology to describe the state of an embryo or foetus.

In vitro: In cells in test tubes or other condition of nature.

In vivo: In whole organisms or in nature.

Koc: Organic carbon normalised partition coefficient.

Log K_{ow}: Octanol-water partition-coefficient; it quantifies the liphophilicity of a substance and is assumed to be an index of the ability to pass through biological membranes and to bio-accumulate in living organisms.

Micelles: An aggregate of surfactant molecules dispersed in a liquid colloid. A typical micelle in aqueous solution forms an aggregate with the hydrophilic 'head' regions in contact with the surrounding solvent, sequestering the hydrophobic tail regions in the micelle centre. This type of micelle is known as a *normal phase micelle* (oil-in-water micelle). Inverse micelles have the head groups at the centre with the tails extending out (water-in-oil micelle). Micelles are approximately spherical in shape. Other phases including shapes such as ellipsoids, cylinders and bilayers are also possible. The shape and size of a micelle is a function of the molecular geometry of its surfactant molecules and solution conditions such as surfactant concentration, temperature, pH and ionic strength. The process of forming micellae is known as *micellisation* and forms part of the phase behaviour of many lipids according to their polymorphism.

nd: Not detected

nM: Nanomolar; 10^{-9} molar; 1 molar = In chemistry, molar concentration (also called molarity, amount concentration or substance concentration) is a measure of the concentration of a solute in a solution, or of any molecular, ionic, or atomic species in a given volume.

Non-point source: Non-point source (NPS) pollution is water pollution affecting a water body from diffuse sources, such as polluted run-off from agricultural areas draining into a river or wind borne debris blowing out to sea. NPS pollution can be contrasted with point source (PS) pollution, where discharges occur to a water body at a single location, such as discharges from a chemical factory or from ships at sea.

NP: 4-nonylphenol.

NP1EO: Nonylphenol monoethoxylate.

NP2EO: Nonylphenol diethoxylate.

NPOE: 1-(2-Nitrophenoxy) octane, also known as nitrophenyl octyl ether.

Organogenesis: The formation and development of the organs of an organism from embryonic cells.

Ovotestis: An ovotestis is a gonad with both testicular and ovarian aspects. In humans, ovotestes are an anatomical abnormality associated with gonadal dysgenesis.

Pallial sex glands: Accessory sex glands in molluscs.

PCBs: Polychlorinated biphenyls (PCBs). Historically used as fire retardants in power transformers and capacitors.

Phytoestrogen: Defined as any plant compounds structurally and/or functionally similar to ovarian and placental estrogens and their active metabolites. Phytoestrogens exists in many plants such as soybeans, fruits and cabbages.

Prosobranch: The majority of marine gastropods are prosobranch, as are a few land snails and freshwater snails. In other words, these snails have their gill, mantle cavity and anus situated in front of the heart (in contrast to the arrangement in the opisthobranchs). Most prosobranchs also have separate sexes. The prosobranch gastropods include the majority of marine snails, among them conches, cones, cowries, limpets, murexes, periwinkles, volutes and whelks, as well as numerous freshwater groups and some land snails with an operculum.

pK_a: An acid dissociation constant, K_a, (also known as acidity constant, or acid-ionization constant) is a quantitative measure of the strength of an acid in solution. It is the equilibrium constant for a chemical reaction known as dissociation in the context of acid-base reactions.

Skewed sex ratio: Imbalance of sex ratio, rather skewed to either males or females.

Sn/l: Stannum/l; Tin is a chemical element with the symbol Sn (Latin: *Stannum*) and atomic number 50. It is a main group metal in group 14 of the periodic table.

Steroid: Any of the numerous compounds containing the carbon ring system of the sterols and including the sterols and various hormones and glycosides.

Sterol: Are important classes of organic molecules. They occur naturally in both plants and animals with the most familiar type of animal sterol being cholesterol, which has been shown to contribute to high blood pressure and heart disease. In the past decade, interest in plant sterols as a dietary supplement has increased due to studies showing that they can contribute to lower cholesterol levels.

Surfactants: Surfactants are usually organic compounds that are amphiphilic, meaning they contain both hydrophobic groups (their 'tails') and hydrophilic groups (their 'heads'). Therefore, they are soluble in both organic solvents and water e.g. detergents.

Teratogenic: Of or relating to substances or agents that can interfere with normal embryonic development.

Vas deferens: The vas deferens (plural: vasa deferentia) also called ductus deferens (Latin: "carrying-away vessel") is part of the male anatomy of some species, including humans.

Vitellogenin (Vtg): An egg yolk precursor protein expressed in female fish and on other vertebrates (amphibia, reptiles and birds) and invertebrates (oysters, mussels, snails and crabs), dormant in male fish and female insects. In the presence of estrogenic endocrine disruptive chemicals (EDCs), males can express the Vtg gene in a dose dependent manner. Vtg gene expression in male fish can be used as a molecular marker of exposure to estrogenic EDCs.

WWTPs: Wastewater treatment plant.

Xenobiotics: Agent foreign to an organism.

11.1: Introduction

Endocrine disrupting chemicals (EDCs) are heterogenous group of substances (either natural or synthetic) that may interfere with the endocrine system of organisms. According to Soto *et al.* (1995) EDCs can:

- mimic the sex steroid hormones estrogens and androgens by binding to hormone receptors or influencing cell signalling pathways;
- block, prevent and alter hormonal binding to hormone receptors or influence cell signalling pathways;
- alter production and breakdown of natural hormones; and
- modify levels and function of hormone receptors.

The US Environment Protection Agency (USEPA 1997) defines endocrine disrupting chemicals (EDCs) as 'xenobiotics that interfere with the synthesis, secretion, transport, binding, action or elimination of natural hormones in the body that are responsible for the maintenance of homeostasis, reproduction, development and /or behaviour'. Various natural and synthetic chemical compounds have been identified that mimic or induce estrogen like responses including natural estrogens/androgens, synthetic estrogens/androgens, alkyl phenols, phthalates, pesticides, heavy metals and phytoestrogens. These chemicals are also called as 'estrogenic EDCs' or in short e-EDCs. EDCs are contaminants that may be hormonally active at very low concentrations (parts per billion to parts part trillion) and are emerging as a major concern for water quality and disturbances of wildlife reproduction and development.

11.2: Endocrine systems

The endocrine systems in humans and animals are made up of internal glands, which manufacture and secrete hormones. These hormones carry messages from one part of a cell to another or from one part of the body to another that trigger or control specific parts/activity of organisms. Hormones help in the development, growth, reproduction, energy use and maintenance of the internal environment (including blood pressure and heart beating rate) and behaviour of animals and humans. The human endocrine system includes hypothalamus, pituitary, parathyroids, thyroid, adrenals, pancrease, ovaries and testes (Figure 11.1).

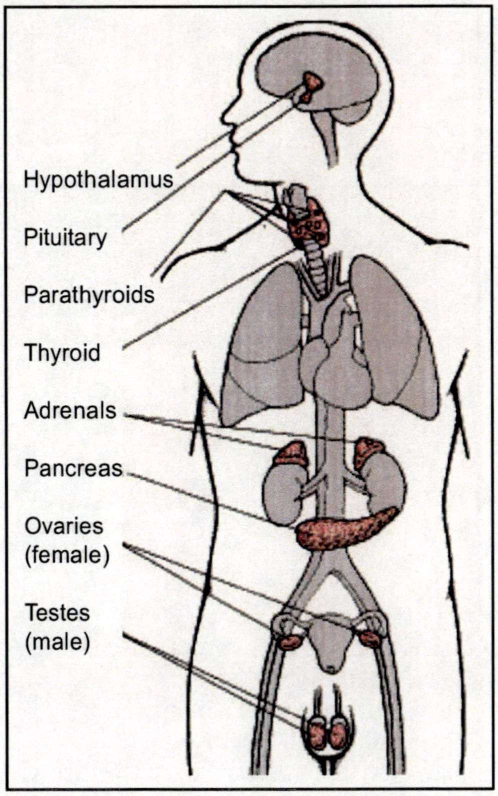

Figure 11.1: ***Human endocrine system.*** *[source: European Commission Research_Endocrine Disruptors http://www.cushings-help.com/the_body.htm].*

11.3: Categories of EDCs

More than 562 chemicals (man-made, synthetic and natural) have been known or suspected to be EDCs (see EEC 2001; http://ec.europa.eu/environment/docum/01262_en.htm) that can affect or disrupt the endocrine system of humans and animals. Many more chemicals are suspected to have similar effects. Chemicals that have been found or suspected to be capable of disrupting the endocrine systems fall into the following chemical categories: synthetic hormones/estrogens, industrial chemicals (alkylphenols; phenols; phthalates; antifouling agents/organotins; pesticides, polyaromatic hydrocarbons; dioxins; polychlorinated biphenyls), metals and phytoestrogens (Table 11.1).

Table 11.1: ***Some suspected/ known endocrine disrupting chemicals (EDCs), their sources and uses*** *[http://www.ourstolenfuture.org/basics/chemlist.htm; Ying and Kookana 2002; Williams et al. 2007; Jackson and Sutton 2008; Matozzo et al. 2008; Soares et al. 2008; Zeng et al. 2008a and 2008b; Barnabe et al. 2009; Liu et al. 2009a].*

Examples of suspected EDC Compounds	Potential sources	Uses, EDC properties and other information
I: Natural hormones/estrogens *Natural androgens:* Androsterone (A); Dihydrotestosterone (DHT); Testosterone (T) *Natural estrogens:* 17β-Estradiol (E2); Estriol (E3); Estrone (E1)	WWTP effluent	
II: Pharmaceuticals/artificial synthetic estrogens or androgens Diethylstilbestrol (DES); 17α-ethynylestradiol (EE2); Mestranol (Me EE2-b); Norgestrel (N); Tamoxifen, Trenbolone (Tr)	WWTP effluent, ingestion	Contraceptive pills; see chapter 12 for other pharmaceuticals.
III: Alkyl phenols Butylphenols; Nonylphenols (NPs); Nonylphenols ethoxylates (NPEOs); Octylphenols (OP); Octylphenol ethoxylates; Pentaphenols	Wastewater treatment plant (WWTP) effluent, point sources.	Alklyphenols (AP) are the degradation products of the widely used non-ionic surfactants (detergents) alkyl phenols ethoxylates (APE). APEs are used in personal care products (deodorants make up, hair products), detergents and cleaners, lubricants, defoamers, emulsifiers, paints, textiles, pulp and paper industries, pesticide formulations, oil recovery and mining's. Among APEs, nonylphenol ethoxylates (NPOEs) and octylphenol ethoxylates (OPEOs) account for approximately 80% and 20% respectively of the worldwide production. NPs and NPEOs are highly toxic compounds. OP, NPs and nonylphenol monoethoxylates (NP1EO) are endocrine disruptors toward many fish species. Due to extensive use of nonylphenol ethoxylates, they reach sewage treatment plants in substantial amounts where they are incompletely degraded to nonylphenol (Ferrara*et al.* 2008; Soares *et al.* 2008).
IV: Phenols Bisphenol A; Bisphenol F	WWTP effluent	Bisphenol A (BPA, (2, 2-bis-(4-hydroxyphenyl)-propane) is one of the key raw materials used to manufacture polycarbonate plastics and epoxy resins. It is used to manufacture various consumer products, including food and water containers, baby bottles, baby toys, computers, CDs, DVDs, cell phones, lining of food and beverage metalcans, medical tubing, epoxy resins, flame retardants and dental fillings. These compounds have been shown to leach into food and water from containers, particularly after heating or as plastic ages (Samital 2008). BPA is a commercially important chemical with worldwide production capacity of 3.7 million metric tons per year of which 95% is used in the production of polycarbonate plastics and epoxy resins BPA has an estrogenic activity with a potency of four to six orders of magnitude less than 17β-estradiol.
V: Phthalates Butyl benzyl phthalate (BBP); Di-n-butyl phthalate (DBP); Di-2-(ethylhexyl)-phthalate (DEHP); Di-ethyl phthalate; Di-*n*-pentyl phthalate; Di-hexyl phthalate; Dicyclohexyl phthalate; Di-propyl phthalate	Leaching from plastics, ubiquitous	Phthalate esters (PAEs) are a class of manufactured chemicals widely used as plasticizers in industrial products and households. The global production of PAEs has reached a level of ~150 million tonnes per year, employed in a broad range of applications such as films, shopping and garbage bags, fluid containers, clothing, toys, household, industrial and building materials, common component of nail polish,

Contd...

		flooring, transportation, food packaging, medicinal products and insect repellents. More than 80% of the plasticisers used worldwide are phthalates, with DEHP representing more than 80% of those. Most phthalates have endocrine disrupting potencies six to seven orders of magnitude lower than 17β-estradiol. They are predominantly fat soluble. PAEs are commonly detected in air, water, sediment/soil, biota and human tissues. The USEPA and several other countries classified PAEs such as BBP, DEHP and DBP as priority pollutants and often suspected as endocrine disrupting effects on vertebrates and humans.
VI: Organotins Tributyltin (TBT); Dibutyltin (DBT); Monobutyltin (MBT); Triphenyltin (TPhT)	Antifouling paint	Tributyltin (TBT) used as an antifouling agent in boat paints is a common contaminant of marine and freshwater ecosystems. Under favourable conditions, TBT degrade s to produce progressively less toxic dibutyltin (DBT) and monobutyltin (MBT) and ultimately inorganic tin. Once released from an antifouling coating, TBT is rapidly absorbed by organic materials such as bacteria and algae or adsorbed onto suspended particles in the water. It is readily incorporated into the tissues of filter feeding zooplankton, grazing invertebrates, and higher organisms such as fish, water birds and mammals, where it accumulatesExposures to organotins (TBT) result in imposex, the abnormal induction of male sex characteristics in female gastropod molluscs.
VII:Phytoestrogens (plant compounds) β-sitosterol; Coumestrols; Genistein; Isoflavonoids; Lignans; Zearalenone **others**	PPM effluent, ingestion	Phytoestrogen exit in many plants with high concentrations are reported in soybeans, fruits and some vegetables.
VIII: Polyaromatic hydrocarbons (PAHs) Anthracene; Benzo(*a*)pyrene; Benz(*a*) anthracene; Benzo(*b/h*) fluoranthene; 6-hydrooxy-chrysene; *n*-butyl benzene; Pyrene; Phenanthrene	Pulp and paper mill (PPM) effluent, ubiquitous	By-products of incomplete combustion of fossil fuels, wood and mineral oil. They are known to be toxic and carcinogenic and also accumulate in soils.
IX: Metals Arsenic; Cadmium; Lead; Mercury; Selenium; Zinc	Point sources, ubiquitous	See chapter 8 for detail
XI: Persistent Organo halogens Dioxins; furans; PBBs; polychlorinated biphenyls (PCBs); PBDEs; 2,4-Dichlorophenol	PPM effluents, ubiquitous	PCBs used in dielectric fluids of transformers and as plasticisers, component of cement, hydraulic lubricants, flame retardants, plastics, paint and adhesives. PCBs exhibit agonistic and antagonistic behaviour in estrogenic systems. Dioxins/furans are by-products of a wide range of industrial processes including metal processing, production of chlorinated compounds. Both dioxins/furans are human carcinogen and potential EDCs for humanand wildlife (see chapter 10 on dioxins, furans and PCBs for more information).
XII: Pesticides 2,4-D; Amitrol; Alachlor; Atrazine; Benomyl; Carbaryl; Chlordane; Cypermethrin; DDT and DDT metabolite, p, p'-DDE; Dicofol; Dieldrin/ Aldrin; Diquat dibromide; Diuron; Endosulfan; Endrin; Fipronil; Heptachlor; Hexachlorobenzene (HCB); Iprodione; Kepone (chlordecone); Lindane; Malathion; Mancozeb; MCPA; Methomyl; Methoxychlor; Metribuzin; Mirex; Pendimethalin; Permethrin; Simazine; Thiram; Toxaphene; Trifluralin; Vinclozolin; Ziram	Agriculture, WWTP effluent, ubiquitous	Insecticides, herbicides; see chapter 9 for detail.

11.4: Sources of EDCs

Both point and non-point sources are major contributors of many EDCs in the environment (see Figure 11.2). Discharge of treated waste water or effluent from sewage treatment plants is one of the common sources of EDCs in the environment. Intensive agriculture, industrial waste and mining activity and landfill are other sources of EDCs (see Figure 11.2). For example, run-off and waste water discharge from intensive dairy, beef-cattle, poultry pigs and aquaculture farms are important contributors of EDCs, in particular estrogenic compounds such as Estrone (E1), 17β-estradiol (E2). Livestock feed lots could also be a potential sources of estrogenic compounds from excretion of hormones in manure and urine (see Table 11.2). Steroidal hormones, such as estradiol, estrone, progesterone and testosterone are naturally excreted by humans (Table 11.3). Genistein and daidzein have been frequently detected in waste water and this may be correlated with high urinary excretion rate of these two phytoestrogens observed (see section 11.6.6). In general, urinary phytoestrogens excretion rate are much high than that of most frequently detected natural estorgen E1.

Pesticides used in intensive agriculture and animal husbandry could be another important sources of EDCs in the environment via run-off, spray drift and drainage discharges. Effluents discharged from pulp mills may contain chlorinated organic compounds (dioxins), surfactants (alkyl phenols ethyoxylates) and natural phytosterols (β-sitosterol; see Table 11.1 for examples of EDCs, their potential sources and uses). Application of poultry manure on land may increase the background concentrations of estradiol and testosterone. Biosolids applied for agronomic, silvicultural and land restoration activities may increase the levels of NP (see section 11.6.2).

Table 11.2: ***Estimated yearly steroid hormone excretion by farm animals*** *[Lange et al. 2002].*

USA			European Union		
Head	**Estrogens (tons)**	**Androgens (tons)**	**Head**	**Estrogens (tons)**	**Androgens (tons)**
Cattle (98 million)	45	1.9	Cattle (82 million)	26	4.5
Pigs (59 million)	0.83	0.35	Pigs (122 million)	3.0	1.0
Sheep (7.7 million)	0.092	-	Sheep (112 million)	1.3	-
Chickens (1,816 million)	2.7	1.1	Chickens (1,002 million)	2.8	1.6

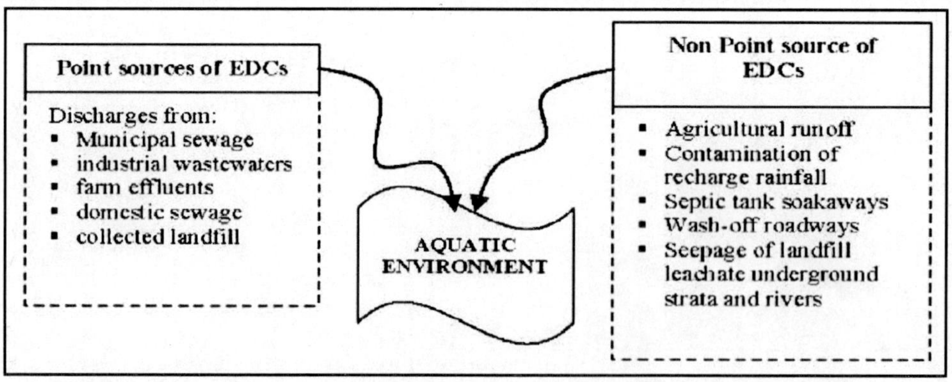

Figure 11.2: ***Sources of endocrine disruptor compounds*** *[Bolong et al. 2009].*

Table 11.3: ***Daily excretion (mg) of estrogenic steroids in humans*** *[data compiled by Ying et al. 2002].*

Category	**Estrone (E1)**	**17β-Estradiol (E2)**	**Estriol (E3)**	**17α-ethynylestradiol (EE2)**
Males	3.9	1.6	1.5	-
Menstruating females	8	3.5	4.8	-
Menopausal females	4	2.3	1	-
Pregnant women	600	259	6000	-
Women	-			35

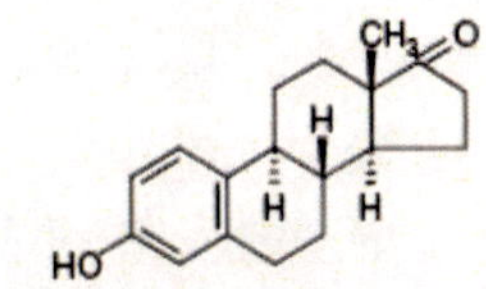 Estrone (E1) $C_{18}H_{22}O_2$ Natural estrogen	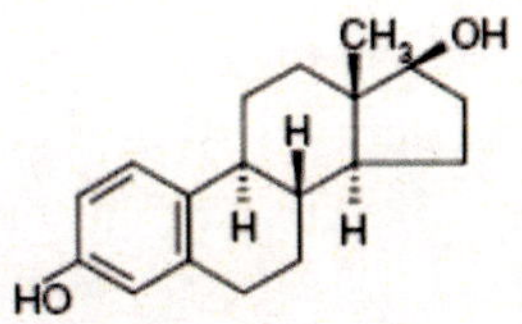 17β-Estradiol (E2) $C_{18}H_{24}O_2$ Natural estrogen	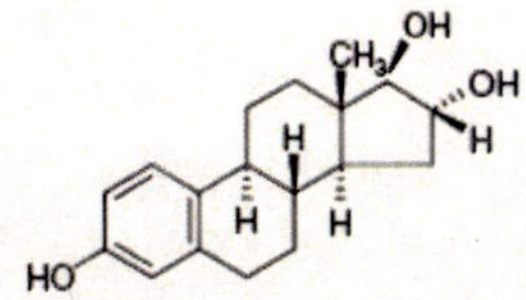 Estriol (E3) $C_{18}H_{24}O_3$ Natural estrogen	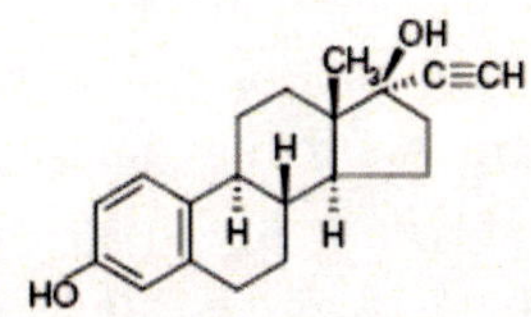 17α-Ethynylestradiol (EE2) $C_{20}H_{24}O_2$ Synthetic estrogen
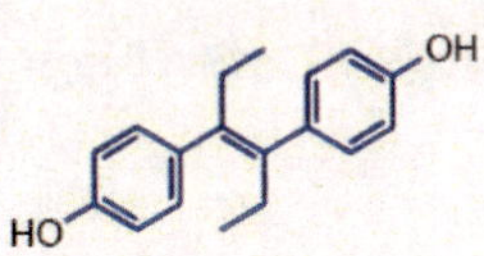 Diethylstilbestrol (DES) $C_{18}H_{22}O_2$ Synthetic estrogen	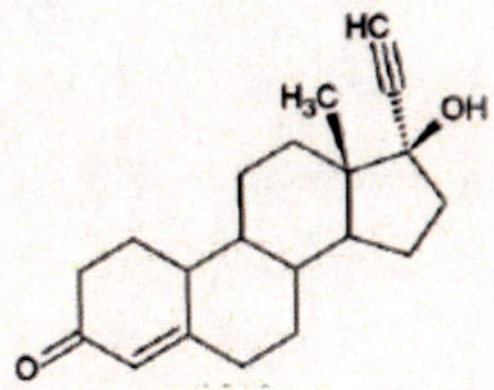 Norethindrone $C_{20}H_{26}O_2$ Synthetic estrogen	 4-n-NP (4-*n*-nonylphenol) $C_{15}H_{24}O$ Industrial chemical-alkylphenol	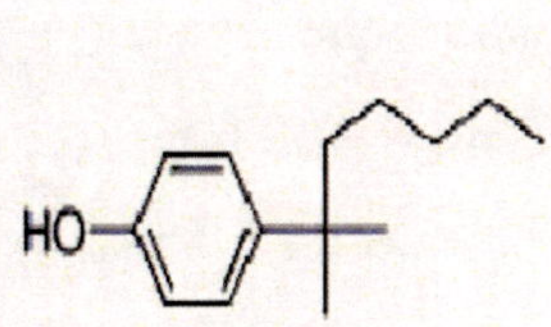 4-t-OP (4-tert-octylphenol) $C_{14}H_{22}O$ Industrial chemical-alkylphenol
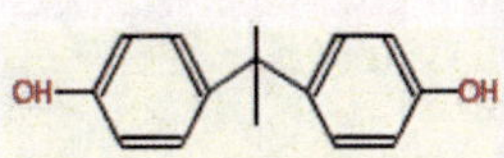 Biphenol-A (BPA) $C_{15}H_{16}O_2$ Industrial chemical	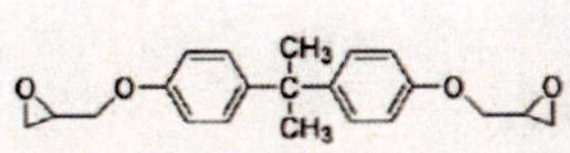 Bisphenol A diglycidyl ether (BADGE) $C_{21}H_{24}O_4$ Industrial chemical	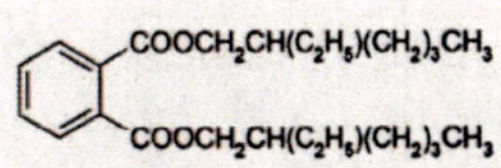 Di-2-(ethylhexyl)-phthalate (DEHP)-phthalate $C_{24}H_{38}O_4$ Industrial chemical	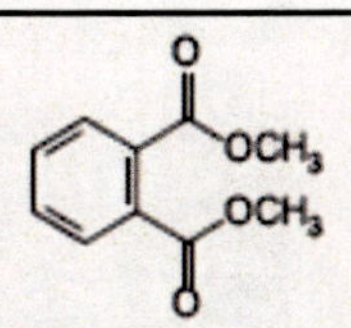 Dimethyl phthalate $C_{10}H_{10}\ O_4$ Industrial chemical
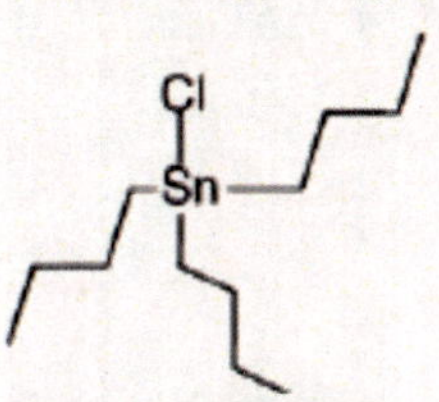 Oranotin-tributyltin $C_{12}H_{28}Sn$ Industrial chemical	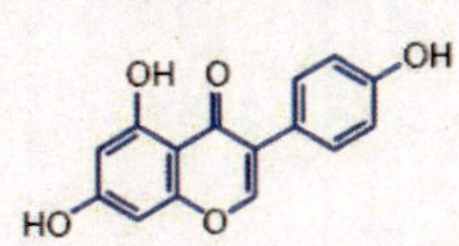 Genistein $C_{15}H_{10}\ O_5$ Phytoestrogen(isoflavones)	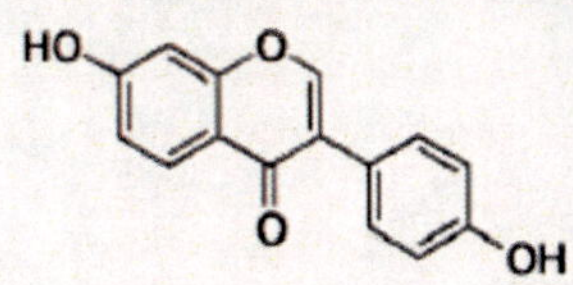 Daidzein $C_{15}H_{10}\ O_4$ Phytoestrogen (isoflavones)	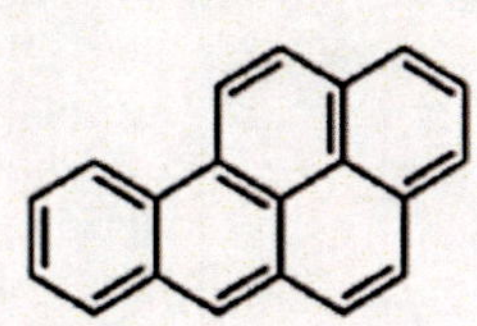 Benzo(*a*)pyrene $C_{20}H_{12}$ Polycyclic Aromatic Hydrocarbon

Figure 11.3: *Structure of some EDC molecules* *[structure based on Barnabe et al. 2009; Cao et al. 2009; http://www.emolecules.com/cgi-bin/more?vid=475094].*

11.5: Physico-chemical properties, fate and transport of EDCs

[based on Ying et al. 2002; Campbell et al. 2006; Crain et al. 2007; Matozzo et al. 2008; Neale et al. 2009].

EDCs are ubiquitous and persistent in the environment. This may be related with the lipophilic nature of EDCs as demonstrated by its high log K_{ow} values (Table 11.4). Owing to their high lipophilicity and persistence, most of xenoestrogens and their metabolites may bio-accumulate and magnify, i.e. EDCs can accumulate in the fatty tissues of organisms (fish, molluscs) and increase in concentrations as they move up through the food chain. The K_{oc} (the partition coefficients between aqueous and solid phases) shows that EDCs have moderate to high log K_{oc} values (i.e. they are likely to be sorbed to sediments or suspended organic materials). The EDCs that are adsorbed in the sediments have the potential for biological uptake and the water column can be an exposure pathway of EDCs in solution to humans and wildlife from consumption of water or biomass. The solubility values (Table 11.4) suggest that EDCs are non- mobile or less mobile (i.e. would generally not remain in solution). However, many EDCs [e.g. estrogens (E1, E2, E3, EE2)] and industrial chemicals (nonylphenols, bisphenol) have been identified in water samples throughout the world (Figure 11.3 for structures of EDCs and section 11.6.1.1 on detection of EDCs in environmental compartments) including surface water, drinking water, groundwater, aquatic life, sediment and wastewater. This fact suggests that some type of soluble transport characteristics may exist with EDCs, which Campbell *et al.* (2006) hypothesised as follows:

- more soluble precursors or metabolites experience transport (e.g. nonylphenol carboxylic);
- colloidal facilitated transport;
- enhanced solubility through elevated pH (many e-EDCs have a pKa around 10) (see Table 11.4) and
- formation of micelles which can greatly enhance the stability of a compound as well as facilitate the stability of other low solubility e-EDCs in solution.

Table 11.4: ***Physico-chemical properties of some selected e-EDCs*** *[based on Kuster et al. 2004; Petrovic et al. 2004; Campbell et al. 2006; Williams et al. 2007; Oehlmann 2008; Liu et al. 2009a].*

Class	Compound	Estrogenic potency (YES or E-screen)	Log K_{oc} (l/kg)	Log K_{ow}	Water solubility (mg/l)	pKa	RBA* (ER)
Natural hormones/ estrogens	Estrone (E1)	0.01-0.38	2.45-3.34	3.13-4.01	2.1-30	10.3-10.8	0.44
	17β-Estradiol (E2)	1.0	3.10-4.01	3.94-4.01	0.8-12.4	10.71	1
	17α-Estradiol (αE2)	0.075-0.3	3.10	3.67-4.01	3.2	10.5-10.71	0.801
	Estriol (E3)	0.001-0.3	2.13-2.62	2.45-3.85	13-441	10.4	0.33
	2-Hydroxyestrone (2-OHE1)	2.6e-3	-	-	-	-	0.04
	16α-Hydroxyestrone (α-OHE1)	-	-	-	-	-	0.065
	Equilin	-	-	3.35	1.14	-	0.289
	17α-dihydroequilin (D-equilin)	-	-	-	-	-	0.18
	Testosterone (Te) MW:288.4	e-5		3.32	30	17.4	-
	Androstenedione (An)	-	-	-	58	-	-
	Androsterone (A)	-	-	3.69	12	-	-
	Progesterone (PR)	-	-	3.87	2.05-8.84	-	-
Pharmaceuticals/artificial estrogens	17α-Ethynylestradiol (EE2) MW:296.4	1.03-1.50	2.91-3.04	3.67-4.15	3.1-44	10.4	1.4
	Diethylstilbestrol	-	-	5.07	12	-	-
	Mestranol MW:310.4	-	4.21	-	-	-	-
	Norethindrone (N)	2.8e-4		2.97	7.04		1.53e-3
	Tamoxifen	<2e-8; 4e-5					0.47
	17β-trenbolone (β-Tr)	-	-	-	-	-	-
	17α-trenbolone (α-Tr)	-	-	-	-	-	-

Contd...

Industrial chemicals	Bisphenol A (BPA) 4-*tert*-octylphenol (OP)	0.0001 0.0004	2.50-6.60 3.54-5.18	2.20-3.32 4.12	109-300 12.6	9.6-11.3 NA	 1.24e-3
	4-*n*-nonylphenol (4-n-NP)	0.000009	3.56-5.67	-	4.9-7.0	NA	1.4e-3
	4-nonylphenol (4-NP)	0.00014	4.48-5.39	5.66	NA	10.2	-
	Nonylphenol monoethoxylates; Nonylphenol diethoxylates (NP_1EO, NP_2EO)	0.000006	3.91-5.64 4.17-5.46	4.2-5.76 4.17-4.21	3.02-7.65 3.02-3.38	10.28	-
	Phthalate esters					-	-
	Di-butyl phthalate (DBP)	$1.0x10^{-7}$-1.0-$x10^{-5}$	3.80	4.57	11.2		
	Di-ethylexyl phthalate (DEHP)	$2.5x10^{-5}$ -	5.77 5.45	7.5 8.80	0.003 0.0002	- -	- -
	Di-isodecyl phthalate (DIDP)	- $1.0x10^{-7}$-1.0-$x10^{-5}$	5.45 1.9-2.6	8.80 1.6	0.0006 4000	- -	-
	Di-isononyl phthalate (DINP)						
	Di-methyl phthalate						
	Tributyl tin (TBT)	-	-	4.4	10	6.25	
Phytoestrogens	Biochanin-A	-	-	-	-	-	1.26e-5
	Coumestrol	6.7e-4	-	-	-	-	2.64e-3
	Daidzein	2e-6	-	-	-	-	1.8e-3
	Equol	2e-5	-	-	-	-	-
	Formononetin	-	-	-	-	-	2.65e-4
	Genistein	0.002	-	2.84	258	-	12e-3

*RBA = Relative binding affinity determined by estrogen receptor ligand competitive binding assay (ER-binding assay); Log k_{oc} = organic carbon sorption coefficient; Log K_{ow} = organic water partition coefficient; Relative estrogenic potency as determined by different bioassays - 17β-estradiol equivalency factor.

Biodegradation of EDCs generally depends on important factors, such as, bacteria, sludge-retention times, aeration, temperature and pH. Laboratory studies conducted by Ying *et al.* 2003 on sorption and degradation of selected EDCs in aquifer materials showed half-lives of 17α–Ethynylestradiol (EE2) and 17β-Estradiol (E2) under aerobic conditions as 81 days and 2-7 days respectively with little change in Bisphenol A (BPA) and Octylphenols (OP). However, there was no degradation observed of these EDCs under anaerobic conditions. Similarly, Auriol *et al.* 2006 (cited from a Danish study - DEPA 2004) concluded that under anaerobic conditions, the degradation rates for EDCs like E1 and EE2 were 10-20 times lower compared to that under aerobic conditions. Furthermore, sorption of steroidal hormones (estradiol, estrone, progesterone and testosterone) to organic matter (see Neale *et al.* 2009) was generally found to be strongest in acidic solutions, whereas in alkaline pH conditions sorption decreased considerably. Estrogens are largely excreted from humans as inactive conjugates of sulphuric acid and glucuronic acids (see also chapter 12, section 12.6.3 for detail on excretion of other pharmaceutical products). Because the conjugates have stronger polarity, higher solubility and low sorption affinity with soils or sediments, the mobility of these compounds is higher than their parent compounds. Due to the presence of micro-organisms in raw sewage and sewage treatment plants, the inactive conjugates are cleaved and active steroid hormones are released to the environment (Ying *et al.* 2002; Pan *et al.* 2008). The half-lives of estrogenic steroids were estimated to be between 2-9 days in water and sediment (E1: 2- 3 days; E2: 2-9 days; EE2: 4-6 days). In some lab. studies, it was shown that E2 was more readily mineralised than EE2 and E1 and that synthetic estrogens (EE2) exhibited more resistance to the active sludge process. Micro-organisms in English river water samples were able to transform E2 to E1 with half-lives of 0.2 to 9 days when incubated at 20°C and E1 was then degraded at similar rates.

The industrial chemical nonylphenol is a semi-volatile compound capable of water/air exchange which can be transported to aquatic and terrestrial ecosystems by wet deposition. The main source of nonylphenol to the environment is the degradation of nonylphenol ethoxylates. Nonylphenol is reported to be more persistent to degradation than nonylphenol ethoxylate and mean sediment dissipation times (DT_{50}) vary between a few weeks and many months. It has been demonstrated that temperature has a strong influence on the biodegradation rate of nonylphenol ethoxylates (biodegradation was 2-10 times higher at 22.5°C than at 13°C). Nonylphenol can decrease its concentration due to photolysis induced by sunlight and through the action of micro-organisms but in sediments it has an estimated half-life of more than 60 years (see review by Soares *et al.* 2008 on nonylphenol). Bisphenol-A (BPA) has an environmental half-life of between 4.5 and 4.7 days in aerobic environments (degraded primarily by bacteria), however, under anaerobic conditions, the degradation of BPA is limited, leading to higher BPA accumulation in aerobic sediments such as estuaries.

The potency of EDCs (relative to bench mark compounds) is generally assessed based on their potential to bind to biological receptors. Recently, most research on EDCs has focused on estrogenic compounds which are likely to cause the greatest risk once released into the environment. Among the estrogenic compounds, the steroidal hormone 17α–ethynylestradiol or EE2 is often used as a bench mark for the estrogenic potential of a compound and EE2 is 25 times more potent than E2 (Williams *et al.* 2007). In terms of potency, steroidal estrogens (EE2, E2, and E1) are more potent than xenoestrogens (e.g. alkyl phenols and bisphenol) (see Table 11.4 for estrogenic potency of different EDC compounds). The relative binding affinities of EE2, E2 and E1 to receptors are also high compared to industrial chemicals and phytoestrogens (see Table 11.4 for RBA values).

11.6: Environmental and biological impacts of EDCs

11.6.1: Natural and synthetic estrogens

[based on Purdom et al. 1994; Manning 2005; Andrew et al. 2008; Matozzo et al. 2008; Jugan et al. 2009; Kawahara et al. 2009].

11.6.1.1: Occurrence

Natural and synthetic estrogens that have been detected in different environmental compartments from Australia, Belgium, Canada, China, France, Germany, Japan, Israel, Italy, Spain, the Netherlands, UK, and USA (Table 11.5) are as follows:

- Surface waters: E1 (nd-262.5 ng/l); E2 (nd- 93 ng/l); E3 (nd-73.6); EE2 (nd-35.6 ng/l) (Table 11.5).
- Sediments: E1 (nd-7.38 ng/g); E2 (nd- 2.48 ng/g); EE2 (nd-2.18 ng/g) (Table 11.5).

In addition, they have been reported from groundwater, drinking water and rain water (Vethaak *et al.* 2005; Campbell *et al.* 2006; Liu *et al.* 2009a and 2009b; Zhang *et al.* 2009):

- groundwater (E1: nd-120 ng/l; E2: nd-80 ng/l; E3: nd-0.16 ng/l and EE2: nd-0.94 ng/l).
- drinking water (E1: 0.2-0.6 ng/l; E2: nd-2.6 ng/l; and EE2: 0.15-0.5 ng/l).
- rain water (E1: <0.6 ng/l; E2: <1.5 ng/l; E3: <0.3 ng/l and EE2: <0.3 ng/l).

11.6.1.2: Ecotoxicity

The acute and chronic toxicity values of natural and synthetic estrogens are given in Table 11.6, which show that the toxicity data for algae and fish are generally lacking. The acute toxicity of most potent EE2 are in the range of 0.84-5.7 mg/l, however, chronic toxicity test showed that EE2 can cause adverse effects on some invertebrates at very low concentrations of ~1.0 ng/l or less (Schulte-Oehlmann *et al.* 2004). The chronic toxicity of EE2 to fish *Pimephales promeals* is 1.0 ng/l and for snail *Marisa cornuarietis* < 1.0 ng/l (Länge *et al.* 2001; Schulte-Oehlmann *et al.* 2004).

Table 11.5: *Comparison of estrogenic hormones (E1, E2, E3 and EE2) concentrations (ng/l) in rivers, estuaries and oceans [data compiled by Kuster et al. 2004; Petrovic et al. 2004; Auriol et al. 2006; Gong et al. 2008; Lei et al. 2009; Zhang et al. 2009; Ying et al. 2009].*

Country	Environment	Estrone (E1)	17β-estradiol (E2)	Estriol (E3)	17β-estradiol (EE2)	Others-sex hormones
Surface waters						
Australia	Sewage treatment plants-effluents	9.12-32.22 ng/l	1.37-5.69 ng/l	-	0.11-1.20 ng/l	-
Belgium	Estuary - Scheldt	0.37-10 ng/l	nd	-	nd	-
Canada	River water	-	-	-	-	2-67 ng/l (steroids)
China	River - Beitang	4.29-49.8 ng/l	2.51-21.2 ng/l	nd-15.4 ng/l	1.64-24.4 ng/l	-
China	Lake – Tai Hu	-	5.5-15.5 ng/l	nd	5.7-30.8 ng/l	-
China	River –Yong ding New	0.64-20.2 ng/l	nd-13.6 ng/l	nd-10.8 ng/l	nd-12 ng/l	
China	River - Dagu River	5-55.3 ng/l	0.93-33.4 ng/l	2.26-46.4 ng/l	nd-35.6 ng/l	-
China	South (ZR and DR)	<LOQ-8.2 ng/l	<LOQ-82	-	-	-
China	North (Haihe River, Tianjin)	19.1-106 ng/l	-	-	-	-
China	East (5 rivers, Jiaozhou Bay)	29.2-262.5 ng/l	-	-	-	-
China (Taiwan)	River -Dan-Shui	22.4-66.2 ng/l	1.4-33.9 ng/l	12.4-73.6 ng/l	7.53-27.4 ng/l	-
France	River - Paris area	1.1-3.0 ng/l	1.4-3.2 ng/l	1.0-2.9 ng/l	1.1-2.9 ng/l	-
France	Paris seven River	0.8-3.9 ng/l	0.8-3.6 ng/l	0.6-3.1 ng/l	0.6-3.5 ng/l	-
France	River - Jalie d'Eysines	nd-30 ng/l	nd-1.5 ng/l	nd-1.8 ng/l	nd-2 ng/l	
Germany	Rivers and creeks	0.10-4.1 ng/l	0.15-3.6 ng/l	-	0.10-5.1 ng/l	-
Germany	River-Rhine, Ahr and Agger	1.3-9.2 ng/l	-	-	0.3-1.0 ng/l	
Germany	River water	-	-	-	-	<0.1-5.1 ng/l (natural and synthetic estrogens)
Italy	River - Tiber	1.5 ng/l	0.11 ng/l	0.33 ng/l	0.04 ng/l	-
Japan	River - Tama	8.7-19.7 ng/l	4.2-5.3 ng/l	-	nd	-
Japan	Lake Kasumigaura	0.2-0.8 ng/l	nd	-	nd	-
The Netherlands	Coastal/ Estuarine and Rivers	0.1-3.4 ng/l	0.3-5.5 ng/l	-	0.1-4.3 ng/l	-
The Netherlands	Dutch surface water	0.3-7.2 ng/l	0.8-1.0 ng/l	-	0.3-0.4 ng/l	-
Spain	River water	-	-	-	-	0.2-71.1 ng/l (estrogens and progestogens)
UK	River water	-	-	-		<0.2-5.1 ng/l (natural and synthetic estrogens)

Contd...

UK	River water	-	-	-	-	2-17 ng/l (synthetic estrogens/ progestogens)
USA	139 streams	nd-112 ng/l	nd-93 ng/l	51	-	-
USA	River water	-	-	-	-	5-2,000 ng/l (steroids and hormones)
Sediments						
Australia	Sediments Ocean (deep ocean outfall)	0.16-1.17 ng/g	0.22-2.48 ng/g	-	<0.05-0.5 ng/g	
China	Sediments (Xiamen Bay)	nd-7.38 ng/g	nd-2.35 ng/g	-	nd-2.18 ng/g	nd-8.79 (DES)
China	Sediments-Pearl River Estuary, South China Sea	nd	nd	-	-	-
Germany	Sediments - River Elbe and its tributaries	<2 ng/g	<2 ng/g	-	<2 ng/g	-
Germany	Activated and digested sludge	-	-	-	-	<2-49 ng/g (natural and synthetic estrogens-E1, E2, EE2, MES)
Japan	Sediments - Tokyo Bay	0.05-3.60 ng/g	nd-0.59 ng/g	-	-	-
Israel	Activated sludge	-	-	-	-	19-64 ng/g (estrogen)
Spain	Sediment (River)	-	-	-	-	0.05-22.8 ng/g (estrogens/ progestogens-E1, E3, EE2, DES, norethindrone, levonorgestrel)
UK	River sediment	-	-	-	-	<0.04-0.3888 ng/g (progesterone)
Drinking water						
Germany	-	-	--	-	-	<0.1-2.1 ng/l (natural & synthetic estrogens)
UK	-	-	-	-	-	<1-10 ng/l (synthetic estrogens/ progestogens
Biota						
Sweden	Rainbow trout bile	-	-	-	-	<0.1-2.5 µg/g (natural and synthetic estrogens)

11.6.1.3: Effects on biota

Estrogenic compounds are suspected to be responsible for increased disturbances in wildlife reproduction and development. The best characterised pathways for endocrine disruption is binding to the estrogens receptor (ER) of chemicals that can either block (antagonists) or mimic (agonists) natural hormone function (Jugan *et al.* 2009). In the aquatic environment, exposure of organisms to estrogens has been linked to endocrine effects in male fish such as vitellogenin (Vtg) induction and feminised reproductive organs. The compounds E2 and EE2 have been found to be the most potent at feminising male gonads in contrast to alkyl phenol, bisphenol-A, *o, p'DDE* and some PCBs (Hecker *et al.* 2006).

Table 11.6: ***Acute and chronic toxicity of natural and synthetic estrogens*** *[Roepke et al. 2005; Crane et al. 2006; Barbosa et al. 2008; Oehlmann et al. 2008; Soares et al. 2008; Sanderson and Thomson 2009; Mihaich et al. 2009; Sánchez-Avila 2009].*

Name of EDCs	Algae	*Daphnia*/invertebrates	Fish
Natural and synthetic			
4-Hydroxyandrostendione (4-OHA)	-	***Acute toxicity*** Daphnia, *Daphnia magna* EC_{50}: 4.26 -6.16 mg/l ***Chronic toxicity*** *Daphnia magna*: 0.84 mg/l (increased mortality of the neonates) *Daphnia magna*: 0.31-2.48 mg/l (reduced fecundity and fertility)	-
DES (Diethylstilbestrol)	-	***Chronic toxicity*** Copepod, *Nitocra spinepes*: 0.003 mg/l Copepod, *Tisbe battagliai*: 0.01 mg/l Water flea, *Daphnia magna*: 0.062 mg/l	***Chronic toxicity*** Fish *Oryzias latipes*: 10 ng/l
E1 (Estrone)	-	***Acute*** Sea urchin embryos EC_{50}: 604.4 ng/ml	-
E2 (17β-Estradiol)	-	***Acute toxicity*** Sea urchin embryos EC_{50}: 14.2 ng/ml	-
E3 (Estriol)	-	***Acute toxicity*** Sea urchin embryos EC_{50}: 1,515.4 ng/ml	-
EE2 (17α – ethynylestradiol)	***Acute toxicity*** Algae LC_{50}: 0.84 mg/l ***Chronic toxicity*** Algae EC_{10}: 0.054 mg/l	***Acute toxicity*** Daphnid LC_{50}: 5.7 mg/l ***Chronic toxicity*** Amphipod, *Hyalella azteca*: 0.0001 mg/l Copepod, *Nitocra spinepes*: 0.05 mg/l Snail, *Bithynia tentaculata*: <0.125 ng/l Snail, *Lymnaea stagnalis*: <1.25 ng/l Snail, *Marisa cornuarietis*: <1.0 ng/l Water flea, *Daphnia magna*:0.01 mg/l	***Acute toxicity*** Fish LC_{50}: 1.6 mg/l ***Chronic toxicity*** Fish, *Oncorhynchus mykiss*: <0.1-<0.3 ng/l Fish, *Oryzias latipes*: 10 ng/l Fish, *Pimephales promelas*: 1 ng/l
Mestranol	Algae LC_{50} : >100	Daphnid: LC_{50} - >100> mg/l	Fish LC_{50}: >100 mg/l
Norethindrone (N)	Algae LC_{50}:1.39 mg/l	Daphnid: LC_{50}>100 mg/l	-
Progesterone (PR)	--	Sea urchin embryos: EC_{50}-546.6 ng/ml	-
Testosterone (Te)	-	***Acute toxicity*** Daphnia LC_{50}: 1.95 mg/l Daphnia, *Daphnia magna*: EC_{50}-6.20 mg/l (no effect) ***Chronic toxicity*** *Daphnia magna*: 0.31-2.48 mg/l (reduced fecundity and fertility)	-

Vtg is a glycolipophospoprotein and produced in the liver under the influence of estrogens, particularly 17β-estradiol (E2), which is a precursor of the yolk reserves in the egg. Vtg is produced in sexually mature females, and Vtg production in males and juvenile fish are normally very low. Although males and juvenile fish have the estrogen receptor and Vtg producing function, they scarcely produce estrogenic substances inside the body. Therefore, the production of Vtg in male fish is an indication of endocrine disruption and has become a useful biomarker for exposure of estrogenic substances (Sumpter and Jobling 1995).

Steroidal compounds as low as 0.1 ng/l of 17α-ethynylestradiol (EE2-synthetic estrogen) and 1.0 ng/l of estradiol (E2-natural estrogen) can cause increased level of Vtg in fish (Hansen *et al.* 1998; Metcalfe *et al.* 2001). Exposure of EE2 (1-25 ng/l) resulted in a dose-dependent increase in vitellogenin in zebrafish staring at 2.0 ng/l (Figure 11.4). Vitollogenesis and feminisation in male fish have been observed in UK Rivers, which are attributed to the presence of estrogenic compounds (E2, E1, E3, EE2 and MeEE2). When rainbow trout (*Oncorhynchus mykiss*) were kept close to the discharges of wastewater treatment plant effluents, vitellogenin synthesis was induced in the male fish. Elevated levels of vitellogenin were also reported in male carp (*Cyprinus carpio*), roach (*Rutilus rutilus*) and flounder (*Platichthys flesus*) caught near WWTP discharges. The other effects of natural and synthetic estrogens include reduced egg production, all males with testis-ova or more females than males or development of testicular oocytes in male fish (Table 11.7).

A recent study observed that sediments, periphyton and invertebrates (fish) can accumulate natural and synthetic estrogens and thereby through diet human can get accumulation of natural and synthetic estrogens through food chain (Viganó *et al.* 2008).

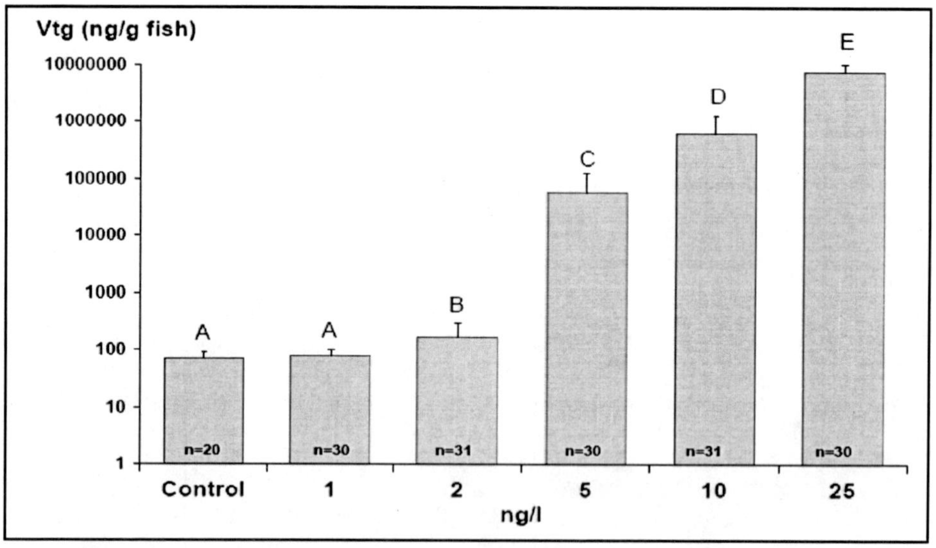

Figure 11.4: ***Mean vitellogenin concentrations at 38 days post-hatch in zebrafish (60 days) exposed to different concentrations of EE2.*** *The exposure was initiated at 20 days post-hatch and terminated at 60 days. Different letters indicate significant differences (P<0.005) between the groups. The number of individuals analysed per group are given at the base of each column [Örn et al. 2003 page 404].*

Table 11.7: ***Effects of steroids and endocrine-active compounds on fish and amphibia*** *[compiled by Mills and Chichester 2005; Hecker et al. 2006; Hogan et al. 2008; Gyllehammar et al. 2009].*

EDC	Fish and amphibia	LOEC* range (ppb)	Effect
17β-Estradiol (E2)	Fish-Medaka	0.01-5000	- No egg production; high mortality; reduced egg production; decreased male sexual behaviour; all female fish; males with testis-ova; reduced male GSI, egg production and fertility; decreased egg hatch; more phenotypic females than males.
	Fish-Fathead minnow	0.12	- Inhibition of egg production.
	Fish-Goldfish	1 -10000	- Reduced testicular growth; reduced GSI and courting behaviour; fewer males with milt.
	Fish-Guppies	10	- No adverse effects on reprodudion; complete inhibition of male sexual behaviour.
	Fish-Common carp	1-9	- Reduced GSI; no milt production; all juvenle males developed into females.

Contd...

	Amphibia-*Xenopus laevis; Rana pipiens*	-	- Development of testicular oocytes in males; ovotestis; occurrence of separate ovary and testis in the same individual
17α-Ethynyl estradiol (EE2)	Medaka	0.01-0.488	- All males with testis ova; more females than males; high mortality with only females surviving; reduced egg production and high male mortality.
	Fathead minnow	0.0001-0.1 (0.1 ng/l-100 ng/l)	- All fish with female gonads; severe physical deformities; increased or reduced or no egg production ; reduced egg fertilisation.
	Rainbow trout	0.002	- Reduced testicular growth.
	Zebra fish	0.00167-0.0154 -	- Delayed maturation; altered mating behaviour; reduced egg production and fertilisation; all females; more females than males.
	Amphibia-*Rana Pipiens*	-	- Development of testicular oocytes in males;ovotestis; occurrence of separate ovary and testis in the same individual; development of oviduct in testis.
	Amphibia-Frogs/Tadpoles	0.0015-0.181 (1.5-181 ng/l)	- Sex reversal at concentrations as low as 1.8 ng/l; increased proportion of phenotypic females in *Xenopus tropicalis*; strong female-biased sex ratio in *Rana pipens.*
Estrone (E1)	Medaka	1	- All males with testis-ova; more females than males.
	Fathead minnow	0.3177	- Reduced testicular growth (GSI).
Estriol (E3)	Medaka	0.01-10	- More males than females; all males with testis-ova.
Testosterone	Amphibia-*Rana clamitans*		- Shrinking of ovary and degeneration of oocytes
Nonlylphenol	Medaka	0.54-0.77	- Increased egg production; more males than females.
	Fathead minnow	NA	- Reduced egg production; reduced males.
	Rainbow trout	1-54.3	- Reduced testicular growth (GSI; reduced egg viability; offsprings with disrupted hormonal balance.
	Amphibia- *Rana Pipiens*	-	- Development of testicular oocytes in males; ovotestis.
Octylphenol	Medaka	10-50	- Significant developmental abnormalities in 2nd generation; reduced courting activity and fertilisation by males; low incidence of testis-ova in males.
	Rainbow trout	4.8-30	- Reduced testicular growth.
	Amphibia -*Rana catesbeiana*		- Shorter time to gonadal differentiation in males and females.
	Guppies	26	- No adverse effects on reproduction or young observed; inhibition of male sexual behaviour.
Bisphenol A	Medaka	10-2282.2	- Reduced egg production and reduced hatch; few testis-ova; morphological changes in testis.
	Zebra fish	375-1500	- Altered gonad histology; delayed maturation; altered mating behaviour; reduced egg production and fertilisation.
	Fathead minnow	640-1280	- Reduced GSI and egg hatchability; reduced egg production.
Atrazine	Amphibia- *Xenopus laevis; R. pipiens*	-	- Development of testicular oocytes in males; ovotestis; decreased testicular development.
PCBs	Amphibia- *Xenopus*	-	- Development of testicular oocytes in males
*o,p'*DDE	Fish-Medaka *Oryzias latipes*	-	- Development of testicular oocytes in males.
DEHPhthalate	Medaka	NA	- No adverse effects observed.
Butyl Benzyl Phthalate	Fathead minnow	NA	- No adverse effects observed.

*LOEC= Lowest tested concentration at which noted effect occurred.

11.6.2: Alkylphenols (nonylphenol)

[based on Ying et al. 2002; Ferrara et al. 2008; Fielder et al. 2007].

11.6.2.1: Occurrence

Nonylphenols (NP) have been detected in sewage treatment plant effluents, surface waters and sediments from Australia, Brazil, Canada, China, Germany, Holland, Italy, Japan, Korea, Norway, Portugal, Spain, Switzerland, UK, and USA in the following ranges (see also Table 11.8):

- Sewage treatment plant effluent: 0.1-63 μg/l.
- Sewage treatment plant sludge: 0.0195-500 μg/kg.

- Surface waters: 0.007-644 mg/l.
- Sediments: <0.01-3520 mg/kg.

NP has also been detected in drinking water (2.5-16 ng/l), groundwater (200-760 ng/l), rain water (<410 ng/l) and in the air (<0.001-81 ng/m^3) (Campbell *et al.* 2006; Liu *et al.* 2009a; Zhang *et al.* 2009). Because of their ecotoxicological properties and occurrence, NP has been included among the priority substances for surface water monitoring in the European Union.

Table 11.8: ***Distribution of nonylphenol in rivers, lakes, estuaries and oceans in different countries*** *[data compiled by Soares et al. 2008; Ying et al. 2009].*

Country	Sewage treatment plants	Water environment River or Lake or Estuaries or Ocean or Surface waters	Sediments River or Lake or Sea
Australia	614-2991 ng/l (effluents)	-	-
Brazil	-	-	1-10 µg/kg reservoirs (total alkylphenol)
Canada	-	- 0.001-0.92 µg/l, LakeandRiver - <0.092 µg/l, River	- <37 mg/kg, Lake - 0.0403-72 mg/kg, River
China	1.5 µg/l: final effluent	1.9-32.8 µg/l, Lake	3.5-32.4 mg/kg, Lake
Germany	- 80-500 mg/kg: sludge aerobically stabilized - 128.2 mg/kg: sludge - 0.25-2.3 µg/l: effluent	- 0.0007-1.22 µg/l, River - <0.03 µg/l, Estuaries	- 0.01-0.153 mg/kg, Marine - <0.01-0.055mg/kg, Sea
Holland		- 0.031-0.934 µg/l, Estuary	- 0.0004-1.08 mg/kg, Estuary
Italy	- 308 mg/kg: sludge anaerobically stabilized - 242 mg/kg sludge before anaerobic digestion	- 158 µg/l (1994), River - 8.8 µg/l (1996), River	-
Japan	- nd: primary effluent - 0.1-0.9 µg/l: raw influent 0.3 µg/l: secondary effluent 0.2-1.1 µg/l: final effluent	- 0.051-1.08 µg/l, River	- 0.5-13 µ g/kg, River
Korea		0.0232-0.1876 µg/l, River	0.0254-0.932 mg/kg, River
Norway	0.0195 mg/kg: final sludge 0.05-1.31 µg/l: final effluent		-
Portugal		<10 µg/l, River and Ocean	-
Spain	<1.5 µg/l: raw effluent 1-2.3 µg/l: primary effluent 5.5-6.6 µg/l: final effluent	15 µg/l, River <LOD-644 µg/l, surface waters 0.15-4.1 µg/l, Ocean	0.022-0.645 mg/kg, River 0.008-1.05 mg/kg, harbour
Switzerland	450-2,530 mg/kg: sludge anaerobically stabilized 120-650 mg/kg: sludge aerobically stabilized 43.5 µg/l: primary effluent	0.015-2.25 µg/l, River <LOD-26 µg/l, surface waters	3,520 mg/kg, River
Taiwan		1.8-10 µg/l, surface waters	-
UK	<0.02-0.1 µg/l: raw effluent 32-63 µg/l: final effluent	<0.03-53 µg/l, surface waters	-
USA	754-1,800 mg/kg: sludge anaerobically stabilized 496 mg/kg: heated treated sludge 470 mg/kg: lime treated sludge 64 mg/kg: composted sludge	<1 -0.5 µg/l, River	0.075-<3,000 mg/kg, River

11.6.2.2: *Ecotoxicity*

NP has moderate to high toxicity to algae, *Daphnia* and fish with acute toxicity values ranging between 25 and 1,400 μg/l (Table 11.9). The detected environmental concentrations of NP in water (0.007 to 644 mg/l) are within the acute toxicity threshold values that cause adverse effects on biota living in the aquatic environment.

Table 11.9: ***Acute toxicity of nonylphenols and octylphenol*** *[Soares et al. 2008].*

Type	Algae	*Daphnia and invertebrates*	Fish
Nonylphenol (NP)	- Algae LC_{50}: 25-750 μg/l - Algae (*Pseudokirchneriella subcapitata*)EC_{50}: 05-0.53 mg/l	- *Daphnia magna* EC_{50}: 93-470 μg/l - *Daphnia magna* EC_{50}: 0.3 mg/l; - Invertebrates LC_{50}: 20-1,590 μg/l - Zooplankton NOEC: 0.001 mg/l	- Fish LC_{50}: 130-1,400 μg/l - Kill fish, *Fundulus heteroclitus*: EC_{50}: (0.21-0.26 mg/l (larvae); - Kill fish, *F. heteroclitus*: EC_{50}:5.4 mg/l (embryos) - Marine fishes, bivalves, crustaceans EC_{50}: 17-195 μg/l
Nonylphenol monoethoxilathed	-	- *Daphnia magna* EC_{50}: 100 μg/l	-
Nonylphenol diethoxilathed	-	- *Daphnia magna* EC_{50}: 100 μg/l	-
Octylphenol (OP)	-	- *Daphnia magna* EC_{50}: 100 μg/l - Sea urchin embryos: EC_{50}-0.174 ng/l	-
Octylphenol (OP) and octylphenol ethoxylates	-	- *Daphnia magna* EC_{50}: 90-140 μg/l	-

11.6.2.3: *Effects on biota*

Nonylphenol has endocrine disruption effects as it induced vitellogenin in male rainbow trout at 20.3 mg/l (Jobling *et al.* 1996), and medaka at 0.1 mg/l (Tabata *et al.* 2001). Dose dependent vitellogenin induction was also observed in crabs (*Carcinus aesluarii*) and mussels (*Mytilus edulis*) (Ricciardi *et al.* 2008). In addition, NP reduced egg production in female zebra fish and developed testicular oocytes in males (Jackson and Sutton 2008; Bolong *et al.* 2009). Other effects of nonylphenol on fish are summarised in Table 11.7.

Owing to its relatively high lipophilicity and persistence, NP may bio-accumulate and bio-magnify. For example, bio-accumulation of NP has been reported in algae, fish, oysters and aquatic birds. The algae contained highest concentrations of NP reported to be between 1.5 and 38 mg/kg with bio-concentration factors from 200 to 10,000. In fish, the concentrations varied from 0.03 to 1.59 mg/kg with bio-concentration factors from 13 to 408. The bio-concentration factor for NP in fathead minnows, *Pimephales promelas* was found to be 245-380 after 42 days exposure in a flow through system in which the concentration of NP was 0.33-2.36 μg/l which indicates that long-term exposures at even low concentrations can lead to accumulation of NP in the fish. The concentrations of NP in wild ducks ranged between 0.03 and 1.2 mg /kg with higher values in muscle (Ahel *et al.* 1993; Soares *et al.* 2008). High accumulation of NP was also observed in Sydney rock oyster, *Saccostrea glomerata* tissue as compared to 17α-ethynylestradiol (EE2) for an eight week exposure period in experimental aquaria. Estradiol equivalents (EEQ) (which provide an indication of the relative concentration of accumulated xeno-estrogen/or endogenous estrogen) were highest with NP exposure (28.7±2.3 ng/g tissue EEQ) than for EE2 which was 2.2 ±1.5 ng/g tissues (Andrew *et al.* 2008). An investigation conducted in Italy (Ferrara *et al.* 2008) found that almost all of the tested seafood (anchovies, hakes, mackerels, mantin shrimp, mullets, octopus, sea breams, swordfish and tuna) contained high concentrations of NP. NPs alone are accounted for some 80% on an average of the total seafood contamination, of which tuna exhibited very high concentrations of total alkylphenolic compounds (Table 11.10).

Table 11.10: ***Concentration ranges of the measured estrogenic chemicals (nonylphenols, bisphenol A, phthalates) in seafood*** *[adapted from Vethaak et al. 2005 and Ferrara et al. 2008].*

EDCs	Bream muscle (ng/g-wet weight)	Flounder muscle (ng/g -wet weight)	*Other seafood (ng/g wet weight)**
Phenols			
Bisphenol-A	<0.18-1.4	1.2-2.6	
Alkylphenols			
Octylphenol	<10-80	<10	Anchovies: 0.2-0.9 Hakes: 1.1-4.8 Mackerels: 0.3-2.8 Mantis shrimp: 1.4-15.7 Mullets: 1.1-2.3 Octopus: 0.8-14.4 Sea breams: 7.9-18.7 Swordfish: 4.3 Tuna: 27.9
Octylphenol ethoxylates	<10	10	
Nonylphenols	<10-160	<10-10	Anchovies: 5-10 Hakes: 10-30 Mackerels: 7-75 Mantis shrimp: 7-1220 Mullets: 12-41 Octopus: 6-58 Sea breams: 5-503 Swordfish: 45 Tuna: 45
Nonylphenol ethoxylates	<20-520	<10-20	
Total alkylphenolic compounds (APE)			Anchovies: 6-37 Hakes: 36-34 Mackerels: 12-121 Mantis shrimp: 37-1255 Mullets: 44-55 Octopus: 19-24 Sea breams: 27-526 Swordfish: 234 Tuna: 889
Phthalates			
Butylbenzylphthalate (BBP)	<0.89-299	<0.2-33	
Di-(2-ethylhexyl) phthalate (DEHP)	70-1503	<2.2-144	
Dimethyl phthalate (DMP)	<0.22-3.2	0.14-5.4	
Di-*n*-butyl phthalate (DBP)			

*Proposed tolerable daily intake (TDI) of NP is 300 mg/day for an adult of 60 kg (Ferrara *et al.* 2008)

11.6.2.4: Effects on humans

Xenoestrogens such as alkylphenols, phthalates, pesticides and PCBs are suspected to increase the risk of breast cancer. In 1991, researchers from Tufts University discovered that a chemical leaching from polystyrene tubes was causing breast cancer cells to grow in the absence of estrogen. This chemical was found to be *p*-nonylphenol, which is a common additive in plastics. *In vitro* studies conducted in Texas, USA in 2005 found increased risk of breast cancer in mice exposed to 4-nonylphenol compared to mice exposed to equivalent doses of estradiol (Dey *et al.* 2009). Table 11.11 provides a list of EDCs suspected to cause breast cancer from long-term exposure of women to these chemicals. NP has also been detected in human blood (nd-268 ng/ml), breast milk, urine and adipose tissue (10-567 ng/g) (see Table 11.12) confirming the persistent and bio-accumulative nature of NP.

Table 11.11: ***Some EDCs linked to breast cancer*** *[Dey et al. 2009].*

Chemical class	Potential sources	Example
Alkylphenols	Detergent, plastic, pesticide formulations	Nonylphenol
Banned pesticides	Historical pesticide used in/near the home.	DDT, dieldrin, chlordane
Flame retardants	Furniture foam and stuffing, carpets and drapes, electronic equipment (TVs, computers).	Polybrominated diphenyl ether (PBDE 47)
Phthalates	Plastic, nail polish and other cosmetics	Dibutyl phthalate
Polychlorinated biphenyls (PCBs)	Older electric equipment.	PCB 52
Polycyclic aromatic hydrocarbons (PAH)	Stoves and heaters, cigarette smoking, outdoor air pollution, auto exhaust, combustion sources.	Benzo(a)pyrene
Currently used pesticides	Recent pesticide used in/near the home.	Chlopyrifos, pyrethrin
Other phenols and miscellaneous	Disinfectants, polycarbonate plastics, cosmetics.	0-phenyl phenol, bisphenol-A, parabens

Table 11.12: ***Reported concentration of NP in human samples (compared to octylphenol) from different countries*** *[data compiled by Lopez-Espinosa et al. 2009].*

Country	Population	Sample	Nonylphenol (min-max.)	Octylphenol (min-max.)
Spain (southern)	Women (24-81 yrs)	Adipose tissue	10-567 ng/g	nd-8.6 ng/g
Switzerland	15 males, 10 females (3-100 yrs)	Adipose tissue (cadavers)	19.8-84.4 ng/g	0.58-4.07 ng/g
China	Healthy people	Blood (serum)	-	0.2-0.2 ng/ml
Japan	Healthy people	Blood (plasma)	0.5-1.0 ng/ml	nd-<0.25 ng/ml
Japan	Healthy people (24-25 yrs)	Blood (plasma)	0.2-0.3 ng/ml	0.1-0.2 ng/ml
Malaysia	Women	Cord blood (plasma)	nd-15.2 ng/ml	nd-1.15 ng/ml
Taiwan (North)	Women	Cord blood (plasma)	nd-211 ng/ml	-
Taiwan	Housekeepers	Blood (plasma)	12.1-28.5 ng/ml	12.68-23.9 ng/ml
Taiwan	Expectant women (18-40 yrs)	Blood (plasma)	nd-268 ng/ml	-
Taiwan	Women (18-40 yrs)	Cord blood (plasma)	nd-100 ng/ml	-
Taiwan (Central)	Women	Cord blood (plasma)	nd-182 ng/g	-
Italy	Women	Breast milk	13.4-56.3 ng/ml	nd-0.2 ng/ml
Japan	Women	Breast milk	0.65-1.4 ng/ml	-
USA	Women	Breast milk	-	nd-7.6 ng/ml
USA	People painting their homes	Urine	-	0.4-13.9 ng/ml
Japan	Healthy people (20-24 yrs)	Urine	-	-
China	Males (21-29 yrs)	Urine	nd-2.3 ng/ml	-
China	Females (21-29 yrs)	Urine	nd-0.14 ng/ml	-
USA	> 6 yrs (NHANES)	Urine	-	-
USA	House keepers	Urine	nd-71.9 ng/ml	nd-7.02 ng/ml
Taiwan	Textile workers, pre-shift	Urine	nd-71.9 ng/ml (median 21 ng/ml)	nd-7.02 ng/ml
Taiwan	Textile workers, post-shift	Urine	- (median 38 ng/ml)	nd-7.02 ng/ml
Japan	Healthy people	Urine	2-20 ng/ml	nd-0.05 ng/ml
Japan	Healthy people, 22-25 yrs	Urine	<1.1-2.1 ng/ml	-
USA	> 6 yrs (NHANES)	Urine	-	(median 0.3 ng/ml)

11.6.3: Phenols - Bisphenol-A (BPA)

[based on Crain et al. 2007; Samital 2008; Ben-Jonathan et al. 2009].

11.6.3.1: Occurrence

Levels of BPA in the environment are reported from various countries including Australia, Germany, Italy, Japan, Spain, Sweden, the Netherlands, UK and USA with the highest concentrations found in landfill leachate and anaerobic sediments (Table 11.13). Concentrations in streams/river water samples are typically less than 21 μg/l, whereas concentrations in landfill leachate are typically upto 17.2 mg/l (Table 11.13). BPA has also been detected in ground water (3-1,410 ng/l), drinking water (nd-930 ng/l) and in rain water (<15-57 ng/l) (Campbell *et al.* 2006; Liu *et al.* 2009a; Zhang *et al.* 2009).

Table 11.13: ***Maximum concentrations of BPA from different countries*** *[compiled by Crain et al. 2007; Ying et al. 2009].*

Country	Surface waters	Non-surface waters
Australia	-	13-44 ng/l (STP) effluents
Austria	0.60 μg/l, surface water	0.93 μg/l, groundwater
Germany	0.0160 μg/l, river water 0.020 μg/l, drinking water	0.70 μg/l, sewage treatment plant
Italy	0.494 μg/l, river water 0.030 μg/l, lagoon water	-
Japan	0.058-19 μg/l, river	5400-17200 μg/l, landfill leachate 5.1 μg/l, treated leachate
Spain	-	0.27-1.5 μg/l, sewage treatment influent
Sweden	-	0.49 μg/l, sewage treatment influent
The Netherlands	21 μg/l, river water 0.330 μg/l, estuary	-
UK	-	0.046 μg/l, crude waste water 0.93 μg/l, treated wastewater
USA	12 μg/l, stream water 0.158 μg/l, canal water 0.147 μg/l, river water	1.7 μg/l, untreated sewage 1.4 μg/l, landfill plume

11.6.3.2: Ecotoxicity

The acute toxicity values of BPA to algae, invertebrate and fish appears to be low and are in the range of 1-1.36 mg/l, 1.1-16 mg/l and 3-15 mg/l respectively (Table 11.14). The available chronic toxicity data indicates considerable species and taxon-specific differences regarding BPA sensitivity. Fish (fathead minnow) and molluscs (snail, *Marisa cornuarietis*) are the most sensitive groups of aquatic animals with NOEC and EC_{10} values for BPA of 1.0 and 0.015 μg/l respectively (Oehlman *et al.* 2008).

11.6.3.3: Effects on biota

[based on Crain et al. 2007].

BPA acts both as a teratogen and endocrine disruptor to various wildlife. The following teratogenic effects were observed in amphibia and fish:

- Salmon fry (*Salmo salar*) exposed to 1.0 mg/l of BPA for 6 days had yolk-sac edema and haemorrhage.
- Exposure of more than 4.6 mg/l of BPA induced short body length, abnormal gut coiling and edema in the amphibian, *Xenopus laevis* whereas exposure in excess of 5.7 mg/l caused development of scoliosis, head malformations and organogenesis suppression.

- In amphibia, *Rana nigromaculata* an environmentally high concentration of 200 μg/l of BPA induced tail flexure malformations after 45 days exposure.

Table 11.14: ***Acute toxicity of BPA*** *[data compiled by Staples et al. 1998; Oehlmann et al. 2008; Mihaich et al. 2009].* NOEC= no observed effect concentrations

Algae and plants	Daphnia/invertebrates	Fish/vertebrates
Acute toxicity - Algae *Pseudokirchneriella subcapitata* NOEC/EC_{10}: 1.36 mg/l - Green algae: EC_{50} :1-3 mg/l	***Acute toxicity*** - Amphipod *Corophium volutator* EC_{50}: 1.1-1.3 mg/l. - Chironomoid *Chironomus tentans* EC_{50}: 2.7 mg/l. - Chironomoid *Chironomus tentans* NOEC: 1.4 mg/l. - Copepod *Tigriopus japonicus* EC_{50}: 4.32 mg/l. - Copepod *Tigriopus japonicus* NOEC: 3.5 mg/l. - Hydra *Hydra vulgaris* EC_{50}: 6.9 mg/l. - Mysid *Mysidopsis bahia* EC_{50}: 1.1 mg/l. - Mysid *Mysidopsis bahia* NOEC: 0.51 mg/l. - Rotifer *Brachionus calyciforus* 48h NOEC: 1.8 mg/l. - Snail *Marisa cornuarietis* EC_{50}: 2.24 ->4.03 mg/l. - Snail *Marisa cornuarietis* NOEC: 1.18 mg/l. - Water flea *Daphnia magna* EC_{50} : 10.2-16 mg/l. - Water flea *Daphnia magna* NOEC: 4.1 mg/l.	***Acute toxicity*** - Fathead minnow (*P. promelas*) LC_{50}: 4.6-4.7 mg/l. - Rainbow trout (*O. mykiss*) LC_{50}: 3-7 mg/l. - Madaka (*O. latipes*) LC_{50}: 15 mg/l. - Sheepshead minnow *Cyprinodon variegatus* LC_{50} :; 7.5 mg/l.
Chronic toxicity - Duckweed, *Lemna gibba* EC_{50}: 20-32 mg/l. - Duckweed, *Lemna gibba* NOEC: 7.8 mg/l. - Green algae, *Selenastrum capricornutum* EC_{50}: 2.7-8 mg/l. - Green algae, *Selenastrum capricornutum* NOEC: 1.2 mg/l. - Marine diatom, *Skeletonema costatum* EC_{50}: 1.0-2.5 mg/l. - Marine diatom, *Skeletonema costatum* NOEC: 0.4 mg/l.	***Chronic toxicity*** - Amphipod *Hyalella azteca* LOEC: 1.1 mg/l. - Amphipod *Hyalella azteca* NOEC: 0.49 mg/l. - Chironomoid *Chironomus riparius* LOEC: 1.0 mg/l - Chironomoid *Chironomus riparius* NOEC: 0.1 mg/l. - Cnidaria *Hydra vulgaris*: NOEC/EC_{10} : 42 μg/l. - Cnidaria *Hydra vulgaris* LOEC: 0.460 mg/l - Crustaceans *Hyalella azteca* NOEC/EC_{10}: 0.490 μg/l - Hydra *Hydra vulgaris* LOEC: 0.460 mg/l. - Hydra *Hydra vulgaris* NOEC: 0.042 mg/l. - Insects *Chironomous riparius* NOEC/EC_{10}: 0.1 mg/l - Rotifer *Brachionus calyciflorus* LOEC: 3.6 mg/l. - Rotifer *Brachionus calyciflorus* NOEC: 1.8 mg/l. - Snail *Marisa cornuarietis*: 0.015 μg/l. - Snail *Marisa cornuarietis* NOEC: 0.025 mg/l - Snail *Marisa cornuarietis* LOEC: 0.640 mg/l - Sponge *Heteromyenia* spp. LOEC: 160 mg/l. - Sponge *Heteromyenia* spp. NOEC: 1.6 mg/l. - Water flea: *Daphnia magna*: NOEC: ≥3.6 mg/l	***Chronic toxicity*** - Fathead minnow (*P. promelas*) NOEC/EC_{10}: 1.0 μg/l. - Amphibians *Xenopus laevis* NOEC/EC_{10}: 2.23 μg/l.

The following endocrine disrupting effects generally occur at environmentally relevant concentrations:

- Alteration of sex determination from exposure during gonadal organogenesis (for example, high concentrations of BPA cause abnormal sex determination in some fishes, amphibians, reptiles and birds) (see Table 11.7).
- Alteration of gonadal functions from exposure both during and after gonadal organogenesis BPA has been shown to adversely affect gonadal functions in female fish. For example, BPA caused female brown trout to delay ovulation of 2 weeks when exposed at 2.4 μg/l and at 5 μg/l BPA eliminated ovulation altogether. The water flea, *Ceriodaphnia dubia* exposed to BPA in excess of 1.0 mg/l had reduced reproduction. Similarly, exposure of environmentally relevant concentrations

of BPA elevated fecundity in three species of prosobranch snails (*Potamopyrgus antipodarum*, *Marisa cornuarietis* and *Nucella lapillus*) (Crain *et al.* 2007).

- Induction of hepatic vitellogenin production following exposure of fully organised individuals (fathead minnow, *P. promelas;* medaka fish, *O. latipes;* rainbow trout, *O. mykiss;* zebra fish, *D. rerio;* amphibians, *X. laevis* and freshwater mussels, *Mytilus edulis)* show elevated Vtg but only at concentrations greater than the typically measured in surface waters (Levy *et al.* 2004; Crain *et al.* 2007).

Ferrara *et al.* 2008 reported that exposure to BPA (10-50 mg/l) was able to produce morphological and physiological effects on the growth of seedlings of broad bean, tomato and durum wheat and lettuce during a 21 days growth experiment. Tomato and broad bean seedlings absorbed BPA to various extents and it bio-accumulated in roots and shoots of tomato. BPA has also been found to accumulate in bream (<0.18-1.4 ng/g) and flounder (1.2-2.6 ng/g) muscle (Vethaak *et al.* 2005).

11.6.3.4: Effects on humans

Human can be exposed to BPA via hydrolysis of polycarbonate plastics and epoxy resins, which results in low concentrations of free BPA in food and liquids. Moreover, BPA can leach from polycarbonate plastic when exposed to heat. Studies conducted in USA, Europe and Japan has documented widespread human exposures to BPA with detected levels ranging from 0.3 to 5.0 ng/ml (approximately 1-20 nM) in serum and breast milk). BPA also accumulates in human fat. Workers who spray epoxy resins had higher levels of both Bisphenol A diglycidyl ether (BADGE) (which is a component of epoxy resins) and BPA in urine and plasma. BPA has also been detected in food cans and dental cement (Ben-Jonathan *et al.* 2009). Canned food contributed about 60% of the total daily intake of BPA among adults in Japan and Europe. Fish, shell-fish, meat and vegetables are important sources of BPA in canned foods (Ye *et al.* 2009).

BPA has low acute toxicity with an oral LD_{50} of 3,250 mg/kg in rats (http://en.wikipedia.org/wiki/bisohenol_A). It is a hormone-mimicking chemical that can disrupt the endocrine disrupting system at very low concentrations. Studies conducted with animals have found that low doses of BPA caused a variety of adverse health effects (reduced sperm counts, impaired immune system functioning, increase prostate tumour proliferation, altered prostate and uterus development, insulin resistance, alteration of brain chemistry, early puberty and behavioural changes). Though there is some evidence that BADGE is a rodent carcinogen, unfortunately data for humans is lacking. Male workers exposed to BADGE through spraying epoxy resins have associated depressed gonadotrophic hormones (Samital 2008). Further studies have demonstrated developmental toxicity, carcinogenic effects and possible neurotoxicity at low doses from exposure to BPA (Table 11.15).

Table 11.15: ***Selected studies on low dose bisphenol A exposure in animals*** *[http://en.wikipedia.org/wiki/bisohenol-A].*

Dose (μg/kg/d)	Effects (measured in studies with mice or rats)	Publication year
0.025	Permanent changes to genital tract.	2005
0.025	Changes in breast tissue that predispose cells to hormones and carcinogens.	2005
2.0	Increased prostate weight 30%.	1997
2.0	Lower body weight, increase of anogenital distance in both genders, signs of early puberty and longer estrus.	2002
2.4	Decline in testicular testosterone.	2004
10.0	Prostate cells more sensitive to hormones and cancer.	2006
10.0	Decreased maternal behaviours.	2002
30.0	Reversed normal sex differences in brain structure and behaviour	2003
50.0	US human exposure limit - a guideline set by USEPA/ BPA USEPA safe levels	1998

11.6.4: Phthalates
(based on Zeng et al. 2008a and 2008b)
11.6.4.1: Occurrence

Phthalates (BBP, DEHP, DEP, DiBP, DMP, DnBP) have been detected in surface waters (<0.0005-16.6 mg/l), sediments (<0.0002-6.3 mg/g), soils (nd-29.4 µg/g) and rain water (8-1700 ng/l) from China, Denmark, Germany, Italy, Malaysia, the Netherlands (Table 11.16).

11.6.4.2: Ecotoxicity

The acute toxicity values of phthalates are summarised in Table 11.17. Toxicity data for algae and fish is only available for DBP and DEHP and are not available for other phthalates. Comparing the acute toxicity of *Daphnia magna,* DEHP appears to be the most toxic phthalate to aquatic organisms (lowest EC_{50} is 0.1 mg/l) (see Table 11.17). The level of DEHP detected in surface waters is in the range of 0.32-16.6 mg/l (Table 11.16) which exceeded the lowest acute toxicity value of DEHP (0.1 mg/l, Table 11.17).

11.6.4.3: Effects on biota

Studies with rodents showed that some phthalate esters such as DEHP, DnBP, and BBP and their metabolites are estrogenic and exhibit adverse reproductive effects. Phthalates inhibit estrogen receptors in fish and bio-accumulated in the aquatic plant, *Elodea Canadensis*; the calm, *Sphaerium striatinum* and the water flea *Daphnia magna* (Jackson and Sutton 2008; Bolong *et al.* 2009). BBP, DEHP DMP have been detected in bream and flounder muscle (Vethaak *et al.* 2005, Table 11.10).

11.6.4.4: Effects on humans
[based on Jackson and Sutton 2008].

As phthalate plasticisers are not chemically bound to PVC, they can leach, migrate or evaporate into indoor air and atmosphere, food stuffs, other materials. Consumer products containing phthalates can result in human exposure through direct contact and use, indirectly through leaching into other products, or through general environmental contaminations. Moreover, humans can be exposed to phthalates in the following ways:

- via food wrapping materials and food processing.
- high exposure via medical treatment, i.e. via use of medical devices containing DEHP or pharmaceutical drugs containing DBP phthalate in their coatings.
- via carpeted houses which have more phthalate dust than uncarpeted houses and consequently will lead to more phthalate uptake by humans (Table 11.18).

A recent study in China found that agricultural soils were contaminated with phthalate esters (TΣPAE) (0.195-3.6 g/g dry-weight) (Zeng *et al.* 2008a). Excessive accumulation of PAEs in agricultural soils may not only cause environmental contamination but may lead to elevated uptake by crops and thereby contamination of vegetables and food chains, leading to direct and indirect human exposures.

Table 11.16: ***Comparisons of concentrations of major phthalate esters (PAEs) in soil, sediment and water from the world*** [compiled by Zeng *et al.* 2008b (soil); Zeng *et al.* 2009; Zhang *et al.* 2009].

Environmental matrices	Country	BBP	DEHP	DEP	DiBP	DMP	DnBP	DnOP
Soils (μg/g dw)	China (urban soil, Guangzhou)	nd-1.797	0.892-264	0.003-0.198	-	0.009-0.348	0.206-30.1	nd-2.31
	China (urban soil, Beijing)	nd-0.060	0.170-6.49	nd-0.252	0.149-0.936	nd-0.067	0.279-3.82	nd-0.172
	China (soil from different regions including East, South, West, North, Northeast and Northwest China)	-	0.20-5.98	nd-2.61	-	nd-0.2	0.14-1.38	-
	China (Greenhouse soils, Beijing)	-	0.22-0.74	<0.01-0.05	0.31-0.94	<0.01-0.02	0.34-1.66	<0.09
	China (vegetable soil, Pearl River Delta)	nd-1.48	2.82-25.11	nd-1.77	-	nd-0.068	nd-20.55	nd-0.92
	China (agricultural soils, Guangzhou)	nd-1.58	0.107-29.4	0.001-0.178	0.017-1.63	0.001-0.157	0.009-2.74	nd-0.084
	Denmark (soil)	0.00006-0.00038	0.025-0.40	<0.0001	-	<0.0001	0.0005-0.00016	0.0015-0.0043
	Denmark (sewage sludge - amended soil)	0.00001-0.029	0.012-1.111	<0.0001	-	<0.0001	0.0003-0.439	0.00061-0.052
	Netherlands (soil)	-	0.0318	-	-	-	0.006	-
	Netherlands (uncultured soil)	0.00009	0.016	<0.0001	-	<0.0001	0.0021	0.0012
	UK (soil)	0.0002-0.008	0.0222-0.0758	0.0002-0.0009	-	0.0001	0.079-0.080	0.0115-0.0137
	UK (sewage sludge - amended soil)	0.0024-0.0029	0.3162-0.5499	0.0005	-	0.0002-0.0003	0.0113-0.0118	0.0056-0.0068
Sediments (mg/g)	China (Guanghou)	0.034	1.30	0.13	5.80	0.039	0.28	na
	China (Taiwan)	0.20	4.6	0.20	-	-	6.3	na
	Germany (Berlin)	-	0.70	-	-	-	0.45	na
	Italy (Rieti district)	0.002	0.069	0.002	0.013	<0.0002	0.008	na
	Malaysia (Klang River basin)	-	1.63	0.003	0.056	0.003	0.25	na
	Netherlands (Dutch coast)	0.014	0.60	0.114	-	0.014	0.39	na
	Netherlands (Freshwater)	-	4.3	-	-	-	0.088	na
Water (mg/l)	China (Guanghou)	<0.005	0.17	0.031	0.43	0.009	1.99	na
	China (Taiwan)	<0.1	9.3	0.50	-	-	4.9	na
	Germany (Berlin)	-	2.27	-	-	-	0.50	na
	Italy (Rieti district)	<0.010	4.3	<0.010	0.30	<0.012	1.6	na
	Malaysia (Klang River basin)	-	16.6	<0.05	0.25	<0.05	1.6	na
	Netherlands (Dutch coast)	0.077	0.32	0.43	-	0.017	0.25	na
	Netherlands (Freshwater)	-	0.33	-	-	-	0.21	na
Fish muscle (mg/kg dw)	*Oreochromis niloticus niloticus*		33.6					na
	Liza subviridis		61.8					na
								na

nd = not detected; na = not available

Table 11.17: ***Acute toxicity of phthalates to freshwater and marine organisms*** *[data compiled by van Wezel et al. 2000; Sánchez-Avila 2009].*

Phthalate	Algae (LC_{50}/EC_{50})	Invertebrates (LC_{50}/EC_{50})	Fish (LC_{50}/EC_{50})
BBP (Benzylbutyl phthalate)		*Daphnia magna*: 0.96 mg/l	
DBP (Di-n-butyl phthalate)	*FW algae*: 0.40-9 mg/l *MW algae*: 0.0034-0.28 mg/l	***FW crustacean*** *Daphnia magna*: 3-5.2 mg/l *Gammarus* spp.: 2.1 mg/l ***MW crustacean*** *Artemia salina*: 8 mg/l ***FW insecta*** *Chironomus* spp.: 0.76-6.3 mg/l	***FW fish*** *Brachydanio rerio*: >0.32->100 mg/l *Lepomis macrochirus*: 1.6->6.5 mg/l *Oncorhynchus mykiss*: 1.2-6.5 mg/l *Pimephales promelas*: 0.85-4 mg/l ***MW fish*** *Ictalurus punctatus*: 0.46-2.91 mg/l
DEHP (Di(2-ethylhexyl) phthalate	*FW algae*: >0.1-130 mg/l	***FW crustacean*** *Daphnia magna*: >0.1-2 mg/l *Gammarus* spp.: >0.40-32 mg/l ***MW crustacean*** *Palaemonetes pugio*: > 1 mg/l ***FW insecta*** *Chironomus* spp.: >10->18 mg/l ***MW mollusca*** *Mytilus edulis*: >0.042 mg/l	***FW fish*** *Brachydanio rerio*: >0.32->100 mg/l *Ictalurus punctatus*: >0.2->100 mg/l *Lepomis macrochirus*: 0.2->250 mg/l *Oncorhynchus mykiss*: >0.32->100 mg/l *Pimephales promelas*: >0.16->10 mg/l ***MW fish*** *Cyprinodon variegatus*: >0.17-550 mg/l
DEP (Diethyl phthalate)		*Daphnia magna*: 860 mg/l	
DMP (Dimethyl phthalate)		*Daphnia magna*: 459 mg/l	

FW = freshwater; MW = marine

Table 11.18: ***Amounts of phthalates (mg/kg) in house dust and TDI of five phthalates*** *[based on various studies and Abb et al. 2009].*

	BBP	DBP	DEHP	DIDP	DINP	References
Phthalates in house dust	13-29.7	29-87.4	480-970	31-60	29-65	Abb et al. 2009
TDI (tolerable daily intake) g/kg body weight	200	100	37	250	150	Plastics Europe 2006

Exposure to high levels of phthalates has been reported to cause miscarriages and pregnancy complications in humans (Bolong *et al.* 2009). Some PAEs (DnBP, DEHP) and their metabolites are suspected to have link to breast cancer (Dey *et al.* 2009). Furthermore, exposure to phthalates may cause profound and irreversible changes in the development of reproductive tract, especially in males and could be the leading cause of the reproductive disorders in humans including reduced sperm counts, histological changes in testes and reduced fertility (Latini *et al.* 2006; Heudorf *et al.* 2007). IARC (2000) classified phthalates (Table 11.19) as 'not classifiable as to carcinogenicity in humans' however, in recent years toxicological concerns have risen regarding the possible endocrine disruption potency of phthalates (Heudorf *et al.* 2007).

Table 11.19: ***Classification and labelling of phthalate plasticizers*** [PAC 2002].

	Cancer	Fertility	Developmental
BBP	none	category 3	category 2 proposed
DBP	none	category 3	category 2
DEHP	none	category 2	category 2
DIDP	none	none	none
DINP	none	none	none

Category 1 = substances known to cause effects in humans; Category 2 = substances to be regarded as if they cause effects in humans; Category 3 = substances of concern for humans.

11.6.5: Organotins-Tributyl tin (TBT)

[based on Antizar-Ladislao 2008; Ketata et al. 2008].

11.6.5.1: Occurrence

TBT has been detected in seawater (upto 500 ng sn/l), sediments in harbours and shipping lines (upto 16,800 ng sn/g) and biota (bivalves upto 1,610 ng sn/g) from a wide range of countries including Australia, Bangladesh, Canada, China, Denmark, Egypt, France, Greece, Hong Kong, India, Italy, Japan, Korea, Malaysia, the Netherlands, Poland, Portugal, Spain, Sweden, Taiwan, UK, USA, and Vietnam (Table 11.20). Therefore, marine food could be an important route of exposure of TBT to humans since marine fishery product contains high TBT concentrations.

11.6.5.2: Effects on biota

It is one of the earliest recognised EDCs which led to the banning use of TBT as antifouling agent in many countries. TBT pollution has been related specifically to the development of imposex, i.e., a masculination of prosobranch gastropod females by the development of a vas deferens and /or a penis in addition to their female sex characteristics (because TBT causes a build-up of testosterone). Imposex is defined as a pseudohermaphroditic condition characterised by the development and superimposition of non-functional male sexual characteristic (penis, vas deferens and/ or seminiferous tubules) in female individuals. Imposex can be induced by TBT at ambient concentration as low as 1.0 ng TBT-sn/l (Strand *et al.* 2009). The highest levels of both TBT and imposex in the molluscs were found in the vicinity of harbours and marinas (in areas with a high ship density and a likely high exposure to antifouling agents like TBT).

Nucella lapillus, Ilyanassa obsolete, Thais emarginata and *Ocenebra erinacei* were the first discovered (female) penis-bearing neogastropods. Imposex has been described in more than 240 gastropod species worldwide and has been recognised as a valuable biomarker for baseline surveys and monitoring of biological effects of contaminants in the environment. Imposex is considered as the best and most relevant example of chemically-induced endocrine disruption in wildlife. The imposex effects have direct consequences on the reproductive capability of the population limiting the numbers of juveniles and their recruitment to the adult population. As a consequence, gastropods with low fecundity, limited mobility and hatchlings that are not widely dispersed may be eliminated from the environment.

The acute toxicity (LC_{50}) and chronic toxicity values for aquatic organisms ranged between 0.42-1800 µg/l and 0.002-320 µg/l respectively. For example, the acute toxicity (LC_{50}) for larvae and adults of the Pacific oyster *Crassostera gigas* are 1.6 and 1800 µg/l for TBT, while chronic larval mortality occurs at 0.05 µg/l and shell deformities occur at 0.02 µg/l (see Sekizawa *et al.* 2003 for a review).

Table 11.20: ***Butyltin compounds in seawater, sediments and biological tissues reported from several regions of the world*** *[data compiled by Antizar-Ladislao 2008].*

Organotins	Seawater (ng sn/l)	Sediments (ng sn/g dw)	Biota (ng sn/g dw)
Tributyltin	Canada (<dl-500) France (<0.0015-0.12) Greece (<dl-70) Japan (3.9-27) Korea (<dl-4.5) Spain (0.4-196.6)	Australia (<dl-1275) Canada (<dl-5,100) Denmark (100-5,000) Egypt (1-2,076) France (<dl-89) India (4.5-16,816) Japan (1.2-14,000) Italy (3-27) Malaysia (2.8-1,100) Portugal (1.2-130) Spain (0.6-7,673) Taiwan (2.4-8,548) UK (1-60) USA (1.5-14,000) Vietnam (8.3-51)	Bangladesh (fish: 0.47-3)[a, b] Canada (mussel: 20-1440)[a, b] China, HK (gastropods: <dl-170) Denmark (bivalve: 200-300)[b] Greece (bivalve: <dl-109) Italy (*H. trunculus*: <dl-91) Japan (fish: <dl-240)[a, b] Japan (mussel: 0.8-287)[a] Korea (bivalves: 16-1610) Korea (starfish: 7-323) Malaysia (fish: 2.4-190) the Netherlands (fish: 9.2-6.7)[a, b] Poland (mussel: 2.2-3.9)[a] Portugal (mussel:11-789) Spain (deep sea fish:1-52)[a]

Contd...

			Spain (*H. trunculus*: 48) Spain (mussel: 52.8-96) Spain (oyster: 269) Sweden (bivalve: 200-300)[b] Taiwan (oyster: <3.8-417) Vietnam (clam: 3.8-15)
Dibutyltin	Canada (<dl-270) France Greece (<dl-159) Japan (2.1-13) Korea (<dl-22.3) Spain (0.3-33.7)	Australia (<dl-71) Canada (<dl-1100) Denmark (na) Egypt (0.1-379) France (<dl-87) India (<dl-469) Japan (<dl-3400) Italy (<dl) Malaysia (3.8-310) Portugal (<5.3-65) Spain (0.5-3519) Taiwan (na) UK (12-219) USA (59-350) Vietnam (8.1-42.7)	Bangladesh (fish: <0.36-15)[a,b] Canada (mussel: <dl-1062) China, HK (gastropods: <dl-197) Denmark (-) Greece (bivalve: <dl-366) Italy (*H. trunculus*:<dl-316) Japan (fish: <dl-50) [a,b] Japan (mussel: 0.8-3.1)[a] Korea (bivalves: 23-699) Korea (starfish: 8-139) Malaysia (fish: <1.3-13) the Netherlands (fish: 13-183) [a, b] Poland (mussel: <1.4-24)[a] Portugal (mussel:<2.5-18) Spain (deep sea fish: 4-67) [a] Spain (*H. trunculus*: 85) Spain (mussel: 20.2-25.7) Spain (oyster: 7.6-441) Sweden (-) Taiwan (oyster: <3.9-281) Vietnam (clam: 4.4-2.7)
Monobutyltin	Canada (<dl-460) France Greece (<dl-19) Japan (2.5-23) Korea (<dl-13.4) Spain (<dl-51)	Australia (<dl-161) Canada (<dl-1100) Denmark (na) Egypt (0.1-186) France (25-74) India (<dl-470) Japan (<dl-3,300) Italy (<dl) Malaysia (5-360) Portugal (5.2-78) Spain (0.7-2,870) Taiwan (na) UK (12-172) USA (21.3-320) Vietnam (3.9-30)	Bangladesh (fish: <5.6-170)[a, b] Canada (mussel: <dl-708) China, HK (gastropods: <dl-336) Denmark (2.5-15) Greece (bivalve: <dl-151) Italy (*H. trunculus*:<dl-167) Japan (fish: <dl-20) [a, b] Japan (mussel: 0.8-2.9)[a] Korea (bivalves: 23-461) Korea (starfish: 8-2860) Malaysia (fish: <2.3-7.4) the Netherlands (fish: 23-41)[a,b] Poland (mussel: <1.4-4.7)[a] Portugal (mussel: <7.9-41) Spain (deep sea fish: <dl-54)[a] Spain (*H. trunculus*: 63) Spain (mussel: 52.8-96.1) Spain (oyster: 0.4-12.9) Sweden (2.5-15) Taiwan (oyster: <3.3-407) Vietnam (clam: 2.8-18)

a = wet weight; b = ng organotin instead of sn; dl = below detection limit; na =no data available; sn = tin.

11.6.5.3: Effects on humans

[http://seagrant.uaf.edu/nosb/papers/2004/soldotna-tbt.html; http://www.lenntech.com/elements-and-water/tin-and-water.htm].

The effects of TBT on humans are not known, however, shipyard workers exposed to TBT dust and vapours while repairing submarine developed breathing problems, irritated skin, headaches, colds, flu, fatigue, dizziness and stomach ache. In addition, exposure of TBT can irritate the eye, skin, and mucous membranes and prolonged exposure may cause liver and kidney damage (Cornell University 1993). Human intake of organotins from seafood, especially from fish farmed in TBT-treated cages could lead to exceeding the tolerable daily intakes based on more sensitive immunotoxicological end-points for some products retailed in North America, Europe and Asia (Santillo 2001). Trimethyltin and triethyltin are mildly toxic to humans. TBT may cause hormonal disorder in humans, which may ultimately result in infertility.

11.6.6: *Phytoestrogens*

[based on Kiparissis et al. 2003; Clotfelter and Rodriguez 2006; Waring and Harris 2005; Liu et al. 2009b].

11.6.6.1: *Occurrence*

Phytoestrogens are plant compounds structurally and/ or functionally similar to ovarian and placental estrogens (17β-estradiol) and their active metabolites. They fall into three main categories: flavones/isoflavones, coumestans and lignans. Phytoestrogen exists in many plants with high concentrations such as soybeans, fruits and some vegetables (Table 11.21). They were detected with high frequency in municipal wastewater treatment plants, the main source of which are attributed to daily excretions of urine and faeces by humans. The excretion rates were related to diets, being highest with soybeans and isoflavones (Table 11.22).

Table 11.21: ***Phytoestrogen content (µg/100 g wet-weight) of some selected fruits and vegetables (with high amount of phytoestrogen)*** *[Kuhnle et al. 2009].*

Food (taxonomic name)	Phytoestrogens	Isoflavones*	Lignans	Coumestrol
Apricot, *Prunus armenica* (dried)	443	12	431	<1
Asparagus, *Asparagus officinalis* (cooked)	154	2	152	-
Beans, butter beans, *Phaseolus limensis* (dried)	196	51	143	2
Beans, France beans, *Phaseolus vulgaris* (cooked)	159	48	109	2
Beans, Haricot beans, *Phaseolus vulgaris* (dried)	132	21	106	5
Beans, Kidney beans, *Phaseolus vulgaris* (dried)	172	73	89	10
Beans, Runner beans, *Phaseolus coccineus* (trimmed & strung)	201	164	26	11
Beans sprout, *Vigna radiata* (pre-washed)	798	351	86	361
Black berries, *Rubius* sp. stewed from fresh	221	<1	220	<1
Blackcurrant, *Ribes nigrum* (fresh)	109	2	107	<1
Carrots, *Daucus carota*	125	4	121	<1
Chestnuts, *Castanea sativa*	217	2	214	<1
Chick peas, *Cicer arietinum* (dried)	609	607	2	<1
Dates, *Phoenix dactylifera* (dried)	599	14	584	<1
Dates, *Phoenix dactylifera* (stones removed)	192	35	157	<1
Fig, *Ficus* spp.	389	12	376	<1
Fig, *Ficus* spp. (dried)	129	14	114	1
Kiwi, *Actinidia delicosa*	111	<1	111	<1
Goose berries, *Ribes uva-crispa*	121	<1	121	<1
Mung beans, *Vigna radiata* (dried)	323	32	289	2
Parsley, *Pteroselium crispum* (leaves)	197	59	137	<1
Plum, yellow - *Prunus domestica* (cooked)	152	2	150	<1
Pomegranate, *Punica granatum* (flesh and seeds)	304	<1	304	<1
Prune, *Prunus domestica* (dried)	363	6	357	<1
Prune, *Prunus domestica* (cooked from dried)	108	2	106	<1
Prune, *Prunus domestica* (semi-dried)	284	3	281	<1
Pumpkin, *Curcurbita* sp.	154	<1	<1	<1
Soya bean, *Glycine max* (cooked)	17,556	17,544	11	<1
Soya bean, *Glycine max* (frozen, cooked)	10,687	10,621	64	2
Soya flour, *Glycine max*	124,227	124,381	345	<1
Sweet potato, *Ipomoea batatus*	259	1	258	<1

* *Isoflavones = sum of daidzein, genistein, glycitein, biochanin A and formononetin*

Table 11.22: *Daily excretion rates of phytoestrogens from human urine (μg/d) [data compiled by Liu et al. 2009b].*

Name of EDCs	Sex	Daily average excretion rates (min-max. range, μg/d))	Comments
Coumestrol	Women	8.6-9.5	Diet containing highest concentrations of isoflavones resulted highest average excretion rates in women
Daidzein	Men Women	211-1,093.9 94.1-3,686.4	Soya protein diet and diet with highest isoflavones concentrations resulted in highest average excretion rates in women.
Dihdrodaidzein	Women	183.5-1,891.9	Soya proteins diet resulted in highest average excretion rates in women.
Enterolactone	Men Women	1,020.4 389.6-1,851.2	Normal and soya protein diet resulted in highest average excretion rates in women.
Equol	Men Women	18.2-72.1 nd-271.0	Soya protein diet and diet with highest isoflavones concentrations resulted in highest average excretion rates in women.
Genistein	Men women	78.4-477.9 45.9-3,553.1	Diet containing highest concentrations of isoflavones and soya protein diet resulted in highest average excretion rates in women.
Glycitein	Men women	51.3 83.3-690.8	Diet containing highest concentrations of isoflavones resulted in highest average excretion rates in women
Kaempferol	Men and women	51.3	
Naringenin	Women	653.8	
ODMA	Men women	48.1-179.0 12.7-2,687.3	Diet containing highest concentrations of isoflavones resulted in highest average excretion rates in women,
Quercetin	Men and women	23.4	
Phloretin	Men and women	84.8	

nd = not detected

11.6.6.2: Effects on biota

Phytoestrogens have positive effects on human health but most phytoestrogens also act as endocrine disruptors and may cause intersex in aquatic organisms and affect behaviour of fishes. Both reporter gene assay (GA) and Yeast estrogen screen (YES) showed that most phytoestrogens are weaker estrogenic substances, however, the transient gene expression assay (TGEA) found that some of phytoestrogens have estrogenic potencies which are at the same level, or even greater than that of E2 (TGEA showed coumestrol potencies as 0.98-1.02; genistein as 1.82-1.98). In addition, some phytoestrogens also possess androgenic activities (Liu *et al.* 2009b). Table 11.4 provides a comparison of estrogenic potencies of different EDCs. Phytoestrogens (e.g. equol, coumestrol) from clover have long been recognised as the cause of reproductive dysfunction and infertility in sheep. Exposure to paper mill effluents have contributed to the negative reproductive effects in fish which researchers believed to be linked to β-sitosterol. Due to the possible negative effects of phytoestrogens, current research is mainly limited to the occurrence and fate of phytoestrogens in wastewater and surface waters (Table 11.23). The detected concentrations are as follows:

- River water nd-143,400 ng/l;
- Drainage water nd-2,000 ng/l
- Sediments 16.1-36,142 ng/g.

11.7: Drinking and environmental water quality guidelines for EDCs

Australia has recommended drinking water (Table 11.24) and ecosystems (Table 11.25) protection guidelines for EDCs to protect human and environmental health. These guidelines can be used for any environmental risk assessment. Additional guideline values relevant to maximum contaminant level and health advisory levels are provided in section 11.8.2. - EDCs in USA streams.

Table 11.23: ***Occurrence and fate of phytoestrogens*** *[data compiled by Liu et al. 2009b].*

Phytoestrogens	Influent (Inf); Effluent (Eff) (countries where reported) (ng/l)	River water (ng/l)	Drainage water (ng/l)	River sediments (ng/g)	Seawater (ng/l)	Drinking water (ng/l)	Air (pg/m³)
Daidzein	Inf: 75-12,000; Eff: nd-81 (Germany, Italy, Japan)	2-42,900	5-30	-	<0.43		-
Daidzein	Inf: 9-25; Eff:18-74 (Italy)	nd	-	-	-		-
Genistein	Inf: 160-13,100; Eff: 7-10,100 (Canada, Italy)	<2.6-143,400	14	-	<0.61		-
Genistein	Inf: 11-62; Eff: nd-19 (Italy)	nd-<400	-	-	-	-	
Glycitein	Inf: 6-25; Eff: nd-16 (Italy)	nd--6	-	-	-	-	-
Formononetin	Inf: nd-10; Eff: nd (Italy)	nd-21	44-157	-	-	-	-
Coumestrol	Inf: 9-23; Eff: nd-19 (Italy)	1-2	-	-	-	-	-
Biochanin-A	Inf: 8-76; Eff: nd-21 (Italy)	1-3	7-22	-	-	-	-
Equol	Inf: 9.4e-1.1e6; Eff: 170-410 (Japan)	nd-23	34-121	-	-	-	-
Zearlenone	Inf: 9.4e-1.1e6; Eff: 170-410 (Italy)	2-5	<30	-	-	-	-
α-Zearalanol	Inf: nd-10; Eff: nd-7 (Italy)	nd-3	nd	-	-	-	-
α-Zearalenol	Inf: Eff: 20 or 30 (Italy)	<10	-	-	-	-	-
β-Zearalnol	Inf: nd-10; Eff: nd-7 (Germany)	nd-3	nd	-	-	-	-
β-Sitosterol	Inf: 4,194-49,920; Eff: 60-50,165 (Canada, France, Germany, Italy, Switzerland, USA)	<930-6,000	2,000	16.1-36,142	63.4-3,156	nd	321-1,765
Enterolactone	Inf: 600-1,223; Eff: 659 (Australia, Finland)	5	-	-	26	13	-

nd = not detected

Table 11.24: ***Australian drinking water guideline values for EDCs*** *[NWQMS 2007].*

Class	Compound	Drinking water guideline value
Natural and synthetic hormones	***Estrogenic hormones***	
	Estrone (E1)	300 ng/l (0.300 μg /l)
	17β-estradiol (E2)	175 ng/l
	17α-estradiol (αE2)	175 ng/l
	Estriol (E3)	500 ng/l
	17α-ethynylestradiol (EE2)	15 ng/l
	Mestranol	25 ng/l
	Norethindrone	2,500 ng/l
	Progesterone	105 μg /l
	Androgens	
	Androsterone	14 μg /l
	Progesterone	105 μg /l
Industrial chemicals-phenols	Bisphenol A (BPA)	200 μg /l
Industrial chemicals-alkylphenols	4-tert-octylphenol (OP)	50 μg /l
	4-nonylphenol (NP)	500 μg /l

Contd...

Industrial chemicals-organotins	Tributyltin (TBT) Dibutyltin (DBT) Monobutyltin (MBT)	1,000 ng/l 2 μg /l 700 ng/l
Industrial chemicals-phthalates	Di-n-butyl phthalate Phthalate anhydride	35 μg /l 7 ng/l
Industrial chemicals-PAH	Anthracene Benzo(a)pyrene Naphthalene Phenanthrene Pyrene	150 μg /l 10 ng/l 70 μg/l 150 μg/l 150 μg /l

The UK Environment Agency, the European Chemicals Bureau and the Australian Environmental protection Authority (Williams *et al.* 2007; Goonan 2008) have proposed the predicted no effect concentrations (PNEC) for following steroidal estrogens for the protection of aquatic life:

- 3-5 ng/l for estrone (E1).
- 1 ng/l for 17β-estradiol (E2).
- <0.1 ng/l for 17α-ethynylestradiol (EE2.)
- 330 ng/l for nonylphenol and octylphenol.
- 1,600 ng/l for bisphenol-A.
- 0.4 ng/l for tributyltin.

Table 11.25: *ANZECC water quality trigger values (TV) for EDCs for the protection of aquatic ecosystems [ANZECC and ARMCANZ 2000].*

EDCs	TV for freshwater	TV for marine water
Natural and synthetic estrogens- No ANZECC guideline values		
-	-	-
Alkylphenols- nonylphenol		
Nonylphenol	ID	ID
Phenols-BPA - No ANZECC guideline values		
-	-	-
Phthalates		
Dimethylphthalate	3,000 μg/l for 99% species protection 3,700 μg/l for 95% species protection 4,300 μg/l for 95% species protection	ID
Diethylphthalate	900 μg/l for 95% species protection 1,000 μg/l for 95% species protection 1,100 μg/l for 95% species protection	ID
Dibutylphthalate	9.9 μg/l for 99% species protection 26 μg/l for 95% species protection 40.2 μg/l for 95% species protection	ID
Di-2-(ethylhexyl)-phthalate (DEHP	ID	ID
Oragnotins		
Tributyltin	ID	0.0004 μg/l -99% species protection 0.006 μg/l -95% species protection 0.02 μg/l -90% species protection
Polycyclic aromatic hydrocarbons (PAH)		
Anthracene	ID	ID
Benzo(a)pyrene	ID	ID
Fluoranthene	ID	ID
Naphthalene	16 μg/l for 95% species protection	50 μg/l for 99% species protection
Phenanthrene	ID	ID

ID = Insufficient data to derive a reliable trigger value

11.8: Case studies - EDCs in Asian mussels, USA streams and Australian rural environment

11.8.1: Phenolic EDCs in Asian seafoods (mussels)

As part of the Asian Mussel Watch Project, green mussel (*Perna viordis*) samples were collected from 48 locations in Cambodia, India, Indonesia, Singapore, Malaysia, the Philippines, Thailand, and Vietnam during 1994-1999. Mussel samples were analysed for PAHs, nonylphenol, octylphenol, bisphenol A. The objective of the study was to examine the regional distribution and pollution status of phenolic EDCs (nonylphenol, octylphenol and bisphenol A) and PAHs in south and south east Asia using mussels as a sentinel monitoring media. The study found highest levels of NP in Malaysia, Singapore, the Philippines and Indonesia (range 18-663 ng/g) and, the results were compared to those of industrialised nations. However, octylphenol (OP) concentrations in mussels (0.3-16 ng/g) were found much lower which may be due to much less usage of OP as surfactants in Asian countries and as well as due to less production of OP in Asia. BPA was in the range of 0.51-13.7 ng/g, with high concentrations found in India and Indonesia. The monitoring results showed the bio-accumulation potential of phenolic EDCs in aquatic biota in the marine environment. Mussels from countries such as Cambodia and Vietnam did not show elevated concentrations of EDCs (Table 11.26), which is probably related to the low extent of industrialization of those two nations. PAHs ranged from 11 to 1,133 ng/g, which can be categorised as low (0-100 ng/g dw) to moderate levels (100-1,000 ng/g) of pollution.

Table 11.26: ***Distribution of phenolic EDCs in mussels (ng/g dry tissue) from South and South East Asian countries*** *[Isobe et al. 2007]*. Bold figures indicate detection of high concentrations from those countries.

Country	Octylphenol	Nonylphenol	Bisphenol A	PAHs
Cambodia (6 sites) *Koh Kang (2), Kandal (2), Sihanouk ville (1), Koh Kong (1)*	0.3-16	27-228	0.96-4.54	21-156
India (5 sites) *Maharashta (2), West Bengal (1), Tamil Nadu (1), Kerala (1)*	2-8	72-202	**1.3-13.7**	**67-1133**
Indonesia (6 sites) *Surabaya (1), Jakarta (2), Ujung Pandarg (1) Medan (1), Cirebon (1)*	3-16	**72-643**	**0.32-6.3**	**58-191**
Malaysia (12 sites) *Langkawi (2), Penang (2), Port Dickson (1), Malacca (1), Bahru (4), Sabah (2)*	6-16	**18-663**	0.51-4.2	**11-451**
Philippines (6 sites) *Cavite, Bataan, Metro Manila (1from each), Samar (1), Bulacan (1), Capiz (1)*	1-14	**21-578**	1.1-4.7	**75-266**
Singapore (1 site) *Changi airport (1)*	7	**605**	3.3	na
Thailand (6 sites) *Samut Prakarn (1), Phuket (1), Pattani (1), Samut Sakhon (1), Chanthaburi (1), Chonburi (1)*	1-16	81-228	0.96-4.54	84-156
Vietnam (5 sites) *Thua Thien Hue (1), Thanh Hoa Province (1), Ky Anh Prævince (1), Binh Dinh (1), Phan Ri (1)*	1-7	71-121	0.58-1.3	34-110

11.8.2: EDCs in USA streams

The US Geological Survey conducted a nationwide investigation of organic wastewater contaminants in 139 streams across 30 states during 1999 and 2000. The survey found 95 organic wastewater contaminants including several EDCs, hormones (eleven), alkylphenols (five), phthalates (six) and PAH (six). Measured concentrations were low and rarely exceeded drinking water guidelines, drinking water health advisories or aquatic life criteria. However, many compounds do not have such established guidelines (Kolpin *et al.* 2002; Table 11.27). Detergent metabolites, plasticisers and steroids contributed almost 80% of the total measured concentrations indicating that such compounds survive wastewater treatment and biodegradation. Collection of such information provides vital information on the fate and transport of chemicals including EDCs in the hydrological systems and baseline information on the occurrence of organic wastewater across a range of hydrological settings.

Table 11.27: ***Organic waste water contaminants (EDCs) detected in US streams*** *[Kolpin et al. 2002].*

Chemical groups	Compounds	Max. Conc. Detected (µg/l)	Most sensitive species	Lowest LC_{50} of most sensitive species (µg/l)	*MCL/ HAL/DWG (µg/l)
PAH	Anthracene (1999-2000)	4.7	Fathead minnow *(Pimephales promelas)*	5.4	-
	Benzo(a)pyrene (1999-2000)	0.24	Water flea *(Daphnia magna)*	1.5	0.2
	Fluoranthene (1999-2000)	1.2	Fathead minnow *(Pimephales promelas)*	74	-
	Naphthalene (1999-2000)	0.08	Rainbow trout (*Oncorhynchus mykiss*)	910	-
	Phenanthrene	0.53	Water flea *(Daphnia magna)*	590	20
Plasticizer/ Phthalates	Bis (2-ethylexyl) adipate	10	Water flea *(Daphnia magna)*	480	400
	Bis (2-ethylexyl) phthalate	20	Water flea *(Daphnia magna)*	7,500	6
	Bisphenol A	12	Fathead minnow *(Pimephales promelas)*	3,600	-
	Diethylphthalate	0.42	Rainbow trout (*Oncorhynchus mykiss*)	12,000	-
	Ethanol,2-butoxy-phosphate	0.51	Fathead minnow *(Pimephales promelas)*	10,400	-
	Triphenyl phosphate	0.22	Rainbow trout (*Oncorhynchus mykiss*)	280	-
Alkylphenols/ Detergent metabolite	4-Nonylphenol	40	Fathead minnow *(Pimephales promelas)*	130	-
	4-Nonylphenol monoethoxylate	20	Water flea *(Daphnia magna)*	14,450	-
	4-Nonylphenol diethoxylate	9	Water flea *(Daphnia magna)*	5,500	-
	4-Octylphenol monoethoxylate	2	-	-	-
	4-Octylphenol diethoxylate	1	-	-	-
Hormones	Equilenin	0.278	-		-
	Equilin	0.147			-
	17α-ethynyl estradiol (EE2)	0.831	-	-	-
	17α-estradiol (? E2)	0.074	-	-	-
	17β-estradiol (E2)	0.2	-	-	-
	Estriol (E3)	0.051	-	-	-
	Estrone (E1)	0.112	-	-	-
	Mestranol	0.407	-	-	-
	19-norethisterone	0.872	-	-	-
	Progesterone	0.199	-	-	-
	Testosterone	0.214	-	-	-

* MCL= maximum contaminant level; HAL= health advisory levels

11.8.3: EDCs in Australian rural environment

EDCs in the Australian riverine environment, a collaborative work in November 2007 between Australia's Land and Water Innovation program and CSIRO reported that streams across Australia carry EDCs at concentrations that can stimulate changes to the sexual cycles of native fauna (see http://products.lwa.gov.au/files/PR071403.pdf). The study indicated that the presence of livestock industries (cattle, dairies, piggeries and poultry) in rural Australia is likely to contribute to the presence of steroidal estrogens within streams in particular catchments. Higher concentrations of estrone (E1) and17β-estradiol (E2) were always found in the aquatic systems receiving effluents from dairies. The report found that there are very limited monitoring on EDCs in waterways and no systematic survey has been conducted in Australia to date though EDCs exposure occur from both urban facilities and rural industries.

The study was a joint three year pilot project to gain better understanding on the occurrence and risks of selected EDCs in the Australian environment. The EDCs investigated were natural and synthetic hormones (E1, E2, EE2) and a number of phenolic xenoestrogens. The findings of this study are now available on line at http://downloads.lwa2.com/downloads/publications_pdf/PR071403.pdf. The study found that some EDCs have a moderate to high affinity for organic carbon. The degradation rates of EDCs were found to be rapid in aerobic conditions and slower with low levels of organic carbon (even under aerobic conditions).

As part of this study, a survey was conducted from WWT effluents in Queensland, Australian Capital Territory and South Australia during wet and dry seasons to assess whether seasonal variations could have an influence on concentrations of target compounds. Concentrations of the selected EDCs were found to be constant over the two sampling periods both within and between WWTPs. The concentrations of NPIEO, NP2EO and NP were high ng/l to μg/l, while the xenoestrogens and steroidal estrogens were in the low ng/l range in eleven surveyed WWTPs.

A number of streams in rural regions within South Australia were also surveyed, where dairy farming, stock grazing and horticultures are practised. E1 and E2 were measured from sites receiving inputs from dairy farming, stock grazing and horticulture and the National park (reference sites). The results of this survey is summarised below:

- E1 concentrations ranged form 0.17 ng/l at National park (reference site) to 38.5 ng/l in dairy cattle effluents coming from a milking shed.
- E2 concentrations ranged from 0.5 ng/l at National park (reference site) to 8.6 ng/l also from dairy cattle effluents coming from a milking shed
- The concentrations of E1 were always found to be higher than E2 at each sampling site.
- Higher concentrations of E1 and E2 were always present in the aquatic systems receiving effluents from dairies.
- Lower concentrations of E1 and E2 were found from areas influenced by urban inputs and stock grazing at slightly lower levels.
- Areas with horticulture practices had levels of E1 and E2 comparable with reference sites.
- Overall, the results suggest that animal sources such as dairy farming, piggeries, poultry and other concentrated animal operations could be a potential source of E1 and E2 to receiving aquatic systems.

Previous research conducted by Mispagel *et al.* (2005) and Allinson *et al.* (2008) reported both estrogenic and androgenic activity of effluent of municipal waste water treatment plants in Victoria. A study was conducted during winter (2006) and summer (2007) periods to assess hormones in municipal wastewater treatment plant (WWTP) discharges across the state of Victoria (see Allinson *et al.* 2007), the study reported both testosterone (< 3 ng/l) and androstenedione (< 10 ng/l) in WWTPs. The estradiol (nd-32 ng/l) and estrone (nd-18.5 ng/l) concentrations varied between winter and summer periods being higher levels found during the summer periods.

11.9: Conclusion

EDCs are widespread in the environment, water and food chain. EDCs have the potential to accumulate in fish and wildlife to levels that have significant hormonal and reproductive affects on fish and wildlife. EDCs can cause neurological, hormonal, immune and reproductive effects on humans.

Recent studies in Canadian lakes treated with the synthetic steroids, 17α-ethynyl-estradiol (EE2) led to extinction of fish stocks of fathead minnows (*Pimephales promelas*) within two years (Kidd et al. 2007). Several UK studies were able to show a relationship between various reproduction endpoints of freshwater fish and feminization of resident species, the causative agents of which were 17β-estradiol (E2), estrone (E1), and 17α-ethynyl-estradiol (EE2). Abnormally elevated levels of plasma vitellogenin, a biomarker response for estrogens exposure, as well as a decrease in gonadal development, reduced fecundity and /or fertility, are the most commonly observed effects.

Among EDCs, nonylphenol and octylphenol deserve special attention not only because of their disrupting properties but due to their wide spread and high concentrations found in the environment.

The bio-accumulation of NP in seafood (fish, oysters) may be an exposure route of these chemical residues to human. Furthermore, human exposure to NP can also occur after contact with contaminated drinking water and food such as vegetables, crops, milk and meat or during occupational exposure. Food can be contaminated with NP from use of cleaning agents in the food processing industries, application of pesticides and wax coats on vegetables and leaching of NP from food packaging materials. NP was found to be ubiquitous in food packaged with concentrations ranging from 0.1 to 19.4 g/kg (see Soares et al. 2008). An investigation conducted in Italy found the presence of alkylphenols such as nonylphenol (32 ng/ml), octylphenol (0.12 ng/ml), and octylphenol ethoxylates (0.07-0.16 ng/ml) in human breast milk. The study found a positive correlation between fish consumption and levels of nonylphenol in breast milk in Italy (Ademollo *et al.* 2008; Soares *et al.* 2008).

Sewage treatment plants (STP) produce tonnes of sludge each year and during STP treatment. An accumulation of NP, OP and BPA generally occurs in sewage sludge. Therefore, the use of sludge in agriculture could be a potential source of alkylphenols and BPA to the terrestrial environment. Application of poultry manure to land may increase the background concentrations of estradiol and testosterone. BPA which may accumulate in soils and through crops may enter into the food chain. NP and OP have been identified in groundwaters as a result of leaching of sludge fertilized fields. NP is highly toxic to aquatic life.

To reduce or minimise the impacts of EDCs it would be advisable to have restrictions on the use of most harmful EDCs, or deregistration of some chemicals, or to encourage using some alternative chemicals wherever possible. Proper measures should be taken in order to prevent or reduce the release of EDCs into the environment. For most endocrine disruptors the dose-response relationships have yet to be established from species to species i.e. the concentration of chemicals that can cause an effect on people or wildlife. Regular monitoring of fish and other food for EDCs would help to reduce risks of these chemicals to humans. Though there is evidence that EDCs in some situations can cause some significant effects on aquatic organisms or terrestrial vertebrates. There are still some knowledge gaps on the exact concentration of chemicals that can cause an effect on humans or wildlife.

EDCs are micro-pollutants with estrogenic or androgenic activities at very low concentrations and are emerging as a major concern for water quality maintenance. Most conventional wastewater treatment plants do not have advanced treatment process/technology and therefore these emerging micro-pollutants are not removed but are released to the receiving natural waters. Therefore, further research to develop advanced treatment technologies is required so that discharge of EDCs in the environment can be minimised and wastewater can be reused for beneficial purposes.

In Australia, on 23rd November 2007, a workshop assembled Australian researchers, policy makers, regulators, water suppliers and research investors at Black Mountain, Canberra to discuss the range of issues relating to EDCs in Australian waters. The workshops' outcome was (The Black Mountain Declaration on Endocrine Disrupting Chemicals in Australian Waters 2007) (http://products.lwa.gov.au/files/PN21411.pdf) recommending the following measures as a future action plan:

- Identifying current impacts of EDCs on native Australian wildlife.
- Developing methods for monitoring EDCs in various exposure matrices.
- Minimising unnecessary discharge of EDCs to the environment.
- Minimising unnecessary exposure of EDCs to humans via food, water and air.
- Desk-top based modelling of EDCs exposure to wildlife and humans.
- Quantitative understanding of the relationship between exposure to EDCs and their effects toward understanding what levels of exposure can be considered as 'safe' in wildlife and humans.
- Understanding the significance of complex mixtures and interactions of EDCs and other chemicals.
- Communication (increased awareness) to enhance understanding of the issue by a broader audience.

On September 2008, a second workshop was held in Sydney which was attended by Australian and UK researchers and policy makers and agreed to set up collaboration on EDC research between the UK and Australia.

11.10: References

Abb, M., Heinrich, T., Sorkau, E. and Lorenz, W. 2009. Phthalates in house dust. *Environmental International*.Doi.10.1016/j.envint.2009.04.007.

Ademollo, N., Ferrara, F., Delise, M., Fabietti, F. and Funari, E. 2008. Nonylphenol and octylphenol in human breast milk. *Environmental International* 34: 984-987.

Ahel, M., McEvoy, J and Giger, W. 1993. Bio-accumulation of the lipophilic metabolites on non-ionic surfactants in freshwater organisms. *Environmental Pollution* 79: 243-248.

Allinson, G., Allinson, M., Salzman, S., Shiraishi, F., Myers, J., Theodoropoulos, T., Hermon, K., and Wightwick, A. 2007. Hormones in treated sewage effluent. Final Report. Department of Primary Industries, Queenscliff, Australia. 54 pp.

Allinson G, Allinson M., Shiraishi F., Salzman S. A., Myers J. H., Hermon K. M., Theodoropoulos, T. 2008. Androgenic activity of effluent from forty-five municipal waste water treatment plants in Victoria, Australia. In. *Second International Conference on Environmental Toxicology –Environmental Toxicology II* WIT Transaction. p. 293-304. WIT Press, Southampton, UK

Andrew, M. N., Dunstan, R. H., O'Connor, W. A., Zwieten, L. V., Nixon, B. and MacFarlane, G .R. 2008. Effects of 4-nonylphenol and 17a-ethynylestradiol exposure in the Sydney rock oyster, *Saccostrea glomerate*: Vitellogenin induction and gonadal development. Aquatic Toxicology, Volume 88, Issue 1, 2 June 2008, Pages 39-47.

Antizar-Ladislao, B. 2008. Environmental levels, toxicity and human exposure to tributyltin (TBT)-contaminated marine environment. A review. *Environmental International* 34: 292-308.

ANZECC and ARMCANZ 2000. Australian and New Zealand Environment and Conservation Council (ANZECC) and Agriculture and Resource Management Council of Australia and New Zealand (ARMCANZ). Australian and New Zealand Guidelines for Fresh and Marine Water Quality, Volume 1, The Guidelines Chapters 1-7). *http://www.mincos.gov.au/publications/australian_and_new_zealand_guidelines_for_fresh and_marine_water_quality*

Auriol, M., Filali-Meknassi, Y., Tyagi, R. D., Adams, C. D. and Surampalli, R. Y. 2006. Endocrine disrupting compounds removal from wastewater, a new challenge. *Process Biochemistry* 41: 525-539.

Barbosa, I. R., Nogueira, A. J. A., Soares, M. V. M. 2008. Acute and chronic effects of testosterone and 4-hydroxyandrostenedione to the crustacean *Daphnia magna*. *Ecotoxicology and Environmental Safety* 71 (3): 757-764.

Barnabe, S., Brar, S. K., Tyagi, R. D., Beauchesne, I. and Surampalli, R. Y. 2009. Pre-treatment and bioconversion of wastewater sludge to value-added products-fate of endocrine disrupting compounds-Review. *Science of the Total Environment* 407: 1471-1488.

Ben-Jonathan, N., Hugo, E. R. and Brandebourg, T. D. 2009. Effects of bisphenol A on adipokine release from human adipose tissue: implications for the metabolic syndrome. *Molecular and Cellular Endocrinology* 304: 49-54.

Bolong, N., Ismail, A. F., Salim, M. R. and Matsuura, T. 2009. A review of the effects of emerging contaminants in wastewater and options for their removal. *Desalination* 239: 229-246.

Campbell, C. G., Borglin, S. E., Green, F. B., Grayson, A., Wozei, E. and Stringfellow, W. T. 2006. Biologically directed environmental monitoring, fate, and transport of estrogenic endocrine disrupting compounds in water: A review. *Chemosphere* 65: 1265-1280.

Cao, Q., Garib, V., Yu, Q., Connell, D. W. and Campitelli, M. 2009. Quantitative structure – property relationships (QSPR) for steroidal compounds of environmental importance. *Chemosphere* doi.: 10.1016/j.chemosphere.2009.03.045.

Clotfelter, E. D. and Rodriguez, A. C. 2006. Behavioural changes in fish exposed to phytoestrogens. *Environmental Pollution* 144: 833-839.

Cornell University. 1993. Web site, Extension Toxicology Network, 5123 Comstock Hall, Cornell University, Ithaca, New York 14853-0901, 9/1993.

Crain, D. A., Eriksen, M., Iguchi, T., Jobling, S., Laufer, H., LeBlanc, G. A. and Guillette, L. J. 2007. An ecological assessment of bisphenol-A: evidence from comparative biology. *Reproductive Biology* 24: 225-239.

Crane, M., Watts, C. and Boucard, T. 2006. Chronic aquatic environmental risks from exposure to human pharmaceuticals. *Science of the Total Environment* 367: 23-41.

DEPA (Danish Environmental Protection Agency) 2004. Degradation of estrogen in sewage treatment processes. Environmental Project No. 899. Danish Environmental Protection Agency, Danish Ministry of the Environment.

Dey, S., Soliman, A. S. and Merajver, S. D. 2009. Xenoestrogens may be the cause of high and increasing rates of hormone receptor positive breast cancer in the world. *Medical Hypothesis* 72: 652-66.

EEC (European Economic Council) 2001. Community strategy for endocrine disruptors - a range of substances suspected of interfering with the hormone systems of humans and wildlife. COM 262. Bruxelles. 46 pp.

Ferrara, F., Ademollo, N., Delise, M., Fabietti, F. and Funari, E. 2008. Alkylphenols and their ethoxylates in seafood from the Tyrrhenian Sea. *Chemosphere* 72: 1279-1285.

Fiedler, S., Foerster, M. And Glaser, B. And Zech, W. 2007. Alkylphenols in sediments of the Atlantic rainforest south-west of São Paulo, Brazil. *Chemosphere* 66: 212-218.

Gyllenhammar, I., Holm, L., Eklund, R. and Berg, C. 2009. Reproductive toxicity in *Xenopus tropicalis* after developmental exposure to environmental concentrations of ethynylestradiol. *Aquatic Toxicology* 91: 171-178.

Gong, J., Ran, Y., Chen, D. and Yang, Y. 2008. Occurrence and environmental risk of endocrine-disrupting chemicals in surface waters of the Pearl River, south China. *Environmental Monitoring and Assessment*. Doi 10.1007/s10661-008-0474-4.

Goonan, P. 2008. Risk from endocrine disrupting substances in the South Australian aquatic environment. Environment Protection Authority, Adelaide, South Australia. 36 pp. available online: http://www.epa.sa.gov.au/pdfs/risks_endocrine.pdf

Hansen, P., Dizer, H., Hock, B., Marx, A., Sherry, J. and Mcmaster, M. 1998. Vitellogenin-a biomarker for endocrine disruptors. *Trends in Analytical Chemistry* 17: 448-451.

Hecker, M., Murphy, M. B., Coady, K. K., Villenevue, D. L. 2006. Terminology of gonadal abnormalities in fish and amphibians resulting from chemical exposure. *Review Environmental Contamination and Toxicology*. 187: 103-131.

Heudorf, U., Mersch-Sundermann, V. and Angerer, J. 2007. Phthalates: Toxicology and exposure. *International Journal of Hygiene and Environmental Health* 210: 623-634.

Hogan, N. S., Duarte, P., Wade, M. G., Lean, D. R. S. and Trudeau, V. L. 2008. Estrogenic exposure affects metamorphosis and alters sex ratios in the northern leopard frog (*Rana pipiens*): identifying critically vulnerable periods of development. *General and Comparative Endocrinology* 156: 515-523.

IARC (International Agency for Research on Cancer) 2000. Some industrial chemicals. IARC monographs 77: 41-148.

Isobe, T., Takada, H., Kanai, M., Tsutsumi, S., Isobe, K. O., Boonyatumanond, R. and Zakaria, M. P. 2007. Distribution of polycyclic aromatic hydrocarbons (PAHs) and phenolic endocrine disrupting chemicals in South and Southeast Asian mussels. *Environmental Monitoring and Assessment* 135: 423-440.

Jackson, J. and Sutton, R. 2008. Sources of endocrine-disrupting chemicals in urban wastewater, Oakland, CA. *Science of the Total Environment* 405: 153-160.

Jobling, S., Sheahan, D., Osborne, A., Mathiessen, P. and Sumpter, J. P. 1996. Inhibition of testicular growth in rainbow trout (*Oncorhynchus mykiss*) exposed to estrogenic alkylphenolic chemicals. *Environmental Toxicology and Chemistry* 15: 194-202.

Jugan, M. L., Oziol, L., Bimbot, M., Huteau, V., Tamisier-Karolak, S., Blondeau, J. P. and Lévi, Y. 2009. *In vitro* assessment of thyroid and estrogenic endocrine disruptors in wastewater treatment plants, rivers and drinking water supplies in the greater Paris area (France). *Science of the Total Environment* 407: 3579-3587.

Kawahara, S., Hrai, N., Arai, M. and Tatarzako, N. 2009. The effect of *in vivo* co-exposure to estrone and AhR-ligands on estrogenic effect to vitellogenin production and EROD activity. *Environmental Toxicology and Pharmacology* 27: 139-143.

Ketata, I., Denier, X., Hameza-Chaffai, A. and Minier, C. 2008. Endocrine-related reproductive effects in molluscs-review. *Comparative Biochemistry and Physiology*, part C 147: 261-270.

Kidd, K. A., Blanchfield, P. J., Mills, K. H., Palace, V. P., Evans, R. E., Lazorchak, J. M. and Flick, R .W. 2007. Collapse of a fish population after exposure to a synthetic estrogen. *Proceedings of the National Academy of Sciences* 104: 8897-8901.

Kiparissis, Y., Balch, G. C., Metcalfe, T. L. and Metcalfe, C. D. 2003. Effects of the isoflavones genistein and equol on the gonadal development of Japanese medaka (*Oryzias latipes*) *Environmental Health Perspectives* 111: 1158-1163.

Kolpin, D. W., Furlong, E. T., Meyer, M. T., Thurman, E. M., Zaugg, S. D., Barber, L. B. and Buxton, H. T. 2002. *Environmental Science and Technology* 36: 1202-1211.

Kuhnle, G.G.C., Dell'Aquila, C., Aspinall, S.M., Runswick, S.A., Joosen, A.M.C.P., Mulligan, A.A. and Bingham, S.A. 2009. Phytoestrogen content of fruits and vegetables commonly consumed in the UK based on LC-MS and ^{13}C-labelled standards. *Food Chemistry* 116 : 542-554.

Kuster, M., López de Alda., M., Barceló, D. 2004. Analysis and distribution of estrogens and progestogens in sewage sludge, soils and sediments. *Trends in Analytical Chemistry*. 23 (10-11): 790-798.

Länge, R., Hutchinson, T. H., Croudace, C. P., Siegmund, F., Schweinfurth, H., Hampe, P. 2001. Effects of the synthetic estrogen 17 alpha ethynylestradiol on the life cycle of the fathead minnow (*Pimephales promelas*). *Environmental Toxicology and Chemistry* 20: 1216-1227.

Lange, I. G., Daxenberger, A., Schiffer, B., Witters, H., Ibarreta, D. and Meyer, H. H. D. 2002. Sex hormones from different livestock production systems: fate and potential disrupting activity in the environment. Review. *Analytica Chemica Acta* 473: 27-37.

Latini, G., Vecchio, A. D., Massaro, M., Verrotti, A. and Felice, C. D. 2006. Phthalate exposure and male infertility. Review.*Toxicology* 226: 90-98.

Lei, B., Huang, S., Zhou, Y., Wang, D. and Wang, Z. 2009. Levels of six estrogens in water and sediment from three rivers in Tianjin area, China. *Chemosphere* doi.:10.1016/j.chemosphere.2009.02.035

Levy, G., Lutz, I., Krüger, A. and Kloas, W. 2004. Bisphenol-A induces feminisation in *Xenopus laevis* tadpoles. *Environmental Research* 94: 102-111.

Liu, Z-H., Kanjo, Y. and Mizutani, S. 2009a. Removal mechanism for endocrine disrupting compounds (EDCs) in wastewater treatment-physical means, biodegradation, and chemical advanced oxidation: A review. *Science of The Total Environment* 407: 731-748.

Liu, Z-H., Kanjo, Y. and Mizutani, S. 2009b. A review of phytoestrogens: their occurrence and fate in Environment. *Water Research.* doi.:10.1016/j.watres.2009.03.025.

Lopez-Espinosa, M. J., Freire, C., Arrebola, J. P., Navea, N., Taoufiki, J., Fernandez, M. F., Ballesteros, O., Prada, R. and Olea, N. 2009. Nonylphenol and octylphenol in adipose tissue of women in southern Spain. *Chemosphere*doi.:10.1016/j.chemosphere.2009.03.063.

Manning, T. 2005. Endocrine disrupting chemicals: A review of the state of the science. *Australasian Journal of Ecotoxicology* 11 (1): 1-52.

Matozzo, V., Gagné, F., Marin, M. G., Ricciardi, F. and Blaise, C. 2008. Vitellogenin as biomarker of exposure to estrogenic compounds in aquatic invertebrates: A review. *Environmental International* 34: 531-545.

Metcalfe, C., Metcalfe, T., Kiparissis, Y., and Koenig, B. 2001. Estrogenic potency of chemicals detected in sewage treatment plant effluents as determined by *in vivo* assays with Japanese medaka (*Oryzias latipes*). *Environmental Toxicology and Chemistry* 21: 297-308.

Mihaich, E. M., Friederich, U., Caspers, N., Hall, T., Klecka, G. M., Dimond, S. S., Staples, C. A., Ortego, L. S. and Hentges, S. G. 2009. Acute and chronic toxicity testing of bisphenol A with aquatic invertebrates and plants. *Ecotoxicology and Environmental Safety* doi.:10.1016/j.ecoenv.2009.02.005.

Mills, L. J. and Chichester, C. 2005. Review of evidence: Are endocrine disrupting chemicals in the aquatic environment impacting fish populations? *Science of the Total Environment* 343: 1-34.

Mispagel, C., Shiraishi., F, Allinson, M., and Allinson, G. 2005. Estrogenic activity of treated municipal effluent from seven sewage treatment plants in Victoria, Australia. *Bulletin of Environmental Contamination & Toxicology*, 74: 853-856.

NWQMS (National Water Quality Management Strategy) 2007. Australian guidelines for water recycling: Managing health and environmental risks (phase 2). Augmentation of drinking water supplies. Natural Resource Management Ministerial Council. Environment Protection and Heritage Council. Australian Health Ministers Conference. Biotex Pty. Ltd., Canberra; 140 pp.

Neale, P. A., Escher, B. I., Schafer, A. I. 2009. pH dependent of steroid hormone-organic matter interactions at environmental concentrations. *Science of the Total Environment* 407: 1164-1173.

Oehlmann, J., Oetken, M. and Schulte-Oehlmann, U. 2008. A critical evaluation of the environmental risk assessment for plasticizers in the freshwater environment in Europe, with special emphasis on bisphenol A and endocrine disruption. *Environmental Research* 108: 140-149.

Örn, S., Holbech, H., Masdsen, T. H., Norrgren, L. and Peterson, G. I. 2003. Gonadal development and vitellogenin production in zebrafish (*Danio rerio*) exposed to ethynylestradiol and methyltestosterone. *Aquatic Toxicology* 65: 397-411.

PAC (Plastic Additive and Compounding) 2002. Health and environmental impact of phthalates. *Plastic Additives and Compounding*, June. pp. 28-29.

Pan, B., Ning, P. and Xing, B. 2008. Part V-sorption of pharmaceuticals and personal care products. *Environmental Science and pollution Research.* doi.: 10.1007/s11356-008-0052-x.

Petrovic, M., Eljarat, E., Lopez de Alda, M. J. and Barceló, D. 2004. Endocrine disrupting compounds and other emerging contaminants in the environment: a survey on new monitoring strategies and occurrence data. *Anal Bioanal Chemistry* 378: 549-562.

Plastics Europe 2006. Kunststoff und Phthalate. http://www.plasticseurope.org/DocShareNoFrame/docs/4/NCNFAIOCFMEIMIMIMPAJCIPCKKVKB1BOHYPDE3PDBNWY/PlasticsEurope/docs/DLS/Phthalate1206-20061207-011-DE-v1.pdf

Purdom, C. E., Hardiman, P. A., Bye, V. J., Eno, N. C., Tyler, C. R. and Sumpter, J. P. 1994. Estrogenic effects of effluents from sewage treatment works. *Chemical Ecology* 8: 275-285.

Ricciardi, F., Matozzo, V. and Marin, M. B. 2008. Effects of 4-nonylphenol exposure in mussels (*Mytilus galloprovincialis*) and crabs (*Carcinus aestuarii*) with particular emphasis on vitellogenin induction. *Marine Pollution Bulletin* 57: 36-372.

Roepke, T. A., Snyder, M. J. and Cherr, G. N. 2005. Estradiol and endocrine disrupting compounds adversely affect development of sea urchin embryos at environmentally relevant concentrations. *Aquatic Ecotoxicology* 71: 155-173.

Samital, T. 2008. Acute and chronic effects of emerging contaminants. *Hdb Environmental Chemistry* 5, part S/1: 105-142.doi. 10.1007/698_5_105.

Sánchez-Avila, J., Bonet, J., Velasco, G. and Lacorte, S. 2009. Determination and occurrence of phthalates, alkyl phenols, bisphenol A, PBDEs, PCBs and PAHs in an industrial sewage grid discharging to a municipal wastewater. *Science of the Total Environment* 407: 4157-4167.

Sanderson, H. and Thomsen, M. 2008. Comparative analysis of pharmaceuticals versus industrial chemicals acute toxicity classification according to the United Nations classification system for chemicals. *Toxicology Letters*, doi.:10.1016/j.toxlet.2009.02.003.

Santillo, David. 2001. "Tributyltin (TBT) Antifoulants: A Tale of Ships, Snails and Imposex", European Environmental Agency, Kgs. Nytorv 6, DK–1050 Copenhagen K, Denmark, copyright 1993-2003.

Schulte-Oehlmann, U., Oetken, M., Bachmann, J. and Oehlmann, J. 2004. Effects of ethynylestradiol and methyltestosterone in prosobranch snails. *In*: Kümmer, K (ed.). *Pharmaceuticals in the Environment: Sources, Fate, Effects and Risks*. Second edition. Berlin, Germany: Springer. pp. 233-247.

Sekizawa, J., Suter, G. and Brinbaum, L. 2003. Integrated human and ecological risk assessment: A case study of tributyltin and triphenyltin compounds. *Human and Ecological Risk Assessment*. 9(1) : 325-342.

Soares, A., Guieysse, B., Jefferson, B., Cartmell, E. and Lester, J. N. 2008. Nonylphenol in the environment: A critical review and treatment in wastewaters. *Environmental International* 34: 1033-1049.

Soto, A. M., Sonnenschein, C., Chung, K. L., Fernandez, M. F., Olea, N. and Serrao, F. O. 1995. The E-SCREEN assay as a tool to identify estrogens: an update on estrogenic environmental pollutants. *Environmental Health Perspectives* 103: 113-122.

Staples, C. A., Dorn, P. B., Klecka, G. M., O'Block, S. T. and Harris, L. R. 1998. A review of the environmental fate, effects, and exposure of bisphenol A. *Chemosphere* 36 (10): 2149-2173.

Strand, J., Jøgensen, and Tairova, Z. 2009. TBT pollution and effects in molluscs at US Virgin Islands, Caribbean Sea. *Environmental International* 35: 707-711.

Sumpter, J.P. and Jobling, S. 1995. Vitellogenin as a biomarker for estrogenic contamination of the aquatic environment. *Environmental Health and Perspectives* 103: 173-178.

Tabata, A., Kashiwada, S., Ohnishi, Y., Ishikawa, H., Miyamoto, N., Itoh, M. 2001. Estrogenic influences of estradiol-17 beta, *p*-nonylphenol and bis-phenol-A on Japanese medaka (*Oryzias lattipes*) at detected environmental concentrations. *Water Science and Technology* 43: 109-116.

USEPA (United States Environment Protection Agency) 1997. Special report on environmental endocrine disruption: an effects assessment and analysis. Washington, DC: Office of Research and Development.

Vethaak, A. D., Lahr, J., Schrap, S. M., Belforid, A. C., Rijs, G. B. J. 2005. An integrated assessment of estrogenic contamination and biological effects in the aquatic environment of the Netherlands. *Chemosphere* 59: 511-524.

van Wezel, A. P., Van Vlaardingen, P., Posthumus, R., Crommentuijin, G. H. and Sijm, D. T. H. M. 2000. Environmental risk limits for two phthalates, with special emphasis on endocrine disruptive properties. *Ecotoxicology and Environmental Safety* 46: 305-321.

Viganó, L., Benfenati, E., Van Cauwenberge, A., Eidem, J. K., Erraticio, C. 2008. Estrogenicity profile and estrogenic compounds in river sediments by chemical analysis, ELISA and yeast assays. *Chemosphere* 73: 1078-1089.

Waring, R. H. and Harris, R. M. 2005. Endocrine disruptors: A human risk. *Molecular and Cellular Endocrinology* 244: 2-9.

Williams, M., Woods, M., Kumar, A., Ying, G., Shareef, A., Karkkainen, M. and Kookana, R. 2007. *Endocrine disrupting chemicals in the Australian Riverine Environment*. A pilot study on estrogenic compounds. A joint project between CSIRO and Land and Water Australia. November 2007. CSIRO, Glen Osmond, South Australia. 90 pp.

Ye, X., Pierik, F. H., Angerer, J., Meltzer, H. M., Jaddoe, W. V., Tiemeier, H., Hoppin, J. A. and Longnecker, M. P. 2009. Levels of metabolites of organophosphates pesticides, phthalates, and bisphenol A in pooled urine specimens from pregnant women participating in the Norwegian mother and child cohort study (MoBA). *International Journal of Hygiene and Environmental Health* doi.:10.1016/j/ijheh.2009.0.004.

Ying, G-G. and Kookana, R.S. 2002. Endocrine disrupting: An Australian perspective. *Australian Water Association Journal* 29(9): 42-45.

Ying, G-G., Kookana, R. S. and Ru, Y-J. 2002. Review article-occurrence and fate of hormone steroids in the environment. *Environmental International* 28: 545-551.

Ying, G-G., Kookana, R. S. and Dillion, P. 2003. Sorption and degradation of selected five endocrine disrupting chemicals in aquifer material. *Water Research* 37: 3785-3791.

Ying, G-G., Kookana, R.S., Kumar, A. and Mortimer, M. 2009. Occurrence and implications of estrogens and xenoestrogens in sewage effluents and receiving waters from southeast Queensland. *Science of the Total Environment*. Doi:10.1016/j.scitotenv.2009.002

Zeng, F., Cui, K., Xie, Z., Wu, L., Liu, M., Sun, G., Lin, Y., Luo, D. and Zeng, Z. 2008a. Phthalate esters (PAEs): Emerging organic contaminants in agricultural soils in peri-urban areas around Guangzhou, China. *Environmental Pollution* 156: 425-434.

Zeng, F., Cui, K., Xie, Z., Wu, L., Liu, M., Sun, Lin, M., Li, Y., Lin, Y., Zeng, Z. and Li, F. 2008b. Occurrence of phthalate esters in water and sediment of urban lakes in a subtropical city, Guangzhou, South China. *Environmental International* 34: 372-380.

Zeng, F., Cui, K., Xie, Z., Wu, L., Liu, Luo, D., Chen, L., Lin, Y., Liu, M. and Sun, G. 2009. Occurrence of phthalate esters in water and sediment of urban lakes in a subtropical city, Guangzhou, South China. *Environmental International* 34: 372-380.

Zhang, X., Li, Q., Li, G., Wang, Z. and Yan, C. 2009. Levels of estrogenic compounds in Xiamen sediment, China. *Marine Pollution Bulletin*, doi,:10.1016/j.marpolbul.2009.03.011.

Websites

http://www.ourstolenfuture.org/basics/chemlist.htm - widespread pollutants with endocrine-disrupting effects- a list of endocrine-disrupting chemicals accessed on 3 July 2007

http://www.ea.gov.au/industry/chemicals/international/endocrine.html - Australian Government EDC site

http://website.lineone.net/~mwarhurst/pesticides.html - Endocrinic ally active chemicals in the environment

http://ehp.niehs.nih.gov/who- Global assessment of the state of the science of endocrine disruptors (http://www/foe.uk/resource/consultation_responses/endocrine_disrupting_foe.html

http://www.epa.gov/ORD/webpubs/endocrine/endocrine.pdf

http://www/iatp.org/edrc/hormonedisrupters082599.htm

http://products.lwa.gov.au/files/PR071403.pdf

http://www.setac.org - Society of Environmental Toxicology and Chemistry (2000). Endocrine disruptor and modulators. Florida, USA

http://www.who.int/pcs/emerg_main.html.- Global Assessment of the State-of-the Science of Endocrine disruptors. International Programme on Chemical safety.

http://en.wikipedia.org/wiki/Bisphenol_A

□□□

12 Chemical Impacts - Pharmaceuticals (Human and Veterinary)

Key facts

- Biotoxins are naturally occurring toxic compounds produced by some freshwater cyanobacteria (blue-green algae) and marine algal species (dinoflagellates, diatoms).
- Tonnes of pharmacologically active substances are used annually in human and veterinary medicines, some of these substances are excreted (human and animal) non-metabolised or an active metabolites and enter into the environment since waste water treatment plant (WWTP) cannot degrade them.
- In Europe, around 4,000 different pharmaceutically active compounds are used as human and veterinary drugs and the annual consumption is about several hundred tonnes.
- European Union and the United States laws require an assessment of potential risks to the environment as part of license application processes for new medicinal products.
- Pharmaceutical products are generally polar, water soluble and low volatile molecules, it means that they have high mobility in the environment.
- About 20 most commonly detected pharmaceuticals in environmental waters are: atenolol (*beta blocker*), benzafibrate (*lipid regulator*), carbamazepine (*antiepileptic*), clofibric acid (*metabolites of clofibrate-lipid regulator*), ciprofloxacin (*antibiotic*), diazepam (*antidepressant*), diclofenac (*analgesic/ Non-steroidal anti inflammatory drug,* (*NSAID*), erythromycin (*antibiotic*), gemfibrozil (*lipid regulator*), ibuprofen (*analgesic/NSAID*), iopromide (*contrast media*), ketoprofen (*analgesic/NSAID*), mefenamic acid (*analgesic/NSAID*), metoprolol (*beta blocker*), norfloxacin (*antibiotic*), propranolol (*beta blocker*), roxithromycin (*antibiotic*), salicyclic acid (*metabolite of aspirin*), sulfamethoxazole (*antibiotic*), trimethoprim (*antibiotic*).
- A total of about 93 pharmaceuticals (comprising 49 antibiotics and 17 NSAID/analgesic) of human and veterinary medicines have been reported from surface waters and 26 pharmaceuticals from drinking water from around the world. Five pharmaceuticals (benzafibrate, carbamazepine, gemfibrozil, ibuprofen and sulfamethoxazole) have been found in all environmental compartments (WWTP, surface waters, groundwater and drinking water).
- Pharmaceuticals have been detected in surface waters in Australia, Canada, Finland, France, Germany, India, Poland, Spain, Switzerland, Italy, UK, USA and Vietnam.
- The impact of pharmaceuticals on environment include negative effects on fish, crustaceans (daphnids), algae, bacteria, earthworms, plants and dung invertebrates. Based on various trophic levels toxicity experiments, algae and bacteria were found to be in general more sensitive to pharmaceuticals than zooplankton and fish.
- The spread of antibiotics in the environment can contribute to bacterial resistance or transfer of drugs through food chain, biomagnification. Antibiotics were recently classified as a priority risk group due to their high toxicity to algae and bacteria at low concentrations and their potential cause of resistance amongst natural bacterial populations.
- The decline of Asian vulture population in Bangladesh, India and Pakistan were reported to be due to renal failure related to analgesic pharmaceutical diclofenac used in cattle treatment. The lowest observed effect concentrations (LOEC) of diclofenac were reported to be 0.007 mg/kg in vultures.
- Pregnant women and their babies, as well as children, are at particular risk from consumption of drinking water contaminated with pharmaceuticals. Some of the drugs that are of high risk during pregnancy and breast feeding and are detected in drinking water are aspirin, atenolol, carbamazepine, clofibric acid, cyclophosphamide, diazepam, gemfibrozil, oxytetracycline, tetracycline and sulfamethoxazole.

- Relatively little is known about the presence of pharmaceuticals in the environment in particular implications for human health and ecosystems from chronic toxicity and continuous exposure to multiple compounds. Safety threshold values for pharmaceutical compounds are often related to single compound-single organism toxicity studies though the impact of a mixture of chemicals could be more toxic than that of the individual compound alone.
- To reduce pharmaceuticals impact on ecological health and environment, approaches are required that may include improvement of disposal systems, segregating sources of pharmaceuticals and changes in husbandry and treatment processes.

Contents

List of Tables

List of Figures

List of Boxes

Acronyms and Glossary

Acute toxicity: Any poisonous/harmful effect produced from a single or short exposure (24 to 96 hours) resulting in severe biological harm or death.

ACR: Acute-chronic ratio (ACR) means a standard measure of the acute toxicity of a material divided by an appropriate measure of the chronic toxicity of the same material under comparable conditions.

Antacid: An antacid is any substance, generally a base or basic salt, which counteracts stomach acidity. In other words, antacids are stomach acid neutralizers.

Analgesic: Pain killer.

Antibiotic: An antibiotic is a substance or compound (also called chemotherapeutic agent) that kills or inhibits the growth of bacteria. Antibiotics belong to the group of antimicrobial compounds used to treat infections caused by micro-organisms, including fungi and protozoa. Antibiotic and antimicrobial, these two terms are now regarded as synonymous to each other and used in this book.

Antimicrobial: An antimicrobial is a substance that kills or inhibits growth of micro-organisms such as bacteria, fungi, or protozoans, as well as destroying viruses. Antimicrobial drugs either kill microbes (microbicidal) or prevent the growth of microbes (microbistatic). Disinfectants are anti-microbial substances used on non-living objects.

Anticonvulsant: The anticonvulsants are a diverse group of pharmaceuticals used in the treatment of epileptic seizures

Antineoplastics: Any of several drugs that control or kill neoplastic cells used in chemotherapy to kill cancer cells and all have unpleasant side effects that may include nausea, vomiting, hair loss and suppression of bone marrow function.

Bank filtration: A type of filtration that works by passing water to be purified for use as drinking water through the banks of a river or lake. It is then drawn-off by extraction wells some distance away from the waterbody. The process may directly yield drinkable water or be a relatively uncomplicated way of pre-treating water for further purification.

Beta blockers: Sometimes written as â-blocker, are a class of drugs used for various indications, but particularly for the management of cardiac arrhythmias, cardioprotection after myocardial infarction (heart attack) and hypertension. Propranolol was the first clinically useful beta adrenergic receptor antagonist. Invented by Sir James W. Black, it revolutionized the medical management of angina pectoris and is considered to be one of the most important contributions to clinical medicine and pharmacology of the 20th century

Carcinogenic: The term carcinogen refers to any substance, radionuclide or radiation that is an agent directly involved in the promotion of cancer or in the increase of its propagation. This may be due to the ability to damage the genome or to the disruption of cellular metabolic processes.

CAS Number: CAS Number is a numeric designation given to a specific chemical compound by the Chemical Abstract Service.

Chronic toxicity: Is the ability of a substance or mixture of substances to cause harmful effects over an extended period, usually upon repeated or continuous exposure, sometimes lasting for the entire life of the exposed organism.

Conjugates: A chemical compound that has been formed by the joining of two or more compounds.

COX: Cyclooxygenase (COX) is an enzyme that is responsible for formation of important biological mediators called prostanoids, including prostaglandins, prostacyclin and thromboxane. Pharmacological inhibition of *COX* can provide relief from the symptoms of inflammation and pain. Non-steroidal anti-inflammatory drugs, such as aspirin and ibuprofen, exert their effects through inhibition of COX. The names "prostaglandin synthase" and "prostaglandin synthetase" are still used to refer to COX.

Cytotoxic: Toxic to cells

EC_{50}: Is the concentration of chemical that affects 50% of the population exposed (usually refers to an effect other than death).

ECOSAR: Program available from the U.S. Environmental Protection Agency (EPA) that uses structure-activity relationships to predict aquatic toxicity of chemicals based on the similarity of their structure to chemicals with previously measured aquatic toxicity.

ECOSAR Data: Lowest concentration causing toxicity for each organism group as predicted using the ECOSAR program (Sanderson *et al.* 2004).

Emerging pollutants: Compounds that are not currently covered by existing water quality regulations, have not been studied before, and are thought to be potential threats to environmental systems, human health and safety.

EROD: 7-ethoxyresorufin-*o*-deethylase; a cytochrome P-450 enzyme capable of activating procarcinogenic polycyclic hydrocarbons and halogenated aromatic hydrocarbons into mutagenic compounds. Ethoxyresorufin acts as a substrate for CYP1A1 and measurement of ethoxyresorufin o-deethylase provides a more direct method of detection for this enzyme.

LC_{50}: The concentration of a chemical/substance that is lethal for 50% of the population exposed.

Genotoxic: Genotoxicity describes a deleterious action on a cell genetic material affecting its integrity. Genotoxic substances are known to be potentially mutagenic or carcinogenic, specifically those capable of causing genetic mutation and contributing to the development of tumors. This includes both certain chemical compounds and certain types of radiation.

Glucuronidation: The reaction of a biological substance with glucuronic acid, which is an uronic acid (carboxylic acid) derived from glucose. Glucuronic acid is highly soluble in water. In the animal body, glucuronic acid is often linked to the xenobiotic metabolism of substances such as drugs, pollutants, bilirubin, androgens, estrogens, mineralocorticoids, glucocorticoids, fatty acid derivatives, retinoids and bile acids. These linkages involve glycosidic bonds and this linkage process is known as glucuronidation.

Half-life: The time required for a chemical to degrade to half of its initial concentration.

Hepatocyte: Hepatocytes make up 70-80% of the cytoplasmic mass of the liver. These cells are involved in protein synthesis, protein storage and transformation of carbohydrates, synthesis of cholesterol, bile salts and phospholipids, and detoxification, modification and excretion of exogenous and endogenous substances. The hepatocyte also initiates the formation and secretion of bile.

Hypostome: The apical portion of the polyps of hydrozoan cnidarians (like *Hydra*).

Isophosphoramide mustard (IPM) : Cytotoxic alkylating metabolite of ifosfamide (IFOS).

K_{oc}: Contaminant organic carbon partition coefficient, proportion sorbed to carbon in soil

K_{ow}: Contaminant octanol-water partition coefficient; octanol/water ratio.

Log: The logarithm of a number.

Metabolite: By-products; metabolites are the intermediates and products of metabolism. The term metabolite is usually restricted to small molecules.

ìM: Micromolar.

Moiety: In physical organic chemistry moiety is generally used to signify part of a molecule.

Molecular weight: The sum of the relative atomic masses of the constituent atoms of a molecule.

Mutagenic: Is a physical or chemical agent that changes the genetic material (usually DNA) of an organism and thus increases the frequency of mutations above the natural background level. Many mutations cause cancer hence mutagens are typically also carcinogens.

Mutation: In biology, mutations are changes to the nucleotide sequence of the genetic material of an organism.

Nanogram (ng): One billionth (1/1,000,000,000) gram. Conversion of nanogram to microgram: 1 nanogram = 0.001 micrograms or 1 microgram = 1000 nanogram.

Necrosis: The death of living cells or tissues. Necrosis can be due, for example, to ischemia (lack of blood flow).

NOEC: No observed effect concentration. No effect concentration (NEC) is a risk assessment parameter that represents the concentration of a pollutant that will not harm the species involved, with respect to the effect that is studied. It is often the starting point for environmental policy.

NSAID: Non-steroidal antiinflammatory drugs

Opioid analgesic : Also known as narcotic analgesics are pain relievers that act on the central nervous system.

Oxidative stress: Adverse effects occurring when the generation of reactive oxygen species (ROS) in a system exceeds the system's ability to neutralize and eliminate them; excess ROS can damage a cell's lipids, protein or DNA. It is caused by an imbalance between the production of reactive oxygen and a biological system's ability to readily detoxify the reactive intermediates or easily repair the resulting damage. All forms of life maintain a reducing environment within their cells. This reducing environment is preserved by enzymes that maintain the reduced state through a constant input of metabolic energy. Disturbances in this normal redox state can cause toxic effects through the production of peroxides and free radicals that damage all components of the cell, including proteins, lipids and DNA.

Peroxisome: Organelles from the microbody family and are present in almost any eukaryotic cell. They participate in the metabolism of fatty acids and many other metabolites. Peroxisomes harbour enzymes that rid the cell of toxic peroxides. Peroxisomes are bound by a single membrane that separates their contents from the cytosol (the internal fluid of the cell) and contain membrane proteins critical for various functions, such as importing proteins into the organelles and aiding in proliferation.

pH: p (potential) of H (hydrogen); provides a measure on a scale from 0 to 14 of the acidity or alkalinity of a solution.

Photosensitize: In pharmacology, a photosensitizer is a drug, food or other chemical that increases photosensitivity of the organism. This may be an adverse effect of many drugs or a desired effect in photodynamic therapy of tumours. In photochemistry, a photosensitizer is a chemical compound that readily undergoes photo excitation and then transfers its energy to other molecules, thus making the reaction mixture more sensitive to light.

Prostaglandin A prostaglandin is any member of a group of lipid compounds that are derived enzymatically from fatty acids and have important functions in the animal body. Every prostaglandin contains 20 carbon atoms, including a 5-carbon ring. They are mediators and have a variety of strong physiological effects. Although they are technically hormones, they are rarely classified as such.

Sludge: Is the residual semi-solid material left from industrial or wastewater treatment processes.

Solubility: The quantity of a particular substance that can dissolve in a particular solvent.

Subtle: Difficult to understand or distinguish.

TCEP: (*tris*(2-carboxyethyl)phosphine). A reducing agent frequently used in biochemistry and molecular biology applications.

Teratogenic: Of or relating to substances or agents that can interfere with normal embryonic development.

Toxicity: The degree to which a substance can harm humans or animals. Toxicity can be acute, subchronic or chronic. *Acute toxicity* involves harmful effects in an organism through a single or short-term exposure. *Subchronic toxicity* is the ability of a toxic substance to cause effects for more than one year but less than the life time of the exposed organism. *Chronic toxicity* is the ability of a substance or mixture of substances to cause harmful effects over an extended period, usually upon repeated or continuous exposure, sometimes lasting for the entire life of the exposed organism.

WWTP: Waste water treatment plants.

Xenobiotics: Any substance foreign to living systems. Xenobiotics include drugs, pesticides, and carcinogens. Detoxification of such substances occurs mainly in the liver.

12.1: Introduction

Pharmaceuticals can be referred to as medicines or drugs or a group of xenobiotic compounds used in the treatment, cure, prevention or diagnosis of disease or used to otherwise enhance physical or mental well-being. Pharmaceutically active compounds are molecules with different functionalities, physico-chemical and biological properties. These chemicals can be organic micro pollutants. Tonnes of pharmacologically active substances are used annually in human and veterinary medicines. Some of these substances are excreted (via human and animal excretion) non-metabolised or an active metabolites and enter into the environment directly or in sewage effluent.

Pharmaceutical compounds are now referred to as 'emerging pollutants or emerging environmental contaminant groups' (Table 12.1) since they are not currently covered by existing water quality regulations, and have not been studied extensively, and are thought to be potential threats to environmental ecosystems, human health and safety (Faràe *et al.* 2008). Research on environmental impacts of pharmaceuticals receiving much attention in recent times and is now recognised as an important field throughout the world. This is due to the occurrence and detection of human and veterinary pharmaceuticals in the European, Japanese, and North American and more recently in Australian environment including rivers, groundwater, lakes and drinking water. Pharmaceuticals such as estrogenic and androgenic pharmaceuticals (such as 17a-ethynylestradiol) are discussed in chapter 10 as endocrine disrupting chemicals (EDCs) and will not be repeated in this chapter.

Table 12.1: ***Classes of emerging compounds*** *[Farré et al. 2008].*

Compound	Examples
Drugs of abuse	Amphetamine, cocaine, tetrahydrocannabinol.
Flame retardants	C10-C13 chloroalkanes; hexabromocyclododecane; polybrominated diphenyl ethers; tetrabromo bisphenol A; tris (2-chloroethyl) phosphate.
Gasoline additives	Dialkyl ethers; methyl-t-butyl ether.
Industrial additives and agents	Chelating gents (EDTA), aromatic sulfonates.
Personal care products (Fragrances; insect repellents; soaps; antimicrobials; sun-screen agents)	Bezophenone; N,N-diethyltoluamide; methylbenzylidene; nitro, polycyclic and macrocyclic musks; triclosan.
Pharmaceuticals (see Table 12.3 for classification)	Acetaminophen (paracetamol); acetylsalicylic acid (aspirin); benzafibrate; carbamazepine; diclofenac; diazepam; iopromide; iopamidol.
Steroid hormones (see Chapter 11 for classification and effects)	Estradiol; estriol; estrone.
Surfactants and surfactant metabolites (see chapter 11 for effects)	Alkylphenol ethoxylates; alkylphenols (nonylphenol and octylphenol).
New classes	Nanomaterials; 1-4 dioxane; swimming-pool-disinfectant by-products.

12.2: Pharmaceutical industry

Pharmaceutical production is a significant global industry with worldwide sales exceeding US$ 400 billion in 2006 (Table 12.2). The industry is growing worldwide at about 8% a year. In 1998, pharmaceutical companies in the United States accounted for about 39% of all pharmaceutical products marketed worldwide followed by Japan (15.4%), Germany (7.2%) and France (5.6%). According to Berry and Rondinell (2000), the growth of this industry in the early years of 21st century is projected to be 9.8% in North America, 10.6% in the Middle East, 9.8% in Australasia, and 11% in Southeast Asia and China. The forces driving growth in the global pharmaceutical industry include a rapidly increasing elderly population in post-industrial countries, rising living standards in developing countries and expanding research and development leading to new products. Though pharmaceuticals are most effective and beneficial for human and animal health, the US EPA reported that pharmaceutical industries are also responsible for releasing a significant proportion of pollutants (air, soil and water pollutants) into the environment. Furthermore, pharmaceutically active compounds, used as human and veterinary drugs, are suspected to reach every environmental compartment.

Table 12.2: ***List of world's largest pharmaceutical and biotech companies (top 50).***
[http://en.wikipedia.org/wiki/List_of_pharmaceutical_companies]

Revenue ranking 2006	Company	Country	Total revenue (US$ million)	Healthcare R&D 2006 (US$ million)	Net income/loss 2006 (US$)	Employees 2006
1	Johnson and Johnson	US	53,324	7,125	11,053	138,000
2	Pfizer	US	48,371	7,599	19,337	122,200
3	Bayer	Germany	44,200	1,791	6,450	106,200
4	GlaxoSmithKline	UK	42,813	6,373	10,135	106,000
5	Novartis	Switzerland	37,020	5,349	7,202	102,695
6	Sanofi-Aventis	France	35,645	5,565	5,033	100,735
7	Hoffmann-La Roche	Switzerland	33,547	5,258	7,318	100,289
8	AstraZeneca	UK/Sweden	26,475	3,902	6,063	98,000
9	Merck & Co.	US	22,636	4,783	4,434	74,372
10	Abbott Laboratories	US	22,476	2,255	1,717	66,800
11	Wyeth	US	20,351	3,109	4,197	66,663
12	Bristol-Myers Squibb	US	17,914	3,067	1,585	60,000
13	Eli Lilly and Co.	US	15,691	3,129	2,663	50,060
14	Amgen	US	14,268	3,366	2,950	48,000
15	Boehringer Ingelheim	Germany	13,284	1,977	2,163	43,000
16	Schering-Plough	US	10,594	2,188	1,057	41,500
17	Baxter International	US	10,378	614	1,397	38,428
18	Takeda Pharmaceutical Co.	Japan	10,284	1,620	2,870	15,000
19	Genentech	US	9,284	1,773	2,113	33,500
20	Procter & Gamble	US	8,964	n/a	10,340	29,258
21	Teva Pharmaceutical Industries	Israel	8,408	495	546	26,670
22	Astellas Pharma	Japan	7,850	1,435	1,122	23,613
23	Daiichi Sankyo	Japan	7,158	1,459	671	20,100
24	Novo Nordisk	Denmark	6,520	1,063	1,086	15,358
25	Eisai	Japan	5,583	926	604	14,993
26	Merck KGaA	Germany	5,175	772	1,258	13,900
27	Alcon	US	4,897	512	1,348	13,500
28	Akzo Nobel	Netherlands	4,694	741	1,449	13,000
29	UCB	Belgium	4,426	1,024	492	12,741
30	Nycomed	Switzerland	4,264	n/a	-105	10,533
31	Forest Laboratories	US	3,442	941	454	9,649
32	Solvay	Belgium	3,268	533	1,026	9,000
34	Allergan	US	3,063	1,056	-127	8,423
35	Gilead Sciences	US	3,026	384	-1,190	6,772
36	CSL	Australia	2,788	161	454	6,400

Contd...

37	Chugai Pharmaceutical Co.	Japan	2,787	467	328	5,962
38	Biogen Idec	US	2,683	718	218	5,907
39	Bausch & Lomb	US	2,292	197	15	5,830
40	Taiho Pharmaceutical Co.	Japan	2,069	244	132	5,756
41	King Pharmaceuticals	US	1,989	254	289	5,191
42	Watson Pharmaceuticals	US	1,979	131	445	5,126
43	Mitsubishi Pharma	Japan	1,945	403	208	5,111
44	Shire	UK	1,797	387	278	4,958
45	Cephalon	US	1,764	403	145	4,913
46	Dainippon Sumitomo Pharma	Japan	1,763	350	193	3,750
47	Kyowa Hakko Kogyo	Japan	1,698	268	108	2,895
48	Shionogi & Co.	Japan	1,640	320	159	2,868
49	Mylan Laboratories	US	1,612	104	217	2,800
50	H. Lundbeck	Denmark	1,552	329	186	2,515

12.3: Classification of pharmaceuticals

Pharmaceuticals can be classified in various ways, e.g. by chemical properties, mode of administration, biological system affected or therapeutic effects (beneficial effects). However, most acceptable classifications are according to their anatomical therapeutical purposes (Table 12.3). On the other hand, classification according to chemical structure is used mainly for the sub-groups of the active substances. For example, antibiotics are sub-grouped into aminoglycosides, b-lactams, macrolides, sulfonamides, tetracyclines, lincosamides and fluoroquinolone (Table 12.4).

Table 12.3: ***Classification of pharmaceuticals (human and veterinary) based on therapeutic uses*** *[Mompelat et al. 2008] - Note: highlighted (in bold fonts*) pharmaceuticals are those most investigated according to Miége et al. 2009; italic compounds (shown in italic fonts) are metabolite of parent compound].*

Therapeutic class	Examples
Antacid -histamine/antiulcer	cimetidine; ranitidine
Antianginal	*dehydronifedipine* (nifedipine metabolite)
Antiasthmatic	albuterol
Antibiotic	amoxicillin; ceftriaxone; cefuroxime; chlortetracycline; **ciprofloxacin**; **clarithromycin**; **erythromycin**; erythromycin H_2O (erythromycin metabolite); flumequine; lincomycin; **norfloxacin**; oxolonic acid; oxytetracycline; penicillin G1/2 -benzathine salt; penicillin V potassium salt-demeclocycline; **roxithromycin**; **sulfamethoxazole**; tetracycline; **trimethoprim**; tylosin, virginiamycin
Anticoagulant	warfarin
Anticonvulsant/antiepileptics	**carbamazepine**; dilantin; gabapentin; primidone; *10,11-dihydro-10-hydroxycarbamazepine* (Carbamazepine metabolite)
Antidepressant-antianxiety-antipsychotic	Amitryptiline; **diazepam**; fluoxetine; gabapentin; primidone
Anti-diabetic	glibenclamide; metformin
Anti-fungal	clotrimazole

Contd...

Antihelminthic	ivermectin
Antihistamine	diphenhydramine
Antihypertensive	diltiazem; enalapril; *enalaprilat* (= enalapril maleate metabolite); valsartan
Antiinflammatory substances	See NSAID
Antineoplastic	cyclophosphamide; epirubicin/doxorubicin; ifosafamide; methotrexate; tamoxifen
Beta blocker/antihypertensor	acebutolol; **atenolol;** celiprolol; **metoprolol;** nadolol; pindolol; **propanolol**; sotalol
Birth control	ethynylestradiol (synthetic steroid hormone or EE2); 6á-methyl-17á-hydroxyprogesterone acetate
Bronchodilator	clenbuterol; salbutamol; terbutaline
Cardiac stimulant	digoxin; *digoxigenin* (digoxin metabolite)
Contrast media	diatrizoate; gadolinium (Gd-DTPA: GD-diethylenetriamine pentaacetic acid); **iopromide**-X-ray; iomeprol; iohexol; iopamidol
Diuretic	bendroflumethiazide; furosemide
Lipid regulator	atorvastatin; **benzafibrate**; clofibrate; ***clofibric acid*** (active metabolite of clofibrate); fenofibrate; *fenofibric acid* (active metabolite of fenofibrate); **gemfibrozil**; *o-and p-hydroxyatorvastatin* (atorvastain metabolite); pravastain; simvastain; *simvastain hydroxy acid* (simvastain metabolite)
Non-steroidal anti-inflammatory drugs (NSAID) and analgesic	5-aminosalicyclic acid; aminopyrine; aspirin (acetyl salicylic acid or ASA); **diclofenac**; DMAA (dimethylaminophenazone); fenoprofen; flufenamic acid; flurbiprofen; **ibuprofen**; indomethacin; **ketoprofen**; meclofenamic acid; **mefenamic acid;** naproxen; nimesulide; **paracetamol** (acetaminophen); phenazone; propyphenazone; ***salicylic acid*** (ASA active metabolite); sulfapyridine; sulfasalazine; tolfenamic acid;
NSAID: DMAA metabolites/degradation products	AAA *(acetaminoantipyrine)*, ADMOPH *(1-acetyl-1-methyl-2-dimethyloxamoyl-2-phenylhydrazine)*; AMPH *(1-acetyl-1-methyl-2 – phenylhydrazine)*; FAA *(formylaminoantipyrine)*
NSAID: phenazone degradation products	DP *(1,5-dimethyl-1,2-dehydro-3-pyrazolone)*; DMADP *(4-(N,N-dimethyl)-amino-1,5-dimethyl-1,2-dehydro-3-pyrazolone)*
NSAID: prophyphenazone degradation products	PDP *(4-(2-methylethyl)-1,5-dimethyl-1,2-dehydro-3-pyrazolone)*
Opioidanalgesic	codeine; fentanyl (synthetic); morphine; methadone (synthetic); tramadol; *6-acetylmorphine* (6-AM, morphine metabolite); EDDP, *norcodeine* (codeine metabolite); *normorphine* (morphine metabolite)
Psycho-stimulant	amphetamine; caffeine; *1,7-dimethylxanthine* (caffeine metabolite)
Steroid hormones	androsterone, estradiol (E2); estriol *(E3)*; estrone *(E1)*; progesterone, testosterone (see chapter 10 on EDC)
Vasodilatator	**pentoxyfilline**

*most investigated (highlighted pharmaceuticals) are those compounds which are highly prescribed, continuously discharged in the environment and could be potentially toxic.

Table 12.4: ***Antibiotic classes and important antibiotics in human and animal medicines*** *[Serrano 2005; Kemper 2008]*

Antibiotic classes	Human medicine	Veterinary medicine
Aminoglycosides	Sisomycin Gentamycin/gentamicin	Apramycin – pigs only Gentamycin – all animals Kanamycin - cattle, dogs, horses, pigs Neomycin - all animals including aquaculture Spectinomycin - cattle, sheep, pigs, poultry
β-Lactams: penicillins	Azlocillin Flucloxacillin Methicillin Mezlocillin	Amoxicillin – all animals including aquaculture Ampicillin – all animals including aquaculture Benzylpenicillin – all animals Cloxacillin – cattle

Contd...

	Nafcillin Piperacillin Phenoxymethylcillin Penicillin-G	Dicloxacillin –cattle Oxacillin - cattle
Cephalosporines	Cefalotin Cefazolin Cefotaxim Cefotiam	Cefalexin – dogs Ceftiofur – cattle, pigs Cefquinom – cattle, pigs
Fenicoles	-	Chloramphenicol – cats, dogs (banned for aquaculture) Florfenicol premix – aquaculture/fish
Fluoroquinolones/fluorochinolones	Ciprofloxacin Flumequin Ofloxacin	Enrofloxacin – all animals Marbofloxacin – all animals
Lincosamides	Clindamycin	Clindamycin – dogs Lincomycin – pigs, cats, dogs, cattle
Macrolides	Azithromycin Clarithromycin Erythromycin Roxithromycin Vancomycin	Erythromycin – cattle, chicken, aquaculture/fish Spiramycin – all animals including aquaculture Tylosin -animals
Quinolones	-	Enrofloxacin – aquaculture Flumequine – aquaculture Oxolinic acid - aquaculture
Sulfonamides	Sulphanilamide Sulphamethoxazole Sulphathiazole	Sulphadimethoxine - cattle, chicken, pigs, aquaculture Sulphadimidine - -cattle, chicken, sheep Sulphapyridine - pigs
Trimethoprim	-	In combination with sulphonamides
Tetracyclines	Deoxycycline Oxytetracycline Tetracycline	Chlortetracycline – cattle, pigs, aquaculture Deoxycycline – cats , dogs, aquaculture Oxytetracycline - cattle, sheep, pigs, aquaculture/fish Tetracycline - horse, sheep, pigs
Sarafloxacin	-	Aquaculture/fish

12.4: Usage of pharmaceuticals *[Based on Mompelat et al. 2008]*

The consumption of pharmaceutical products in North American countries (Canada, USA, and Mexico) is highest, followed by European Union (see Table 12.5 for top prescription medicines used in Australia, Canada, UK and the USA). In Europe, around 4,000 different pharmaceutically active compounds are used as human and veterinary drugs and the annual consumption is several hundred tons. For instance, NSAID like aspirin (acetylsalicylic acid), paracetamol, ibuprofen, and diclofenac were produced in Germany in 2001 respectively at 836, 622, 345 and 86 t, while metformin (antidiabetic) and carbamazepine (anticonvulsant) were produced respectively at 517 and 88 t.

Among all the pharmaceuticals, antibiotics are the main ones that are consumed by humans, and are also used in poultry production and fish/aquaculture farming (Table 12.4). As a result, antibiotic consumption can reach upto several thousand tons per year because of their extensive use in livestock farming and human medicine. For example, in 1999 the EU and Switzerland used more than 13,000 t of antibiotics (65% as human use), France being the highest consumer in Europe. A comparison of human and animal microbial usages in France (1999-2005) showed that β-lactams and macrolides are predominantly used in human medicines and tetracyclines and sulfonamides are predominantly used in veterinary medicines.

Table 12.5: *Top prescription medicines (classes or actual drugs) in some selected countries.*

Australia[1] *PBS/RPBS [prescription medicines distributed through community pharmacies, 2001]*	**Canada**[2] *[Based on prescriptions (000s) dispensed in 2007]*	**UK**[3] *[Based on defined daily dose, 2007]*	**USA**[4] *[by number of prescriptions dispensed in 2005]*
1. Atorvastatin (*cholesterol lowering*) (5,189) 2. Simvastatin (*cholesterol lowering*) (4,984) 3. Paracetamol (*analgesic*) (4,775) 4. Salbutamol (*asthma treatment*) (3,585) 5. Amoxicillin (*antibiotic*) (2,393) 6. Codeine with paracetamol (*strong analgesic*) (2,985) 7. Ranitidine hydrochloride (*peptic ulcer treatment*) (3,324) 8. Celecoxib (*osteoarthritis and rheumatoid arthritis*) (3,770) 9. Atenolol (*blood pressure reduction*) (2,744) 10. Temazepam (*for sleep*) (2,787)	1. Cardiovasculars (65,697) 2. Psychotherapeutics (53,268) 3. Gastrointestinal/Genitourinary (27,315) 4. Cholesterol agents (26,588) 5. Hormones (24,611) 6. Anti-infectives systemic (24,239) 7. Analgesics (20,410) 8. Diabetes therapy (18,141) 9. Diuretics (16,351) 10. Neurological disorders (16,127)	1. Simvastatin (32,967 kg) (*lipid regulator)* 2. Ramipril (3,475 kg) 3. Aspirin (1,147 kg) *(NSAID)* 4. Atorvastain (10,625 kg) *(lipid regulator)* 5. Folic acid (357 kg) 6. Bandroflumethiazide (1,967 kg) 7. Amiodipine (3,502 kg) 8. Salbutamol (484 kg) 9. Omeprazole (8,851 kg) 10. Lisinopril (4,076 kg) 11. Atenolol (29,919 kg) *(beta blocker)* 12. Furosemide (15,172 kg) 13. Levothyroxine (53 kg) 14. Lansoprazol (9,636 kg) 15. Metformin (573,837 kg) 16. Diclofenac sodium (28,572 kg) *(NSAID)* 17. Fluticasone propionate (161 kg) 18. Ferrous sulfate (52,094 kg) 20. Paracetamol (769,498 kg) *(NSAID)*	1. Paxil (paroxetine hydrochloride) *(antidepressant drugs* 2. Lexapro (escitalopram oxalate) antidepressant drugs) 3. Lortab (hydrocodone bitartrate and acetaminophen) *(opioid analgesic and antitussive)* 4. Xanax (CIV alprazolam) *(central nervous system-active compounds)* 5. Ultram (tramadol hydrochloride) *(centrally acting analgesic)* 6. Vicodin (hydrocodone bitartrate and acetaminophen) *(opioid analgesic and antitussive)* 7. Lyrica (pregabalin) capsules, CV 8. Roxicodone (oxycodone hydrochloride) 9. Prinivil (Lisinopril) *(angiotensin converting enzyme inhibitor)* 10. Cymbalta (duloxetine hydrochloride) Delayed-release Capsules

1. Commonwealth Department of Health and Ageing-*http://www.aihw.gov.au/publications/phe/sdua02/sdua02-c05.pdf*
2. 2008 IMS health Canada- *http://www.aihw.gov.au/publications/phe/sdua02/sdua02-c05.pdf*
3. Richman and Castensson 2008
4. *http://www.xrlist.com/script/main/art.asp?articlekey=7909 (top 300)*

12.5: Environmental sources of pharmaceuticals

[based on Zuccato et al. 2006; Hernando et al. 2006; Lin et al. 2008; Faràe et al. 2008; Mompelat et al. 2008; Richman and Castensson 2008].

Pharmaceuticals can enter into the environment (surface water, groundwater) from treatment of humans via different routes, for example, from households, hospitals, schools, retirement facilities. Improper disposal of unused or expired drugs thrown in garbage, toilets or sewerage may end up in landfills, which are the local points of contaminations. Audit has shown that a significant percent of waste medicines (unused or expired) are disposed-off in dustbins/domestic garbage (> 45%) and 10-28% are disposed directly to sewerage (Table 12.6 and 12.7).

The pharmaceutical ingredients, metabolites and conjugates are mainly excreted in urine or faeces, from where they enter municipal sewage treatment plants where they can be degraded, adsorbed to sewage sludge or eventually diluted into surface water (Figure 12.1). One of the most important sources of pharmaceutical products (PPs) in aquatic environments is the effluents from wastewater treatment plants (WWTPs) since PPs are not completely removed during treatment. For example, Hernando *et al.* (2006) reported that sewage treatment plants (STP) partially removed carbamazepine, atenolol, metoprolol, trimethoprim or diclofenac (<10% for most of them and 10-39% for diclofenac). Therefore, those compounds that can not be removed by sewage treatment can be found in surface water (see also section 12.10 management of pharmaceuticals).

Table 12.6: ***Pharmaceutical disposal practices in the USA*** *[Kotchen et al. 2009].*

Disposal method	Respondent (%)
Trash	45.2
Toilet/sink	28.0
Pharmacy	5.9
Hazardous waste centre	5.1
Store at home	11.8
Other	4

Table 12.7: ***Disposal characteristics of different drug types in the UK*** *[Bound et al. 2006].*

Drug type	Bin	Sink/toilet	Return to Pharmacy	other
Painkiller	69.6	10.9	18.5	1
Antihistamine	75.3	9.1	14.3	1.3
antibiotic	71.4	3.6	14.3	10.7
Antiepileptic	100	-	-	-
b-blocker	66.7	16.7	16.7	-
Hormone	75	-	25	-
Lipid regulator	66.7	-	-	33.3
Antidepressant	66.7	-	33.3	-

Veterinary pharmaceuticals can enter into the environment via different pathways. These include emissions during manufacture, formulation, treatment processes, and/or disposal of unused medicines and disposal of containers. The most important routes of entry into the environment are direct discharge of fish/aquaculture products (via use of medicines or as fish farm effluents), excretion of substances in urine and faeces of livestock animals, and the wash-off of topical treatments from livestock animals. It can enter indirectly via the application of manure containing excreted products (manure, slurry) to land (Figure 12.1). The main by-products from cattle feed lots are manure and liquid effluents. Run-off from animal feed-lots could also increase the amount of livestock chemicals input into surface waters and groundwater (from leaching). Livestock direct access to open land and water would also contribute additional chemicals to the environment by urine and faeces. Aquaculture is an increasingly significant contribution to veterinary antibiotic transfer to the environment.

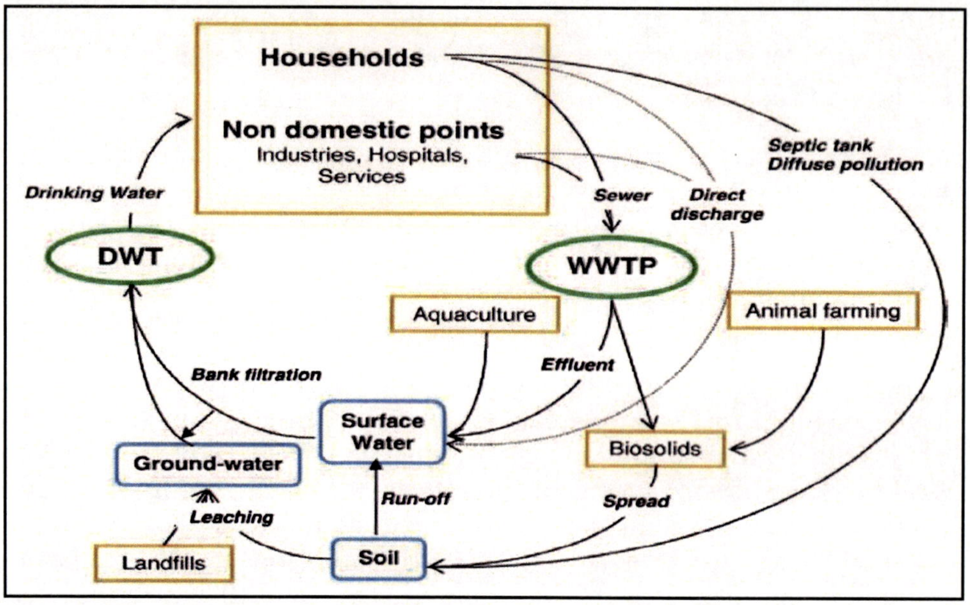

Figure 12.1: ***Origin and routes of human and veterinary pharmaceutical products contaminating the environment*** *[Mompelat et al. 2008]*. DWT= drinking water treatments; WWTP = wastewater treatment plants

Compounds that adsorbed to sludge can reach the terrestrial environment when sludge is used as agricultural fertilizer. In addition, leakage from sewer networks can be an additional source of pharmaceuticals reaching the aquatic environment. In a recent study in Taiwan (Lin *et al.* 2008) documented that hospitals accounts for 37.5% of human pharmaceutical contamination sources, followed by animal husbandry (27.9%) and drug production facilities (23.4%). Antibiotics were the predominant pharmaceuticals in all hospital and animal husbandry effluents which are as follows:

- Hospital effluents: antibiotics 39.3% and NSAIDs 37.7%; psycho-stimulants 11.2% and estrogens 7.2%.
- Animal husbandry effluents: antibiotics 88.5%; estrogens 8.2%.

12.6: Physico-chemical properties and fate of pharmaceuticals

[Based on Halling-Sørensen 1998; Zuccato et al. 2006; Díaz-Cruz et al. 2003; Mompelat et al. 2008]

12.6.1: Fate of pharmaceuticals

Pharmaceutical products (PPs) and their by-products undergo biotic, chemical and physico-chemical transformation in water, though they are designed to resist microbial degradations and to be chemically stable. Abiotic PPs and their residues are most likely eliminated in surface water via direct and indirect photo-degradation. The photo-degradation of numerous PPs include solar absorption or indirectly through radicals (OH) generated by the solar irradiation of photosensitizers like nitrate and humic acids. However, photo-degradation also depends on the intensity of solar irradiation, eutrophic conditions and depth of watercourse and composition of organic matter, latitude and season. Fluoxetine, diclofenac, triclosan, albuterol, diazepam, fenofibrate, iodoarene and fluoroquinolones are photo-reactive substances. Substances like tetracyclines can not be photo degraded because of its adsorption onto sediments. In fact, photo-degradation as well as biodegradation may be limited by the adsorption of PPs in sediments or the complexation with dissolved organic matter (DOM), sorption into the shallow layer of the soil or during bank filtration. A number of pharmaceuticals are either non-degradable or remain in the environment after extended period of time (see Table 12.8 for fate of medical substances in the environment). Sorptions of pharmaceutical products depend on physico-chemical properties (see Table 12.9 for physico-chemical properties and Figure 12.2 for chemical structure of pharmaceuticals) and also geology, soil texture (e.g. sand, clay), sediment and dissolved organic matter (DOM). Generally, the degradation of most xenobiotic is faster and more complete under aerobic as compared to anaerobic conditions.

Table 12.8: ***Fate of some selected human and veterinary medical compounds in the environment (soils, manure and sewage)*** *[compiled from Halling-Sørensen et al. 1998; Thiele-Bruhn 2003]; $T_{1/2}$ = environmental half life; Vet = veterinary medicine].*

Pharmaceuticals (active ingredient)	Drug category	Concentrations	Fate/degradation	Conditions
Ampicillin	antibiotic	-	48% biodegradable	sewage treatment
Aspirin	analgesic (pain killer)	-	readily biodegradable	sewage treatment
Avermectin B1 (vet)	antiparasitic agent	-	$T_{1/2}$ = 14-28 days	soils
Bacitracin (vet)	antibiotic-growth promoter	-	$T_{1/2}$ = 22.5 days (20°C)	soils, faeces
Caffeine	stimulants	-	readily biodegradable	sewage treatment
Ceftiofur sodium (vet)	promoter antibiotic (µ-lactams)	-	$T_{1/2}$ 22.2 days 50% degraded; 22.2 days	aerobic degradation in soils; clay loam, sand and silty clay loam clay loam
Chlortetracycline (vet)	antibiotic-growth promoter (sulfonamides)	- 5.6 µg/g 5.6 µg/g 4.7 µg/kg	-at 30°C after 30 days 44% remained; -at 20°C and 4°C 88% and 100% persistent after 30 days 24% degraded; 84 days 88% degraded; 30 days 0% degraded	- Cattle manure Sandy loam. pH 6.1, and manure soil; ca 180 days
Ciprofloxacin	antibiotic (fluorqinolones)	10 µg/g 250 µg/g	2.2-35.3% degraded; 56 days 50% degraded; 1.6-2.5 days	mineral., media+wood rotting, fungi sewage sludge
Clofibrate	lipid lowering agent	-	non-degradable	sewage treatment
Codeine phosphate	narcotic analgesic	-	non-degradable	sewage treatment
Cyclophosphamide	antineoplastic agent (chemotherapy)	-	not degradable	sewage treatment
Enrofloxacin	antibiotic (fluorqinolones)	10 µg/g 10 µg/g	30.3% degraded 0.1-12.8% degraded; 56 days	sandy loam cattle manure + basidomycetes
Erythromycin	antibiotic (macrolides)	 5.6 µg/g	non-degradable 25% degraded; 30 days	sewage treatment sandy loam; pH 6.1+ manure
Flumequine	antibiotic	-	$T_{1/2}$ = 150 days	surface sediment
Ibuprofen	analgesic (pain killer)	-	inherently biodegradable	sewage treatment
Ifosfamide	antineoplastic agent	-	not degradable	sewage treatment
Ivermectin (vet)	antiparasitic agent	-	persistence in dung (<6 days)	field conditions
Monesin (vet)	antibiotic-growth promoter (polyether)	- -	persistent under anaerobic conditions-60-70% unchanged after 10 weeks 30-40% degraded; 70 days	- manure, anaerobic

Contd...

Naproxen	analgesic (pain killer)	-	not degradable	sewage treatment
Oxolinic acid (vet)	antibiotic-feed additive		$T_{1/2}$ = 150-1,000 days	sediment
Oxytetracycline (vet)	antibiotic-feed additive fish farms (sulfonamides)	-	$T_{1/2}$ = 30 to 142 days 0% degraded;180 days 50% degraded, 43.8 days	surface sediment soil and manure sediment slurry and aerobic
Paracetamol	analgesic	-	readily degradable	sewage treatment
Streptomycin	antibiotic (aminoglycosides)	 5.6 μg/g	complete adsorption to clay fraction of the soil 0% degraded; 30 days	faeces (5%)/ soil matrix sandy loam/pH 6.1 + manure
Penicillin	β-lactams	5.6 μg/g	0% degraded; 30 days	sandy loam/pH 6.1+manure
Sulfanilamide	antibiotic (sulfonamides)	250-1000 μg/l 250 1000 μg/l	-0% degraded; 14 days -0%/50% degraded	-various soils -manure/sludge
Sulfadiazine	antibiotic (sulfonamides)		0%/50% degraded; 28 days/3.8 days	manure/sludge
Sulfamethoxazole	antibiotic (sulfonamides)	-	not degradable	sewage treatment
Sulfasalazine	antibiotic	-	not degradable	sewage treatment
Tetracycline	antibiotic (tetracyclines)	- 50-300 μg/kg 10 μg/kg 10 μg/kg 20-100 μg/l 10 μg/kg 10 μg/kg	not degradable 0% degraded; ca 180 days 65% degraded; ca 180 days 100% degraded; 84 days 50% degraded; 55-105 days -50% degraded; 4.5 days -50% degraded; 9 days	sewage treatment soil poultry manure manure + soil pig manure -pig manure with aeration -pig manure with aeration
Theobromine	hypertension	-	Readily degradable after acclimatisation	sewage treatment
Theophylline	psychomotor stimulants		Readily degradable	sewage treatment
Tolbutamide	hypoglycaemic agent		not degradable	sewage treatment
Trimethoprim	antibiotic (sulfonamides)	500 μg/l	50%degraded; 22-41 days	sewage sludge
Tylosin (vet)	antibiotic-growth promoter (macrolides)	- 5.6 μg/g 100 μg/g	40% of the compound remain active after 30 days of incubation (4°C) 0% degraded; 30 days -50% degraded; 4.2 days -50% degraded; 5.7 days	faeces (5%)/soil matrix sandy loam/pH 6.1 + manure -slurry+ sand; pH 6.3; 1.4% OC -slurry+ sandy loam; pH 6.8; 1.6% OC

Pharmaceutical products are generally polar, water soluble and low volatile molecules. They have high mobility in environment and moderate retention on sediments and sludges. The mobility of organic compounds such as, pharmaceuticals can be classified in 3 log classes of the n-octanol water distribution coefficient (log K_{ow}):

- Low sorption: log K_{ow} <2.5
- Medium sorption: log K_{ow} between 2.5 and 4
- High sorption: log K_{ow} >4.0 (see Table 12.10)

Few pharmaceuticals have a log K_{ow} values >4 (Table 12.9), many pharmaceuticals are weak acids/ bases (Table 12.9) therefore, subject to ionisation and exist as ionised moiety under conditions of ambient pH. There is a great variability between PPs as regards to their pKa values (see Table 12.9). NSAIDs, lipid regulators are mostly acidic compounds, whereas beta-blockers, bronchodilators, hormones, psycho-stimulants, opioidanalgesic, paracetamol, antibiotic phenols and macrolides are mostly basic compounds.

12.6.2: *Persistence of pharmaceuticals* [*Díaz-Cruz et al. 2003; Zuccato et al. 2006*].

According to Díaz-Cruz *et al.* (2003), three factors can determine whether a pharmaceutical compound is likely to concentrate in the aquatic, terrestrial or atmospheric environment, these are:

- Octanol/water partition coefficient
- Dissociation constants
- Vapour pressure/Henrys Law of constant

Furthermore, the persistence of drug in a sediment or soil mostly depends on its photo-stability, binding and partition coefficient, dissociation constants and adsorption capability, degradation rate, and leaching in water. Strongly adsorbed pharmaceuticals tend to accumulate in soil or sediment, whereas highly mobile pharmaceuticals tend to leach into groundwater and be transported with groundwater, drainage water and surface run-off to surface waters. In the terrestrial environment, the persistence of antibiotics ranges from less than one day to weeks or months. Despite the polar nature, some drugs do persist in the environment for more than a year (*e.g.* clofibric acid- the main active metabolite of the blood lipid regulators; clofibrate can persist for 21 years). Clofibric acid sometimes referred to as persistent organic pollutants or persistent bio-accumulative toxicants. The environmental persistence of erythromycin, cyclophosphamide, naproxen, sulfamethoxazole or sulfasalazine is longer than one year.

Samuelsen *et al.* (1994) conducted laboratory studies to determine the persistence of pharmaceuticals (antibiotics) in an artificial marine environment, the findings of the research are summarised below:

- Oxytetracycline (OTC) found highly persistent in the marine sediments (^{3}10 months).
- Oxolinic acid (OXA), flumequine (FLU) and sulfadiazine (SDZ) retain their antimicrobial activity in the sediment for at least 180 days.
- OTC, OXA, FLU and SDZ do not degrade.
- Sulfadimethoxine (SMX) and ormethoprim (OMP) are short lived (<21-62 days).

Yamamoto *et al.* (2008) conducted laboratory based experiments with eight pharmaceuticals namely acetaminophen, atenolol, carbamazepine, ibuprofen, ifenprodil, indomethacin, mefenamic acid and propranolol to examine persistence and partitioning in the aquatic environment. Propranolol, indomethacin and ifenprodil were easily photo-degraded, whereas other five pharmaceuticals were relatively persistent against sunlight. This suggests that some pharmaceuticals may remain intact once released in the environment.

12.6.3: *Excretion of pharmaceutical products* [*Mompelat et al 2008*]

Pharmaceutical by-products (metabolites) are excreted via urine or faeces. Once in the human/ animal body, the parent molecule or ingested molecule of a pharmaceutical substance undergoes a set of biochemical reactions, of which the elimination of pharmaceutical products in urine or faeces is driven by hepatic metabolism (steps include oxidation, reduction and hydrolytic cleavage) leading to more polar molecules than the parent compounds. A second conjugation step followed consists of transferring polar group on parent compounds or metabolites, such as glucuronidation, sulphoration and acetylation, allowing metabolites to become more hydrophilic and water soluble and more easily eliminated through urine and/or faeces.

Table 12.9: ***Pharmaceuticals and their physico-chemical properties*** *[Daughton and Ternes 1999; Thiele-Bruhn 2003; Jjemba 2006; Chenxi et al. 2008; Gibs et al. 2007; Pan et al. 2008; Sanderson and Thomsen 2008; Zorita et al. 2009].*

Pharmaceuticals (Active ingredient) MW (molecular weight)	Drug category	Environmental half life (t 1/2)	Log K_{ow}	Log K_{oc} l/kg	pKa*	Solubility in water
Amoxicillin	Antibiotic	-	0.97	-	2.4/7.4/9.6	3.43 mg/ml
Aspirin/acetylsalicylic acid MW:180.2	Analgesic (pain killer)/NSAID	-	1.13	-	3.5	
Atenolol (basic)	Beta blocker/ Cardiovascular	1.88 days	-0.03	-	9.6	13.3 mg/ml
Atrovastatin MW:1155.4	Anticholesterol	-		-	4.46	-
Benzafibrate (acidic) MW:361.8	Lipid regulator	-		-	3.29	
Caffeine	Respiratory/stimulant	1.5±0.4 days	-0.09	--	10.4	21.6 mg/mg
Carbamazepine (basic) MW:236.3	Anticonvulsant/ antiepileptic	-	2.25		7-13.9	17.66 mg/l
Ceftriaxone	Antibiotic	-		-	3/3.2/4.1	400 mg/ml
Chlortetracycline	Antibiotic	-	-0.62	-	3.3/3.3/7.4/9.3	630 mg/l
Ciprofloxacin MW : 331.3	Antibiotic	1.6-2.5 days	0.28	1127-61000	3.01/6.14/8.7/10.58	0.7 g/l
Clarithromycin	Antibiotic	-	3.18	-	8.7	-
Clofibrate (acidic) MW:242.70	Lipid regulator	≥365 days	3.62	-	3	Soluble in common solvents but not in water
Clofibric acid MW:214.66	Lipid regulator	-	2.72	-	3.18	1000 g/l
Cyclophosphamide MW:261.09	Cytotoxic/antineoplastic chemotherapy	-	0.2-1.77	-		10-50 g/l
Diazepam MW:284.7	Antidepressant-antianxiety	-	2.58	-	3.4	-
Diclofenac MW:296.2	Analgesic/NSAID	-	4.51	-	4.15	8.4 g/l
Diclofenac sodium (acidic)	Central nervous system (analgesic)	-	0.7	-	4	2.43 mg/ml
Doxycycline	Antibiotic	-	-0.02	-	3.0/7.97/9.15	Soluble in water
Erythromycin (basic)	Antibiotic	-	3.06	-	8.9	
Erythromycin base (basic)	Antibiotic	≥365 days	3.06	-	8.90	0.00144 mg/ml
Fenofibrate MW:360.84	Lipid regulator	-	5.19	-	-	Insoluble in water
Fluoxetine MW:345.8	Antidepressant	-	4.35	-	10.5	2.8 g/l
Gemfibrozil (acidic) MW:250.34	Lipid regulator	-	4.77	-	4.75	-

Contd...

Ibuprofen (acidic) MW:206.3	Analgesic/NSAID	-	3.79	-	4.4/5.2/4.9	21 g/l
Indomethacin MW:357	Analgesic/NSAID	-	4.27	-	4.5	0.000937 mg/ml
Ivermectin	Antihelminthic	-		-	-	-
Meclofenamic acid MW:295	Anti-inflammatory	-	5.12	-	4.2	-
Metoprolol MW:267.37	Beta blocker	-		-	9.17	-
Metformin hydrochloride	Hormone	-	-1.25	-	12.4	Soluble in water
Monensin MW:692	Growth promoter	-	2.75-3.89	-	6.65	
Naproxen (acidic) MW:230.26	Analgesic/NSAID	> 365 days	3.18	-	4.15	4.6 g/l
Ofloxacin	Antibiotic	-	0	-	-	Soluble in aqueous solution
Oxytetracycline	Antibiotic)	101-364 days	-2.08	195-93320	3.27/7.32/9.11	313 mg/l
Paracetamol/acetaminophen MW:151.2	Analgesic/NSAID	-	0.51	-	9.4	-
Propranolol MW:259	Beta blocker	-	2.6	-	9.5	-
Ranitidine MW:314.4	Antacid/antiulcer	160 hours	0.29	-	-	Soluble in water
Sulfamethoxazole MW:253.2	Antibiotic	≥365 days	0.89	-	6/5.6	0.61 mg/ml
Tetracyclines MW:444.4	Antibiotic	-	-1.33	-	3.32/7.78/9.6	230-52000 mg/l
Tetracycline hydrochloride	Antibiotic	> 365 days	-1.4	-	3.32/7.78/9.6	-
Tramadol hydrochloride	Central nervous system (analgesic)	-		-	9.41/8.3/9.42	Readily soluble
Triclosan MW:289.54		-	4.8	110-7990	8.1	-
Tylosin	Antibiotic	-	1.05	550-7990	-	5000 mg/l
Trimethoprim (Basic) MW:290.3	Antibiotic	20-100 days	0.64	205	7.2/6.6	0.3-0.4 mg/ml
Warfarin	anticoagulant	-	-0.84	-	-	

* pKa = An acid dissociation constant; K_a, (also known as acidity constant, or acid-ionization constant), is a quantitative measure of the strength of an acid in solution. It is the equilibrium constant for a chemical reaction known as dissociation in the context of acid-base reactions.

Based on excretion rates, the pharmaceutical products can be classified into four excretion rates:

- Low excretion or unchanged form (<5%),
- Moderate excretion (6-39%),
- Relatively high excretion (40-69%)
- High excretion (³70%) (see Table 12.11)

Most antibiotics do not accumulate in the organism and after medication; they are mostly excreted as parent compounds (70-90%) (e.g. amoxicillin, ciprofloxacin, doxycycline tetracycline, trimethoprim), whereas antidepressants (e.g. diazepam, diltiazem) are lowly excreted (£5%) and NSAIDs (e.g. ibuprofen, paracetamol) poorly excreted (£15%). As a result, municipal sewage treatment networks become one of the most important sources of pharmaceuticals emission to the environment.

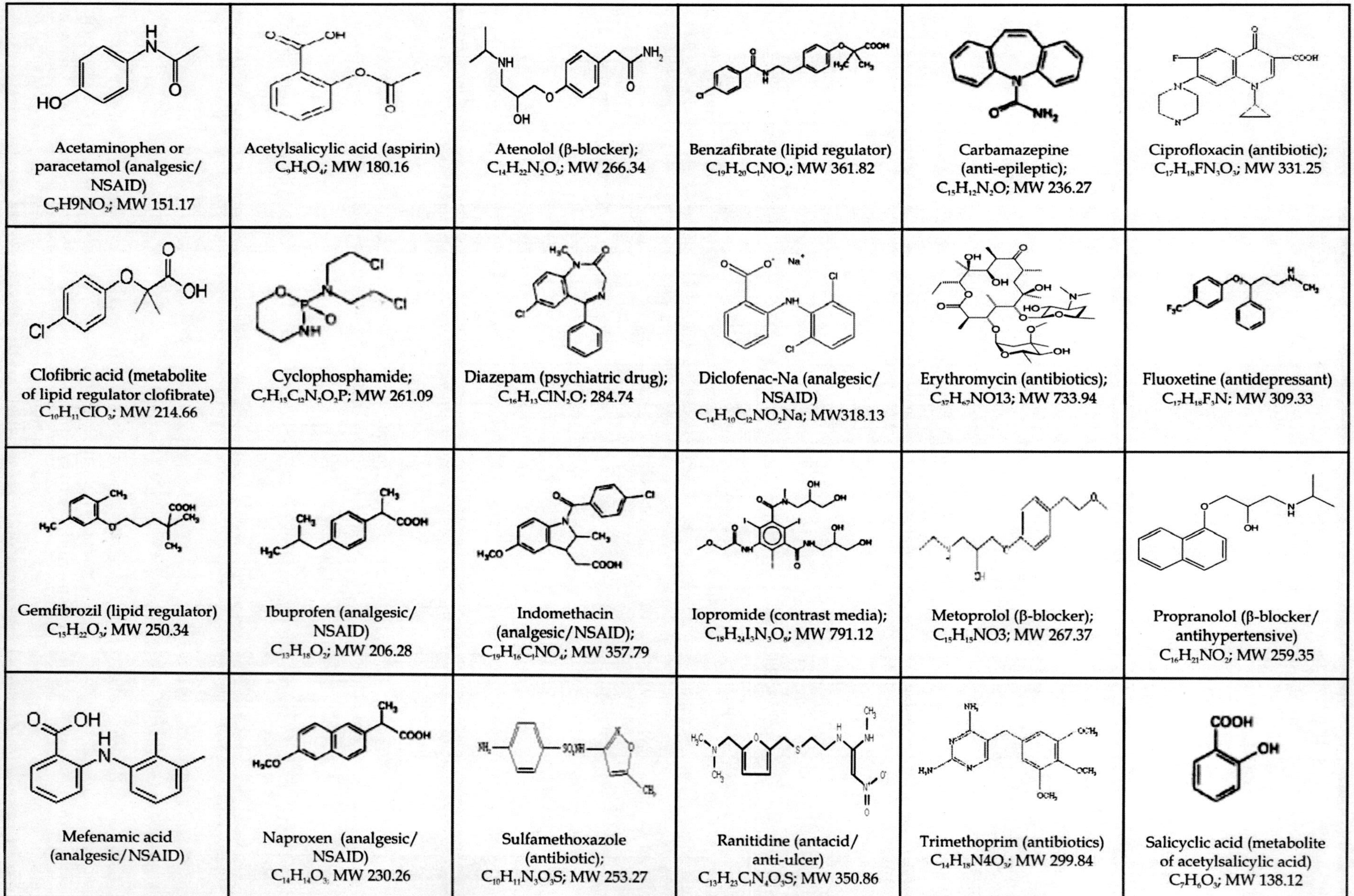

Figure 12.2: ***Structures of some selected pharmaceuticals*** *[Daughton and Ternes 1999; Yamamoto et al. 2008; Feitosa-Felizzola and Chiron 2009; Lavén et al. 2009; http://commons.wikimedia.org/wiki/File:Iopromide.png]*

Table 12.10: ***Classification of pharmaceutical products according to log K_{ow}*** *[Mompelat et al. 2008; chemlDplus-http://chem.sis.nlm.gov/chemidplus].*

Therapeutic class	Log $K_{ow} \leq 2.5$	Log K_{ow} 2.5 - 4	Log K_{ow} >4
Antacid -histamine	cimetidine; ranitidine	-	-
Antibiotic classes	beta-lactams; cyclines fluoroquinolones; lincosamide; quinolones; phenicol; pyrimidines; sulfonamides, others	macrolides	phenol (triclosan)
Anticoagulants	-	warfarin	-
Anticonvulsants	dilantin, gabaentin	carbamazepine	-
Antidepressants	meprobamate; oxazepam	diazipam	amitryptiline
Antihelminthics	-	ivermectin	-
Antihistamines	-	diphennydramine	-
Antihypertensives	-	diltiazem	valsartan
Antineoplastics	5-fluorouracil; cyclophosphamide; doxorubicin	-	-
β-blocker	acebutolol; atenolol; metoprolol; pindolol; sotaol	propanolol	-
Bronchodilatators	clenbuterol; salbutamol; terbutaline	-	-
Contraceptives	-	ethynylestradiol	-
Cardiac stimulant	digoxin	-	-
Contrast media	iopamidol; iopromide X-ray; lomeprol	-	-
Diuretics	bendroflumethiazide	-	-
Lipid regulator	-	clofibric acid	benzafibrate; gemfibrozil
NSAIDs and analgesic	5-aminosalicyclic acid; aminopyrine; aspirin; paracetamol; phenazone; propyphenazone	diclofenac; fenoprofen; ibuprofen; ketoprofen; naproxen; tramadol	indomethacin; tolfenamic acid
Opioanalgesics	amphetamine; caffeine; codeine	tramadol	-
Steroid hormones	estriol	androstenedione; estrone; testosterone	estradiol; progesterone
Vasodilatator	pentoxyfilline	-	-

12.7: Environmental risk assessment of pharmaceuticals

[Carlsson et al. 2006; EMEA 2006; Richman and Castensson 2008; Ågerstrand and Rudén 2009].

European Union and the US laws require an assessment of potential risks of pharmaceuticals to the environment as part of license application processes for new medicinal products. European Medicines Evaluating Agency (EMEA) for Environmental Risk Assessment (ERA) came into effect on 1st December 2006. The EMEA as well as the Food and Drug Administration (FDA, 1998) of the USA employ tiered approach (two tiered: tier-I and tier-II) for environmental risk assessment (ERA) of new and existing chemicals. ERA procedures for pharmaceuticals follow the same principles of ERAs as applied for chemicals. It is a step-wise tiered procedure starting with rough estimates and progressing to more elaborate refined methods if a potential risk cannot be excluded. The assessment may be terminated either when sufficient information is available indicating that the product/compound is unlikely to represent an environmental risk or when a risk has been identified and sufficiently characterised. A risk quotient or hazard quotient (RQ/HQ) is usually calculated from a predicted environmental concentration (PEC) or measured environmental concentration (MEC), respectively and a predicted no effect concentration (PNEC). The RQ or HQ is the basic principle internationally accepted and adopted in the development of environmental risk assessment (ERA) guidelines.

The PEC value is calculated by a formula following the basic principles of the EMEA guideline for environmental risk assessment of pharmaceuticals (EMEA 2006). For PEC derivation several factors, including the predicted market volume, the water consumption of the target population or a dilution factor accounting for dilution of effluent when reaching the surface waters are some of common parameters for this approach. The first-tier procedure also uses the levels of concern or measured environmental concentrations (MEC), the concentration range at which a chemical produces adverse effects on organisms. An example of PEC calculation is given below following Ågerstrand and Rudén (2009):

$$PEC = \frac{A \times 1\ 000\ 000\ 000 \times (100\text{-}R)}{365 \times P \times V \times D \times 100}$$

where,

A = total API (active pharmaceutical ingredients) sales in the country, this should include products containing the same API;

R = removal rate during sewage treatment;

P = number of inhabitants in the country;

V = volume of wastewater per capita and per day;

D = dilution factor of wastewater by surface water flow.

The factor 1 000 000 000 converts the quantity used from kg to µg

Table 12.11: ***Class of pharmaceutical products based on their excretion rates – parent compounds*** *[Mompelat et al. 2008].*

Excretion rate (%)	Compound	Therapeutic use
Low (£5%)	amitryptilline (little)	antidepressants-anti-anxiety
	amphetamine (1-74)	psycho-stimulants
	carbamazepine (1-2)	anticonvulsant
	cyclophosphamide (£5-20)	antineoplastics
	diazepam (1)	antidepressants-anti-anxiety
	diltiazem (2-4)	antidepressants-anti-anxiety
	erythromycin (5-10)	antibiotics
	Ibuprofen (1-10)	NSAIDs and analgesics
	ifosafamide (upto 15)	antineoplastics
	metoprolol (3-30)	beta-blockers
	paracetamol (acetaminophen) (<5)	NSAIDs and analgesics
	propranolol (<0.5)	beta-blockers
Moderately low (6-39%)	5-aminosalicyclic acid (<12)	NSAIDs and analgesics
	5-fluorouracil (11-20)	antineoplastics
	acebutolol (39)	beta-blockers
	amphetamine (1-74)	psycho-stimulants
	chloramphenicol (5-10 or 8-12)	antibiotics
	clarithromycin (25)	antibiotics
	cyclophosphamide (£5-20)	antineoplastics
	diclofenac (15)	NSAIDs and analgesics
	enalapril (30-36)	anti-hypertensives
	epirubicin/doxorubicin (biliary excretion : 9-11 within 48 h)	antineoplastics
	erythromycin (5-10)	antibiotics
	fluorouracil	antineoplastics
	ifosfamide (upto 15)	antineoplastics
	metronidazole (40 or 20)	antibiotics
	metoprolol (3-30)	beta-blockers

Contd...

-	norfloxacin (30)	antibiotics
	primidone (15-40)	anticonvulsant
	salbutamol (30)	bronchodilatators
	simvastatin (10-15)	lipid regulator
	spiramycin (20)	antibiotics
	sulfamethoxazole (15 or 30)	antibiotics
	tramadol (15-35)	opioanalgesics
Relatively high (40-69%)	amphetamine (1-74)	psycho-stimulants
	atenolol (50-90)	beta-blockers
	benzafibrate (45-50)	lipid regulators
	cimetidine (48-75)	antacids
	ciprofloxacin (83.7 or 50-70)	antibiotics
	digoxin (50-70)	cardiac stimulants
	epirubicin/doxorubicin (40 biliary excretion within 72 h)	antineoplastics
	furosemide (40-90)	diuretics
	ranitidine (30-70)	antacids
	testosterone (60)	steroid hormones
High (^{3}70)	amoxicillin (80-90 or 60-80)	antibiotics
	amphetamine (1-74)	psycho-stimulants
	atenolol (50-90)	beta-blockers
	ceftriaxone (70)	antibiotics
	cimetidine (48-75)	antacids
	ciprofloxacin (83.7 or 50-70)	antibiotics
	codeine (70 free or as conjugates)	opioanalgesics
	doxycycline (70)	antibiotics
	furosemide (40-90)	diuretics
	gabapentin (100)	antibiotics
	morphine (71.6)	opioanalgesics
	sotalol (>75)	beta-blockers
	tetracycline (80-90)	antibiotics
	trimethoprim (60 or 80)	antibiotics
	valsartran (80)	anti-hypertensives

In second-tier procedure, PNEC for the aquatic environment is estimated from EC_{50} or LC_{50} values obtained with acute toxicity tests of the most sensitive species (algae, daphnia and fish i.e. species from three trophic levels) and by application of an assessment factor. Whether a substance presents a risk to organisms in the environment is based on a comparison of PEC or MEC of the substance with its PNEC to organisms in ecosystems (see Table 12.12). When a risk is suspected from this comparison on a case-by-case basis is subjected to third tier-procedure, including chronic toxicity tests, micro-organism specific test, bio-accumulation study.

Table 12.12: ***Risk assessment of pharmaceuticals based on PEC/PNEC (HQ) values*** *[Richman and Castensson 2008].*

Risk categories	PEC/PNEC	Comments
No risk	< 0.1 (0.01-0.1)	use of the medicine will result in insignificant environmental risk
Low risk	0.1-1.0	use of the medicine will result in low or minimal environmental risk
Medium risk	1-10	use of the medicine will result in moderate environmental risk
High risk	>10	use of medicine will result in high environmental risk

The PNEC in water compartments ($PNEC_{water}$) can be determined following the formula:

$$PNEC_{water} = \frac{EC_{50} \text{ or } LC_{50}}{1000}$$

Table 12.13 provide examples of risk assessment of selected pharmaceuticals based on measured environmental concentrations in influent and effluent samples, predicted no effect concentrations (PNEC) and risk quotients (MEC/PNEC) of each pharmaceutical compound.

Table 12.13: *Examples of risk assessment of selected pharmaceuticals based on measured environmental concentrations in influent (MECinf) and effluent (MECefl) samples, predicted no effect concentrations (PNEC) and risk quotients (MEC/PNEC) of each pharmaceutical compound* [Santos et al. 2007].

Compound	MECinf (μg/l)	MECeffl (μg/l)	PNEC (μg/l)	MEC inf/PNEC	MEC eff/PNEC	Comments
Caffeine	11.44	3.20	182	0.06	0.02	No risk or low risk (risk quotients in influent and effluent wastewaters is <1)
Carbamazepine	2.15	1.29	6.36	0.34	0.20	No risk or low risk (risk quotients in influent and effluent wastewaters is <1)
Ibuprofen	373.11	48.24	9.1	41.00	5.3	Risk ratios were 41 and 5.32 for influent and effluent respectively, indicating though an important decrease of the concentration occurred due to wastewater treatments, an ecological risk is till present, risk quotients >1
Ketoprofen	3.59	1.50	15.6	0.23	0.10	No risk or low risk (risk quotients in influent and effluent wastewaters is<1)
Naproxen	27.04	4.28	21.2	1.28	0.20	The risk quotient of naproxen was higher than one in influent but lower than one in effluent wastewater, therefore any ecological risk due to naproxen has been considerably decreased

12.8: Pharmaceuticals in the environment

[based on Boxall 2004; Jones et al 2005; Fent et al. 2006; Hernando et al. 2006; Collier 2007; Barnes et al. 2008; Mompelat et al. 2008; Richman and Castensson 2008; Spongberg and Witter 2008; Zhou et al. 2008; Kümmerer 2009; Miége et al. 2009; New Scientist 2009; Titz and Döll 2009;Watkinson et al. 2009].

Pharmaceuticals generally are biologically active compounds that are intended not to be easily biodegradable and are often water soluble and therefore can be found in wastewaters and can easily end up in natural waters. Pharmaceutical ingredients have been reported to be ubiquitously present in sewage treatment effluents, other wastewaters (hospital wastewaters), surface waters and groundwaters, sediments and drinking water (Box 12.1) reservoirs in the ng/l to μg/l range. In total 109 pharmaceutical residues have been detected in different environmental compartments (Table 12.14) including: antibiotics, analgesic and antiinflammatory, antacid, antidepressants, antidiabetics, antiepileptics and cytostatic, bronchodilator, beta or β-blockers, lipids regulators, psychiatric and x-ray contrast media as summarised below. Benzafibrate (lipid regulator), carbamazepine (antiepileptic/analgesic), gemfibrozil (lipid regulator), ibuprofen (analgesic/antiinflammatory) and sulfamethoxazole (antibiotic) have been detected in all environmental waters. Pharmaceuticals and their products so far detected from various environmental compartments are summarized below:

- *WWTP:* 35 pharmaceuticals have been detected from WWTP (Table 12.14), these include the following pharmaceutical groups: 14 antibiotics; 8 NSAID and analgesics; 4 beta blockers; 4 lipid regulators; 2 antiepileptics; 1 x-ray contrast media; 1 stimulant and 1 disinfect.

- *Surface waters:* 93 pharmaceuticals have been detected in surface waters (Table 12.14) from Australia, Brazil, Finland, France, Germany, Japan, Italy, Poland, Spain, Switzerland, UK, USA, and Vietnam. These include the following pharmaceutical groups: 49 antibiotics; 17 NSAID and analgesic; 5 beta blockers; 4 lipid regulators; 3 bronchodilator; 3 x-ray contrast media; 2 antacids; 2 antineoplastics; 1 antianginal; 1 opiodanalgesic; 1 antidepressants; 1 antidiabetics; 1 disinfect; 1 stimulant; 1 antiepileptics and 1antihypertensive.
- *Groundwater:* 22 pharmaceuticals have been detected in groundwater (see Table 12.14). These include the following pharmaceutical groups: 9 antibiotics; 2 NSAID and analgesics; 2 lipid regulators; 2 x-ray contrast media; 1 antianginal; 2 antidepressants; 1 stimulant; 1 antiepileptics; 1 antihypertensive and 1 beta blocker.
- *Drinking water:* 26 pharmaceuticals have been detected in drinking water (see Table 12.14). These include the following pharmaceutical groups: 3 antibiotics; 8 NSAID and analgesics; 3 lipid regulators; 1 x-ray contrast media; 3 antidepressants; 1 stimulant; 3 antiepileptics;1 antineoplastics; 1 disinfect; 1 opiodanalgesic and 1 stimulant.

Box 12.1: Disturbing pharmaceutical levels in water *[WWEE 2009; http://www.redorbit.com/news/health/1628532/pharmaceutical_waste_found_in_indias_streams; http://redgreenandblue.org/2009/01/25/record-levels-of-pharmaceuticals-found-in-indias-water/].*

Pharmaceuticals and hormonally active chemicals have been detected at low concentrations in the drinking water supply for 28 million Americans according to a survey carried out by the Southern Nevada Water Authority (NSWA), in Las Vegas between 2006 and 2007. The NSWA analyzed drinking water from 10 water utilities for 51 compounds and reported that 11 of the most frequently detected compounds are: Atenolol – a beta blocker, Atrazine – organic herbicide banned in the European Union, Carbamazepine - a mood stabilizing drug, Estrone - an oestrogen hormone, Gemfibrozil – an anti-cholesterol drug, Meprobamate – a tranquilizer, Naproxen – a painkiller, Phenytoin – an anti-convulsant, Sulfametho-xazole and Trimethoprim – two antibiotics and TCEP – a reducing agent used in molecular biology.

In contrast, researchers have discovered that the huge pharmaceutical industry in India has led to severely contaminated wastewater downstream from drug factories. How severe? Surface water downstream of a plant that treats effluents from 90 pharmaceutical factories in Patancheru, India contained 150 times the highest levels detected in the USA, according to the Associated Press. Most researchers agree that this is the highest level of pharmaceuticals ever found in drinking water. Twenty-one different pharmaceuticals could be in that water at any given moment. Much of India's poor are left to drink and use that water as it flows downstream and results of scientific testing show that some of those chemicals, even in trace amounts, can cause serious damage in both humans and other species in the wild. When mixed with sewage in water, bacteria can be even more potent.

India is one of the world's leading exporters of pharmaceuticals and the US – which spent US$ 1.4 billion on Indian made drugs in 2007. 'If you take a bath there (in stream near Patancheru, India) then you have all the antibiotics you need for treatment' – chemist and an expert on drug resistance in the environment Professor Klaus Kümmerer, from Environmental Chemistry and Environmental Hygiene at the University of Freiburg, Germany, told the AP. 'If you just swallow a few gasps of water, you're treated for everything' – told Joakim Larsson, an environmental scientist at the University of Gothenburg, Sweden who studied levels of pharmaceuticals in the stream near Patancheru Enviro Tech Ltd., India. It is reported last year; the AP found that, human cells fail to grow normally in the laboratory when exposed to trace concentrations of certain pharmaceuticals. Researchers in India noted that tadpoles that were exposed to water from the plant were 40% smaller than those living in clean water.

But Larsson's findings could spell out more trouble as continued exposure to drugs could lead to mutation of bacteria, making them more resistant to powerful treatments. In recent years, many bacteria have developed resistance to the drug, leaving it significantly less effective. 'We are prescribing these drugs, and the disease is not being cured – there is resistance going on there', said Dr. A. Kishan Rao, a Medical doctor and environmental activist who has treated people for more than 30 years near the drug factories. Dr. Rao called it 'a Global concern'.

All these findings reveal disturbing trends in the effects of high consumption, unregulated production and contamination of natural waters by pharmaceuticals worldwide.

Table 12.14: ***Pharmaceutical's concentrations in environmental waters***

1. *WWTP effluent*: *data compiled by Hernando et al. 2006; Fent et al. 2006; Spongberg and Witter 2008; Zhou et al. 2008; Kümmerer2009;Miége et al. 2009.***2. *Surface waters*:** *data compiled by Boxall 2004; Hernando et al. 2006; Mompelat et al. 2008; Kümmerer2009; Watkinson et al. 2009.* **3. *Groundwater*** *: Barnes et al. 2008; Richman and Castensson 2008; Kümmerer 2009.* **4. *Drinking water*:** *data compiled by Jones et al 2005; Mompelat et al. 2008; New Scientist 2009.*

Pharmaceuticals	Therapeutic class	WWTP (Effluent)[1] (max. levels or range - ng/l)	Surface waters[2] (max. levels or range - ng/l)	Ground water[3] (max. levels or range - ng/l)	Drinking water (ng/l)[4]
AAA (acetoaminoantipyrine)	*NSAID-DMMA metabolites or degradation products*	-	1000 (Germany)	-	-
1,7-dimethylxanthine	*caffeine metabolite*	-	-	-	-
Acetaminophen/ paracetamol	*NSAID and analgesic*	-	5-10000 (France, Poland, UK, USA)	57	210.1 (France)
Acetylsalicylic acid/ aspirin	*NSAID and analgesic*	600	340 (UK)	-	-
AMDOPH (*1-acetyl-1-methyl-2-dimethyloxamoyl-2-phenylhydrazide)*	*Dimethylaminophenazone (DMAA) metabolite or degradation products*	-	300 (Germany)	-	900-1250 (Germany)
Aminopyrine	*analgesic-antiinflammatory*	-	340	-	-
Amitriptyline	*antidepressant, anti-anxiety*	-	-	-	1.4 (France)
Amoxicillin	*antibiotic*	-	30-200 (Australia, UK)	-	-
AMPH (*1-acetyl-1-methyl-2-phenylhydrazide)*	*Dimethylaminophenazone (DMAA) metabolite or degradation products*	-	160 (Germany)	-	-
Atenolol	*beta blocker*	154	145	-	-
Azithromycin	*antibiotic*	138	upto 3	-	-
Benzafibrate	*lipid regulator*	816-2353	1100-3100	Detected	27 (Germany)
Betaxolol	*beta blocker*	190	28	-	-
Bleomycin	*antineoplastics*	-	-	-	13 (UK)
Bisprolol	*beta blocker*	-	2900	-	-
Caffeine	*stimulant*	2-9900	6000	130	22.9-119 (France, USA)
Carazolol	*Beta blocker*		110	-	-
Carbamazepine	*anticonvulsant/antiepileptic/analgesic*	674-3117 (Canada, Sweden, Wales)	1100	Detected	24-530 (Canada, France, Germany, USA)
Cefaclor	*antibiotic*	-	200 (Australia)	-	-
Cephalexin	*antibiotic*	-	100 (Australia)	-	-
Cephalexin	*antibiotic*	-	200 (Australia)	-	-
Chloramphenicol	*antibiotic*	68-580	10-60 (Switzerland)	-	-
Chlortetracycline	*antibiotic*	-	nd -690 (Australia, Italy, USA)	-	-

Contd...

Cimetidine	*antacid*	-	580	-	-
Ciprofloxacin	*antibiotic*	7-72.3 (Sweden & other places)	nd-1300 (Australia, Finland, Italy, USA)	-	-
Clarithromycin	*antibiotic*	710	0.5-260 (Italy)	-	-
Clenbuterol	*bronchodilator*	-	50	-	-
Clindamycin	*antibiotic*	upto 110	10 (Australia)	-	-
Clofibrate	*lipid regulator*	-	40	-	-
Clofibric acid	*metabolite of clofibrate*	150-361	270	-	5.3-270 (Germany)
Codeine	*opioidanalgesic*	-	1000	-	30 (USA)
Danofloxacin	*antibiotic*	-	10 (France)	-	-
Dehydronifedipine	*Antianginal (nifedipine metabolite)*	-	30	22	-
Demeclocycline	*antibiotic*	-	120-325 (USA)	-	-
Diatrizoate	*X ray contrast media*	-	100000	-	1200 (Germany)
Diazepam	*antidepressant*	-	-	Detected	10 -23.5 (Italy, UK)
Diclofenac	*analgesic-antiinflammatory*	70-1276 (Sweden, Wales)	225-1200 (France, Germany, UK, USA)	-	2.5-35 (France, Germany)
Dilantin	*antiepileptics/anticonvulsants*	-	-	-	1.3 (USA)
Diltiazem	*antihypertensive*	-	49	28	-
Doxycline	*antibiotic*	-	400 (Australia)	-	-
DP (1,5-dimethyl-1,2-dehydro-3-pyrazolone)	*NSAID-phenazone degradation products*	-	50 (Germany)	-	1.10 (Germany)
Enrofloxacin	*antibiotic*	-	nd-10 (Australia, France, USA)	-	-
Eoxacin	*antibiotic*	-	10 (France)	-	-
Erythromycin	*antibiotic*	212-886	nd-1022 (Australia, Italy, UK)	-	-
Erythromycin H_20	*metabolite of erythormycin*	up to 6000	nd-1700 (Australia, Japan, Italy,, UK, USA Vietnam)	-	-
FAA (formylaminoantipyrine)	*Dimethylaminophenazone (DMAA) metabolite or degradation products*	-	1000 (Germany)	-	-
Fenofibrate	*lipid regulator*	110	-	-	-
Fenoprofen	*NSAID*	-	10 (Germany)		-
Fenoterol	*bronchodilator*	-	61	-	-
Flucloxacillin	*antibiotic*	-	7		-
Flumequine	*antibiotic*	-	10-30 (France)	-	-
Fluorchinolone	*antibiotic*	upto 100	upto 19	-	-
Fluoxetine	*antidepressant*	-	12	56	-
Gemfibrozil	*lipid regulator*	564-2366	510-790	Detected	70 (Canada)
Ibuprofen	*analgesic-antiinflammatory*	1960	226-3400 (France, Germany, Italy, Spain, UK, USA)	3110	0.6-1350 (Finland, France, Germany, USA)

Contd...

Indomethacin	*NSAID and analgesic*	9-35 (England)	10-200 (Germany, Spain)	-	-
Iopamidol	*X ray contrast media*			Detected	-
Iopromide	*X ray contrast media*	5680	-	Detected	-
Ketoprofen	*analgesic-antiinflammatory*	325-733	120 (France, Germany, UK)	-	3-8 (Finland, France)
Lincomycin	*antibiotic*	-	nd-730 (Australia, France, Italy, USA)	320	-
Markolide	*antibiotic*	upto 700	upto 20	upto 2	-
Mefenamic acid	*analgesic-antiinflammatory*	1140	5-68 (UK)	--	-
Meprobamate	*antidepressant, anti-anxiety*	-	-	-	5.9-14000 (USA)
Metformin	*antidiabetic*	-	150	-	-
Metoprolol	*beta-blocker*	338-777	2200	-	-
Monesin	*antibiotic*	-	150 (Australia)	-	-
Nalidixic acid	*antibiotic*	-	750 (Australia)	-	-
Naproxen	*analgesic-antiinflammatory*	1847-1890	266-390 (France, UK, USA)	-	-
Norfloxacin	*antibiotic*	-	30-1150 (Australia, Brazil, Finland, USA)	-	-
Ofloxacin	*antibiotic*	upto 82	30-60 (France)	-	-
Oleandomycin	*antibiotic*	-	9.79-74.2 (Italy)	-	-
Ornidazole	*antibiotic*	-	10-40 (France)	-	-
Oxolonic/oxolinic acid	*antibiotic*		12 (France)	-	-
Oxytetracycline	*antibiotic*	-	nd-340 (Australia, Italy, USA)	-	-
PDP (4-(2-methylethyl)-1,5-dimethyl-1,2-dehydro-3-pyrazolone)	*NSAID (prophyphenazone degradation product)*	-	-	-	0.24 (Germany)
Penicillin G	*antibiotic*	-	250 (Australia)	-	-
Penicillin V	*antibiotic*		10 (Australia)	-	-
Phenazone	*analgesic-antiinflammatory*		950 (Germany)	-	250-400 (Germany)
Piperacillin	*antibiotic*	-	48		-
Primidone	*antiepileptics/anticonvulsants*	-	-	-	40 (Germany)
Propyphenazone	*NSAID and analgesic*	-	220 mean (Germany)	-	80-240 (Germany)
Propranolol	*beta-blocker*	20-676 (England, Sweden, Wales)	25-590	-	-
Ranitidine	*antacid*		10	-	-
Ronidazol	*antibiotic*	-	-	upto 10	-
Roxithromycin	*antibiotic*	49.6-1000	560 (Australia, USA)	upto 26	-
Salbutamol	*bronchodilator*	-	35	-	-
Salicylic acid	*acetylsalicylic acid metabolite (NSAID)*	10-59600	15-30 (UK)	-	-
Salinomycin	*antibiotic*	-	150 (Australia)	-	-
Sarafloxacin	*antibiotic*	-	20 (France)	-	-

Contd...

Sotalol	*beta blocker*			Detected	-
Spiramycin	*antibiotic*	-	9.79-74.2 (Italy)		-
Sulfadiazin/sulfadiazine	*antibiotic*	-	-	upto 17 (Germany)	-
Sulfadimethoxine	*antibiotic*	-	nd-15000 (USA)		-
Sulfadimidin	*antibiotic*	-	upto 7	upto 23 (Germany)	-
Sulfamethazine	*antibiotic*	-	nd-220 (Switzerland, USA)	360	-
Sulfamethizole	*antibiotic*	-	130 (USA)		-
Sulfapyridine	*antibiotic*		10-25 (Poland, UK)		-
Sulfamethoxazole	*antibiotic*	nd-2000 (Sweden, Wales)	nd-2000 (Australia, France, Poland, USA, Vietnam)	1110 (Germany)	1200 (USA)
Sulfamethoxine	*antibiotic*	-	70 (USA)	-	-
Sulfasalazine	*antibiotic*		30 (Australia)	-	-
Sulfathiazole	*antibiotic*	-	nd-80 (Australia, USA)	-	-
Sulphonamide	*antibiotic*	upto 1000	upto 40	Up to 20	-
Tamoxifen	*antineoplastic, cytostatic*	0.2-238 (England)		-	-
Tetracycline	*antibiotic*	282	nd-300 (Australia, USA)	-	-
Timolol	*beta blocker*	-	10	-	-
Triclosan	*disinfect*	150	150	-	734 (USA)
Trimethoprim	*antibiotic*	38-660	nd-710 (Australia, France, Germany, Italy, UK, USA Vietnam)	-	-
Tylosin	*antibiotic*	-	0.29-280 (Australia, Italy, USA)	-	1.7 (Italy)

12.9: Environmental and biological impacts of pharmaceutical residues

[based on Diaz-Cruz et al. 2003; Boxall 2004; Kümmerer 2004; Jones et al. 2005; Besse and Garric 2008; Copper et al. 2008; Fountoulakis et al. 2008; Watkinson et al. 2009].

12.9.1: General effects

Pharmaceuticals have been developed to produce biological effects, so their residues in the environment can cause different ecotoxicological effects. Although, detected environmental concentrations of pharmaceuticals are generally at ultra trace levels (ng/l to µg/l), researchers found that these concentration can induce toxic or chronic effects on non-target organisms. The impacts of pharmaceuticals on environment include negative effects on fish, crustaceans (daphnids), algae, bacteria, earthworms, plants and dung invertebrates. For example, pharmaceuticals may cause subtle effects on aquatic and terrestrial organisms that are not detected in standard studies. These impacts include effects on oocytes and testicular maturation, impacts on insect physiology and behaviour, inhibition or stimulation of growth in aquatic plant and algal species, and the development of antibacterial resistance in soil microbes (Table 12.15).

Table 12.15: ***Reported subtle effects of pharmaceutical compounds on aquatic and terrestrial organisms*** *[Based on Boxall 2004].*

Substance(s)	Medicine class	Reported effect
Amlodipine	Calcium channel blocker	Inhibition in the ability of dissected polyps from the cnidarian *Hydra vulgaris* to regenerate a hypostome, tentacles and a foot.
Carbamazepine	Analgesic	Inhibition of basal 7-ethoxyresorfun-o-deethylase (EROD) activity in cultures of rainbow trout hepatocytes; inhibition of emergence of *Chironomous riparius.*
Clofibrate	Lipid regulator	Inhibition of basal EROD activity in cultures of rainbow trout hepatocytes.
Diazepam	Antianxiety drug	Inhibition in the ability of dissected polyps from the cnidarian *Hydra vulgaris* to regenerate a hypostome, tentacles and a foot.
Diclofenac	Analgesic	Inhibition of basal EROD activity in cultures of rainbow trout hepatocytes.
Digoxin	Cardiac glycoside	Inhibition in the ability of dissected polyps from the cnidarian *Hydra vulgaris* to regenerate a hypostome, tentacles and a foot.
Erythromycin	Antibacterial	Inhibition of growth of cyanobacteria and aquatic plants.
Fenfluramine	Anorexic	Enhance release of serotonin (5-HT) in crayfish which in turn triggers the release of ovary-stimulating hormone resulting in large oocytes with enhances amounts of vitellin; In feeder crabs, stimulates the production of gonad-stimulating hormone accelerating testicular maturation (endocrine disruption).
Fenofibrate	Lipid regulator	Inhibition of basal EROD activity in cultures of rainbow trout hepatocytes.
Ibuprofen	Anti-inflammatory	Stimulation of growth of cyanobacteria and inhibition of growth of aquatic plants.
Propanolol	Beta blocker	Weak EROD inducer in cultures of rainbow trout hepatocytes.
Sulfamethoxazole	Antibacterial	Inhibition of basal EROD activity in cultures of rainbow trout hepatocytes.
Tetracycline	Antibacterial	Inhibition of growth of cyanobacteria and aquatic plants
Tetracyclines, macrolides and streptomycin	Antibacterials	Antibacterial resistance in soil bacteria.
Tylosin	Antibacterial	Impacts on the structure of soil microbial communities.

12.9.2: Bacterial resistance

Antibiotics have been detected in environmental waters including waste water, surface water, ground water, sediments and soils since either they are not fully eliminated during the sewage purification process and are not biodegradable. As a result there is increasing concern about the growing resistance of pathogenic bacteria in the environment and their ecotoxic effects. Resistance genes as well as resistant bacteria in the environment are increasingly seen as an ecological problem. The transfer of resistant bacteria (Table 12.16) to humans may occur via water or food if plants are watered with surface water or sewage sludge, if manure is used as a fertiliser or if resistant bacteria are present in the meat.

Tables 12.16: ***Examples of resistant bacteria and genetic material associated with resistance found in different environmental compartments*** *[Kümmerer 2004].*

Source	Resistant bacteria/resistance gene detected
Hospital effluent (e.g. treatment of infections and direct release into the aquatic environment)	*vanA, mecA*
Waste water (e.g. various sources)	*vanA, ampC,* gentamicin resistance genes, *P.aeruginosa, Acinetobacter* spp., *Escherichia coli, Campylobacter* spp., *Pseudomonas* spp., *Enterococci*
Sediments (e.g. application of antibiotics in aquaculture/fish farms)	*E.coli*
Surface water (e.g. effluent of sewage treatment plants)	*vanA, ampC*
Groundwater (e.g. leaching from fields fertilised with animal slurry)	*Mycobacterium* spp., *Streptomyces* spp.

12.9.3: Effects on treatment process

Due to low biodegradability and hydrophobic nature of some pharmaceuticals, they tend to concentrate in sludge and may cause adverse effects on the anaerobic digestion process, commonly applied for sewage sludge stabilization. Some common antibiotics used as feed additives for veterinary treatment have been found to cause the complete inhibition of biogas production. Pharmaceuticals such as propranolol hydrochloride, diclofenac sodium, and floxacin have caused a reduction in the methane production. Antibiotics such as, erythromycin, clarithromycin and amoxicillin at concentrations of 100 mg/l (though not likely to occur in the environment) significantly decreases denitrification rates. Furthermore antibiotics present in sewage can also negatively impact bacteria associated with biological processes within sewage treatment plants.

12.9.4: Effects on terrestrial biota

Animal wastes used to fertilise fields contain drugs that can accumulate in the soil and become persistent soil pollutants. The effects of such contamination on terrestrial biota include altered post germinative development of plants and the growth of roots, hypocotyls and leaves. Sulfonamides and tetracyclines have high persistence and adsorption capability, therefore repeated application of manure containing such residues to agricultural field can risk contamination of the food chain. Ivermectin used as veterinary antiparasitic have affected non-target organisms such as, dung beetles and dung-flies.

The catastrophic declines in the three species of vultures, *Gyps bengalensis*, *G. indicus* and *G. tenuirostris* in Bangladesh, India, and Pakistan were believed to be due to exposure of diclofenac during feeding on the carcasses of domestic cattle treated with a normal veterinary dose of the drug shortly before death. The drug was regularly used for veterinary medication and residues entered the vultures as they fed on dead domestic livestock causing renal failure and resulting in an over 95% decline in some populations since the early 1990s. The lowest observed effect concentrations (LOEC) were reported to be 0.007 mg/kg in vultures. Since loss of vultures has caused major ecological effects over the subcontinent and as a consequence the Indian Government banned use of diclofenac in September 2005. The collapse of the vulture populations has benefited the feral dog (*Canis familiaris*) and rat (*Rattus* spp.) due to the increases in carcasses usually preyed upon by the *Gyps* spp.

12.9.5: Effects on aquatic biota

Ivermectin (veterinary pharmaceutical) is extremely toxic to the freshwater crustacean, *Daphnia magna* (LC_{50} 0.000025 mg/l) (Boxall *et al.* 2003). Cardiovascular pharmaceuticals are reported to be sensitive to amphibians and fish at macromolecular levels. Low concentrations (10 μg/l) of the pharmaceuticals diazepam, digoxin and amlodipine inhibited the ability of freshwater cnidarian *Hydra vulgaris* to regenerate the dissected body parts. Similarly, low, sublethal concentrations of simvastatin (0.16 μg/l-50 μg/l) were found to affect the development time, body length, RNA content and growth rate of hepacaticoid copepods. Antineoplastics such as, cyclophosphamide and ifosafamide are not degraded in sewage treatment plant and they may cause mutation in non-target organisms. Fluoroquinolone antibiotics were shown to exert genotoxic effects for the genetically modified bacterial strain *Salmonella typhimurium* at concentration as low as 5 μg/l for norfloxacin and 25 μg/l for ciprofloxacin. Ciprofloxacin also caused a consistent reduction of algal population diversity (more information on aquatic toxicity of algae, bacteria, crustacean and fish are given in section 12.9.7 under ecotoxicological effects).

12.9.6: *Effects on human*

Pregnant women and their babies, as well as children are at particular risk if drinking water is contaminated with pharmaceuticals. The detection of drug residues in drinking water can expose new born human babies from very low age before the blood brain barrier and the detoxifying systems are fully developed. The possible impacts of life long exposure of humans are not yet available. However, the exposure of the pregnant mother or her foetus to cytotoxic drugs (antineoplastic) and chemotherapy drugs via drinking water could be very harmful. As shown in Table 12.13, 26 drugs have been detected in drinking water, of which, a number of drugs are not recommended for children. If children are exposed to these pharmaceuticals the relative risk is up to eight fold higher than adults according to Collier (2007). Several drugs are teratogenic and are not recommended during breast feeding. Some of the drugs that are of high risk during pregnancy and breast feeding and are detected in drinking water (Table 12.14) are aspirin, atenolol, carbamazepine, clofibric acid, cyclophosphamide, diazepam, gemfibrozil, oxytetracycline, tetracycline and sulfamethoxazole. Antibiotics present in the environment can produce resistance in microbial assemblages, which can have potentially drastic effects upon human health.

12.9.7: *Ecotoxicological effects*

[Based on Webb 2004; Fent et al. 2006; Hernando et al 2006; Nunes et al. 2008; Khalaf et al. 2009]

12.9.7.1: *Ecotoxicity classification of pharmaceuticals*

Acute toxicity of pharmaceuticals has been assessed by standard tests according to established guidelines based on bacteria, algae, invertebrates and fish. According to EU directive 93/67/EEC (CEC 1996), the ecotoxicity of pharmaceutical substances are based on measured effective concentrations (EC_{50} value) in a scheme with different risk classes (Table 12.17). Based on various trophic levels toxicity experiments, algae and bacteria were found to be in general more sensitive to pharmaceuticals than zooplankton and fish (see Tables 12.18).

Table 12.17: *Ecotoxicity classification of pharmaceuticals [CEC 1996; Kim et al. 2008].*

EC_{50}	Ecotoxicity classification	Hazard classification
<0.1 mg/l	Extremely toxic to aquatic organisms.	Very high
0.1-1.0 mg/l	Very toxic to aquatic organisms.	High
1-10 mg/l	Toxic to aquatic organisms.	Medium
10-100 mg/l	Harmful to aquatic organisms.	Harmful
>100 mg/l	Non-toxic to aquatic organisms.	Low

The following section provides an account of acute toxicity and toxic effects of some selected pharmaceutical groups on aquatic organisms:

Table 12.18: ***Acute toxicity data of most commonly detected pharmaceuticals on algae, daphnia, invertebrates, fish and other biota*** *[based on Carlsson et al. 2006; Enick and Moore 2007; Grung et al. 2008; Pounds et al. 2008; Sanderson and Thomsen 2009] PNEC = predicted no effect concentration; NA = not available*

Compound	Test species	Test type	Ecotoxicity
Amitriptyline (*antidepressant*)	Algae Crustacean - Daphnid Fish	LC_{50} 24 h LC_{50} 48 h LC_{50} 96h	NA 0.78 mg/l 0.78 mg/l
Amoxicillin (*antibiotic*)	Algae Algae-*Lemna gibba* Algae/bacteria *Microcystis aeruginosa* Crustacean - Daphnid Fish-rainbow trout	LC_{50} 24 h LOEC 7d chlorophyll a/b EC_{50} growth LC_{50} 48 h EC_{50} 24h-hepatocyte cytotoxicity	2.2 >1 mg/l 0.0037 mg/l NA >182.7 mg/l
Aspirin or acetylsalicylic acid (*analgesic/NSAID*)	Algae – *Desmodesmus subspicatus* Algae Crustacean-Daphnid Fish	NOEC LC_{50} 24 h LC_{50} 48 h LC_{50} 96h	3.2 mg/l 106.7 mg/l 88 mg/l 150 mg/l
Atenolol (*beta blocker*)	Green algae Algae Crustacean-Daphnid *Hydra vulgaris* (cnidaria) Fish	PNEC, ECOSAR LC_{50} 24 h LC_{50} 48 h Acute toxicity, 7 days effect LC_{50} 96h	77.7 μg/l 620 mg/l 33.4 mg/l 1 mg/l -10 mg/l NA
Benzafibrate (*lipid regulator*)	Algae Crustacean - Daphnid Fish	LC_{50} 24 h LC_{50} 48 h LC_{50} 96h	>100 mg/l >100 mg/l >100 mg/l
Caffeine (*stimulant*)	Algae Crustacean - Daphnid Fish	LC_{50} 24 h LC_{50} 48 h LC_{50} 96h	NA 182 mg/l 87 mg/l
Carbamazepine (*antiepileptic*)	Algae/bacteria *Synechococcus leopolensis* Crustacean - Daphnid Crustacean – *Ceriodaphnia dubia* Fish – rainbow trout (*Oncorhynchus mykiss*)	LC_{50} 24 h LC_{50} 48 h NOEC 7 days reproduction LC_{50} 48 h cell function	33.6 mg/l 13.8 mg/l 0.025 mg/l 111.8 mg/l
Chloramphenicol (*antibiotic*)	Algae/bacteria *Selenastrum capricornutum* Algae/bacteria *Microcystis aeruginosa* Crustacean - Daphnid Fish	EC_{50} 72 h EC_{50} 7 d LC_{50} 48 h LC_{50} 96h	3.1 mg/l 0.05 mg/l 305 mg/l 305 mg/l
Chlortetracycline (*antibiotic*)	Algae Crustacean - Daphnid Fish	LC_{50} 24 h LC_{50} 48 h LC_{50} 96h	NA 128 mg/l NA
Cimetidine (*antacid*)	Algae Crustacean - Daphnid Fish	LC_{50} 24 h LC_{50} 48 h LC_{50} 96h	NA 740 mg/l > 1000 mg/l

Contd...

Ciprofloxacin *(antibiotic)*	Algae - *Microcystis aeruginosa* Algae Algae - *Lemna gibba* (duckweed) Crustacean - *Daphnia magna* Fish - *Brachydanio rerio*	EC_{50} EC_{50} EC_{50} 7 day EC_{50} EC_{50}	0.005 mg/l 0.05-18.7 mg/l 0.69 mg/l 0.61 mg/l >60 mg/l
Clarithromycin *(antibiotic)*	Algae Crustacean - Daphnid Fish	LC_{50} 24 h LC_{50} 48 h LC_{50} 96h	0.002 mg/l 18.7 mg/l NA
Clofibrate *(lipid regulator)*	Algae Crustacean - Daphnid Fish	LC_{50} 24 h LC_{50} 48 h LC_{50} 96h	12 mg/l 17.7 mg/l 7.7 mg/l
Clofibric acid *(lipid regulator)*	Algae Crustacean - Daphnid Fish Fish –zebra fish *Brachydanio rerio* Fish - *Danio rerio*	LC_{50} 24 h LC_{50} 48 h LC_{50} 96h LC_{50} 48 h NOEC 10 days mortality	86 mg/l 72 mg/l 89 mg/l 8.6 mg/l 7 mg/l
Cyclophosphamide *(antineoplastic)*	Algae Crustacean - Daphnid Fish	LC_{50} 24 h LC_{50} 48 h LC_{50} 96h	>100 mg/l >100 mg/l NA
Diazepam *(antidepressant)*	Algae Crustacean - Daphnid *Daphnia magna* Fish	LC_{50} 24 h LC_{50} 48 h EC_{50} 24 h LC_{50} 96h	3.1 mg/l 4.3 mg/l 14.1 mg/l - 24.6 mg/l 12.7
Diclofenac *(analgesic/NSAID)*	Algae – Green *Pseudokirchneriella subcapitata = S. capricornutum* Algae – Green *Pseudokirchneriella subcapitata = S. capricornutum* Aquatic vascular plant, *Lemna minor* Bacteria, *Vibrio fischeri* Bird, vulture *Gyps bengalensis, Gyps indicus* and *Gyps tenuirostris* Crustacean - *Daphnia magna* Crustacean- *Ceriodaphnia dubia* Crustacean - *Ceriodaphnia dubia* Crustacean - *Ceriodaphnia dubia* Rotifer- *Brachionus calyciflorus* Rotifer- *Brachionus calyciflorus* Zebra fish, *Danio rerio* Zebra fish, *Danio rerio* Fish – rainbow trout (*Oncorhynchus mykiss*)	NOEC 96 h, growth LOEC 96 h, growth EC_{50}, 7 days, growth EC_{50} LOAEL, dietary intake, renal failure EC_{50}, 48 h EC_{50}, 48 h NOEC, 7 days, reproduction LOEC, 7 days, reproduction NOEC, 48 h, reproduction LOEC, 48 h, reproduction NOEC, 10 days, ESL LOEC, 10 days, ESL EC_{50} 24 h hepatocyte cytotoxicity	10 mg/l 20 mg/l 7.5 mg/l 11.45 mg/l - 13.7 mg/l 0.007 mg/l 68 mg/l-224.3 mg/l 22.7 mg/l-224.3 mg/l 1 mg/l 2 mg/l 12.5 mg/l 25 mg/l 4 mg/l 8 mg/l 6.0 mg/l

Contd...

Erythromycin *(antibiotic)*	Algae – *Lemna gibba* Algae Crustacean - Daphnid Fish –lake trout *Salvelinus namaycush* Fish	LOEC 7d LC_{50} 24 h LC_{50} 48 h LC_{50} 96 h LC_{50} 96h	>1 mg/l 0.02 mg/l 0.94 mg/l 410 mg/l 80 mg/l
Flumequine *(antibiotic)*	Algae Crustacean - Daphnid Fish	LC_{50} 24 h LC_{50} 48 h LC_{50} 96h	>5000 mg/l >96.4 mg/l NA
Fluoxetine *(antidepressant)*	Algae Crustacean - Daphnid Crustacean- *Ceriodaphnia dubia* Fish – rainbow trout (*Oncorhynchus mykiss*) Fish- fathead minnow (*Pimephales promelas*)	LC_{50} 24 h LC_{50} 48 h LC_{50} 48 h EC_{50} 24 h –cytotoxicity LC_{50} 48 h	0.024 mg/l 0.51 mg/l 0.23 mg/l 1.5 mg/l 0.71 mg/l
Gemfibrozil *(lipid regulator)*	Algae Crustacean - Daphnid Fish	LC_{50} 24 h LC_{50} 48 h LC_{50} 96h	NA >100 mg/l NA
Ibuprofen (*analgesic/NSAID*)	Algae - *Skeletonema costatum* Algae vascular plant – *Lemna gibba* Bacteria - *Vibrio fischeri* Bacteria - *Vibrio fischeri* Crustacean - *Daphnia magna* Crustacean - *Daphnia magna* Mollusc – *Planorbis carinatus* Mollusc – *Planorbis carinatus* Fish –bluegill sunfish, *Lepomis macrochirus* Fish - *Lepomis macrochirus*	EC_{50}, 96 h NOEC, 7 days, growth EC_{50}, microtox EC_{50}, Tox Alert 100, 15 min LC_{50}, 48 h NOEC, 48 h LC_{50}, 72h, survival NOEC 21d LC_{50}, 96 h NOEC, 96 h	7.1 mg/l >1 mg/l 12.1 mg/l 19.1 mg/l 9.06 mg/l 3.37 mg/l 17.2 mg/l 5.36 mg/l 7.1 mg/l 10 mg/l
Iopromide *(X ray contrast media)*	Algae Crustacean - Daphnid Fish	LC_{50} 24 h LC_{50} 48 h LC_{50} 96h	962 mg/l 137 mg/l 4.5
Ketoprofen *(analgesic/NSAID)*	Algae Bacteria – *Vibrio fischeria* Bacteria – *Vibrio fischeria* Crustacean-Daphnid Daphnid Fish	EC_{50}, ECOSAR EC_{50}, microtox EC_{50},ToxAlert LC_{50} 48 h EC_{50}, ECOSAR EC_{50}, ECOSAR	164 mg/l 19.3 mg/l 15.6 mg/l 64 mg/l 248 mg/l 32 mg/l
Linomycin *(antibiotic)*	Algae Crustacean - Daphnid Fish	LC_{50} 24 h LC_{50} 48 h LC_{50} 96h	0.07 mg/l 0.68 mg/l NA
Metformin *(antidiabetic)*	Algae Crustacean - Daphnid Fish	LC_{50} 24 h LC_{50} 48 h LC_{50} 96h	NA 1345 mg/l NA

Contd...

Metoprolol *(betablocker)*	Algae - *D. subspicatus* Crustacean – *Ceriodaphnia magna* Crustacean - *Daphnia magna* Crustacean – *T. platyurus* Fish, Japanese ricefish – *Oryzias latipes*	EC_{50} LC_{50}, 48 h LC_{50}, 48 h EC50 LC_{50}, 48 h	7.9 mg/l 8.8 mg/l 63.9 mg/l 77.5 mg/l >100 mg/l
Naproxen *(analgesic-antiinflammatory)*	Algae Crustacean - Daphnid Fish	LC_{50} 24 h LC_{50} 48 h LC_{50} 96h	21 mg/l 37 mg/l 560 mg/l
Ofloxacin *(antibiotic)*	Algae Crustacean - Daphnid Fish	LC_{50} 24 h LC_{50} 48 h LC_{50} 96h	1.44 mg/l 17.41 mg/l NA
Oxytetracycline *(antibiotic)*	Algae/cyanobacteria - *M. aeruginosa* Algae Algae - *Selenastrum capricornutum* Crustacean - Daphnid Crustacean – *D. magna* Fish	EC_{50} growth LC_{50} 24 h EC_{50} growth LC_{50} 48 h EC_{50} 21 days reproduction LC_{50} 96h	0.207 mg/l 0.17 mg/l 4.5 mg/l 18.7 mg/l 46.2 mg/l 62.5 mg/l
Paracetamol (acetaminophen) *(analgesic/NSAID)*	Algae – *P. subcapitata* Algae Crustacean - *Daphnia magna* Invertebrates (fairy shrimp)- *Streptocephalus proboscideus* (most sensitive of 5 invertebrates) Fish – *O. mykiss* Fish – *B. rerio*	EC_{50} LC_{50} 24 h EC_{50}, 48 h LC_{50} 24h EC_{50}, 24 h LC_{50}	> 91 mg/l 134 mg/l 9.2 mg/l 29.6 mg/l >120 mg/l 378 mg/l
Propranolol *(beta blocker)*	Algae Crustacean - Daphnid Fish	LC_{50} 24 h LC_{50} 48 h LC_{50} 96h	0.7 mg/l 0.8 mg/l 1.87 mg/l
Ranitidine *(antacid/antiulcer)*	Crustacean - *Daphnia spp.*	EC_{50}	650 mg/l
Salicyclic acid *(Acetyl salicyclic acid metabolite)*	Algae Crustacean - Daphnid Fish	LC_{50} 24 h LC_{50} 48 h LC_{50} 96h	>100 mg/l 118 mg/l 37 mg/l
Sulfadiazine *(antibiotic)*	Algae Crustacean - Daphnid Fish	LC_{50} 24 h LC_{50} 48 h LC_{50} 96h	NA 112 mg/l NA
Sulfamethoxazole *(antibiotic)*	Algae - *P.subcapitata, Cyclotella meneghiniana, Synechococcus leopoliensis* Crustacean – *D. magna, C. dubia, B. calciflorus* Fish - *B rerio*	EC_{50} EC_{50} LC_{50}	0.027 mg/l -2.4 mg/l 15.5->100 mg/l >1000 mg/l

Contd...

Tetracycline (*antibiotic*)	Algae – *Nitzschia closterium, Selenastrum capricornutum, Microcystis aeruginosa* Algae Crustacean - Daphnid Crustacean - *Daphnia magna* Fish – *Salvelinus namaycush* (lake trout) Fish – *Morone saxatilis* (striped bass)	EC_{50} LC_{50} 24 h LC_{50} 48 h EC_{50} LC_{50} 24/96 h LC_{50} 24, 48, 96 h	0.09 -16 mg/l 2.2 mg/l 0.165 mg/l >340 mg/l 220 mg/l >182 mg/l
Tricloson (*disinfect*)	Algae Crustacean - Daphnid Fish	LC_{50} 24 h LC_{50} 48 h LC_{50} 96h	0.0045 mg/l 0.13 mg/l 0.26 mg/l
Trimethoprim (*antibiotic*)	Algae - *Pseudokirchneriella subcapitata, Microcystis aeruginosa, R. salina* Crustacean Fish	EC_{50} EC_{50} LC_{50}	16 mg/l-112 mg/l 123 mg/l >100 mg/l
Tylosin (*antibiotic*)	Algae Algae/bacteria- *Microcystis aeruginosa* Crustacean - Daphnid Fish	LC_{50} 24 h EC_{50} 7 days LC_{50} 48 h LC_{50} 96h	NA 0.034 mg/l 680 mg/l NA
Warfarin (*anticoagulant*)	Invertebrates (fairy shrimp) - *Streptocephalus proboscideus* (most sensitive of 5 invertebrates)	EC_{50}	67 mg/l

ESL = ecological structure activity relationship value; ECOSAR = ecological structure-activity relationships

12.9.7.2: Antidepressants and antiepileptics
[based on Ferrari et al. 2003; Fent et al. 2006].

Antiepileptic (carbamazepine) and antidepressants (diazepam and fluoxetine) can be classified as harmful to aquatic organisms, since the acute toxicity data for algae, daphnids and fish are below 100 mg/l (see Table 12.17). Carbamazepine is very toxic to invertebrates (NOEC 0.025 mg/l in *Ceriodaphnia dubia*; and non-toxic to fish (LC_{50} in rainbow > 111 mg/l) (Table 12.17)). Diazepam is toxic to aquatic organisms; with acute toxicity range of 3.1- 12.7 mg/l (see Table 12.17). Diazepam at 10 μg/l inhibited polyp regeneration in *Hydra vulgaris*. Euryhaline fish, *Gambusia holbrooki* exhibited behavioural changes when exposed to diazepam including dark pigmentation, lethargy and anomalous movement (Nunes *et al.* 2008). Fluoxetine has been one of the most toxic human drugs to aquatic species with acute toxicity for algae (LC_{50} 48 h) occurring as low as 0.024 mg/l (see Table 12.18). Fluoxetine has also been detected in tissues of fish (*Lepomis macrochirus, Ictalurus punctatus, Cyprinus carpio, Pomoxius nigromaculatus)* residing in a municipal effluent- dominated stream in north Texas at levels of at least 0.1 ng/g. Laville *et al.* (2004) tested the effects of nine pharmaceuticals (amoxicillin, clofibrate, carbamazepine, diclofenac, fluoxetine, propranolol, sulfamethoxazole and gadolinium chloride) on fish hepatocytes and found that all pharmaceuticals were cytotoxic except amoxicillin at concentration upto 500 μM.

12.9.7.3: *Analgesics and non-steroidal anti-inflammatory drugs (NSAIDs)*
[based on Fent et al. 2006; Hernando et al. 2006; Farré et al. 2008; Sanderson and Thomsen 2008].

Acetylsalicylic acid (aspirin) affected reproduction of *Daphnia magna* and *D. longispina* at concentration around 1.8 mg/l. In addition, the metabolite of paracetamol has been shown to be highly toxic to domestic cats at relatively low doses. Diclofenac is one of the most commonly detected pharmaceuticals in the environment and seems to be the compound having the highest acute toxicity within the class of NSAIDs (compare the acute toxicity of diclofenac, ibuprofen, naproxen and ketoprofen in Table 12.17). Diclofenac can bio-accumulate in fish (rainbow trout) and can cause histopathological organ lesions (severe damage of the capillary walls at the gills (LOEC = 5 μg/l) following long term exposure. Impairment of renal and gill function can occur in fish after long term exposure of diclofenac. Diclofenac, ibuprofen, naproxen and ketoprofen have been found to be harmful to bacteria (EC_{50} < 20 mg/l) but less harmful to algae and invertebrates.

12.9.7.4: *Beta blockers (β-blockers)*
[based on Fent et al. 2006; Hernando et al. 2006; Farré et al. 2008; Sanderson and Thomsen 2008].

The acute toxicity of beta blockers has not been extensively studied. The propranolol and metoprolol showed acute and chronic toxicity to aquatic organisms, of which propranolol was reported to be more toxic to algae and daphnia (EC_{50}<0.8 mg/l) (See Table 12.17). Long term exposure of propranolol reduced reproduction in *Ceriodaphnia dubia* at 250 μg/l. Propranolol have caused chronic toxicity on the cardiovascular system and on reproduction of *C. dubia* with NOEC and LOEC on reproduction of propranolol were 125 and 250 μg/l respectively.

12.9.7.5: *Lipid regulators*
[based on Hernando et al. 2006; Sanderson and Thomsen 2008].

The lipid regulators, gemfibrozil and clofibric acid are very toxic to bacteria on the basis of EC_{50} values (EC_{50} range 0.018-0.09 mg/l) (Hernando *et al.* 2006). Other lipid regulators such as benzafibrate are not harmful to daphnid, algae and fish (LC_{50} >100 mg/l) (see Table 12.17). On the other hand, fenofibrate is harmful to invertebrates (LC_{50} 50 mg/l). Clofibric acid showed high toxicity on bacteria (EC_{50} 0.091 mg/l) (Hernando *et al.* 2006) however the toxicity to algae, invertebrates and fish are less than 70 mg/l (see Table 12.17). Clofibrate showed LC_{50} of 7.7 mg/l for fish *Gambusia holbrooki*. Fish also showed oxidative based alternations in fish *G. holbrooki* after exposure to clofibrate and clofibric acid, with modifications of several enzymatic activities (Nunes *et al.* 2008).

12.10: Management of pharmaceuticals in the environment
[based on Boxall 2004; Roig et al. 2009]

As indicated in section 12.5, that pharmaceuticals are not completely removed in WWTPs as evidenced from effluent data (Table 12.13 and Figure 12.3). The removal efficiency is generally computed as the percentage of reduction between the dissolved aqueous phase concentration of the contaminant in the influent and the dissolved aqueous phase concentration of the contaminant in the effluent. In order to reduce emissions of pharmaceutical products to surface waters there is a need to improve the performance of existing conventional treatment processes or implementation of novel and more effective techniques. Though activate sludge process believed to be one of the most efficient methods for the removal of chemicals in waste water treatment plant, however, such treatment process have low removal efficiency for certain pharmaceuticals such as, atenolol, carbamazepine, clofibric acid, diclofenac, mefenamic acid, metoprolol, trimethoprim (see Figure 12.3). Furthermore, to reduce impacts on ecological health and environment, measures are required that could reduce the release of the substances to the environment. Such measures may include the following:

- Improvement of disposal systems for out-of-date medicines and waste containers;
- Segregating sources of pharmaceuticals such as, hospital wastewater, which is likely to be heavily contaminated with pharmaceuticals and antibiotic resistance bacteria

Apart from the above approaches, there are a number of physical processes that can be used to remove the pharmaceuticals such as, sorption, volatilisation, biological degradation or chemical reactions (treatment with ozone). However, the suitability of different options is highly specific for each substance. For example, the antibiotic ciprofloxacin is removed by strong sorption onto suspended solids of sewage sludge whereas diclofenac and 17α-ethinylestradiol (EE2) undergo significant bio-degradation in aged activated sludge. Many of the treatment methods, whilst removing the pharmaceuticals, may also produce transformation products that are more persistent and mobile than the parent compounds, some of which may also have similar or enhanced toxicity. Despite that, several approaches could be taken to reduce overall pharmaceutical impacts on environment which may include source control, source separation and changes in husbandry and treatment process (see Table 12.19).

Table 12.19: ***Approaches to reduce amounts of human and veterinary pharmaceuticals released to the environment*** *[adapted from Ternes et al. 2002 and 2004; Daughton, 2003 a and b; Boxall 2004; Titz and Döll 2009].*

Source control	***Human and veterinary medicine:*** Medicine labels provide information on possible environmental impact allowing physicians and veterinary surgeons to consider potential environmental impact when prescribing; develop an environmental classification for pharmaceuticals.
Source separation	***Human medicine:*** Separation of hospital and nursing homes waste from other wastes to allow targeted treatment; separation of urine from solid material; ***Veterinary medicine:*** Separation of treated animals from untreated animals.
Changes in treatment and husbandry practices	***Veterinary medicine:*** Advice to user on usage and husbandry methods to reduce environmental exposure, for example, pasture animals should not be allowed near water bodies for X number of days after administration of veterinary medicines.
Treatment of water and slurry	***Human medicine:*** Wastewater treatment; it is known that sorption removes fluorquinolones; bio-degradation removes benzafibrate, ethynylestradiol, diclofenac, ibuprofen, sulfamethoxazole; and ozonation removes antibiotics, antiphlogistics, beta blockers, and estrone; advanced drinking water treatment. ***Veterinary medicine:*** slurry and sludge storage; for example, tylosin degrades very rapidly during slurry storage (a mechanism to reduce amounts released to soils).
Good farming practice	***Human and veterinary medicine:*** sludge and slurry should be applied to land as good farming practice and they then reduce the potential for drugs to contaminate surface and groundwater.
Disposal	***Human and veterinary medicine:*** Appropriate disposal of unused medicines and containers including print advice for proper disposal on packaging and instruction leaflet; operate a communication campaign for proper disposal of unused/outdated pharmaceuticals
Research	Initiate specific research programmes for the development of sustainable pharmaceuticals such as development of environment-friendly pharmaceuticals.

12.11: Environmental water quality guidelines for pharmaceuticals

Currently there are no drinking water standards and health advisories for veterinary pharmaceuticals and prescription and non-prescription drugs in the USA (see Focazio *et al.* 2008). However, Australia has recommended drinking water guideline for pharmaceuticals to manage health and environmental risks, the guideline can be used for any environmental risk assessment (Table 12.20).

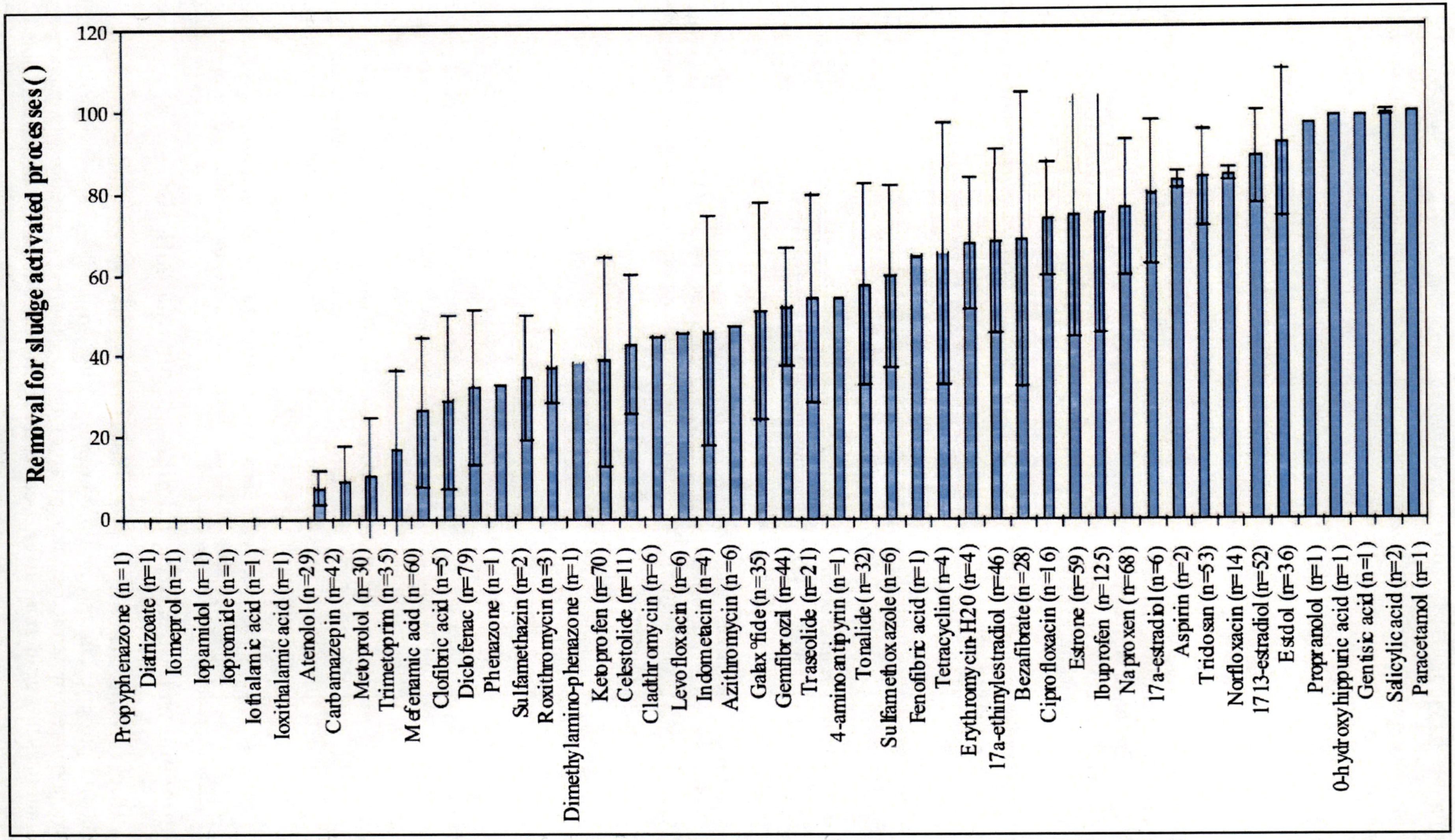

Figure 12.3: ***Mean removal efficiency (%) and relative standard deviation for pharmaceuticals and personal care products in wastewater treatment plants with activated sludge processes*** *[Miége et al. 2009; page 1724 data are based on 1724 international scientific publications; Note: High removal contaminants: 17β-estradiol, estriol, triclosan, norfloxacin; Low removal contaminants: atenolol, carbamazepine, clofibric acid mefenamic acid, metoprolol, trimethoprim].*

Table 12.20: ***Australian drinking water quality guideline values for pharmaceuticals*** *[NWQMS 2007].*

Pharmaceuticals	Australian recommended drinking water guideline (μg/l)	Pharmaceuticals	Australian recommended drinking water guideline (μg/l)	Pharmaceuticals	Australian recommended drinking water guideline (μg/l)
Antibiotics		*Nonsteroidal antiinflammatory drugs (NSAIDs)*		Other pharmaceuticals	
Amoxicillin	7,000	Aspirin (acetylsalicylic acid)	290	Cimetidine	2,000
Azithromycin	40	Diclofenac	18	Clenbuterol	150
Chloramphenicol	17,500	Dipyrone	5,250	Clofibric acid	7,500
Chlortetracycline	105	Fenoprofen	4,500	Codeine	500
Ciprofloxacin	2,500	Ibuprofen	4,000	Cotinine	100
Clarithromycin	2,500	Indomethacin	250	Cyclophosphamide	35
Clindamycin	3,000	Ketoprofen	35	Dehydronifedipine	200
Doxycycline	105	Tolfenamic acid	175	Diazepam	25
Enrofloxacin	220	β-androgenic blockers		Diltiazem	600
Erythromycin	175	Betaxolol	100	Enalaprilat	12.5
Lincomycin	35,000	Bisoprolol	6.25	Fluoxetine	100
Monensin	350	Carazolol	3.5	Gemfibrozil	6,000
Nalidixic acid	10,000	Metoprolol	250		
Oxytetracycline (Terramycin)	105	Nadolol	200	Iohexol	7,200
Penicillin G	15	Propranolol	400	Iopamidol	4,000
Penicillin V	15	Timolol	100	Iopromide	7,500
Roxithromycin	1,500	***Other pharmaceuticals***		Isophosphamide	35
Sulfadimethoxine	350	Acetaminophen (paracetamol)	1,750	Metformin	2,500
Sulfamethazine	350	Antipyrine	2.5	Methotrexate	0.05
Sulfamethiazole	350	Atrovastatin	50	Salbutamol	30
Sulfamethoxazole	350	Benzafibrate	750	Salicyclic acid	105
Tetracycline	105	Carbamazepine	3,000	Stigmastanol	10,000
Trimethoprim	700	Cefaclor	1,000	Sulfasalazine	5,000
Tylosin	10,500	Cephalexin	350	Temazepam	50

12.12: Conclusion

Several thousand tonnes of medicinally active substances are used every year in the world for humans and animals and excreted as parent compound or active metabolites, escaping degradation in STPs and entering freshwater. Pharmaceuticals are biologically active compounds that are intended not to be easily degradable and are often water soluble and therefore can be found in wastewaters and in natural waters. As methods for detection improves it is becoming more evident that pharmaceuticals are ubiquitous pollutants in both ground and surface waters and even drinking water. Furthermore, the presence of human and veterinary pharmaceuticals in the environment is of increasing global concern due to their effects on ecological receptors. Potentially, they could have similar function or cause side effects in non-target organisms as they do to their intended users.

The spread of antibiotics in the environment contributes to bacterial resistance or transfer of drugs through food chain, biomagnification. Antibiotics were recently classified as a priority risk group due to their high toxicity to algae and bacteria at low concentrations and their potential cause of resistance amongst natural bacterial populations. Antibiotics in surface waters have the potential to disrupt key bacterial cycles/processes critical to aquatic ecology (nitrification/denitrification) or agriculture (soil fertility) and animal production (rudimentary processes). Ecological disasters such as the decline of vultures in Bangladesh, Pakistan and India due to diclofenac toxicity are one of the most important ecological effects of pharmaceuticals.

The toxicity data of many pharmaceutical compounds are limited to acute effects of individual compounds only, though it is now apparent that the chronic toxicity data and information are more important. Acute toxicity of pharmaceuticals to aquatic organisms may not occur at measured environmental concentrations as acute effect concentrations of many pharmaceuticals are 100-1,000 times higher than residues found in the aquatic environment. For example, the lowest acute effect concentration of fluoxetine was reported to be 20 μg/l, whereas the highest estimated environmental concentration was detected in the environment at 0.01μg/l. Many pharmaceuticals have not yet been studied as extensively as others and reliable toxicity data are limited to acute effects only. The chronic toxicity data are important to assess long-term effects from exposure of pharmaceuticals to human and other biota.

Relatively little is known about the presence of pharmaceuticals in the environment in particular implications for human health and ecosystems from chronic toxicity and continuous exposure to multiple compounds. Safety threshold values for pharmaceutical compounds are often related to single compound-single organism toxicity studies though the impact of a mixture of chemicals could be more toxic than that of the individual compound alone (Chévre *et al.* 2008). Quinn *et al.* (2009) exposed freshwater cnidarian *Hydra attenuate* to a mixture of eleven pharmaceuticals (benzafibrate and gemfibrozil, ibuprofen, naproxen carbamazepine), antibiotics (novobiocin, oxytetracycline, sulfamethizole, sulfapyridine, trimethoprim) and caffeine (stimulant). Quinn *et al* (2009) concluded that pharmaceuticals act additively having sub-lethal effects observed at environmentally relevant concentrations. Cleuvers (2003) has shown that clofibric acid in combination with carbamazepine as well as diclofenac in combination with ibuprofen exerted a much stronger toxic effect to *Daphnia* spp. and algae than that could be expected from the very weak toxic effects seen after exposure to each compound individually. Therefore, combined effects of pharmaceuticals on aquatic organisms require further investigation since each compound could have little effect but when acting together they may pose a hazard, which may be underestimated by focusing on individual compound.

12.13: References

Ågerstrand, M. and Rudén, C. 2009. The Swedish environmental classification and information system for pharmaceuticals – an empirical investigation of the motivations, intentions and expectations underlying its development and implementation. *Environment International*, doi: 10.1016/j.envint.2008.12.001.

Barnes , K. K., Kolpin, D. W., Furlong, E. T., Zaugg, S. D., Meyer, M. T. and Barber, L. B. 2008. A national reconnaissance of pharmaceuticals and other organic wastewater contaminants in the United States. 1 Groundwater. *Science of the Total Environment* 402: 192-200.

Berry, M. A. and Rondinelli, D. A. 2000. Environmental management in the pharmaceutical industry: integrating corporate responsibility and business strategy. *Environmental Quality Management* spring 2000: 21-35.

Besse, J-H and Garric, J. 2008. Human pharmaceuticals in surface waters. Implementation of a prioritisation methodology and application to the French situation. *Toxicology Letters* 176: 104-123.

Boxall, A.B.A., Kolpin,D.W., Halling- Sørensen,B and Tolls, J. 2003. Are veterinary medicines causing environmental risks? *Environmental Science and Technology* 37 : 286-294A.

Boxall, A. B. A. 2004. The environmental side effects of medication. *EMBO Reports* 5(12): 1111-1116.

Bound, J. P., Kitsou, K., Voulvoulis, N. 2006. Household disposal of pharmaceuticals and perception of risk to the environment. *Environmental Toxicology and Pharmacology*.21: 301-307.

Carlsson, C., Johansson, A-K., Alvan, G., Bergman, K. and Kühler, T. 2006. Are pharmaceuticals potent environmental pollutants? Part 1: Environmental risk assessments of selected active pharmaceutical ingredients. *Science of the Total Environment* 364: 67-87.

CEC 1996. Commission of the European Communities (CEC). Technical guidance document in support of commission directive 93/67/EEC on risk assessment for new modified substances. Part II. Environmental risk assessment. Luxembourg: Office for official publication of the European Communities.

Chenxi, W., Spongberg, A. L and Witter, J. D. 2008. Determination of the persistence of pharmaceuticals in biosolids using liquid-chromatography tandem mass spectrometry. *Chemosphere* 73: 511-518.

Chévre, N., Maillard, E., Loepfe, C. and Becker-van Slooten, K. 2008. Determination of water quality standards for chemical mixtures: Extension of a methodology developed for herbicides to a group of insecticides and a group of pharmaceuticals. *Ecotoxicology and Environmental Safety* 71: 740-748.

Cluvers, M. 2003. Aquatic ecotoxicity of pharmaceuticals including the assessment of combination effects. *Toxicology Letter*. 142: 185-194.

Collier, A.C. 2007. Pharmaceutical contaminants in potable water: potential concerns for pregnant women and children. *EcoHelath* 4: 164-171.

Copper, E. R., Siewicki, T. C. and Phillips, K. 2008. Preliminary risk assessment database and risk ranking of pharmaceuticals in the environment. *Science of the Total Environment*. 398: 26-33.

Daughton, C. G. 2003a. Cradle-to-cradle stewardship of drugs for minimizing the deposition whilst promoting human health. I. Rationale for and avenues toward green pharmacy. *Environmental health Perspective* 111: 757-774.

Daughton, C. G. 2003b. Cradle-to-cradle stewardship of drugs for minimizing the deposition whilst promoting human health. II. Drug, disposal, waste reduction and future directions. *Environmental health Perspective* 111: 775-785.

Daughton, C.G. and Ternes, T.A. 1999. Pharmaceuticals and personal care products in the environment: agents of subtle change? *Environmental Health Perspectives* 107 (6): 907-937.

Díaz-Cruz, M. S., López de Alda, M. J and Barceló, D. 2003. Environmental behaviour and analysis of veterinary and human drugs in soils, sediments and sludge. *Trends in Analytical Chemistry*, 22(6): 340-351.

EMEA (European Medicine Agency) 2006. Committee for medicinal products for human use (CHMP), Guideline on the environmental risk assessment of medicinal products for human use: 2006. Ref. EMEA/ CRMP/ SWP/4 447/00.

Enick, O. V. and Moore, M. M. 2007. Assessing the assessments: pharmaceuticals in the environment. *Science Direct* 27: 707-729.

Faràe, M. L., Pérez, S., Kantiani, L. and Barceló. D. 2008. Fate and toxicity of emerging pollutants, their metabolites and transformation products in the aquatic environment. *Trends in Analytical Chemistry* 27 (11): 991-1007.

FDA (Food and Drug Administration of USA) 1998. Guidance for Industry-Environmental assessment of human drugs and biological applications. FDA, CDER/CBER, CMC 6, rev.1.

Fent, K., Weston, A. A. and Caminada, D. 2006. Review- Ecotoxicology of human pharmaceuticals. *Aquatic Ecotoxicology* 76: 122-159.

Ferrari, B., Paxeus, N., Lo Guidice, R., Pollio, A and Garric, J. 2003. Ecotoxicological impact of pharmaceuticals found in treated wastewaters: study of carbamazepine, clofibric acid, and diclofenac. *Ecotoxicology and Environmental safety* 55 (3): 359-370.

Feitossa-Felizzola, J. and Chiron, S. 2009. Occurrence and distribution of selected antibiotics in a small Mediterranean stream (Arc River, Southern France). *Journal of Hydrology* 364: 50-57.

Focazio, M. J., Kolpin, D. W., Barnes, K. K., Furlong, E. T., Meyer, M. T., Zaugg, S. D., Barber, L. B. and Thurman, M. E. 2008. A national reconnaissance of pharmaceuticals and other organic wastewater contaminants in the United States. II. Untreated drinking water sources. *Science of the Total Environment* 402: 201-216.

Fountoulakis, M. S., Stamatelatou, K. and Lyberatos, G. 2008. The effect of pharmaceuticals on the kinetics of methanogenesis and acetogenesis. *Bioresource Technology* 99: 7083-7090.

Gibs, J., Stackelberg, P. E., Furlong, E. T., Meyer, M., Zaugg, S. D., and Lippincott, R. L. 2007. Persistence of pharmaceuticals and other organic compounds in chlorinated drinking water as a function of time. *Science of the Total Environment* 373: 240-249.

Grung, M., Källqvist, T., Sakshaug, S., Skurtveit, S and Thomas, K. V. 2008. Environmental assessment of Norwegian priority pharmaceuticals based on EMEA guideline. *Ecotoxicology and Environmental Safety* 71: 328-340.

Halling-Sørensen, B., Nielsen, S. N., Lanzky, P. F., Ingerslev, F., Lützhøft and Jørgensen, S. E. 1998. Occurrence, fate and effects of pharmaceutical substances in the environment – A review. *Chemosphere* 36 (2): 357-393.

Hernando, M. D., Mezcua, M., Fernández-Alba, A. R and Barceló, D. 2006. Environmental risk assessment of pharmaceutical residues in wastewater effluents, surface waters and sediments. *Talanta* 69: 334-342.

Jjemba, P.K. 2006. Excretion and ecotoxicity of pharmaceutical and personal care products in the environment. *Ecotoxicology and Environmental Safety* 63 : 113-130.

Jones, O. A., Lester, J. N. and Voulvoulis, N. 2005. Pharmaceuticals: a threat to drinking water?. *Trends in Biotechnology*. 23 (4): 163-167.

Kemper, N. 2008. Veterinary antibiotics in the aquatic and terrestrial environment. *Ecological Indicators*. 8: 1-13.

Khalaf, H., Salste, L., Karlsson, P., Ivarsson, P., Jass, J. and Olsson, P-E. 2009. In vitro analysis of inflammatory responses following environmental exposure to pharmaceuticals and inland waters. *Science of the Total Environment* 407: 1452-1460.

Kim, Y., Jung, J., Kim, M., Park, J., Boxall, A. B. A and Choi, K. 2008. Prioritizing veterinary pharmaceuticals for aquatic environment in Korea. *Environmental Toxicology and Pharmacology* 26: 167-176.

Kotchen, M., Kallaos, J., Wheeler, K., Wong, C. and Zahller, M. 2009. Pharmaceuticals in wastewater: Behaviour, preferences, and willingness to pay for a disposal program. *Journal of Environmental Management* 90: 1476-1482. doi: 10.1016/j.envman.2008.10.002

Kümmerer, K. 2004. Resistance in the environment. *Journal of Antimicrobial Chemotherapy*. 54: 311-320. doi:10.1093/jc/dkh325.

Kümmerer, K. 2009. Antibiotics in the aquatic environment- A review-part I. *Chemosphere* doi:10.1016/j.chemosphere.2008.11.086.

Lavén, M., Alsberg, T., Yu, Y., Adolfsson-Erici, M. And Sun, H. 2009. Serial-mixed-mode cation- and anion-exchange solid –phase extraction for separation of basic, neutral and acidic pharmaceuticals in wastewater and analysis by high-performance liquid chromatography-quadrupole time-of –light mass spectrometry. *Journal of Chromatography A*. 1216: 49-62.

Laville, N., Aït-Aïssa, S., Gomez, E., Casellas, C. ad Porcher, J. M. 2004. Effects of human pharmaceuticals cytotoxicity, EROD activity and ROS production in fish hepatocytes. *Toxicology* 196: 41-55.

Lin, A. Y-C., Yu, T-H., Lin, C-F. 2008. Pharmaceutical contamination in residential, industrial and agricultural waste streams: Risk to aqueous environments in Taiwan. *Chemosphere* 74: 131-141.

Miége, C., Choubert, J. M., Ribeiro, L., Eusébe, M. and Coquery, M. 2009. Fate of pharmaceuticals and personal care products in wastewater treatment plants-conception of a database and first results. *Environmental Pollution*, doi.10.1016/j.envpol.2008.11.045.

Mompelat, S., LeBot, B. and Thomas, O. 2008. Occurrence and fate of pharmaceutical products and by–products, from resource to drinking water. *Environment International*. Doi.10.1016/j.envint.2008.10.008.

New Scientist. 2009. Top 11 compounds in US drinking water. 12 January. http://www.newscientist.com/article/dn16397-top-11-compounds-in-us-drinking-water.html

Nunes, B., Gaio, A. R., Carvalho, F. and Guilhermino, L 2008. Behaviour and biomarkers of oxidative stress in *Gambusia holbrooki* after acute exposure to widely used pharmaceuticals and a detergent. *Ecotoxicology and Environmental Safety* 71: 341-354.

NWQMS (National Water Quality Management Strategy) 2007. Australian guidelines for water recycling: Managing heath and environmental risks (phase 2) - Augmentation of drinking water supplies. Natural Resource Management Ministerial Council. Environment Protection and Heritage Council. Australian Health Ministries Conference. Biotex Pty Ltd., Canberra.140 pp.

Pan, B., Ning, P and Xing, B. 2008. Part V. sorption of pharmaceuticals and personal care products. *Environmental Science and Pollution Research,* doi 10.1007/s11356-008-0052-x.

Pounds, N., Maclean, S., Webley, M., Pascoe, D. and Hutchinson, T. 2008. Acute and chronic effects of ibuprofen in the mollusc *Planorbis carinatus* (Gastropoda: Planorbidae). *Ecotoxicology and Environmental Safety* 70: 47-52.

Quinn, B., Gagné, F and Blaise, C. 2009. Evaluation of the acute, chronic and teratogenic effects of a mixture of eleven pharmaceuticals on the cnidarian, *Hydra attenuata. Science of the Total Environment* 407: 1072-1079; doi:10.1016/ j.scitotenv.2008.10.022.

Richman, C. and Castensson, S. 2008. Impact of waste pharmaceuticals: an environmental hazard or greenwash". *The Pharmaceutical Journal* 280: 335-342.

Roig, B., Greenwood, R. and Barcelo, D. 2009. Editorial - An international conference on "pharmaceuticals in the environment" in a frame of EU Knappe project. *Environment International.* Doi:10.1016/j.envint.2009.01.012.

Samuelsen, O. B., Lunestad, B. T., Ervik, A and Fjelde, S. 1994. Stability of antibacterial agents in an artificial marine aquaculture sediment studied under laboratory conditions. *Aquaculture* 126 (3-4): 283-290.

Sanderson, H., Brain, R. A., Johnson, D. J., Wilson, C. J. and Solomon, K. R. 2004. Toxicity classification and evaluation of four pharmaceuticals classes: antibiotics, antineoplastics, cardiovascular, and sex hormones. *Toxicology* 203: 27-40.

Sanderson, H. and Thomsen, M. 2008. Comparative analysis of pharmaceuticals versus industrial chemicals acute toxicity classification according to the United Nations classification system for chemicals. *Toxicology Letters;* doi:10.1016/j.toxlet.2009.02.003.

Santos, J. L., Aparicio, I. and Alonso, E. 2007. Occurrence and risk assessment of pharmaceutically active compounds in wastewater treatment plants: a case study: Seville city (Spain). *Environment International* 33: 596-601.

Serrano, P. H. 2005. *Responsible use of antibiotics in aquaculture. Food and Agriculture Organisation of the United Nations (FAO),* Fisheries Technical paper 469. Rome. 97p.

Spongberg, A. L. and Witter, J. D. 2008. Pharmaceutical compounds in the wastewater process stream in Northwest Ohio. *Science of the total Environment.* 397: 148-157.

Ternes, T. A., Meisenheimer,M., McDowell, D., Sacher, F., Brauch,H.J., Haist-Guide, B., Preuss, U., Wilme, U and Zulei-Seibert, N. 2002. Removal of pharmaceuticals during drinking water treatment. *Environment Sience and Technology* 36 : 3855-3863.

Ternes, T. A., Joss, A., Siegrist, H. 2004. Scrutinizing pharmaceuticals and personal care products in wastewater treatment. *Environmental Science Technology* 38: 393-398.

Titz, A. and Döll, P. 2009. Actor modelling and its contribution to the development of integrative strategies for management of pharmaceuticals in drinking water. *Social Science and Medicine* 68: 672-681.

Thiele-Bruhn, S. 2003. Pharmaceutical antibiotic compounds in soils- a review. *Journal of Plant Nutrition and Soil Science* 166: 145-167.

Watkinson, A. J., Murby, E. J., Kolpin, D. W. and Costanzo, S. D. 2009. The occurrence of antibiotics in an urban watershed: from wastewater to drinking water. *Science of the Total Environment; doi:10.1016/j.scitoenv.2008.11.059.*

Webb, S. F. 2004. A data based perspective on the environmental risk assessment of human pharmaceuticals. II. Aquatic risk characterisation. *In*: Kümmerer, K. (ed.) *Pharmaceuticals in the environment, sources, fate, effects and risks.* Springer-Verlag. Berlin, pp. 319-343.

WWEE (World Water and Environment Engineering) 2009. Surveys reveal disturbing pharmaceutical levels in water. January/February, Volume 32(1): 8.

Yamamoto, H., Nakamura, Y., Moriguchi, S., Nakamara, Y., Honda, Y., Tamura, I., Hirata, Y., Hayashi, A. and Sekizawa, J. 2008. Persistence and partitioning of eight selected pharmaceuticals in the aquatic environment: Laboratory photolysis, biodegradation, and sorption experiments. *Water Research;* doi:10.1016/J.waters.2008.10.039.

Zorita, S., Mårtensson, L and Mathiasson, L. 2009. Occurrence and removal of pharmaceuticals in a municipal sewage treatment system in the south of Sweden. *Science of the Total Environment* 407: 2760-2770.

Zuccato, E., Castiglioni., Fanelli, R., Rietano, G., Bagnati, R., Chiabrando, C., Pomati, F., Rossetti, C and Clamari, D. 2006. Pharmaceuticals in the environment in Italy: causes, occurrence, effects and control –review. ESPR-*Environmental Science and Pollution Research* 13 (1): 15-21.

Zhou, J. L., Zhang, Z. L., Banks, E., Grover, D. and Jiang, J. Q. 2008. Pharmaceutical residues in wastewater treatment work effluents and their impact on receiving river water. *Journal of Hazardous Materials;* doi:10.1016/j.hazmat.2008.11.070.

Website

http://10.10.2.119:8080/00123f24cda4/comfort-file/58b7_9384d2b6_e11b_11dd_b279_00123f24cda4/a0282e00.pdf

http://redgreenandblue.org/2009/01/25/record-levels-of-pharmaceuticals-found-in-indias-water/

http://www.redorbit.com/news/health/1628532/pharmaceutical_waste_found_in_indias_streams/

□□□

13 Chemical Impacts - Biotoxins

Key facts

Freshwater biotoxins

- Biotoxins are naturally occurring toxic compounds produced by some freshwater cyanobacteria (blue-green algae) and marine algal species (dinoflagellates, diatoms).
- Cyanobacteria are now recognised as a serious water quality problem with regard to drinking water supply and recreational use.
- Environmental conditions that can promote cyanobacterial/blue-green algal blooms are: excessive nutrients (phosphorus and nitrogen); warm water temperatures (15-30°C); sunlight for photosynthesis; and quiescent or stagnant water.
- Outbreak of human poisoning attributed to cyanobacteria are neurotoxins (e.g. anatoxin-a, anatoxin-a(s), saxitoxins) and hepatotoxins (e.g. microcystins, nodularin) produced by some strains of the cyanobacteria including *Microcystis, Anabaena, Aphanizomenon* spp.
- Cyanotoxins are intracellular toxins and released to the surrounding waters during cell senescence, death and lysis or water treatment processes. Toxins in the water bodies at the time of a bloom will be present in both the water (free, dissolved or extracellular toxins) and the cyanobacterial cells (intracellular).
- Microcystin toxins are the most commonly reported cyanotoxins world wide.
- Rape, rice and lettuce plants exposed to different levels of microcystins showed an accumulation of toxins. The cyanotoxins such as microcystins can be taken up by plants and when, present in sufficient concentrations can induce various metabolic and morphological changes such as inhibition of photosynthesis, germination and growth inhibition, decrease in roots development and leaf necrosis.
- Cyanobacterial toxins such as microcystins may be accumulated in organs of freshwater fish, mussels and shrimps. Investigations carried out in China, Brazil, Egypt and Portugal reveals that microcystin levels in edible parts of freshwater seafoods (e.g. fish/shrimp/mussel muscle and other organs) have exceeded the WHO recommended limit for the tolerable daily intake (TDI) in many occasions for human consumption of 0.04 µg/kg per day of MC-LR.
- Toxic blue-green algae can affect water quality by causing peculiar tastes and odours, discolouration and unsightly scums. Geosmin (E1, 10-dimethyl-E-9-decalol) and MIB (2-methyl isoborneol) accounts for the majority of off-flavour odour.
- Animals killed due to cyanobacterial toxins are birds, cats, cattle, chicken, dogs, ducks, fish, horses, human, lambs, monkeys, sheep, squirrels, pigs, rhinoceros, rodents and invertebrates.
- The most serious human poisoning due to cyanobacterial toxins was the death of more than 50 patients due to liver failure received contaminated dialysis water in Brazil in 1996.
- The recommended WHO guideline for microcystin for the protection of human health is 1.0 µg/l.

Marine biotoxins

- Marine biotoxins are caused by marine micro-algal species such as dianoflagellates and diatoms, which produce potent toxin called 'phycotoxins'. About 109 species of microalgae are known to produce toxins, of which 70 are dinoflagellates.

- The most important marine algal toxins come under the following categories: amnesic shellfish poisoning (ASP), diarrhetic shellfish poisoning (DSP), neurotoxic shellfish poisoning (NSP), paralytic shellfish poisoning (PSP) and ciguatera fish poisoning (CFP).
- Humans are exposed to marine toxins through consumption of shellfish (oysters, mussels, scallops) and some tropical reef fish (barracuda, coral grouper, red snapper, sword fish, parrot fish).
- PSP is a life threatening syndrome with a mortality rate of approximately 15-20% in humans of those intoxicated.
- Monitoring of the marine environment (for the occurrence of toxic algal species) and fishery products (for toxin levels) and appropriate regulations may prevent shellfish poisoning incidents.
- Regulatory authorities in many countries have set the maximum permitted levels (MPL) of marine toxins in food to protect human health which are as follows: APS (domoic acid) at 2,000 μg/ 100 g shellfish flesh; DSP (okadaic acid) at 16-20 μg/ 100 g shellfish flesh; , NSP (brevetoxins) at 80 μg/ 100 g shellfish flesh); PSP (saxitoxins) at 80 μg/ 100 g shellfish flesh and CFP (ciguatoxins) at 3.5 μg/ 100 g fish flesh.

Contents

List of Tables (Freshwater biotoxins)

List of Figures (Freshwater biotoxins)

List of Boxes (Freshwater biotoxins)

List of Tables (Marine biotoxins)

List of Figures (Marine biotoxins)

Acronyms and glossary

Action potential: In neurophysiology, the action potential is a self-regenerating wave of electrochemical activity that allows nerve cells to carry a signal over a distance. It is the primary electrical signal generated by nerve cells and arises from changes in the permeability of the nerve cell's axonal membranes to specific ions. Action potentials (also known as nerve impulses or spikes) are pulse-like waves of voltage that travel along several types of cell membranes. The best-understood example of an action potential is that which is generated on the membrane of the axon of a neuron but also appears in other types of excitable cells, such as cardiac muscle cells and even plant cells.

Aerosol: A gaseous suspension of fine solid or liquid particles.

Agonist: Drug that binds to a receptor of a cell and triggers a response by the cell. An agonist often mimics the action of a naturally occurring substance. An agonist produces an action. It is the opposite of an antagonist which acts against and blocks an action. Agonists and antagonists are key agents in the chemistry of human body and important players today in pharmacology.

Alga: Primitive, mainly aquatic, one celled or multicellular plants that lack true, stems, roots and leaves containing green pigment - chlorophyll.

Algal bloom: Prolific growth of algae.

Alkaloid: Classes of over 3,000 nitrogen-containing chemicals that are produced by plants and have effects in humans and animals.

Alzheimer's diseases: Alzheimer's (pronounced ALTS-hih-merz) disease (AD) is the most common form of dementia (pronounced dih-MEN-sha). Dementia is a disease of the nervous system characterized by loss of certain mental abilities. This loss is severe enough to interfere with normal activities and lasts at least six months. AD is not present at birth but usually develops during old age. It is marked by a decline in mental functions such as memory, reasoning and the ability to plan.

Amnesic shellfish poisoning: Human illness caused by eating shellfish that are contaminated with domoic acid from diatom *Pseudo-nitzschia* spp.

Amphoteric: In chemistry, an amphoteric substance is one that can react as either an acid or base.

Anal sphincter: The circular band or ring of muscle that controls the release of faeces from the anus.

Anatoxin-a: Is a secondary, bi-cyclic amine alkaloid and cyanotoxin with acute neurotoxicity. The toxin is produced by at least four different genera of cyanobacteria and has been reported in North America, Europe, Africa, Asia and New Zealand. Symptoms of anatoxin exposures are loss of coordination, muscular fasciculations, convulsions and death due to respiratory paralysis.

Antagonist: In biochemistry, an antagonist acts against and blocks an action. For example, insulin lowers the level of glucose (sugar) in the blood, whereas another hormone called glucagon raises it; therefore, insulin and glucagon are antagonists. An antagonist is the opposite of an agonist which stimulates an action. Antagonists and agonists are key players in pharmacology and in the chemistry of human body.

ASP: Amnesic shellfish poisoning

Asphyxiation (suffocation) : The condition of being deprived of oxygen (as by having breathing stopped).

Ataxia: Inability to coordinate voluntary muscle movements; unsteady movements and staggering gait.

Ballast water: Water taken up or released by a ship to stabilize it, or to raise/lower it in the water column; seawater is used by vessels to provide stability, to adjust the vessel's draft for various loading conditions and to position the ship to on-load or off-load amphibious vehicles. Generally seawater is provided to the ballast tanks by a combination of flooding from the sea through valves in the tanks and from the fire main.

Barcoo fever: Is an illness once common in the Australian outback but is now virtually unknown. It is postulated that the disease is due to ingestion of cyanobacterial (blue-green algal) toxins, in particular cylindrospermopsin, a toxin from *Cylindrospermopsis raciborskii* and other cyanobacteria, which is a hepatotoxin. The symptoms of the disease are consistent with a hepatitis or liver disorder and Cylindrospermopsis is known to be widespread in inland Australian water sources.

Benthic: Mussels or fish which lives at great depths, such as sea floor preferring a cold water environment.

Bradycardia: Slowness of the heart rate, usually fewer than 60 beats per minute in an adult human.

Bio-accumulation: Accumulation of a substance in a living organism as a result of its intake both in its food and also from the environment.

Biotoxins: A toxin (poison) which originates from a living thing (a plant, animal, fungi, bacteria, etc).

Biovolume: A measure of the volume of space occupied by a biological individual such as blue-green algae or a group of individual. Biovolume is used as a quantitative measure of the volume of cell material of algae (here cyanobacteria) in an environmental sample.

Blue-green algae: Primitive single or multi-celled algae, characterized by the absence of a nucleus and other membrane bound organelles and often by their blue-green colour.

Brevitoxins: Lipophilic; 10 and 11 ring polyether chemicals which can cause neurotoxic shellfish poisoning (NSP).

Calcium channel: A calcium channel is an ion channel which displays selective permeability to calcium ions. It is sometimes synonymous as voltage-dependent calcium channel.

Chlorophyll-a: Chlorophyll a is a form of chlorophyll. Chlorophyll a is mainly used in light reactions and used in photosynthesis.

Ciguatera: Human illness caused by eating certain coral reef fish that are contaminated with ciguatoxin and maitotoxin from dinoflagellate, *Gambierdiscus toxicus*.

Ciguatoxins: Large, heat soluble, polyether produced by certain strains of *Gambierdiscus toxicus* found in tropical and sub-tropical water around the world and are responsible for the poisoning syndrome known as ciguatera.

Congeners: A member of the same kind, class, or group.

Cognitive: Is a psychiatric diagnosis that describes a mental disorder characterized by abnormalities in the perception or expression of reality.

Corkscrew: Moving in a spiral or zigzag course.

Cyanobacteria: Also known as blue-green algae, blue-green bacteria or Cyanophyta, is a phylum of bacteria that obtain their energy through photosynthesis. The name 'cyanobacteria' comes from the colour of the bacteria (Greek: κυαυοζ (kyanós) = blue).

Cyanotoxins: A cyanotoxin is a toxin produced by certain cyanobacteria (blue-green algae). Cyanotoxins are biotoxins and responsible for acute and (sub) chronic poisoning of wild/domesticated animals and humans.

Cyanophage: Cyanophage is a virus that infects cyanobacteria.

Cytoskeleton: The cytoskeleton (also CSK) is a cellular 'scaffolding' or 'skeleton' contained within the cytoplasm.

Cytotoxic: The degree to which something is toxic to living cells.

Cylindrospermopsin: A cyclic alkaloid produced by cyanobacteria, *Cylindrospermopsis raciborskii* that can be very toxic for plants and animals including humans.

DA: Domoic acid, a type of biotoxin produced by marine diatom.

Daltons: A unit of mass equal to $^1/_{12}$ the mass of the most abundant isotope of carbon, carbon 12, which is assigned a mass of 12. Also called Dalton; a unit of molecular mass. Another name for atomic mass unit. One hydrogen atom weighs one dalton.

Depolarization: In biology, depolarization is a decrease in the absolute value of a cell's membrane potential. Thus, changes in membrane voltage in which the membrane potential becomes less positive or less negative are both depolarizations. In neurons and some other cells, a depolarization large enough may result in an action potential. Hyperpolarization is the opposite of depolarization and inhibits the rise of an action potential. The rising and falling phases of such an action potential are often imprecisely also called depolarization and hyperpolarization, respectively.

Diarrhoea: Frequent and watery bowel movements; can be a symptom of infection or food poisoning or colitis or a gastrointestinal disorder.

Diastereomer: One of two or more compounds that have the same chemical composition and that have the same atoms joined to each other, but that are not mirror images of each other.

Diatoms: Singled-celled golden-brown algae, with siliceous cell walls consisting of two overlapping symmetrical parts.

Dinoflagellates: Singled-celled aquatic organisms bearing two dissimilar flagella and having characteristics of both plants and animals, golden-brown algae, characteristically having a large nucleus with clearly visible chromosomes with one flagellum (whip-like expansions of the cells used for locomotion) protruding from the girdle groove and another from the sulcus groove.

Dinophysis: Toxic marine algae (dinoflagellates) that can be found within some shellfish.

Dinophysis toxins: Heat-stable polyether and lipophilic toxic compounds isolated from dinoflagellates.

Diode: A diode is a specialized electronic component with two electrodes called the anode and the cathode. Most diodes are made with semi-conductor materials such as silicon, germanium or selenium. Some diodes are comprised of metal electrodes in a chamber evacuated or filled with a pure elemental gas at low pressure. Diodes can be used as rectifiers, signal limiters, voltage regulators, switches, signal modulators, signal mixers, signal demodulators and oscillators.

Domoic acid: A water soluble toxic amino acid mimic produced by the marine diatoms. *Pseudo-nitzschia* spp. It is responsible for Amnesic shellfish poisoning (ASP).

DSP: Diarrhetic shellfish poisoning caused by dinoflagellates of the genus *Dinophysis*. Some also referred to as diarrhoeic shellfish poisoning.

Dysarthria: Impaired articulatory ability resulting from defects in the peripheral motor nerves or in the speech musculature. Dysarthria is a motor speech disorder resulting from neurological injury, characterised by poor articulation (cf. aphasia: disorder of the content of speech). Any of the speech subsystems (respiration, phonation, resonance, prosody, articulation and movements of jaw and tongue) can be affected.

EC_{50}: The concentration of material in water that is estimated to be effective in producing some lethal response in 50% of the test organisms.

Edema: Swelling from excessive accumulation of watery fluid in cells, tissues, or serous cavities.

Embayment: An area of water protected by land forming a bay such as Maumee Bay.

Epidemiology: The study of the distribution and determination of health/disease states in human populations.

Eutrophication: High levels of nutrients, usually phosphorus and nitrogen that can cause increases in natural biological production in rivers, lakes and reservoirs or degradation of water quality due to enrichment by nutrients (nitrogen and phosphorus), resulting in excessive algal growth and decay and often with low dissolved oxygen in the water.

Excitatory: An excitatory substance that excites the nerve cell and makes it able to receive critical information.

Gastroenteritis: Inflammation of the stomach and intestines causing abdominal pain, diarrhoea and vomiting. It may be caused by infection, food poisoning or allergy. Severe cases can cause dehydration.

Glutathione S-transferase and glutathione peroxidase activities : An enzyme. Enzymes of the glutathione S-transferase (GST) family are composed of many cytosolic, mitochondrial, and microsomal proteins. GSTs catalyse a variety of reactions and accept endogenous and xenobiotic substrates. They are members of MAPEG family of transmembrane proteins.

Glutamate: An amino acid precursor. Glutamate is the most common excitatory neurotransmitter in the brain and spinal cord.

Glutamyltransferase (-GT): An enzyme that catalyses reversibly the transfer of a glutamyl group from a glutamyl-peptide and an amino acid to a peptide and a glutamyl-amino acid. It is often elevated above normal in hepatitis and other conditions that cause chemical liver damage.

Guanidine: Is a crystalline compound of strong alkalinity formed by the oxidation of guanine. It is used in the manufacture of plastics and explosives. It is found in urine as a normal product of protein metabolism.

Haemorrhage: In general term, bleeding, the flow of blood from a ruptured blood vessel.

Hepatotoxins: Toxins that cause degeneration of liver cells.

Hepatopancreas: The hepatopancreas is an organ of the digestive tract of arthropods, gastropods (shellfish) and fish. It provides the functions which in mammals are provided separately by the liver and pancreas.

Hyperplasia: Or 'hypergenesis' is a general term referring to the proliferation of cells within an organ or tissue beyond that which is ordinarily seen (e.g. constantly dividing cells). Hyperplasia may result in the gross enlargement of an organ, the formation of a benign tumour or may be visible only under a microscope. Hyperplasia is considered to be a physiological response to a specific stimulus and the cells of a hyperplastic growth remain subject to normal regulatory control mechanisms.

Hydrophilic: Having affinity for water (polar).

Hyperosmotic: Of relating to or characterized by increased osmosis. Hypertonic, containing a higher concentration of salts or other dissolved materials than normal tissues.

Hyperphosphorylation: Hyperphosphorylation occurs when a biochemical with multiple phosphorylation sites is fully saturated. Hyperphosphorylation is one of the signalling mechanisms used by the cell to regulate mitosis. When these mechanisms fail, developmental problems or cancer are a likely outcome. The mechanism appears to be largely conserved throughout eukaryote species.

Hypotension: Abnormally low blood pressure.

Hypovolemic shock: Shock caused by severe blood or fluid loss.

Intoxication: Is the state of being affected by one or more psychoactive drugs. It also refers to the effects caused by the ingestion of poison or by the over consumption of normally harmless substances.

Iminium: An iminium salt or cation in organic chemistry, has the general structure $[R^1R^2C = NR^3R^4]+$ and is as such a protonated or substituted imine.

i.p: Intraperitoneal injection into the gut or peritoneum, a common method for injecting drugs into extra cellular fluid for gradual absorption into the blood stream.

Intraperitoneal: Intraperitoneal injection or IP injection is the injection of a substance into the peritoneum (body cavity).

Irritation: An observable physiological reaction by the body (i.e., skin, eyes, nose and throat) to a stimulus or substance.

Kainate: Any salt or ester of kainic acid agonist for the K-type excitatory amino acid receptor. Kainic acid is a natural marine acid present in some algae. It is a specific agonist for the kainate receptor used as an ionotrophic glutamate.

LC_{50}: Lethal concentration of material in water that is estimated to be lethal to 50% of the test organisms. The LC_{50} is usually expressed as a time-dependent value, e.g. 24 hour or 96 hour; LC_{50}, the concentration estimated to be lethal to 50% of the test organisms after 24 or 96 hours of exposure.

LD_{50}: Appropriate to use with test animals such as rats, mice and dogs. It is rarely applicable to aquatic organisms because it indicates the quantity (dose, the lethal dose) of a material introduced directly into the body by injection or ingestion rather than the concentration of the material in water in which aquatic organisms are exposed during toxicity tests.

Lipophilic: Having an affinity for lipids; literally 'fat-loving'. Applied to molecular entities (or parts of molecular entities) having a tendency to dissolve in fat-like (e.g. hydrocarbon) solvents.

Lipopolysaccharide: Also known as lipoglycans, are large molecules consisting of a lipid and a polysaccharide joined by a covalent bond; they are found in the outer membrane of Gram-negative bacteria,

Lysis: Destruction.

Macrolides: The macrolides are a group of drugs (typically antibiotics) whose activity stems from the presence of a macrolide ring, a large macrocyclic lactone ring to which one or more deoxy sugars, usually cladinose and desosamine, may be attached. The lactone rings are usually 14, 15 or 16-membered. Macrolides belong to the polyketide class of natural products. In more technical terms, the macrolides are a group of antibiotics produced by various strains of Streptomyces (spore forming bacteria that grow slowly in soil or water as a branching filamentous mycelium similar to that of fungi) and have a complex chemical (macrocyclic) structure. They act by inhibiting protein synthesis, specifically by blocking the 50S ribosomal subunit. They are broad spectrum antibiotics.

Macrophages: A type of white blood cell that assists in the body's fight against bacteria and infection by engulfing and destroying invading organisms by white blood cell, or leucocyte, found in all vertebrate animals. Macrophages specialize in the removal of bacteria and other micro-organisms, or of cell debris after injury. Like phagocytes they engulf foreign matter but they are larger than phagocytes and have a longer life span. They are found throughout the body but mainly in the lymph and connective tissues and especially the lungs, where they ingest dust, fibres and other inhaled particles.

Microalgae: Microscopic algae.

Microcystin-LR: One of the groups of toxins produced and released by cyanobacteria is called *microcystin* named from the species *Microcystis aeruginosa.* The microcystins are a group of cyclic heptapeptide hepatotoxins produced by a number of cyanobacterial genera. The peptide ring is made up of two protein amino acids and five non-protein amino acids. It is the two protein amino acids that distinguish microcystin types from each other, while the other amino acids are relatively constant. By using amino acid single letter code classification, each microcystin is designated a name depending on the variable amino acids which complete their structure. For instance, one of the most common toxins found in water supplies around the world, microcystin-LR contains the amino acids Leucine (L) and Arginine (R) in these variable positions.

Microcystis: A free floating single celled cyanobacterium genus that can form large dense blooms with some species producing the toxin microcystin.

***Microcystis aeruginosa*:** A species of Microcystis which was historically the first to be identified as producing microcystin.

Micromolar: A concentration of 1/1,000,000 (one millionth) molecular weight per liter (mol/l).

Moiety: In physical organic chemistry moiety is generally used to signify part of a molecule.

Mydriasis: Prolonged abnormal dilatation of the pupil of the eye caused by disease or a drug.

Mouse bioassay: Mouse bioassay is used to assess the effects of marine biotoxins and is based on intraperitoneal administration of toxic mussel extracts and monitoring of survival time within a 24-h period. Here toxic effects on mice were examined for extracts of mussel samples. About 100 g of shellfish meat is macerated in a blender, gently boiled for 5 minutes with 100 ml 0.1 N HCl and 1 ml of the clarified extract (pH adjusted to 2-4) injected intraperitoneally into a 20 g test mouse.

MU: Mouse units; 1 mouse unit is the amount of toxin required to kill a mouse weighing 20 g in 15 minutes upon intraperitoneal injection.

Nanomolar (nM): In chemistry, molar concentration (also called molarity, amount concentration or substance concentration) is a measure of the concentration of a solute in a solution, or of any molecular, ionic, or atomic species in a given volume. Nanomolar (nM) = 10^{-9} molar.

Nausea: The condition or state that precedes vomiting.

Necrosis: The localized death of plant and animal tissue (as from infection or the interruption of blood supply). An affected area is described as being 'necrotic'.

Neuron cell: Nerve cell.

Neurotoxins: Toxins that disrupt the normal functioning of the human nervous system or a toxin that acts specifically on nerve cells or neurons, usually by interacting with membrane proteins and ion channels and can cause paralysis.

ng: Nanogram (a measure of weight), one billionth (1/1,000,000,000) of a gram.

NHMRC: National Health and Medical Research Council of Australia.

Nodularins: Cyclic nonribosomal peptides produced by cyanobacteria that can be very toxic for plants and animals including humans.

Numbness: Partial or total lack of sensation in a part of the body; a symptom of nerve damage or dysfunction.

OA: Oakadic acid.

Paralytic shellfish poisoning: Human illness caused by eating shellfish that are contaminated with saxitoxins from dinoflagellate *Alexandrium* spp., *Gymnodinium* spp.

Paresthesia: An abnormal sensation of the skin, such as numbness, tingling, pricking, burning, or creeping on the skin that has no objective cause. Paresthesia is the usual American spelling and paraesthesia the preferred English spelling.

Peptides: Molecules that hydrolyze into amino acids and form the basic building blocks of proteins.

Phosphatase: Any of a group of enzymes that act as a catalyst in the hydrolysis of organic phosphates. A common phosphatase in many organisms is alkaline phosphatase. Phosphatases can be categorised into two main categories: Cysteine-dependent phosphatases (CDPs) and metallo-phosphatases (which are dependent on metal ions in their active sites for activity).

Phosphorylation: Is the addition of a phosphate (PO_4) group to a protein or other organic molecule.

Phycotoxins: Marine biotoxins that accumulate in fish and shellfish from their diet (algae including diatoms and dianoflagellates) causing ciguatera, shellfish poisoning when the fish are eaten, as distinct from toxins naturally present (tetramine poisoning).

Photosensitizer: The act or process of inducing photosensitivity or to make (an organism, a cell, or a substance) photosensitive.

pKa: An acid dissociation constant, Ka, (also known as acidity constant, or acid-ionization constant) is a quantitative measure of the strength of an acid in solution. It is the equilibrium constant for a chemical reaction known as dissociation in the context of acid-base reactions. pKa is the measurement of how acidic a molecule is (how easily it gives up a proton). The lower the pKa of a molecule the more acidic it is.

Polyether: A polymer in which the repeating unit contains two carbon atoms linked by an oxygen atom, i. e. compounds with more than one ether group. While the term generally refers to polymers like polyethylene glycol and polypropylene glycol, low molecular compounds such as the crown ethers may sometimes be included.

Polyketides: Are secondary metabolites from bacteria, fungi, plants and animals. Polyketides are biosynthesized by the polymerization of acetyl and propionyl subunits in a similar process to fatty acid synthesis (a Claisen condensation). They are the building blocks for a broad range of natural products or are further derivatized. Polyketides are structurally a very diverse family of natural products with diverse biological activities and pharmacological properties. Polyketide antibiotics, antifungals, cytostatics, anticholesterolemics, antiparasitics, coccidiostatics, animal growth promoters and natural insecticides are in commercial use.

Prefecture: The town in which the administration of a department is located, the district administered by a prefect (as in France or Japan or the Roman Empire).

Protein phosphatase: Are enzymes that remove phosphate groups that have been attached to aminoacid residues of proteins by protein kinases.

PSP: Paralytic shellfish poisoning.

Quiescence: A period in which a cell is not increasing its mass or going through the cell.

Red tide: Dense concentrations of dinoflagellates (*Gynodinium breve*) that can colour the sea water red.

Roe: Fish eggs or egg-filled ovary.

Sabah: A region of Malaysia in north-eastern Borneo.

Schizophrenia: Is a psychiatric diagnosis that describes a mental disorder characterized by abnormalities in the perception or expression of reality.

Saxitoxins: Saxitoxin (STX) is a neurotoxin naturally produced by certain species of marine *dinoflagellates* (*Alexandrium* sp., *Gymnodinium* sp., *Pyrodinium* sp.) and cyanobacteria (*Anabaena* sp., some *Aphanizomenon* spp., *Cylindrospermopsis* sp., *Lyngbya* sp., *Planktothrix* sp.). The term saxitoxin originates from the butter clam (*Saxidomus giganteus*) in which it was first recognized.

Scums: A filmy layer of extraneous or impure matter that forms on or rises to the surface of a liquid or body of water. Or a green slime found in stagnant water.

Seizure: A sudden occurrence (or recurrence) of a disease.

Serine/threonine: A group of enzymes that catalyze the phosphorylation of serine or threonine residues in proteins with ATP or other nucleotides as phosphate donors.

Shellfish: Mussels, oysters, scallops, shrimps, crabs.

Sodium channels: Are integral membrane proteins that form ion channels, conducting sodium ions (Na^+) through a cell's plasma membrane. They are classified according to the trigger that opens the channel for such ions, i. e. either a voltage-change (voltage-gated sodium channels) or binding of a substance (a ligand) to the channel (ligand-gated sodium channels). In excitable cells such as neurons, myocytes and certain types of glia, sodium channels are responsible for the rising phase of action potentials.

Superoxide dismutase: A zinc and copper or manganese containing enzyme which reacts with superoxide radicals to convert them to less dangerous chemical entities.

Synaptic: Pertaining to or affecting a synapse or synapsis. Synapse is the point of connection usually between two nerve cells. Specifically, a synapse is a specialized junction at which a nerve cell (a neuron) communicates with a target cell. The neuron releases a chemical transmitter (a neurotransmitter) that diffuses across a small gap and activates specific specialized sites called receptors situated on the target cell. The target cell may be another neuron or a specialized region of a muscle cell or a secretory cell (a cell that can make and secrete a substance). Neurons can also communicate through direct electrical connections (electrical synapses).

Tingling: To have a prickling, stinging sensation, as from cold, a sharp slap or excitement causing or experiencing a painful shivering feeling as from many tiny pricks, 'a prickling blush of embarrassment'.

Toxicology: The study of poisons, their effects, antidotes and detection.

Tricarboxylic acid: A tricarboxylic acid is an organic carboxylic acid whose chemical structure contains three carboxyl functional groups (-COOH). The best-known example of a tricarboxylic acid is citric acid.

Toxigenic organisms: Producing toxic substances.

Tropane: Is a nitrogenous bicyclic organic compound. It is mainly known for a group of alkaloids derived from it (called tropane alkaloids), which include among others, atropine and cocaine. Both alkaloids contain tropinone from which tropane is a derivate. Tropane alkaloids occur in plants of the families Erythroxylaceae (including coca) and Solanaceae (including mandrake, henbane, deadly nightshade, datura, potato, tomato).

Tunicates: Also known as urochordata, tunicata (and by the common names of urochordates, sea squirts and sea pork) is the sub-phylum of a group of underwater sac-like filter feeders with incurrent and excurrent siphons, that are members of the phylum Chordata. Most tunicates feed by filtering sea water through pharyngeal slits but some are sub-marine predators such as the *Megalodicopia hians*. Like other chordates, tunicates have a notochord during their early development but lack myomeric segmentation throughout the body and tail as adults.

Tumour-promoting: A non-carcinogenic substance that enhances tumour production in a tissue previously exposed to a sub-carcinogenic doses of a carcinogen.

Viscera: The internal organs of the body, specifically those within the chest (as the heart or lungs) or abdomen (as the liver, pancreas or intestines). The singular of 'viscera' is 'viscus' meaning in Latin 'an organ of the body'.

13: Introduction

Biotoxins are naturally occurring toxic compounds produced by certain living organisms in fresh and marine waters. Both human activities and natural forces cause pollution of the environment and alterations in habitats that can support toxin producing organisms. In freshwater, the blue-green algae (also known as cyanobacteria) produce potent toxins called cyanotoxins and in marine waters many (about 5,000 micro-algal species, e.g. dinoflagellates) produce toxins called phycotoxins. Though algae are in general harmless, a few species are highly toxic to humans and wildlife. This chapter has been divided into two sections: freshwater biotoxins and marine biotoxins.

13.1: Freshwater biotoxins

Freshwater biotoxins is caused by blue-green algae (BGA), which have the characteristics of both algae and bacteria and is now classified as Cyanobacteria. Cyanobacteria or cyanoprokaryota (cyano = dark blue) are one of the oldest forms of life on earth, which is about 2.7-3.5 billion years old. Cyanobacteria or the blue-green algae occur worldwide. Cyanobacteria are gram-negative bacteria and the class cyanobacteria include 150 genera and about 2,000 species (van Apeldoorn *et al.* 2007).

They may grow as filaments or colonies (Figure 13.1) and occurs widely in lakes, reservoirs, ponds and slow-flowing rivers. Algae inhabit all natural waters and become a problem only when they increase to excessive numbers (water blooms). Concern about the effects of cyanobacteria has grown in many countries due to poisoning attributed to toxic cyanobacteria bloom (Håkanson *et al.* 2007) as well as due to bad taste and smell of the water. Cyanobacteria are now recognised as a serious water quality problem with regard to drinking water supply and recreational use. The earliest recorded cyanobacterial bloom occurred in 1878 in Lake Alexandrina at the mouth of the Murray River in South Australia (caused by brackish water cyanobacteria *Nodularia spumigena*) and since then reported from many countries around the world including Algeria, Australia, Bangladesh, Brazil, Canada, Checz Republic, China, Denmark, Egypt, England, France, Finland, Germany, Greece, Hungary, India, Ireland, Israel, Italy, Japan, Latvia, Lithuania, Morocco, New Zealand, Norway, Philippines, Poland, Portugal, Scotland, Saudi Arabia, South Africa, South Korea, Spain, Sweden, Switzerland, Turkey, UK and USA (see Figure 13.2 and Table 13.1).

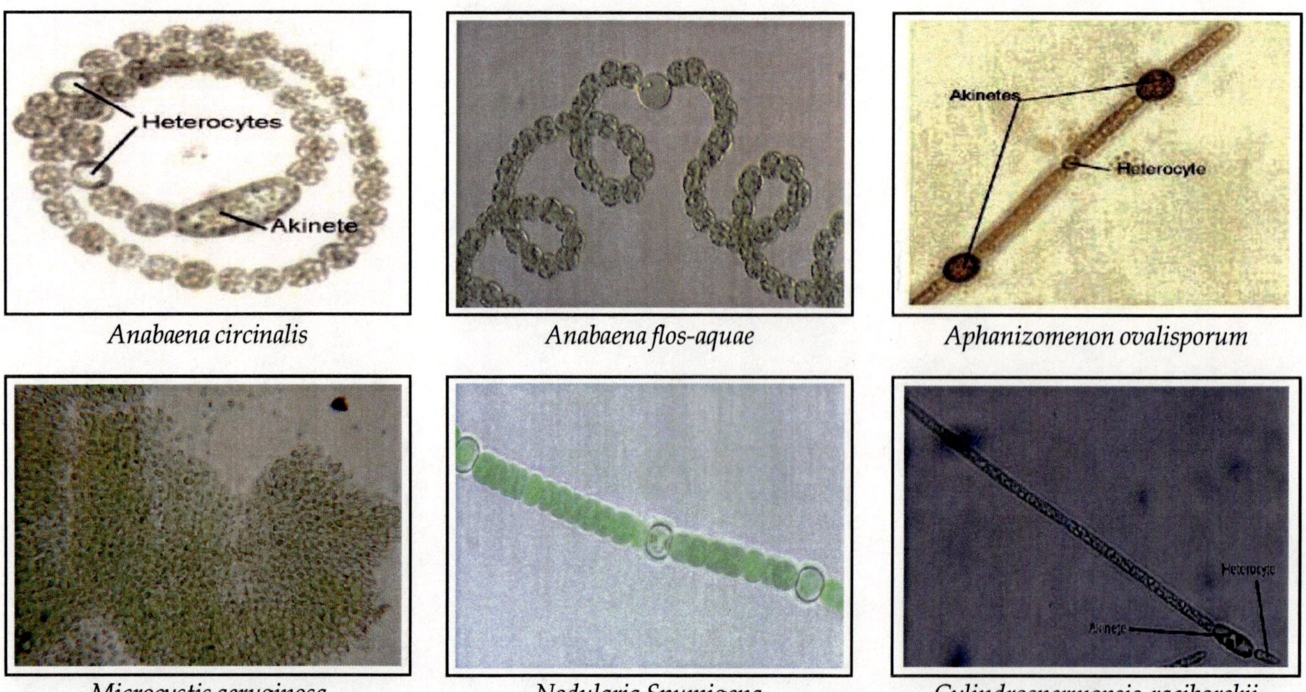

Anabaena circinalis — *Anabaena flos-aquae* — *Aphanizomenon ovalisporum*

Microcystis aeruginosa — *Nodularia Spumigena* — *Cylindrospermopsis raciborskii*

Figure 13.1: ***Commonly occurring blue-green algae or cyanobacteria in inland waters*** [photo sources: *www.waterquality.crc.org.au; http://microbewiki.kenyon.edu/index.php*].

Toxins and their toxicity (effects) are often used synonymously, but they are not. Clarification regarding toxins and toxicity are given in Box 13.1.

Countries Exhibiting One or More Documented HAB Outbreaks

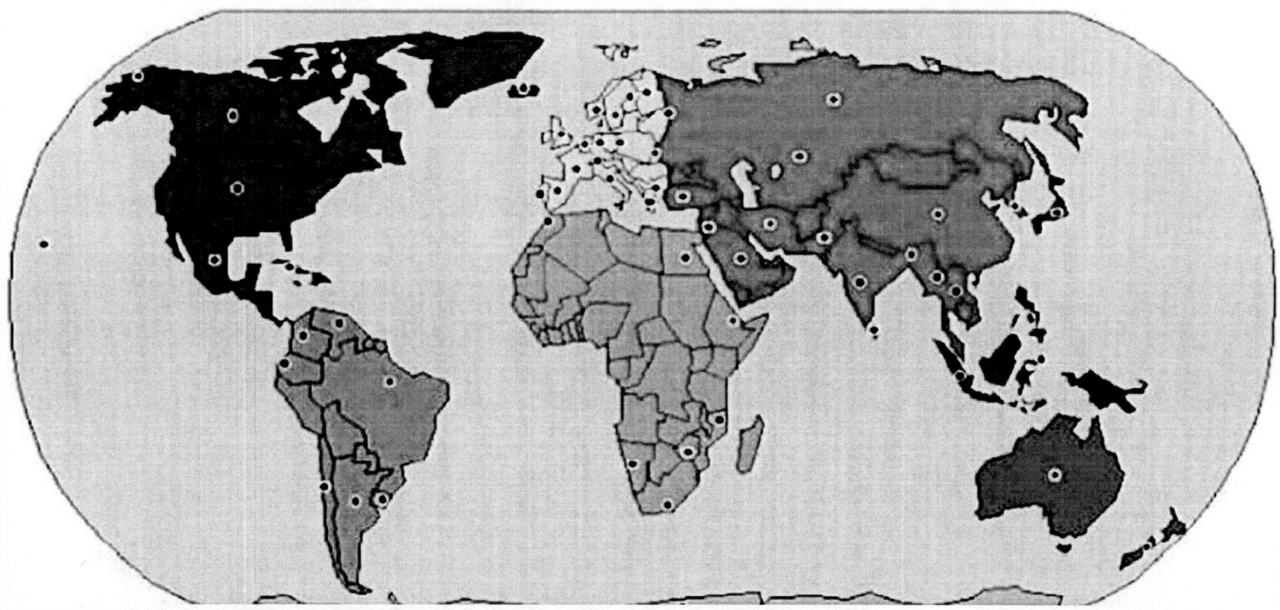

Figure 13.2: ***Countries where cyanobacteria water bloom producing cyanotoxins have been documented*** *[Figure source: Carmichael 2008; page 109]. Note: dots do not indicate numbers of cyanobacterial bloom occurrences, in most cases there have been multiple occurrences but not every occurrence has had a poisoning event documented.*

13.1.1: Conditions favouring cyanobacterial blooms

Environmental conditions that can promote cyanobacterial/blue-green algal blooms are :

- excessive nutrients (phosphorus and nitrogen);
- warm water temperatures (15-30°C; mostly> 20°C);
- sunlight for photosynthesis;
- stagnant water (note: turbulence and high flows are unfavourable to the growth of cyanobacteria).

In addition to growth rates, light and temperature also influence toxin production for many species of cyanobacteria, for example, several studies have shown that toxin production in *Microcystis aeruginosa* increases with irradiance under light-limited conditions. Microcystins composition of *Planktothrix agrdhii* changed toward a higher proportion of a more toxic variant with increased light intensity (see review by Paul 2008). Temperature has been shown to influence the type of toxin produced with high temperatures (>25°C) enhancing microcystin MC-RR production and lower temperatures favouring microcystin MC-LR synthesis (Rapala and Sivonnen 1998 and Håkanson *et al.* 2007). Cyanobacterial abundance can increase during warmer year's suggesting that global warming may cause more frequent harmful algal blooms (see chapter 2, sections 2.4.4; 2.8.2.1 on climate change impacts on freshwater resources).

13.1.2: Physico-chemical characteristics of cyanotoxins

[based on Chorus and Bartram 1999; Duy et al. 2000; Kankaanpää et al. 2005; van Apeldoorn et al. 2007; Pegram et al. 2008]

13.1.2.1: Cyanobacterial toxins

Many species and strains of cyanobacteria are able to produce cyanotoxins (Table 13.1). The World Health Organisation (WHO) recommends that public health management be focussed on the cyanotoxins and cyanobacteria in drinking water be managed as a chemical problem. Outbreaks of human poisoning attributed to toxic cyanobacteria are mainly of two groups: the neurotoxins (*neuro* = nerve) and hepatotoxins (*hepa* = liver). According to global data and information on cyanobacteria, the hepatotoxins (those causing liver damage) occur most frequently, although there have been blooms producing neurotoxins that have lead to animal deaths.

Table 13.1: ***Toxic cyanobacteria species reported from around the world.***

[de Figueiredo et al. 2004; New Zealand Government 2005; van Apeldoorn et al. 2007; http://www.health.vic.gov.au/environment/;http://www.moh.govt.nz/moh.nsf/0/5A25BF765B400911CC25708F0002B5A8/$File/09-cyanobacteriacompliance.doc].

Toxic species	Cyanotoxin	Location where toxin identified	Toxic species	Cyanotoxin	Location where toxin identified
Anabaena bergii	cylindrospermopsin	Australia	*Cylindrospermum* sp.	anatoxin-a	Finland
Anabaena blooms	anatoxin-a	China, Germany, New Zealand, South Korea	*Haphalosiphon (hibernicus)* (soil isolate)	microcystins	USA
A. circinalis	microcystins	France	*Lyngbya wollei*	saxitoxins	USA
A. circinalis	saxitoxins	Australia, New Zealand	*Microcystis aeruginosa*	Microcystins	Worldwide
A. flos-aquae	microcystins	Canada, Norway, New Zealand, Scotland	*M. flos-aquae*	microcystins	Australia
A. flos-aquae	anatoxin-a	Canada	*Microcystis* spp.	anatoxin-a (minor amounts)	Japan Germany, Morocco
A. flos-aquae	anatoxin-a(S)	Canada, Latvia, New Zealand,	*M. viridis*	microcystins	Japan
A. plantonica	anatoxin-a	Italy	*M. wesenbergii*	microcystins	Japan
Anabaena spp.	microcystins	Egypt	*M. botrys*	microcystins	Denmark
Anabaena spp.	anatoxin-a	Finland, Ireland	*Nodularia spumigena*	nodularins	Australia, Baltic Sea, New Zealand
Anabaena spp.	anatoxin-a (minor amounts)	Japan	*Nostoc* spp.	microcystins	Finland, England
Anabaena spp. (*A. flos-aquae, A. lemmermannii, A. circinalis*)	microcystins	Finland	*Oscillatoria limosa*	microcystins	Switzerland

Contd...

A. lemmermannii	anatoxin-a(S)	Denmark	*Oscillatoria* spp. benthic	anatoxin-a	Scotland, South Korea
Anabaena spp.	microcystins	Denmark	*Oscillatoria* spp.	anatoxin-a	China, Ireland, Morocco, New Zealand,
Anabaenopsis millenii	microcystins	Greece	*Phormidium* sp.	microcystins, anatoxin-a	New Zealand, Switzerland
Aphanizomenon	anatoxin-a	Germany	*Planktothrix agardhii*	microcystins	Checz Republic China, Denmark, Finland, Germany Norway
A. flos-aquae	saxitoxins	USA	*P. formosa*	homoanatoxin-a	Norway
A. ovalisporum	cylindrospermopsin	Israel, Australia, New Zealand	*Planktothrix* spp.	anatoxin-a	Finland
Aphanizomenon sp.	anatoxin-a	Finland	*Planktothrix* spp.	saxitoxins	Italy
Aphanizomenon spp.	anatoxin-a, cylindrospermopsin, LPS, saxitoxins, microcystins	New Zealand	*P. mougeotii*	microcystins	Denmark
Aphanocapsa cumulus	microcystins	Brazil	*Pseudanabaena* spp.	microcystins	Morocco
Aphanocapsa cumulus	microcystins	New Zealand	*Raphidiopsis curvata*	cylindrospermop sin	China
Arthrospira sp.	microcystins	New Zealand	*Synechocystis* spp.	microcystins	Morocco
Cylindrospermopsis raciborskii	cylindrospermopsin	Australia, Hungary, New Zealand	*Snowella lacustris*	microcystins	Norway
C. raciborskii	saxitoxins	Brazil	*Umezakia natans*	cylindrospermop sin	Japan

Examples of neurotoxins and hepatoxins are listed below:

- Neurotoxins are anatoxin-a, anatoxin a(s), saxitoxin, neosaxitoxin-a (see Figure 13.3 for the structure).
- Hepatoxins are microcystins, cylindrospermopsin, nodularin (see Figure 13.3 for the structure).

The Cyanobacterial toxins are classified by how they affect the human body. For example, hepatotoxins (which affect the liver) are produced by some strains of the cyanobacteria such as *Microcystis, Anabaena, Oscillatoria, Nodularia, Nostoc, Cylindrospermopsis* and *Umezakia*. Neurotoxins (which affect the nervous system) are produced by some strains of *Aphanizomenon* and *Oscillatoria*. Cyanobacteria, *Cylindroapermopsis raciborski* may also produce toxic alkaloids, causing gastrointestinal symptoms or kidney disease in humans. Table 13.2 provides an overview of the most important cyanotoxins so far known and their mode of acute actions. A good review on cyanobacterial toxins including their chemical nature and mechanisms of toxicity can be found in Chorus and Bartram (1999) and Duy *et al.* (2000).

13.1.2.2: Chemical nature of cyanotoxins

[based on Chorus and Bartram 1999; Duy et al. 2000; Cazenave et al. 2005; van Apeldoorn et al. 2007; Pergram et al. 2008].

Anatoxin-a is a tropane alkaloid toxin. It is a bicyclic secondary amine, 2-acetyl-9 azabicyclo [4-2-1]non-2-ene, and is a structural analogue of cocaine (see Figure 13.3) having a molecular weight of 165 daltons. Anatoxin-a is produced by *Anabaena flos-aquae,* and species of *Oscillatoria, Aphanizomenon* and others (Table 13.1). It has a pka of 9.4 and exists in the protonated form at physiological pH. Anatoxin-a(s) is a phosphorylated cyclic N-hydroxy guanine with a structure and action similar to organophosphate pesticides with a molecular weight of 252 dalton (Figure 13.3). It is produced by *Anabaena flos-aquae.* Saxitoxin and neosaxitoxin are carbamate alkaloid neurotoxins and have a common basic structure but unrelated to other neurotoxins. Saxitoxins are similar to that produced by marine dinoflagellates (see section of marine biotoxins 13.2.6 on paralytic shellfish poisoning or PSP). Saxitoxins have been found in the cyanobacteria *Aphanizomenon flos-aquae, Anabaena circinalis, Lyngbya wollei.*

Mycrocystins are a family of peptide toxins, primarily from species of *Microcystis, Anabaena, Oscillatoria, Nostoc,* and *Planktothrix*. There are about 80 different MCs which differ mainly in the two L-aminoacids at positions 2 and 4 (see Figure 13.3). The most common form of MCs are MC-LR, MC-RR and MC-YR. These compounds were first isolated from the cyanobacterium *Micorcystis aeruginosa* from the Rideau River in Ottawa, Canada and therefore the toxins were named as microcystins (Chorus and Bartram 1999). Microcystins are cyclic heptapeptides with a molecular weight of about 1,000 daltons. Microcystins contain seven aminoacid peptides. The general structure of microcystin is *cyclo-(D-Alanine-X-D-MeAsp-Z-Adda-D-Glutamate-Mdha)* where X and Y are variable L amino acids. D-MeAsp is D-*erythro*-β-methylaspartic acid; and Mdha is *N*-Methyldehydroalanine. The aminoacid Adda, (2S, 3S, 8S, 9S)-3-amino-9-methoxy-2,6,8-trimethyl-10-phenyldeca-4,6-dienoic acid, is the most unusual structure in this group of cyanobacterial cyclic peptide toxins (Figure 13.3). Structural variations have been reported in all the seven amino acids, but most frequently with substitution of L-amino acid *X* (mostly *X* is leucine, arginine or tyrosine but also homotyrosine, alanine, phenylalanine, hoophenylalanine, methionine-*S*-oxide or tryptophan is possible) and Y (*Y* is arginine, or alanine but it might also be also aminoisobutyric acid, homoarginine or methionine-S-oxide and demethylation of D-MeAsp and/or Mdha. Most of the microcystins contain β-methylaspartic acid, glutamic acid and alanine with methylamine attached to the glutamic acid. In some microcystins, Mdha is replaced by L-serine and D-Ala by D-serine. The Adda chain determines the biological activity of the toxins. Any structural modification to the Adda region or acylation of the glutamate renders microcystins less toxic or even non-toxic (van Apeldoorn *et al.* 2007).

Nodularins produced by the cyanobacteria, *Nodularia spumigena* are monocyclic pentapeptides, with the two terminal aminoacids of the linear peptide being condensed (joined) to form cyclic compound (Figure 13.3). The general structure of the nodularin is *cyclo-(D-MeAsp-L-Arg-Adda-D-Glutamate-Mdhb),* in which L-Arg is L-Arginine, D-MeAsp is D-*erythro*-β-methylaspartic acid and Mdhb is *N*-(Methyldamino)-2-dehydrobutyric acid. The aminoacid Adda (2S, 3S, 8S, 9S)-3-amino-9-methoxy-2,6,8-trimethyl-10-phenyldeca-4,6-dienoic acid, is the most unusual structure in this group of cyanobacterial cyclic peptide toxins (Figure 13.3). A few (seven) natural occurring variations of nodularin have been found; two dimethyl variants, one with D-Asp instead of D-MeAsp, other with DMAdda instead of Adda.

Other chemical information about anatoxin-a, anatoxin-a(S), saxitoxin, microcystins, cylindrospermopsin and nodularin are summarised in Table 13.2

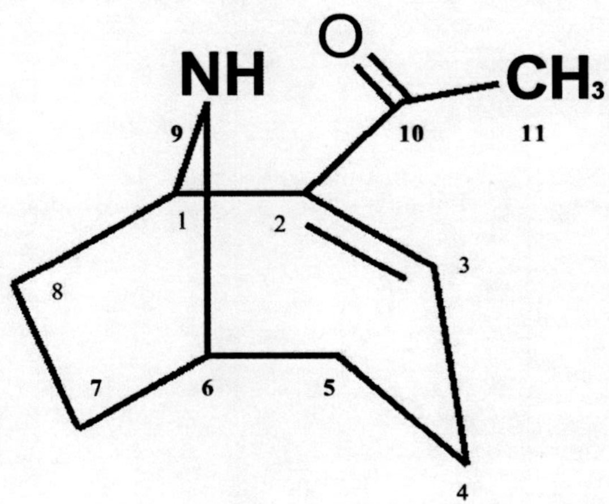

Anatoxin-a

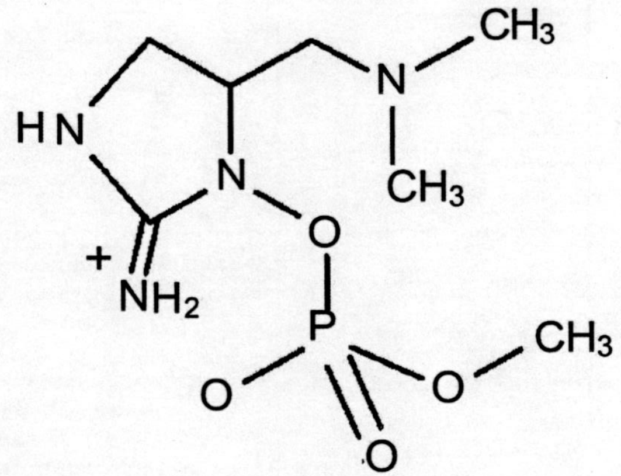

Anatoxin-a(S)

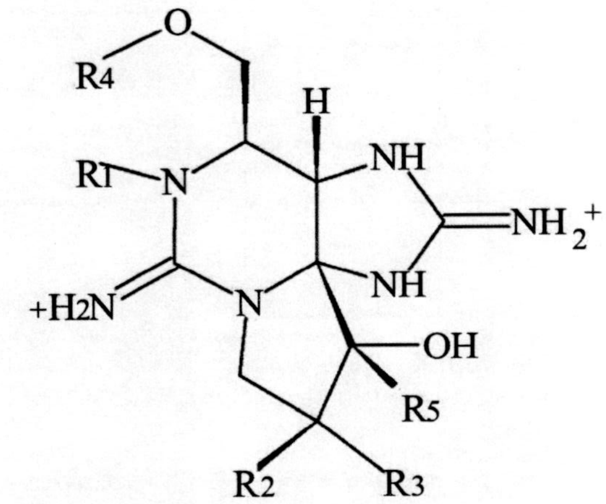

Saxitoxins

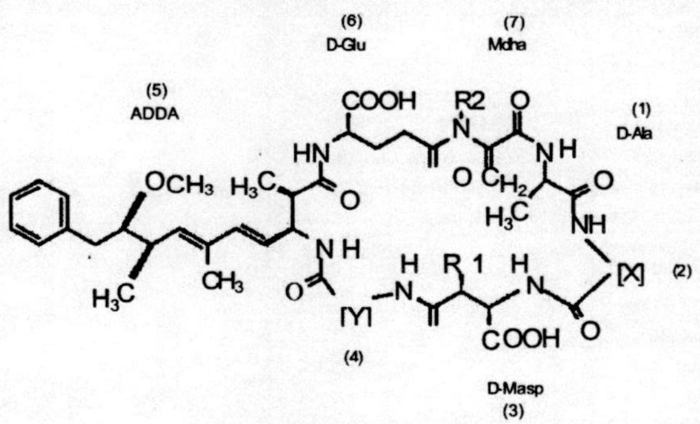

Microcystin

Some common microcystins

Microcystin-LR: R= $(CH(CH_3)_2$;

Microcystin-RR: R= $CH_2CH_2NHC(NH_2) = NH$;

Microcystin-YR: R= C_6H_4-p-OH

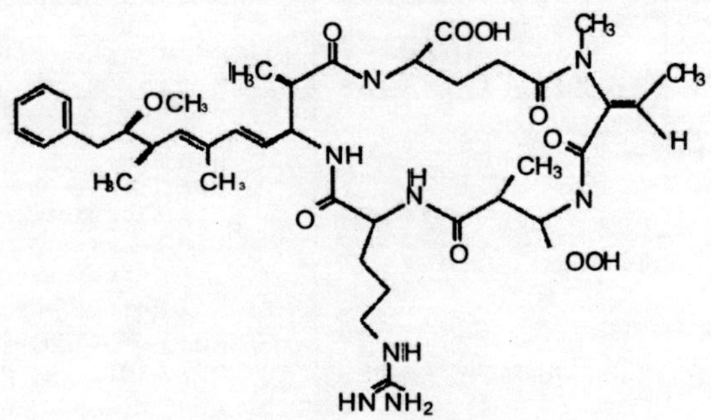

Nodularin

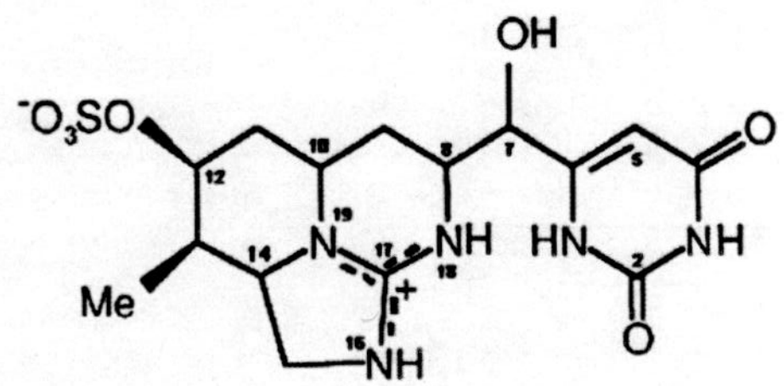

Cylindrospermopsin

Figure 13.3: ***Structures of different cyanotoxin groups*** *[Carmichael and Falconer 1993; Sadler 1994; Duy et al. 2000 http://www.epa.gov/cyano_habs_symposium/monograph/Ch16.pdf-Humpage 2008]..*

Table 13.2: ***Cyanobacterial toxins, their chemical nature and their mode of action*** *[Chorus and Bartram 1999; Duy et al. 2000; Camargo and Alonson 2006; van Apeldoorn et al. 2007; Hudnell 2009; Messineo et al. 2009; http://www.who.int/water_sanitation_health/bathing/srwe1-chap8.pdf].*

Toxins type	Cyanotoxins	Chemical nature	LD_{50} (i.p. mouse) of pure toxin (µg/kg bw)	Taxa known to produce the toxin(s)	Site and mode of action
Neurotoxins	Anatoxin-a (alkaloid)	- tropane related alkaloid toxin; - anataoxin-a is rapidly absorbed; when taken by the oral route; - it has a pka of 9.4	Anatoxin- a: 200-250	*Anabaena circinalis; A. flos-aquae; Anabaena* spp.; *Oscillatoria* spp., *Aphanizomenon* spp.	- nervous system by mimicking the action of acetyl choline, and overstimulating muscle cells
	Anatoxin-a(s) (unique organophosphate)	- N-hydroxyguanidine methyl phosphate ester; molecular weight of 252 daltons	Anatoxin-a(s): 20-40	*Anabaena flos-aquae; A. lemmermanii*	- nervous system by inhibiting acetyl-cholinesterase and overstimulating muscle cells
	Saxitoxins (carbamate alkaloids)	- toxins produced similar to marine dinoflagellates (PSP); - tricyclic molecules with hydropurine rings.	Saxitoxins: 10-30	*Anabaena circinalis Aphanizomenon flos-aquae; Lyngbya wollei; Cylindrospermopsis raciborskii*	- nervous system by binding to sodium channels (site 1), and disrupting nerve conduction
Hepatoxins/protein -phosphatase-blocker	Microcystins in general (about 80 known congeners)	- cyclic heptapeptides with molecular weight of 1,000 daltons; - most of the microcystins	45->1,000	*Microcystis aeruginosa; M. viridis; Planktothrix agardhii; Anabaena circinalis; Aphanizmenon flos-aquae,*	- liver, hepatopancreas by shrinking the cytoskeleton, distorting cells and causing hepatic haemorrhages and by inhibiting protein phosphatases, and causing tumour promotion.
	Microcystin-LR (MC-LR)	contain β-methylaspartic acid, glutamic acid, and alanine with methylamine attached to	MC-LR: 25-150 (50-60 accepted)		
	Microcystin-YR (MC-YR)	glutamic acid; - contain seven aminoacid peptides;	MC-YR: 25-125 (60 accepted)		
	Microcystin-RR (MC-RR)	- most known -microcystins have molecular weight between 909 and 1,115 daltons.	MC-RR: 300-600 (500 accepted)		
	Nodularin	- structure similar to that of microcystin; - contains five aminoacid peptides; molecular weight of 825 daltons; - contains the aminoacids D-glutamic acid, D-β-methylaspartic acid, L-arginine, D-*N*-methyl-dehydrobutyrine, and adda.	30-50	*Nodularia spumigena*	- liver, hepatopancreas by shrinking the cytoskeleton, distorting cells and causing hepatic haemorrhages, and by inhibiting protein phosphatases, and causing tumour promotion.
Cytotoxins	Cylindrospermopsin (alkaloid)	- tricyclic alkaloid, with a tricyclic guanidine moiety combined with hydroxymethyluracil (Figure 13.3); - molecular weight of 415 daltons.	2100 in 1 day 200 in 5-6 days	*Cylindrospermopsis raciborskii*	-blocks protein synthesis; substantial cumulative toxicity; liver and kidney most affected.

[LD_{50} = lethal dose 50 (the dose of a chemical that will kill 50% of a group of experimental animals); i.p = intraperitoneal; see Box 13.1 for the difference between the terms toxicity and toxin]

13.1.2.3: Persistence and degradation of cyanotoxins

[based on Chorus and Bartram 1999; van Apeldoorn wt al. 2007; Lemes et al. 2008; Messineo et al. 2009].

Cyanotoxins are intracellular toxins (i.e. contained within the cyanobacterial cells) and released to the surrounding waters during cell senescence, death and lysis or water treatment processes (e.g. per-chlorination or algicide application). At the time of a bloom toxins will be present in both the water (free, dissolved or extracellular toxins) and the cyanobacterial cells (intracellular).

Laboratory studies showed rapid decomposition of anatoxin-a in the presence of sunlight, generally at elevated pH. Anatoxin-a in pure solution in the absence of pigments undergoes rapid photochemical degradation in sunlight. Breakdown is further accelerated at alkaline conditions. The half-life of photochemical breakdown is 1-2 hour, whereas under normal day and night light conditions at pH 8 or 10, the half-life was reported to be about 14 days. Anatoxin-a(S) decomposes rapidly in basic solutions but is relatively stable under neutral and acidic conditions. Saxitoxins are reported to degrade slowly with half-lives ranging from 1-10 weeks, and more than three months often required for greater than 90% breakdown.

Microcystins are intracellular toxins and degrade slowly. Microcystins are only released into the water by senescence or cell death or through water treatment processes (perchlorination or algicide application). Once released into the water, microcystins can persist for a relatively longer period before being removed by biodegradation or photolysis. In full sunlight, microcystins undergo slow photochemical breakdown and isomerization, with the reaction rate being enhanced by the presence of water-soluble cell pigments (presumed by phycobiliproteins). On the other hand, a more rapid breakdown under sunlight can occur in the presence of humic substances (which can act as photosensitizers). In deeper and muddy waters, the rate of breakdown is reported to be lower. In natural waters, aquatic bacteria, *Sphingomonas* spp. and *Burkholderia* spp. (previously known as *Pseudomonas*) are capable of degrading microcystins with degradation rates dependent on temperature (Chorus and Bartram 1999; Lemes *et al.* 2008).

Scums of *Microcystis aeruginosa, Planktothrix* spp. blooms that dry on shores of lakes may contain high concentrations of microcystins for several months and these toxins are released back into the water body when re-immersed. Studies carried out found that dissolved microcystin is more persistent than particulate toxin (e.g. 30 and 15 days respectively required for 90% degradation) (Chorus and Bartram 1999; Messineo *et al.* 2009).

In environmental waters, microcystins are neutral or anionic. They are readily soluble in water, the solubility of MC-LR being higher than 1 g/l. Owing to their hydrophobicity and their polar functions (contain polar groups such as carboxylic acids), microcystins remain in the aqueous phase rather than being adsorbed on sediments or on suspended particulate matter (Rivasseau *et al.* 1998).

Cylindrospermopsin is relatively stable in the dark, with slow breakdown occurring at elevated temperature (>50°C), however, in sunlight and in the presence of cell pigments, breakdown occurs quite rapidly ($t^{1/2}$ within 2-3 days). In natural waters in which plant pigment levels were low, $t^{1/2}$ was found to be 11-15 days.

Box 13.1: ***Toxins and toxicity: what's in a name?*** *[Chorus and Bartram 1999]*

There is often misunderstanding of the terms toxicity or toxin content in regard to cyanobacteria or water samples that contain cyanotoxins. In particular, the terms are often taken to be synonymous, which they are not. ***Toxicity*** *refers only to animal testing data and is expressed as the amount of cyanobacteria lethal to an animal (e.g. per kilogram of body weight). The commonly used* LD_{50} *value is a measure of toxin potency and it is the amount of cyanobacteria or pure toxin needed (per kilogram of body weight) to kill 50% of experimental trial animals. The lower the* LD_{50} *values the more potent is the cyanobacterial sample or pure toxin.* ***Toxin concentration*** *and* ***toxin content*** *refers either to the amount of toxin per litre of water or the amount of toxin per mass of cyanobacterial bloom material.* ***Toxin quota*** *refers to the amount (mass or moles) of toxin per cyanobacteria cell.*

13.1.3: Environmental and biological impacts of freshwater biotoxins

13.1.3.1: Concentrations of cyanobacterial toxins in surface waters

A compiled measured concentrations of cyanotoxins in surface waters are given in Table 13.3. The table reveals that microcystin toxins are the most commonly reported cyanotoxins world wide. The range and the published highest concentrations of cyanotoxins from cyanobacterial bloom samples are highlighted below:

- Anatoxin-a: range 0.02-4,400 μg/g dw; highest concentrations (4,400 μg/g dw) reported from Finland.
- Anatoxin-a (S): range 4-3,300 μg/g dw; highest concentrations (3,300 μg/g dw) reported from the USA.
- Saxitoxins: range 5-3,400 μg/g dw; highest concentrations (3,400 μg/g dw) reported from Australia.
- Microcystin: range 0.004-25,000 μg/g dw; highest concentrations (25,000 μg/g dw) from Germany.
- Nodularin: range 3,000-18,000 μg/g dw; highest concentrations (18,000 μg/g dw) from the Baltic Sea.
- Cylindrospermopsin: 5,500 μg/g dw from Australia.

13.1.3.2: Effects on crops and plants

[based on McElhiney et al. 2001; Babica et al. 2006; Chen et al. 2004; van Apeldoorn et al. 2007; Crush et al. 2008].

Cyanobacteria can grow fast, shade and foul rice crops if they occur in bays which may lead to yield loss. Food crop plants can uptake microcystins; for example, several investigation results reveal that microcystins can affect plant growth and germination at a much higher concentrations (see Table 13.4). Irrigation with cyanobacteria contaminated water may cause accumulation of toxins on the external surfaces of edible plant material. The toxins are deposited when the water dries on the plant surface between irrigation periods or when the water becomes trapped in the centres of, for example, salad plants. As some cyanobacterial toxins could be toxic in a dry form, therefore continued application of heavily affected waters on pastures (e.g. from dairy waste recycling dams) may lead to significant toxin build-up on foliages. Long-term feeding of stocks with contaminated pastures may cause poisoning of stocks. Rape, rice and lettuce plants exposed to different levels of microcystins showed an accumulation of toxins (see Table 13.4) that were correlated to the toxin levels in the water. Therefore, consumption of edible plants exposed to microcystins via irrigation water might have health risks. This is an area for concern, especially in countries with a high reliance on the irrigation for high value crops, and with a high incidence of eutrophication and blooms like Australia.

The cyanotoxins such as microcystins can be taken up by plants and when, present in sufficient concentrations can induce various metabolic and morphological changes such as inhibition of photosynthesis, germination, growth inhibition, decrease in roots development and leaf necrosis (Babica *et al.* 2006). In a laboratory bioassay experiment (McElhiney *et al.* 2001), microcystin-LR (1.12 μg/ml) showed necrosis in leaves of runner bean (*Phaseolis vulgaris*) after 25 days of exposure, the effects followed by the desiccation and loss of older leaves in exposed plants. In addition, MC-LR impaired the development of the roots of the exposed plants. When potato (*Solanum tuberosum*) plants were exposed to solutions containing 1-5,000 μg microcystin-LR/l, no significant growth effect was seen at 1 μg/l, but at concentrations of 5, 10, 50, 100 and 500 μg/l significant growth inhibitions was recorded. Dose dependent decreases in chlorophyll content were observed in potato plants at concentrations of 50, 100, 500, 1,000; 2,500 and 5,000 μg/l. In addition, necrosis in shoot and leaf tissue at various concentrations was also observed (McElhiney *et al.* 2001). Studies conducted by Chen *et al.* 2004, found that microcystin significantly inhibited the elongation of primary roots of rape and rice seedlings (see Table 13.4).

Table 13.3: ***Toxin concentrations reported in cyanobacterial bloom or water samples from surface waters.***
[data compiled by Chorus and Bartram 1999; Cook et al. 2004; Nasri et al. 2004; Welker et al. 2004; Kardinaal and Visser 2005; Addico et al. 2006; Carrrasco et al. 2006; Madhavi et al. 2006; Ueno et al. 2006; Wood et al. 2006; Bláhova et al. 2007; Wang et al. 2007; Carvalho et al. 2008; Mohamed 2008; Tango and Butler 2008; Bajpal et al. 2009; Bláhova et al. 2009; Gurbuz et al. 2009; Messineo et al. 2009; Mohamed and Shehri 2009; Paldavièiene et al. 2009].

Country and location	Year (monitoring year or publication year)	Toxins identified	Concentrations µg/g dw unless otherwise stated	Remarks (nos. of toxic samples and analysis methods)
Anatoxins				
Denmark	1993-1995	Anatoxin-a(S)	4-3,300	9 toxic samples (out of 10); chE inhibition assay
Finland	1985-1987	Anatoxin-a	10-4,400	13 toxic samples (out of 30); GC/MS
Finland and Japan	na	Anatoxin-a	0.4-2,600	3 toxic samples (out of 3); LC/MS
Japan	1988-1992	Anatoxin-a	0.3-16	9 toxic samples (out of 10); TSP-LC/MS
Germany	1995-1996	Anatoxin-a	0.02-0.36 (given as µg/l)	10 toxic samples (out of 45); GC-ECD
Italy	1989-2006	Anatoxin-a	nd-100 µg/g dw	Data from 28 lakes; GC-MS
Ireland	1995	Anatoxin-a	10-100	2 toxic samples (out of 2); HPLC (fluorometric)
Saxitoxins				
Australia	1990-1992	Saxitoxins	85-2,040	11 toxic samples (out of 11); HPLC and FAB-MS
Australia	1992-1994	Saxitoxins	50-3,400	24 toxic samples (out of 31); HPLC
USA	1994	Saxitoxins	5-60 µg STX equivalent	7 toxic samples (out of 8); HPLC/AOAC
Microcystins				
Algeria	2000-2001	Microcystin (MC-LR)	3-29,163 µg/l	Lake Oubeira; protein phosphate type 2A assay
Argentina	2003 - publication	Microcystin	5.8-2400	97% toxic (35 samples); San Roque reservoir
Australia	1991	Microcystins, 24 unidentified	2100-4100*	4 toxic samples, HPLC
Bangladesh	2002	Microcystins (total)	<0.1->1000 µg/l	Ponds; 50% of positive samples contained > 10 µg/l; highest frequency and highest concentrations detected in ponds of Mymensingh district
Brazil	2004	Microcystin	28-45 µg/l (in reservoir water) 138-223 µg/g (freeze dried cyanobacterial cell)	Monjolinho Reservoir; HPLC
Brazil	1995-1996	Microcystins	0.003-10 µg/l	Surface water in Parna; 18% toxic (50 samples); ELISA
Canada (Alberta)	1990	MC-LR	4-610	37 toxic samples (out of 50), HPLC
Canada (Alberta, 3 lakes)	1990-1993	MC-LR	1-1,550	168 toxic samples (out of 226 samples); HPLC

Contd...

Canada				
China	1998	MC-LR, MC-RR	200-7,300	5 toxic samples (out of 10); HPLC
China	2004	MC-LR MC-RR	0.045-2.73 μg/l 0.045-2.73 μg/l	Drinking water reservoirs in Zhjiang province
China	2003-publication	Microcystins	24.5-97.3	Surface water from Meiliang Bay; 100% toxic (6 samples)
Chile	1999-2000-publication	Microcystins	8-130	Chilean lake; 100% toxic (2 samples)
Czech Rep.	2004	Microcystins (extracellular)	>0.125 μg/l in 145 water samples (70% samples); 0.67 μg/l median	ELISA; 206 water samples from 94 localities;
Czech and Slovak Republic	1995-1996	MC-LR	4-6,836	63 toxic samples; HPLC
Denmark	1992-1994	MC-RR, MC-LR	3-2,800	HPLC
Denmark	1993-1995	Microcystins	5-1,900	198 toxic samples (out of 296); HPLC
Denmark	-	Microcystins (superficial scums)	11-737 μg/g dw	-
Egypt	1993	MYCSTs	18.2-40 mg/kg	Mouse bioassay
Egypt	1993	MYCSTs	300 μg/g	River Nile, ELISA
Finland	1994-1995	MC-LR	<10-800	17 toxic samples (out of 18); HPLC
France	1994	Microcystins	70-3,900	16 toxic samples (out of 30); HPLC
France	1998-1999	MYCSTs	0.270 μg/g (HPLC0; 0.014-0.074 (PP2A)	Saint-Caprais reservoir; HPLC and PP2A
France	1999-2000	MYCSTs	0.05-4.1 μg/l (PP2A); 0.2-5.2 μg/l (HPLC)	Saint-Caprais reservoir; HPLC and PP2A
France	2002-publication	Microcystins	0-5.2 μg/l	Lake; 72% toxic (25 samples)
Germany	1992	MC-LR	36-360	8 toxic samples (out of 15); HPLC
Germany	1996 publication	Microcystins	14700	Bautzen reservoir; 1005 toxic (1 sample)
Germany	2001-publication	Microcystins	500-6500	Lake Ammerrsee; 100% toxic (11 samples)
Germany	1993	Microcystins	0.15-36* (given as μg/l)	17 toxic samples (out of 18); ELISA
Germany	1995-1996	Microcystins	1-5,000	385 toxic samples (out of 533); HPLC
Germany	1997	Microcystins, several	1-25,000	34 samples; HPLC
Germany	2000-publication	Microcystins	0.14-119 μg/l concentration)	Berlin water; 100% toxic (129 samples)
Ghana	2006-publication	Microcystins	3.21 mu g/l (highest)	Weija and Kpong reservoirs
Greece	1987-2000	Microcystins	50.3-1638 μg/g scum dw	33 freshwaters
Hungary	2000-publication	Microcystins	1710-6600	Lake Valencei; 100% toxic (2 samples)
India	2009-publication	Microcystins	25.2	Pond; ELISA; *Nostoc* sp BHU001
Italy	1989-2006	Microcystins (sum of all variants)	nd-226.16 ng/ml	Data from 28 lakes; 12 out of the 51 samples exceeded the WHO guideline value of 1 μg/l; HPLC and LC-MS/MS

Contd...

Italy	1989-2006	Microcystins (superficial scums)	107-1160x10³ ng/g	Data from 28 lakes; 12 out of the 51 samples exceeded the WHO guideline value of 1 μg/l; HPLC and LC-MS/MS
Japan	1980-1991	MC-RR, MCYR, MC-LR	30-2,100	13 samples; HPLC
Japan	1986-1988	MC-RR, MCYR, MC-LR	100-860	4 toxic samples (out of 4); HPLC
Japan	1988-1992	MC-RR, MCYR, MC-LR	70-1,1610	11 toxic samples (out of 19); HPLC
Japan	1990	MC-RR, MCYR, MC-LR	160-950	12 toxic samples (out of 14); HPLC
Japan	1992-1995	MC-RR, MCYR, MC-LR	0.04-480	18 toxic samples (out of 22); HPLC
Japan	1993-1995	Microcystins	0.05-1,300* (given as μg/l)	46 toxic samples (out of 57); ELISA
Japan	1993-1994	Microcystins	0.06-94* (given as μg/l)	12 toxic samples (out of 17); ELISA
Japan	1989-1994	Microcystins	300-19,500*	10 toxic samples (out of 10); ELISA
Japan	1994-1995	Microcystins	52-52000 pg/ml	ELISA
Kenya	2003-publication	Microcystins	310-19822	Lake Baringo; 100% toxic (4 samples)
Korea	2001-publication	Microcystins	0-0.2 μg/l	Daechung reservoir; 100% toxic (25 samples)
Lithuania	2006-2007	Microcystins (MC-LR, MC-RR, MC-LY, MC-YR)	<0.1-134.2 μg dm³	Curonian Lagoon; HPLC, ELISA, protein phosphate inhibition assay (PPI)
Morocco	2002-publication	Microcystins	0-8800	100% toxic (8 samples)
Morocco	1994-1998	Microcystins	700-8800	Lalla Takerkoust reservoir; 89% toxic (9 samples)
Morocco	1999	Microcystins	0.1-0.78 μg/g dw (ELISA); 502-1924 mg/kg (mouse bioassay)	Qued Mellah lake
Morocco	1994-1999	Microcystins	2.2-944 μg/g (ELISA); 26.8-1844 μg/l (freshwaters); 28-350 mg/kg (mouse bioassay)	Seven freshwaters
New Zealand	2001-2004	Microcystins	36.5 mg/l (highest microcystin concentration)	266 water and cyanobacterial mat samples from 277 water bodies; ELISA and LC/MS and neuroblastoma assays; microcystins/nodularin found in 102 samples; anatoxin-a in 3 samples and saxitoxin in 48 samples
Netherlands	2000 publication	Microcystins	0-2420	96% toxic (48 samples)
Philippines	2003-publication	Microcystin	0-1344	Laguna de Bay; 55% toxic (11 samples)
Philippines	2002-publication	Microcystin	649-4019	Laguna de Bay; 100% toxic (3 samples)
Poland	-	Microcystin s (superficial scums)	23-1687 μg/g	-

Contd...

Portugal	1989-1992	Microcystin LR, plus six known and three unidentified microcystins	1,000-7,000	12 toxic samples (out of 12); HPLC
Portugal	1994-1995	Microcystins	0.1-37* (given as μg/l)	28 toxic samples (out of 29); ELISA
Saudi Arabia	2006	Microcystins (in cyanobacterial mats)	468-512.5 μg/g	Hot springs; ELISA
Saudi Arabia	2007	Microcystins (in *Anabaenopsis arnoldi*)	Upto 364 μg/g dw (in bloom) 1.6-8.3 μg/l (in raw water) 0.33-1.6 μg/l (in treated water)	Mountain lake; ELISA
South Africa	1985-1986	Microcystin-FR, LR, YR, LA-LAba	5-420	28 toxic samples (out of 29); HPLC
South Africa	1988-1989	Microcystin-FR, LR, YR, LA-LAba	40-630	HPLC
Spain	1997-1999	Microcystins	1-150 μg/l	Madrid reservoirs
Spain	2002-2003	Microcystins	> 1 μg/l in four reservoirs and upto 70 μg/l in one reservoir	Seven water reservoirs
Thailand	1993-1995	Total microcystins	Mean 161 pg/ml; highest 354 pg/ml	Chiang Mai; 6 out of 10 samples were positive; ELISA
Turkey	2005	Microcystin (MC-LR)	0.73-48.5 μg/g	Lake Kovada
UK	1992	Microcystins, 3	17-131* (given as μg/l)	3 toxic samples (out of 3); HPLC
Uruguay	2001 publication	Microcystins	101-1074	La plata River; 100% toxic (9 samples)
USA	1993	Microcystins	1,900-12,800*	9 samples; ELISA
USA	2000-publication	Microcystins	1200 (maximum)	Lake; 5 samples
USA	2000-2006	Microcystins	2.9×10^{-2} to 6.58×10^{2} μg/ L	Chesapeake Bay; opportunistic samples collected from cyanobacteria blooms.
USA	2003-2004	Microcystins (MC-LR)	Av. conc. range: -0.46-1.78 μg/l (Buffalo lake); -0.44-1.08 μg/l (Ransom lake)	Buffalo Sprink Lake (BSL) and Lake Ransom Canyon (LSC); concentration range include spring, summer, fall and winter data
Nodularin				
Australia (Tasmania)	1992-1993	Nodularin	2,000-3,500	7 toxic samples (out of 9); HPLC
Baltic sea	1985-1987	Nodularin	300-18,000	17 toxic samples (out of 23); HPLC
Baltic sea	1990-1991	Nodularin	2,000-3,500	6 toxic samples (out of 16); HPLC
Lithuania	2006-2007	Nodularin	<0.1-284.6 μg dm^3	Curonian Lagoon; HPLC, ELISA, protein phosphate inhibition assay (PPI)
Cylindrospermopsin				
Czech Republic	2007	Cylindrospermopsin	0.01-4 μg/l	Lake waters; ELISA, LC/MS
Israel	1994	Cylindrospermopsin	465-2000	Lake Kinneret; HPLC

** = microcystin LR used as standard*

AOAC = mouse bioassay done according to the Association of Official Analytical Chemists; GC-ECD = gas chromatography-electro capture detection; GC/MS = gas chromatography/mass spectrometry; ELISA = enzyme linked immunosorbent assay; HPLC = high pressure liquid chromatography; FAB-MS = fast atom bombardment-mass spectrometry; LC/MS = liquid chromatography/mass spectrometry; PP2A=protein-phosphatase2A-inhibition assay; TSP-LC/MS = thermospray-liquid chromatography/mass spectrometry

Table 13.4: ***Effects of microcystins (MCs) on selected terrestrial plants (crops)*** *[information compiled by Babica et al. 2006].*

Crops	Investigated variant of MC	Exposure concentrations (µg/l) and exposure duration (days)	Observed effects	References
Brassica napans (Rape/oil seed)	Cyanobacterial extract containing MC-LR, -RR, -YR	24-3,000 (sum of MCs), 10 days	- growth inhibition and plant browning (≥ 120 µg/l) - inhibition of germination (≥ 600 µg/l) - inhibition of superoxide dismutase and increase of peroxide activities (≥ 24 µg/l) - accumulation of MCs upto 0.065 µg/g fresh weight	Chen *et al.* 2004
Oryza sativa (Rice)	Cyanobacterial extract containing MC-LR, -RR, -YR	24-3,000 (sum of MCs), 10 days	- growth inhibition (especially roots ≥ 120 µg/l) - no effects on germination and peroxide activity (≥ 24 µg/l) - accumulation of MCs up to 0.005 µg/g fresh weight	Chen *et al.* 2004
Lepidium sativum (Garden cress)	Pure MC-LR, cyanobacterial extract	10, 2-6 days	- growth inhibition, increase of glutathione S-transferase and glutathione peroxidase activities	Gehringer *et al.* 2003
Solanum tuberosum (Potato)	MC-LR	1-5,000; 16 days	- growth inhibition (≥ 5 µg/l) - decrease in number of roots (≥ 10 µg/l) - necrosis and decrease in chlorophyll content (≥ 50 µg/l)	McElhiney *et al.* 2001
Phaseolis vulgaris (Runner bean seeds)	MC-LR	1.12 µg/ml, 18 days	- accumulation of MCs after 3 days of exposure (11.55 µg/ml of plant extract) - growth inhibited	McElhiney *et al.* 2001
Lactuca sativa (commercial lettuce)	MC-LR (spray irrigation with water containing *Microcystis)*		- absorption of microcystin by plants - MC-LR was present at 2.5 mg/kg (dw) in the central leaves; 0.833 mg/kg (dw) in the distal zones of mature leaves; 0.094 mg/kg (dw) in the basal zone of mature leaves.	Codd et al. 1999
Lolium perenn (Rye grass), *Trifolium repens* (clover), *Brassica napans*(rape) and *L. sativa* (lettuce)	Microcystins (water containing microcystin	Each plant was exposed to 170 µg of a mixture of 10 different microcystins	- microcystins were nor present at measurable levels in shoots of rye grass or rape but were present in lettuce (0.79 mg/kg dw) and clover 90.20 mg/kg dw) - microcystins in roots were highest in clover (1.45 mg/kg dw) and intermediate in lettuce (0.68 mg/kg dw) and low in ryegrass (0.12 mg/kg dw).	Crush et al. 2008

13.1.3.3: Effects on aquatic organisms

Cyanobacteria are regular components of the diet of the cichlid, cyprinid and other fishes. For example, phytoplanktivorous fish, silver carp directly consume cyanobacteria, whereas zooplanktivorous fish like stickle backs and smelt feed on zooplankton that directly consumes cyanobacteria (Ibelings and Havens 2008). Many fish species are not able to avoid ingestion of these toxic organisms in lakes, dams or ponds during cyanobacterial blooms. As a result fish can be exposed to cyanotoxins during normal feeding, drinking or gill respiration. Fish that do not feed on cyanobacteria can still receive toxins through the food web such as from feeding on zooplankton or shellfish that have been contaminated with microcystins. Cyanobacterial toxins can be toxic to aquatic organisms such as zooplankton. Toxin can affect the survival rate or may even cause death of crustaceans and fish. Furthermore, during cyanobacterial blooms accumulation of mycrocystins in the gut, liver, kidney, blood, gills and muscle of fish can occur that may cause damage of the vital organs leading to the death of fish (see Table 13.5). de Figueiredo *et al.* 2004, van Apeldoorn *et al.* 2007 and Ibelings and Havens 2008 reviewed and summarised the effects of cyanotoxins on aquatic organisms which are as follows:

- Exposure of *Ceratophylum demersum* for 6 weeks to MC-LR levels of 0.1-5.0 μg/l caused a significant growth reduction. Photosynthesis of the aquatic macrophytes (*Elodea canadensis, Myriophyllum spicatum* and *Phragmites australis*) and macroalgae (*Cladophora* spp.) have been affected at exposure of 0.5 μg microcystin-LR/L.
- Macrophytes such as *Phragmites australis, Ceratophylum demersum and Elodea canadensis* have been shown to absorb MC-LR. In case of *P. australis* the higher values for absorbed MC-LR were found in stem and rhizome.
- Natural *Daphnia* populations subjected to blooms of *Cylindrospermopsis raciborskii* experienced high mortality, reduced individual body growth and negative effects on fecundity because of the toxic metabolites present in *C. raciborskii.*
- Microcystins are toxic to brine shrimp, *Artemia salina,* rotifers and copepods. Cladocerans were reported to be killed by microcystins.
- Exposure of zebra fish (*Danio rerio*) embryos to 400 μg of purified anatoxin-a/L under laboratory conditions caused an altered heart rate in different developmental stages.
- Zebrafish embryos exposed to saxitoxin (10, 50, 100 and 500 μg STX/l), at 10 μg/l and above concentrations delayed hatching and led to malformations and mortalities at 500 mg/l.
- In 1992, about 2,000 brown trouts were killed in Loch Leven and the River Leven, Scotland believed to have been caused by toxins of *Anabaena flos-aquae.* The killed fish had liver and gill damaged.
- In the USA, more than 100,000 catfish died from consumption of *M. aeruginosa* during a bloom.
- In Brazil (2000), the Caragalheiras reservoir suffered a massive bloom of *Pseudanabaena schimdlei* (with the presence of *M. aeruginosa*) which led to the death of all most all fish in the reservoir (Chellapa *et al.* 2000).
- In Denmark, death of thousands of fish and a cow were related to hepatotoxic blooms of *Anabaena flos-aquae* and *P. agardhi* (Henriksen 2001).
- According to Ibelings and Havens (2008), fish kills during cyanobacterial blooms could be due to combination of stress factors that occur concurrently during blooms of toxic cyanobacteria (high temperature and pH, enhanced level of ammonia, low oxygen and cyanobacterial toxins). Several fish organs can be affected (liver, kidney, heart, gonads) but liver is the organ often found to be most affected during blooms.
- Anatoxin-a(S) was responsible for the death of ducks in the USA and waterfowl in Canada.

Table 13.5: ***Effects of cyanobacteria toxins on aquatic organisms.***
pg/ml=picograms per millilitre

Species	Toxins	Effects	Reference
Crustaceans			
Daphnia pulicania	MC	Reduced filtering rates; reduced survival of new born daphnids	Bell and Codd 1996
Rotifers, copepods and cladocerans	MC	Death	Penaloza *et al.* 1990
Fish			
Brown trout	Toxins of *Anabaena flos-aquae*	Liver and gill damage resulting in fish kills	Codd 1994
Medaka fish embryos, *Oryzias latipes*	MC-LR (1, 10, 100 pg/ml; 1, 10, 100 ng/l; 1, 10 μg/l)	Survival rates reduced up to 90% at higher concentrations (10 and 1 μg/l); liver hypertrophy and hepatic haemorrhage observed	Jacquet *et al.* 2004
Rainbow trout, *Oncorhynchus mykiss*	*Microcystis aeruginosa* toxin; 6,600 μg of microcystin/kg body weight	Death within 96 hour due to acute liver damage	Tencalla *et al.* 1994
Tilapia, *Oreochromis* spp.	MC-LR	Liver most affected, followed by kidney and gills	Jos *et al.* 2005
Tilapia, *Oreochromis* spp.	MC-LR; 60 μg MC-LR/fish per day	Damage of hepatopancreas, kidney, gills and gastrointestinal tract	Molina *et al.* 2005
Zebra fish, *Danio rerio*	0.5, 5, 15 and 50 μg/l microcystin-LR (MC-LR)	Increased concentration reduces the spawning activity	Baganz *et al.* 1998

13.1.3.4: Bio-accumulation of cyanotoxins in aquatic organisms

[based on Chorus and Bartram 1999; Ibelings and Chorus 2007; van Apeldoorn et al. 2007; Ibelings and Havens 2008].

Literature review reveals that the majority of cyanotoxins accumulation studies dealt with microcystin (an abundant and highly toxin protein). ***Microcystins*** have been found in phytoplankton, zooplankton, gastropods (*Lymnaea stagnalis, Helisoma trivolvis* and *Physa gynna*) as well as in copepods and crab larvae (Chorus and Bartram 1999; Duy *et al.* 2000). Of several fish species investigated, microcystins were found to accumulate mainly in the liver, gut, kidneys and gonads, and much less in muscle tissue (Ibelings and Havens 2008; see also Table 13.6). Microcystins are also found in fish faeces in substantial amounts, as a consequence the benthic community can be exposed to cyanobacterial toxins.

Several studies reported accumulation of microcystins in freshwater seafoods that are eaten by humans including fish, mussels and shrimps (see Table 13.6). Therefore, the transfer of the microcystins to human through the food web is easily possible. Furthermore, investigations carried out in China, Brazil, Egypt and Portugal reveals that microcystin levels in edible parts of freshwater seafoods (e.g. fish/shrimp/mussel muscle and other organs) have exceeded the WHO recommended limit for the tolerable daily intake (TDI) in many occasions for human consumption of 0.04 μg/kg per day of MC-LR (see Ibelings and Chorus 2007).

Redclaw crayfish (*Cherax quadricarinatus)* and rainbow fish (*Melanotaenia eachamensis*) were able to accumulate ***Cylindrospermopsin*** from an aquaculture pond infested by *Cylindrospermopsis raciborskii* (pond water contained 589 μg cylindrospermopsin/l; 93% in cells, 7% in water). The toxins accumulated in crayfish hepatopancreas tissue (4.3 mg/kg cylindrospermopsin/kg freeze dried) and muscle tissue

(0.9 mg/kg freeze dried muscle tissue) whereas rainbow fish viscera contained 1.2 μg/g freeze dried (Saker and Eaglesham 1999).

Nodularins (produced by *N. spumigena)* are reported to accumulate in mussels (*Mytilus edulis*), clams (*Macoma balthica)* and in some Baltic fish (e.g. flounder, Atlantic cod and three-spine stickleback), prawn (Australia) (see review by van Apeldoorn *et al.* 2007, see Table 13.6).

Table 13.6: ***Maximum or range of accumulation of cyanobacterial toxins in aquatic animals in natural systems*** *[data compiled by Smith et al. 2008; Ibelings and Havens 2008; Nyakairu et al. 2009].*
Afdw = ash freed ry weight; Dw = dry weight; fw = fresh weight; n/r = not reported; MC = microcystin; NOD = nodularin

Group	Country	Species	Maximum accumulation levels in tissue
Mollusc	Australia	Mussels	Whole : 2.5 μg NOD/g fw
	Baltic sea	*Mytilus edulis,* blue mussel *Mytilus edulis,* blue mussel *Macoma balthica,* clam	Whole : 2.15 μg NOD/g dw Whole : 1.49 μg NOD/g dw Whole : 0.13 μg NOD/g dw
	Canada	*Lymnaea stagnalis,* freshwater snail *Anodonta grandis simpsoniana,* giant floater mussel	Whole : 140 μg MC/g dw Whole : 1.35 μg MC/g dw
	China	*Bellamya aeruginosa,* freshwater snail	Hepatopancreas : 7.42 μg MC/g fw (range 1.06-7.42) Digestive tract : 4.54 μg MC/g fw (range 0.8-4.54) Gonad : 2.62 μg MC/g fw (range 0-2.62) Foot : 0.06 μg MC/g fw (range 0.01-0.06)
		Anodonta woodiana, Hyriopsis cumingii	Intestine : 20.65 μg MC/g dw
		Cristaria plicata, freshwater pearl mussel	Hepatopancreas : 38.48 μg MC/g dw
		Lamprotula leai, freshwater mussels	Visceral mass : 1.7 μg MC/g dw Gills : 0.64 μg MC/g dw Foot : 0.58 μg MC/g dw
	Japan	*Unio douglasiae*	Hepatopancreas : 420 μg MC/g dw
		Cristaria plicata, cockscomb pearl mussel	Hepatopancreas : 297 μg MC/g dw
	Netherlands	*Dreissena polymorphia,* zebra mussel	Whole : 30 μg MC/g afdw
	Portugal	Mussels	Not reported : 16 μg MC/g fw
Crustacean	Australia	Prawn	Viscera : 6.400 μg NOD/g fw Muscle : 0.022 μg NOD/g fw
	Brazil	Shrimp Crab (unspecified)	Muscle : 0.011 μg MC/g fw Muscle : 1.0 μg MC/g fw (range 0.02-1.0)
	China	*Palaemon modestus,* freshwater shrimp	Stomach : 4.53 μg MC/g dw Hepatopancreas : 4.29 μg MC/g dw Eggs : 2.34 μg MC/g dw Gonads : 1.17 μg MC/g dw Gills : 0.51 μg MC/g dw Muscle : 0.13 μg MC/g dw
		Procambrus clarkii, red swamp crayfish	Stomach : 9.97 μg MC/g dw Gonads : 0.93 μg MC/g dw Hepatopanacreas : 0.08 μg MC/g dw Gills : 0.27 μg MC/g dw Muscle : 0.05 g MC/g dw

Contd...

		Macrobrachium nipponensis, freshwater shrimp	Stomach : 2.92 μg MC/g dw Hepatopancrease : 0.53 μg MC/g dw Gonads : 0.48 μg MC/g dw Eggs : 0.27 μg MC/g dw Gills : 0.05 μg MC/g dw Muscle : 0.04 μg MC/g dw
	Portugal	Crayfish	Not reported : 2.7 μg MC/g dw
Fish	Argentina	*Odontesthes bonariensis*, silver side	Liver : 1.01 μg MC/g fw
		Odontesthes bonariensis, silver side	Gills : 0.10 μg MC/g fw
		Odontesthes bonariensis, silver side	Muscle : 0.34 μg MC/g fw
	Australia	Flounder, snapper, tailor	Muscle : 0.003 μg NOD/g fw
		Garfish	Viscear : 0.152 μg NOD/g fw
	Baltic sea	*Platichthys flesus*, flounder	Liver : 0.14 μg NOD/g dw
		Gadus morhua, cod	Liver : 0.05 μg NOD/g dw
		Platichthys flesus, flounder	Liver : 0.41 μg NOD/g dw
		Platichthys flesus, flounder	Liver : 0.637 μg NOD/g dw
		Platichthys flesus, flounder	Liver : 0.399 μg NOD*/g dw
	Brazil	Fish	Muscle : 0.039 μg MC*/g fw
		Tilapia rendalii, red breast tilapia	Liver : 31.1 μg MC/g fw (range 0-31.1)
		Tilapia rendalii, red breast tilapia	Viscera : 67.8 μg MC/g fw (range 0-67.8)
		Tilapia rendalii, red breast tilapia	Muscle : 0.026 μg MC/g fw (range 0.003—0.026)
	China	*Carassius auratus*, red carpeted oranda	Muscle : 3.19 μg MC/g fw
	China	*Various fish species:* *Hypopothalmichthys molitrix*, phytoplanktivorous *Parabramis pekinensis*, herbivorous *Carassius auratus*, omnivorous *Culter ilishaeformis*, carnivorous	Intestines : 22 μg MC/g dw Blood : 14.4 μg MC/g dw Liver : 7.8 μg MC/g dw Kidney : 5.8 μg MC/g dw Muscle : 1.8 μg MC/g dw
	Netherlands	*Osmerus eperlanus*, smelt	Liver : 874 μg MC/g afdw (range 59-874)
		Gymnocephalus cernua, ruffle	Liver : 194 μg MC/g afdw (range 9-194)
		Perca fluviatilis, perch	Liver : 51 μg MC/g afdw (range 17-51)
	Portugal	*Cyprinus carpio*, carp	Muscle : 0.3 μg MC*/g fw
		Barbus sp. barbel	Muscle : 0.1 μg MC*/g fw
		Lisa sp, grey mullet	Muscle : 0.1 μg MC/g fw
	Uganda	*Oreochromis niloticus*	Gut : 1312.08-1479.24 ng/g Liver : 48.07-73.10 ng/g Muscle : 9.65-208.65 ng/g
	Uganda	*Lates niloticus*	Gut : 27.78 ng/g Liver : 3.74 ng/g Muscle : 1.86 ng/g

* = indicates ELISA results were not confirmed by another method

13.1.3.5: Effects on livestock and wildlife

[based on de Figueiredo et al. 2004; van Apeldoorn et al. 2007; Stewart et al. 2008].

Cyanobacteria toxins can affect both domesticated animals and wildlife. The effects of cyanobacterial toxins were first reported as stock deaths at Lake Alexandrina, South Australia in 1878. Since then there have been reports of animal poisoning in Australia and other countries such as Canada, South Africa, USA, and UK. Animals killed due to cyanobacterial toxins are birds, cats, cattle, chicken, dogs, ducks, fish/shrimp, horses, lambs, monkeys, sheep, squirrels, pigs, rhinoceros, rodents and invertebrates. Livestock including cattle, sheep and poultry may be affected as a result of drinking water contaminated with cyanobacteria. Table 13.7 summarizes a world wide account of cyanobacterial poisoning of livestock and wild animals since 1878. Summary of the effects of cyanotoxins on livestock and wild animals are as follows:

- Anatoxin-a cause death within minutes to a few hours depending on the species, amount of toxin ingested, and amount of food in the stomach. Documented cases of animal fatalities due to consumption of water with anatoxin-a are cattle in Finland and Canada, and dogs in the USA, New Zealand, Scotland and Ireland.
- Anatoxin-a(S) was responsible for the death of dogs, pigs and ducks in the USA. Animal poisoning due to anatoxin-a(S) was also reported from Denmark.
- Three cows and ten calves died on a cattle property located near a farm dam at McKinlay in northwest Queensland, Australia resulting from toxic effects of *Cylindrospermopsis raciborskii.*
- Microcystins were responsible for repeated poisoning in wild and domesticated animals, death of cattle, geese, sheep, pigs, horses, dogs, cats, squirrels, poultry, waterfowl and birds. (Table 13.7).
- In the UK (1989), *Microcystis* blooms in lakes and reservoirs have been associated with the death of sheep and dogs with human illness in the past decade (WHO 2003).
- *Planktothrix agardhi* blooms caused intoxication of pets and 121 persons in Sweden (WHO 2003)
- The episodic mass mortalities of bird, Lesser Flamingos (*Phoeniconaias minor*) at Kenyans Rift Valleys during 1990s have been linked to microcystin-LR, microcystin-RR and anatoxin-a (ANA). Death of birds may have been caused from feeding on floating cyanobacterial scum (note: scum contains high levels of toxins such as microcystin). Research conducted found these three cyanotoxins (MC-LR, MC-RR and anatoxin-a) in dead flamingo livers, stomach (microcystin 1.96 μg/g dw; anatoxin-a 43.4 μg/g dw, intestine (microcystin 0.36 μg/g dw); anatoxin-a 7.62 μg/g dw) and faecal pellets (microcystin 0.48 μg/g dw; anatoxin-a 2.45 μg/g dw) from Lake Bogoria and Nakuru. The MC-LR in the bird livers varied from 0.21 to 0.93 eq/g fresh weight and anatoxin-a 1.06-5.82 μg/g fresh weight, and these concentrations may have been sufficient to have caused the birds death (van Apeldoorn *et al.* 2007)
- In the summer of 1985, in Alberta, Canada, over 1,000 bats were found dead at a lake covered with a thick white scum containing anatoxin-a.
- In New South Wales, Australia (summer of 1971) almost total mortalities of bees from 84 hives was associated with the insects watering from a lake affected by a wind borne scum of *Anabaena circinalis.*
- In South Africa (1994-1995), livestock poisoning outbreak was related to contamination of drinking water by the presence of *M. aeruginosa* (MC-LR).
- In the USA, hepatotoxic blooms of *M. aeruginosa* were related to deaths of 24 young cows.

Possible symptoms that livestock show when poisoned by cyanobacteria are convulsions, paralysis, liver damage, constipation and abortion and skin sensitivities. Milder cases of algal poisoning in livestock's can cause loss of appetite and may cause a consequent decline in milk yield in dairy cattle. In extreme cases, stock deaths can occur following ingestion of high concentrations of potentially toxic blue-green algae. Dogs are particularly susceptible to blue-green algal poisoning because scums can attach to their coats and be swallowed during self-cleaning.

13.1.3.6: Effects on drinking water quality

Algal blooms of toxic blue-green algae can affect the water quality by causing peculiar tastes and odours, discolouration and unsightly scums. Cyanobacteria have the ability to produce earthy and musty smelling compounds or odour metabolites (compounds that are not essential for primary cell metabolism). Over 200 compounds have been identified as being produced by algae, of which filamentous forms produce more than 25% of all known off-flavour compounds. Geosmin (E1, 10-dimethyl-E-9-decalol) and MIB (2-methyl isoborneol) accounts for the majority off-flavour odour. The cyanobacterial genera that are known to produce geosmin are *Anabaena, Aphanizomenon, Lyngbya, Microcystis, Oscillatoria, Phormidium, Schizothrix* and *Symploca* (see Table 13.8). The presence of objectionable odour has caused an estimated 30% of the potential revenue loss in the US channel catfish industry alone. Off-flavours have also impacted Nile tilapia, Atlantic salmon, rainbow trout, and shrimp aquaculture industry. Moreover, the presence of cyanobacteria in water can cause clogging of irrigation and water delivery systems including filters, meters, valves and sprinklers, thereby increases the costs of water treatment.

Table 13.7: ***Historical records of animal poisoning incidents or health impacts on aquatic organisms associated with cyanobacterial toxins*** *[NHMRC 1994; Duy et al. 2000 and Landsberg 2002].*

Country	Location where occurred	Animals affected	Species producing toxins
Argentina	Guyena	1984 – 72 cows.	Hepatotoxins from *Microcystis aeruginosa*
Australia	Darling River; Lake Mokoan, Victoria (1 time); Lake Alexandriana; New England; NSW (4 times); QLD (2 times); Western Australia.	1878 -sheep, dogs, horses, pigs & cattle; 1959 -300 sheep, 5 cattle, , and a horse; 1965 - 20 lambs; 1966 - 16 sheep died & 50 were sick; 1971 - death of honeybees; 1974 to 75- 34 sheep & 52 lambs; 1977 – Turkeys; 1981 – 25 sheep died; 1984 - death of >30 cattle; 1991 – 1600 sheep & cattle died, fish killed associated with anoxic conditions; 1992 – sheep; 1994 – 14 sheep; 1998 – 13 cows & calves.	*Anabaena circinalis;* *Microcystis aeruginosa;* *Nodularia spumigena.*
Belgium		1995 – approx. 30 herons and ducks.	*Microcystis* spp.
Bermuda		1945 - Cattle deaths	*Aphanizomenon flos-aquae*
Brazil		1996 – fish (long whiskered catfish)	*Microcystis* spp.
Canada	Alberta (7 times); Ontario (2 times); Manitoba (3 times); Saskatchewan (9 times).	1967 – 25 pigs; 1971 to 1918 – cattle, Hogs, horses, poultry, wild bird; 1924 – 20 cattle; 1945 – a horse, several calves, 2 pigs, and a cat; 1948-1949 – cattle deaths; 1950 – a cow, horses, pigs, dogs, turkeys, geese, chickens, and wild birds; 1949-1951 – heavy mortality of wild ducks; 1951 – a horse, 9 spaniel dogs; 1954 – pigs dies, cattle affected; 1950 – 14 beef cattle; 1959 – 300 dogs, 1 goose, horse, and cattle; 1959 – 30 dogs, 1 goose, horse, and cattle; 1961 – 20 dogs, 3 cattle, perch, and wild ducks; 1962 – 1 horse, 8 cows died, 60 cows sick; 1962 – 3 dogs;	*Anabaena* spp.; water bloom; *Anabaena flos-aquae;* *Aphanizomenon;* *Aphanizomenon flos-aquae;* *Microcystis aeruginosa;* *M. flosaquae* *Anabaena* spp.; *Microcystis;* Blue-green algae; algae; *Anabaena circinalis.*

Contd...

		1964 – 5 dogs; 1964 – 20 calves sick, 1 died moderate fish mortalities; 1965 – 17 cattle; 1966 – 2 calves and 1 dog; 1966 – 2 calves and 1 dog; 1967 – 25 pigs; 1972 – 3 calves; 1972 – 12-15 cattle; 1975 – 20 lambs, 34 cattle; 1985 – 1000 bats, hoary bat, mallards, & American wigeon; 1989 – 16 cattle died.	
Chile		1984?- Fish- silverside, common carp	*Microcystis aeruginosa*
Denmark		1975 - 30 dogs, were sick, 20 died; 1993-94 - birds	*Nodularia spumigena;* *Anabaena lemmemanni*
England		1989 – 20 sheep, 15 dogs; 1978 – 3 cows	*Microcystis* spp.; *Planktothrix agardhii*
East Indian Archipelago		1873 – fish	*Tricodesmium erythraeum*
Finland	Lake Vesijarvi; Coast of the Baltic sea; Winhelmhaven; Lake Saakjarvi.	1928 – about 40 cattle; 1982 – fish roach; 1984 - water fowl, fish, muskrat; 1 dog and 3 puppies; 1985 – cows; 1990 – 2 dogs sick and sacrificed; 1997 – 2 dogs.	*Anabaena lemmemanni;* *Anaboenopsis frogilissima;* *Nodularia spumigena;* *Anabaena circinalis;* *Planktothrix agardhii.*
France-Forez		1988 – farmed fish	*Microcystis aeruginosa.*
Georgia		1964-Fish; 1977 – cattle; 1997 – fish.	*Nodularia spumigena;* *Microcystis* spp.; *Rhizosolena forgillssima;* *Anabaenopsis elenkinii;* *Rhizosolenia calcaravis.*
Germany		1963 - About 400 ducks; 1983 – 16 young cattle; 1990 - 2 dogs;	*Nodularia spumigena.*
Holland		1942 - Fish, frogs, newts.	*Aphanizomenon flos-aquae*
Hungary		1914 – Fish.	*Anabaena circinalis.*
India		1973 - Pearl oyster.	*Tricodesmium thiebautii.*
Ireland		1992-94 – dogs.	Benthic species
Italy		1997 - Fish and bivalves.	*Planktothrix* spp.
Japan		1995 - 20 spot -billed ducks.	*Microcystis* spp.
Latvia		1982-1987 – fish-pike, pike perch.	*M. aeruginosa.*
New Zealand		1966- Lambs.	*M. aeruginosa.*

Contd...

Norway		1978 – 4 heifers	*M. aeruginosa*
Russia		1953 – horses, livestock, ducks, geese, cats, dogs, waterfowl; 1956-1959 – common carp, catfish, shellfish, roach, bream	*M. aeruginosa*
Scotland		1990-91 – 4 dogs 1992 - 1000 brown trout (fish); 1992 – dogs; 1994 – 6 young cattle; 1995 – 2 cattle.	*Aphanizomenon flos-aquae;* *Microcystis aeruginosa;* *Planktothrix agardhii;* *P. agardhii* *Oscillatoria* spp.
South Africa	Free state & Transvaal; Hartbeespoort Dam; Lake Bonney; Vaal Dam.	1913-1943 – thousands of sheep, cattle, horses, mules, donkeys, dogs, hares, turkeys, ducks, fish; Sheep; 1927 – livestock, rabbits and wild birds; 1959 – 300 sheep, 5 cattle, and 1 horse; 1973-1974 – cattle deaths; 1977 – 3 lambs; 1979 – 3 rhinoceroses; 1980 – cattle; 1993-94 – cattle & sheep; 1994-11 sheep; 1996 – 200 dairy cows, 1996 – 18 goats; 1996 – 290 dairy cows, 1996 -18 goats.	Algae including blue green algae; Benthic spp.; *Microcystis aeruginosa;* *Microcystis* spp.; *Nodularia spumigena*
Spain		1989 – 600,000 rainbow trout (fish); 1983 – fish in ponds.	*Anabaena* spp.
Sri Lanka	Kursadi island	1942- a variety of marine fauna including fish eel, milk fish, butterfly fish, Emperor angelfish, large scale mullet - associated poor water quality, low dissolved oxygen.	*Tricodesmium*
Sweden		1940-1942- fish, mergansers, divers, cats; 1982 – 9 dogs died; 1983 – 9 dogs.	*Anabaena* spp.; *Nodularia spumigena*
Thailand		1991 – shrimp	
UK		1978 - 3 cows; 1989 – death of 20 sheep, and 14 dogs; 1990 – 3 dogs died; 1991 – 1 dog died.	*Oscillatoria* spp.; *Oscillatoria agardhii;* *Microcystis aeruginosa*
USA	Colorado; Indiana; Iowa (4 times); Kentucky; Illinois (2 times); Minnesota (8 times); Montana (3 times); New	1882-1884 - cattle, horse, hogs; 1900 - several cattle; 1914 – cattle; 1918 - 1 sheep, 7 hogs, and about 30 chicken;	*Anabaena* spp.; *Anabaena flosaquae;* *Anabaena lemmemanni;* *Anabaena spiroides;* *Aphanizomenon flos-aquae;*

Contd...

	Hampshire (2 times); North Dakota; Mississippi; Oklahoma (2 times); South Dakota; Texas; Washington; Wisconsin.	1930 - 9 cattle; 1931-1933 – fish; 1933 - more than 21 sheep and chickens; 1933 - 45 turkey, 4 ducks, 2 geese, cow, pigs, horse, poultry; 1933- 3 cattle; 13 sheep, 8 lambs, and numerous chicken; 1939 – 4 ducks, wild birds, carps, snakes, salamanders, a calf; 1943 – sheep; 1944-45 – 37 hogs, 4 sheep, 2 cattle, 3 horses, several dogs, cats, squirrels, chickens, turkeys, and song birds; 1946 – cattle and deer; 1946 – carp, northern pike, yellow pike, perch, black crappies, bluegills, suckers, black bullheads, hog suckers, ells; 1948 –79 hogs, 2 horses, many ducks, chicken, cats, wild animals, few dogs; 1948 – horses, a dog, and wild birds, buffalo fish, common carp, black bullhead; 1952 – thousands of Franklins gulls; 560 ducks, 400 coots, 200 pheasants, 50 fox squirrels, 18 muskrats, 15 dogs, 4 cats, 2 hogs, 2 hawks, 1 mink; 1956 – 58-fish, frogs, chickens, ducks, turkeys and cattle died or became ill; 1956-58 – fish, frogs, chickens, ducks, turkey, and cattle; 1966 – tons of fish died after $CuSO_4$ treatment; 1976 – 4 dogs died, 7 dogs, 1 horse, 1 cow were sick; 1977 – 8 dogs, and 30 cattle; 1977 – several cattle; 1980 – toxic to tadpoles; 1981 – tadpoles died; 1981 – 10 sows; 1984 – 11 cattle; 1985 – 9 cows died, 11 were sick; 1985 – 9 dogs; 1986 – 5 ducks, and 13 pigs died; 1987 – 5 cows were sick; 1987 – 20 hogs died; 1988 – 4 cows, 6 calves, 18 pigs, and 7 ducks died; 2 dogs died; 1989 – 18 pigs, 7 mallard ducks; 1991 – 2 dogs; 1992- 3 Holstein heifers; 1997 – cattle; 1998 – cattle.	*Coleosopharium Kuetzzingianum; Gleotrichia echinulata;* *Microcystis* spp.; *Microcystis aeruginosa;* *M. flos-aquae;* *Nostoc rivulae;* *Oscillatora* spp.; algae including blue-green algae; hepatotoxins from water bloom.
USSR	Lake Juksa; Lake Semehivichi.	1933- cats; 1953 – death of cats, dogs, and water fowl	Blue-green algal bloom; *Microcystis aeruginosa*

13.1.3.7: Effects on mammals and human

13.1.3.7.1: Toxicity and toxic effects

[based on Duy et al. 2000; Codd et al. 2005; van Apeldoorn et al. 2007; Pergram et al. 2008; Hudnelll 2009]

Cyanotoxins are among the most potent toxins known; only botulinum and dioxins are among few toxic substances known to be more potent than cyanotoxins. The USEPA listed microcystins, cylindrospermopsin and anatoxin-a as highest priority cyanotoxins, and saxitoxin and anatoxin-a(S) as medium to high priority. Human poisoning due to cyanobacteria was mainly caused by microcystins, cylindrospermopsin and anatoxin-a (Hudnell 2009).

Anatoxin-a are nicotionic acetylcholine receptor having a LD_{50} of 200 µg/kg. Typical symptoms recorded in mice are loss of muscle coordination, gasping, convulsions and death within minutes from respiratory arrest. Anatoxin-a(s) is a potent acetylcholinesterase inhibitor with a LD_{50} (i.p mouse of 20 µg/kg). The saxitoxin and neosaxitoxin are fast acting neurotoxins that inhibit nerve connections by blocking sodium channels. The LD_{50} (i.p. mouse) is reported to be 10 µg/kg for saxitoxin.

Table 13.8: ***Cyanobacteria species known to produce off-flavour compounds*** *[information compiled by Smith et al. 2008].*

Odour metabolites	Cyanobacteria species
Geosmin (E1,10-dimethyl-E-9-decalol)	*Anabaena circinalis, A. Crassa, A. laxa, A. lemmermanii, A. macrospore, A. solitaria, A. viguieri; Aphanizomenon flos-aquae, A. gracile; Fisheriella muscicola; Geitlerinema spendidium (= Oscillatoria splendida); Leibleinia subtilis (= Lyngbya subtilis); Lyngbya aestuarii, L. cryptovaginata; Oscillatoria amphibia; Phormidium amoeneum, P. breve, P. calcicola, P. cortianum, P. formosum, P. simplissimum, P. uncinatim, P. viscosum, Phormidium.* spp.; *Planktothrix aghardhii, P. prolifica; Pseudanabaena catenata; Schizothrix muellerii; Symploca muscorum; Tychonema borneti, T. granulatum.*
MIB (2-methylisoborneol)	*Hyella* spp.; *Jagerinema genimatum (= Oscillatoria geminate); Lyngbya wollei; Oscillatoria curiceps, O. limosa, O. tenuis, O. variabilis; Phormidium autumnale, P. breve, P. calcicola, P. favosum, P. tenue, Phormidium* spp.; *Planktothrix aghardhii, P. cryptovaginata, P. attenuata, P. perornata, P. prolifica; Porphyrosiphon martensianus; Pseudanabaena articulate, P. catenate, P. limnetica; Synechococcus cedrorum, Synechococcus* spp.; *Tychonema granulatum.*
Ketones, ionines	*Calothrix parietina; Rivularia* spp.; *Tolypothrix distorta*

Microcystins are genotoxic, protein phosphatase inhibiting toxins responsible for liver failure and death in humans, wild animals, livestock and aquatic life. Microcystins can cause adverse effects by rapidly binding and inhibiting the activity of serine/threonine protein phosphatase 1 and 2A. The inhibition is irreversible. MC-LR has a LD_{50} (i.p mice, 24 hours) of 60 µg/kg, whereas i.p LD_{50} for MC-RR is about ten times higher. The primary acute effect of protein phosphate inhibition is hyperphosphorylation of many cellular proteins including hepatocellular cytoskeleton, which causes loss of cell to cell contacts and thus intra-hepatic haemorrhage. Death is due to hypovolemic shock. Microcystins have also been implicated in tumour promotion in both the liver and colon. In addition it can produce gastrointestinal and allergic reactions.

Nodularin is a peptide toxin and the mechanism of toxicity is quite similar to that of microcystins. The acute toxicity and mode of action of different cyanotoxins is summarised in Table 13.2.

Cylindrospermopsin causes wide spread necrotic injury in mammals (liver, kidneys, lungs, spleen and intestine). It is also genotoxic and can cause chromosomal loss and DNA

The possible effects of cyanotoxins to humans are summarised below:

- Hepatotoxins (e.g. mycrocystin - most potential for human injury including liver damage and tumour promotion).
- Neurotoxins (e.g. saxitoxins class of compounds or Paralytic Shellfish poison-PSP's) - Sodium channel blocking agent - acute poisoning results death by paralysis and respiratory failure.
- Cytotoxin (e.g. cylindrospermopsin) - Cytotoxic; liver, kidney and other organ damaged.

- Endotoxins (e.g. lipopolysaccharides (LPS) of outer cell wall of cyanobacteria) - Gastro-intestinal disorders; skin, eye irritation; skin rashes; respiratory distress.

13.1.3.7.2: *Exposure routes of cyantoxins to humans*

People may be exposed to cyanobacterial toxins via different routes as highlighted below according to Codd *et al.* 2005:

- *Oral or ingestion*: drinking water or recreational contact or through showering or bathing in contaminated water or via food (shellfish, finfish if toxin accumulation has occurred), plant foods (after irrigation with water containing cyanobacterial toxins), dietary supplementary (pills, capsules if containing dried cyanobacterial cells with toxins).
- *Pulmonary (inhalation, aspiration)*: water, aerosols, spray during recreation, work, showering.
- *Dermal (skin, mucosal contact):* water during recreation, work, showering.
- *Haemodialysis*: water used for haemodialysis.

13.1.3.7.3: *Human health risks and illness*

Surface scums of cyanobacteria, where they occur, represent a specific hazard to human health because of their particularly high toxin contact. Cyanotoxins in recreational waters pose human health risks even without direct water contact. For example, cyanobacteria toxins may become airborne due to wind and wave action. Human's miles away from affected water bodies may inhale cyanotoxins that may cause acute respiratory distress in asthmatic and other susceptible populations, and may contribute to the chronic and delayed health effects associated with exposure to algal toxins (Hudnell 2005 and 2009).

Disease due to cyanobacterial toxins varies according to the type of toxin and the type of water or water-related exposure (drinking, skin contact, etc.). Humans are affected with a range of symptoms including skin irritation, stomach cramps, vomiting, nausea, diarrhoea, fever, sore throat, headache, muscle and joint pain, blisters of the mouth and liver damage. Swimmers in water containing cyanobacterial toxins may suffer allergic reactions, such as asthma, eye irritation, rashes, and blisters around the mouth and nose.

The first human illness was due to an outbreak of gastroenteritis associated with cyanobacterial toxins in the Ohio River, USA in 1931, however, the most serious human poisoning due to cyanobacterial toxins was the death from liver failure of more than 50 patients received contaminated dialysis water in Brazil in 1996 (see Box 13.2). Illness caused by toxins has also been reported from other countries such as Australia, Brazil, Canada, China, UK, USA and Zimbabwe (see Table 13.9).

Table 13.9: ***Reported human illness associated with exposure of cyanobacterial toxins*** *[Duy et al. 2000; Azevedo et al. 2009; http://www/whowater_sanitation_health]*

Country	Location	Affected population	Exposure route and species
Australia	Palm Island (Queensland); Armidale (New South Wales); Lake Alexandrina (Central Australia); Northern and Central Australia; River Murray town (South Australia)	1975 - 140 children (2-16 yr) and 10 adults experience hepatitis-like syndrome with headache, abdominal pain, vomiting, anorexia, tender hepatomology; increase in the level of Y-glutamyltransferase; 1991 - 5 adults and 3 children with skin rashes;	Drinking water from open reservoir; water contact; River water; rain water tanks *Anabaena* sp.; *Cylindrospermopsis raciborskii; Microcystis* sp.; *Nodularia* sp.

Contd...

		1992 – 'barcoo fever' disease characterised by nausea and vomiting at the sight of food; 1992 – 26 cases, aged 1-84 yr with skin, systematic, and multiple symptoms	
Brazil	Caruaru, Pernambuco state	1996 – 116 to 136 patients at a dialysis centre experienced visual disturbances, nausea, and vomiting; 63 patients died	Receiving dialysis water contaminated by microcystins
Canada	Saskatchewan	1959 – 12 people with acute gastroenteritis	Lake water contact, swimming *Anabaena* spp.
China		1994 – High rate of hepatocellular carcinoma cancer in population	Consuming pond/ditch and shallow well
UK	Staffordshire	1989 - cases of pneumonia, 16 cases with sore throat headache, abdominal pain, vomiting, dry cough, diarrhoea, vomiting, and blistered mouths	Reservoir-canoeing, swimming in water *Microcystis aeruginosa*
USA	West Virginia; Allegheny Country, Pennsylvania (2 times)	1930-1931 – 9,000-60,000 population affected with acute gastroenteritis; 1975 – 62% of the 8,000 population affected by acute gastroenteritis; 1980-1981 – over 100 persons affected by skin and eye irritation, earache, hay-fever like symptoms, and acute gastroenteritis	Drinking water from River, reservoir; contact with water from lake (swimming, water skiing) *Anabaena* sp.and *Aphanizomenon* sp. and *Schizothrix calcicola*
Zimbabwe	Harare	1960-1961 - acute gastroenteritis among children	Consuming water

Box 13.2: ***First human intoxication by microcystins in Brazil*** *[Azevedo et al. 2009].*

The first human deaths from cyanobacterial hepatotoxins occurred following intravenous exposure in a dialysis clinic in Caruaru city, Pernambuco State, Brazil. In February 1996, 116 patients (out of 131) began to complain of headache, pain, blurred vision, nausea, vomiting. Subsequently, 100 patients developed liver failure, and 52 had died by December of 1996 with a common syndrome called 'Caruaru syndrome'. On investigation, it was found that the referral clinic was not receiving treated municipal water during the 1996 summer drought. Instead, the clinic received water (from Tabocas reservoir which is water supply for Caruaru city) without complete treatment delivered by tank trucks from the municipal water treatment plant. Occasionally, the truck driver was instructed to add chlorine to the water in his truck when the visual turbidity of water was high. After arriving at the clinic, water was treated in its in house water treatment system (a process that involved sand, carbon, and cation/anion exchange filtrations, followed by a micropore filtration, before being used for hemodialysis). The analysis of phytoplankton from Tabocas reservoir revealed that cyanobacteria represented 99% of the total density of phytoplankton community, the most common genera were Microcystis, Anabaena (which are known as microcystins producers) and Cylindrospermopsis (known to produce cylindrospermopsin). Analysis of carbon, sand and cation/anion resin filters showed microcystin in the range of 0.5-2.1 μg/g and cylindrospermopsin in the carbon extract of 19.7 μg/g. All victims' serum and liver tissue were positive for microcystins (the average microcystin concentration of liver samples who died during February December 1996 was 223 μg/g). The chemical analysis of live samples identified microcystins - MC-LR, MC-YR and MC-AR but did not confirm for cylindrospermospin. Histopathology examination of liver tissue obtained from 16 victims revealed a uniform pattern with disruption of liver plates, liver cell deformity, necrosis, apoptosis, cholestasis, cytoplasmic vacuolization, mixed leukocyte infiltration and multinucleated hepatocytes. An estimated microcystin concentration of 19.5 µg/l of water used during dialysis treatment from 13 to 17 February 1996. The available biological and chemical evidence as cited above suggest that water being a major factor in the deaths of dialysis patients. The result further suggests that intoxication of patients may have been caused by microcystin released from cyanobacterial cells they were lysed by chlorination in the tank trucks. The first outbreak of human poisoning from cyanotoxins signifies the importance of monitoring of water quality (chemicals) to safeguard the human health.

13.1.4: Regulations and guidelines for cyanobacteria and cyanotoxins

[based on Burch 2008; Izdorczyk et al. 2009].

A number of countries have developed regulations or guidelines for cyanobacteria and cyanotoxins in drinking water, or in water used for recreational activity and agriculture. The main focus has been upon microcysins produced by *Microcystis aeruginosa* and *Planktothrix agardhii* since these species are regarded as the most significant potential source of human injury from cyanobacteria on a world-wide scale (Burch 2008).

13.1.4.1: Alert levels for cyanobacteria

Alert levels are public warnings of the risk posed by cyanobacteria to users of public waters. It is based on identification of cyanobacteria species, their cell numbers and size, toxin concentration and other prevailing conditions. The NHMRC (National Health and Medical Research Council, Australia (see <http://www.nhmrc.gov.au>) and other authors (Hallegraeff 1991; Burch 1993, 2002 and 2008) proposed a number of cyanobacterial alert levels (see Table 13.10) to protect the human health, as listed below:

- Alert level 1: low probability of adverse health effects.
- Alert level 2: moderate probability of adverse health effects.
- Alert level 3: high probability of adverse health effects.

13.1.4.2: Guideline values for cyanotoxins

[WHO 2006; van Apeldoorn et al. 2007].

13.1.4.2.1: Guidelines for drinking water

The recommended WHO guideline for microcystin for the protection of human health is 1 µg/l. See Box 13.3 for calculation of WHO guidelines for microcystin. Many counters have adopted WHO provisional guideline for microcystin (MC-LR) for drinking water including Brazil, Czech Republic, China, France, Italy, Japan, Korea, New Zealand, Norway. However there are no guidelines for other cyanobacterial toxins, except in New Zealand (see Table 13.11).

13.1.4.2.2: Guidelines for food

The general guideline ('health alert' level) for Australian seafood for nodularin and cylindrospermopsin (Van Buynder *et al.* 2001; Saker *et al.* 2004) are as follows:

- *Nodularin*: for fish, prawn and mussels are 214, 943 and 1286 µg/kg food respectively
- *Cylindrospermopsin*: for fish, prawn and mussels are 158, 720 and 933 µg/kg food respectively.

The WHO recommended tolerable daily intake (TDI) for human exposure to microcystin (MC-LR) is 0.04 µg/kg per day in food and most countries apply a tolerance level of 80 mg STX eq/100 g mussel meat. For cyanobacterial food supplements the proposed guideline for MC-LR is 10 µg/g.

Table 13.10: *Comparison of different threshold definitions of alert levels for cyanobacteria proposed by different authors.*

Alert Level	Burch (1993) (based on cell nos.)	Falconer *et al.* (1994) (based on microcystin concentration)	Bartram *et al.* (1999); Chorus and Bartram (1999) (based on cell numbers and chlorophyll-a concentration)	House *et al.* (2004) (based on cell numbers)	NHMRC (2008) (based on cell numbers and microcystin concentration)	Izydorczyk (2009) (based on chlorophyll-a concentration determined by algae on line (AOA)	Comments
Alert level 1 Or *Green level surveillance mode*	500 cells/ml	0.4 µg/l microcystin	20,000 cells/ml or 0.2 mm³/l or 1.0 µg chl a/l	2,000 cells/ml	≥ 500 to 5,000 cells/ml or >0.04 to <0.4 mm³/l (*Microcystis aeruginosa* or biovolume equivalent for the combined total of all cyanobacteria)	1.9 µg/l chlorophyll-a	- Stage typical of non bloom-conditions and the water is generally free of visible colour. - Low probability of adverse health effects.
Alert level 2 or *Amber level alert mode*	2,000 cells/ml	1.0 µg/l	100,000 cells/ml or 10 mm³/l or 50 µg chl a/l	5,000 cells/ml or 1 mm³/l	> 5000 to 50,000 cells/ml or ≥ 0.4 to <4 mm³/l (*Microcystis aeruginosa* or biovolume equivalent for the combined total of all cyanobacteria where a known toxin producer is dominant in the total biovolume or ≥ 0.4 to <10 mm³/l (for the combined total of all cyanobacteria where known toxin producers are not present	4.9 µg/l chlorophyll-a	- Stage when algae/cyanobacteria are multiplying, water may be of green tinge, or may be musty or organic tastes and odours; scums may form at cell numbers exceeding 5,000 cells/ml. -Moderate probability of adverse health effects.
Alert level 3 or *Red level action mode*	15,000 cells/ml	10.0 µg/l	Presence of scums	50,000 cells/ml or 10 mm³/l	Level 1 guideline ≥ 50,000 cells/ml toxic *Microcystis aeruginosa* or ≥ 10 µg/l total microcystins or biovolume equivalent of ≥ 4 mm³/l for the combined total of all cyanobacteria where a known toxin producer is dominant in the total biovolume Level 2 guideline >10 mm³/l for total biovolume of all cyanobacterial material or cyanobacterial scums are consistently present	49.4 µg/l chlorophyll-a	-Stage of bloom conditions, water is green, have strong musty taste and odour; toxin must be present and water is unsafe for human and stock consumption and contact and recreational activities; scums often present. -high probability of adverse health effects.

13.1.4.2.3: Guidelines for livestock drinking water

Australia has developed livestock drinking water guidelines for cyanobacteria (ANZECC and ARMCANZ 2000). Algal blooms should be treated as possible toxic event and the water source should be withdrawn from stock until the algae are identified and the level of toxin determined. An increasing risk to livestock health is likely when cell counts of *Microcystis* exceed 11,500 cells/ml and/or concentrations of microcystins exceed 2.3 μg/l expressed as microcystin-LR toxicity equivalents.

Table 13.11: *Guideline standards for cyanobacterial toxins in drinking water for various countries [compiled by Burch 2008].*

Country	Guideline value/standard	Comments
Argentina	Under review	
Australia	Microcystin (MC-LR): 1.3 μg/l Saxitoxins: 3 μg STX eq/l	MC-LR calculation is based on 70 kg average body weight of Australian adulthuman.
Brazil	Cylindrospermopsin: 15 μg/l Microcystins: 1.0 μg/l Saxitoxins: 3.0 μg/l	Use of algicides prohibited and toxicity testing/ toxin analysis is required when cell counts exceed 10,000 cells/ml or 1mm^3/l biovolume.
Canada	Microcystin (MC-LR): 1.5 μg/l	
Czech Republic	Microcystin (MC-LR): 1μg/l	Follows WHO provisional guideline for microcystin (MC-LR).
China	Microcystin (MC-LR): 1μg/l	Follows WHO provisional guideline for microcystin (MC-LR).
France	Microcystin (MC-LR): 1μg/l	Drinking water ruling
Italy	Total microcystins: 0.84μg/l	Follows WHO provisional guideline for microcystin (MC-LR).
Japan	Total microcystins: 1.0μg/l	Follows WHO provisional guideline for microcystin (MC-LR).
Korea	Total microcystins: 1.0μg/l	Follows WHO provisional guideline for microcystin (MC-LR).
New Zealand	Anatoxin-a: 6.0 μg/l Anatoxin-a(S): 1.0 μg/l Cylindrospermopsin: 1.0μg/l Homoanatoxin-a: 2.0 μg/l Microcystin (MC-LR): 1.0 μg/l Nodularin:1.0 μg/l Saxitoxins (STX-eq): 3.0 μg/l	Maximum acceptable values (MAV) are applied for micro-organisms; MAVs are based on the WHO guidelines for drinking water quality.
Norway	Microcystin (MC-LR): 1.0 μg/l	Follows WHO provisional guideline for microcystin (MC-LR).
Poland	Microcystin (MC-LR): 1.0 μg/l	National legislation for guideline value in drinking water.
South Africa	Microcystin (MC-LR): 0-0.8 μg/l	Guideline levels for microcystins in potable water as a "Target water quality range".
Spain	Microcystin (MC-LR): 1.0 μg/l	Maximum permissible amount in drinking water
Thailand	No guideline yet	Awareness of the need for guidelines.
USA	No guideline yet	Cyanobacteria and their toxins are listed as microbiological contaminants on the contaminant candidate list, currently recognised as unregulated contaminants.
Uruguay	Under review	
World Health Organisation (WHO)	Microcystin (MC-LR): 1.0 μg/l	Based on 60 kg average body weight of adult (see Box 13.3).

Box 13.3: ***Calculation of WHO guideline for microcystin*** *[WHO 2006].*

WHO Guideline value has been calculated as $= \dfrac{TDI \times bw \times P}{L}$

A guideline value for life time consumption of a chemical contaminant of drinking water is calculated by applying TDI or tolerable daily intake. TDI is the amount of potentially harmful substance that can be consumed daily over a life time with negligible risk of adverse health effects. The life time consumption of a chemical contaminant of drinking water is usually calculated by applying the derived TDI to a typical daily water intake in litres (l) by an individual of a given body weight (bw). The proportion (P) of daily intake of the contaminant that is ingested from the drinking water is considered since some intake may come from food or inhalation from air. Calculation of WHO guideline for microcystin 1.0 µg/l is as follows:

Average adult body weight (bw) is considered as 60 kg

Average water intake for adult is considered as 2 litres per day

WHO TDI is 0.04 µg/kg of body weight per day

The proportion (P) of TDI allocated to drinking water assumed to be 0.8

Therefore; WHO Guideline value $= \dfrac{TDI\ (0.04) \times bw\ (60) \times P\ (0.8)}{L(2)}$

$= 0.04 \times 60 \times 0.8/2 = 0.96$ *µg/l or 1 µg/l microcystins*

13.2: Marine biotoxins

According to FAO (2004) marine biotoxins are produced by a number of harmful micro-algal species which have the capacity to produce potent toxins (called phycotoxins). There are over 5,000 species of marine algae, but only 109 species of micro-algae are reported to produce toxins (out of 109 toxic algae, 7) are dinoflagellates; see Table 13.12 for the list of toxic micro-algae). Approximately 20% of all food borne disease outbreaks in the USA result from seafood consumption, half of which are associated with micro-algal toxins. Worldwide, marine algal toxins cause more than 60,000 poisoning events annually with an associated mortality rate of 1.5%.

Toxic algae may cause harm through the production of toxins or by their accumulated biomass, which can affect co-occurring organisms and alter food-web dynamics. Impacts include human illness and mortality following consumption of or indirect exposure to toxins and substantial economic losses to coastal communities, commercial fisheries and associated fish, bird and mammal mortalities. These potent toxins can find their way through the food chain (e.g. molluscs and finfish) to human consumers causing a variety of gastrointestinal and neurological illness. The toxins are probably used by their producers as a way of competing for space to fight predation or as a defense against the overgrowth of other organisms.

The following are the three main algal groups that produce toxin:

- *Cyanobacteria*: Single-celled, blue-green algae without a distinct nucleus; produces neurotoxins and most commonly hepatotoxins (covered in section 13.2).
- *Diatoms*: Single-celled algae with cell wall composed of silica; produces domoic acid associated with amnesic shellfish poisoning (section 13.3.3).
- *Dinoflagellates*: Single-celled algae with two flagella or long thin organelles extending from the surface of the organism; produces okadaic acid, ciguatoxins (section 13.3.4-13.3.7).

13.2.1: Causes of marine algal blooms

Most of the harmful marine algal species have a restricted distribution pattern, whereas some have a worldwide distribution. In recent time, the frequency, intensity and geographic distribution of harmful algal blooms has increased and this may be related with a number of causes as listed below:

- Increasing utilization of coastal waters for aquaculture,
- Transfer of shellfish stock from one place to another,
- Increasing input of nutrients from domestic, industrial and agricultural wastes, and
- Possible introduction of exotic species into waterways via ballast water containing algae, cyst or spores.

13.2.2: Types of phycotoxins

The most important marine phycotoxins at present are shellfish toxins and ciguatoxins, as listed and described below (see also Table 13.3.1):

Table 13.12: ***Taxonomic reference list of toxic marine micro-algae*** *[Intergovernmental Oceanographic Commission- http://www.bi.ku.dk/ioc/default.asp].*

Diatoms (13 species) *Amphora coffeaeformis, Nitzschia navis-varingica, Pseudo-nitzschia australis, P. calliantha, P. cuspidata, P. delicatissima, P. fraudulenta, P. galaxiae, P. multiseries, P. multistriata, P. pungens, P. seriata, P. turgidula*	**Dinoflagellates; Order: Prorocentrales** (13 species) *Prorocentrum arabianum, P. arenarium, P. belizeanum, P. borbonicum, P. cassubicum, P. concavum, P. emarginatum, P. faustiae, P. hoffmannianum, P. lima, P. maculosum, P. minimum, P. rhathymum*
Dinoflagellates; Order: Dinophysiales (11 species) *Dinophysis acuminata, D. acuta, D. caudata, D. fortii, D. miles, D. mitra, D. norvegica, D. rapa, D. rotundata, D. sacculus, D. tripos*	**Dinoflagellates; Order: Gymnodiniales** (19 species) *Amphidinium carterae, A. operculatum, A. operculatum var. gibbosum, Cochlodinium polykrikoides, Gymnodinium catenatum, G. corsicum, Karenia bicuneiformis, K. brevis, K. brevisulcata, K. concordia, K. cristata, K. digitata, K. mikimotoi, K. papilionacea, K. selliformis, K. umbella, Karlodinium armiger, K. veneficum, Takayama cladochroma*
Dinoflagellates **Order: Gonyaulacales** (23 species) *Alexandrium acatenella, A. andersonii, A. balechii, A. catenella, A. fundyense, A. hiranoi, A. minutum, A. monilatum, A. ostenfeldii, A. tamarense, A. tamiyavanichii, Coolia monotis, Gambierdiscus australes, G. pacificus, G. polynesiensis, G. toxicus, G. yasumotoi, Ostreopsis lenticularis, O. mascarenensis, O. ovata, O. siamensis, Protoceratium reticulatum, Pyrodinium bahamense*	**Haptophytes** (9 species) *Chrysochromulina leadbeateri, C. polylepis, Phaeocystis globosa, P. pouchetii, Prymnesium calathiferum, P. faveolatum, P. parvum, P. patelliferum, P. zebrinum*
Dinoflagellates Order: Peridiniales (5 species) *Azadinium spinosum; Heterocapsa circularisquama, Pfiesteria piscicida, P. shumwayae, Protoperidinium crassipes*	**Raphidophyceans** (7 species) *Chattonella antiqua, C. globosa, C. marina, C. subsalsa, C. verruculosa, Fibrocapsa japonica, Heterosigma akashiwo*

13.2.2.1: Shellfish toxins

- Amnesic shellfish toxins causing amnesic shellfish poisoning (ASP).
- Diarrhoeic or diarrhetic shellfish toxins causing diarrhoeic shellfish poisoning (DSP)
- Neurotoxic shellfish toxins causing neurotoxic shellfish poisoning (NSP)
- Paralytic shellfish toxins causing paralytic shellfish poisoning (PSP)

13.2.2.2: Finfish toxins

- Ciguatoxins causing ciguatera fish poisoning (CFP)

13.2.3: Amnesic shellfish poisoning (ASP)

[Based on ANZFA 2001; FAO 2004; Landsberg et al. 2005; Ciminiello and Fattorusso 2006; Camacho et al. 2007; Zaccaroni and Scaravelli 2008 and http://www.whoi.edu/redtide/page.do?pid; http://www.whoi.edu/redtide/illness/illness.html]

ASP is the only shellfish poisoning caused by a diatom that produce the toxin domoic acid. Several members of the chain forming diatoms (*Pseudo-nitzschia* spp.) are responsible for ASP. *Pseudo-nitzschia* species (Figure 13.4) are distributed worldwide (Table 13.13). The disease was first identified on Prince Edward Island, Canada in 1987 where more than 100 people were poisoned after consumption of blue mussels harvested from the Islands. Three people died and the survivor suffered memory loss.

Recently another closely related diatom genus, *Nitzschia*, originating from a Vietnamese shrimp pond was also added to the new list of domoic acid producing taxa (FAO 2004).

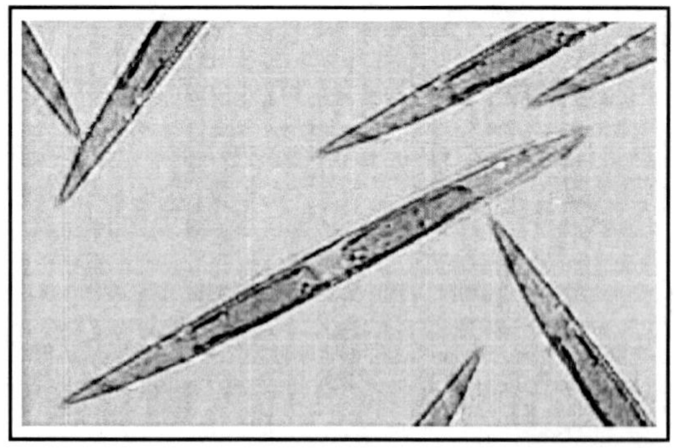

Figure 13.4: ***Domoic acid producing Pseudo-nitzschia pungens*** *[Zaccaroni and Scaravelli 2008 page 120].*

Table 13.13: ***Some examples of Pseudonitzschia*** **spp.** ***and their distribution*** *[FAO 2004; http://www.fao.org/docrep/007/y5468e/y5486e00.htm].*

Diatom species possessing toxin domoic acid	Distribution
Pseudo-nitzschia multiseries	***Atlantic*** - North America; Europe; South America. ***Pacific***- North America; Northeast Asia; Japan.
Pseudo-nitzschia pseduodelicatissima	***Atlantic*** - Europe; Africa (Denmark strait to Northwest Africa); Norwegian and Danish coastal waters, Skagerrak, Kiel Bay), South America (Argentina), USA (Arctic to Gulf of Mexico). ***Pacific***- California, British Columbia, Coastal waters of Penn Cove, Washington.
Pseudo-nitzschia delicatissima	***Atlantic*** - Norwegian coastal waters; Danish waters; Northwest Africa; Rhode island and Argentina and coastal waters of USA. ***Pacific***- California.
Pseudo-nitzschia australis	***Atlantic*** - Coastal waters of Spain, Portugal, Southwest Africa and Argentina ***Pacific*** - coastal waters of Peru, Chile, New Zealand, Mexico (Gulf of California), West Coast of North America from San Diego to Puget Sound, Washington, British Columbia.
Pseudo-nitzschia pungens	***Atlantic*** - Coastal waters of Argentina and coastal waters of USA ***Pacific***- Coastal waters of USA and Mexico; New Zealand coastal waters ***Sea of Japan*** - Peter the Great Bay, the Russian Federation.

Domoic acid can enter the marine food chain via uptake by molluscan shellfish (e.g. mussels) that filter their food from the water. The water can contain both diatoms themselves and the toxin. The toxins accumulate in the digestive gland and certain tissues of shellfish. Fish such as anchovies can also contain domoic acid in their guts by feeding on toxic *Pseudo-nitzschia* spp.

Shellfish identified containing ASP toxins are listed below:

- Blue mussels (*Mytilus edulis*), Canada 1987
- Soft shell clams (*Mya arenaria*), Canada, 1988
- Razor calm (*Siliqua patula*), Oregon and Washington, USA 1991

Other marine organisms

- Dungeness crabs (*Cancer magister*), Washington and Oregon states (USA) 1991.

13.2.3.1: Chemical nature of domoic acid

The toxin responsible for ASP is domoic acid (DA), a naturally occurring compound belonging to the kainoid class of compounds (Figure 13.5). DA is the major toxin present in plankton or shellfish. It is crystalline thermo stable, water- soluble excitatory tricarboxylic acid which acts as a glutamate antagonist on the kainite receptors of the central nervous system. DA is absorbed by the gastrointestinal tract and slowly reaches central nervous system. Several congeners (ten) of DA have been identified of which three geometrical isomers, isodomoic acids D, E and F and the C5′-diastereomer were found in small amounts in both the diatom and in shellfish tissues (Figure 13.5).

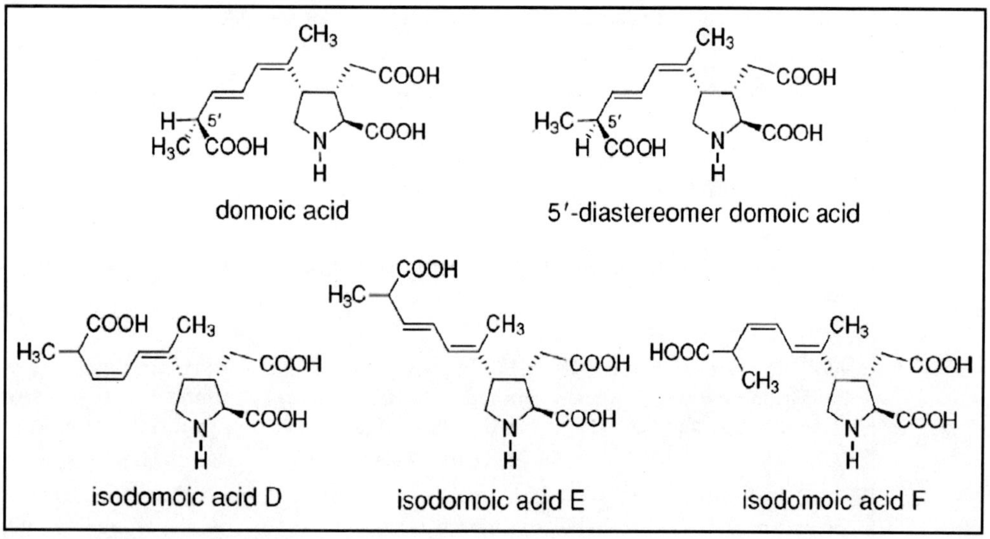

Figure 13.5: ***Chemical structure of domoic acid and some of its congeners*** *[Ciminiello and Fattorusso 2006; page 66].*

13.2.3.2: Toxic effects to aquatic organisms and human
13.2.3.2.1: Toxicity to aquatic organisms

Mortalities of sea lions, sea birds and sea otters, were found to be linked with domoic acid and *Pseudo-nitzschia* spp. In 1998 (May and June), over 400 Californian sea lions (*Zalophus californianus*) death was related to feeding on planktivorous northern anchovies (*Engraulis mordax*) contaminated with DA. Similarly in September 1991, brown pelicans (*Pelecanus occidentalis*) and cormorants (*Phalacrocorax penicillatus)* died in Monterey Bay, California (USA) after eating anchovies (*E. mordax)* contaminated with DA. Domoic acid related mortality also occurred with benthic feeding sea otters along the US West Coast. Most recently, high domoic acid levels were detected in dead humpback whales in George Bank off the northeast US coast. Table 13.14 provides specific examples of ASP (as cases and outbreaks) in different countries over the years.

13.2.3.2.2: Toxicity to human

DA is a potent neurotoxin and it binds to specific receptor proteins in neuronal cells causing continual depolarization of the neuronal cell until cell rupture occurs. DA is toxic to both the central and peripheral nervous systems of humans. Amnesic shellfish poisoning or ASP can be a life threatening syndrome and may cause both gastrointestinal and neurological disorders. The clinical symptoms of ASP include abdominal cramps, vomiting and diarrhoea, incapacitating headaches, disorientation and short-term memory loss. The memory loss associated with extreme cases of human intoxication lead to the description of the syndrome as ASP. In the most severe case of poisoning, patients are victim to seizure, coma, profuse respiratory secretion, unstable blood pressure and death. The loss of memory in patients associated with mussel toxin (ASP) is reported to be similar to patients with Alzheimer's disease, however, loss of memory in mussel-intoxicated patients is not affected by the age of the of patients, whereas symptoms of Alzheimer's disease's intensity increases with advancing age. Information relating to different marine biotoxins (ASP, DSP, NSP, PSP and CFP) including their causative algal species, principal toxins and biochemical effects are summarised in Table 13.15.

Table 13.14: *Cases and outbreaks of ASP in some selected countries* [*ANZFA 2001; FAO 2004*].

Country	Cases and outbreaks
North America	
Canada	- 1987 (November and December): food poisoning from cultured blue mussels (*Mytilus edulis*) from the Cardian Bay, Prince Edward Island, 3 deaths and 105 acute intoxication due to toxin DA sources from diatom *Pseudo-nitzschia pungens, P. nitzschia multiseries.*
	- 1988 (October): DA levels detected (20 mg/g) in soft shell clam (*Mya arenaria*) and blue mussels (*Mytilus edulis*) from the Bay of Fundy, New Brunswick resulting closure of some shellfish harvesting areas.
	- 1995 (April and May): sea scallops on Georges Bank showed DA levels in their digestive glands in excess of 1,300 mg/g and up to 150 mg/g in the roe.
USA	*USA-Alaska* - 1992: no severe problems with DA have existed, but the highest DA values was 11.1 mg/g (for a razor clam)
	USA-West Coast - 1991 (September): more than 100 brown pelicans and cormorants in Monterey Bay, Central California died which were attributed to the neurotoxin DA sourced from a bloom of the pennate diatom *P. australis.* DA was reported in viscera (up to 190 mg/g) and flesh (up to 40 mg/g) of local anchovies, a principal food source of seabirds.
	- 1991 (October-November): razor clams (*Siliqua patula*) from the Pacific coats of Washington and Oregon contained DA in the edible parts (foot, siphon and mantle) at levels as high as 154 mg/g (wet weight), 24 cases of illness in humans reported in the State of Washington. Recreational and commercial harvest of clams was forbidden; other shellfish (oysters and mussels) were never toxic; DA was also found in the viscera of Dungeness crabs (*Cancer magister*) in coastal waters of California, Oregon and Washington.
	- 1994: a bloom of *P. pungens, P. multiseries and P. australis* were reported in western Washington; mussels contained ~10 mg DA/g (wet weight) and the phytoplankton ~14 mg/g (wet weight).
	- 1997: a bloom of *Pseudo-nitzschia* spp. was observed (four species: *P. pungens, P. multiseries, P. australis* and *P. pseudodelicatissima*) with highest *Pseudo-nitzschia* concentrations of 13 x 10^6 cells/litre (28 July 1997).
	- 1998: elevated DA levels prompted a coast-wide closure of the Washington razor clam fishery.
	- 1998 (May and June): over 400 Californian sea lions died along the central California coast; contaminated anchovies with DA were the causative agent sourced from diatom *P. australias.*
	- 2002: sardines, anchovies, crabs and shellfish along the Californian Coast contained high levels of *DA* (upto 380 µg/g found in mussels); a number of marine mammals were sick or died (200 sea lions and 200 seabirds have became sick); no human illness.
Asia	
Japan	- 1991-1994: DA has not been detected in shellfish, or in diatoms.
	- 1994: DA was detected in the red alga, *Chondria armata* in Kagoshima prefecture; in this area xanthin crab (*Ategatis floridus*) contained 10 mg DA/kg.
Oceania	
Oceania (Australia, Tasmania and New Zealand)	- 1994: traces of DA have been detected in some scallop viscera (< 20 mg/g of shellfish meat). - 1992-1993: DA was detected at low levels in phytoplankton samples (Otago to Northland) in New Zealand; shellfish samples contained low levels of DA (maximum level was 16.5 mg/g).

	- 1993 (August): Scallop viscera from the lake Entrance, Victoria showed positive to domoic acid (26 µg/g for one sample and others < 20 µg/g). -1994 (December): maximum levels of domoic acid detected in New Zealand have been up to 187 µg/g in green mussels from Marborough Sounds; 72.4 µg/g in scallops from Tauranga harbour; 600 µg/g in scallop gut from Doubtless Bay.

13.2.3.3: Preventive measures

- Monitoring of the marine environment (for the occurrence of toxic algal species such as *Pseudo-nitzschia* spp.) and fishery products (for ASP toxin levels).
- A guideline value of 20 mg DA/kg is valid for total ASP toxin content in the edible parts of molluscs (shellfish meat) in Australia, Canada, the European Union and USA.
- Closures of the shellfish fishery if the DA levels exceed 20 mg/kg.

Table 13.15: ***Marine biotoxins affecting humans*** *[Daranas et al. 2001; Camacho et al. 2007; NHMRC 2008; Zaccaroni and Scaravelli 2008; http://www.whoi.edu/redtide/page.do?pid=12779]*

Syndrome	Causative Organisms (*highlighted organism are the main species*)	Main aquatic species containing toxins	Principal toxins and (chemical form)	Biochemical target	Affected areas
Shellfish toxins (ASP, DSP, NSP, PSP)					
Amnesic shellfish poisoning (ASP)	*Amphora coffaeformis*, ***Pseudo-nitzschia* spp.**, *Pseudo-nitzschia pungens, P. multiseries, P. australis, P. delicatissima, P. seriata, Nitzschia actydrophila*	Shellfish	Domoic acid and its congers (amino acids like compounds, hydrophilic)	Binds to sub-types of the glutamate receptor resulting both in gastrointestinal and neurological effects.	Mainly North America, Atlantic Canada, Australia, Chile, New Zealand, USA and U.K.
Diarrhoetic shellfish poisoning (DSP)	*Coolia* spp., ***Dinophysis*** **spp.**, ***Prorocentrum*** **spp.**, *Prorocentrum lima, P. maculosum, Protoceratium reticulatum*	Shellfish (mussels, scallops, clams, gastropods)	Okadaic acid; Dinophysistoxins (DTXs), Yessotoxins (YTXs), Pectenotoxins (PTXs). (Polyether, lipophilic)	Inhibit serine/threonine protein phosphatases 1, 2A and at high concentrations 2B, resulting in diarrhoea, impaired balance and loss of fluids. Tumour promoters	Europe, Japan, Atlantic Canada, South Africa, Chile, Thailand, New Zealand and Australia
Neurotoxic shellfish poisoning (NSP)	*Gymnodinium breve*, ***Karenia brevis***, *K. papilonacea, K. sellformis, K. bicuneiformis*	Shellfish (oysters, clams, mussels, cockles, whelks). (Aerosol)	Brevetoxins (PBTxs) (Transfused polyether, lipophilic)	Binds to site 5 on the voltage-dependent sodium channel, affecting the nervous system. Aerosol linked to respiratory problems	U.S. Gulf coast, New Zealand

Contd...

Paralytic shellfish poisoning (PSP)	***Alexandrium* spp. *Gymnodinium catenatum*,** *Karenia* spp., *Pyrodinium bahamense*	Shellfish	Saxitoxins (STXs) (freshwater cyanophyta also contain STX) (purine derivatives, hydrophilic)	Binds to site 1 on the voltage-dependent sodium channel, affecting the nervous system.	Worldwide: Asia, Australia, Brazil, Canada, Chile, Europe, New Zealand, South Africa, USA
Fish toxins (CFP)					
Ciguatera fish poisoning (CFP)	*Amphidinium carterae*, *Coolia monotis*, ***Gambierdiscus toxicus***, *Ostreopsis* spp., *Prorocentrum* spp., *Thecadinium* spp.	Reef fish	Ciguatoxins (CTXs), maitotoxin (MTX), (palytoxin, Gambierol) (transfused polyether, lipophilic)	Binds to site 5 on the voltage-dependent sodium channel, affecting the nervous system.	Tropical and sub-tropical areas: Hawaii, Florida Keys, Puerto Rico, Caribbean, Australia and many Pacific islands

13.2.4: *Diarrhoetic or diarrhetic shellfish poisoning (DSP)*

[based on ANZFA 2001; Daranas et al.2001; van Egmond 2004; FAO 2004; Landsberg et al. 2005; Ciminiello and Fattorusso 2006; Camacho et al. 2007; Christian and Luckas 2008; Zaccaroni and Scaravelli 2008; http://www.whoi.edu/redtide/page.do?pid; http://www.whoi.edu/redtide/illness/illness.html; http://www.fao.org/DOCREP/006/Y4743E/y4743eod.htm]

DSP toxins are produced usually by dinoflagellates that belong to the genera *Dinophysis* spp. (Figure 13. 6). Other dinoflagellate genus *Prorocentrum* spp. is also known to produce DSP toxins. The ingestion of contaminated bivalves such as mussels, scallops, oysters or clams can cause DSP in humans.

DSP is widespread in its distribution (South America, South Africa, New Zealand, Australia and Thailand) with most seasonal occurrence in Europe and Japan. The first incidence of DSP occurred in Japan in late 1970s when the dinoflagellate, *Dinophysis fortii* was identified as the causative organism. Since then several other *Dinophysis* species such as *D. acuminata* (in Europe), *D. acuta, D. norvegica* (in Scandinavia), *D. fortii* (in Japan), *D. mitra, D. rotundata, D. tripos* and benthic dinoflagellates *Prorocentrum lima, P. concavum* and *P. redfieldi* were also identified to produce DSP toxins. In addition, *D. caudata, D. hastaae* and *D. sacculus* are also suspected. DSP from North America (Nova Scotia, Canada) was confirmed in 1990 from a benthic dinoflgellate, *Prorocentrum lima.*

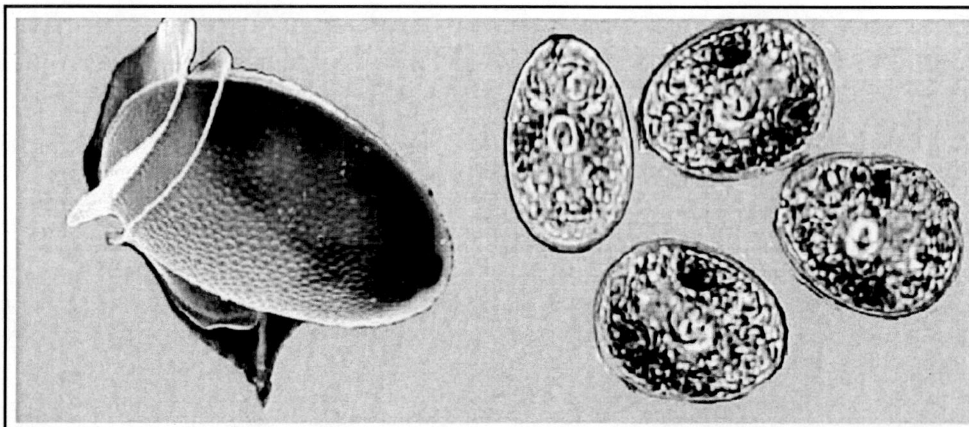

A B

Figure 13.6 ***Diarrhetic toxin producing organisms:* A. *Dinophysis* spp.; B. *Prorocentrum* spp.** *[Zaccaroni and Scaravelli 2008; page 106]*

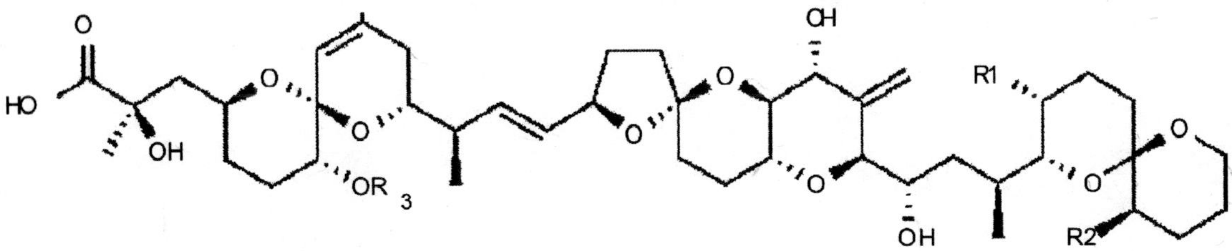

R1	R1	R3	
CH_3	H	H	okadaic acid
CH_3	CH_3	H	dinophysistoxin 1 (DTX1)
H	CH_3	H	dinophysistoxin 2 (DTX2)
CH_3	CH_3	Acyl	dinophysistoxin 3 (DTX3)

Figure 13.7: ***Chemical structure of main DSP toxins (okadaic acid group)*** *[Ciminello and Fattorusso 2006; page 59].*

13.2.4.1: Chemical nature of okadaic acid

The DSP toxins are produced by okadaic acid (OA) isolated from the black sponge, *Halichondria fortii*. All DSP toxins are relatively non-polar (lipophilic) polyether compounds with molecular weights higher than 500 daltons and accumulate in the fatty tissue of shellfish. Subsequent research revealed the presence of a number of okadaic acid derivatives including dinophysis toxins (DTXs), pectenotoxins (PTXs after the scallop genus *Pecten*) and yessotoxins (YTXs after the scallop *Patinopecten yessoensis*) in shellfish. This toxin class consists of eight congeners of which OA, DTX1 and dinophysis toxin 2 (DTX2) are the primary congeners involved in shellfish poisoning (Figure 13.7). The toxins are heat stable and survive normal cooking.

13.2.4.2: Toxic effects to aquatic organisms and human

13.2.4.2.1: Toxicity to aquatic organisms

OA inhibits growth of a variety of non-DSP producing micro-algae at micromolar concentrations (e.g. diatom, *Thalassiosira weissflogi*, EC_{50} 2.2 and 1.0 mM for OA-diol and OA respectively). Killifish (*Oryzias latipes*) exposed to yessotoxin (YTX) (1 or 0.5 mg/l) did not die within 24 hours whereas fish exposed to bisdesulfated YTX (dsYTX) (0.5 mg/l) died after six hours.

13.2.4.2.2: Toxicity to human

OA and its derivates being lipophilic accumulate in the fatty tissue of shellfish. These compounds are potent serine/threonine phosphatases inhibitors. Protein phosphatases are a critical group of enzymes closely linked with many crucial metabolic processes, ion balance, neurotransmission and cell cycle regulation. Diarrhoea associated with DSP is most likely due to the hyperphosphorylation of proteins, including ion channels, in the intestinal epithelia, resulting in impaired water balance and loss of fluids. The toxin has been classified as DSP toxins due to their production of symptoms of diarrhoea. Among the DSP toxins, OA, DTX1 and DTX3 (see Figure 13.7) are the most important in causing diarrhoea in humans. Other DSP toxins such as pectenotoxins (PTXs) have a low diarrhoetic potential and yessotoxins (YTXs) did not induce diarrhoea in rodents. However, both PTXs and YTXs are lethal to mice at i.p injection and they exert toxicity to liver and heart respectively. Epidemiological data from Japan (1976-1977) indicated that as little as 12 MU (mouse unit) was enough to induce a mild form of poisonings (12 MU corresponds to 43.2 μg and 1 MU corresponds to approximately 3.2 μg DTX1 and 4 μg of OA). DSP symptoms are diarrhoea, nausea, vomiting and abdominal pain starting 30 minutes to few hours after ingestion of the toxic shellfish with complete recovery within three days. The clinical symptoms of

DSP often may have been mistaken for those of bacterial gastric infections and the problem may be much more widespread than currently thought. However, no human fatalities have been reported. OA and DTX1 are also believed to be potential as tumour promoters. Table 13.16 provides specific examples of DSP (as cases and outbreaks) in different countries over the years.

13.2.4.3: Preventive measures

- The European Commission in 2002 laid down the following rules to protect human health: Maximum level of Okadaic acid (OA) + dinophysis toxins (DTXs) + pectenotoxins (PTXs) (together) in edible tissues (whole body or any part of edible separately of molluscs, echinoderms, tunicates and marine gastropods) shall be 16 μg OA equivalents/kg. The regulatory level in shellfish in Australia and NZ is 16-20 μg OA equivalents/100 g shellfish.
- A regular shellfish testing and dinoflagellates monitoring (*Dinophysis* spp. and *Prorocentrum* spp.) program should be carried out, such as weekly shellfish testing for toxins and phytoplankton analysis.
- Closure of fishery product harvesting areas should be imposed when DSP toxin levels in shellfish exceed tolerable levels (i.e. > 0.2 mg/g meat = 5 MU/100 g meat).

Table 13.16: *Cases and outbreaks of DSP in some selected countries [FAO 2004].*

Country	Cases and outbreaks
Europe	
Belgium	- 2003 (February): 403 cases of DSP reported after consumption of blue mussels imported from Denmark; presence of OA, DTXs, YTXx, PTXx were confirmed; 229 μg free OA/kg meat was found in the mussels.
Denmark	- 1990: 170 mg OA/100 g detected in mussels on the north Danish coasts imported from France, poisoning 415 persons. - 1999: three toxic events, domestic production areas were closed because of the presence of DSP toxins in blue mussels. - 2001: DSP toxins exceeded the regularity limits in three production areas of commercially fished blue mussels. - 2002: several production areas were closed for several weeks or months for DSP toxins.
France	- 1984 and 1985: mussels raised in France caused DSP symptoms in 10,000 and 2,000 people respectively. - 1998: *Dinophysis* spp. has been found to produce DSP; 27 production areas on the Atlantic and 2 production areas on the Mediterranean coast were closed due to presence of DSP.
Germany	- 1986-1989: *Dinophysis acuminata* was found regularly along the coasts of German Bight. More than 1,000 *Dinophysis* cells/litre (max. 25,000 cells/litre) were detected; mussel's beds were closed. - 2000: one case of DSP intoxication reported. - 2001: 5.9 x 103 cells per litre of *D. acuminata* was observed at the East Frisian coast waters, - DSP toxins in mussels increased resulting a ban on mussel harvesting in the contaminated area.
Greece	- 1999: 10 positive DSP samples were reordered; *D. acuminata* were detected at a concentration of 1,500 cells/litre. - 2000: one production area was closed due to presence of DSP toxins.
Ireland	- 1980: the first events of DSP poisoning. - 1984, 1987-1991: mussels showed variable toxicity levels, 200 mg OA/100 g and 25 mg DTX1/100 g of mussel meat. - 1994: OA levels 0.7 mg/g and DTX2 levels 6.3 mg/g hepatopancreas were reported.

Contd...

	- 2000: 30 areas were closed for harvesting of bivalve shellfish; people developed symptoms of DSP. - 2001: 17% of samples surveyed were found positive for DSP. - 2002: seven percent of sampled mussels contained OA or DTX2 above the regulatory limit.
Norway	- 1979: DSP outbreak in the Oslo Fjord due to bloom of *Prorocentrum minimum, P. lima,* (19,000 cells/l counted). - 1984-1989: DSP was an annual events in southeast and Western Coast; widespread intoxication; *Dinophysis* spp. were regularly found; 300-400 cases of DSP were recorded; *D. acuta, D. norvegica, D. acuminata,* and *D. acuta* blooms recorded in 1984, 1985 and 1986 in Skagerrak; DSP levels in mussels exceeded toxic thresholds in 1989 and 1990; closures of harvesting areas due to DSP in 1994; toxins detected were DTX1 and YTX. - 1999: 135/473 samples were positive for DSP and YTX respectively and YTX was the dominant toxin (33% cases); closures of production areas. - 2000: 45% of 414 samples gave positive results for DSP. - 2001 (until October): 26% of 915 samples were positive for DSP.
Portugal	- 1988: molluscs were contaminated by DSP toxins (north east of Portugal); the species involved was *D. acuta.* - 1989: *D. acuminata, D. sacculus* and *D. caudata* (around 1,600 cells/litre) caused DSP contamination on the Algarve coast; OA and DTX2 were responsible. - 2001 (summer): all shellfish species (except oysters) in Aveiro region contained OA and its esters above the regulatory limits (57-170 µg/100 g). - 2002: OA and DTX2 were detected at high concentration in the Aveiro Lagoon where the green crab (a shellfish predator) also accumulated these toxins; several people were sick. - 2002: blooming of *D. acuminata* led to prolonged closures of wild intertidal mussel and benthonic bivalve harvesting areas; recreational harvest of rock mussels and cockles caused several human poisoning; maximum level of 1,860 µg total OA/ 100 g in wild mussels recorded (Povoa do Varzim area); dozen of people had severe gastroenteritis, 18 percent of 738 samples were positive.
Spain	- 1981: *D. acuminata* and *D. acuta* were associated with DSP events causing 5,000 cases gastroenteritis throughout Spain. - 1989-1990: *D. acuta* was the causative species involved in DSP outbreaks. - 1993, 1995, 1996, 1997, and 2002: shellfish bed closing occurred due to DSP toxins OA and DTX2.
Sweden	- 1993: DSP was observed among people consuming mussels. - 1989-1990: harvesting of farmed mussels was stopped due to high OA concentrations. - 1997: OA and DTX2 were found. - 1999: closure of production areas (September- March) due to a peak in DSP toxicity.
UK and Northern Ireland	- 1997: 49 patients showed symptoms after consuming mussels in two London restaurants. - 1999 (England and Wales): DSP events have been more frequent and prolonged; toxins found in cockles from the southeast England and south Wales. - 2002: banned harvesting of shellfish (queen scallops) in several parts of England. - 2002 (April) 2003 (March) (Scotland): shellfish samples from Scotland examined for DSP; 66 out of 931 samples were found positive. - 2001 (Ireland): 25 oyster samples, 10 mussel samples, and 24 cockle samples were positive for DSP.
Asia	
China	- 1986 and 1987: 26 out of 89 samples contained DTX1 or DA; six samples contained levels above the regulatory limits (20 µg/100 g soft tissues); the highest level of 84 µg/100 g found in *Perna viridis* from Shenzhen.

Contd...

Japan	- 1976 and 1986: 164 persons suffered severe vomiting and diarrhoea; 1,300 DSP cases; dinoflagellate, *D. fortii* was responsible.
India	- 1984-1986: DA level in shellfish ranged around 0.37-1.5 MU/g hepatopancreas; no reports of DSP episodes in the general population.
The Philippines	- 1995: five species of *Dinophysis* detected but no human poisoning reported.
The Russian Federation	-1993: *Dinophysis acuminata, D. acuta, D. fortii, D. norvegica* have been identified in the far eastern coastal waters but no human poisoning have been reported.
Oceania	
Oceania (Australia and New Zealand)	Australia - 1995: dinoflagallate, *Prorocentrum lima* producing OA and methyl-OA was isolated at three locations on Heron Island, Australia. - 1997: pipi shellfish (*Donax delatoides*) poisoning reported (56 cases of hospitalisation) in New South Wales in December. -2000: an elderly women became seriously ill after eating pipies from a local beach, high levels of pectenotoxins PTX-2SA found in the pipies. New Zealand -1998: the presence of YTXs and related compounds in shellfish and phytoplankton was reported. - September 1994-July 1996 : 0.7% of shellfish samples collected (weekly basis) around the coastline showed a DSP toxin level above the regulatory limit (maximum 96 μg/100 g mussels).

13.2.5: Neurotoxic shellfish poisoning (NSP)

[based on ANZFA 2001; Daranas et al. 2001; FAO 2004; Ciminiello and Fattorusso 2006; Camacho et al. 2007; Christian and Luckas 2008; Zaccaroni and Scaravelli 2008; http://www.whoi.edu/redtide/page.do?pid; http://www.whoi.edu/redtide/illness/illness.html; http://www.fao.org/DOCREP/006/Y4743E/y4743eod.htm]

The NSP toxin is caused by polyether brevetoxins produced by the dinoflagellate *Gymnodinium breve* (Figure 13.8) (since 2000 it is called *Karenia brevis*). Blooms of *G. breve* are also known as red-tide that causes the sea water red in colour.

The first record of NSP was in 1880 from the West coast of Florida. Until 1992-93, NSP was considered to be endemic to the Gulf of Mexico and the east Coast of Florida. *G. breve* has also been reported from the West Atlantic, Spain, Portugal, Greece, Japan and New Zealand. In early 1993, more than 180 human shellfish poisoning were reported from New Zealand from an organism similar to *G. breve*. Similar dinoflagellates have also been identified in low concentrations from Victorian, South Australian and West Australian waters.

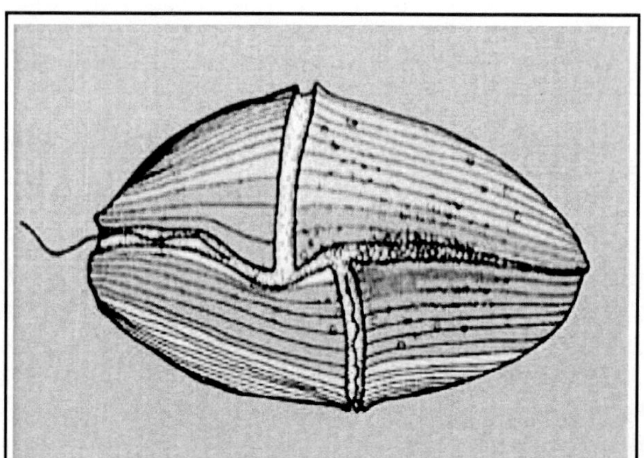

Figure 13.8: ***Gymnodinium breve*** *[Zaccaroni and Scaravelli 2008; page 116]*

An unusual feature of *G. breve* is the formation of toxic aerosols by wave action which can lead to asthma-like symptoms in humans. The lack of an external shell in *G. breve* make this micro-alga easily lysed in turbulent waters. The lysis allows the toxin to be released in water making aerosol and droplets potentially toxic.

Brevetoxin-like toxins were also produced by four other algal species belonging to class Raphidiophyceae *Chattonella* spp. (e.g. CaTx1; CaTx-II, CaTX-iii).

13.2.5.1: Chemical nature of brevetoxins

The NSP toxins are brevetoxins called "PbTXs". The toxins are tasteless, odourless, heat and acid stable, lipid-soluble, ladder-like cyclic polyether. The molecular structure of bervetoxins consists of 10 to 11 transfused rings (Figure 13.9) and their molecular weights are around 900 daltons. Ten brevetoxins have been isolated and identified from the field blooms and *G. breve* cultures (Figure 13.9).

Brevetoxins congeners fall into two types, based on backbone structure:

- the brevetoxin B backbone (type 1);
- brevetoxin A backbone (type 2).

The ring systems in the middle of the molecules differ somewhat; type 1 and type 2 toxins share a lactone in the 'A ring' (head of the molecule) and a conserved structure on the 'tail ring', both of which are required for their toxicity.

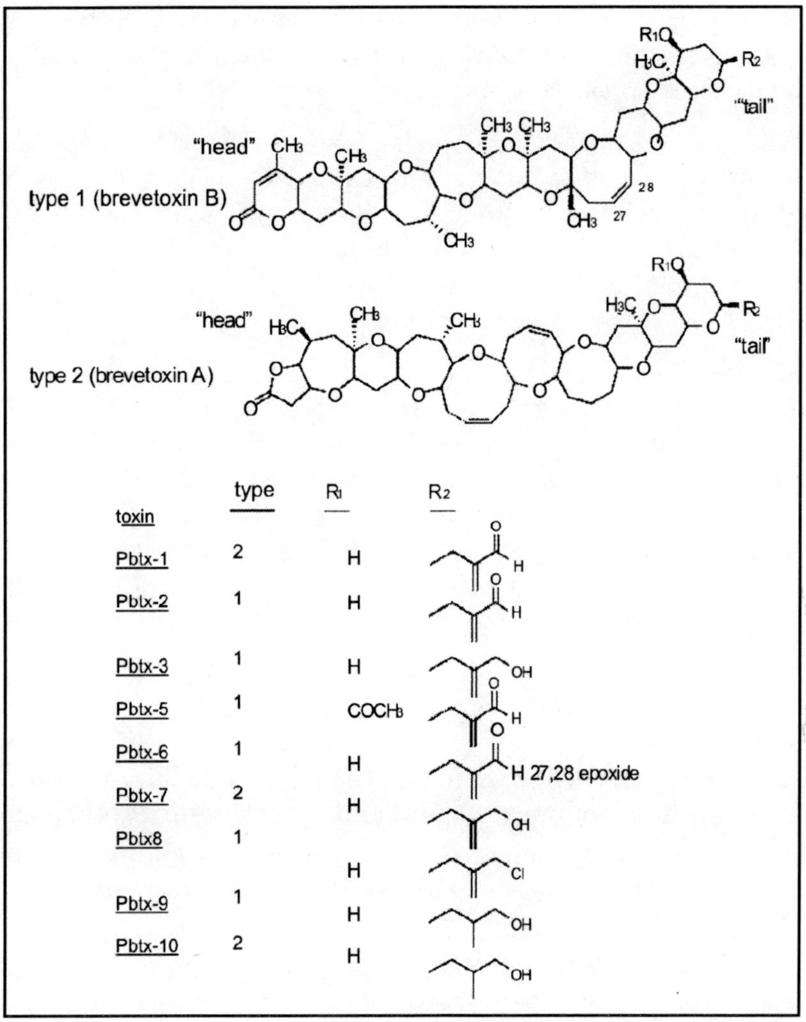

Figure 13.9: *Structure of brevetoxins* *[Ciminiello and Fattorusso 2006; page 65].*

13.2.5.2: Toxic effects to aquatic organisms and humans
13.2.5.2.1: Toxicity to aquatic organisms

The brevitoxins are toxic to fish, marine mammals, birds and humans, but not to shellfish. Fish kills may occur when *G. breve* counts reach the range between 100,000 and 250,000 cells/litre. The ichthyotoxic symptoms from bervetoxins effect include violent twisting, corkscrew swimming and loss of equilibrium, respiratory paralysis and death.

Medaka fish (*Oryzias latipes*) eggs injected with doses of 0.1- 8.0 ng PbTx-1 showed inhibition of hatching (maximum effect at 3 ng/egg) and larval survival (maximum effect at 4.5 ng/egg). Hatching abnormalities were also seen at doses of 2 ng/egg, while dose of 4.1 ng/egg produced embryos which failed to hatch. Toxicity experiments conducted with mosquito fish showed that PBTx-1 is the most potent toxin (see Table 13.17).

Table 13.17: ***Toxicity of brevetoxins on female mosquito fish, Gambusia affinis*** *[Rein et al. 1994].*

Brevetoxins	LC_{50} (24 hours) (nM)
PBTx-1	2.57
PBTx-2	14.3
PBTx-3,	15.8
PBTx-6	77.7
PBTx-9	31.4

In April and May 1996, an estimated 1,700 tonnes of cultured blue fin tuna (*Thunnus maccoyi*) were killed in South Australia after a bloom of *Chattonella marina* (Table 13.18). High levels of breve-like toxins (up to 142 mg/100 g) were found in livers of farmed blue fin tuna. Plankton samples revealed a bloom of the Raphidophyta *Chattonella marina* (upto 66,000 cells/l).

The death of 149 manatees (*Trichechus manatus latirostris*) in 1996 along the southwest Coast of Florida, USA is believed to be due to exposure to *G. breve* during a bloom via inhalation and oral ingestion (Table 13.18). Severe nasopharyngeal, pulmonary, hepatic, renal and cerebral congestion was observed in all manatees. Mortality among the double-crested cormorant (*Phalarocorax auritus*) caused by brevetoxins was observed along the Florida Gulf coast.

13.2.5.2.2: Toxicity to human

G. breve is a 'naked' organism having no outer shell of polysaccharide plates like other dinoflagellates; therefore it is easily broken open in the rough surf releasing the toxins. As a result during swimming direct contact with the toxic blooms can take place causing eye and the nasal membrane irritation. Furthermore, inhalation of brevetoxins may cause respiratory distress. In addition, during an active in-shore red tide, the aerosol of contaminated salt spray will contain the toxins and organisms fragments both in the droplets and attached to salt particles which can be carried in inland depending on wind and other environmental conditions.

Brevetoxins and their derivatives exert their toxic effects by specific bindings to site 5 of voltage-sensitive sodium channels. Since brevetoxins are lipophilic therefore it can accumulate in fatty tissues of shellfish. Consumption of the raw or cooked shellfish can cause NSP. The symptoms occur within 30 minutes to three hours (after mussel's ingestion) and can last few days and so far no mortality has been recorded. The symptoms include both neurological and gastrointestinal problems:

- Neurological symptoms are paresthesia of area around the mouth, face and throat, muscular ache, ataxia, inversion of thermal perception, bradycardia and mydriasis.
- Gastrointestinal symptoms are abdominal pain, nausea and diarrhoea, nausea, vomiting, diarrhoea, chills, sweats, reversal of temperature, hypotension, numbness, cramps, paralysis, seizures and coma, tingling and paresthesia's of lips, face and extremities. The symptoms are similar to PSP and ciguatera intoxication but less severe.

Respiratory problems in humans occur at about 10^5-10^6 cells/litre, while fish mortality occurs at > 10^6 cells/litre (ANZFA 2001). The LD_{50} value in rats being in the order of 520-6,600 µg/kg, pathogenic dose for human is in the order of 42-72 mouse units (MU) (http://www.fao.org/DOCREP/006/Y4743E/y4743eod.htm). Table 13.18 provides specific examples of NSP (as cases and outbreaks) in different countries over the years.

Table 13.18: ***Cases and outbreaks of NSP in some selected countries*** *[FAO 2004].*

Country	Cases and outbreaks
North America	
Canada	- 2001: Red-tides *Heterosigma akashiwo* (Raphidophyceae) have been reported causing mortality of cultured fish.
USA	*USA- East Coast* - 1987-1988: brevitoxin associated mortality in bottlenose dolphins (*Tursiops truncatus); G. breve* was identified (6 x 10 ^{6} cells/l) in North Carolina over 600 km northward; 48 cases of NSP reported in humans and 1,480 km^2 of shellfish (oyster and clam) harvesting waters were closed. - 2001: Red-tides *H. akashiwo* caused cultured fish mortality. *USA- Florida and the Gulf of Mexico* - 1946-1947 and 1987-1988: bottlenose dolphins *(T. truncates)* mortality were due to berevetoxins and a bloom of *G. breve.* - 1996 (June): three patients diagnosed with NSP for eating clams (*Chione cancellata*) and whelk (unidentified species). - 1996 (March-April): 149 manatees (*Trichechus manatus latirostris*) died in the southwest coast of Florida believed to be related to toxin producing *G. breve* bloom (23.3 x 10^6 cells/l) present in the area during the same time. - 1999: Brevetoxins have been detected in menhaden and mullet from the coast of Florida. *USA-West Coast* - 2001: Red-tides *H. akashiwo* caused mortality of cultured fish.
Asia	
China, Hong Kong	- 1987: The first blooms of Red-tides due to *H. akashiwo* (Raphidophyceae) have been reported from Hong Kong causing mortality of cultured fish; other algae identified were *Chattonella antiqua.*
Japan	- 1972-1973: Red-ides due to *Fibrocaspa japonica* have been reported (Ehime Prefecture, Atsumi Bay, the Seto Island sea) causing mortality of cultured fish (yellow tail, *Seriola quinqueradiata*). - 2001: Red-tides due to *H. akashiwo* (Raphidophyceae) have been reported causing mortality of cultured fish (yellow tail, *Seriola quinqueradiata*). - 2001: Red-tides due to *Chattonella antiqua* reported from the coastal regions killing large number of cultured fish (caged yellow tail).
The Republic of Korea	- 2001: Red-tides due to *H. akashiwo* (Raphidophyceae) have been reported in embayment causing mortality of cultured fish.
Malaysia	- 2001: Red-tides due to *H. akashiwo* (Raphidophyceae) have been reported in embayment causing mortality of cultured fish.
Oceania	
Australia	- 1994 (January): Mussels from Tamboon inlet (Gippsland coast of Victoria) contained a NSP toxin level of 27.5 MU/100 g in association with *G. breve* type outbreak. - 1996 (April and May): An estimated 1,700 tonnes of cultured blue fin tuna (*Thunnus maccoyi*) were killed in South Australia after a bloom of *Chattonella marina.* High levels of breve-like toxins (up to 142 mg/100 g) were found in livers of farmed blue fin tuna; plankton samples revealed a bloom of the Raphidophyta *Chattonella marina* (up to 66,000 cells/l).
New Zealand	- 1993: During shellfish poisoning outbreak, NSP toxin level reached 592 MU/100 g of edible shellfish. - 1994 (September)-1996 (July): 10 NSP events (maximum level 26 MU/100 g of various shellfish) reported; one human NSP poisoning involving 186 cases.

Contd...

	- 1998 (Mid February-April): Red-tides due to *H. akashiwo* (Raphidophyceae) have been reported in embayment causing mortality of cultured fish. - 1998 (Mid February-April): *Gymnodinium mikimotoi* bloom killed Eels, flounders, other pelagic fish and marine invertebrates; caused illness to 87 people (respiratory, dry cough, a severe sore throat, running nose, skin and eye irritations).

13.2.5.3: Preventive measures

- Monitoring programs relying on the microscopic identification and enumeration of harmful taxa in water samples may be sufficient for prevention of human intoxication.
- In the USA, a level of 80 mg PBTx-2/100 g of shellfish tissue (0.9 mg/kg or 20 MU/100 g or 4 mg/mouse) analysed by mouse bioassay in shellfish triggers regulatory action by US Food and Drug Administration (USFDA).
- The Florida Department of Natural Resources has run a general control program where closures of shellfish beds are made when *G. breve* concentrations exceeds 5,000 cells/l.
- In New Zealand, a mouse bioassay is in force and 20 MU/100 g is employed as an acceptable level.
- Local anaesthetics, calcium administration and hyperosmotic solutions are useful in poisoning treatment, as they counteract brevetoxins action on membrane. In case of aerosolized Red-tide toxins respiratory irritation, the use of particle filter masks or retreat to air conditioned environment may anecdotally provide relief from the air borne irritation.

13.2.6: Paralytic shellfish poisoning (PSP) toxins

[based on ANZFA 2001; FAO 2004; Landsberg et al. 2005; Ciminiello and Fattorusso 2006; Camacho et al. 2007; Christian and Luckas 2008; Zaccaroni and Scaravelli 2008; http://www.whoi.edu/redtide/page.do?pid; http://www.whoi.edu/redtide/illness/illness.html; http://www.fao.org/DOCREP/006/Y4743E/y4743eod.htm]

Incidents of paralytic shellfish poisoning (PSP) have been recorded throughout the world (Figure 13.10) and are associated with both temperate and tropical dinoflagellates of the Genera *Alexandrium, Gymnodinium* and *Pyrodinum* (Figure 13.11) and other species. *Alexandrium* occurs along the Atlantic and Pacific coast lines where it may grow in great quantities, particularly during the summer period. Toxins produced by *Alexandrium* were named PSP toxins due to the observation that their consumption caused symptoms of poisoning in warm blooded species similar to paralytic phenomena such as cramps, signs of paralysis and blocking of respiration.

Though historically, PSP endemic areas have been in temperate waters of the northern hemisphere but over the last quarter century PSP has expanded to southern hemisphere including Australia, New Zealand, Chile, Argentina and South Africa (see Figure 13.10). In the case of Australia, this expansion is attributed to the transport of dinoflagellate *G. catenatum* cysts from Japanese waters via ballast water of ships to Tasmanian waters which was favourable to their growth. In tropical waters, the causative species is *Pyrodinum bahamense,* which has become a significant human health issue in south-east Asia, for example in the Philippines over 100 fatal poisoning have been recorded since 1980.

The dinoflagellate species recorded from different waters are listed below:

- Temperate waters (northeast and west coast of North America): *Alexandrium tamarensis, A. fundyense, A. catenella* and *Gymnodinium catenatum.*
- Europe: *Alexandrium tamarensis, A. catenella* and *Gymnodinium catenatum.*
- Asia: *Alexandrium catenella, Gymnodinium catenatum* and *Pyrodinum bahamense.*
- Australia: *Alexandrium catenella, A. minutum, A. tamarense, Gymnodinium catenatum, Pyrodinium bahamense var. compressum.*
- New Zealand: *Alexandrium angustitabulatum, A. catenella, A. minutum, A. ostenfeldii* and *A. tamarense.*

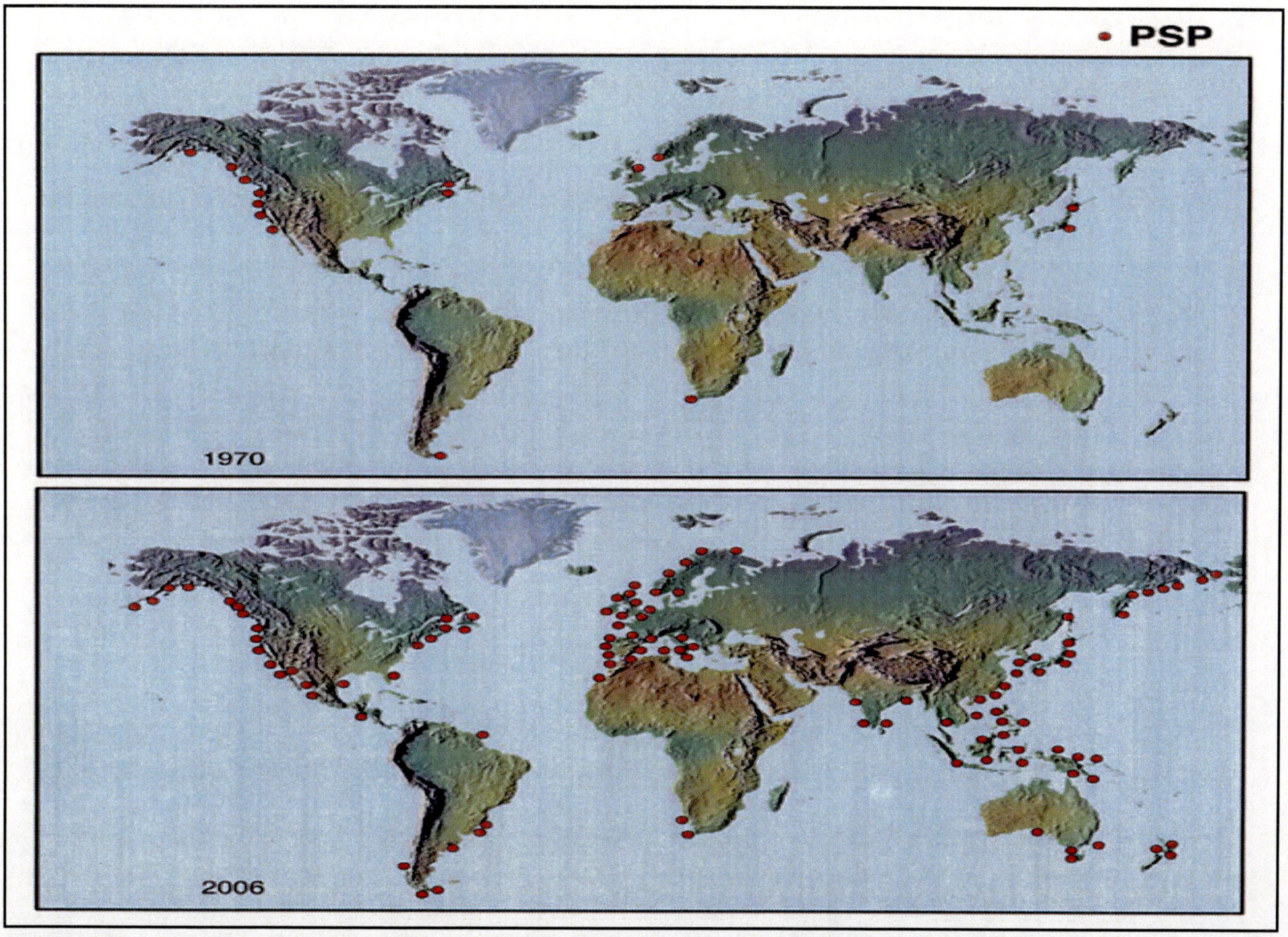

Figure 13.10: ***Global distribution of PSP toxins recorded as of 1970 (top) and 2006 (bottom).*** *[http://www.whoi.edu/redtide/page.do?pid=14899].*

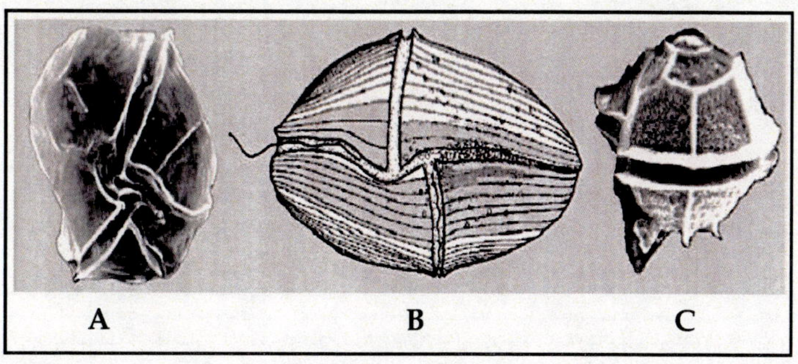

Figure 13.11: ***Saxitoxins producing organisms. A. Alexandrium*** **spp.,** ***B. Gymnodinium*** **spp.,** ***C. Pyrodinium*** **spp.** *[Zaccaroni and Scaravelli 2008; page 110].*

Climatic and environmental conditions favouring blooming of dinoflagellates include changes in salinity, rising water temperatures, increased nutrients and sunlight that triggers cyst/spore germination.

13.2.6.1: Chemical nature of PSP toxins - saxitoxins

In 1957, a PSP toxin was isolated from *Saxidomus gigantenus* (butter clams) from Alaska and the chemical was assigned as saxitoxin (STX). Since then, a number of toxin fractions have been identified from plankton and shellfish.

The toxin responsible for PSP are a suite of heterocyclic guanidines collectively called saxitoxins, of which there are currently over 29 known congeners (Figure 13.12). Saxitoxins are hydro-soluble, thermostable and resistant to acidic environment. Their structures vary having different combinations of hydroxyl and sulfate substitutions at four sites on the molecule. Based on substitutions at R4, the saxitoxins can be subdivided into following four groups:

- Neurotoxic and highly potent carbamate toxin (non-sulfated saxitoxin (STX) and neosaxitoxin (NEO) and gonyautoxins (GTX1-GTX4) (see Table 13.19).
- Weakly toxic *N*-sulfo-carbamoyl-11-hydroxysulfate toxins (B1, B2, C1-C4). Of all the PSP toxins GTX 5-6; C1-C4 are least toxic to mammals (see Table 13.19).
- Decarbamoyl (dc-) analogs, which are thought to arise from the metabolism of dinoflagellate toxins within the shellfish (dcSTX, dcNEO, dcGTX2-3, dcGTX1,4), STX, dcGTX1-4).
- Deoxydecarbamoyl (do-) toxins, have been detected only in Australian populations of *G. catenatum* (doSTX, doGTX 2-3).

The freshwater cyanophyta has also been reported to contain saxitoxins and neosaxitoxins.

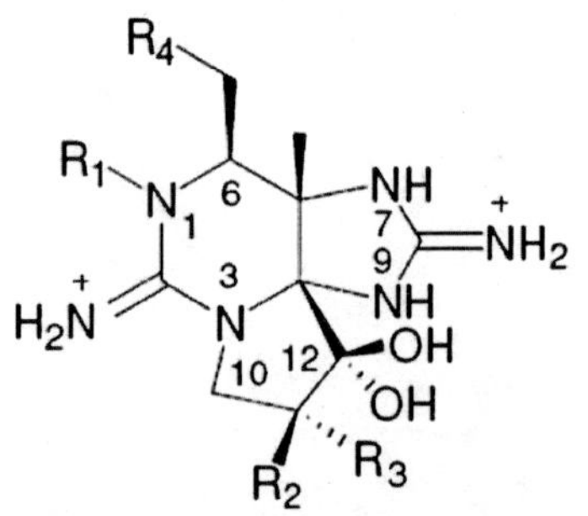

R_1	R_2	R_3	carbamate ions	N-sulfocarbamoyl toxins	decarbamoyl toxins	deoxydecarbamoyl toxins
H	H	H	**1** STX	**11** GTX, B1	**17** dcSTX	**27** doSTX
OH	H	H	**2** NEO	**12** GTX6, B2	**18** dcNEO	
H	H	OSO_3^-	**3** GTX2	**13** C1	**19** GTX2	**28** doGTX2
H	OSO_3^-	H	**4** GTX3	**14** C2	**20** dcGTX3	**29** doGTX3
OH	H	OSO_3^-	**5** GTX1	**15** C3	**21** dcGTX1	
OH	OSO_3^-	H	**6** GTX4	**16** C4	**22** dcGTX4	
H	H	OH	**7** 11αOH-STX		**23** 11αOH-dcSTX	
H	OH	H	**8** 11βOH-STX		**24** 11βOH-dcSTX	
OH	H	H	**9** 11αOH-NEO		**25** 11αOH-dcNEO	
OH	OH	H	**10** 11βOH-NEO		**26** 1βOH-dcNEO	

Figure 13.12: ***Structure of paralytic shellfish toxins showing 29 congeners*** *[Ciminiello and Fattorusso 2006; page 58].*

13.2.6.2: Toxic effects to aquatic organisms and human
13.2.6.2.1: Toxicity to aquatic organisms

In addition to human health effects, saxitoxins (STX) have been implications in the deaths of fish, seabirds and marine mammals. For example, blooms of *Alexandrium* can cause fish kills (adult herring) with a lethal dose of 10-20 mg STX eq. Milkfish (*Chanos chanos*) when exposed to toxic *Alexandrium minutum* at a density of 1.4 x 10^4 cell/ml ~ 3.0 x 10^4 cells/ml or toxic algal extract of 5.13 x 10^3 cells/ml ~ 2.05 x 10^4 cells/ml showed edema, hyperplasia and necrosis of secondary lamellae in the gill and a higher oxygen consumption rate. On the contrary, milkfish exposed to non-toxic *A. minutum* did not cause any damage to the gills nor a rise in oxygen consumption rate or critical oxygen demand. In another experiment, copepods grazing on toxic *Alexandrium* were fed to larval mummichogs (*Fundulus heteroclitus)* showed reduced swimming and prey capture performances suggesting that toxic

Alexandrium species can impact on both fish and their prey thereby facilitating transfer of the toxin through the food web.

Table 13.19: ***Relative toxicity of PSP toxins in mouse bioassay*** *[FAO 2004 original from Usleber et al. 1997]. [a] = a/b epimeric mixture*

Toxin	Toxin group	Relative toxicity
STX	Carbamate toxin	1
neoSTX	Carbamate toxin	0.5-1.1
GTX 2/3[a]	Carbamate toxin	0.39/1.09-0.48/0.76
GTX 1/4[a]	Carbamate toxin	0.8/0.33-0.9/0.9
dcSTX	Decarbamoyl toxin	0.43
dcneoSTX	Decarbamoyl toxin	0.43
B1	*N*-sulfo-carbamoyl	0.07-0.17
B2	*N*-sulfo-carbamoyl	0.07-0.09
C1 to C4	*N*-sulfo-carbamoyl	<0.01-0.14

Saxitoxins were responsible for the mortality of 14 humpback whales (*Megaptera novaeangliae*) in Cape Cod Bay, Massachusetts, USA during *Alexandrium tamarense* bloom (November 1997-January 1988) with mackerel probably serving as vector. Similarly, saxitoxins were a key suspect in the mortality of over 100 highly endangered Mediterranean monk seals (*Monachus monachus*) in the coast of Mauritania, West Africa during May and June 1997 (Table 13.20).

However, there are no reports that PSP toxins caused any mortality in shellfish, such as bivalves, which feed on toxic algae. It may be related with the fact that many of the bivalves, though filter feeders, are relatively insensitive to the PSP toxins because many of them have nerves and muscles operated mainly by voltage-gated calcium channels. Saxitoxins and other PSP toxins block only the voltage-gated sodium channels with great potency. This mechanism enables shellfish to accumulate elevated levels of toxins in body parts while feeding on toxic dinoflagellates. Table 13.20 provides specific examples of PSP (as cases and outbreaks) in different countries over the years.

Table 13.20: ***Cases and outbreaks of PSP in some selected countries*** *[FAO 2004]. http://www.fao.org/docrep/007/y5486e/y5486e00.HTM].*

Country	Cases and outbreaks
Europe	
Denmark	- 1987-1990: PSP was found in mussels caused by *Alexandrium tamarensis* and *A. ostenfeldii.*
France	- 1992: PSP toxins found in mussels from the Atlantic coast.
	- 1998: *A. tamarensis* was detected at concentrations of upto 350,000 cells/l with closure of clams, oysters and mussels production areas for two months.
	- 2000: Two areas in Brittany were closed due to the presence of PSP toxins.
Germany	- 1987: Three cases of PSP occurred in Lower Saxonia contributed by the consumption of canned Spanish mussels.
Ireland	- 1992 (July): PSP event was associated with *Alexandrium tamarensis.*
	- 1996 (June): GTX2 and GTX3 were identified in wild mussels (*Mytilus edulis*) and cultivated oysters (*Crassostrea gigas*) following a bloom of *A. tamarensis.* Maximal levels were 360 and 88 µg STX eq/100 g meat in *M. edulis* and *C. gigas* respectively.
	- 1999: Cysts of *A. tamarensis* were detected in surface sediments in Cork harbour.
	- 2002: PSP toxins were detected above the regulatory limit in both mussels and oysters from Cork harbour.

Contd...

Italy	- No cases of PSP reported until 1995. - 1994: some mussel samples showed PSP more than 80 μg STX eq/100 g meat but no public health problems were reported. - 2001: PSP toxins at a level < 40 μg STX eq/100 g of meat associated with *Alexandrium minutum* in Friuli Venezia Giulia. - 2002 (May and early June): PSP toxins were detected in Sardinia due to the presence of *A. catenella* and *A. minutum;* STX levels were higher than 80 μg STX eq/100 g and a maximum of 25,210 μg STX eq/100 g were reported.
Norway	- PSP has been detected in many localities and cases of human intoxication reported in 1901, 1939, 1959, 1979, 1981, 1991 and 1992 with a total of 32 victims including two fatalities. - 1999: 11 percent of surveyed samples were tested positive for the presence of PSP toxins; cases of human illness from mussels imported from Sweden were reported.
Portugal	- 1986-1990: PSP outbreaks along the Portuguese coast north of Roca Cape were reported. - 1992: 100-500 μg STX eq/100 g of shellfish occurred off the south coast of Lisbon and along the Algarve coast. - 1993: PSP recorded almost all year round covering the entire coast (113 mg STX eq/100 g of shellfish in December at the Northern Coast upto 9,145 mg STX eq/100 g in the Lisbon coast in September). - 1994: PSP outbreaks off the south coast of Portugal, off the Algarve coast and the Abides Lagoon were due to *Gymnodinium catenatum, Alexandrium lusitanicum.* Nine patients (6 women and 3 men) suffered after consumption of molluscs, *Mytilus edulis; Gymnodinium catenatum* concentrations of >3,000 cells/l was observed in the Algarve area.
Spain	- 1976: Mussel aquaculture in the Galician Rias has been seriously affected by PSP incidents; Spanish mussels exported to Germany, France, Switzerland and Italy have caused food poisoning; 120 people were affected following consumption of Spanish mussels *Mytilus edulis.* - 1989: PSP toxins were detected in different molluscan species (*Venus verrucosa, Venerupis rhomboides, Callista choine, Acanthocardia tuberculatum*). - 1993: Dinoflagellate, *Gymnodinium catenatum* caused PSP incidents with PSP toxins; B1, STX, GTX2 and GTX3 were identified in mussels (concentrations were higher than 80 μg STX eq/100 g meat). - 1994 (October-November): *G. catenatum* caused a PSP outbreak on the Atlantic shores. - 1995: Toxic dinoflagellate, *Alexandrium minutum* reaching concentrations upto 45 x 10^6 cells/l; natural population of bivalve molluscs (*Mytilus galloprovincialis, Chamelea gallina and Lithopaga lithopaga*) showed positive for PSP toxins with a maximum of 1170 STX eq/100 g tissue. 2002: Short toxic events related to PSP toxins in Galicia due to *A. minutum*, and *A. catenella;* production areas (for *Acanthocardia tuberculatum* and scallops) in Andalusia were closed due to the presence of *G. catenatum.*
Sweden	- 1985-1998: PSP toxins have been detected in mollusc meat with highest level between 1986 and 1987 (1,000 MU/100 g meat). The species involved was *Alexandrium excavatum.*
UK and the Northern Ireland	- 1968: The first cases of PSP intoxication after consumption of mussel, *Mytilus edulis* (600-6,000 mg STX eq/kg) from the north-eastern coast of England; 78 people hospitalised but there were no deaths. - 1972, 1973: PSP were detected very often (< 400 MU/100 g shellfish). - 1977: Toxicity reached 1,792 MU/100 g. - 1978-1981: PSP events were annual occurrences. - 1990: First cases of PSP were recorded along the west coast of Scotland with maximum toxin concentrations of 16,480 MU/100 g in mussels and scallops. - 2000-2001: Low levels of PSP toxins were detected in some areas of England and Wales; in Scotland, PSP toxins were detected in early May in mussels from aquaculture sites along the west coast and in scallop gonad tissues from scallop fishing grounds in Orkney; restrictions on fishing activity were imposed.

Contd...

	- 2002 (April) -2003 (March), Scotland: Out of 1,690 samples analysed (collected from 76 primary and 36 secondary production areas respectively); 21 samples were positive for PSP.
Africa	
Morocco	- 1994: Blooms of *Gymnodinium catenatum* in October to November caused a PSP outbreak (also observed in Spain and Portugal at the same time); the maximum PSP levels in bivalve mollusc *Mytilus galloprovincialis* was 6,000 µg STX eq/100 g meats.
South Africa	- 1969-1979: 17 cases of PSP were reported after consumption of local black mussels (*Chloromytilus meridionalis*) containing upto 84,000 µg STX eq/kg, two fatalities were reported; *Gonyaulax catenella* has caused several human deaths. - 1994: *Alexandrium tamarensis* bloom was reported; mussels and oysters contained PSP toxins beyond the safe limit of human consumption.
Tunisia	- 1998: 700 tons of cultured sea bass, sea bream and other wild fish were found dead in the lagoons of Burger and Gar-el-Melh probably caused by PSP toxins produced by *Alexandrium minutum* and *A. tamarensis*.
Asia	
China (main land)	- 1967-1979: 40 events resulting 23 fatalities and 423 non-fatal intoxication; vector was the marine snail, *Nassarius succinstus* (Zhejiang province). - 1986 (November): One fatality and 136 hospitalisation; vector was the clam, *Ruditapes phillipenensis* (Fujian province). - 1989 (November): Four fishermen showed PSP symptoms, one died; vector was the clam, *Pinna pectinata* (Fujian province). - 1991 (March): Two human fatalities out of four illness; vector was the mussel, *Perna virdis* (Daya Bay). - 1990-1992: Survey in Guandong province found PSP toxins in 33 edible marine organisms up to 1,000 mg STX eq/100 g meat. - 1994 (June): Five illness, one death; vector was the snail, *N. succinstus.*
China (Hong Kong)	- 1992: 3 outbreaks. -1996 (September): *Chlamys nobilis* contained 320 mg STX eq/100 g soft tissue. -1998 (March-April): Within 40 days after red-tide events levels of PSP toxins went upto 79 mg STX eq/100 g in *Perna virdis,* but no human poisoning reported.
China (Taiwan)	- 1986: Two people died out of 30 illnesses after consuming *Soletellina diphos.* - 1991 (February): Eight people were ill after eating *Soletellina diphos.* -1995 (August) – 1997 (March): PSP and TTX (tetrodotoxin) found in shellfish *Niotha clathrata, Natica lineata* and *Natica vitellus.*
India	1981: 98 cases of PSP reported, vector was the local mussels, one fatality.
Japan	- 1992: First infestation of shellfish with PSP reported; bivalves were scallop and oysters; toxic dinoflagellates were *Alexandrium catenella, A. tamarense* and *Gymnodinium catenatum.* - 1996 (July) -1996 (October): Six out of 30 species of shellfish (scallop, mussels) contained PSP toxins from Fukue Island, Nagasaki Prefecture. - 1997 (March): 20 people poisoned from eating oysters (Fukue island). - 1999: bloom of *Alexandrium tamiyavanichii* occurred in Seto island and affected several oysters' farms.
Malaysia	- 1977: 201 cases recorded four fatalities after eating local clams. - 1990: PSP confined to the west coast of Sabah in Borneo, the dinoflagellate was *Pyrodinium bahamense.* - 1991: Three people became ill after eating farmed mussels caused by *Alexandrium tamiyanavichii* (Sebatu, Straits of Malacca).

Contd...

	- 2001 (September): Six people became ill after eating clams, one died, high density of *Alexandrium tropicale* and unidentified *Alexandrium* sp. were reported from the area (Coastal lagoon, Kelatan).
The Philippines	- 1983-1998: 2,000 cases of PSP reported.
	- 1992: Marine puffers (*Arothron* sp.) contained considerable amounts of PSP toxins.
	- 1992: Cultured mussels caused PSP in many people.
Thailand	- 1983: 62 cases of PSP reported after consumption of local mussels, *Mytilus* sp., and one fatality.
	- 1993: freshwater puffer (*Tetraodon fangi*) contained saxitoxin (STX).
Oceania	
Australia	- 1987-1997: Port Philip Bay, Victoria: Maximum PSP toxins recorded was 10,000 µg/ STX eq/100 g edible shellfish.
	- 1994-July 1996: Tasmania; Max levels in mussels, oysters and scallops were 18,429, 699 and 83 g/ STX eq /100 g edible shellfish.
New Zealand	- 1993 (January): An outbreak of PSP poisoning found in shellfish (pipi, scallop, mussel and tuatua) caused by *Alexandrium minutum;* more than 180 cases registered.
	- 1993 (January)-1996 (July): 11 PSP events (maximum level 920 µg/ STX eq/100 g mussel).

13.2.6.2.2: Toxicity to human

PSP is the most widespread shellfish poisoning and outbreaks that are occurring worldwide (Figure 13.10). One of the most predominant toxins within this PSP group is STX and it is so potent that upto 50 human can be poisoned from the level of poison contained in just one contaminated shellfish. PSP is a neurotoxic syndrome with a mortality rate of approximately 15-20% of those intoxicated. The sulfo-carbamoyl compounds are considerably less toxic than any other groups of PSP toxins (Table 13.19), however, they might be converted to the more toxic carbamates under acidic conditions.

Symptoms of PSP intoxication develop fairly rapidly within 0.5-2.0 hours after ingestion of the contaminated shellfish. In humans, the peripheral nervous system is affected with symptoms including tingling and numbness of extremities, muscular non-coordination, respiratory distress and muscular paralysis leading to death by asphyxiation. According to Zaccaroni and Scaravelli (2008) the PSP symptoms can be classified into mild, severe and extreme:

- *Mild*: Mouth paresthesia which can spread to the whole face and neck, to fingers and ears. Other symptoms are nausea, headache and vomiting.
- *Severe*: Paresthesia spreading to arms and legs; general sensation of numbness, muscular weakness and floating sensation. Other symptoms are altered speech, dysarthria, severe ataxia, motory in coordination and some respiratory problem appears.
- *Extreme*: Peripheral paralysis, including respiratory paralysis which can lead to death, if not rapidly treated.

The Australia and New Zealand Food Authority reported that 120-180 µg PSP toxins can produce moderate symptoms in humans and 400 to 1,060 µg can cause death and 2,000 to 10,000 µg is more likely to constitute a fatal dose. Globally, PSP is responsible for some 2,000 cases of human poisoning per year (15% mortality) from temperate waters of Europe, North America and Japan to the Southern Hemisphere in South Africa, India, Australia, New Zealand, Thailand, Brunei, Sabah (Malaysia), the Philippines and Papua New Guinea. In 1987, an outbreak of PSP with 187 cases and 26 deaths was reported after consumption of a clam (*Amphichaena kinderman*) soup. The fatalities were the highest among young children (50%) compared to 7% in adults. The estimated lethal dose in this incident was estimated to be 25 µg STX eq/kg bw for children and 86-788 µg STX eq/kg bw for adults who died. Table 13.20 provide specific examples of PSP (as cases and outbreaks) in different countries over the years.

13.2.6.3: *Preventive measures*

Until now, no antidote has been found for PSP treatment. However, outbreaks of PSP intoxications are preventable by large-scale monitoring programs including assessing toxin levels in shellfish (mussels, oysters, scallops, clams) and rapid closure of shellfish harvesting beds suspected or demonstrated toxic areas. Worldwide regulations for PSP toxins have been set at 400 MU/100 g or 80 μg STX eq/100 g to protect human health.

13.2.7: *Ciguatera fish poisoning (CFP)*

[based on Lehane and Lewis 2000; ANZFA 2001; FAO 2004; Landsberg et al. 2005; Ciminiello and Fattorusso 2006; Camacho et al. 2007; Christian and Luckas 2008;; Zaccaroni and Scaravelli 2008; http://www.whoi.edu/redtide/page.do?pid; http://www.whoi.edu/redtide/illness/illness.html]

Ciguatera fish poisoning (CFP) (or ciguatera) is an illness caused by consumption of tropical fish that are contaminated with toxins (ciguatoxins or CTX) produced by a benthic micro-algae (dinoflagellate), *Gambierdiscus toxicus* (Figure 13.13). The dinoflagellate, *G. toxicus* was identifed in the late 1970s near the Gambier islands, French Polynesia. Tuamotu Archipelago. Other toxic benthic dinoflagellates that have also been suspected to cause CFP are *Prorocentrum, Ostreopsis, Coolia* spp. Ciguatera fish poisoining has been known for centuries and reported in the West Indies in 1511. The term 'cigua' means intoxication caused by the ingestion of coral reef fishes (originally dervied from spanish trival name of an univalve mollusc, 'cigua' *Turbo pica* which caused intoxication following ingestion).

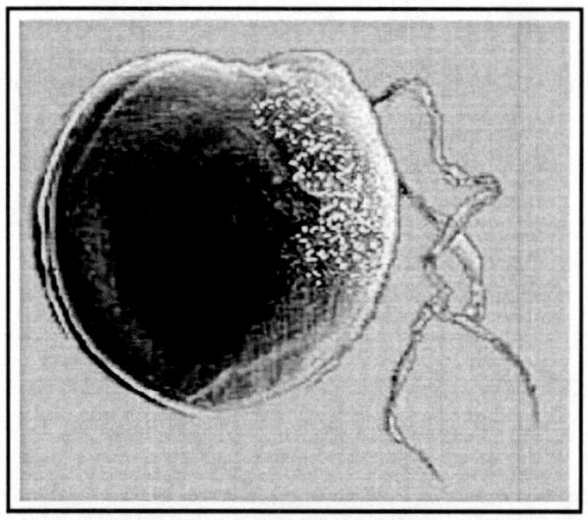

Figure 13.13: ***Gambierdiscus toxicus*** *[Zaccaroni and Scaravelli 2008; page 124].*

CFP are endemic to the tropical and sub-tropical Pacific and Indian Oceanic insular regions and the tropical Carribbean including coral reef areas (Figure 13.14). Currently, ciguatera is the most common type of marine food poisoining worldwide and an estimated 10,000 to 50,000 people worldwide suffering from the disease annually.

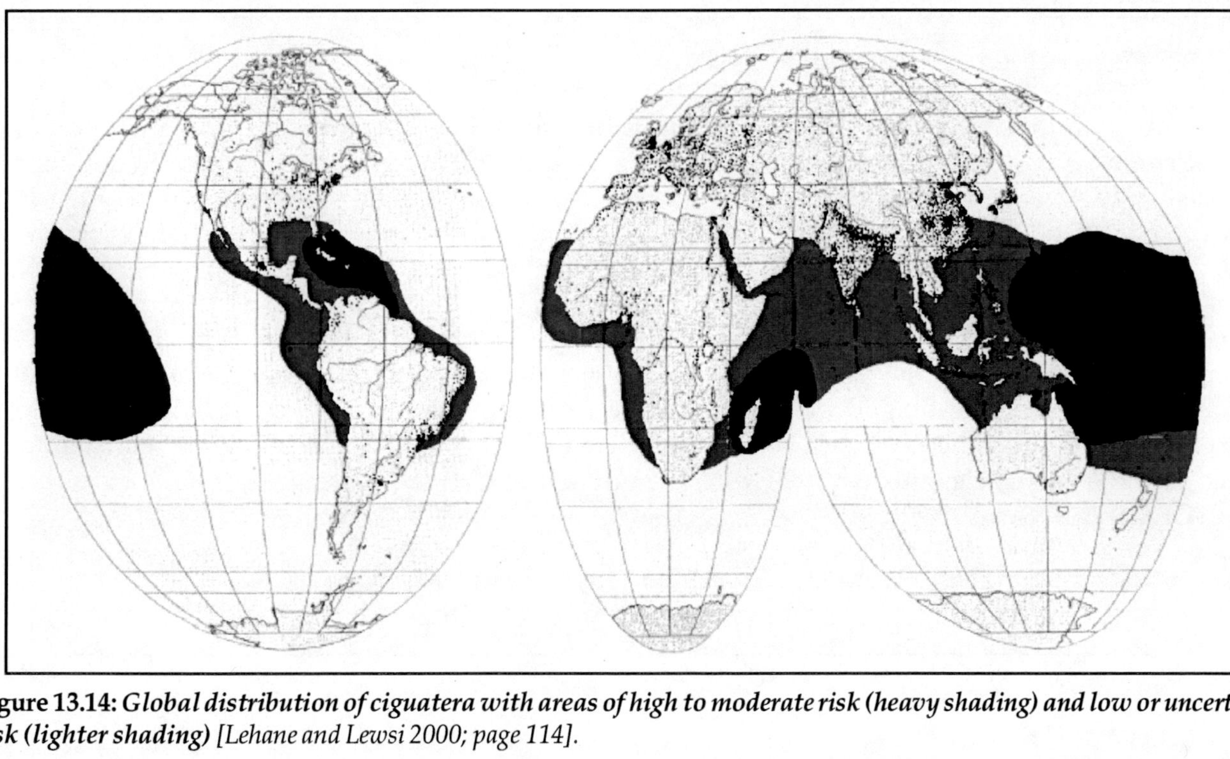

Figure 13.14: ***Global distribution of ciguatera with areas of high to moderate risk (heavy shading) and low or uncertain risk (lighter shading)*** *[Lehane and Lewsi 2000; page 114].*

G. toxicus lives in epiphytic association with bushy red, brown and green seaweeds and is naturally found in subtropical and tropical reef systems. The dead coral and marine algae thriving in tropical and sub-tropical reef systems are eaten by herbivorous fish and these hebivorous fish accumulate and concentrate the toxins (ciguatoxins) produced by the dinoflagellate. The herbivorous fish are eaten by larger carnivorous fish and the toxins are then transfererd to carnivorous fish. At least 400 fish species have been implicated in CFP with the largest carnivores such as moray eels, groupers, snappers, spanish mackerels and barracudas are the most toxic. Some examples of fish species associated with ciguatera is presented in Table 13.21.

Table 13.21: ***Examples of fish associated with ciguatoxins*** *[Compiled from Fastard and Chow 2001].*

Species	Distrbution
Blue parrotfish (*Sparus coeruleus*)	Western Atlantic
Bonefish (*Albula vulpes*)	World wide warm seas
Chinaman fish (*Symphorus nematophorus*)	Western Pacific
Gaucereye porgy (*Calamus calamus*)	Western Atlantic
Giant moray (*Gymnothorax javanicus*)	Indo-Pacific
Gray triggerfish (*Balistes carolinensis*)	Atlantic, Gulf of Mexico
Great Barracuda (*Sphyraena barracuda*)	Indo-Pacific, Western Atlantic
Heavybeak parrotfish (*Chlorus gibbus*)	Indo-Pacific
Hog fish (*Lachnolaimus maximus*)	Western-Atlantic
Horse-eye jack (*Caranx latus*)	Atlantic
Humphead parrotfish (*Cheilinus undulatus*)	Indo-Pacific
Lesser amberjack (*Seriola fasciata*)	Western Atlantic
Lined surgeon (*Acanthurus linearis*)	Indo-Pacific
Narrowhead gray mullet (*Mugil capurri*)	East Central Atlantic
Northern red snapper (*Lutjanus campechanus*)	Western Atlantic, Gulf of Mexico
Red groupper (*Epinephelus morio*)	Western-Atlantic
Spanish mackerel (*Scomberomoous maculatus*)	Western Atlantic
Spotted coral grouper (*Plectropomus maculatus*)	Western Pacific
Sword fish (*Xiphias gladius*)	Mediterranean
Tarpon (*Megalops atlanticus*)	Eastern Atlantic
Whitelip shark (*Carcharinus longimanus*)	Worldwide
Yellowtail snapper (*Ocyurus chrysurus*)	Western Atlantic

13.2.7.1: Chemical nature of ciguatoxins

G. toxicus is the source of two types of marine toxins, the water-soluble maiotoxins (MTXs, produced by all strains of *G. toxicus*) and the fat-soluble ciguatoxins (CTXs, produced by certain strains of *G. toxicus*). MTXs are principally found in the gut of herbivorius fishes and have no proven role in CFP. On the otherhand, ciguatoxins are found in the liver, muscles, skin and bones of large carnivorous fishes and are the principal casue of CFP in humans. Ciguatoxins are lipid-soluble polyether compounds consisting of 13-14 rings fused by linkages into a most rigid ladder–like structure (Figure 13.15). They are highly stable toxins, as cooking, freezing and salting do not reduce their toxicity. Ciguatoxins arise from biotransformation in the fish of precursor gambiertoxins. The study of molecular structure of the toxins led to the identification of two different toxins, one from the Pacific area (P-CTX-1) and the other from the Caribbean (C-CTX-1). Caribbean (and Indian Ocean) ciguatoxins is less polar and 10 fold less toxic than the Pacific ciguatoxins (P-CTX-1). The number of C in the molecule of P-CTX1 is 60 and 62 in C-CTX10 (see Table 13.22).

Table 13.22: ***Comparison of ciguatoxins (C-CTX and P-CTX)*** *[Zaccaroni and Scaravelli 2008].*

	C-CTX	P-CTX
Molecular formula	$C_{62}H_{92}O_{19}$	$C_{60}H_{86}O_{19}$
LD_{50}	3.6 μg/kg	0.35 μg/kg
Number of cycle ethers	14	13

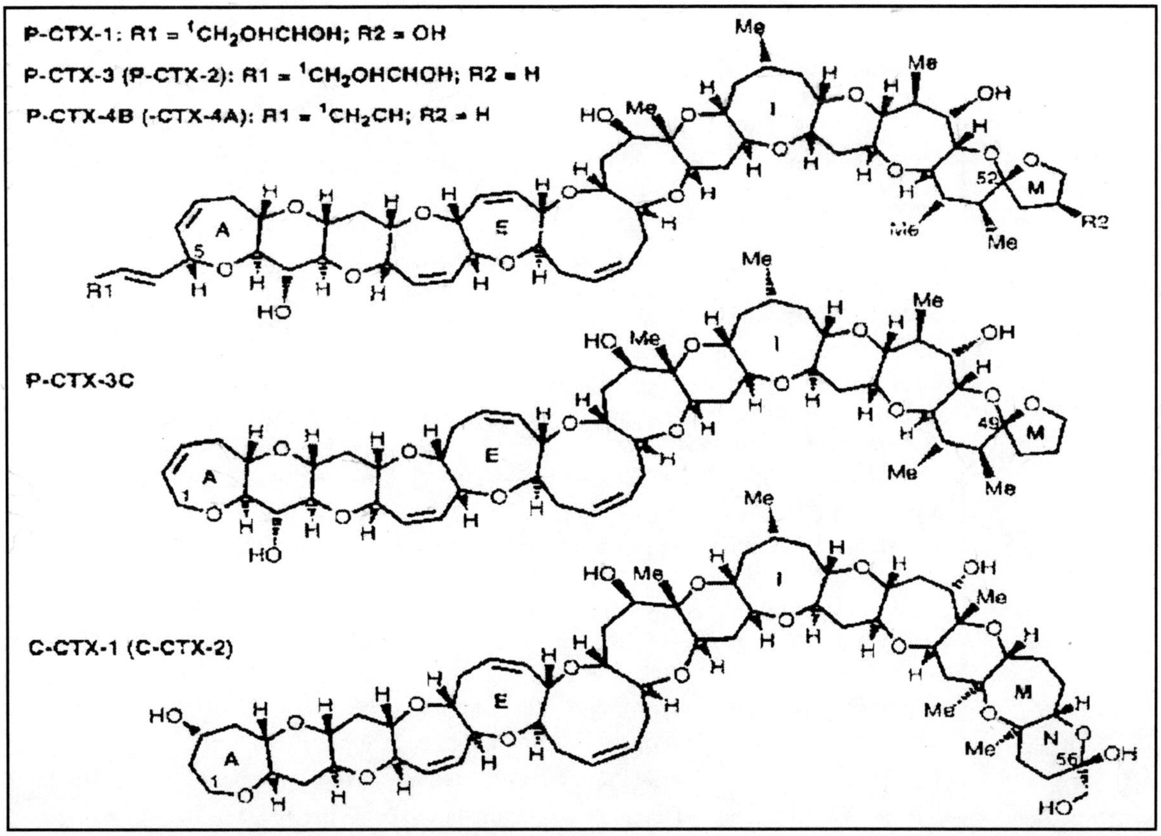

Figure 13.15: ***Structure of ciguatoxins*** *[source Lehane and Lewis 2000; Camacho et al. 2000; Note: The Pacific ciguatoxins is shown as P-CTX and the Caribbean as C-CTX, for more explanation see Lehane and Lewis 2000; page 94].*

13.2.7.2: Toxic effects to aquatic organisms and human

13.2.7.2.1: Toxicity to aquatic organisms

Experiments conducted have shown that ciguatoxins can be lethal to fish when administered orally, by intraperitoneal injection or when dissolved in aquarium in which fish were swimming. Fish exposed to ciguatoxins behaved abnormally, exhibiting erratic movement, disorientation, inactivity, quiescence and loss of equilibrium as well as experiencing physiological distress.

13.2.7.2.2: Toxicity to human

Ciguatoxins (CTX-1) is one of the most lethal natural substances known with LD_{50} in mice around 0.25-0.36 µg/kg i.p. oral intake of as little as 0.1µg ciguatoxin can cause illness in the adult human. After consumption of ciguatoxin contaminated fish, the onset of first symptoms can be as short as 30 minutes for severe intoxication. In milder cases onset can be delayed upto 24-48 hours. The first symptoms can be either gastrointestinal such as, vomiting, dirrahoea, nausea, abdominal pain (>50% cases) or neurological such as, tingling of the lips, hands and feet, unusual temperature perception where cold objects may feel warm and warm or hot objects may feel cold and a severe localized itch of the skin (>70% cases). In the Pacific Ocean, neurological symptoms predominate, while in the Caribbean Sea, gastrointestinal symptoms are a dominant feature of the disease. These differences in symptoms may confirm the existence of different ciguatoxins in the Pacific and Caribbean waters (see chemical nature section 13.2.7.1). Other symptoms include a profound feeling of fatigue (90% cases) throughout the illness as well as muscle joint and teeth aches can occur at varying extents. In severe cases, hypotension with bradycardia, respiratory difficulties and paralysis can involve but deaths are uncommon (less than 1 percent). Ciguatoxins share their mechanism of action with brevetoxins because they bind to site 5 of voltage-dependent sodium channel. Table 13.23 provides specific examples of CFP (as cases and outbreaks) in different countries over the years.

Table 13.23: ***Cases and outbreaks of CFP in some selected countries [FAO 2004].***

Country	Cases and outbreaks
Africa	
Madagascar	- 1993 (November): Severe outbreak of ciguatera due to consumption of a shark in Manakara; 98 out of 500 poisoned people died.
Asia	
China (main land)	not available
China (Hong Kong)	- 1998-1992: 47 outbreaks involving 397 people reported of which snapper had accounted for 59.6%. - No date: CFP incidence reported after consumption of a mangrove snapper, symptoms were gastrointestinal and neurological features (nausea, abdominal pain, diarrhoea, paresthesia and numbness of extremities). - No date: 8 family members showed sign of CFP after consuming a grouper of which eight people were hospitalised.
Fiji	- 1984: 925 cases of CFP reported after eating local snapper, barracuda, grouper and emperor, 1 person died.
French-Polynesia -New Caledonia (South Pacific)	- 1960-1984: 24,000 patients were reported. - 1979: 3,009 people affected by CFP by eating local fish (surgeon fish, parrot fish, grouper, snapper, carrang, emperor and barracuda). - 2001: Four adult tourists developed CFP after eating contaminated fish in Vanuatu.
Indian Ocean	- No date: Annual incident rate was estimated to be 0.78 per 10,000 residents. - 1978: 367 people from La Reunion were affected after eating snapper from Salya de Malha.
South Pacific Islands	- 1979-1983: CFP reported incident was estimated to be 97 per 100,000. - 1987: Annual mean CFP incident 217 per 100,000 people. - 1985-1990: the Pacific Islands (Kiribati, Tokelau, Vanuatu), reported 90-100 cases per 10,000 population per year; in French Polynesia, Vanuatu, Marshall Islands and Cook Islands the reported cases varied from 35 to 50 per 10,000 population per year; less than 20 cases per 10,000 population per year from Fiji, Northern Marianas, New Caledonia, Wallis and Futuna, American and Western Samoa, Niue, Guam, Nauru, Fed St. of Micronesia, Palau, Tonga and Papua New Guinea.
Central and South America	
Anguilla	- 2001: Incidence varied between 2-5 cases per 1,000 inhabitants.
The Bahamas	- 1982: 14 members of crew of an Italian freighter showed CFP poisoning after eating a local barracuda. - 1997: 17 crew members of Norwegian cargo ship showed symptoms of CFP after eating a contaminated barracuda caught from the Cay Sal Bank. Other fish such as red barracuda, red snapper caught were tested positive for ciguatoxin.
Cuba	- Ten cases recorded officially except in 1974 (when 174 cases reported). - 1978: 100 cases reported after eating local moray eel and Spanish mackerel, no mortality. - 1987: 57 cases reported, most people ate barracuda.
Dominican Republic	- 1989: 81 CFP cases reported.
Guadeloupe	- Three cases per 10,000 inhabitants reported per year. - 1980-1985: 255 cases reported.
Haiti	- 1985: 2 cases reported. - 1995: Six US soldier became ill after eating local amber jack (*Seriola dumerlli*). Analysis of cooked fish showed 20 ng Caribbean ciguatoxin- (C-CTX-1)/g flesh.
Jamaica	- 1978: 250 people showed CFP symptoms after eating local grouper and barracuda.
Martinique	- 1983: Annual incidence estimated at 41 per 10,000 inhabitants.
Mexico	- 1974: 24 people on-board of a ship showed CFP symptoms after eating barracuda from the Gulf of Mexico; no mortality.

Contd...

	- 1984: 200 cases of ciguatera intoxication occurred in La Paz Baja California Sur in Mexico after consuming contaminated snapper (*Lutjanus* spp.).
	- 1993 (May): The entire crew of a fishing boat became ill after eating fish *Serranidae* and *Labridae*. Ciguatera like toxins was confirmed in the suspected fish.
	- 1994 (July): 10 cases of CFP occurred on the Isla de Mujeres after consuming barracuda.
Puerto Rico	- 1980-1982: 100 outbreaks involving 215 persons recorded; an annual incidence of 90 per 10,000 inhabitants estimated.
Venezuela	- 2001: 200 cases of CFP on a cruise ship resulted in several fatalities.
Virgin Islands (USA)	- An annual incident rate of 730 per 10,000 people reported.
	- 1981: 14 outbreaks of CFP involving 65 patients reported after eating black fin snapper.
	- 1982 and 1983: 33 and 55 people showed CFP symptoms after eating local carrang and or snapper.
Oceania	
Australia	- 1987: Major outbreaks in Sydney, 63 people were affected from consumption of Spanish mackerel from Queensland; outbreak of CFP was reported from eating Coral cod, *Plectropomus* spp. captured from Arafura sea (northern Australia) causing 20 poisoning events.
	- 1988: Compiled record produced in Queensland showed 617 cases from 225 outbreaks; the annual incidence of CFP in Queensland is 1.6 cases per 100,000 populations, outbreaks of CFP occurred from consumption of fish caught along the tropical coast of eastern Australia.
	- 1997 (July) – 1998 (August): 3 small outbreaks in inner Sydney caused by reef fish: 1st outbreak (6 cases) from Fiji caused by coral trout, 2nd outbreak (10 cases) caused by coral trout from Queensland and 3rd outbreak (10 cases) caused by spotted cod from Queensland.
	- 1997 (September): CFP outbreak in outer Melbourne caused by 16.2 kg Maori Wrasse fish imported from Trunk reef in Queensland, 30 people who ate the fish at an Asian restaurant had the symptoms and 18 had neurological symptoms.
	- 1999-2000: 132 CFP cases in 10 outbreaks registered in Australia (excluding Queensland).
New Zealand	- 1997: One person became ill three weeks after returning from Fiji where the person ate barbecued fish (required respiratory support).

13.2.7.3: Preventive measures

Avoidance from eating of any tropical reef fish/ large predatory fish may help prevent intoxication of CFP since there is currently no easy way to routinely monitor ciguatoxin in any seafood product prior to consumption. The avoidance of consuming viscera (liver, spleen, kidney) and roe of reef fish would also reduce risk from ciguatoxins since the concentration of ciguatoxins are highest in the liver, spleen and kidney and gonads of affected fish. It is also advisable to avoid fish caught in the area known for frequent occurrence of ciguatoxins intoxication.

13.2.8: Regulatory limits of marine biotoxins

Shellfish and finfish are a delicious and nutritional food. They can, however, accumulate toxins which can cause food poisoning and other health effects including deaths. One health risk posed by the consumption of shellfish is due to accumulation of algal biotoxins. As a consequence, regulatory authorities in many countries have set the maximum permitted levels (MPL) of various marine toxins in food to protect human health as listed below:

- *ASP*: A maximum permitted level (MPL) of 2,000 µg 'domoic acid' per 100 g shellfish has been adopted by many countries.
- *DSP*: Regulations for this group limit the toxin content to below 16-20 µg 'okadaic acid equivalents' per 100 g shellfish flesh.

- *NSP*: Regulatory authorities have set the maximum permitted level of toxin as 80 μg 'brevetoxin' per 100 g shellfish flesh, equivalent to 20 standards mouse units (MU)/100 g flesh (Table 13.24).
- *PSP*: Regulations restrict sale or harvest of shellfish containing PSP levels above 80 μg 'saxitoxin equivalents' /100 g shellfish flesh.
- *CFP*: Because of the high toxicity (see Table 13.24 for LD_{50}), the consumer safety limit for this toxin have been set at 3.5 μg/100 g fish flesh.

Table 13.24: ***Toxin properties: toxicity and regulatory action levels of phycotoxins*** *[Garthwaite 2000].*

	LD_{50} (i.p. mouse μg/kg)	MPL (μg/ 100 g flesh)	MPL (mouse unit)
ASP (domoic acid)	120	2,000 (shellfish flesh)	Sensitivity vey low
DSP (okadaic acid)	200	16 or 20 (shellfish flesh)	5
NSP (brevetoxins)	200	80 (shellfish flesh)	20
PSP (saxitoxins)	10	80 (shellfish flesh)	400
CTX (ciguatoxin)	0.25-0.9	3.5 (fish flesh)	Sensitivity too low

13.3: Conclusion

Harmful algal blooms pose serious threats to human and animal health, aquatic ecosystem sustainability and economic viability. A conservative estimate shows that it costs the US economy between 2.2 and 4.6 billion dollars annually (Hundnell 2009). Globally, the incidence of algal blooms has increased in recent time and projected to increase in the future due to the effect of global warming or climate change (Delpa *et al.* 2009).

It is apparent that most of harmful freshwater and marine algal blooms are associated with increased nutrient inputs into natural waterways. For example, in China alone, 40 to 50 red-tides (PSP) are now reported every year, which seems to be linked to increased pollution in China's coastal waters. Surveys conducted worldwide reveals that toxic cyanobacteria are internationally ubiquitous; however there is not much information from tropical and sub-tropical regions.

The major effects of fresh water and marine biotoxins are: the loss of wild and domesticated animals; toxicity of domestic animal and humans; effects on aquatic species and increased expense for water treatment facilities (Svreck and Smith 2004) to remove toxigenic cells and monitoring for cyanobacteria and marine algae. Furthermore, irrigation water contaminated with cyantoxins (such as microcystins) may inhibit germination and plant growth causing crop failures. The exposure of edible crop plants to microcystins is a concern for human health, as the toxins can accumulate in plant tissues and might be transferred into the food chain. Human poisoning due to cyanobacteria was mainly caused by microcystins, cylindrospermopsin and anatoxin-a and occurred through exposure to contaminated drinking water supplies, recreational waters and medical dialysis. Biotoxins are bio-accumulative in seafood (shellfish, finfish) and there is a high likelihood of transfer of biotoxins to human consumers via seafood. The consumption of seafood causes an increasing number of human intoxication around the world. Therefore, regular monitoring for toxin levels in water, food plants and seafood may be an important human health safety procedure. This procedure would help to check whether seafood is within the government regulatory limit standards for shellfish and fish toxins. Lastly, it is important to manage causes of algal blooms rather than monitoring toxins in seafood which is very expensive.

13.4: References

Addico, G., Hardege, J., Komarek, J., Babica, P. and de Graft-Johnson, K.A.A. 2006. Cyanobacteria species identified in the Weija and Kpong reservoirs, Ghana, and their implications for drinking water quality with respect to microcystin. *African Journal of Marine Science* 28(2):451-456.

ANZECC (Australian and New Zealand Environment and Conservation Council) and ARMCANZ (Agriculture and Resource Management Council of Australia and New Zealand). 2000. *Australian and New Zealand Guidelines for Fresh and Marine Water Quality*, Volume 1, The Guidelines Chapters 1-7). http://www.mincos.gov.au/publications/australian_and_new_zealand_guidelines_for_fresh and_marine_water_quality

ANZFA (Australia and New Zealand Food Authority). 2001. Shellfish toxins in food. A toxicological review and risk assessment. Technical report series no. 14. Australia and New Zealand Food Authority. Canberra. 21 pp. http://www.anzfa.gov.au

Azevedo, S.M.F.O., Carmichael, W.W., Jochimsen, E.M., Rinehart, K.L., Lau, S., Shaw, G.R. and Eaglesham. 2009. Human intoxication by microcystins during renal dialysis treatment in Caruaru-Brazil. *Toxicology* 181-182: 441-446.

Babica, P., Bláha, L. and Maršálek, B. 2006. Exploring the natural role of microcystins - A review. *Journal of Phycology* 42: 9-20.

Baganz, D., Staaks, G., Steinberg, C. 1998. Impact of the cyanobacteria toxin, microcystin-LR induced behavior of zebra fish, *danio rerio*. *Water Research* 32: 948-952.

Bajpal, R., Sharma, N. K., Lawton, L. A., Edwards, C. and Rai, A. K. 2009. Short communication- Microcystin producing cyanobacterium *Nostoc* sp. BHU001 from a pond in India. *Toxicon* 53: 687-590..

Bartram, J., Burch, M., Falconer, I. R., Jonem, G., Kuiper-Goodmand, T. 1999. *Situation assessment, planning and management. In: I. Chorus., and J. Bartram, (eds.), Toxic cyanobacteria in water. A guide to their public health consequences. Monitoring and management.* E&N Spon on behalf of the World Health Organisation, London, 179-209 pp.

Bell, S. G. and Codd, G. A. 1996. *Detection, analysis, and risk assessment of cyanobacterial toxins*: *In*: R. M. Harrison, and R.E Hester (eds). Agricultural Chemicals and the Environment. Royal Society of Chemistry, Information Service, London, 109-122 pp.

Bláhova, L., Babica, P., Maršálková, E., Maršálek, B and Bláha. 2007. Concentrations and seasonal trends of extracellular microcystins in freshwaters of the Czech Republic-Results of the national monitoring program. *CLEAN-Soil, Air and Water* 35(4): 348-354.

Bláhova, L., Oravec, M., Maršálek, B., Šejnohova, L., Šimek, Z. and Bláha, L. 2009. The first occurrence of the cyanobacterial alkaloid toxin cylindrospermopsin in the Czech Republic as determined by immunochemical and LC/MS methods. *Toxicon* 53: 519-524.

Burch, M. D. 1993. *The development of an alert levels and response framework for the management of blue-green algal blooms.* Australian centre for water quality research and the University of Adelaide symposium on blue-green algal blooms- New developments in research and management, February 17, Adelaide, South Australia.

Burch, M. D. 2002. *Cyanobacterial toxins-The Australian perspective on guidelines and management.* pp.15- 22. *In*: Blue-green algae: their significance and management within water supplies. Occasional paper 4- blue-green algae. CRC for Water Quality and Treatment, Adelaide. South Australia. www.waterquality.crc.org.au.

Burch, M. D. 2008. Effective doses, guidelines and regulations. In: *Hudnell, H.K* (editor). Cyanobacterial harmful algal blooms: State of the science and research needs. *Advances in Experimental Medicine and Biology* 619: chapter 36. Springer Press, New York, 831-853 pp.

Camacho, F. G., Rodgiguez, J. G., Mirón, A. S., García, M. C. C., Belarbi, E. H., Chisti, Y. and Grima, E. M. 2007. Biotechnological significance of toxic marine dinoflagellates-research review paper. *Biotechnology Advances* 25: 176-194.

Camargo, J. A. and Alonson, A. 2006. Review article. Ecological and toxicological effects of inorganic nitrogen pollution in aquatic ecosystems: A global assessment. *Environmental International* 32: 831-849.

Carmichael, W. W and Falconer, I. R. 1993. *Diseases related to freshwater blue-green algal toxins and control measures.* In: I.R Falconer (ed.) *Algal toxins in seafood and drinking water*. Academic Press, London, 187-209 pp.

Carmichael, W. 2008. A world review-one hundred-twenty -seven- years of research on toxic cyanobacteria-where do we go from here? In: Hudnell, H.K (ed.). Cyanobacterial harmful algal blooms: State of the science and research needs. *Advances in Experimental Medicine and Biology* 619: chapter 4. Springer Press, New York, 105-125 pp.

Carrasco, D., Moreno, E., Sanchis, D., Wormer, L., Paniagua, T., Cueto, A. D. and Quesada, A. 2006. Cyanobacterial abundance and microcystin occurrence in Mediterranean water reservoirs in Central Spain: microcystins in the Madrid area. *European Journal of phycology* 41(3): 281-291.

Carvalho, G., Dellamano-Oliveira, M. J. and Rocha, O. 2008. Occurrence and toxicity of an *Anabaena* bloom in a tropical reservoir (southeast Brazil). *Harmful Algae* 7(5): 590-598.

Cazenave, J., Wunderlin, D. A., Bistoni, M. D. L., Ame, V. V. and Krause, E. 2005. Uptake, tissue distribution and accumulation of microcystin-RR in *Corydoras paleatus, Jenysia multidentate and Odontesthes bonariensis* – A field and laboratory study. *Aquatic Ecology* 75: 178-190.

Chellappa, N. T., Costa, M. A. A. M. and Marinho, I. R. 2000. Harmful cyanobacterial blooms from semi-arid freshwater ecosystems of Northeast Brazil. *Australian Society of Limnology* (Newsletter) 38: 45-49.

Chen, J. Z., Song, L. R., Dai, E., Gan, N. Q. and Liu, Z. L. 2004. Effects of microcystins on the growth and the activity of superoxide dismutase and peroxidase of rape (*Brassica napus* L.) and rice (*Oryza sativa* L.). Toxicon 43(4): 393-400.

Chorus, I. and Bartram, J. (eds.) 1999. *Toxic cyanobacteria in water. A guide to their public health consequences, monitoring and management.* Published by WHO, E and FN Spon Press, London. 416 pp.

Christian, B. and Luckas, B. 2008. Determination of marine biotoxins relevant for regulations: from the mouse bioassay to coupled LC-MS methods. *Anal Bioanal Chem* 391: 117-134.

Ciminiello, P. and Fattorusso, E. 2006. *Bivalve molluscs as vectors of marine biotoxins involved in seafood poisoning.* pp. 54-82. In: Progress in Molecular and Subcellular Biology. Subseries Marine Molecular Biotechnology. G. Cimino, M. Gavagnin (eds.): Molluscs. Springer-Verlag Berlin Heidelberg.

Codd, G. A. 1994. *Biological aspects of cyanobacterial toxins.* In: D. A. Steffensen and B. C. Nicholson (eds.). Toxic cyanobacteria, current status of research and management: International workshop, 22-26 March 1994, Adelaide, S.A. Proceedings, *Australian Centre for Water Treatment and Water Quality Research, Salisbury, S.A.*

Codd, G. A., Metcalf, J. S. and Beattie, K. A. 1999. Retention of *Microcystis aeruginosa* and microcystin by salad lettuce (*Lactuca sativa)* after spray irrigation with water containing cyanobacteria. *Toxicon* 37: 1181-1185.

Codd, G.A., Morrison, L.F. and Metcalf, J.S. 2005. Cyanobacterial toxins: risk management for health protection. *Toxicology and Applied Pharmacology* 203: 264-272.

Cook, C.M., Vardaka, E. and Lanaras, T. 2004. Toxic cyanobacteria in Greek freshwaters, 1987-2000: occurrence, toxicity, and impacts in the Mediterranean region. *Acta Hydrochim. Hydrobiol* 32(2); 107-124.

Crush, J. R., Briggs, L. R., Sprosen, S. N. 2008. Effect of irrigation with lake water containing microcystin content and growth of rye grass, clover, rape, and lettuce. *Environmental Toxicology* 23(2): 246-252.

Daranas, A. H., Norte, M. and Fernández, J. J. 2001. Toxic marine algae-review. *Toxicon* 39: 1101-1132.

DCNR (Dept of Conservation and Natural Resources) 1995. Brochure on blue-green algae, September Victoria, Australia.

Delpa, I., Baures, J. E., Clement, M. and Thomas, O. 2009. Review article- Impacts of climate change on surface water quality in relation to drinking water production. *Environmental International* doi:10.1016/j/envint.2009.07.001.

de Figueiredo, D. R., Azeiteiro, U. M., Estves, S. M., Goncalves, F. J. M. and Perieira, M. J. 2004. Microcystin -producing blooms - a serious global public health issue. *Ecotoxicology and Environmental Safety* 59: 151-163.

Duy, T. N., Lam, P. K. S., Shaw, G. R., Connell, D. W. 2000. Toxicology and risk assessment of freshwater cyanobacteria (blue-green algal) toxins in water. *Reviews of Environmental Contamination and Toxicology* 163: 113-186.

Falconer, I. R., Burch, M. D., Steffensen, D. A., Choice, M. and Coverdale, O. R. 1994. Toxicity of the blue-green alga (cyanobacterium) *Microcystis aeruginosa* in drinking water to growing pigs, as an animal model for human injury and risk assessment. *Environmental Toxicology and Water Quality*: An international Journal 9: 131-139.

Farstad, D. J. and Chow, T. 2001. A brief case report of ciguatera poisoning. *Wilderness Environmental Medicine.* 12: 263-269.

FAO (Food and Agricultural Organization of the United Nations) 2004. *The marine biotoxins.* Food and Agriculture Organization of the United Nations. FAO, Food and Nutrition Paper 80. http://www.fao.org/docrep/007/y5486e/y5486e00.HTM.

Garthwaite, I. 2000. Keeping shellfish safe to eat: brief review of shellfish toxins, and methods for their detection. *Trends in Food Science and Technology* 11: 235-244.

Gehringer, M. M., Kewada, V., Coates, N. and Downing, T. G. 2003. The use of *Lepidium sativum* in a plant bioassay system for the detection of microcystin-LR. *Toxicon* 41: 871-876.

Gurbuz, F., Metclaff, J. S., Karahan, A. G. and Codd, G. A. 2009. Analysis of dissolved microcystins in surface water samples from Kovada Lake, Turkey. *Science of the Total Environment* 407: 4038-4046.

Hallegraeff, G. M. 1991. Aquaculturists guide to harmful Australian microalgae. CSIRO and Fishing Industry Training Board of Tasmania Inc. Australia. 111 pp.

Håkanson, L., Bryhn, A. C and Hytteborn, J. K. 2007. On the issue of limiting nutrient and predictions of cyanobacteria in aquatic systems. *Science of the Total Environment* 379: 89-108.

Henriksen, P. 2001. Toxic freshwater cyanobacteria in Denmark. In: I. Chorus (ed.) Cyanotoxins-occurrence, causes, consequences. Springer, Berlin, 104-115 pp.

House, J., Ho, L., Newcombe, G., Burch, M. 2004. *Management strategies for toxic blue-green algae: Literature survey.* A report for the CRC for water quality and treatment. 2.0.2.4.1.2, 61 pp.

Hudnell, H. K. 2005. Chronic biotoxin-associated illness; Multiple-system symptoms, a vision deficit and effective treatment. *Neurotoxicology and Teratology*. 27: 733-743.

Hudnell, H. K. 2009. The State of US freshwater harmful algal blooms assessments, policy and legislation. Doi:10.1016/ *j.toxicon*.2009.07.021.

Ibelings, B. W. and Chorus, I. 2007. Accumulation of cyanobacterial toxins I. freshwater "seafood" and its consequences for public health: A review: *Environmental Pollution* 150: 177-192.

Ibelings, B. W. and Havens, K. E. 2008. Cyanobacterial toxins: a qualitative meta-analysis of concentrations, dosage and effects in freshwater, estuarine and marine biota. In: Hudnell, H.K (ed.). Cyanobacterial harmful algal blooms: State of the science and research needs. *Advances in Experimental Medicine and Biology* 619: chapter 12. Springer Press, New York, 675-732 pp.

Izdorczyk, K., Carpentier, C., Mrówczyñskli, J, Wagenvoort, A., Jurczak, T and Tarczyñska, M. 2009. Establishment of an alert level framework for cyanobacteria in drinking water resources by using the algae online analyser for monitoring cyanobacterial chlorophyll a. *Water Research*. (in press) doi. 10.1016/j.watrs.2008.11.048.

Jacquet, C., Thermes, V., de Luze, A., Puisex-Dao, S., Bernard, C., Joly, J-S., Bourrat, F. and Edery, M. 2004. Effects of microcystin-LR on development of medeka fish embryos (*Oryzias latipes*). *Toxicon*. 43: 141-147.

Jos, A., Pichardo, S., Prieto, A., Repetto, G., Vazquez, C. M., Moreno, I. and Camean, A. M. 2005. Toxic cyanobacterial cells containing microcystins induce oxidative stress in exposed tilapia fish (*Oreochromis* sp.) under laboratory conditions. *Aquatic Ecotoxicology*, 72: 261-271.

Kankaanpää, H.T., Holliday, J., Schröder, H., Goddard, T. J., Fister, R. V. and Carmichael, W.W. 2005. Cyanobacteria and prawn farming in northern New South Wales, Australia - a case study on cyanobacteria diversity and hepatotoxin accumulation. *Toxicology and Applied Pharmacology*. 203: 243-256.

Kardinaal, W.E.A. and Visser, P.M. 2005. *Dynamics of cyanobacterial toxins. In* : J. Huisman, H.C.P. Matthijs and P.M.Visser (Eds). Harmful cyanobacteria, p. 41-63. Springer. Netherlands.

Landsberg, J. H. 2002. The effects of harmful algal blooms on aquatic organisms. *Review in Fisheries Science*. 10(2): 113–390.

Landsberg, J. H., Van Dolah, F., Doucette, G. 2005. Marine and estuarine harmful algal blooms: impacts on human and animal health. In: *Ocean and Health: Pathogens in the marine environment.* Belkin and Colwell (eds.), Springer, New York.

Lehane, L. and Lewis, R. 2000. Ciguatera: recent advances but the risk remains-review. *International Journal of Food Microbiology* 61: 91-125.

Lemes, G. A. F., Kersanach, R., Pinto, L. D. S., Dellagostin, O. A., Yunes, J. S. and Matthiensen, A. 2008. Biodegradation of microcystins by aquatic *Burkholderia* sp. From a south Brazilian coastal lagoon. *Ecotoxicology and Environmental Safety* 69: 358-365.

Madhavi,B., Tang, L., Cai, Q., Sandeep, M., Guan, H., Wang, P.,Wang, Z., Christopher, T., Ronald, K. and Jia-Sheng, W. 2006. Seasonal variations in the concentration of microcystin-LR in two lakes in Western Texas, USA. Environmental toxicology and chemistry .25(2): 349-355.

McElhiney, J., Lawton, L. A., and Leifert, C. 2001. Investigations into the inhibitory effects of microcystins on plant growth, and the toxicity of plant tissues following exposure. *Toxicon.* 39: 1411-1420.

Messineo, V., Bogialli, S., Melchiorre, S., Sechi, N., Lugliè, A., Casiddu, P., Mariani, M.A., Padedda, B.M., Corcia, A.D., Mazza, R., Carloni, E. and Bruno, M. 2009. Cyanobacterial toxins in Italian freshwaters. *Limnologica* 39: 95-106.

Mohamed, Z. A. 2008. Toxic cyanobacteria in public hot springs in Saudi Arabia. *Toxicon* 51: 17-27.

Mohamed, Z. A. and Al Shehri, A.M. 2009. Microcystin-producing blooms of *Anabaenopsis Arnoldi* in a potable mountain lake in Saudi Arabia. *FEMS Microbiology Ecology* 69 (1): 98-105.

Molina, R., Moreno, I., Pichardo, S., Jos, A., Moyano, R., Monterde, J. G. and Camean, A. 2005. Acid and alkaline phosphatase activities and pathological changes induced in tilapia fish (*Oreochromis* spp.). Exposed subchronically to microcystins from toxic cyanobacterial blooms under laboratory conditions. *Toxicon.* 46 (7): 725-735.

Nasri, A-B., Bouaïcha, N. and Fastner, J. 2004. First report of a microcystin-containing bloom of the cyanobacteria *Microcystis* spp. in Lake Oubeira, Eastern Algeria. *Archives of Environmental Contamination and Toxicology* 46: 197-202.

NHMRC (National Health and Medical Research Council) 1994. *Health effects of toxic cyanobacteria (Blue-green algae).* National Health and Medical Research Council, Commonwealth of Australia.

NHMRC (National Health and Medical Research Council) 2008. Guidelines for managing risks in recreational water. National Circuit, Canberra. Australian Government 215p. http://www.nhmrc.gov.au/publications/synopses/eh38.htm.

NZG (New Zealand Government) 2005. *Draft Guidelines for Drinking-water Quality Management* for New Zealand, October 2005.

http://www.moh.govt.nz/moh.nsf/0/5A25BF765B400911CC25708F0002B5A8/$File/09-cyanobacteriacompliance.doc (accessed on 18 Jan 2008).

Nyakairu, G. W. A., Nagawa, C. B. and Mbabazi, J. 2009. Assessment of cyanobacteria toxins in freshwater fish: A case study of Murchison Bay (Lake Victoria) and Lake Mburo, Uganda. *Toxicon* doi:10.106/j.toxicon.2009.07.024.

Paldavièiene, A., Mazur-Marzec, H. and Razinkovas, A. 2009. Toxic cyanobacteria blooms in the Lithuanian part of the curonian lagoon. *Oceanologia* 51 (2): 203-216.

Paul, V. J. 2008. Global warming and cyanobacterial harmful algal blooms. In: Hudnell, H.K (editor). Cyanobacterial harmful algal blooms: State of the science and research needs. *Advances in Experimental Medicine and Biology* 619: chapter 11. Springer Press, New York, 239-258 pp.

Pegram, R. A., Humpage, A. R., Neilan, B. A., Runnegar, M. T., Thacker, R. W., Pflugmacher, S., Etheridge, S. M and Love, A. H. 2008. Chapter 15. Cyanotoxins workgroup report. pp. 317-181. In: *Advances in Experimental Medicine and Biology. Volume 619. Book series - Cyanobacterial harmful algal blooms-State of the science and research Needs.* H. Kenneth Hundnell (ed.). Springer New York. Doi 10.1007/978-0-387-75865-7.

Penaloza, R., Rojas, M., Vila, L. and Zambrano, F. 1990. Toxicity of a soluble peptide from *Microcystis* sp. to zooplankton and fish. *Freshwater Biology* 24: 233-240.

Rapala, J. and Sivonnen, K. 1998. Assessment of environmental conditions that favour hepatotoxic and neurotoxic *Anabaena* spp. strains cultured under light limitation at different temperatures. *Microbial Ecology* 36: 181-192.

Rein, K. S., Lynn, B., Gawley, R.E., and Baden, D.G. 1994. Chemical modifications synaptosome binding, toxicity and an unexpected conformational effect. *Journal of Organic Chemistry* 59: 2107-2113.

Rivasseau, C., Martins, S. and Hennion, M-C. 1998. Determination of some physicochemical parameters of microcystins (cyanobacterial toxins) and trace level analysis in environmental samples using liquid chromatography. *Journal of Chromatography* A 799: 155-169.

Sadler, R. 1994. Report on cyanobacteria in water. Government Chemical Laboratory, Coopers Plains, Queensland, Australia.

Saker, M. L. and Eaglesham, G.K. 1999. The accumulation of cylindrospermopsin from the cyanobacterium *Cylindrospermopsis raciborskii* in tissue of the Redclaw crayfish *Cherax quadricarinatus* Toxicon. 37: 1065-1077.

Saker, M. L., Metcalf, J. S., Codd, G. A., Vasconcelos, V. M. 2004. Accumulation and depuration of the cyanobacterial toxin cylindrospermopsin in the freshwater mussel *Andonta cygnea*. Toxicon. 43: 185-194.

Smith, J. L., Boyer, G. L. and Zimba, P. V. 2008. A review of cyanobacterial odours and bioactive metabolites: impacts and management alternatives in aquaculture. *Aquaculture* 280: 5-20.

Stewart, I., Seawright, A.A. and Shaw, G.W. 2008. Cyanobacterial poisoning in livestock, wild mammals and birds-an overview. In: Hudnell, H.K (editor). Cyanobacterial harmful algal blooms: State of the science and research needs. *Advances in Experimental Medicine and Biology* 619: chapter 28. Springer Press, New York, 613-637 pp.

Svrcek, C. and Smith, D. W. 2004. Cyanobacteria toxins and the current state of knowledge on water treatment options, a review. *Journal of Environmental Engineering and Science* 3: 155-185.

Tango, P. J. and Batler, W. 2008. Cyanotoxins in tidal waters of Chesapeake Bay. *Northeastern Naturalist* 15(3): 403-416.

Tencalla, F. Dietrich, D. and Schlatter 1994. Toxicity of *Microcystis aeruginosa* peptide toxin to yearling rainbow trout (*Oncorhynchus mykiss*). *Aquatic Toxicology* 30: 583-595.

Ueno, Y., Nagata, S., Tsutsumi, T., Hasegawa, A., Yosshida, F., Suttajit, M., Mebs, D., Putsc, M. and Vasconcelos, V. 2006. Survey of microcystins in environmental water by a highly sensitive immunoassay based on monoclonal antibody. *Natural Toxins* 4(6);271-276.

Usleber, E., Donald, M. and Straka, M. 1997. Comparison of enzyme immunoassay and mouse bioassay for determining paralytic shellfish poisoning toxins in shellfish. *Food Add. Contam.* 14(2): 193-198.

van Apeldoorn, M. E., van Egmond, H. P., Gerrit, J. A. S. and Bakker, G. J. I. 2007. Review. Toxins in cyanobacteria. *Mol. Nutr.Food Res.* 51: 7-60.

van Buynder, P. G., Oughtred, T., Kirkby, B., Phillips, S., Eaglesham, G., Thomas, K., and Burch, M. 2001. Nodularin uptake by seafood during a cyanobacterial bloom. *Environmental Toxicology* 16: 468-471.

van Egmond, H. P. 2004. Natural toxins: risk, regulations and the analytical situation in Europe. *Anal. Bioanal Chem.* 378:1152-1160.

Wang, J., Pang, X., Ge, F. and Ma, Z. 2007. An ultra performance liquid chromatography-tandem mass spectrometry method for determination of microcystins occurrence in surface water in Zhejiang province, China. *Toxicon.* 49: 1120-1128.

Welker, M., Chorus, I. and Fastner, J. 2004. Occurrence of cyanobacterial toxins (microcystins) in surface waters of rural Bangladesh-pilot study. WHO collaborating centre for research on drinking-water hygiene at the Federal Environmental Agency. WHO/SW/WSH/04.06. 23 pp. http://www.who.int/water_sanitation_health/emerging/wsh0406/en/index.html.

WHO (World Health Organization) 2003. Algae and cyanobacteria in freshwater. *In* : Guidelines for safe recreational water environments. Vol. I: Coastal and freshwaters. World Health Organisation, Geneva, Switzerland, 136-158 pp.

WHO (World Health Organisation) 2006. *Guidelines for drinking water quality, incorporating first addendum. In:* recommendations, third ed., vol. 1. World Health Organisation, Geneva, Switzerland, pp. 407-408.

Wood, S. A., Holland, P. T., Stirling D. J., Briggs, L. R., Sprosen, J., Ruck, J. G. and Wear, R. G. 2006. Survey of cyanotoxins in New Zealand water bodies between 2001 and 2004. *New Zealand Journal of Marine and Freshwater Research.* 40: 585-597.

Zaccaroni, A. and Scaravelli, D. 2008. *Toxicity of sea algal toxins to humans and animals.* pp. 91-158. In: V. Evangelista (ed.). *Algal toxin: Nature, occurrence, effect and detection.* Springer Science.

Websites

WHO 2007: Harmful algae. http://www.whoi.edu/redtide/illness/illness.html. updated October 2007.

http://www-cyanosite.bio.purdue.edu/cyanotox/toxins.html

http://www.nwfsc.noaa.gov/research/divisions/ec/ecotox/fishneurobiology/devconsequences.cfm

http://www.g-mwater.com.au/environment/waterquality/b-g_algae (blue-green algae) Goulburn-Murray Water

http://www.worldwaterday.org/2001/disease/cyano.html

http://www.whoi.edu/redtide/page.do?pid=9257

http://www.whoi.edu/redtide/species/alexandrium_images.html

http://www.whoi.edu/redtide/page.do?pid=12779

http://www.zooplankton-online.net/gallery.html

http://www.whoi.edu/redtide/illness/ASP_chemstructure.gif

http://www.waterquality.crc.org.au/dwfacts/techfact_cyanobacteria.pdf

□□□

Index

A

B

C

D

E

F

G

H

I

N

T

U

V

□□□